Lecture Notes in Computer Science   16536

Founding Editors

Gerhard Goos
Juris Hartmanis

Editorial Board Members

Elisa Bertino, *Purdue University, West Lafayette, IN, USA*
Wen Gao, *Peking University, Beijing, China*
Bernhard Steffen, *TU Dortmund University, Dortmund, Germany*
Moti Yung, *Columbia University, New York, NY, USA*

The series Lecture Notes in Computer Science (LNCS), including its subseries Lecture Notes in Artificial Intelligence (LNAI) and Lecture Notes in Bioinformatics (LNBI), has established itself as a medium for the publication of new developments in computer science and information technology research, teaching, and education.

LNCS enjoys close cooperation with the computer science R & D community, the series counts many renowned academics among its volume editors and paper authors, and collaborates with prestigious societies. Its mission is to serve this international community by providing an invaluable service, mainly focused on the publication of conference and workshop proceedings and postproceedings. LNCS commenced publication in 1973.

Hyungsoo Jung · Tianzheng Wang ·
Masashi Toyoda · Hyuk-Yoon Kwon ·
Jae-woong Lee
Editors

Database Systems
for Advanced Applications

31st International Conference, DASFAA 2026
Jeju, South Korea, April 27–30, 2026
Proceedings, Part II

Springer

Editors
Hyungsoo Jung
Seoul National University
Seoul, Korea (Republic of)

Masashi Toyoda
The University of Tokyo
Tokyo, Tokyo, Japan

Jae-woong Lee
Kangwon National University
Chuncheon-si, Korea (Republic of)

Tianzheng Wang
SFU Computing Science, ASB10070
Simon Fraser University
Burnaby, BC, Canada

Hyuk-Yoon Kwon
SeoulTech
Seoul, Korea (Republic of)

ISSN 0302-9743   ISSN 1611-3349 (electronic)
Lecture Notes in Computer Science
ISBN 978-981-92-0365-9   ISBN 978-981-92-0366-6 (eBook)
https://doi.org/10.1007/978-981-92-0366-6

Preface

It is our great pleasure to present the proceedings of the 31st International Conference on Database Systems for Advanced Applications (DASFAA 2026), held on April 27–30, 2026, in South Korea. DASFAA is a leading annual international conference in the database field, showcasing state-of-the-art research and development in database systems and advanced applications. It serves as a premier forum for technical presentations and discussions among researchers, developers, and users from both academia and industry.

This year, DASFAA received 834 research paper submissions. Following the long-standing DASFAA tradition, we conducted a rigorous double-blind review process. To support this process, we assembled a large and dedicated Program Committee consisting of 31 Senior Program Committee (SPC) members and 245 Program Committee (PC) members, including emergency reviewers. Each valid submission was reviewed by mostly three PC members and meta-reviewed by one SPC member. Based on the recommendations of the SPC members, we, the PC co-chairs, carefully examined each submission together with its reviews before making the final decisions. As a result, 228 full papers were accepted, yielding an acceptance rate of 27.3%. The review process was supported by the Microsoft CMT system.

During the three main conference days, the accepted full papers were presented across 44 research sessions. The dominant keywords among the accepted papers included language models, graph, recommendation, knowledge, efficient, data, framework, network, prediction, detection, neural, and federated, reflecting several of the major research trends currently shaping the database community. In addition to the research track, the conference program also included 16 industry papers, 11 demo papers, and 5 tutorials.

To highlight important emerging directions in the field, DASFAA 2026 featured four invited keynote presentations by Victor Leis (Technical University of Munich), Wenjie Zhang (University of New South Wales), Meihui Zhang (Beijing Institute of Technology), and Wook-Shin Han (Pohang University of Science and Technology).

Five tutorials were selected by the Tutorial Co-chairs:

1. Continual Recommender Systems, by Seunghan Lee (Korea University), Seunghyun Baek (Korea University), Dojun Hwang (Korea University), Hyunsik Yoo (University of Illinois Urbana-Champaign), and SeongKu Kang (Korea University);
2. Efficient Compression and Queries of Large Graphs, by Fan Zhang (Guangzhou University), Qingshuai Feng (Great Bay University), and Kai Wang (Shanghai Jiao Tong University);
3. Advances in Real-Time Processing of Longitudinal Data: From Statistical and Deep Learning Methods to Applications, by Ying-Ren Chien (National Taipei University of Technology), Pavel Loskot (Zhejiang University-University of Illinois Urbana-Champaign Institute), and Yu Gao (Midea Group);
4. Trustworthy Foundation Models with a Data-Centric Approach, by Wenjie Fang (University of Science and Technology of China), Dan Li (Sun Yat-sen University), and Jian Lou (Sun Yat-sen University);

5. Reliable Visual Analytics with Dimensionality Reduction: Quality Evaluation and Interpretation of Projections, by Jeon Hyeon (Seoul National University), Takanori Fujiwara (University of Arizona), and Rafael M. Martins (Linnaeus University).

Two workshops were selected by the Workshop Co-chairs to be held in conjunction with DASFAA 2026: the 10th International Workshop on Graph Data Management and Analysis (GDMA 2026) and the 12th International Workshop on Big Data Management and Service (BDMS 2026). The papers accepted to these workshops are included in a separate proceedings volume, also published by Springer in the Lecture Notes in Computer Science (LNCS) series.

We are deeply grateful to the General Chairs, Sang-Won Lee (Seoul National University, South Korea), Bin Cui (Peking University, China), and Zi (Helen) Huang (University of Queensland, Australia) for their outstanding leadership and support. We also sincerely thank all SPC members, PC members, and external emergency reviewers for their tremendous dedication, time, and expertise. Finally, we would like to express our appreciation to all members of the Organizing Committee and the many volunteers for their invaluable contributions to the success of DASFAA 2026.

April 2026

Hyungsoo Jung
Tianzheng Wang
Masashi Toyoda

Organization

Steering Committee Members

Chair

Lei Chen — Hong Kong University of Science and Technology (Guangzhou), China

Vice Chair

Stéphane Bressan — National University of Singapore, Singapore

Treasurer

Yasushi Sakurai — Osaka University, Japan

Secretary

Kyuseok Shim — Seoul National University, South Korea

Members

Aamir Cheema	Monash University, Australia
Arnab Bhattacharya	IIT Kanpur, India
Atsuyuki Morishima	University of Tsukuba, Japan
De-Nian Yang	Academia Sinica, Taiwan
Eun-Jun Hwang	Korea University, South Korea
Guoliang Li	Tsinghua University, China
Sang-Won Lee	Seoul National University, South Korea
Sourav Bhowmick	Nanyang Technological University, Singapore
Yang-Sae Moon	Kangwon National University, South Korea
Zhanhuai Li	Northwestern Polytechnical University, China
Zhiyong Peng	Wuhan University, China
Zi Huang	University of Queensland, Australia

Organizing Committee

General Co-chairs

Bin Cui	Peking University, China
Zi (Helen) Huang	University of Queensland, Australia
Sang-Won Lee	Seoul National University, South Korea

Program Chairs

Hyungsoo Jung	Seoul National University, South Korea
Tianzheng Wang	Simon Fraser University, Canada
Masashi Toyoda	University of Tokyo, Japan

Industry PC Chairs

Lei Chen	Hong Kong University of Science and Technology, China
Min-Soo Kim	KAIST, South Korea

Demo Chairs

Byungchul Tak	Kyungpook National University, South Korea
Siqiang Luo	Nanyang Technological University, Singapore

Local Chairs

Jongwuk Lee	Sungkyunkwan University, South Korea
Buru Chang	Korea University, South Korea

Panel Chair

Seung-won Hwang	Seoul National University, South Korea

Workshop Chairs

Joyce Jiyoung Whang	KAIST, South Korea
Wenjie Zhang	University of New South Wales, Australia

Tutorial Chairs

Shenghua Liu   Chinese Academy of Science, China
Yasushi Sakurai   Osaka University, Japan
Kijung Shin   KAIST, South Korea

Publicity Chairs

Jaeyoung Do   Seoul National University, South Korea
Yi Cai   South China University of Technology, China
Atsuyuki Morishima   Tsukuba University, Japan

Web/Slack Chairs

Suan Lee   Semyung University, South Korea
Sun-Young Ihm   Pai Chai University, South Korea

Publication Chairs

Jae-woong Lee   Kangwon National University, South Korea
Hyuk-Yoon Kwon   Seoul National University of Science and
                                    Technology, South Korea

Registration Chair

Hyunsouk Cho   Ajou University, South Korea

Treasurer

Jonghyeok Park   Korea University, South Korea

Sponsorship Chairs

Kyongha Lee   Korea Institute of Science and Technology
                                    Information, South Korea
Kyoungsook Kim   AIST, Japan
Xin Wang   Tianjin University, China

SC Liaison

Yang-Sae Moon   Kangwon National University, South Korea

Advisory Board

Kyuseok Shim	Seoul National University, South Korea
Yang-Sae Moon	Kangwon National University, South Korea
Young-Koo Lee	Kyunghee University, South Korea
Ha-Joo Song	Pukyong National University, South Korea

Program Committee

Research Track

Chairs

Hyungsoo Jung	Seoul National University, South Korea
Tianzheng Wang	Simon Fraser University, Canada
Masashi Toyoda	University of Tokyo, Japan

Senior Program Committee

Alkis Simitsis	Athena Research Center, Greece
Byung Suk Lee	University of Vermont, USA
Cheng Long	Nanyang Technological University, Singapore
Da Yan	Indiana University Bloomington, USA
Dong Wen	University of New South Wales, Australia
Hideyuki Kawashima	Keio University, Japan
Jianliang Xu	Hong Kong Baptist University, China
Jieming Shi	Hong Kong Polytechnic University, China
Jun Miyazaki	Institute of Science Tokyo, Japan
Ke Wang	Simon Fraser University, Canada
Kyoung-Sook Kim	AIST, Japan
Kyuseok Shim	Seoul National University, South Korea
Leong Hou U	University of Macau, China
Nikos Bikakis	Hellenic Mediterranean University and ATHENA Research Center, Greece
Sang-Wook Kim	Hanyang University, South Korea
Satoshi Oyama	Nagoya City University, Japan
Sebastian Link	University of Auckland, New Zealand
Sen Wang	University of Queensland, Australia
Sibo Wang	Chinese University of Hong Kong, China
Steven E. Whang	KAIST, South Korea

Verena Kantere	University of Ottawa, Canada
Wei-Shinn Ku	Auburn University, USA
Wen Hua	Hong Kong Polytechnic University, China
Wenjie Zhang	University of New South Wales, Australia
Xiangyu Ke	Zhejiang University, China
Xin Cao	University of New South Wales, Australia
Xin Wang	Tianjin University, China
Yanfeng Zhang	Northeastern University, China
Yixiang Fang	Chinese University of Hong Kong, Shenzhen, China
Yon Dohn Chung	Korea University, South Korea
Zeke Wang	Zhejiang University, China

Program Committee

Alexander Zhou	Hong Kong Polytechnic University, China
Anne Laurent	University of Montpellier, France
Aziz Nasridinov	Chungbuk National University, South Korea
Beibei Li	Chongqing University, China
Bin Wang	Northeastern University, China
Bo Xu	Dalian University of Technology, China
Buru Chang	Korea University, South Korea
Byoungwook Kim	Gangneung-Wonju National University, South Korea
Byungchul Tak	Kyungpook National University, South Korea
Cai Xu	Xidian University, China
Changdong Wang	Sun Yat-sen University, China
Chao Zhang	Renmin University of China, China
Chen Shen	Megagon Labs, USA
Chengcheng Mai	Nanjing Normal University, China
Chengliang Chai	Beijing Institute of Technology, China
Chenhao Ma	Chinese University of Hong Kong, Shenzhen, China
Chenxu Wang	Xi'an Jiaotong University, China
Chenyang Wang	Aalto University, Finland
Cheqing Jin	East China Normal University, China
Chihyun Park	Kangwon National University, South Korea
Chongjun Wang	Nanjing University, China
Christos Doulkeridis	University of Piraeus, Greece

Dan He	University of Queensland, Australia
Dan Li	Sun Yat-sen University, China
Debarati B. Chakraborty	University of Hull, UK
Derong Shen	Northeastern University, China
Di Yao	Institute of Computing Technology, Chinese Academy of Sciences, China
Dimitris Kotzinos	ETIS/CY Cergy Paris University, France
Divya Saxena	Indian Institute of Technology Jodhpur, India
Dongha Lee	Yonsei University, South Korea
Dongjin Yu	Hangzhou Dianzi University, China
Dong-Kyu Chae	Hanyang University, South Korea
Eenjun Hwang	Korea University, South Korea
Faming Li	Northeastern University, China
Fei Guo	Central South University, China
Gengrui Zhang	Concordia University, Canada
Gong Cheng	Nanjing University, China
Guanjie Zheng	Shanghai Jiao Tong University, China
Guixian Zhang	China University of Mining and Technology, China
Ha-Myung Park	Kookmin University, South Korea
Hanbing Zhang	Fudan University, China
Hanchen Wang	University of Technology Sydney, Australia
Hang Yu	Shanghai University, China
Hao Xin	Hong Kong University of Science and Technology, China
Haobing Liu	Ocean University of China, China
Haofen Wang	Tongji University, China
Haoyang LI	Hong Kong Polytechnic University, China
Hayato Yamana	Waseda University, Japan
Hengyu Liu	Aalborg University, Denmark
Hiroaki Ohshima	University of Hyogo, Japan
Hogun Park	Sungkyunkwan University, South Korea
Hongzhi Wang	Harbin Institute of Technology, China
Huaijie Zhu	Sun Yat-sen University, China
Hui Li	Xidian University, China
Huiqi Hu	East China Normal University, China
Hyun Ji Jeong	Kongju National University, South Korea
Hyunsouk Cho	Ajou University, South Korea
Hyunwoo Park	Seoul National University, South Korea
Jaebum Kim	Konkuk University, South Korea
Jaemin Yoo	Seoul National University, South Korea
Jaesoo Yoo	Chungbuk National University, South Korea

Jae-woong Lee	Kangwon National University, South Korea
Jay-Yoon Lee	Seoul National University, South Korea
Jen-Wei Huang	National Cheng Kung University, Taiwan
Ji Zhang	University of Southern Queensland, Australia
Jiali Mao	East China Normal University, China
Jianbin Qin	Shenzhen University, China
Jianqiu Xu	Nanjing University of Aeronautics and Astronautics, China
Jianxin Li	Beihang University, China
Jianxiong Guo	Beijing Normal University, China
Jie Shao	University of Electronic Science and Technology of China, China
Jilin Hu	Aalborg University, Denmark
Jin Wang	Arizona State University, USA
Jingya Zhou	Soochow University, China
Jinhong Jung	Soongsil University, South Korea
JinYeong Bak	Sungkyunkwan University, South Korea
Jithin Vachery	National University of Singapore, Singapore
Jonghyeok Park	Korea University, South Korea
Jongik Kim	Chungnam National University, South Korea
Jongwuk Lee	Sungkyunkwan University, South Korea
Joonseok Lee	Google Research & Seoul National University, South Korea
Joyce J. Whang	KAIST, South Korea
Ju Fan	Renmin University of China, China
Jungeun Kim	Inha University, South Korea
Junghoon Kim	Ulsan National Institute of Science and Technology, South Korea
Jun-Gi Jang	Daegu Gyeongbuk Institute of Science and Technology, South Korea
Junhua Fang	Soochow University, China
Junjie Yao	East China Normal University, China
Jun-Ki Min	Korea University of Technology and Education, South Korea
Junya Arai	Nippon Telegraph and Telephone Corporation, Japan
Kai Zhang	University of Science and Technology of China, China
Kaixi Hu	Wuhan Textile University, China
Kangzheng Liu	Huazhong University of Science and Technology, China
Kesheng Wu	Lawrence Berkeley National Laboratory, USA
Ki Yong Lee	Sookmyung Women's University, South Korea

Kijung Shin	KAIST, South Korea
P. Krishna Reddy	IIIT Hyderabad, India
Kristian Torp	Aalborg University, Denmark
Kwanghyun Park	Yonsei University, South Korea
Kyong-Ha Lee	Korea Institute of Science and Technology Information, South Korea
Kyoungsoo Bok	Wonkwang University, South Korea
Kyungyong Lee	Hanyang University, South Korea
Ladjel Bellatreche	ISAE-ENSMA, France
Lanjun Wang	Tianjin University, China
Lei Li	Hong Kong University of Science and Technology (Guangzhou), China
Li Jiajia	Shenyang Aerospace University, China
Libin Zheng	Sun Yat-sen University, China
Lijun Chang	University of Sydney, Australia
Linglin Yang	Peking University, China
Lingwei Chen	Rochester Institute of Technology, USA
Lingyang Chu	McMaster University, Canada
Lizhen Cui	Shandong University, China
Lu Chen	Swinburne University of Technology, Australia
Marwan Hassani	TU Eindhoven, Netherlands
Meng Liu	National University of Defense Technology, China
Mengxuan Zhang	Australian National University, Australia
Mijin An	SAP, South Korea
Ming Zhong	Wuhan University, China
Minghao Zhao	East China Normal University, China
Minghe Yu	Northeastern University, China
Mingyue Cheng	University of Science and Technology of China, China
Min-hwan Oh	Seoul National University, South Korea
Mo Li	Liaoning University, China
Muhammad Haris	Liaoning University, China
Nannan Wu	Tianjin University, China
Ning Liu	Shandong University, China
Ning Wang	Beijing Jiaotong University, China
Ning Yang	Sichuan University, China
Ningning Cui	Chang'an University, China
Norio Katayama	National Institute of Informatics, Japan
O-Joun Lee	Catholic University of Korea, South Korea
Ozan Kahramanogullari	Free University of Bozen-Bolzano, Italy
Peng Cai	East China Normal University, China

Peng Cheng	Tongji University, China
Peng Liu	Guangxi Normal University, China
Pengpeng Zhao	Soochow University, China
Ping Lu	Beihang University, China
Qi Song	University of Science and Technology of China, China
Qiang Qu	Shenzhen Institutes of Advanced Technology, Chinese Academy of Sciences, China
Qiang Yin	Shanghai Jiao Tong University, China
Qianzhen Zhang	National University of Defense Technology, China
Qingyun Sun	Beihang University, China
Qingzhi Ma	Soochow University, China
Qiyao Peng	Tianjin University, China
Quanqing Xu	Independent Researcher, China
Renjie Sun	Zhejiang Gongshang University, China
Rui Zhou	Swinburne University of Technology, Australia
Rui Zhu	Shenyang Aerospace University, China
Ruiyuan Li	Chongqing University, China
Sanghack Lee	Seoul National University, South Korea
Sanghyun Park	Yonsei University, South Korea
Sanjay Kumar Madria	Missouri University of Science & Technology, USA
Savong Bou	University of Tsukuba, Japan
SeongKu Kang	Korea University, South Korea
Shanshan Feng	Wuhan University, China
Shaoxu Song	Tsinghua University, China
Shengan Zheng	Shanghai Jiao Tong University, China
Shiyu Yang	Guangzhou University, China
Shoko Wakamiya	Nara Institute of Science and Technology, Japan
Shuai Xu	Nanjing University of Aeronautics and Astronautics, China
Shunmei Meng	Nanjing University of Science and Technology, China
Shuyuan Li	Beihang University, China
Si Liu	Texas A&M University, USA
Silvestro Roberto Poccia	University of Turin, Italy
Siyuan Chen	Guangzhou University, China
Suan Lee	Semyung University, South Korea
Sungsu Lim	Chungnam National University, South Korea
Sungwon Jung	Sogang University, South Korea
Suprio Ray	University of New Brunswick, Fredericton, Canada

Susik Yoon   Korea University, South Korea
Tae-Sun Chung   Ajou University, South Korea
Taesup Kim   Seoul National University, South Korea
Tangpeng Dan   Wuhan University of Technology, China
Tao Qiu   Shenyang Aerospace University, China
Tao Zhao   National University of Defense Technology,
                              China
Tianzi Zang   Nanjing University of Aeronautics and
                              Astronautics, China
Tieke He   Nanjing University, China
Tiezheng Nie   Northeastern University, China
Tomoki Yoshihisa   Shiga University, Japan
Wei Hu   Nanjing University, China
Wei Li   Harbin Engineering University, China
Wei Liu   Sun Yat-sen University, China
Wei Song   Wuhan University, China
Wei Emma Zhang   University of Adelaide, Australia
Wentao Li   University of Leicester, UK
Wolf-Tilo Balke   TU Braunschweig, Germany
Woohwan Jung   Hanyang University, South Korea
Woong-Kee Loh   Gachon University, South Korea
Xiang Ao   Institute of Computing Technology, CAS, China
Xiang Lian   Kent State University, USA
Xiang Zhao   National University of Defense Technology,
                              China
Xiangguo Sun   Southeast University, China
Xiangmin Zhou   RMIT University, Australia
Xiangtao Li   Jilin University, China
Xiangyu Song   Chang'an University, China
Xiao Pan   Shijiazhuang Tiedao University, China
Xiaofeng Gao   Shanghai Jiao Tong University, China
Xiaoling Wang   East China Normal University, China
Xiaoou Ding   Harbin Institute of Technology, China
Xiaoyang Wang   University of New South Wales, Australia
Xin Huang   Hong Kong Baptist University, China
Xinqiang Xie   Neusoft, China
Xuanhe Zhou   Shanghai Jiao Tong University, China
Yajun Yang   Tianjin University, China
Yan Zhang   Peking University, China
Yang Chen   Fudan University, China
Yang-Sae Moon   Kangwon National University, South Korea
Yanhao Wang   East China Normal University, China

Yanmei Hu	Chengdu University of Technology, China
Yanmin Zhu	Shanghai Jiao Tong University, China
Yeon-Chang Lee	Ulsan National Institute of Science and Technology, South Korea
Yeonsu Park	Kangwon National University, South Korea
Yingxia Shao	Beijing University of Posts and Telecommunications, China
Yishu Wang	Northeastern University, China
Yohan Jo	Seoul National University, South Korea
Yong Zhang	Tsinghua University, China
Yongchao Liu	Ant Group, China
Yonghong Yu	Nanjing University of Posts and Telecommunications, China
Yongpan Sheng	Southwest University, China
Young-Koo Lee	Kyung Hee University, South Korea
Young-Kyoon Suh	Kyungpook National University, South Korea
Yu Hua	Huazhong University of Science and Technology, China
Yu Liu	Huazhong University of Science and Technology, China
Yu Sun	Nankai University, China
Yu Yang	Education University of Hong Kong, China
Yuanbo Xu	Jilin University, China
Yuankai Fan	Institute of Artificial Intelligence (TeleAI), China Telecom, China
Yuchen Li	Singapore Management University, Singapore
Yue Kou	Northeastern University, China
Yunjun Gao	Zhejiang University, China
Yunpeng Chai	Renmin University of China, China
Yunyong Ko	Chung-Ang University, South Korea
Yurong Cheng	Beijing Institute of Technology, China
Yuxiang Zeng	Hong Kong University of Science and Technology, China
Zhao Zhang	East China Normal University, China
Zhaojing Luo	Beijing Institute of Technology, China
Zhaonian Zou	Harbin Institute of Technology, China
Zhengyi Yang	University of New South Wales, Australia
Zhihui Wang	Fudan University, China
Zhiwei Zhang	Beijing Institute of Technology, China
Zhixu Li	Renmin University of China, China
Zhongnan Zhang	Xiamen University, China
Zi Chen	Wuhan University of Technology, China
Zimu Zhou	City University of Hong Kong, China

Zirui Zhuang | Beijing University of Posts and Telecommunications, China

Industry Track

Chairs

Lei Chen | Hong Kong University of Science and Technology, China
Min-Soo Kim | KAIST, South Korea

Demo Track

Chairs

Byungchul Tak | Kyungpook National University, South Korea
Siqiang Luo | Nanyang Technological University, Singapore

Program Committee

Chenhao Ma | Chinese University of Hong Kong, Shenzhen
Chun-Hee Lee | Kyungpook National University, South Korea
Dong-Hyuk Im | Kwangwoon University, South Korea
Ergute Bao | Mohamed bin Zayed University of Artificial Intelligence, UAE
Ki Yong Lee | Sookmyung Women's University, South Korea
Kisung Lee | Louisiana State University, USA
Ling Li | Shanxi University, China
Meng Li | Nanjing University, China
Minsu Cho | Kwangwoon University, South Korea
Peng Fang | Huazhong University of Science and Technology, China
Sahil Suneja | IBM Research, USA
Weiping Yu | Nanyang Technological University, China
Xiaolin Han | Northwestern Polytechnical University, China
Yon Dohn Chung | Korea University, South Korea

Contents

Machine Learning for Database

Cloud Data Management

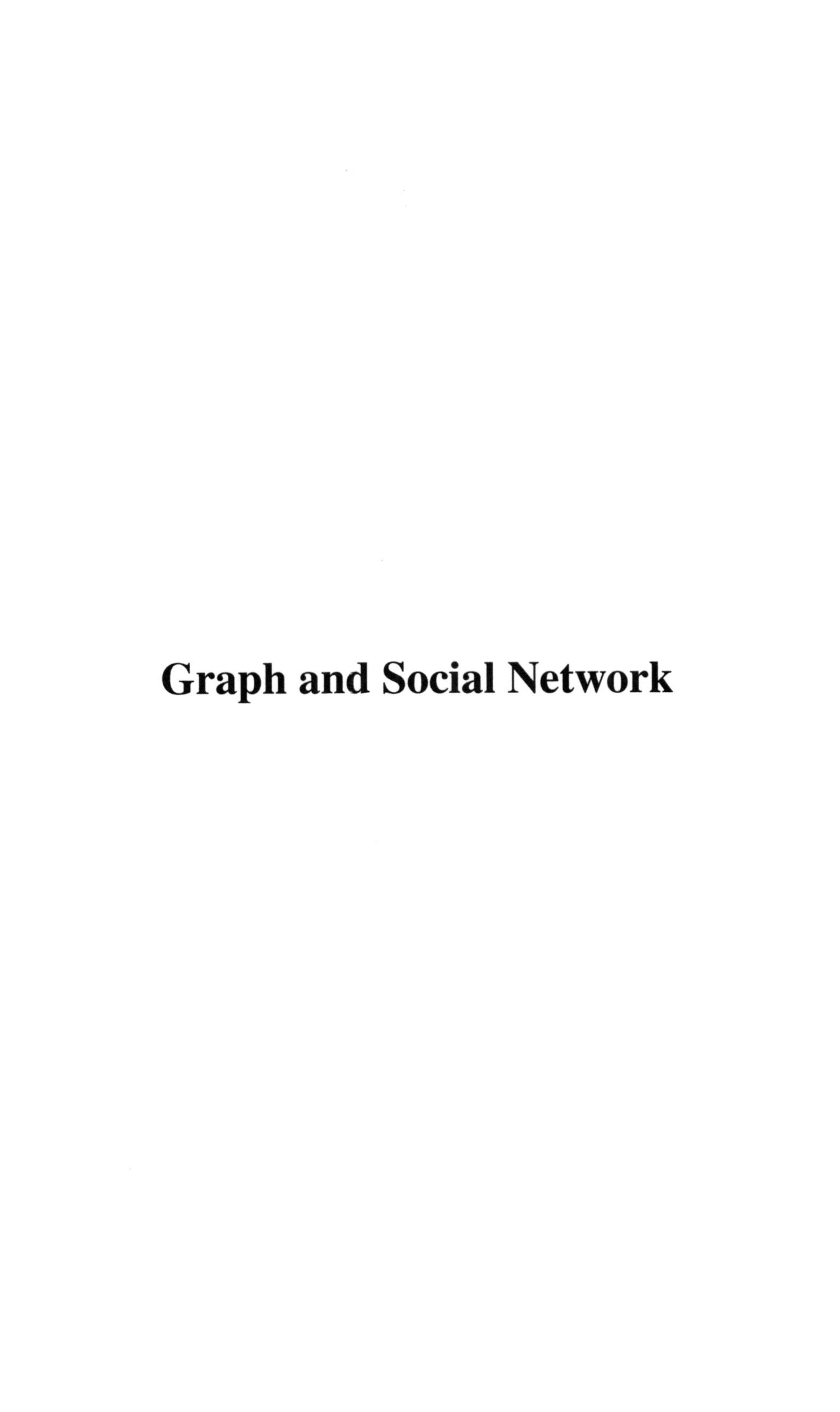

Graph and Social Network

Beyond Similar Information:
A Distinction-Preserving Framework
for Graph Autoencoders

Ge Chen[1] , Yulan Hu[2] , Sheng Ouyang[2] , and Cuicui Luo[1]($\boxtimes$)

[1] University of Chinese Academy of Sciences, Beijing, China
`chenge221@mails.ucas.ac.cn`, `luocuicui@ucas.ac.cn`
[2] Gaoling School of Artificial Intelligence, Renmin University of China,
Beijing, China
`{huyulan,ouyangsheng}@ruc.edu.cn`

Abstract. Graph autoencoders (GAEs), a class of generative self-supervised learning methods, have demonstrated great potential in recent years. Typically, GAEs employ an encoder to map the input graph into a latent representation and a decoder to reconstruct the graph by recovering its characteristics, such as node features or structural information. However, GAEs that rely on feature reconstruction often fail to recover the unique information that differs from neighboring nodes, leading to excessive feature smoothness between neighboring nodes and sub-optimal performance. To address this issue, we propose two complementary strategies applied during the encoding and decoding phases, respectively. At the decoding stage, we develop a simple yet effective approach to preserve the distinctiveness between neighbors in the raw graph. We conceptualize the encoder-decoder architecture of GAEs as a teacher-student framework, where we compute pairwise node dissimilarities in both the original and reconstructed graphs and enforce a Kullback-Leibler divergence constraint to transfer distinctiveness from the input to the output space. At the encoding stage, we introduce a discriminative constraint that encourages decorrelation among similar node pairs, implemented via a covariance-based regularization that jointly considers node and neighborhood embeddings. Based on these strategies, we present ClearGAE, a GAE capable of reconstructing graphs while preserving their essential distinctions. Extensive experiments on three types of graph tasks demonstrate the effectiveness of ClearGAE. Moreover, our strategies are model-agnostic and can be seamlessly integrated as plug-and-play modules into other GAE variants.

Keywords: Graph Representation learning · Graph Self-supervised Learning · Masked Graph Autoencoders

1 Introduction

Generative graph self-supervised learning (SSL), exemplified by GAEs [6,7,9], has attracted increasing attention in recent years. As an emerging learning

© The Author(s), under exclusive license to Springer Nature Singapore Pte Ltd. 2026
H. Jung et al. (Eds.): DASFAA 2026, LNCS 16536, pp. 3–18, 2026.
https://doi.org/10.1007/978-981-92-0366-6_1

paradigm [14], GAEs leverage intrinsic supervisory signals to learn graph representations, which can subsequently be applied to various downstream tasks.

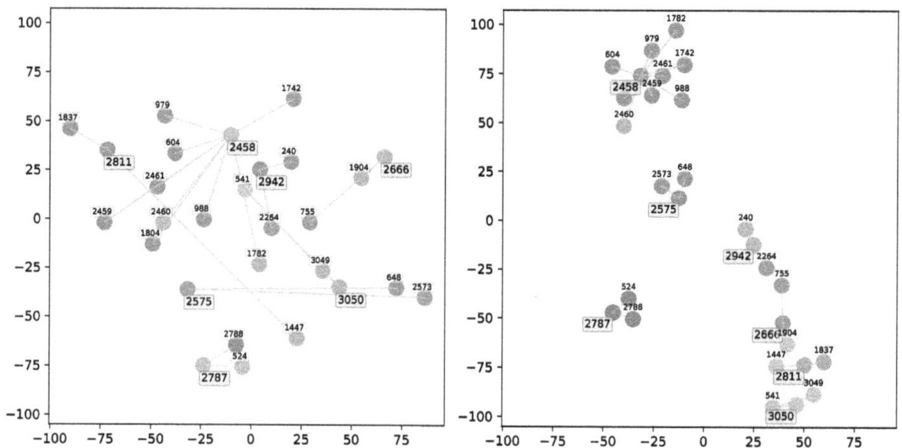

Fig. 1. We selected the nodes that GraphMAE misclassified in the test set, extracted the subgraphs of these nodes, and visualized both the original node representations and the embeddings generated by the encoder, as shown in the figure above. The left side represents the original features, while the right side shows the embedding features. Nodes misclassified in the test set are highlighted in yellow. (Color figure online)

Most existing GAEs focus on reconstructing a single type of graph property, such as node features or structural information. Feature reconstruction-based GAEs are conceptually analogous to pixel-level reconstruction in computer vision (CV), where the goal is to recover detailed input signals. For instance, Graph-MAE [6] and GraphMAE2 [5] reconstruct raw input features either directly or from their encoded counterparts. More recently, GA^{2}E [7] extends this idea by unifying diverse graph tasks through subgraph feature reconstruction. In parallel, another line of research focuses on structure reconstruction. Early works such as GAE and VGAE [9] aim to reconstruct the adjacency matrix, primarily for link prediction tasks. S2GAE [22] further generalizes this idea by formulating several graph learning tasks as missing edge prediction, achieving strong empirical results. Beyond conventional graph learning settings, GAEs have also been adopted in more complex scenarios, such as heterogeneous graph learning [23] and imbalanced node classification [8].

Despite the notable progress made by recent GAE studies, a critical limitation remains. Feature-reconstruction-based GAEs often cause node embeddings to converge with those of their neighbors, reduce representational differences among adjacent nodes, and ultimately forming homogeneous clusters, even when the original data inherently contains greatly distinct characteristics between nodes and their neighbors. This issue primarily stems from an excessive emphasis on preserving local neighborhood similarity, leading to overly similar node representations and impairing node distinguishability in downstream tasks. As a result, adjacent nodes from different classes are often mapped to similar representations, causing them to be incorrectly classified into the same category. To illustrate this issue, we conduct a node classification case study on the Cite-Seer dataset, as shown in Fig. 1. Specifically, we extract a subgraph centered on misclassified nodes (highlighted in yellow). The left part of the figure shows the original node connections, where yellow nodes denote those misclassified after encoding. For example, in the small triangle subgraph formed by nodes 3050, 3049, and 541, nodes 3050 and 3049 share the same class label, while node 541 belongs to a different class. However, as shown on the right side of the figure, all three nodes are encoded to highly similar representations, effectively collapsing onto the same embedding as node 541.

Motivated by this observation, we investigate the root causes of feature over-smoothing in the context of GAEs. **First**, the use of GNN-based encoder and decoder inherently risks representation smooth due to repeated neighborhood aggregation [20]. **Second**, the reconstruction losses operate on a point-to-point basis and the recovered features are derived from common neighboring features in the raw graph. Consequently, standard GAEs tend to overemphasize pairwise common information and generate the local similarity embeddings, hindering node separability and leading to degraded performance in downstream tasks.

In this paper, we propose ClearGAE, a novel GAE framework designed to better reconstruct masked nodes by not only preserving the common information shared with neighboring nodes but also retaining the unique features that differentiate nodes from their neighbors, thereby improving performance on downstream graph tasks. To this end, we introduce two complementary strategies, applied at the encoding and decoding stages, respectively. During the decoding phase, we employ a **distillation-based strategy** to preserve neighborhood-level feature difference in the input graph during reconstruction. Specifically, we treat the original graph as the teacher and the reconstructed graph as the student, and adopt a knowledge distillation technique to transfer feature discriminability between nodes and their neighbors from the teacher to the student.

For each node, we compute the feature distance to its first-order neighbors in both the original and reconstructed graphs, and apply a KL divergence loss to align the two sets of distance distributions. This simple yet effective constraint encourages the decoder to preserve critical node-level distinctions, thereby enabling the model to generate clearer and more meaningful graph reconstructions. During the encoding phase, we incorporate a **decorrelation regularization strategy** into the latent embeddings and the mean embeddings of their 1-hop neighbors to suppress excessive similarity and promote representational diversity. By combining these two strategies, ClearGAE effectively maintains essential pairwise distinctiveness, mitigates feature smoothing, and ultimately leads to improved overall performance. Our contributions can be summarized as follows:

– We identify and analyze a critical limitation in existing GAE studies: reliance on a single reconstruction criterion that emphasizes the recovery of similar

information among nodes and their adjacent neighbors, while disregarding the distinct individual characteristics intrinsic to each node. To the best of our knowledge, this is the first work to explicitly address this issue.
- We propose two simple yet effective strategies, targeting the encoding and decoding stages respectively, that can be seamlessly integrated into existing GAE frameworks as plug-and-play modules.
- We conduct comprehensive experiments across three types of graph tasks, demonstrating both the effectiveness and generality of our approach.

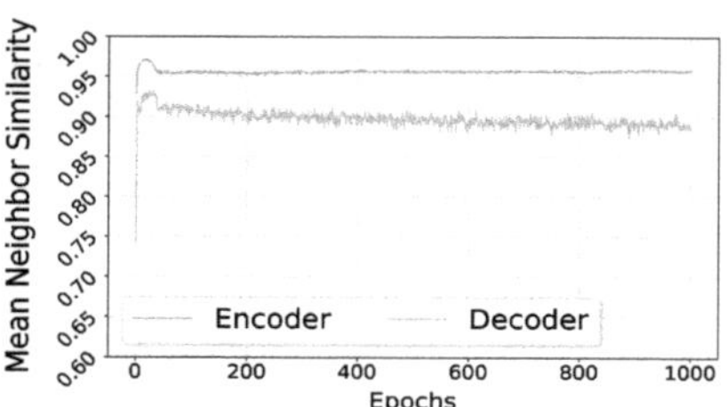
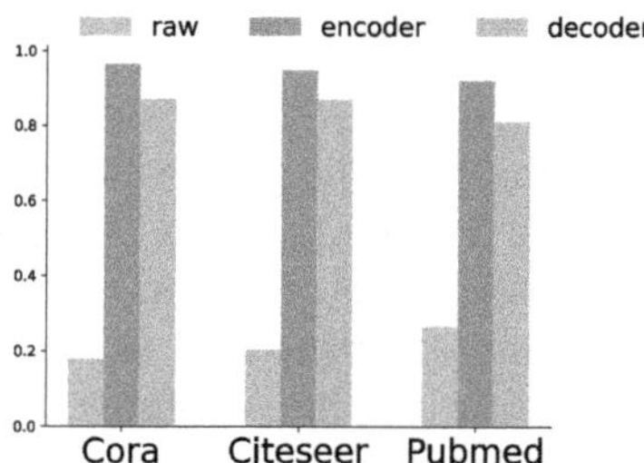

Fig. 2. The left plot illustrates the variations in node neighborhood similarity of the Cora dataset during the training process after encoding and decoding. The right plot compares the pairwise node similarity across the raw graph, the generated embeddings, and the reconstructed graph.

2   Related Works

2.1   Generative Self-Supervised Graph Learning

Generative self-supervised learning (SSL) aims to generate new graphs that closely resemble the given input. In our study, we concentrate on the graph autoencoding family (GAEs) of generative SSL. These GAEs typically involve encoding the inputs into a latent feature space and reconstructing the graphs from these encoded representations. The pioneering work, known as GAE/VGAE [9], leverages a GNN encoder and a dot-product decoder to reconstruct the structural information. For different downstream tasks, later GAEs explore the recovery of various kinds of graph information, which includes features [21], structure [9], or a combination of both [32]. Meanwhile, several works focus on attaining an improved latent space to generate new representations. Since the simplistic prior assumption leads to a suboptimal latent space, SIG-VAE [3] seeks to enhance VGAE by introducing Semi-Implicit Variational Inference. ARGA/ARVGA [17] integrates an adversarial training module into the GAE/VGAE framework, aligning the latent space with the prior distribution seen in actual data.

However, early GAEs without masking performed unsatisfactorily in the classification task [6]. Recently, the outstanding performance of GraphMAE

[6], on par with contrastive learning methods, has spotlighted the "mask-then-reconstruct" scheme in the graph domain. Subsequent GraphMAE series have pivoted towards graph masking modeling and the reconstruction of the graph. GraphMAE [6], GraphMAE2 [5], MaskGAE [11], HGMAE [23], and Bandana [31] mask and reconstruct various components: node features, edges/paths, and portions of information through each edge, respectively. SimSGT [15] suggests that the reconstruction of subgraph-level information improves the efficacy of molecular representation learning. Existing research concentrates on the similarity between reconstructed node features and their original counterparts while ignoring the differences between paired nodes after reconstruction. However, these differences are also crucial for the integrity of the original node.

2.2 Smoothness of Graph Embeddings

GNN and its smoothing operation makes the features of nearby vertices similar, promoting node representations perform well on the classification task. However, stacking more layers can lead to undistinguished output features and performance degradation. This phenomenon is called oversmoothing, initially analyzed by [13]. A series of works have focused on addressing this critical issue through methods such as edge dropping randomly [19], normalization [2], original features or shallow layer information enhancement [10,30]. This seems like a straightforward solution—introducing the aforementioned GNNs into a masked feature framework to address neighbor similarity issues. However, this conflicts with the reconstruction of masked nodes. For example, masked nodes cannot benefit from shallow-layer information enhancement, and randomly dropping edges may leave them with insufficient contextual information. We used a spectral GNN [1] combining high-pass and low-pass filters as the encoder, and we found that it indeed resulted in a decrease in classification performance.

3 Methodology

In this section, we formally propose ClearGAE. By incorporating dual-stream strategies in both the encoding and decoding phases, ClearGAE significantly enhances node discriminability while maintaining topological fidelity. Figure 3 presents an overall depiction of ClearGAE.

3.1 Background

Notations. Given a graph $\mathcal{G} = (\mathcal{V}, \mathcal{E}, \mathcal{X})$, where $\mathcal{V}$ is the set of N nodes and $\mathcal{E} \subseteq \mathcal{V} \times \mathcal{V}$ is the set of edges, and $\mathcal{X} \in \mathbb{R}^{N \times F_s}$ is the input node feature matrix, where F_s denotes the feature dimension. Each node $v \in \mathcal{V}$ is associated with a feature vector $x_v \in \mathcal{X}$, and each edge $e_{u,v} \in \mathcal{E}$ denotes a connection between node u and node v. The graph structure can also be represented by an adjacency matrix $\mathcal{A} \in \{0,1\}^{N \times N}$, where $A_{u,v} = 1$ if $e_{u,v} \in \mathcal{E}$, and $A_{u,v} = 0$ otherwise.

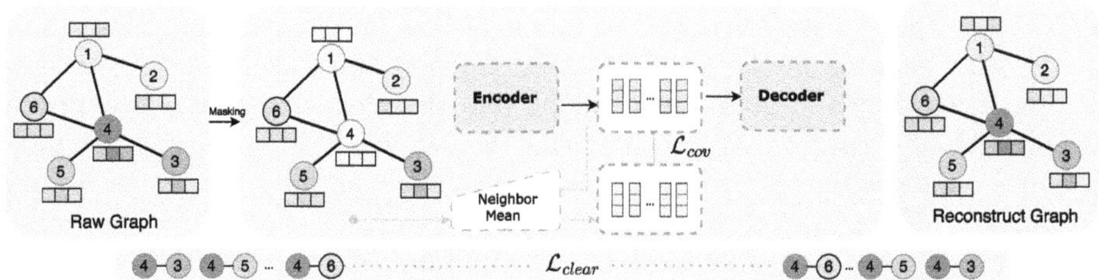

Fig. 3. Illustration of the ClearGAE.

Masked GAEs. The masked GAEs randomly mask a proportion of input graph characteristics, e.g., node features, resulting in a corrupted graph $\tilde{\mathcal{G}} = \{\tilde{\mathcal{X}}, \mathcal{A}\}$. We use $\mathcal{M}$ to denote the masked node patches. Then, we use a single-layer GNN as the encoder f_E to map $\tilde{\mathcal{G}}$ into a latent representation, yielding $\mathcal{H} \in \mathbb{R}^{N \times d_h}$. Another decoder f_D is employed to recover the features from $\mathcal{H}$, producing the reconstructed graph features $\hat{\mathcal{X}}$. This process can be formally expressed as:

$$\tilde{\mathcal{X}} = \mathcal{X} \odot (1 - \mathcal{M}), \quad \mathcal{H} = f_E(\tilde{\mathcal{X}}, \mathcal{A}), \quad \hat{\mathcal{X}} = f_D(\mathcal{H}, \mathcal{A}). \tag{1}$$

Conventional GAEs typically employ MSE or SCE [5,6] as the reconstruction criterion, which is applied to the masked features, and is formulated as:

$$\mathcal{L}_{rec} = \text{Rec}(\mathcal{X}_m, \hat{\mathcal{X}}_m), \tag{2}$$

where $\mathcal{X}_m$ denotes the masked node features in the original graph and $\hat{\mathcal{X}}_m$ denotes the corresponding features in the reconstructed graph.

Problem Description. To illustrate this, we conducted an analytical experiment on the Cora dataset to study the evolution of pairwise node similarity for the reconstructed nodes during the training process. Specifically, for each node v and its set of neighbor nodes $\mathcal{N}_v$, we measure its average similarity with its corresponding neighbors:

$$s_v = \frac{1}{|\mathcal{N}_v|} \sum_{i \in \mathcal{N}_v} \text{sim}(v, i). \tag{3}$$

Here, sim denotes the cosine similarity calculation. The results are presented in Fig. 2. It can be observed that the similarity score quickly rises to a high level at the beginning of training and remains relatively stable throughout the entire training phase. In contrast, the pairwise node similarity in the raw graph is low.

Problem Analysis. We provide a microscopic analysis of how features of a masked node are recovered, thereby demonstrating that minimizing the loss function corresponds to recovering the common inherent information shared between the neighboring nodes and the target node itself in the raw graph.

When the node v is masked, its input feature can be set to $x_v = \mathbf{0}$, and its neighbors can be represented by $\mathcal{N}(v)$. First, we demonstrated the encoding process. The output of a single layer can be represented as $h_v^{(1)} = \sigma\left(W^{(1)} \cdot \mathrm{AGG}\left(x_u \mid u \in \mathcal{N}(v)\right)\right)$, and after K-layer encoder, the latent embeddings are formulated as $h_v^{(K)} = f_\theta\left(x_u \mid u \in \mathcal{N}(v)\right)$. The decoder input includes $h_v^{(K)}$ and the neighbor representations $h_u^{(K)}{}_{u \in \mathcal{N}(v)}$. $h_v^{(K)}$ is entirely determined by its neighbors' information, and the reconstrued node can be represented:

$$\hat{x}_v = g_\phi\left(\mathrm{AGG}\left(f_\theta(x_{\mathcal{N}(v)}), h_u^{(K)}{}_{u \in \mathcal{N}(v)}\right)\right), \tag{4}$$

The loss function of GraphMAE is:

$$\mathcal{L}_{\mathrm{SCE}} = \frac{1}{C}\left(1 - \frac{x_v \cdot \hat{x}_v}{|x_v||\hat{x}_v|}\right)^\gamma, \tag{5}$$

Minimizing the loss is equivalent to:

$$\min_{\theta,\phi}\left|\frac{x_v}{|x_v|} - \frac{\hat{x}_v}{|\hat{x}_v|}\right|^2, \tag{6}$$

$\hat{x}_v$ can be represented as $\hat{x}_v = \Phi\left(x_{u u \in \mathcal{N}(v)}\right) + \epsilon$, where ϵ denotes the higher-order propagation error. The requirement of minimizing reconstruction error expects neighboring nodes of the original node to provide more information that is semantically similar to the original node. The ideal scenario is formulated as:

$$\Phi\left(x_{u u \in \mathcal{N}(v)}\right) \approx x_v. \tag{7}$$

Indeed, only in extremely rare cases can a node's features be fully captured by similar information alone. In the raw graph, nodes are typically connected to both similar and dissimilar neighbors.

3.2   Pairwise Nodes Discrepancy Distillation

As analyzed earlier, the feature recovery of a masked node originates from the shared information between its neighboring nodes and the original node itself. With an increasing number of training epochs, the reconstructed node pairs become progressively more similar. However, the original node's information consists not only of shared components, but also unique attributes that distinguish it from its neighboring nodes. This phenomenon is detrimental to effective representation learning, as excessive similarity between node pairs may eventually cause the entire node set to collapse into nearly identical embeddings. To address this issue, we propose an intuitive strategy applied during the decoding stage to preserve the necessary distinctions among node pairs. The design concept is that the recovery of a node's features should originate not only from the shared information with its neighboring nodes, but also from its unique attributes— that is, the features that differ from its neighbors. This can be interpreted as a form of knowledge distillation, transferring diversity from the raw graph to the reconstructed graph.

The proposed strategy consists of three key steps. **First**, for each node v, we compute the feature-wise distance between v and each of its first-order neighbors $\mathcal{N}_v$, obtaining a set of normalized dissimilarity scores:

$$\mathcal{D}_v = \{\mathrm{Dif}(v, i) \mid i \in \mathcal{N}_v\}, \quad \mathrm{Dif}(v, i) = \sigma\left(\frac{|\mathcal{X}_v - \mathcal{X}_i|}{\tau}\right), \tag{8}$$

where σ denotes the softmax function and τ is a temperature parameter that controls the sharpness of the distribution. **Second**, for each reconstructed node v', we repeat the process to compute the dissimilarity scores with its neighbors, yielding another score set: $\mathcal{D}'_v = \left\{\mathrm{Dif}(v', i), i \in \mathcal{N}_v\right\}$. **Third**, to enforce the preservation of local distinctiveness, we treat the dissimilarity scores from the raw graph as the teacher and those from the reconstructed graph as the student. We then introduce a consistency loss between the two distributions to guide the decoding process:

$$\mathcal{L}_{\mathrm{clear}} = \frac{1}{|\mathcal{E}|} \sum_{v \in \mathcal{V}} \rho(\mathcal{D}_v, \mathcal{D}'_v), \tag{9}$$

where $\rho(\cdot)$ denotes a distance metric (e.g., KL divergence) used to quantify the discrepancy between the teacher and student dissimilarity distributions.

We view this strategy as a form of knowledge distillation, transferring information from the raw input graph to the reconstructed (decoded) graph. Within this teacher-student framework, the raw graph acts as the teacher, preserving unique information that is typically attenuated during reconstruction, while the reconstructed graph serves as the student. By enforcing a constraint on the structural patterns shared between the teacher and student graphs, the decoder is explicitly guided to retain critical discriminative information. This regularization helps produce more informative and discriminative node representations.

3.3   Dimensional Decorrelation Regularization

Despite preserving unique information during the decoding phase, the latent embeddings remain difficult to distinguish when nodes in the raw graph exhibit excessive similarity with their neighboring nodes. Prior work [5] has identified a key limitation of GraphMAE: its reconstruction quality heavily relies on the discriminability of input node features. To further improve embedding quality, we adopt a decorrelation regularization strategy that imposes dimension-wise constraints on the embeddings produced by the encoder. The core rationale behind this regularization design is to enforce that the node embeddings are decorrelated from their neighboring nodes across feature dimensions (i.e., reducing linear dependencies between a node and its neighbors in the feature space), while simultaneously maintaining low correlation among embeddings of distinct nodes, thereby enhancing the model's representation capability for graphs with structurally similar nodes.

Let $\mathcal{H} = [h_1^\top; \ldots; h_N^\top] \in \mathbb{R}^{N \times d}$ denote the latent embeddings, and $\tilde{\mathcal{H}}_{\mathrm{avg}} = [\tilde{h}_1^\top; \ldots; \tilde{h}_N^\top] \in \mathbb{R}^{N \times d}$ denote the mean representations of their respective neighborhoods. The neighbor representations can be computed by:

$$\tilde{\mathcal{H}}_{\text{avg}} = (\mathcal{A} - \mathcal{I})\mathcal{H} \cdot \text{diag}(\frac{1}{(\mathcal{A} - \mathcal{I})\mathbf{1}}) \tag{10}$$

We begin by centering both representations by subtracting their empirical means:

$$\bar{h} = \frac{1}{N}\sum_{i=1}^{N} h_i, \quad \bar{\tilde{h}} = \frac{1}{N}\sum_{i=1}^{N} \tilde{h}_i, \tag{11}$$

$$\mathcal{H}_c = \mathcal{H} - \bar{h}, \quad \tilde{\mathcal{H}}_c = \tilde{\mathcal{H}}_{\text{avg}} - \bar{\tilde{h}}. \tag{12}$$

We then compute the sample covariance matrices of the centered representations:

$$\mathbf{C}_x = \frac{1}{N-1}\mathcal{H}_c^\top \mathcal{H}_c, \quad \mathbf{C}_{xy} = \frac{1}{N-1}\mathcal{H}_c^\top \tilde{\mathcal{H}}_c. \tag{13}$$

Here, $\mathbf{C}_x \in \mathbb{R}^{d \times d}$ captures intra-feature correlations within node representations, while $\mathbf{C}_{xy} \in \mathbb{R}^{d \times d}$ quantifies the cross-covariance between node and neighborhood representations.

To reduce feature redundancy and encourage decorrelated embeddings, we penalize the off-diagonal entries of $\mathbf{C}_x$. Simultaneously, we minimize the squared values of $\mathbf{C}_{xy}$ to reduce cross-feature dependency between node and neighborhood embeddings. The overall covariance regularization loss is defined as:

$$\mathcal{L}_{cov} = \underbrace{\frac{1}{d}\sum_{i \neq j}\left(\mathbf{C}_x^{(i,j)}\right)^2}_{\mathcal{L}_x} + \underbrace{\frac{1}{d}\sum_{i = j}\left(\mathbf{C}_{xy}^{(i,j)}\right)^2}_{\mathcal{L}_{xy}}. \tag{14}$$

Table 1. Summary of the results on node classification.

Dataset	Cora	Citeseer	Pubmed	Photo	Computer	CS	Physics	Ogbn-Arxiv
DGI	82.30 ± 0.60	71.80 ± 0.70	76.80 ± 0.60	91.61 ± 0.22	83.95 ± 0.47	92.15 ± 0.63	94.51 ± 0.52	-
MVGRL	83.50 ± 0.40	73.30 ± 0.50	80.10 ± 0.70	91.74 ± 0.07	87.52 ± 0.11	92.11 ± 0.12	95.33 ± 0.03	70.34 ± 0.16
CCA-SSG	84.00 ± 0.40	73.10 ± 0.30	81.00 ± 0.40	93.14 ± 0.14	88.74 ± 0.28	**93.31 ± 0.22**	95.38 ± 0.06	68.57 ± 0.02
VGAE	71.50 ± 0.40	65.80 ± 0.40	72.10 ± 0.50	92.20 ± 0.11	86.37 ± 0.21	92.11 ± 0.09	75.35 ± 0.14	69.94 ± 0.30
GraphMAE	84.20 ± 0.40	73.40 ± 0.40	81.10 ± 0.40	92.98 ± 0.35	88.34 ± 0.27	93.08 ± 0.17	95.30 ± 0.12	71.75 ± 0.17
GraphMAE2	84.50 ± 0.60	73.40 ± 0.30	81.40 ± 0.50	93.03 ± 0.21	88.50 ± 0.71	92.05 ± 0.04	-	71.89 ± 0.03
MaskGAE	84.30 ± 0.39	73.80 ± 0.81	**83.58 ± 0.45**	93.31 ± 0.13	89.54 ± 0.06	-	-	71.16 ± 0.33
Bandana	84.62 ± 0.37	73.60 ± 0.16	83.53 ± 0.51	93.44 ± 0.11	89.62 ± 0.09	93.10 ± 0.05	95.57 ± 0.04	71.09 ± 0.24
ClearGAE	**85.00 ± 0.30**	74.25 ± 0.25	81.70 ± 0.10	**94.27 ± 0.10**	**90.36 ± 0.00**	93.29 ± 0.01	**95.71 ± 0.06**	**72.11 ± 0.29**

"-" indicates that the corresponding performances are not reported. The best results are marked bold, while the second-best results are underlined.

Table 2. Experimental results for link prediction.

Metrics	Method	Cora	Citeseer	Pubmed	Photo	CS	Physics
AP	DGI	93.60±1.14	96.18±0.68	95.65±0.26	81.01±0.47	92.79±0.31	92.10±0.29
	MVGRL	92.95±0.82	89.37±4.55	95.53±0.30	63.43±2.02	89.14±0.93	-
	GRACE	82.36±0.24	86.92±1.11	93.26±1.20	81.18±0.37	83.90±2.20	82.20±1.06
	GCA	80.87±4.11	81.93±1.76	93.31±0.75	65.17±10.11	83.24±1.16	82.80±4.46
	CCA-SSG	93.74±1.15	95.06±0.91	95.97±0.23	67.99±1.60	96.40±0.30	96.26±0.10
	CAN	94.49±0.60	95.49±0.61	-	96.68±0.30	-	-
	SIG-VAE	94.79±0.71	94.21±0.53	85.02±0.49	94.53±0.93	94.93±0.37	**98.85±0.12**
	GraphMAE	89.52±0.01	74.50±0.04	87.92±0.01	77.18±0.02	83.58±0.01	86.44±0.03
	Bandana	<u>95.25±0.16</u>	<u>97.16±0.17</u>	<u>96.74±0.38</u>	<u>96.79±0.15</u>	<u>97.09±0.15</u>	<u>96.67±0.05</u>
	ClearGAE	**99.34±0.09**	**99.87±0.01**	**98.60±0.00**	**97.05±0.02**	**98.08±0.34**	<u>98.10±0.12</u>
AUC	DGI	93.88±1.00	95.98±0.72	96.30±0.20	80.95±0.39	93.81±0.20	93.51±0.22
	MVGRL	93.33±0.68	88.66±5.27	95.89±0.22	69.58±2.04	91.45±0.67	-
	GRACE	82.67±0.27	87.74±0.96	94.09±0.92	81.72±0.31	85.26±2.07	83.48±0.96
	GCA	81.46±4.86	84.81±1.25	94.20±0.59	70.02±9.66	84.35±1.13	85.24±5.41
	CCA-SSG	93.88±0.95	94.69±0.95	96.63±0.15	73.98±1.31	96.80±0.16	96.74±0.05
	CAN	93.67±0.62	94.56±0.68	-	97.00±0.28	-	-
	SIG-VAE	94.10±0.68	92.88±0.74	85.89±0.54	94.98±0.86	95.26±0.36	**98.76±0.23**
	GraphMAE	90.70±0.01	70.55±0.05	69.12±0.01	77.42±0.02	91.47±0.01	87.61±0.02
	S2GAE	93.52±0.23	93.29±0.49	98.45±0.03	-	-	-
	Bandana	<u>95.71±0.12</u>	<u>96.89±0.21</u>	<u>97.26±0.16</u>	<u>97.24±0.11</u>	<u>97.42±0.08</u>	<u>97.02±0.04</u>
	ClearGAE	**99.39±0.09**	**99.87±0.00**	**98.85±0.00**	**97.54±0.02**	**98.29±0.31**	<u>98.22±0.09</u>

"-" indicates that unavailable code or out-of-memory. The best results are marked bold, while the second-best results are underlined.

Table 3. Experimental results for graph classification.

	Dataset	IMDB-B	IMDB-M	PROTEINS	COLLAB	MUTAG	NCI
Contrastive	GraphCL	71.14±0.44	48.58±0.67	74.39±0.45	71.36±1.15	86.80±1.34	77.87±0.41
	JOAO	70.21±3.08	49.20±0.77	74.55±0.41	69.50±0.36	87.35±1.02	78.07±0.47
	GCC	72.0	49.4	-	78.9	-	-
	MVGRL	74.20±0.70	51.20±0.50	-	-	89.70±1.10	-
	InfoGCL	75.10±0.90	51.40±0.80	-	80.00±1.30	91.20±1.30	80.20±0.60
Generative	GraphMAE	75.52±0.66	51.63±0.52	75.30±0.39	80.32±0.46	88.19±1.26	80.40±0.30
	S2GAE	75.76±0.62	51.79±0.36	<u>76.37±0.43</u>	<u>81.02±0.53</u>	88.26±0.76	80.80±0.24
	ConMAE	<u>75.78±0.23</u>	**52.49±0.45**	-	81.32±0.32	**91.28±0.55**	<u>81.42±0.30</u>
	ClearGAE	**76.73±0.34**	<u>52.11±0.37</u>	**77.15±0.11**	**82.43±0.12**	<u>91.21±0.28</u>	**83.77±0.69**

"-" indicates that unavailable code or out-of-memory. The best results are marked bold, while the second-best results are underlined.

Table 4. The Ablation studies of the proposed loss functions.

Dataset	Cora (NC)	Photo (NC)	Pubmed (LP)	Photo (LP)	PROTEINS (GC)	MUTAG (GC)
ClearGAE	85.00	94.27	98.85	97.54	77.15	91.21
w/o $\mathcal{L}_{clear}$	84.85	93.28	98.84	96.17	76.22	85.85
w/o $\mathcal{L}_{cov}$	84.95	93.19	98.07	94.94	75.89	89.12

3.4   Training Objective

ClearGAE adopts two complementary training strategies: one that enforces distinctiveness among nodes during the decoding phase, and another that regularizes node representations during the encoding phase. These strategies jointly address fundamental limitations of existing GAE approaches. The overall training objective combines the node reconstruction loss $\mathcal{L}_{rec}$, a distillation-based loss $\mathcal{L}_{clear}$ weighted by a coefficient α, and a covariance regularization loss $\mathcal{L}_{cov}$ weighted by a balancing coefficient λ, defined as:

$$\mathcal{L} = \mathcal{L}_{rec} + \alpha\mathcal{L}_{clear} + \lambda\mathcal{L}_{cov} \tag{15}$$

4   Experiment

In this section, we first introduce the fundamental experimental settings. We then evaluate the performance of our model using fourteen publicly available datasets and three downstream tasks. Next, we report and analyze the experimental results of ClearGAE on the different tasks. Following this, we conduct ablation studies and validate the plug-and-play capability.

4.1   Experimental Settings

Datasets. The experiments for node-level tasks were conducted on eight publicly accessible datasets spanning multiple domains and scales: citation networks (Cora, Citeseer, PubMed and Ogbn-Arxiv), co-authorship graphs (Coauthor CS and Coauthor Physics), and co-purchase graphs (Amazon Photo and Amazon Computer). For link prediction, six datasets (Cora, Citeseer, PubMed, Amazon-Photo, Coauthor-CS, and Coauthor-Physics) were used. For graph classification tasks, experiments were conducted on six public datasets: IMDB-B, IMDB-M, PROTEINS, COLLAB, MUTAG, and NCI1.

Baselines. For the node classification comparison, we selected eight state-of-the-art (SOTA) methods within the SSL framework. These include three contrastive methods: DGI [24], MVGRL [4], and CCA-SSG [29], and five generative methods: VGAE [9], GraphMAE [6], GraphMAE2 [5], MaskGAE [12], Bandana [31]. For the link prediction comparison, we selected nine SOTA methods within the SSL framework. These include five contrastive methods: DGI, MVGRL, GRACE [33], GCA [28], and CCA-SSG [29], as well as four generative methods: CAN [16], SIG-VAE [3], GraphMAE, S2GAE [22], Bandana [31]. For the graph classification comparison, we selected seven SOTA methods within the SSL framework. These include five contrastive methods: GraphCL [28], JOAO [27], GCC [18], MVGRL, and InfoGCL [26], as well as three generative methods: GraphMAE, S2GAE [22] and ConMAE [25].

Evaluation. We selected three widely-used and complementary metrics for different tasks: accuracy (Acc), area under the characteristic curve (AUC) and the average precision (AP). To optimize the training process, we applied the Adamw optimizer with a learning rate ranging from 0.0001 to 0.01. The results are reported as the mean $\pm$ standard deviation over ten experimental trials.

4.2 Results of Node Classification

We use Acc as the metric to evaluate node classification. The node classification results are displayed in Table 1. From these results, we observe that ClearGAE outperforms the selected methods in most cases. Compared with the feature reconstruction model, our proposed strategy demonstrates superior performance across all benchmark datasets in node classification tasks, surpassing not only the original GraphMAE but also its enhanced variant, GraphMAE2. Furthermore, these results demonstrate that preserving the different information between nodes during the reconstruction process helps effectively reconstruct the original nodes features, thereby improving node classification accuracy. Recent baselines like Bandana [31] achieve the second-best performance in most datasets. However, while Bandana exhibits suboptimal performance on large-scale graphs such as ogbn-arxiv, our framework demonstrates robust scalability, consistently outperforming existing methods across datasets of varying scales—from small (Cora: 85%) to large (ogbn-arxiv: 72.11%).

4.3 Results of Link Prediction

We use AP and AUC as metrics to evaluate link prediction. The results are reported in Table 2. In the experimental settings, we have outlined the evaluation methodology for link prediction. Although our primary focus lies on effective node feature reconstruction, the performance of ClearGAE in link prediction remains competitive. Our approach is built upon GraphMAE, and its effectiveness can be directly validated through a comparative analysis with GraphMAE. ClearGAE consistently outperforms GraphMAE across all baselines, indicating that preserving node-level differences during embedding generation significantly enhances link prediction performance. Contrary to prevailing assumptions in generative models—where feature reconstruction has often been regarded as suboptimal for link prediction tasks (as seen in structure reconstruction frameworks such as MaskGAE and Bandana)—our method demonstrates that feature reconstruction-based approaches can systematically outperform edge reconstruction models (e.g., Bandana). We conclude that the discriminative capacity of pairwise node representations in mitigating oversmoothing benefits graph property exploration and achieves SOTA performance in link prediction.

4.4 Results of Graph Classification

We use the Acc as the metric to evaluate graph classification performance. The graph classification results are detailed in Table 3. Compared to generative approaches, our proposed method achieves superior performance across all evaluated datasets, confirming its efficacy in optimizing reconstruction quality. The data in Table 3 clearly demonstrates the strong performance of our methodology on bioinformatics datasets, significantly outperforming GraphMAE. The accuracy of PROTEINS, MUTAG, and NCI datasets improved by 1.85%, 3.02%, and

3.37%, respectively. The superior performance of our model on graph classification tasks further validates that generating more discriminative representations enables task-agnostic generalization.

4.5   Analysis and Discussion

Ablation Study. We conducted systematic ablation studies on the core components of ClearGAE to elucidate their operational mechanisms and contributions to task-specific performance. For each of the three fundamental graph learning tasks – node classification (NC), link prediction (LP), and graph classification (GC) – we implemented a controlled experimental protocol. Specifically, we randomly selected two benchmark datasets per task to evaluate both the individual and combined contributions of two critical design strategies. As summarized in Table 4, "w/o $\mathcal{L}_{clear}$" denotes ablation of the pairwise nodes discrepancy distillation (PDD) and "w/o $\mathcal{L}_{cov}$" indicates removal of the dimensional decorrelation regularization(DDR). Both strategies improve performance across all tasks.

The ablation of the pairwise node dissimilarity strategy (w/o $\mathcal{L}_{clear}$) results in an average performance gain of 1.31% across six benchmark datasets. These empirical findings support our hypothesis that preserving local distinctions significantly enhances the quality of learned node embeddings. For link prediction tasks, improving the discriminative capacity between neighboring node embedding leads to more substantial performance gains, particularly evident in metrics like A-Photo (AUC +2.6%).

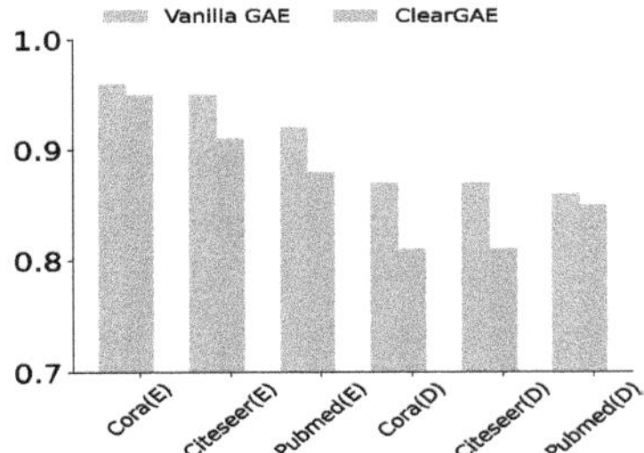

Fig. 4. Similarity changes.

Dataset	Cora	Citeseer	CS
GraphMAE2	84.50	72.93	92.05
GraphMAE2+PDD+DDR	84.87	73.07	92.63
Bandana	82.26	69.35	92.84
Bandana+DDR	83.40	71.31	92.93

Fig. 5. Experimental results for our strategy integrated with other SSL methods.

Results of Pairwise Similarity. To validate the distinctiveness of the node representations, we independently calculated the average neighbor feature similarity for nodes under two scenarios: with and without our strategy applied. For generality, we conducted similarity comparisons across both encoding and decoding phases. We obtained the pre-trained latent representations (encoder outputs) and reconstructed graphs (decoder outputs) generated by GraphMAE and ClearMAE under their best node classification performance. We then calculated the corresponding neighboring node similarity. As shown in Fig. 4, the pairwise similarity of embeddings and reconstructed graph learned by ClearGAE are both lower than that of the vanilla GAE, indicating that our strategy reconstructs nodes with enhanced distinctiveness and generates more discriminative embeddings across all datasets. While parameters tuning can preserve greater dissimilarity in decoder outputs, such adjustments often degrade downstream performance. This validates that feature smoothness plays a critical role: optimal classification performance requires balancing feature smoothness—neither excessively amplifying nor suppressing it. A well-designed embedding should retain sufficient similarity to neighbors while preserving distinctiveness.

Plug-and-Play. To comprehensively evaluate the effectiveness of our method, we extended our experiments beyond GraphMAE to include feature-based GraphMAE2 and edge-based Bandana [31]. We evaluated the node classification performance by applying the corresponding strategies to GraphMAE2 and Bandana, with the experimental results presented in the Fig. 5. The experimental results demonstrate that our method enhances both feature reconstruction and edge reconstruction models, where the improved discrimination of neighboring node embeddings contributes to higher node classification accuracy.

It is worth noting that the results of Bandana and GraphMAE2 in Table 1 differ from those in Fig. 5. Specifically, Table 1 reports the officially published best results for Bandana, whereas in Fig. 5, we reproduce the results using the official implementation and recommended hyperparameters. Subsequently, we sequentially apply each proposed strategy to the two baselines without any additional hyperparameter tuning and report the corresponding performance.

5   Conclusion

In this work, we focus on enhancing reconstruction capability while mitigating excessive similarity among neighboring node embeddings in masked graph autoencoders. Our analysis indicates that GNN-encoder-decoder and relying solely on MSE or SCE as reconstruction criteria may result in the loss of essential distinctiveness between nodes. To address this issue, our approach is centered on adding regularization between nodes and their neighboring nodes. The design and implementation of this idea involve introducing two strategies: incorporating a KL divergence constraint on the pairwise node dissimilarity between the raw graph and the reconstructed graph, and introducing a discriminative loss to decorrelate node representations across the embedding dimensions. We conduct extensive experiments to validate the effectiveness and generality of this strategy. These results demonstrate that our method effectively enhances the quality and discriminative power of graph representations.

Acknowledgment. This work was supported by the National Social Science Fund of China (Major Program, Grant No. 25&ZD161), the National Natural Science Foundation of China (Grant No. 72210107001), and the CAS Pilot Project for Basic and Interdisciplinary Scientific Research (Grant No. XDB1380303).

References

1. Duan, R., et al.: Unifying homophily and heterophily for spectral graph neural networks via triple filter ensembles. In: Proceedings of the 38th International Conference on Neural Information Processing Systems, NIPS 2024. Curran Associates Inc., Red Hook (2025)
2. Guo, X., Wang, Y., Du, T., Wang, Y.: ContraNorm: a contrastive learning perspective on oversmoothing and beyond. In: The Eleventh International Conference on Learning Representations
3. Hasanzadeh, A., Hajiramezanali, E., Narayanan, K., Duffield, N., Zhou, M., Qian, X.: Semi-implicit graph variational auto-encoders. Adv. Neural Inf. Process. Syst. **32** (2019)
4. Hassani, K., Khasahmadi, A.H.: Contrastive multi-view representation learning on graphs (2020)
5. Hou, Z., et al.: GraphMAE2: a decoding-enhanced masked self-supervised graph learner (2023)
6. Hou, Z., et al.: GraphMAE: self-supervised masked graph autoencoders. In: Proceedings of the 28th ACM SIGKDD Conference on Knowledge Discovery and Data Mining (2022)
7. Hu, Y., et al.: Exploring task unification in graph representation learning via generative approach. arXiv preprint arXiv:2403.14340 (2024)
8. Hu, Y., Ouyang, S., Yang, Z., Liu, Y.: VIGraph: self-supervised learning for class-imbalanced node classification. arXiv preprint arXiv:2311.01191 (2023)
9. Kipf, T.N., Welling, M.: Variational graph auto-encoders. arXiv preprint arXiv:1611.07308 (2016)
10. Li, G., Müller, M., Thabet, A.K, Ghanem, B.: DeepGCNs: can GCNs go as deep as CNNs? In: 2019 IEEE/CVF International Conference on Computer Vision (ICCV), pp. 9266–9275 (2019). https://api.semanticscholar.org/CorpusID:201070021
11. Li, J., et al.: What's behind the mask: understanding masked graph modeling for graph autoencoders (2023)
12. Li, J., et al.: What's behind the mask: understanding masked graph modeling for graph autoencoders. In: KDD 2023, pp. 1268–1279. Association for Computing Machinery, New York (2023). https://doi.org/10.1145/3580305.3599546
13. Li, Q., Han, Z., Wu, X.M.: Deeper insights into graph convolutional networks for semi-supervised learning. In: AAAI 2018/IAAI 2018/EAAI 2018. AAAI Press (2018)
14. Liu, Y., et al.: Graph self-supervised learning: a survey. IEEE Trans. Knowl. Data Eng., 1 (2022). https://doi.org/10.1109/tkde.2022.3172903. http://dx.doi.org/10.1109/TKDE.2022.3172903
15. Liu, Z., et al.: Rethinking tokenizer and decoder in masked graph modeling for molecules (2024)
16. Meng, Z., Liang, S., Bao, H., Zhang, X.: Co-embedding attributed networks. In: Proceedings of the Twelfth ACM International Conference on Web Search and Data Mining, pp. 393–401 (2019)
17. Pan, S., Hu, R., Long, G., Jiang, J., Yao, L., Zhang, C.: Adversarially regularized graph autoencoder for graph embedding (2019)
18. Qiu, J., et al.: GCC: graph contrastive coding for graph neural network pretraining. In: Proceedings of the 26th ACM SIGKDD International Conference on Knowledge Discovery & Data Mining, pp. 1150–1160 (2020)

19. Rong, Y., Huang, W., Xu, T., Huang, J.: DropEdge: towards deep graph convolutional networks on node classification. arXiv preprint arXiv:1907.10903 (2019)
20. Rusch, T.K., Bronstein, M.M., Mishra, S.: A survey on oversmoothing in graph neural networks (2023)
21. Salehi, A., Davulcu, H.: Graph attention auto-encoders (2019)
22. Tan, Q., et al.: S2GAE: self-supervised graph autoencoders are generalizable learners with graph masking. In: WSDM 2023, pp. 787–795 (2023)
23. Tian, Y., Dong, K., Zhang, C., Zhang, C., Chawla, N.V.: Heterogeneous graph masked autoencoders (2023)
24. Veličković, P., Fedus, W., Hamilton, W.L., Liò, P., Bengio, Y., Hjelm, R.D.: Deep graph infomax. arXiv preprint arXiv:1809.10341 (2018)
25. Wang, Y., et al.: Generative and contrastive paradigms are complementary for graph self-supervised learning (2025). https://arxiv.org/abs/2025.11776
26. Xu, D., Cheng, W., Luo, D., Chen, H., Zhang, X.: InfoGCL: information-aware graph contrastive learning. Adv. Neural. Inf. Process. Syst. **34**, 30414–30425 (2021)
27. You, Y., Chen, T., Shen, Y., Wang, Z.: Graph contrastive learning automated. In: International Conference on Machine Learning, pp. 12121–12132. PMLR (2021)
28. You, Y., Chen, T., Sui, Y., Chen, T., Wang, Z., Shen, Y.: Graph contrastive learning with augmentations. Adv. Neural. Inf. Process. Syst. **33**, 5812–5823 (2020)
29. Zhang, H., Wu, Q., Yan, J., Wipf, D., Yu, P.: From canonical correlation analysis to self-supervised graph neural networks (2021)
30. Zhang, W., et al.: Model degradation hinders deep graph neural networks. In: KDD 2022. Association for Computing Machinery, New York (2022). https://doi.org/10.1145/3534678.3539374
31. Zhao, Z., Li, Y., Zou, Y., Tang, J., Li, R.: Masked graph autoencoder with non-discrete bandwidths (2024)
32. Zhou, S., et al.: DGE: deep generative network embedding based on commonality and individuality. In: Proceedings of the AAAI Conference on Artificial Intelligence, vol. 34, pp. 6949–6956 (2020). https://doi.org/10.1609/aaai.v34i04.6178
33. Zhu, Y., Xu, Y., Yu, F., Liu, Q., Wu, S., Wang, L.: Deep graph contrastive representation learning. arXiv preprint arXiv:2006.04131 (2020)

NK-GAD: Neighbor Knowledge-Enhanced Unsupervised Graph Anomaly Detection

Zehao Wang[1,2] and Lanjun Wang[2(✉)]

[1] College of Intelligence and Computing, Tianjin University, Tianjin, China
[2] School of New Media and Communication, Tianjin University, Tianjin, China
`wanglanjun@tju.edu.cn`

Abstract. Graph anomaly detection aims to identify irregular patterns in graph-structured data. Most unsupervised GNN-based methods rely on the homophily assumption that connected nodes share similar attributes. However, real-world graphs often exhibit attribute-level heterophily, where connected nodes have dissimilar attributes. Our analysis of attribute-level heterophily graphs reveals two phenomena indicating that current approaches are not practical for unsupervised graph anomaly detection: 1) attribute similarities between connected nodes show nearly identical distributions across different connected node pair types, and 2) anomalies cause consistent variation trends between the graph with and without anomalous edges in the low- and high-frequency components of the spectral energy distributions, while the mid-part exhibits more erratic variations. Based on these observations, we propose NK-GAD, a neighbor knowledge-enhanced unsupervised graph anomaly detection framework. NK-GAD integrates a joint encoder capturing both similar and dissimilar neighbor features, a neighbor reconstruction module modeling normal distributions, a center aggregation module refining node features, and dual decoders for reconstructing attributes and structures. Experiments on seven datasets show NK-GAD achieves an average 3.29% AUC improvement.

Keywords: Unsupervised Graph Anomaly Detection · Graph Mining · Unsupervised Node Classification

1 Introduction

Graph anomaly detection is crucial for identifying irregular or suspicious patterns in graph-structured data, with applications in fields like social networks [34] and financial transactions [28]. Existing works [17] typically categorize anomalies in graph-structured data into two types: 1) contextual anomalies, where node attributes deviate from the normal nodes, and 2) structural anomalies, where irregularities occur in the edges between nodes. However, due to the high cost of obtaining labeled graph data in real-world applications, recent advancements [10, 22] have increasingly focused on unsupervised detection methods.

H. Jung et al. (Eds.): DASFAA 2026, LNCS 16536, pp. 19–34, 2026.
https://doi.org/10.1007/978-981-92-0366-6_2

Recently, the unsupervised graph anomaly detection methods [21] are mainly built on graph neural networks (Gens) [32], which rely on a message-passing mechanism to aggregate information from neighboring nodes and capture the underlying patterns of normal behavior. Prior research has demonstrated that the performance of Graph Neural Networks (Gens) is largely governed by the homophily assumption [19], where connected nodes with the same labels or similar features. However, recent studies [10] reveal the existence of label-level heterophily, where connected nodes belong to different classes (e.g., anomalous vs. normal). This structural characteristic severely undermines the effectiveness of GNN-based models, which rely on neighborhood similarity for message propagation. Motivated by the observed right-shifting effect [8,26], where low-frequency energy gradually shifts toward the high-frequency spectrum as the anomaly degree increases, subsequent works [10,26] have sought to alleviate the negative influence of anomalous nodes on normal ones. These approaches typically modify graph structures or node representations to reduce harmful information transfer between the two node types.

However, the existing unsupervised methods [10,22] overlook the attribute-level heterophily [19,30] present in real-world graph-structured data. In detail, the attribute-level heterophily refers to the dissimilarity between connected nodes in the spatial domain, where the majority of the cosine similarity values fall within the low range of [0, 0.25] (Fig. 1). It is also manifested in the spectral domain of the graphs, where spectral energy is distributed across the mid-frequency components, such as over 90% of the spectral energy in the Weibo dataset is concentrated in the range with eigenvalues between [0.75, 1.25) (Fig. 2).

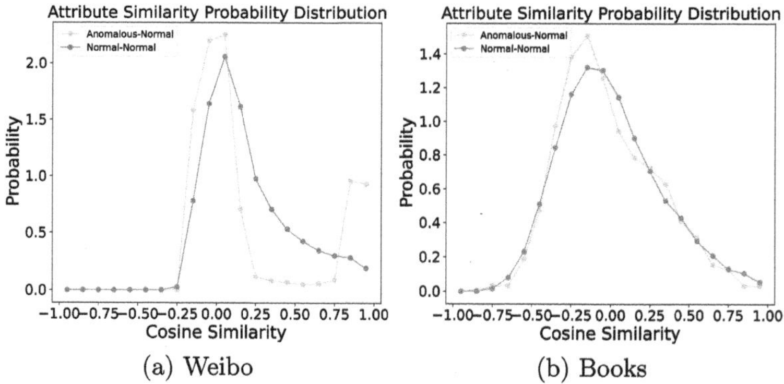

(a) Weibo   (b) Books

Fig. 1. Attribute similarity distributions of anomalous–normal and normal–normal node pairs in the Weibo and Books datasets mostly fall within the low cosine similarity range, indicating strong feature dissimilarity between connected nodes. Moreover, the two distributions largely overlap across pair types.

Furthermore, based on Figs. 1 and 2, we recognize two phenomena that indicate the existing methods are not practical for unsupervised graph anomaly

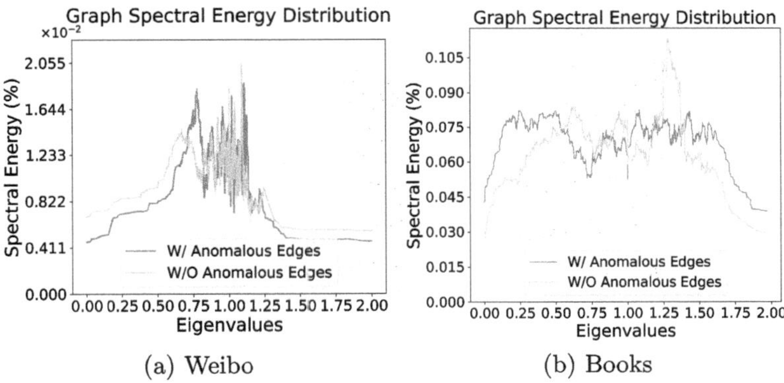

(a) Weibo   (b) Books

Fig. 2. Graph spectral energy distribution for the graphs w/ and w/o anomalous edges in the Weibo and Books datasets. Compared to the graphs w/o anomalous edges, which remove the edges between normal and anomalous nodes [10], the spectral energy distributions of the graphs w/ anomalous edges show consistent variation trends in low- and high-frequency components, while the mid-frequency components exhibit more erratic variations.

detection. Firstly, in the spatial domain (Fig. 1), the attribute similarities of the connected nodes across different connected node pair types (anomalous-normal and normal-normal) display nearly identical distributions. This observation indicates that it is not reasonable to judge whether the edges or attributes are anomalous solely based on the similarity of the connected nodes' attributes. However, existing unsupervised methods [10, 26] modify edges and attributes between connected nodes with highly dissimilar attributes only based on such similarity. Thus, these methods remove anomalous edges and attributes as well as normal ones in attribute-level heterophily graphs, resulting in the loss of information related to the normal patterns between neighbors.

Secondly, in the spectral domain (Fig. 2), anomalies in attribute-level heterophily graphs lead to consistent variation trends in the low- and high-frequency components of the spectral energy distributions [26], while the mid-frequency components exhibit more erratic variations. For instance, the Weibo datasets [17] show increases in spectral energy in the low- and high-frequency ranges after removing abnormal edges. This observation implies that the distinction between normal and anomalous patterns is clearer in the low- and high-frequency components than in the mid-frequency range. However, existing methods [10, 26] neglect the importance of the high-frequency component. For instance, Tang et al. [26] focus on processing the mid-frequency component to extract both the normal and anomalous patterns, while He et al. [10] aim to mitigate the impact of anomalous nodes by reducing high-frequency energy. Thus, they do not consider the dissimilar attributes of neighbors, and it is challenging to distinguish between normal and anomalous patterns in unsupervised settings.

As a result, in this paper, we focus on incorporating neighbor knowledge, e.g. neighbor features and the corresponding distributions, to improve the ability to detect graph anomalies in an unsupervised setting. We design a novel framework, named **N**eighbor **K**nowledge-enhanced unsupervised **G**raph **A**nomaly Detection (NK-GAD), which consists of a joint graph convolutional encoder, a neighbor reconstruction module, a center aggregation module, and two decoders. The joint graph convolutional encoder consists of a low-pass filter and a high-pass filter to parallel extract patterns of both the similar and dissimilar attributes of neighbors. To deal with two types of anomalies, we design two modules for structural and contextual anomalies, respectively. In detail, the neighbor reconstruction module enhances node representations by reconstructing neighboring feature distributions based on the hidden representation in the encoder, aiding in the detection and removal of structural anomalies. The center aggregation module refines node features by leveraging reconstructed neighbor distributions, filtering out contextual anomalies. Finally, two decoders reconstruct node attributes and the adjacency matrix to compute anomaly scores and total loss. The contributions are summarized as follows:

- We present two observations from real-world datasets, analyzed from both spatial and spectral perspectives, that highlight the limitations of existing methods in graph anomaly detection. Building on these insights, we propose a novel framework, NK-GAD, which leverages neighbor knowledge to detect the graph anomaly.
- We introduce a joint graph convolutional encoder that captures both similar and dissimilar neighbor features for extracting normal patterns across diverse attributes.
- To mitigate the impact of different kinds of anomalies on the normal nodes, NK-GAD incorporates a neighbor reconstruction module to extract normal patterns and a center aggregation module to refine the central node's features using reconstructed neighbor distributions.
- Extensive experiments on seven real-world datasets show the effectiveness of NK-GAD, with up to a 3.29% improvement in AUC over SOTA methods.

2   Related Work

2.1   Graph Neural Networks

Graph Neural Networks (Gens) have garnered significant attention for their ability to model graph-structured data effectively across various domains, such as traffic prediction [24] and semantic search [35]. Gens exploit the graph structure by aggregating information from neighboring nodes, enabling them to capture local patterns and dependencies efficiently. A key principle is the homophily assumption [19], where connected nodes share similar attributes or labels. Models such as GCNs [13] and GATs [29] leverage message passing to propagate information along edges, reinforcing node representations for tasks like node classification [33] and link prediction [14]. However, the homophily assumption limits the performance of Gens on attribute-level heterophily graphs [19,30], where connected nodes often have dissimilar attributes or belong to different labels. Recent studies [19,30] show that attribute-level heterophily is not inherently detrimental, as it can provide valuable insights for uncovering hidden patterns in real-world graph-structured data.

Inspired by the findings, we analyze real-world datasets in the graph anomaly detection task and observe two phenomena (Fig. 1 and Fig. 2) that motivate us to combine neighbor knowledge to improve the ability to detect anomalies.

2.2   Unsupervised Graph Anomaly Detection

Graph anomaly detection seeks to identify anomalies in graph-structured data [17], a crucial task with applications in fraud detection [18] and social network analysis [9]. Due to the rarity of anomalies and the high cost of obtaining labeled graph data in real-world applications, recent research has increasingly focused on unsupervised detection methods. Traditional algorithms [16,17] show good performance on certain datasets that own obvious differences between normal and anomalous nodes. Compared to these traditional algorithms, deep learning-based detection models often exhibit better generalization [17]. With the deepening research into graph-structured data, the emergence of Graph Neural Networks (Gens) [13] has revolutionized graph anomaly detection. Gens-based graph anomaly detection methods enable the learning of rich, task-specific representations of nodes and edges [10,22]. Works like ADA-GAD of He. et al. and AnomalyADE of Fan. et al. utilize GNN-based architectures to detect anomalies by reconstructing node attributes and adjacency matrices, identifying discrepancies between reconstructed and observed data. GAD-NR [22] and SmoothGNN [5] focus on leveraging neighbor feature information to detect anomalies.

However, existing studies on unsupervised graph anomaly detection overlook the attribute-level heterophily of real-world graph-structured datasets. Moreover, our observations show that these methods are not suitable for attribute-level heterophily graphs. Thus, we propose NK-GAD, a framework designed for unsupervised graph anomaly detection on attribute-level heterophily graphs.

3   Problem Formulation

A graph with node attributes is formally defined as $\mathcal{G} = (\mathcal{V}, \mathcal{E}, \mathbf{X})$, where $\mathcal{V} = \{v_1, v_2, \ldots, v_{|\mathcal{V}|}\}$ is the set of nodes, $\mathcal{E} = \{e_1, e_2, \ldots, e_{|\mathcal{E}|}\}$ is the set of edges, and $\mathbf{X} \in \mathbb{R}^{|\mathcal{V}| \times dim}$ is the node attribute matrix. Specifically, each edge $e \in \mathcal{E}$ is represented as a tuple (v_i, v_j), where $v_i, v_j \in \mathcal{V}$ are the nodes connected by the edge. The structural relationships captured in $\mathcal{E}$ can alternatively be expressed as a binary adjacency matrix $\mathbf{A} \in \mathbb{R}^{|\mathcal{V}| \times |\mathcal{V}|}$, where $\mathbf{A}_{i,j} = 1$ indicates the presence of an edge $(v_i, v_j) \in \mathcal{E}$, and $\mathbf{A}_{i,j} = 0$ otherwise. In the attribute matrix $\mathbf{X}$, the i-th column is denoted as $\mathbf{x}_i \in \mathbb{R}^{dim}$ which represents the attribute vector of the corresponding node v_i. The degree matrix is defined as $\mathbf{D} = \mathrm{diag}(d_1, \ldots, d_{|\mathcal{V}|})$, where $d_i = \sum_{j=1}^{|\mathcal{V}|} \mathbf{A}_{i,j}$ is the degree of node v_i. The symmetric normalized Laplacian matrix of a graph $\mathcal{G}$ is defined as $\mathbf{L} = \mathbf{D}^{-1/2}(\mathbf{D} - \mathbf{A})\mathbf{D}^{-1/2}$. Since $\mathbf{L}$ is positive semi-definite and symmetric, it can be decomposed as $\mathbf{L} = \mathbf{U}\Lambda\mathbf{U}^\top$, where $\Lambda = \{\lambda_1, \lambda_2, \ldots, \lambda_{|\mathcal{V}|}\}$ are the eigenvalues, and each column $\mathbf{u}_i$ is the unit eigenvector corresponding to eigenvalue λ_i. Additionally, the neighbors of node v_i are denoted by $\mathcal{Z}(i) = \{v_j \mid (v_i, v_j) \in \mathcal{E}\}$. Referring to [22,27], we regard the neighbor feature distribution of the node $v_i \in \mathcal{V}$ as a multi-variate Gaussian distribution $\mathbb{P}_{\mathcal{Z}(i)} \sim \mathcal{N}(\mu_i, \boldsymbol{\Sigma}_i)$, where $\mu_i \in \mathbb{R}^{dim}$ is the vector of mean of each dimension and $\boldsymbol{\Sigma}_i \in \mathbb{R}^{dim \times dim}$ is the covariance matrix.

In this paper, we adapt the unsupervised setting, wherein the model does not have access to node labels during training, following prior studies [10,22]. Given a graph $\mathcal{G} = (\mathcal{V}, \mathcal{E}, \mathbf{X})$, the goal of a reconstruction-based graph anomaly detection model f is to assign an anomaly score s_i to each node $v_i \in \mathcal{V}$ based on the input graph $\mathcal{G}$. Formally, this can be expressed as:

$$f : \mathcal{G} \rightarrow \{s_1, s_2, \ldots, s_i, \ldots, s_{|\mathcal{V}|}\}, \quad s_i \in \mathbb{R}_{\geq 0} \tag{1}$$

where the anomaly score s_i quantifies the degree of abnormality of the node v_i. A higher score indicates a higher likelihood of the node being anomalous, whereas a lower score suggests it is more likely to be normal.

4   Methodology

In this section, we introduce the proposed NK-GAD, and its overview is shown in Fig. 3. NK-GAD consists of a joint graph convolutional encoder, a neighbor reconstruction module, a center aggregation module, and two decoders. The joint graph convolutional encoder combines the low- and high-frequency components to extract patterns from similar and dissimilar attributes between central and neighboring nodes. Next, the neighbor reconstruction module utilizes the extracted patterns to reconstruct the feature distribution of neighbor attributes. Minimizing the loss of the neighbor reconstruction module improves the extraction of normal patterns. Then, the center aggregation module uses these patterns to refine central node attributes. Finally, two decoders reconstruct node attributes and edges, and calculate the anomaly scores.

4.1   Joint Graph Convolutional Encoder

Observations from the spatial domain indicate that distinguishing between normal and anomalous nodes requires further extraction of patterns contained within the features of attribute-level heterophily graphs. Meanwhile, observations from the spectral domain highlight the advantage of leveraging low- and high-frequency components to differentiate between normal and abnormal patterns in an unsupervised setting, where the low- and high-frequency components can be used to represent the similar and dissimilar attributes between neighbors, respectively.

Motivated by these insights, to extract patterns from the features of neighbors, including both the similar and dissimilar attributes, we introduce a joint

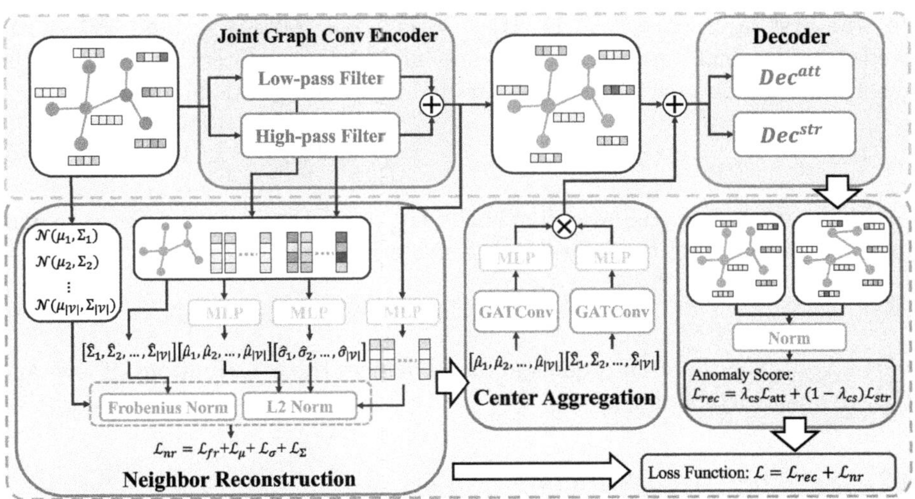

Fig. 3. Overview of NK-GAD. NK-GAD integrates four core components to reconstruct graph data, subsequently computing anomaly scores and total loss for detection. Red, blue, and gray nodes highlight anomalies, normal instances, and unseen labels, respectively. (Color figure online)

graph convolutional encoder that integrates a low-pass filter and a high-pass filter to process both frequency components in parallel.

We first project the node attributes into a hidden representation space using a linear layer, formulated as $\mathbf{H}^0 = \mathrm{Linear}(\mathbf{X})$, where $\mathbf{H}^0 \in \mathbb{R}^{|\mathcal{V}| \times d}$ represents the hidden representation, and d denotes the dimensionality of the hidden space.

The low-pass and high-pass filters in the joint graph convolutional encoder are based on the normalized Laplacian matrix $\mathbf{L}$, facilitating the separation of smooth and sharp transitions in the graph $\mathcal{G}$. Based on the Rayleigh quotient [25], the eigenvalues of $\mathbf{L}$ are bounded in the range $[0, 2]$. These eigenvalues are used to define the graph's frequency components: low-frequency components correspond to eigenvalues in $[0, 1)$, while high-frequency components correspond to eigenvalues in $(1, 2]$. Thus, the low- and high-pass filters are defined as follows:

$$f_{low} = \mathbf{I} + \mathbf{D}^{-\frac{1}{2}}\mathbf{A}\mathbf{D}^{-\frac{1}{2}} = 2\mathbf{I} - \mathbf{L}, \quad f_{high} = \mathbf{I} - \mathbf{D}^{-\frac{1}{2}}\mathbf{A}\mathbf{D}^{-\frac{1}{2}} = \mathbf{L} \tag{2}$$

To process the hidden representations of each node, we adopt a multi-layer architecture that leverages graph convolution operations [6,11]. The representations are separately computed for low- and high-frequency components at each layer as follows:

$$\mathbf{H}^l_{low} = f_{low} *_{\mathcal{G}} \mathbf{H}^{l-1}_{low}\mathbf{W}^l_{low}, \quad l = 1, 2, \ldots, L \tag{3}$$

$$\mathbf{H}^{l'}_{high} = f_{high} *_{\mathcal{G}} \mathbf{H}^{l'-1}_{high}\mathbf{W}^{l'}_{high}, \quad l' = 1, 2, \ldots, L' \tag{4}$$

where $*_{\mathcal{G}}$ denotes the graph convolution operator, $\mathbf{H}^l_{low}$ and $\mathbf{H}^{l'}_{high}$ are the representations of low- and high-frequency components at the l-th layer, respectively.

For the first layer ($l = l' = 1$), $\mathbf{H}^0_{low} = \mathbf{H}^0_{high} = \mathbf{H}^0$. $\mathbf{W}^l_{low}$ and $\mathbf{W}^{l'}_{high}$ are the learnable weight matrices for the low- and high-pass filters. Specifically, the graph convolution operations are computed in the spectral domain as:

$$f_{low} *_{\mathcal{G}} \mathbf{H}^l_{low} = \mathbf{U}\left(2\mathbf{I} - \Lambda\right)\mathbf{U}^T\mathbf{H}^{l-1}_{low}, \quad f_{high} *_{\mathcal{G}} \mathbf{H}^{l'}_{high} = \mathbf{U}\Lambda\mathbf{U}^T\mathbf{H}^{l'-1}_{high} \qquad (5)$$

At the final layer, the representations $\mathbf{H}^L_{\text{low}}$ and $\mathbf{H}^{L'}_{\text{high}}$ are combined to generate the output $\mathbf{Z}$ of the joint graph convolutional encoder as:

$$\mathbf{Z} = (1 - \lambda_{joint})\mathbf{H}^L_{\text{low}} + \lambda_{joint}\mathbf{H}^{L'}_{\text{high}} \qquad (6)$$

where λ_{joint} is a trade-off coefficient that balances the contributions of low- and high-frequency components.

4.2   Neighbor Reconstruction

The patterns hidden in neighbor features can be reflected in the characteristics of neighbor feature distribution. For a central node, similar attributes shared with neighbors represent common patterns, while dissimilar attributes capture variations. Since node feature dimensions are often interdependent [22], their correlations also encode shared patterns. These patterns can be quantified via the neighbor feature distribution's mean, standard deviation, and covariance. Features close to the mean indicate normal patterns, while large deviations reveal structural anomalies.

This helps mitigate structural anomalies by identifying and reducing the influence of neighbors exhibiting anomalous patterns. Therefore, we introduce a neighbor reconstruction module that leverages the characteristics of neighbor feature distributions to improve the detection of structural anomalies.

Self Feature Reconstruction. To ensure the updated representation matrix $\mathbf{Z}$ captures information about each node's neighbors, we employ a simple multi-layer perceptron (MLP) to reconstruct $\mathbf{H}^0$ from $\mathbf{Z}$, which can be formulated as $\hat{\mathbf{H}}^0 = \text{MLP}(\mathbf{Z})$, where $\hat{\mathbf{H}}^0 \in \mathbb{R}^{|\mathcal{V}| \times d}$ denotes the reconstructed representation of $\mathbf{H}^0$. Then, the feature reconstruction loss $\mathcal{L}_{fr}$ is computed as $\mathcal{L}_{fr} = ||\hat{\mathbf{H}}^0 - \mathbf{H}^0||_2$, where $|| \cdot ||_2$ is the L2 norm.

Neighbor Feature Distribution Reconstruction. Based on the hidden representation matrix $\mathbf{H}^0$, we compute the true mean, standard deviation, and covariance matrix of the neighbor feature distribution for each node in the graph.

For a given node v_i, the mean μ_i and standard deviation σ_i of its neighbors' features are calculated as follows:

$$\mu_i = \frac{1}{|\mathcal{Z}(i)|} \sum_{j \in \mathcal{Z}(i)} \mathbf{h}^0_j, \quad \sigma_i = \sqrt{\frac{1}{|\mathcal{Z}(i)| - 1} \sum_{j \in \mathcal{Z}(i)} \left(\mathbf{h}^0_j - \mu_i\right)^2} \qquad (7)$$

where $\mu_i, \sigma_i \in \mathbb{R}^d$, $|\mathcal{Z}(i)|$ is the number of neighbors, and $\mathbf{h}_j^0$ is the feature vector of neighbor v_j in $\mathbf{H}^0$. Then, the covariance matrix $\boldsymbol{\Sigma}_i$ for the neighbor feature distribution of node v_i is computed as:

$$\boldsymbol{\Sigma}_{\mathbf{i}} = \frac{1}{|\mathcal{Z}(i)| - 1} \sum_{j \in \mathcal{Z}(i)} \left(\mathbf{h}_j^0 - \mu_i \right) \left(\mathbf{h}_j^0 - \mu_i \right)^T \tag{8}$$

Next, we predict these characteristics. Since the neighbor feature mean reflects smooth transitions and the standard deviation captures variations influenced by sharp transitions, we employ two MLPs to predict the mean and standard deviation for each node v_i, utilizing the node's low-frequency and high-frequency feature representations, $\mathbf{p}_i \in \mathbf{H}_{\text{low}}^L$ and $\mathbf{q}_i \in \mathbf{H}_{\text{high}}^{L'}$, respectively:

$$\hat{\mu}_i = \text{MLP}_\mu \left(\mathbf{p}_i \right), \quad \hat{\sigma}_i = \text{MLP}\sigma \left(\mathbf{q}_i \right) \tag{9}$$

After calculating the mean and standard deviation for all nodes, the prediction losses for the mean and standard deviation can be formulated as follows:

$$\mathcal{L}_\mu = \frac{1}{N} \sum_{i=1}^{N} \|\hat{\mu}_i - \mu_i\|_2, \quad \mathcal{L}_\sigma = \frac{1}{N} \sum_{i=1}^{N} \|\hat{\sigma}_i - \sigma_i\|_2 \tag{10}$$

We compute the predicted covariance matrix $\hat{\boldsymbol{\Sigma}}_i$ for the neighbor feature distribution of v_i based on $\hat{\mu}_i$ and $\hat{\mathbf{h}}_i^0 \in \hat{\mathbf{H}}_i^0$ as same as Eq. 8.

To avoid gradient explosion when calculating the KL divergence of the covariance matrix, we adopt the Frobenius norm $\|\cdot\|_F$ as the loss function to calculate the loss for the covariance matrix:

$$\mathcal{L}_{\boldsymbol{\Sigma}} = \frac{1}{N} \sum_{i=1}^{N} \|\hat{\boldsymbol{\Sigma}}_i - \boldsymbol{\Sigma}_i\|_F = \frac{1}{N} \sum_{i=1}^{N} \text{Tr}((\hat{\boldsymbol{\Sigma}}_i - \boldsymbol{\Sigma}_i)^T (\boldsymbol{\Sigma}_i - \hat{\boldsymbol{\Sigma}}_i)) \tag{11}$$

By combining the loss from self-feature reconstruction and neighbor feature distribution reconstruction, the loss for the neighbor reconstruction module is defined as follows:

$$\mathcal{L}_{nr} = \mathcal{L}_{fr} + \mathcal{L}_\mu + \mathcal{L}_\sigma + \mathcal{L}_{\boldsymbol{\Sigma}} \tag{12}$$

4.3   Center Aggregation

For each neighbor $v_j \in \mathcal{Z}(i)$ of a central node v_i, the node v_i also serves as a neighbor of v_j. Thus, the normal features of v_i should align with the neighbor feature distributions of all $v_j \in \mathcal{Z}(i)$. Leveraging the mean and covariance of these distributions allows smoothing and updating v_i's representation, mitigating contextual anomalies. Building on this principle, we design the center aggregation module to leverage these neighbor feature distributions and improve the central node's feature representation.

In this module, we utilize the graph attention layer [29] to separately update the central node v_i's hidden representation based on the mean and covariance matrices of its neighbors.

The mean aggregation process is formulated as $\hat{\mathbf{z}}_i^{mean} = \sigma\left(\sum_{j\in\mathcal{N}(i)} \alpha_{ij}\mathbf{W}\hat{\mu}_j\right)$, where $\mathbf{W} \in \mathbb{R}^{d\times d}$ is a learnable weight matrix, $\sigma(\cdot)$ is an activation function, and α_{ij} is the attention coefficient between v_i and v_j, which is computed as:

$$\alpha_{ij} = \frac{\exp\left(\text{LeakyReLU}\left(\mathbf{a}^\top[\mathbf{W}\hat{\mu}_i\|\mathbf{W}\hat{\mu}_j]\right)\right)}{\sum_{k\in\mathcal{N}(i)}\exp\left(\text{LeakyReLU}\left(\mathbf{a}^\top[\mathbf{W}\hat{\mu}_i\|\mathbf{W}\hat{\mu}_k]\right)\right)} \tag{13}$$

where $\|$ is the concatenation operator and $\mathbf{a} \in \mathbb{R}^{2d}$ is a learnable weight vector.

The covariance aggregation follows a similar process to the mean aggregation but operates on the covariance matrix of features. Given the input covariance matrix $\hat{\mathbf{\Sigma}}_i$, it is first flattened into a vector $\hat{\mathbf{z}}_i^{cov} \in \mathbb{R}^{d^2}$ for processing. The aggregation process for covariance is $\hat{\mathbf{z}}_i^{cov} = \sigma\left(\sum_{j\in\mathcal{N}(i)} \alpha'_{ij}\mathbf{W}'\hat{\mathbf{z}}_j^{cov}\right)$, where $\mathbf{W}' \in \mathbb{R}^{d\times d}$ is a learnable weight matrix, α'_{ij} is the attention coefficient between nodes v_i and v_j.

The aggregated representations $\hat{\mathbf{z}}_i^{mean}$ and $\hat{\mathbf{z}}_i^{cov}$ are further transformed into the output space using MLPs:

$$\tilde{\mu}_i = \text{MLP}\left(\hat{\mathbf{z}}_i^{mean}\right), \quad \tilde{\mathbf{\Sigma}}_i = \text{Reshape}_{(d,d)}\left(\text{Norm}\left(\text{MLP}\left(\hat{\mathbf{z}}_i^{cov}\right)\right)\right) \tag{14}$$

where $\tilde{\mu}_i$ is the updated mean and $\tilde{\mathbf{\Sigma}}_i$ is the updated covariance matrix for the node v_i. The function Reshape$(\cdot)$ reshapes the input into a $d \times d$ matrix, and Norm$(\cdot)$ is the normalization function.

Finally, we combine these features to update $\mathbf{z}_i \in \mathbf{Z}$ output from the joint graph convolutional encoder to obtain the updated hidden representation $\tilde{\mathbf{H}}$. Specifically, the $\tilde{\mathbf{h}}_i \in \tilde{\mathbf{H}}$ is calculated as $\tilde{\mathbf{h}}_i = (1 - \lambda_{ca})\mathbf{z}_i + \lambda_{ca}\tilde{\mu}_i\tilde{\mathbf{\Sigma}}_i$, where λ_{ca} is a trade-off coefficient controlling the contribution of the aggregated features.

4.4   Decoder

Finally, we employ two GNN-based decoders, Dec^{att} and Dec^{str}, to reconstruct the node attributes and adjacency matrix, respectively. These decoders take the updated hidden representations $\tilde{\mathbf{H}}$ and the graph adjacency matrix $\mathbf{A}$ as inputs. The reconstruction process is formulated as:

$$\hat{\mathbf{X}} = \text{Dec}^{att}(\tilde{\mathbf{H}}, \mathbf{A}), \quad \hat{\mathbf{A}} = \text{Dec}^{str}(\tilde{\mathbf{H}}, \mathbf{A}) \tag{15}$$

To evaluate the reconstruction quality, we define a loss that balances the reconstruction errors of node attributes and the adjacency matrix using a trade-off coefficient λ_{cs}:

$$\mathcal{L}_{rec} = \lambda_{cs}\mathcal{L}_{att} + (1 - \lambda_{cs})\mathcal{L}_{str} \tag{16}$$

where $\mathcal{L}_{att} = \|\hat{\mathbf{X}} - \mathbf{X}\|_2$, $\mathcal{L}_{str} = \|\hat{\mathbf{A}} - \mathbf{A}\|_F$, and the coefficient $\lambda_{cs} = \frac{\text{std}(\mathbf{A})}{\text{std}(\mathbf{A})+\text{std}(\mathbf{X})}$, following the method used in ADA-GAD [10]. Moreover, $\mathcal{L}_{\text{rec}}$ is also utilized as the anomaly score to quantify the anomaly degree of each node.

Finally, the total loss $\mathcal{L} = \mathcal{L}_{rec} + \mathcal{L}_{nr}$, which combines the reconstruction loss and neighbor reconstruction loss.

5   Experiment

5.1   Experimental Setup

Datasets. We conduct experiments on seven public real-world datasets with organically occurring anomalies, including four domains: social media (Weibo [36], Reddit [15], DGraph [12]), e-commerce (Disney [20], Books [23]), communication (Enron [23]), and financial network (Elliptic [31]). The detailed statistics of all datasets are summarized in Table 1.

Table 1. Statistics of all datasets.

Dataset	Nodes	Edges	Ano.	Ratio
Weibo	8.4k	408k	868	10.3%
Reddit	11.0k	168k	366	3.3%
Disney	124	335	6	4.8%
Books	1.4k	3.7k	28	2.0%
Enron	13.5k	177k	5	0.4%
Elliptic	203k	234k	4,545	9.8%
DGraph	3.7M	4.3M	15.5k	0.4%

Baselines. We compare the proposed NK-GAD with nine anomaly detection methods: DOMINANT [4], DONE [1], AdONE [1], AnomalyDAE [7], GAAN [3], GAD-NR [22], ADA-GAD [10], GADAM [2], and SmoothGNN [5].

Implementation Details. We set the number of epochs to 30, the dropout rate to 0.3, the weight decay to 1e−5, the learning rate to 0.001 for the Reddit and Disney datasets, and 1e−4 for others. The embedding dimension d is set to 32 for Weibo, Disney, and Books, and 16 for the other datasets. We repeat all experiments 10 times using 10 different seeds from 0 to 9. In the joint graph convolutional encoder, the layer depth of the low-pass graph filter is set to 2 for Weibo, Reddit, and Books, and 1 for the others. The depth of the high-pass graph filter is set to 1 for Weibo, Reddit, and Enron datasets, and 2 for the others. For both the contextual and structural decoders, we use GATs for Weibo and GCNs for Reddit. For others, the contextual and structural decoders are GATs and GCNs, respectively. To determine the appropriate values for the coefficients λ_{joint} in the joint graph convolutional encoder and λ_{ca} in the center aggregation, a grid search is conducted for each dataset. The values of λ_{joint} and λ_{ca} are searched over the range from 0.1 to 0.9 in steps of 0.1. The analysis of these hyperparameters is provided in Sect. 5.4.

For large-scale datasets such as Elliptic and DGraph, both the proposed NK-GAD and reconstruction-based baselines (e.g., ADA-GAD and GAAN) adopt a mini-batch scheme [17] to reduce memory consumption. The complexity analysis of NK-GAD is provided in Sect. 5.5.

We use AUC (Area Under the Curve) as the metric to evaluate the performance of all methods [10,17,22].

5.2   Performance Analysis

As shown in Table 2, the proposed NK-GAD consistently achieves superior performance across six of the seven datasets, yielding an average 3.29% AUC

Table 2. AUC (%) results (mean±std) across seven datasets. The best results are highlighted in bold, while sub-optimal results are underlined. OOM denotes out of memory with regard to CPU [17].

Method	Weibo	Reddit	Disney	Books	Enron	Elliptic	DGraph
DOMINANT	89.45 ± 1.76	56.10 ± 0.05	49.36 ± 4.32	57.30 ± 4.53	52.95 ± 3.88	OOM_C	OOM_C
DONE	84.20 ± 1.67	54.90 ± 1.28	42.32 ± 5.27	46.80 ± 6.35	48.71 ± 6.43	OOM_C	OOM_C
AdONE	83.50 ± 3.05	53.89 ± 2.51	48.22 ± 2.64	54.36 ± 1.07	51.79 ± 1.69	OOM_C	OOM_C
AnomalyDAE	89.88 ± 1.32	55.31 ± 1.65	48.45 ± 2.46	58.87 ± 1.42	48.54 ± 2.89	OOM_C	OOM_C
GAAN	89.82 ± 0.03	55.34 ± 0.61	48.02 ± 0.00	58.09 ± 2.57	56.77 ± 5.00	OOM_C	OOM_C
ADA-GAD	$\underline{92.44 \pm 0.46}$	56.89 ± 0.01	66.37 ± 4.83	$\underline{64.43 \pm 3.74}$	69.15 ± 4.26	$\underline{49.85 \pm 2.75}$	48.99 ± 1.16
GAD-NR	73.81 ± 2.01	56.77 ± 2.92	69.83 ± 7.48	58.79 ± 6.87	$\underline{73.44 \pm 9.43}$	44.63 ± 7.33	47.63 ± 1.21
GADAM	25.54 ± 0.85	$\mathbf{57.75 \pm 0.64}$	$\underline{70.72 \pm 1.44}$	59.13 ± 5.40	36.08 ± 1.50	46.89 ± 1.08	45.22 ± 0.82
SmoothGNN	56.97 ± 8.72	53.76 ± 3.39	54.01 ± 10.92	48.16 ± 5.53	51.48 ± 5.47	48.00 ± 5.02	$\underline{50.44 \pm 5.20}$
NK-GAD	$\mathbf{93.70 \pm 0.87}$	$\underline{57.27 \pm 0.07}$	$\mathbf{77.26 \pm 2.25}$	$\mathbf{65.60 \pm 0.72}$	$\mathbf{80.82 \pm 3.10}$	$\mathbf{52.17 \pm 3.41}$	$\mathbf{55.30 \pm 0.16}$

Table 3. AUC (%) results of NK-GAD and its three variants. Best results are in bold; sub-optimal ones are underlined.

Method	Weibo	Reddit	Disney	Books	Enron	Elliptic	DGraph
NK-GAD[†]	89.57	55.32	68.64	60.61	63.16	49.73	48.44
NK-GAD[‡]	91.21	56.21	76.21	64.74	79.88	51.26	49.32
NK-GAD[§]	$\underline{92.71}$	$\underline{56.92}$	$\underline{76.40}$	$\underline{65.18}$	$\underline{80.21}$	$\underline{51.49}$	$\underline{51.44}$
NK-GAD	**93.70**	**57.27**	**77.26**	**65.60**	**80.82**	**52.17**	**55.30**

improvement over the best baselines. Specifically, NK-GAD attains the highest AUC on six datasets, especially the largest datasets, Elliptic and DGraph. These consistent gains demonstrate that NK-GAD effectively enhances its robustness across diverse graph domains. The only marginal decline appears on the Reddit dataset, where NK-GAD slightly underperforms GADAM, due to indistinct high-frequency structural patterns that weaken spectral modeling advantages. Nevertheless, NK-GAD surpasses all other baselines and achieves the highest overall average AUC (68.16% vs. 62.73% for ADA-GAD), confirming its strong generalization capability in graph anomaly detection tasks.

5.3   Ablation Study

To evaluate the contributions of each module in NK-GAD, we design three variants: NK-GAD[†] replaces the joint graph convolutional encoder with a plain GCN-based encoder and removes both the neighbor reconstruction and center aggregation modules; NK-GAD[‡] retains the joint graph convolutional encoder but excludes the neighbor reconstruction and center aggregation modules; NK-GAD[§] removes only the center aggregation module, keeping the joint graph convolutional encoder and neighbor reconstruction module retained. Table 3 highlights the role of each component. The performance gap between NK-GAD[‡] and NK-GAD[†] demonstrates that the joint graph convolutional encoder is effective on datasets with strong high-frequency signals, such as Disney and Enron. Incorporating neighbor reconstruction and center aggregation further enhances detection on graphs with a large amount of edges, like Weibo and DGraph, where numerous neighbors introduce redundant patterns that require refinement.

5.4 Hyperparamter Analysis

We conduct extensive experiments to analyze the impact of the hyperparameters λ_{joint} and λ_{ca} across all datasets, as shown in Fig. 4. For λ_{joint}, NK-GAD shows a performance drop on datasets such as Disney and Enron as the value increases, indicating high sensitivity to high-frequency components. In contrast, performance on Reddit and Books remains stable, suggesting limited dependence on this parameter. These results highlight the importance of balancing high- and low-frequency information in a dataset-specific manner. For λ_{ca}, performance on Disney and Books declines notably beyond intermediate values, reflecting diminishing gains from overemphasizing updated central node features. Conversely, larger graphs like Weibo and DGraph benefit from higher λ_{ca}, where numerous neighbors introduce redundant patterns that require refinement.

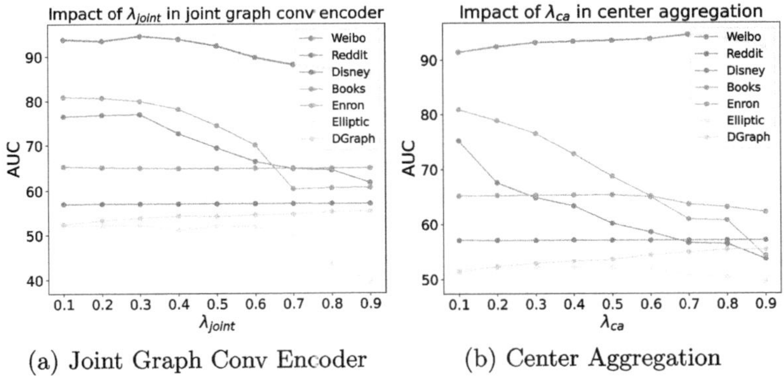

(a) Joint Graph Conv Encoder   (b) Center Aggregation

Fig. 4. Impacts of λ_{joint} in the joint graph convolutional encoder and λ_{ca} in the center aggregation module across datasets.

5.5 Computational Complexity Analysis

We further analyze the computational complexity of the proposed NK-GAD. The joint graph convolutional encoder processes low-pass and high-pass filters over L and L' layers, respectively. Each layer involves sparse graph convolution with complexity $O(|\mathcal{E}|d)$ and linear transformations with complexity $O(|\mathcal{V}|d^2)$, leading to a total encoder complexity of $O\left((L + L') \cdot (|\mathcal{E}|d + |\mathcal{V}|d^2)\right)$, where $|\mathcal{E}|$, $|\mathcal{V}|$, and d denote the number of edges, nodes, and feature dimensions. The neighbor reconstruction module computes self-feature reconstruction ($O(|\mathcal{V}|d^2)$)

and neighbor statistics (mean: $O(|\mathcal{E}|d)$, variance/covariance: $O(|\mathcal{E}|d^2)$), resulting in $O\left(|\mathcal{E}|d^2 + |\mathcal{V}|d^2\right)$. The center aggregation module employs graph attention networks to update node features, requiring $O(|\mathcal{E}|d)$ for attention scores and $O(|\mathcal{V}|d^2)$ for feature updates, with total complexity $O\left(2 \cdot (|\mathcal{E}|d + |\mathcal{V}|d^2)\right)$. For decoders, the attribute decoder incurs $O(|\mathcal{E}|d + |\mathcal{V}|d^2)$, while the structure decoder reconstructs the adjacency matrix with $O(|\mathcal{V}|^2 d)$. The overall complexity is dominated by $O\left(|\mathcal{E}| + |\mathcal{V}|^2\right)$. The complexity is equal to the reconstruction-based methods in baselines like ADA-GAD and GAD-NR.

To adapt to large graphs such as Elliptic and DGraph, NK-GAD is designed to be efficiently deployed on local GPUs through a mini-batch strategy [17]. By sampling node subsets with their k-hop neighborhoods, the model reduces memory complexity from quadratic to near-linear in $|\mathcal{V}|$, enabling scalable learning on million-node graphs. Under this setting, NK-GAD remains superior to full-graph-processed baselines, showing that neighborhood-knowledge aggregation is both effective and deployable on large-scale attribute-heterophily graphs.

6   Conclusion

This paper reveals two phenomena: 1) attribute similarities between connected nodes show nearly identical distributions across different connected node pair types, and 2) anomalies cause consistent variation trends between the graph with and without anomalous edges in the both low- and high-frequency components of the spectral energy distributions, while the mid-part exhibits more erratic variations. Building on these insights, we propose NK-GAD that leverages neighbor knowledge for graph anomaly detection. A joint graph convolutional encoder captures high- and low-frequency patterns, while a neighbor reconstruction module learns normal neighbor distributions to refine detection, complemented by a center aggregation module for feature updating. Extensive experiments on seven real-world datasets demonstrate the effectiveness of NK-GAD.

Acknowledgments. This work was supported in part by the National Natural Science Foundation of China under Grant 62572346.

References

1. Bandyopadhyay, S., Lokesh, N., Vivek, S.V., Murty, M.N.: Outlier resistant unsupervised deep architectures for attributed network embedding. In: Proceedings of the 13th International Conference on Web Search and Data Mining, pp. 25–33 (2020)
2. Chen, J., Zhu, G., Yuan, C., Huang, Y.: Boosting graph anomaly detection with adaptive message passing. In: The Twelfth International Conference on Learning Representations (2024)
3. Chen, Z., Liu, B., Wang, M., Dai, P., Lv, J., Bo, L.: Generative adversarial attributed network anomaly detection. In: Proceedings of the 29th ACM International Conference on Information & Knowledge Management, pp. 1989–1992 (2020)

4. Ding, K., Li, J., Bhanushali, R., Liu, H.: Deep anomaly detection on attributed networks. In: Proceedings of the 2019 SIAM International Conference on Data Mining, pp. 594–602. SIAM (2019)
5. Dong, X., Zhang, X., Sun, Y., Chen, L., Yuan, M., Wang, S.: SmoothGNN: smoothing-aware GNN for unsupervised node anomaly detection. In: Proceedings of the ACM on Web Conference 2025, pp. 1225–1236 (2025)
6. Dong, X., Thanou, D., Toni, L., Bronstein, M., Frossard, P.: Graph signal processing for machine learning: a review and new perspectives. IEEE Signal Process. Mag. **37**(6), 117–127 (2020)
7. Fan, H., Zhang, F., Li, Z.: AnomalyDAE: dual autoencoder for anomaly detection on attributed networks. In: ICASSP 2020-2020 IEEE International Conference on Acoustics, Speech and Signal Processing (ICASSP), pp. 5685–5689. IEEE (2020)
8. Gao, Y., Wang, X., He, X., Liu, Z., Feng, H., Zhang, Y.: Addressing heterophily in graph anomaly detection: a perspective of graph spectrum. In: Proceedings of the ACM Web Conference 2023, pp. 1528–1538 (2023)
9. HC, M.: Retracted: BMADSN: big data multi-community anomaly detection in social networks. Int. J. Electr. Eng. Educ. **60**(1_suppl), 1736–1749 (2023)
10. He, J., Xu, Q., Jiang, Y., Wang, Z., Huang, Q.: ADA-GAD: anomaly-denoised autoencoders for graph anomaly detection. In: Proceedings of the AAAI Conference on Artificial Intelligence, vol. 38, pp. 8481–8489 (2024)
11. Huang, J., Du, L., Chen, X., Fu, Q., Han, S., Zhang, D.: Robust mid-pass filtering graph convolutional networks. In: Proceedings of the ACM Web Conference 2023, pp. 328–338 (2023)
12. Huang, X., et al.: DGraph: a large-scale financial dataset for graph anomaly detection. Adv. Neural. Inf. Process. Syst. **35**, 22765–22777 (2022)
13. Kipf, T.N., Welling, M.: Semi-supervised classification with graph convolutional networks. arXiv preprint arXiv:1609.02907 (2016)
14. Kumar, A., Singh, S.S., Singh, K., Biswas, B.: Link prediction techniques, applications, and performance: a survey. XXPhys. A **553**, 124289 (2020)
15. Kumar, S., Zhang, X., Leskovec, J.: Predicting dynamic embedding trajectory in temporal interaction networks. In: Proceedings of the 25th ACM SIGKDD International Conference on Knowledge Discovery & Data Mining, pp. 1269–1278 (2019)
16. Li, J., Dani, H., Hu, X., Liu, H.: Radar: residual analysis for anomaly detection in attributed networks. In: IJCAI, vol. 17, pp. 2152–2158 (2017)
17. Liu, K., et al.: Bond: benchmarking unsupervised outlier node detection on static attributed graphs. Adv. Neural. Inf. Process. Syst. **35**, 27021–27035 (2022)
18. Liu, Y., et al.: Pick and choose: a GNN-based imbalanced learning approach for fraud detection. In: Proceedings of the Web Conference 2021, pp. 3168–3177 (2021)
19. Luan, S., et al.: When do graph neural networks help with node classification? Investigating the homophily principle on node distinguishability. Adv. Neural Inf. Process. Syst. **36** (2024)
20. Müller, E., Sánchez, P.I., Mülle, Y., Böhm, K.: Ranking outlier nodes in subspaces of attributed graphs. In: 2013 IEEE 29th International Conference on Data Engineering Workshops (ICDEW), pp. 216–222. IEEE (2013)
21. Qiao, H., Tong, H., An, B., King, I., Aggarwal, C., Pang, G.: Deep graph anomaly detection: a survey and new perspectives. arXiv preprint arXiv:2409.09957 (2024)
22. Roy, A., et al.: GAD-NR: graph anomaly detection via neighborhood reconstruction. In: Proceedings of the 17th ACM International Conference on Web Search and Data Mining, pp. 576–585 (2024)

23. Sánchez, P.I., Müller, E., Laforet, F., Keller, F., Böhm, K.: Statistical selection of congruent subspaces for mining attributed graphs. In: 2013 IEEE 13th International Conference on Data Mining, pp. 647–656. IEEE (2013)
24. Shaygan, M., Meese, C., Li, W., Zhao, X.G., Nejad, M.: Traffic prediction using artificial intelligence: review of recent advances and emerging opportunities. Transp. Research Part C Emerg. Technol. **145**, 103921 (2022)
25. Stanković, L., et al.: Vertex-frequency graph signal processing: a comprehensive review. Digit. Signal Process. **107**, 102802 (2020)
26. Tang, J., Li, J., Gao, Z., Li, J.: Rethinking graph neural networks for anomaly detection. In: International Conference on Machine Learning, pp. 21076–21089. PMLR (2022)
27. Tekin, S.F., Kozat, S.S.: Crime prediction with graph neural networks and multivariate normal distributions. SIViP **17**(4), 1053–1059 (2023)
28. Thilagavathi, M., Saranyadevi, R., Vijayakumar, N., Selvi, K., Anitha, L., Sudharson, K.: AI-driven fraud detection in financial transactions with graph neural networks and anomaly detection. In: 2024 International Conference on Science Technology Engineering and Management (ICSTEM), pp. 1–6. IEEE (2024)
29. Veličković, P., Cucurull, G., Casanova, A., Romero, A., Lio, P., Bengio, Y.: Graph attention networks. arXiv preprint arXiv:1710.10903 (2017)
30. Wang, J., Guo, Y., Yang, L., Wang, Y.: Understanding heterophily for graph neural networks. In: Proceedings of the 41st International Conference on Machine Learning. Proceedings of Machine Learning Research, vol. 235, pp. 50489–50529. PMLR (2024)
31. Weber, M., et al.: Anti-money laundering in bitcoin: Experimenting with graph convolutional networks for financial forensics. In: ACM SIGKDD International Conference on Knowledge Discovery and Data Mining (2019)
32. Wu, Z., Pan, S., Chen, F., Long, G., Zhang, C., Philip, S.Y.: A comprehensive survey on graph neural networks. IEEE Trans. Neural Netw. Learn. Syst. **32**(1), 4–24 (2020)
33. Xiao, S., Wang, S., Dai, Y., Guo, W.: Graph neural networks in node classification: survey and evaluation. Mach. Vis. Appl. **33**(1), 4 (2022)
34. Zardi, H., Alrajhi, H.: Anomaly discover: a new community-based approach for detecting anomalies in social networks. Int. J. Adv. Comput. Sci. Appl. **14**(4), 912–920 (2023)
35. Zhang, N., et al.: AliCG: fine-grained and evolvable conceptual graph construction for semantic search at alibaba. In: Proceedings of the 27th ACM SIGKDD Conference on Knowledge Discovery & Data Mining, pp. 3895–3905 (2021)
36. Zhao, T., Deng, C., Yu, K., Jiang, T., Wang, D., Jiang, M.: Error-bounded graph anomaly loss for Gens. In: Proceedings of the 29th ACM International Conference on Information & Knowledge Management, pp. 1873–1882 (2020)

our method uses first-order derivatives to emphasize trend changes while reducing sensitivity to amplitude, and replaces Euclidean distance with Mahalanobis distance for better multivariate correlation capture. We also apply time weighting to emphasize recent data, aligning with the volatile nature of financial series. From the resulting graph structure (see Fig. 1(a)), we use a Gated Recurrent Unit (GRU) to encode each stock's history, and a delay-aggregated multi-head Dynamic Graph Convolution Network (DGCN) to capture evolving inter-stock dependencies. A heterogeneous graph attention mechanism integrates both positive and negative influences to infer stock trends. We validate our model on real-world datasets including CSI300[1], NASDAQ100[2], and S&P500[3], demonstrating state-of-the-art performance. Ablation studies further confirm the effectiveness of each component. Our key contributions include:

- **Temporal Misalignment-Aware Dual-Relational Modeling**: We are the first to model temporally misaligned dependencies between stock pairs as dual (positive and negative) relations, offering a more refined view of inter-stock interactions and market behavior.
- **Unified Framework for Stock Trend Prediction**: To address major limitations in existing methods, we propose DRDGRL, which employs MT-DDTW for temporal-aware dual-relation construction, followed by a delay-aggregated DGCN and heterogeneous graph attention to fuse signals from both relations and the target node.
- **Extensive Experimental Validation**: We rigorously evaluate our framework on diverse real-world datasets, showing how the dual-relational dynamic graph effectively addresses core challenges in stock trend prediction.

2 Preliminaries

We denote the stock set as $S = \{s_1, s_2, \cdots, s_N\}$, where s_i is the i-th stock among N total stocks. For stock s_i on trading day t, its F-dimensional feature vector is $\mathbf{x}_{i,t} = \{x_{i,t}^{(open)}, x_{i,t}^{(close)}, x_{i,t}^{(high)}, x_{i,t}^{(low)}, x_{i,t}^{(volume)}, x_{i,t}^{(turnover)}\}^\top \in \mathbb{R}^F$. The market data on day t is $\mathbf{x}_t = \{\mathbf{x}_{1,t}, \mathbf{x}_{2,t}, \cdots, \mathbf{x}_{N,t}\}^\top \in \mathbb{R}^{N \times F}$, and the entire observation period is $\mathbf{X} = \{\mathbf{x}_1, \mathbf{x}_2, \cdots, \mathbf{x}_T\} \in \mathbb{R}^{T \times N \times F}$. Given $\mathbf{X}$, our goal is to learn a function $f(\mathbf{X}) = \hat{\mathbf{Y}}^{T+1:T+k} \in \mathbb{R}^{N \times 1}$ that predicts the cross-sectional ranking of future stock performance from day T to day $T + k$, where the ground truth label for stock s_i is defined as the average return over this horizon, $y_i^{T+1:T+k} = \frac{1}{k} \sum_{\tau=1}^{k} \frac{x_{i,T+\tau}^{(close)} - x_{i,T}^{(close)}}{x_{i,T}^{(close)}}$, with $k = 5$ in this research.

To model temporal and inter-stock dependencies, we adopt a discrete-time dynamic graph (DTDG) form with a fixed node set $\mathcal{V}$ and time-varying edge sets $\mathcal{E}_t$. A dynamic graph is denoted as $\mathcal{G} = \{\mathcal{G}_1, \mathcal{G}_2, \cdots, \mathcal{G}_T\}$, where each snapshot

---

[1] https://cn.investing.com/indices/csi300.

[2] https://hk.finance.yahoo.com/quote/%5EIXIC/history.

[3] https://hk.finance.yahoo.com/quote/%5EGSPC/history/.

$\mathcal{G}_t = (\mathcal{V}, \mathcal{E}_t, \mathcal{X}_t)$ corresponds to trading day t. Each node $v_i \in \mathcal{V}$ represents a stock, and each edge $e_{i,j} \in \mathcal{E}_t$ captures the evolving relation between stocks s_i and s_j, including both similarity strength and temporal delay. Extending this to a heterogeneous dual-relational form, we define $\tilde{\mathcal{G}}_t = (\mathcal{V}, \{\tilde{\mathcal{E}}_t^{r_1}, \tilde{\mathcal{E}}_t^{r_2}\}, \mathcal{X}_t)$, where $r \in \{pos, neg\}$ denotes positive or negative relationships, as illustrated in Fig. 1(a).

3 Our Proposed Method

3.1 Stock Correlation Graph Generation

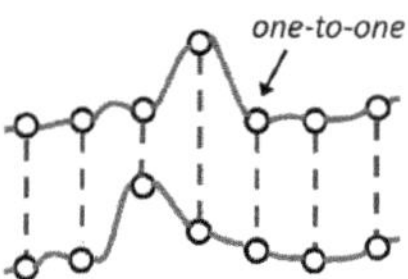

(a) Euclidean Distance

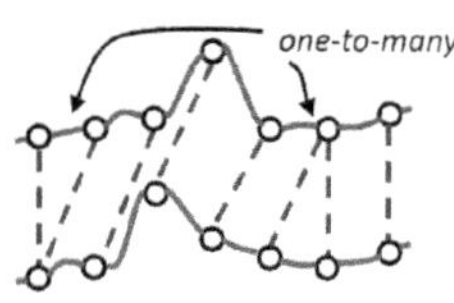

(b) DTW-based Distance

Fig. 2. Comparison of traditional euclidean distance methods and DTW-based methods

We propose a multivariate time-series similarity measurement method, namely MT-DDTW, to compute both temporal similarity and lead–lag relationships between stock pairs, forming dual-relational edges in a dynamic graph, from which both the correlation strength and the corresponding temporal delay between sequences can be jointly derived. Unlike knowledge-driven approaches relying on human-defined or NLP-derived relations, our method directly learns from stock trend signals [19], enabling adaptive and domain-agnostic relational modeling (Fig. 2).

To address limitations of prior correlation-based methods, MT-DDTW extends the Dynamic Time Warping (DTW) framework to multivariate financial data. Given two L-day sequences $\mathbf{x}_i, \mathbf{x}_j \in \mathbb{R}^{L \times F}$, we first apply temporal weighting to emphasize recent market behavior:

$$\mathbf{x}_i' = \mathrm{diag}(\mathbf{W}_k)\mathbf{x}_i, \quad \mathbf{W}_k = \left\{ \frac{1 + (t-1)k}{1 + k\sum_{t=1}^{L-1} t} \mid t = 1, \ldots, L \right\}. \tag{1}$$

A larger k assigns greater importance to more recent days, reflecting recency bias in market dynamics. To mitigate the impact of differing price scales, we adopt the derivative DTW (DDTW) concept by computing first-order derivatives for each feature dimension:

$$x_{i,t}^{(f)''} = \frac{(x_{i,t}^{(f)'} - x_{i,t-1}^{(f)'}) + (x_{i,t+1}^{(f)'} - x_{i,t-1}^{(f)'})/2}{2}, \quad t \in \{2, \ldots, L-1\}, f \in \{1, \ldots, F\}. \tag{2}$$

For robust distance computation, MT-DDTW replaces the Euclidean metric with Mahalanobis distance, capturing feature correlations and preventing singularities:

$$d_M(\mathbf{x}_{i,t_1}'', \mathbf{x}_{j,t_2}'') = \sqrt{(\mathbf{x}_{i,t_1}'' - \mathbf{x}_{j,t_2}'')^\top \mathbf{S}^{-1}(\mathbf{x}_{i,t_1}'' - \mathbf{x}_{j,t_2}'')}, \tag{3}$$

where $\mathbf{S} = \mathrm{Cov}([\mathbf{x}_i''; \mathbf{x}_j'']) + \epsilon\mathbf{I}$ is the regularized covariance matrix (Fig. 3).

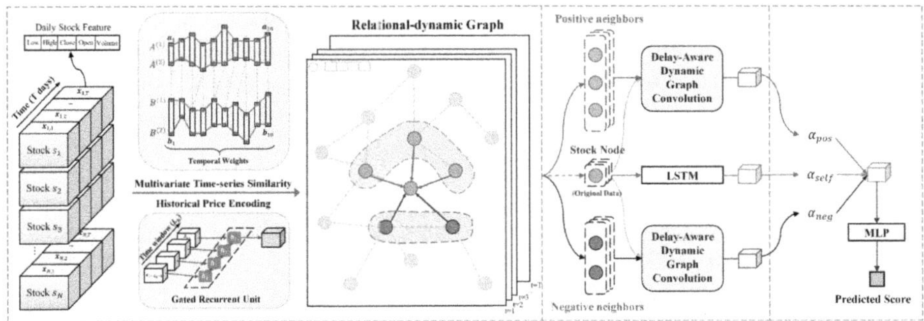

(a) Stock Correlation Graph Generation   (b) Delay-Agg. Dynamic Graph Convolution   (c) Heterogeneous Graph Attention

Fig. 3. The architecture of proposed DRDGRL. Key components include: (a) to capture dual-relational similarity and lead-lag delays between stocks; (b) to aggregate information from leading neighbors at delayed time steps to capture causal dependencies; (c) to produce fused final predictions from multiple sources.

We then compute the cumulative DTW cost matrix Δ_{DTW} with a constrained alignment path. Specifically, we employ the Sakoe–Chiba band constraint such that $|t_1 - t_2| \leq W$, with $W = 3$ in this study, which restricts alignments to a narrow band around the diagonal to prevent excessive temporal distortion and ensure local temporal consistency. The recursive formulation is:

$$\Delta_{DTW}(t_1, t_2) = cost[t_1, t_2] + \min \begin{cases} \Delta_{DTW}(t_1 - 1, t_2), \\ \Delta_{DTW}(t_1, t_2 - 1), \\ \Delta_{DTW}(t_1 - 1, t_2 - 1). \end{cases} \tag{4}$$

The final similarity score between stocks s_i and s_j is obtained as the inverse of the normalized DTW distance:

$$a_{ij} = \mathrm{Sim}_{i,j} = \frac{1}{1 + \mathrm{norm}(DTW(\mathbf{x}_i'', \mathbf{x}_j''))}. \tag{5}$$

Meanwhile, the corresponding lag value $\tau_{i,j}$ is extracted as the average temporal offset along the optimal warping path, representing the lead–lag relationship between the two time series.

After calculating the pair $(a_{i,j}, \tau_{i,j})$ for all stock pairs, we apply inverted min–max normalization to similarity scores to remove the extreme 2% outliers. We then construct two adjacency sets per time snapshot: $\tilde{\mathcal{E}}_t^{pos}$ for positively correlated edges and $\tilde{\mathcal{E}}_t^{neg}$ for negatively correlated edges, where each edge is associated with its similarity and lag values $(a_{i,j}, \tau_{i,j})$. This yields a dual-relational dynamic graph $\tilde{\mathcal{G}}_t = (\mathcal{V}, \{\tilde{\mathcal{E}}_t^{pos}, \tilde{\mathcal{E}}_t^{neg}\}, \mathcal{X}_t)$ that captures evolving inter-stock dependencies and lead–lag structures over time.

3.2   Historical Price Encoding

In this section, we focus on generating the node embeddings $\mathcal{X}_t$ in each graph snapshot $\tilde{\mathcal{G}}_t$. As described above, the input time series feature of stocks on trading day t is $\mathbf{x}_t \in \mathbb{R}^{N \times F}$. We collect data from the previous $L_h - 1$ trading days and concatenate them with $\mathbf{x}_t$ to form $\mathbf{X}_t = [\mathbf{x}_{t-L_h+1}, \mathbf{x}_{t-L_h+2}, \cdots, \mathbf{x}_t]^\top \in \mathbb{R}^{L_h \times N \times F}$. To extract temporal information, we employ a GRU-based encoder to produce time-aware hidden representations:

$$\mathbf{H}_t = GRU(\mathbf{X}_t) \in \mathbb{R}^{L_h \times N \times d_{GRU}}. \tag{6}$$

The final output at the last timestep is then taken as the node embeddings of the graph snapshot:

$$\mathcal{X}_t = \mathbf{H}_t^{(L_h)} \in \mathbb{R}^{N \times d_{GRU}}. \tag{7}$$

By repeating the above process at each time step, we obtain the corresponding subgraph snapshot $\tilde{\mathcal{G}}_t = (\mathcal{V}, \{\tilde{\mathcal{E}}_t^{r_{pos}}, \tilde{\mathcal{E}}_t^{r_{neg}}\}, \mathcal{X}_t)$ and eventually construct the complete dual-relational dynamic graph $\tilde{\mathcal{G}}$ for the entire dataset.

3.3   Delay-Aggregated Dynamic Graph Convolution Mechanism

In this section, we introduce a dynamic graph neural mechanism that leverages temporal delay information derived from MT-DDTW. Starting with dynamic graph modeling, we extend it to handle dual relational graphs with distinct edge types. An overview is shown in Fig. 4. As discussed in Sect. 2, a dynamic graph $\mathcal{G} = (\mathcal{G}_t)_{t \in \mathbb{Z}^T}$ with $\mathbb{Z}^T = \{1, 2, ..., T\}$ consists of undirected snapshots $\mathcal{G}_t = \{\mathcal{V}, \mathcal{E}_t, \mathcal{X}_t\}$. Given nodes $\mathcal{V}$ and edges $\mathcal{E}_t$, The adjacency matrix $\mathbf{A}_t \in \mathbb{R}^{N \times N}$ composed of pairwise similarities a_{ij}, and node features $\mathcal{X}_t \in \mathbb{R}^{N \times d}$ jointly define each time step t.

Delay-Aware Dynamic Graph Convolution. Given adjacency matrix $\mathbf{A} \in \mathbb{R}^{N \times N}$ and vertex features $\mathcal{X} \in \mathbb{R}^{N \times d}$, the Graph Convolution layer with output dimension d_{GC} is:

$$GC_{d_{GC}, \mathbf{A}}^{\mathbf{W}} := \mathrm{ReLU}(\hat{\mathbf{A}} \mathcal{X} \mathbf{W}) \tag{8}$$

where $\hat{\mathbf{A}} := \tilde{\mathbf{D}}^{-\frac{1}{2}} \tilde{\mathbf{A}} \tilde{\mathbf{D}}^{-\frac{1}{2}}$, $\tilde{\mathbf{A}} = \mathbf{A} + \mathbf{I}_N$, and $\tilde{\mathbf{D}}_{k,k} := \sum_l \tilde{\mathbf{A}}_{k,l}$. For sequences $(\mathbf{A}_t)_{t \in \mathbb{Z}^T}$ and $(\mathcal{X}_t)_{t \in \mathbb{Z}^T}$, standard Graph Convolutions are applied across time with shared weights.

To incorporate temporal lead-lag effects, we propose delay-aggregated graph convolution. For each central node i at time t, we aggregate information only from its top-k most similar neighbors j satisfying $\tau_{ij} \leq 0$ (i.e., leading or synchronous). Specifically, for neighbor j leading by τ_{ij} steps, the aggregated feature is approximated using $\mathcal{X}_{t+\tau_{ij}}^j$, assuming that j's state at $t + \tau_{ij}$ can be estimated from historical dynamics, which is implemented via cubic interpolation over recent observations. This yields a temporal-aware convolution process:

$$\text{Delay-GC}_{d_{GC}} : ((\mathcal{X}_{t+\tau_{ij}}^j), \mathbf{A}_t) \mapsto (GC_{d_{GC}, A_t}^{\mathbf{W}}(\mathcal{X}_{t+\tau_{ij}}))_{t \in \mathbb{Z}_T} \tag{9}$$

where each node's embedding depends on aligned (or approximated) neighbor states at shifted time points $t + \tau_{ij}$. Such alignment enhances modeling of causal dependencies across nodes over time.

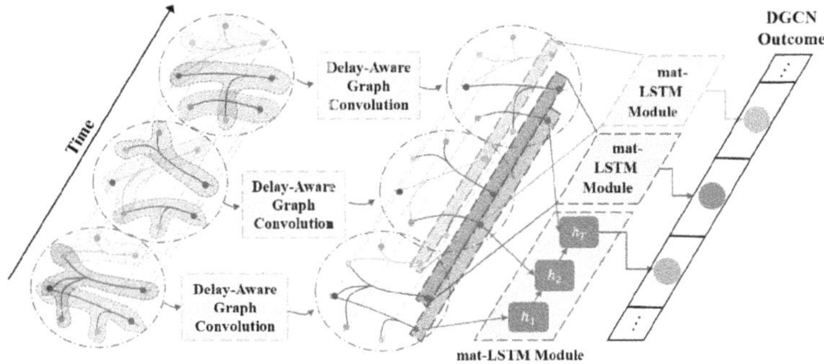

Fig. 4. A simplified illustration of the Delay-Aggregated Dynamic Graph Convolution Mechanism. Stock graphs share fixed nodes but have time-varying edges. Taking the green, red, and blue nodes as examples, delay-aware graph convolution is first applied over time, followed by a mat-LSTM on each node. (Color figure online)

Dynamic Graph LSTM Layer. To capture temporal dependencies in the sequence $(\mathbf{X}_t)_{t \in \mathbb{Z}^T}$, where $\mathbf{X}_t \in \mathbb{R}^{N \times d_{GC}}$, a mat-LSTM with output dimension K is defined as:

$$\text{mat-LSTM}_K : (\mathbf{X}_t)_{t \in \mathbb{Z}^T} \mapsto \mathbf{H}_T \in \mathbb{R}^{N \times K} \tag{10}$$

transforming input from $\mathbb{R}^{T \times N \times d_{GC}} \rightarrow \mathbb{R}^{N \times K}$, with:

$$
\begin{aligned}
\mathbf{H}_t &= \mathbf{O}_t \odot \tanh(\mathbf{C}_t), & \mathbf{F}_t &= \sigma\big(\mathbf{X}_t \mathbf{W}_f + \mathbf{H}_{t-1} \mathbf{U}_f + \mathbf{b}_f\big), \\
\mathbf{C}_t &= \mathbf{J}_t \odot \tilde{\mathbf{C}}_t + \mathbf{F}_t \odot \mathbf{C}_{t-1}, & \mathbf{J}_t &= \sigma\big(\mathbf{X}_t \mathbf{W}_j + \mathbf{H}_{t-1} \mathbf{U}_j + \mathbf{b}_j\big), \\
\mathbf{O}_t &= \sigma\big(\mathbf{X}_t \mathbf{W}_o + \mathbf{H}_{t-1} \mathbf{U}_o + \mathbf{b}_o\big), & \tilde{\mathbf{C}}_t &= \tanh\big(\mathbf{X}_t \mathbf{W}_c + \mathbf{H}_{t-1} \mathbf{U}_c + \mathbf{b}_c\big).
\end{aligned}
\tag{11}
$$

Here, $\odot$ denotes Hadamard product, $\sigma(\cdot)$ is the sigmoid function, and $\mathbf{W}_i, \mathbf{U}_i, \mathbf{b}_i$ are learnable parameters.

Heterogeneous Dynamic Graph Convolution Layer. We further extract $((\tilde{\mathcal{E}}_t^{pos})_{t \in \mathbb{Z}^T}, \mathcal{V})$ and $((\tilde{\mathcal{E}}_t^{neg})_{t \in \mathbb{Z}^T}, \mathcal{V})$ from Sect. 3.2 to construct positive and negative adjacency matrices, denoted as $((\mathbf{A}_t^{pos})_t, (\mathcal{X}_t)_t)$ and $((\mathbf{A}_t^{neg})_t, (\mathcal{X}_t)_t)$, respectively. Each input branch is passed through:

$$DGCN_{d_{GC}, K} : \text{mat-LSTM}_K \circ \text{Delay-GC}_{d_{GC}} \tag{12}$$

where $\circ$ denotes function composition. The resulting outputs $\mathbf{H}_T^{pos}, \mathbf{H}_T^{neg} \in \mathbb{R}^{N \times K}$ capture dynamic embeddings for dual relationship views.

3.4 Heterogeneous Graph Attention Layer

We employ a heterogeneous graph attention mechanism to integrate dual-relational embeddings with the node's intrinsic information. Each node aggregates from three sources: positive neighbors $\mathbf{H}_T^{pos}$, negative neighbors $\mathbf{H}_T^{neg}$, and its own embedding $\mathbf{H}_T^{self} \in \mathbb{R}^{N \times K}$, which can be computed via the mat-LSTM$_K$ in Sect. 3.3. An attention mechanism assigns relevance scores $(\alpha_{\text{self}}, \alpha_{\text{pos}}, \alpha_{\text{neg}})$ to each relation type for final integration.

To estimate each type's contribution, we feed $\mathbf{H}_T^{self}$, $\mathbf{H}_T^{pos}$, and $\mathbf{H}_T^{neg}$ into independent MLPs, generating hidden representations $\mathbf{H}_T^r$ for $r \in \{self, pos, neg\}$. Using a shared attention vector $\mathbf{q}$ and parameters $\mathbf{W}, \mathbf{b}$, we compute a relation-level importance score w_r:

$$w_r = \frac{1}{|\tilde{V}|} \sum_{v^t \in \tilde{V}} \mathbf{q}^T \tanh(\mathbf{W}\mathbf{h}_{v^t,r} + \mathbf{b}) \tag{13}$$

where $\tilde{\mathcal{V}}$ is the set of nodes and $\mathbf{h}_{v^t,r}$ is the v^t-th row of $\mathbf{H}_T^r$. Applying softmax yields attention coefficients:

$$\beta_r = \frac{\exp(w_r)}{\sum_{m \in \{self,pos,neg\}} \exp(w_m)} \tag{14}$$

The final embedding $\mathbf{Z}_T$ is a weighted combination:

$$\mathbf{Z}_T = \sum_{r \in \{self,pos,neg\}} \beta_r \cdot \mathbf{H}_T^r \tag{15}$$

This allows the model to emphasize more informative relations while integrating all three sources into a unified vector for downstream tasks.

3.5 Optimization Objectives

We formulate stock prediction as a semi-supervised node regression task. A fully connected (FC) layer maps node embeddings to predicted stock scores:

$$\hat{\mathbf{Y}} = \tanh(\mathbf{W}\mathbf{Z}_T + \mathbf{b}) \tag{16}$$

Following [4], we employ a hybrid loss combining mean squared error (MSE) and pairwise ranking loss. The regression term aligns predicted and ground truth values:

$$\mathcal{L}_{reg} = \frac{1}{N} \sum_{i=1}^{N} \left\| \hat{\mathbf{Y}}_i - \mathbf{Y}_i \right\|^2 \tag{17}$$

The ranking-aware loss encourages consistency in stock ordering:

$$\mathcal{L}_{rank} = \sum_{i=1}^{N} \sum_{j=1}^{N} \text{ReLU}\left(-(\hat{\mathbf{Y}}_i - \hat{\mathbf{Y}}_j)(\mathbf{Y}_i - \mathbf{Y}_j)\right) \tag{18}$$

To balance both objectives in a data-driven way, we adopt an uncertainty-based aggregation strategy with learnable parameters s_{reg}, s_{rank}:

$$\mathcal{L} = \frac{1}{2}\left(e^{-s_{reg}}\mathcal{L}_{reg} + s_{reg}\right) + \frac{1}{2}\left(e^{-s_{rank}}\mathcal{L}_{rank} + s_{rank}\right) \tag{19}$$

This approach adaptively calibrates the relative importance of prediction accuracy and ranking consistency, improving robustness and convergence.

Table 1. Statistics of the datasets used in this study. Multiple datasets from Chinese and US markets are included ensure comprehensive experimental representativeness.

Market	CSI300	NASDAQ100	S&P500
Stocks (Nodes)	300	80	538
Training Period	01/02/2018–12/29/2023		
Testing Period	01/02/2024–12/31/2024		
Days (Train:Test)	2272 : 242	2263 : 251	2263 : 251

4   Experiments

4.1   Experimental Settings

Datasets. We use the publicly available yfinance API[4] to conduct extensive experiments on real-world datasets of varying scales from both Chinese and American stock markets, covering constituent stocks with complete daily records in the CSI300, NASDAQ100, and S&P500 indices. To balance the requirement of a fixed node set, avoid survivorship bias, and preserve the representative size of each index, we select, for each dataset, stocks that have been included in the respective index between 2018 and 2024 and have uninterrupted daily data (i.e., no trading suspensions). For the CSI300, whose filtered result far exceeds its intended size, we randomly sample 300 such stocks to match the index size, forming the final dataset. We process the sequence data through the following steps. **Step 1**: We select daily indicators in each dataset including opening price, closing price, highest price, lowest price, trading volume for American market entities and additionally turnover indicator for Chinese market entities; **Step 2**: According to the definition of stock trend prediction in Sect. 2, we calculate set the return ratio of a stock as the ground-truth label; **Step 3**: We further perform outlier removal and normalization on the data; **Step 4**: Using the method introduced in Sect. 3.1, we generate a daily stock correlation graph for each date in each dataset using only historical data up to that trading day, ensuring strict temporal causality, and represent it as a daily stock relationship adjacency matrix; **Step 5**: We split the data into training set (2018–2023) and testing set (2024) in chronological order. The detailed dataset statistics is in Table 1.

Parameter Settings. We set the time window T to 20 days of historical data for both model input and daily stock correlation graph generation, with look-back window $L_h = 10$. In the process, the weighing factor k is 2, and the resulting graphs are row-normalized with top-10 neighbors retained for aggregation mechanism. We utilize a two-layer, four-head dynamic graph convolution network to capture positive and negative relationships between stocks. The hidden dimension d_{GRU} of Historical Price Encoding is 50, and the hidden dimension for Delay-Aggregated Dynamic Graph Convolution mechanism and Dynamic Graph

---

[4] https://github.com/ranaroussi/yfinance.

Table 2. Experimental results of all models. We evaluate models using two metric categories: 1) ranking-based metrics (IC, ICIR) to assess consistency between predicted scores and actual labels; and 2) portfolio-based metrics (ARR, AVol, MDD, IR) to measure return and risk performance of the resulting strategies.

Models	CSI 300						NASDAQ 100						S&P 500					
	IC↑	ICIR↑	ARR↑	AVol↓	MDD↓	IR↑	IC↑	ICIR↑	ARR↑	AVol↓	MDD↓	IR↑	IC↑	ICIR↑	ARR↑	AVol↓	MDD↓	IR↑
GRU	0.016	0.193	0.202	**0.223**	−0.142	0.141	0.002	0.017	0.103	**0.192**	−0.150	0.103	0.008	0.081	0.176	0.154	−0.095	0.274
LSTM	0.012	0.090	0.237	0.238	−0.172	0.459	0.011	0.072	0.122	0.220	−0.178	0.264	0.006	0.085	0.123	0.152	−0.092	−0.182
ALSTM	0.009	0.126	0.132	0.248	−0.194	−0.556	0.009	0.068	0.118	0.198	−0.157	0.247	0.002	0.038	0.224	0.163	−0.103	0.565
Transformer	0.015	0.110	0.215	0.253	−0.183	0.304	0.017	0.102	0.173	0.251	−0.191	0.565	0.012	0.148	0.238	0.146	−0.083	0.861
LSTM_GCN	0.006	0.030	0.183	0.357	−0.203	0.141	0.005	0.021	0.068	0.272	−0.173	−0.040	−0.001	−0.014	0.077	0.181	−0.104	−0.664
TGC	0.014	0.075	0.296	0.269	−0.178	0.840	0.011	0.075	0.131	0.236	−0.170	0.368	0.009	0.110	0.203	**0.141**	−0.086	0.530
STHGCN	0.020	0.118	0.299	0.254	−0.145	0.924	0.015	0.092	0.220	0.246	−0.182	0.853	0.010	0.087	0.250	0.170	−0.085	0.575
THGNN	0.027	0.155	0.375	0.297	−0.190	1.362	0.024	0.136	0.204	0.224	−0.141	0.798	0.016	0.121	0.349	0.155	−0.080	1.181
RTGCN	0.028	0.135	0.256	0.261	−0.167	0.528	0.019	0.108	0.166	0.206	−0.145	0.618	0.010	0.124	0.233	0.165	−0.088	0.547
LSR_IGRU	0.034	0.188	0.314	0.282	−0.146	1.037	0.020	0.104	0.279	0.223	−0.146	1.263	0.014	0.107	0.333	0.161	**−0.076**	1.086
MASTER	0.016	0.082	0.238	0.283	−0.192	0.421	0.007	0.033	0.141	0.243	−0.145	0.351	0.004	0.053	0.210	0.161	−0.100	0.013
HGTAN	0.039	0.292	0.421	0.254	−0.154	1.988	0.025	0.140	0.317	0.218	−0.166	1.522	0.020	0.162	0.406	0.166	−0.081	1.418
VGNN	0.033	0.201	0.312	0.248	−0.157	1.220	0.020	0.113	0.271	0.214	**−0.132**	1.327	0.010	0.146	0.268	0.151	−0.088	1.027
MSTNN	0.040	0.242	0.359	0.293	−0.168	1.293	0.021	0.117	0.309	0.207	−0.133	1.589	0.020	0.158	0.329	0.163	−0.087	1.024
DRDGRL	**0.046**	**0.311**	**0.481**	0.267	**−0.124**	**2.205**	**0.031**	**0.164**	**0.330**	0.243	−0.182	**1.688**	**0.022**	**0.207**	**0.447**	0.200	−0.099	**1.936**

LSTM Layer is 90. During model training, we employ a batch size of 32, using the Adam optimizer with an initial learning rate of 0.0001 to optimize the combined loss function. We treat prediction as a regression task and build a daily virtual investment portfolio. Experiments run on a server with NVIDIA GeForce RTX 3090 (24GB), and all models are implemented in PyTorch.

Trading Protocols. Building on [7], we evaluate stock prediction using a Top-K-Drop strategy, which maintains K assets and dynamically replaces the lowest-ranked ones. During the test period (January–December 2024), portfolios are rebalanced daily based on model forecasts.

1. At the close of day t, the model predicts scores for all stocks, producing a daily ranking of expected returns. Current holdings are compared to this ranking to identify candidates for replacement.
2. The N lowest-ranked stocks in the portfolio are sold, and an equal number of top-ranked, unheld stocks are purchased. Additional top-ranked stocks are added if needed to restore K holdings.
3. Stocks retained across days are held and marked-to-market until sold. Realized daily returns of the evolving Top-K portfolio are calculated for cumulative performance and other metrics. Transaction costs are set to 0.03% per trade with a minimum of 5 units.

For each dataset, K is set to 5% of all stocks, with $N = K/T_{\text{label}}$ to ensure full portfolio turnover within $T_{\text{label}} = 5$ days, consistent with the label design.

Compared Baselines. We compare our model with the following state-of-art baseline models, including 1) Non-graph methods of GRU [5], LSTM [14],

ALSTM [18], Transformer [17], and 2) Graph-based models, with LSTM_GCN [2], TGC [4], STHGCN [13], THGNN [19], RTGCN [23], LSR_IGRU [24], MASTER [11], HGTAN [3], VGNN [20] and MSTNN [16].

Evaluating Metrics. To comprehensively evaluate the effectiveness of slope-based stock recommendation, we adopt two categories of ranking-based and portfolio-based metrics that together better reflect the model's predictive quality and real-world profitability.

- **Information Coefficient (IC)**: Spearman rank correlation between predicted scores $\hat{y}_t$ and realized returns y_t at time t, i.e., $\text{IC} = \text{SpearmanCorr}(\hat{y}_t, y_t)$. A higher IC implies more consistent ranking performance.
- **Information Coefficient Information Ratio (ICIR)**: Temporal stability of IC, calculated as $\text{ICIR} = \text{mean}(\text{IC}_t)/\text{std}(\text{IC}_t)$ over time.
- **Annual Return Ratio (ARR)**: Annualized return of strategy, given by $\text{ARR} = (1 + \sum_{t=1}^{n} r_t)^{\frac{252}{n}} - 1$, where r_t is daily return and n is trading days.
- **Annual Volatility (AVol)**: Annualized standard deviation of daily returns, $\text{AVol} = \text{std}(r_t) \times \sqrt{252}$.
- **Maximum Drawdown (MDD)**: Maximum observed loss from peak to trough, $\text{MDD} = \max_{t \in [1,T]} \left[\frac{\max_{i \in [1,t]} (R_i - R_t)}{\max_{i \in [1,t]} R_i} \right]$, where R_t is cumulative return up to day t.
- **Information Ratio (IR)**: Risk-adjusted excess return over a benchmark, $\text{IR} = \text{mean}(r - r_b)/\text{std}(r - r_t)$, where r_b is benchmark daily return.

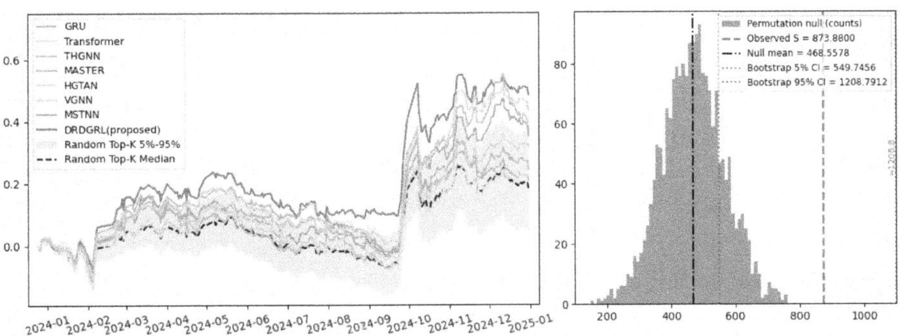

Fig. 5. The test return curves of primary baselines and DRDGRL on CSI300.

4.2 Experimental Results

We systematically compare non-graph models, graph-based baselines, and our DRDGRL on stock trend prediction. As shown in Table 2 and Fig. 5, traditional non-graph architectures exhibit weak ranking capability (low IC/ICIR)

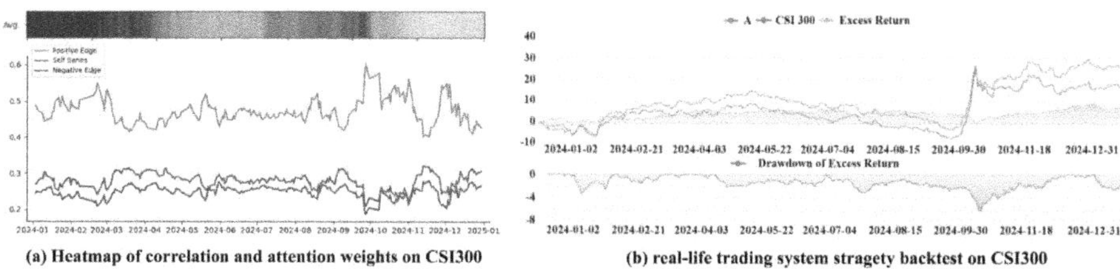

(a) Heatmap of correlation and attention weights on CSI300   **(b) real-life trading system stragety backtest on CSI300**

Fig. 6. (a) Heatmap of average stock correlations over time via the MT-DDTW method (top), and line chart of heterogeneous attention weights over time (bottom); (b) Real-life trading system strategy deployment results.

and inferior portfolio performance (lower ARR/IR and larger drawdowns), indicating their inability to capture inter-stock dependencies. Graph-based baselines improve both predictive and portfolio metrics by incorporating relational structure, yet they still fail to model dynamic and bidirectional correlations effectively.

In contrast, DRDGRL achieves simultaneous gains in both ranking and portfolio metrics across CSI300, NASDAQ100, and S&P500 datasets. It delivers the highest IC (0.046/0.031/0.022) and ICIR (0.311/0.164/0.207), confirming stronger slope-based stock ordering, while also yielding the best portfolio results with ARR of 0.481, 0.330, and 0.447 and IR of 2.205, 1.688, and 1.936, respectively. These improvements are accompanied by lower volatility and controlled maximum drawdown, indicating enhanced risk-adjusted performance. The joint uplift in ranking quality and investability demonstrates that DRDGRL not only identifies profitable assets more accurately, but also converts ranking superiority into superior realized returns, validating its robustness and practicality for real-world portfolio allocation. As shown in Fig. 5 (right panel) and Fig. 6(b), we further conduct a permutation test aligned with the label distribution, and perform real-life monthly backtesting based on a production trading system architecture provided by Shanghai Seek Data Group. The former shows that the predicted scores from DRDGRL consistently lie on the far-right tail of the randomly simulated null distribution. These results confirm that the model's outperformance is statistically significant and not due to random chance.

4.3   Parameter Sensitivity Analysis

We conduct parameter sensitivity analysis on CSI300 and S&P500 datasets, focusing on key hyperparameters: the number of historical days used for prediction T, the hidden dimension of the Historical Price Encoding module d_{GRU}, the dimension of the Dynamic Graph Convolution module d_{GCN}, and the number of GCN layers and attention heads. As shown in Fig. 7(a) and (b), increasing d_{GRU} initially boosts performance, peaking at an optimal size before dropping, indicating a trade-off between capacity and overfitting. A similar trend is seen for d_{GCN} in Fig. 7(c) and (d), where overly large dimensions reduce performance. Figures 7(e) and (f) show that changing T causes only slight variation, suggesting low sensitivity. Lastly, Fig. 7(g) and (h) indicate that both the number of

GCN layers and attention heads require tuning; moderate settings perform best, while excessive complexity impairs accuracy.

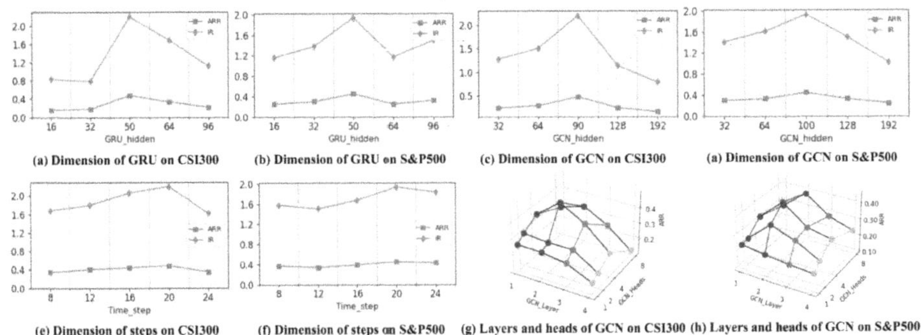

Fig. 7. Parameter sensitivity analysis experiments on CSI300 and S&P500.

4.4   Ablation Study

To assess the contribution of each component, we conduct an ablation study by removing or altering specific modules and relation structures. As shown in Table 3, removing Historical Price Encoding (w/o HPE) causes notable degradation, confirming its role in temporal modeling. Eliminating positive or negative relations (w/o positive/w/o negative), or using only self-node features, also harms performance, proving the necessity of dual relations. Removing self-node features (w/o self) further weakens results, as local context is essential. Using only one relation type (w/o self+neg or w/o self+pos) causes greater drop, and removing all relations (w/o pos+neg) performs worst, highlighting that node features alone are inadequate. Without delay aggregation (w/o delay agg.), performance drops slightly, suggesting potential in refining delay-aware design. Finally, replacing MT-DDTW with static, Pearson, or vanilla DTW under the w/o delay agg. setting consistently performs worse, confirming that conventional similarity metrics fail to capture the dynamic, misaligned characteristics of financial time series, reinforcing the value of temporal encoding and dual-relational modeling.

4.5   Time Complexity Analysis

Offline: Delay–Aware Graph Construction. In the offline stage, the main computational cost lies in constructing the delay–aware dual–relational graph via MT–DDTW. For N stocks, each with historical window length L and feature dimension F, pairwise similarity computation requires $O(N^2 LF)$ operations using Mahalanobis distance and derivative-based DTW within a Sakoe–Chiba band of width W. Since $W \ll L$, the effective cost per stock pair reduces to

Table 3. Performance of ablated or modified models on CSI300 and S&P500. Removing modules or changing relation leads to drops in metrics, highlighting their importance.

Models	CSI 300				S&P 500			
	IC↑	ICIR↑	ARR↑	IR↑	IC↑	ICIR↑	ARR↑	IR↑
w/o HPE	0.019	0.126	0.287	0.899	0.010	0.083	0.252	0.873
w/o self	0.024	0.154	0.297	0.936	0.012	0.144	0.220	0.554
w/o positive	0.032	0.258	0.343	1.201	0.009	0.103	0.201	0.499
w/o negative	0.031	0.247	0.380	1.365	0.013	0.144	0.260	1.003
w/o self+pos	0.012	0.079	0.261	0.503	0.005	0.083	0.126	-0.053
w/o self+neg	0.027	0.205	0.335	1.106	0.006	0.088	0.136	0.026
w/o pos+neg	0.016	0.123	0.241	0.433	0.014	0.109	0.322	1.007
w/o delay agg.	0.042	0.306	0.413	1.877	0.020	0.166	0.410	1.422
Corr-static	0.024	0.165	0.316	0.998	0.009	0.114	0.234	0.587
Corr-pearson	0.039	0.292	0.352	1.237	0.017	0.139	0.364	1.196
Corr-dtw	0.034	0.237	0.399	1.546	0.018	0.157	0.356	1.278
DRDGRL	**0.046**	**0.311**	**0.481**	**2.205**	**0.022**	**0.207**	**0.447**	**1.936**

$O(WLF)$, giving an overall complexity of $O(N^2WLF)$, quadratic in N. In practice, generating all graphs for one trading day on the CSI300 dataset takes approximately 6.55 s.

Online: Delay-Aggregated Dynamic Graph Neural Network. During the online training and inference stage, inference mainly involves temporal encoding via GRU and dynamic graph message passing. The Historical Price Encoding module requires $O(NL_h d_{GRU}^2)$ operations, where L_h is the historical window length. For the Delay-Aware Dynamic Graph Convolution Network, each layer aggregates up to k neighbors, yielding $O(TNkd_{GCN})$ per layer, where T is the number of temporal snapshots. With L_{GCN} layers and H attention heads, the total cost is $O(TNkd_{GCN}L_{GCN}H)$. As k, L_{GCN}, and H are constants, the overall online complexity scales linearly with both N and T. In our setup, training on the CSI300 dataset for one epoch takes about 56.25 s.

5   Related Works

Stock trend prediction is a long-standing topic in finance, driven by complex market factors. Traditional models like AR [10], ARIMA [15], and GARCH [6] capture temporal patterns but fall short in modeling nonlinear dynamics. Machine learning methods, such as SVM [8] and Random Forest [9], improved accuracy by learning data patterns, while deep models like GRU [5] and LSTM [14] further advanced temporal modeling. However, these approaches often treat stocks independently, neglecting industry or supply-chain interactions. To address this, Graph Neural Networks (GNNs) model stocks as nodes with inter-stock relations as edges. Early works used static GCN or GAT on sector-based graphs [4], but fixed edges cannot capture dynamic market changes or differentiate positive and negative effects. More recent models [19,22,24] introduce dynamic graphs for temporal adaptability, yet face limitations: (1) weak handling of dual relations; (2) use of similarity metrics like Pearson correlation that miss lead–lag patterns; and (3) reliance on intra-window relations, ignoring evolving structures. These gaps motivate our dual-relational dynamic framework to reflect evolving and asymmetric dependencies in financial markets.

6 Conclusion

In this work, we propose DRDGRL, a dual-relational dynamic graph representation learning framework for stock trend prediction. Unlike prior methods, it effectively captures dynamic, asymmetric, and dual inter-stock relationships. By introducing MT-DDTW, our framework models both positive and negative dependencies while addressing temporal misalignments like lead-lag effects. Through delay-aggregated graph convolution and heterogeneous attention, DRDGRL learns evolving structures and asymmetric market behaviors. Extensive experiments on real-world datasets show that our model significantly outperforms state-of-the-art baselines. The framework offers a general and extensible approach for modeling complex temporal and relational patterns in financial time series. Future work may explore integrating textual or event-driven data and further examining temporal misalignment in markets.

Acknowledgments. The work is supported by the National Natural Science Foundation of China (Grant No. 6247231.7).

Disclosure of Interests. The authors have no competing interests to declare that are relevant to the content of this article.

References

1. Chan, K.: A further analysis of the lead-lag relationship between the cash market and stock index futures market Rev. Financ. Stud. **5**(1), 123–152 (1992)
2. Chen, Y., Wei, Z., Huang, X.: Incorporating corporation relationship via graph convolutional neural networks for stock price prediction. In: CIKM, pp. 1655–1658 (2018)
3. Cui, C., Li, X., Zhang, C., Guan, W., Wang, M.: Temporal-relational hypergraph tri-attention networks for stock trend prediction. Pattern Recogn. **143**, 109759 (2023)
4. Feng, F., He, X., Wang, X., Luo, C., Liu, Y., Chua, T.S.: Temporal relational ranking for stock prediction. ACM Trans. Inf. Syst. (TOIS) **37**(2), 1–30 (2019)
5. Gupta, U., Bhattacharjee, V., Bishnu, P.S.: Stocknet–GRU based stock index prediction. Expert Syst. Appl. **207**, 117986 (2022)

6. Herwartz, H.: Stock return prediction under GARCH–an empirical assessment. Int. J. Forecast. **33**(3), 569–580 (2017)
7. Hu, Y., et al.: Fintsb: a comprehensive and practical benchmark for financial time series forecasting. arXiv preprint arXiv:2502.18834 (2025)
8. Huang, W., Nakamori, Y., Wang, S.Y.: Forecasting stock market movement direction with support vector machine. Comput. Oper. Res. **32**(10), 2513–2522 (2005)
9. Kumar, M., Thenmozhi, M.: Forecasting stock index movement: a comparison of support vector machines and random forest. In: Indian Institute of Capital Markets 9th Capital Markets Conference Paper (2006)
10. Li, L., Leng, S., Yang, J., Yu, M.: Stock market autoregressive dynamics: a multinational comparative study with quantile regression. Math. Probl. Eng. **2016**(1), 1285768 (2016)
11. Li, T., Liu, Z., Shen, Y., Wang, X., Chen, H., Huang, S.: Master: market-guided stock transformer for stock price forecasting. In: Proceedings of the AAAI Conference on Artificial Intelligence, vol. 38, pp. 162–170 (2024)
12. Sakoe, H.: Dynamic-programming approach to continuous speech recognition. In: 1971 Proceedings of the International Congress of Acoustics, Budapest (1971)
13. Sawhney, R., Agarwal, S., Wadhwa, A., Shah, R.R.: Spatiotemporal hypergraph convolution network for stock movement forecasting. In: 2020 IEEE International Conference on Data Mining (ICDM), pp. 482–491. IEEE (2020)
14. Schmidhuber, J., Hochreiter, S., et al.: Long short-term memory. Neural Comput. **9**(8), 1735–1780 (1997)
15. Shumway, R.H., Stoffer, D.S., Shumway, R.H., Stoffer, D.S.: Arima models. Time Series Analysis and Its Applications: With R Examples, pp. 75–163 (2017)
16. Song, L., et al.: Multi-scale temporal neural network for stock trend prediction enhanced by temporal hyepredge learning. In: IJCAI, pp. 3272–3280 (2025)
17. Vaswani, A., et al.: Attention is all you need. In: Advances in Neural Information Processing Systems, vol. 30 (2017)
18. Wang, Q., Hao, Y.: ALSTM: an attention-based long short-term memory framework for knowledge base reasoning. Neurocomputing **399**, 342–351 (2020)
19. Xiang, S., Cheng, D., Shang, C., Zhang, Y., Liang, Y.: Temporal and heterogeneous graph neural network for financial time series prediction. In: CIKM, pp. 3584–3593 (2022)
20. Xing, R., Cheng, R., Huang, J., Li, Q., Zhao, J.: Learning to understand the vague graph for stock prediction with momentum spillovers. IEEE Trans. Knowl. Data Eng. **36**(4), 1698–1712 (2023)
21. Ying, Z., et al.: Predicting stock market trends with self-supervised learning. Neurocomputing **568**, 127033 (2024)
22. You, M., Cheng, D., Zhang, M., Zhu, P., Liang, Y.: Delay-aware graph neural stochastic differential equations for financial time series modeling and forecasting. In: Proceedings of the ACM Web Conference 2026 (2026). https://doi.org/10.1145/3774904.3792829
23. Zheng, Z., Shao, J., Zhu, J., Shen, H.T.: Relational temporal graph convolutional networks for ranking-based stock prediction. In: 2023 IEEE 39th International Conference on Data Engineering (ICDE), pp. 123–136. IEEE (2023)
24. Zhu, P., Li, Y., Hu, Y., Liu, Q., Cheng, D., Liang, Y.: LSR-IGRU: stock trend prediction based on long short-term relationships and improved GRU. In: CIKM, pp. 5135–5142 (2024)

PVGCL: Graph Contrastive Learning with Purified View Modeling for Spurious Link Detection

Jinfang Xue[1,3,4], Yifan Hong[2], Ruohan Yang[1,3,4], Ling Wang[5], Bo Li[1,3,4], and Huan Wang[1,3,4(✉)]

[1] College of Informatics, Huazhong Agricultural University, Wuhan, China
`{jfxue,ruohanyang}@webmail.hzau.edu.cn,{lio,hwang}@mail.hzau.edu.cn`
[2] Data Space Research Institute, Hefei Comprehensive National Science Center, Hefei, China
[3] Key Laboratory of Smart Farming for Agricultural Animals, Wuhan, China
[4] Engineering Research Center of Intelligent Technology for Agriculture, Wuhan, China
[5] School of Automation and Software Engineering, Shanxi University, Taiyuan, China

Abstract. Spurious link detection (SLD) aims to uncover the observed links but should not exist in a perturbed graph. Current SLD research suffers from the structure uncertainty challenge, where it is harder to identify correct spurious links that appear strong wiggly due to pollution of spurious links. In this paper, we propose a novel framework **PVGCL** to alleviate the widespread uncertainty reduced by spurious links with purified view modeling and contrastive learning. Firstly, we take original perturbed graph as input and Graph Convolutional Network (GCN) as polluted feature encoder to obtain the predecessor for purified view modeling. Secondly, we propose the Residual Swin Window Attention Module (RSWA) to leverage the graph invariant properties (e.g., communities and degree heterogeneity) and jointly aggregate multi-window messages of the attention mechanism. RSWA is capable of purifying already polluted feature and capturing the representative semantics of correspondingly clean graph. Thirdly, to mitigate the problem of structure uncertainty, we approximate the polluted feature and purified feature by graph contrastive learning (GCL). By doing this, the informative information of spurious links could be well preserved while the irrelevant affected features among representations could be eliminated, hence improving the discriminative capacity of the spurious link representation. The experimental results demonstrate that our approach substantially increases AUC with relative improvements reaching as high as 12.5% over ten other competing models. These results highlight the efficacy of our method. The code is available at here.

Keywords: Spurious link detection · Contrastive learning · Structure uncertainty · Graph structure learning

H. Jung et al. (Eds.): DASFAA 2026, LNCS 16536, pp. 51–66, 2026.
https://doi.org/10.1007/978-981-92-0366-6_4

1 Introduction

Spurious links are links observed in a graph that should not exist [44]. Due to limitations in sampling tools, modeling errors, and malicious forgeries by attackers, spurious links sneak into the clean graph [39,40]. These links are harmful, since graph's credibility is significantly weaken, which in turn reduces the performance of downstream tasks, such as link prediction, community detection and node classification [3,4,23,34,41].

One line of research related to spurious links primarily focuses on mitigating their adverse impacts on downstream tasks [6,24]. They label spurious links as noisy links, and term the process of mitigating impacts as network purification/denoising [12,58]. Another line of research deliberately injects spurious links into the raw graph. It aims to explore/enhance the robustness of graph neural networks (GNNs) [7,19]. Only a few studies attempt to detect spurious link detection. Through reasonable consideration, we summarize two reasons why spurious link detection (SLD) has attracted less attention. Firstly, SLD is difficult for graph structure learning. Due to the pollution characteristic of the spurious links [8,16,21], when spurious links and genuine links share the same local structure pattern, GNNs tend to mistakenly classify genuine as spurious links. Secondly, since the damaging nature of spurious links on the network, we usually remove them spur-of-the-moment instead of detecting them. However, SLD is crucial for graph structure learning. On the one hand, SLD is the preliminary step to evaluate the credibility of the network. We need to define reasonably upper bound on the number of spurious links for determining threshold the of the network credibility. Undoubtedly, such a definition requires the detection of spurious links as a foundational prerequisite. On the other hand, SLD contributes to trace the origin of the spurious information. For example, in recommendation systems and short video platforms [49,50,52], it enables the identification of malicious actors behind spurious links, thereby further reducing the spreading risk of fake information.

We summarize existing few methods for SLD, which primarily rely on local structure pattern in the network, including the k-hop neighbors, local subgraphs, or community structures. The first GNN-based detection method [45] utilizes 2-hop neighbors to compute scores for the observed links in the network, with low scores considered to be spurious links. Several studies [28,55] employ structural similarity metrics—such as Common Neighbors [42], Adamic-Adar [55], and Resource Allocation [42]—essentially rely on 1-hop neighbor information for SLD. E-Net [44] introduces a denoising layer into Graph Convolutional Network (GCN) to reduce the spread of noise information (i.e., spurious links) and then performs link prediction and SLD tasks. However, the blocking process will damage important cues that are critical for SLD. LPCA-R [26] first divides the network into communities and then detects spurious links by calculating the probability of link breakage. To the best of our knowledge, LPCA-R is the latest method for SLD. The reason why existing methods focus on local structure pattern is that they assume spurious links and genuine links exhibit heterogeneous local structure patterns. However, real-world networks often display struc-

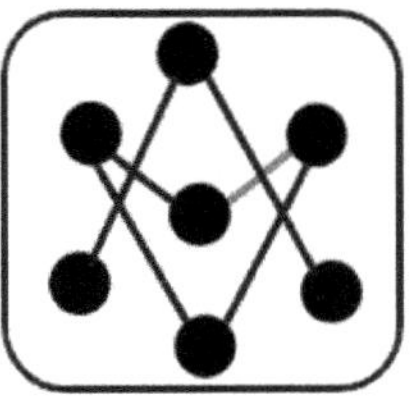

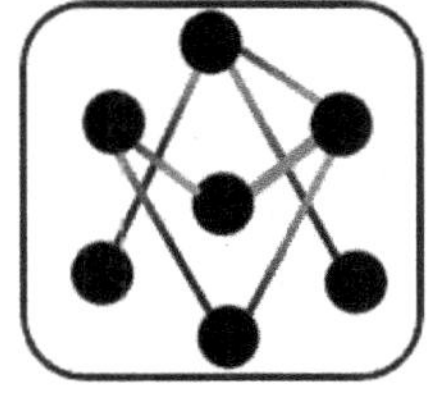

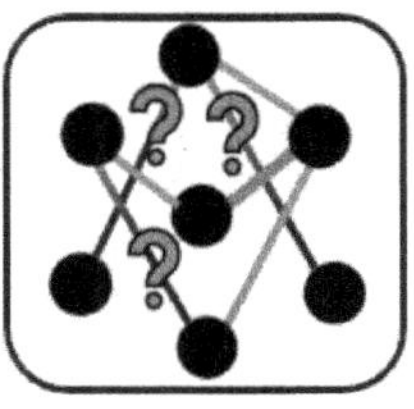

(a) Perturbed Graph   (b) Pollution Spread   (c) Structure Uncertainty

Fig. 1. An example of detecting spurious links without addressing structure uncertainty (•: nodes, —: genuine links, –: spurious links). In (a), the perturbed graph has six links, where red links denote spurious links. Panel (b) shows process of pollution spread that red links contaminate neighbor genuine links. In (c), part of genuine links are predicted as spurious and others exhibit ambiguous semantics, we define this phenomenon as structure uncertainty. Best viewed in color. (Color figure online)

ture uncertainty, that is, different links sharing the same local structure pattern may convey ambiguous semantics [9,10,54]. Consequently, existing methods are prone to misclassifying links into the same category—purely based on their local structure pattern. We present an example in Fig. 1 to illustrate the structure uncertainty. Figure 1(a) shows the perturbed network. In Fig. 1(b), we identify the 1-hop and 2-hop neighbors of spurious link, revealing they have been polluted due to spread of spurious information. In Fig. 1(c), existing methods predict label of spurious link and its neighbors as the same. Given that their neighbors are genuine links, these methods are inclined to classify both targets as spurious.

In light of the insufficiency of existing methods for SLD, it is imperative to explore structure uncertainty based on the already polluted graph, with the aim of capturing subtle discrimination under the local structure pattern. Graph Contrastive learning (GCL) [35,36] constructs complementary views via graph topology augmentation and learns representations by contrasting positive samples against negative ones. Although GCL has achieved promising success in various downstream learning tasks [30,48,56], it is difficult to directly construct the complementary view (i.e., a likely-clean purified view) from the already contaminated graph without prior learning. For this, we identify two key challenges:

Challenge 1—*Precise purified feature learning:* When genuine links are contaminated by neighbor spurious links, they often exhibit highly similar semantics. How to subtly design a mechanism capable of capturing prune structure at the perturbation level.

Challenge 2—*Enabling solution of structure uncertainty in contrastive process:* Polluted and purified semantics provide two distinct yet complementary views for representing spurious links. How to mitigate structure uncertainty in process of contrasting these two views to comprehensively learn the shared representations of spurious links and thereby improve the detection performance.

To address the above challenges, we proposed a **Graph Contrastive Learning** with **P**urified **V**iew Modeling for SLD (PVGCL). For **Challenge 1**, we designed a purified feature encoding module based on the window attention mechanism. This mechanism captures deep prune semantic feature from the view of polluted structure pattern. The uncertainty awareness learning module that differentiates both the polluted and purified views via graph contrastive learning to address **Challenge 2**. In this process, the reliable samples selection mechanism effectively maintains discrimination representation while promoting shared representation of spurious links. For evaluation, we conduct extensive experiments to demonstrate the superiority of PVGCL over ten state-of-the-art methods on PubMed and Cora datasets. In summary, our contributions are as three fold:

- We propose PVGCL, a novel graph contrastive learning framework that models purified view from the polluted structure pattern view and addresses the structure uncertainty for SLD.
- We develop a purified feature encoding module to learn the semantic discrimination between spurious and genuine links from local polluted structure view, then utilize an uncertainty awareness learning module to address the shared feature of spurious links from the polluted pattern view and the purified view.
- We conduct extensive experiments to demonstrate the superior performance of PVGCL over ten state-of-the-art methods on real-world datasets.

2   Preliminary

We consider an attributed and undirected network as $G = (\mathcal{V}, \mathcal{E}, \mathcal{X})$, where $\mathcal{V}$ and $\mathcal{E}$ are the node set and link set respectively, and $\mathcal{X} \in \mathbb{R}^{n \times d}$ represents the original feature matrix. For each link $e \in \mathcal{E}$, we denote its label as $y_e \in \{0, 1\}$ where $y_e = 1$ indicates e is spurious and $y_e = 0$ indicate e is genuine. $\mathcal{Y} = \{y_e \mid e \in \mathcal{E}\}$ denotes a set of link labels. In the context of spurious link detection (SLD), we represent a detector as $f_\theta(\cdot)$ where θ indicates learnable parameters. The predicted scores s_e and the predicted label $\hat{y}_e \in \{0, 1\}$ are calculated by $z_e = f_\theta(G, e)$ and $\hat{y}_e = \mathbf{1}_{\sigma(s_e) > 0.5}$ where $\sigma(\cdot)$ denotes the sigmoid function. We formulate the objective function of the detector as $\max_\theta \sum_{e \in \mathcal{E}} \mathbb{I}(\hat{y}_e = y_e)$ where $\mathbb{I}(\cdot)$ is an indicator function.

3   Methodology

In this section, we detail PVGCL, a self-contrastive learning model for spurious link detection. As shown in Fig. 2, PVGCL consists of four key components: (i) polluted feature encoding module, (ii) purified feature encoding module, (iii) uncertainty awareness learning module, and (iv) optimization and detection module. Specially, polluted feature encoder that automatically learns the underlying infected graph representations (in Sect. 3.1). Purified feature encoder leverages the graph invariant properties to learn deep clean structure (in Sect. 3.2). Once the dual feature is learned, it is used to differentiate intrinsic link representation through uncertainty awareness learning module (in Sect. 3.3). Finally, these updated link embeddings are used by an optimization and detection module to generate final detection (in Sect. 3.4).

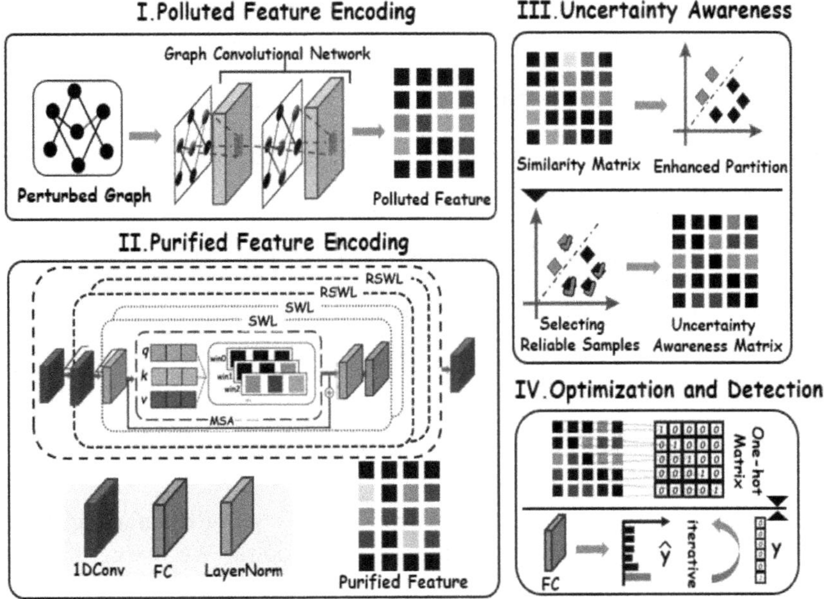

Fig. 2. PVGCL: A Graph Contrastive Learning Framework with Purified View Modeling for Spurious Link Detection.

3.1   Polluted Feature Encoding

Previous studies have demonstrated that most GNNs are vulnerable to spurious links, which can lead to a substantial decline in predictive performance [15]. More recently, [57] revealed that message passing of GNN aggravated pollution induced by spurious links. Conversely, this disadvantage benefits us to learn polluted feature. In this part, we employ Graph Convolutional Network (GCN). Following [11,47], an l-th layer GCN with $\theta = (W_1, W_2, ..., W_l)$ implements f_θ as:

$$f_\theta(X, A) = \hat{A}\,\sigma\left(\left(\hat{A}XW_1\right)W_2\right)\cdots W_l, \tag{1}$$

where $\hat{A} = \tilde{D}^{-1/2}(A + I)\tilde{D}^{-1/2}$ and $\tilde{D}$ is the diagonal matrix of $A + I$ with $\tilde{D}_{ii} = 1 + \sum_j A_{ij}$. σ is the activation function such as ReLU. Note that this is a GCN-like [43] message passing function, any other graph message passing functions [2,17] can also be adopted here. For simplicity, we let f_u to refer to $f_\theta(X, A)$ in the following sections.

3.2   Purified Feature Encoding

With the learned polluted graph structure from Eq. 1, we design a Transformer-like structure that serves as the basic unit for our restored feature encoder. Inspired by [18], our encoder takes polluted feature for nodes as input, and outputs restored node states, through a multi-window messages aggregation with the attention mechanism. To preserve the graph topology patterns, we first aggregate the local projected features for shallow feature learning:

$$\text{shallow feature}: \quad f_{shallow} = W_{conv}f_p, \tag{2}$$

where W_{conv} denotes the 1D convolution weight matrix. Next, we transform shallow node feature to the same d-dimensional space using Residual Swin Window Layer (RSWL) to form deep node feature. The core of RSWL is the sliding window layer (SWL). We independently partition shallow feature into windows for preserving graph invariant properties (e.g., communities and degree heterogeneity) [29,33]. Let F_{win} denotes the features of each window. F_{win} is decomposed into projection matrix Q, K and V, respectively. Let $head_i$ be the attention score for the i-th window:

$$head_i = \text{softmax}(\frac{Q_iK_i^T}{\sqrt{d_k}} + b)V_i, \tag{3}$$

where d_k is scaled factor and $head_i$ denotes i-th head. After calculating the attention score within the attention, we merge all windows. We merge all windows in the same way we split. Let we denote feature as F_{atten}. The entire RSWL process is:

$$F_{rswl} = \text{FC}\left(\text{LayerNorm}\left(F_{atten}\right)\right) + F_{atten}. \tag{4}$$

Our incidence predictor has a simple yet effective structure but other readout functions are also viable [22]. Subsequently, we recursively stack RSWLs, the model gradually mines deeper structural representations to obtain deep feature representations. We denote the deep features of the last RSWL as F_{deep}. To further improve the stability and gradient of the deep representation, we continue to use a residual design to connect the original input features and the output features:

$$f_{merge} = W_{conv}f_{deep} + f_{shallow}. \tag{5}$$

This residual design avoids over-optimization of deep features. We call this process Residual Swin Window Attention Module (RSWA). Progressively stacking multiple RSWA further ensures deep feature mining. For brevity, we let f_r denote final output in the following sections.

3.3   Uncertainty Awareness Learning

Although we have obtained features of the contaminated and the restored graph using dual feature encoder, we have not yet solved the structural uncertainty problem caused by pollution. To address this issue and improve the discriminability of feature, we proposed self-contrastive learning to optimize our training strategy. Unlike relying solely on the cross-entropy branch to directly obtain labels, self-contrastive learning is based on similarity to map the features of both to the same feature space, which reduces the reliance on classification. Even if the structural uncertainty increases which is caused by spurious links in the contaminated graph, or causes the restored graph features to deviate from the raw clean graph, we enforce the features to be mapped to the unified space, thereby achieving the purpose of mutual optimization and restraint. Specifically, the design of our module focuses on the following key insight: screening reliable negative contrast samples, the model can capture the underlying link distinctive semantics. Thus, we ensure that the features of the contaminated and restored graph have the same distribution, which is crucial to alleviating structural uncertainty.

To accomplish this, we initially transform the polluted feature f_u with its augmented feature f_r to project a unified space. We accomplish this through matrix multiplication. Before this, we utilize the concatenation operation to translate feature from node-level into link-level. Formulizelly, we let z_u and z_r denote link-level feature of f_u and f_r, respectively. Hence, the similarity matrix $S_{u,r}$ can be defined as follows:

$$S_{u,r} = \frac{(z_u)^\top (z_r)}{\|z_u\|_2 \|z_r\|_2}, \tag{6}$$

where the subscript represents the L2 normalization [20]. We apply L2 normalization to constrain $S_{u,r}$ within the range of [0,1]. Notably, our similarity calculation function is a simple yet effective way but other metric functions are also viable [53]. As discussed in Sect. 3.2, spurious links can cause structure uncertainty. To address this issue, inspired by [31,51], we employ to screen reliable negative samples. In simple terms, based on the similarity matrix in Eq. 6, we treat the main diagonal elements as positive sample pairs and mask the off-diagonal elements proportionally, so that their contribution to self-contrastive learning is one in a millon. Based on this, the screening process can be expressed as:

$$\tilde{S}_{u,r} = \begin{cases} s_{u,r}, & \text{if } s_{u,r} \in \Phi_k \\ c, & \text{otherwise} \end{cases}, \tag{7}$$

where Φ_k denotes the set of k high-similarity negative link candidates and $s_{u,r}$ is element of similarity metrix $S_{u,r}$. It is noted that term c is negative number(<0), we will conjunction with Eq. 8 to discuss why c is negative later.

Based on $\tilde{S}_{u,r}$, our objective of self-contrastive learning is maximize mutual information among positive pairs while reducing their with negative pairs. To ensure consistent labeling views, both polluted and puried verisions of a link. For facilitate expression, we use s_{ii} refer to positive pairs of $\tilde{S}_{u,r}$ and s_{ij} refer to negative pairs. Hence, we have:

$$\mathcal{L}_{\text{contrast}} = -\mathbb{E}_{i \sim \{1,...,|\mathcal{E}|\}} \left[\log \frac{\exp(s_{ii}/\tau)}{\sum_{j=1}^{|\mathcal{E}|} \exp(s_{ij}/\tau)} \right], \tag{8}$$

where $|\mathcal{E}|$ denotes number of links and τ is the temperature parameter to adjust the contrastive scale. Now we explain why c is negative. As we know, exponential function is monotonically increasing. Exponential value is close to 0 when power falls into the negative range. Further discussion, it means negative term (i.e., $s_{ij} = c$) makes little contribution in $\mathcal{L}_{\text{contrast}}$. Back to task, that indicates we subtly choose reliable negative samples as expected in optization process.

3.4   Optimization and Detection

PVGCL takes an end-to-end supervised training approach that aims to minimize the loss between the ground truth label and the predictions. Link-level features are formed by the node-level polluted and purified features (obtained in Sect. 3.1 and 3.2). The transformation manner like in Sect. 3.3. z_u and z_r denote polluted and purified feature, respectively, aforementioned. We employ a full connected layer as the projection layer, but other structure are viable [22]:

$$\hat{y} = \mathrm{sigmoid}\left[FC(z_u, z_f)\right], \tag{9}$$

where sigmoid is nonlinear activation function to contrain predictions into $[0,1]$. We use the Binary Cross Entropy (BCE) [27] as the classification loss during the model optimization:

$$\mathcal{L}_{\mathrm{class}} = -\mathbb{E}_{e \in \mathcal{E}}\left[y_e \cdot \log(\hat{y}_e) + (1 - y_e) \cdot \log(1 - \hat{y}_e)\right], \tag{10}$$

where $\hat{y}_e$ is predicted spurious score of observed link e. Our final loss function is a combination of the self-contrastive optimization loss and the classification loss using a weight coefficient α:

$$\mathcal{L} = \alpha\mathcal{L}_{\mathrm{class}} + (1 - \alpha)\mathcal{L}_{\mathrm{contrast}}. \tag{11}$$

4   Experiments

In this section, we conduct comprehensive experiments to answer the following research questions:

- RQ1: How does our PVGCL perform compared against existing state-of-the-art methods as baselines?
- RQ2: Can the purified feature encoding module and uncertainty awareness learning actually work?
- RQ3: What is the effect of PVGCL on various ratio of spurious links?

4.1   Experimental Setting

Datasets. We utilize two public datasets, both of which are provided from PyTorch-Geometric[1]. These datasets are PubMed [32] and Cora [1]. An overview of the datasets is shown as Table 1.

Baselines. For comparison with PVGCL, we use ten baselines: (i) AA [55], (ii) CN [55], (iii) PA [55], (iv) RA [55], (v) LP [45], (vi) LPCA-R [26], (vii) PS2 [38], (viii) S2-GAE [37] (ix) DVGAE [5] and (x) TAGNN [14]. AA, PA, RA, LP and LPCA-R methods are designed to detect spurious links. As we now, spurious link detection remains an underexplored area, with only a few available methods. Previous studies have shown that link prediction methods are effective in detecting spurious links [25]. Thus, we select a series of link prediction methods as baselines, including S2-GAE, DVGAE, TAGNN, and PS2.

---

[1] https://pytorch-geometric.readthedocs.io/en/latest/modules/data.html.

Evaluation Metric. For evaluate with PVGCL, we utilize AUC [13], F1_score [46], Precision [46] and Recall [46]. Please noted, since spurious links constitute a highly imbalanced class in real-world datasets, we report Precision and Recall specifically on spurious link samples, denoted as Precision on <u>SL</u> and Recall on <u>SL</u>, to assess the performance of both the baseline methods and our proposed approach on this category.

Experimental Setup. The datasets were randomly divided into training, validation, and test sets, with proportions of 80%, 10%, and 10%, respectively. For AA, CN, PA and RA, the original paper did not specify the fracture probability threshold, we set it to the optimal threshold on AUC in our experiments for consistency with common practice. For other baseline methods, we followed the parameter settings provided in the official code repositories. For GCN, we set $l = 2$ and hidden layer to 512 dimensions and the output layer to 128 dimensions. Correspondingly, the dimensions in the middle of the purified feature encoding module are all 128. The window size is 64 dimensions. In SWL, drop_out is adopted, with a drop_out size of 0.2. In the uncertainty awareness learning module, c and τ are set to -10 and 0.5, respectively. The optimization and detection module contains 2 layers of FC, with the middle layer dimension being 512 and the drop_out being 0.7. The learning rate is set to 0.0001, the training epoch is 100, the Adam optimizer is selected, and GradScaler is used for mixed-precision training. Finally, we set batch size as 4096.

Table 1. Statistics of real-world datasets.

Dataset	#Nodes	#Edges	#Features
PubMed	19,717	88,648	500
Cora	19,793	126,842	8,710

4.2   Results and Analysis

Comparison with Baselines (RQ1). We randomly connect unrelated node pairs in the graph to inject spurious edges. Table 2 reports the performance of all baselines and PVGCL under a spurious-edge injection ratio of 0.1. Please note, Precision on <u>SL</u> denotes Precision on spurious links. The same as Recall on <u>SL</u>. From the results, we make the following key observations:

(i) Overall Performance on AUC anf F1_score: Our proposed method PVGCL consistently achieves the best performance across both datasets under the 0.1 spurious-link ratio setting. Specifically, PVGCL attains the highest AUC and F1_score values (0.9639/0.9471 on PubMed and 0.9563/0.9164 on Cora), significantly outperforming both classical similarity-based indices (AA, CN, PA, RA) and representative GNN-based baselines (S2-GAE, DVGAE, TAGNN).

Table 2. Performance comparison of different methods on real-world datasets

Dataset	Method	AUC	F1_score	Precision on SL	Recall on SL
PubMed	AA	0.6216	0.2515	0.0144	**1.0000**
	CN	0.6216	0.2515	0.0144	**1.0000**
	PA	0.7029	0.7191	0.0255	0.6636
	RA	0.6216	0.2515	0.0144	**1.0000**
	LP	0.9005	0.9188	<u>0.7460</u>	0.4289
	LPCA-R	0.6048	0.5625	0.1545	<u>0.7549</u>
	PS2	<u>0.9224</u>	<u>0.9393</u>	0.7015	0.7030
	S2-GAE	0.8410	0.9000	0.0000	0.0000
	DVGAE	0.5727	0.8185	0.1455	0.1671
	TAGNN	0.5432	0.9000	0.0000	0.0000
	PVGCL	**0.9639**	**0.9471**	**0.7580**	0.6962
Cora	AA	0.7781	0.5640	0.0244	**0.9968**
	CN	0.7779	0.5640	0.0244	**0.9968**
	PA	0.7584	0.8281	0.0307	0.4730
	RA	0.7783	0.5640	0.0244	**0.9968**
	LP	0.7782	0.8929	<u>0.4279</u>	0.1272
	LPCA-R	0.7291	0.5625	0.1878	0.7464
	PS2	<u>0.8311</u>	0.8902	0.3623	0.4281
	S2-GAE	0.7608	<u>0.9000</u>	0.1000	0.0151
	DVGAE	0.5836	0.8532	0.1639	0.1139
	TAGNN	0.5078	<u>0.9000</u>	0.0000	0.0000
	PVGCL	**0.9563**	**0.9164**	**0.4693**	<u>0.7640</u>

These results demonstrate the effectiveness of PVGCL in learning discriminative representations under noisy structural conditions.

(ii) Overall Performance on Precision and Recall on <u>SL</u>: Considering the severe class imbalance in real-world graphs, we further evaluate each method's precision and recall specifically on spurious link (SL) samples. PVGCL achieves superior Precision on <u>SL</u> (0.7580 on PubMed, 0.4693 on Cora) while maintaining competitive Recall on <u>SL</u>, indicating its stronger capability to accurately identify false connections without sacrificing detection sensitivity. This highlights PVGCL's robustness and generalization in distinguishing noisy or spurious structural patterns compared with both traditional heuristics and recent graph learning methods.

Module Effectiveness (RQ2). To further verify whether polluted feature encoding module and purified feature encoding module effectively contribute to

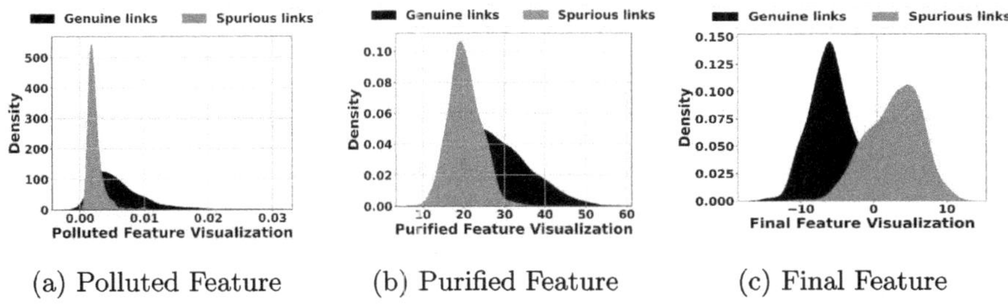

(a) Polluted Feature (b) Purified Feature (c) Final Feature

Fig. 3. Feature visualization of PVGCL on the PubMed dataset.

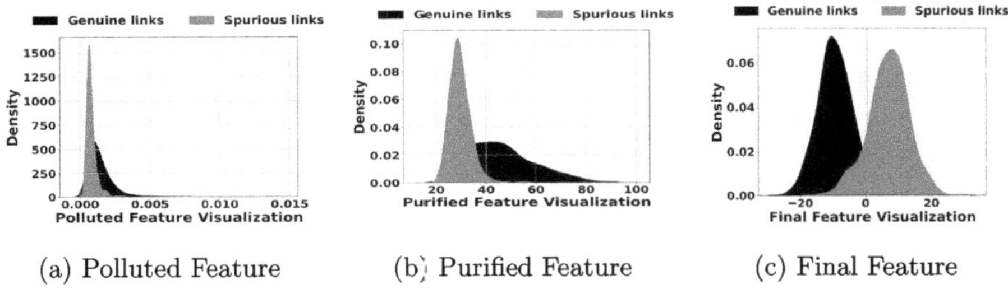

(a) Polluted Feature (b) Purified Feature (c) Final Feature

Fig. 4. Feature visualization of PVGCL on the Cora dataset.

discriminating genuine and spurious links, we visualize the feature distributions before and after purification on both PubMed and Cora datasets.

(i) Feature Distribution Evolution: As shown in Figs. 3 and 4, the polluted features exhibit significant overlap between genuine and spurious links, suggesting that raw graph representations are highly contaminated by structural noise. After applying polluted feature encoding module and purified feature encoding module, the purified features display improved separability, where the density curves of the two categories begin to diverge. Finally, the output features of PVGCL form two clearly distinguishable distributions, demonstrating the model's capability to recover clean structural semantics from noisy graphs.

(ii) Effectiveness of the Designed Modules: Compared across datasets, the improvement trend is consistent—on both PubMed and Cora, the purified and final features show a progressive reduction in overlap between two link types. This confirms that purified feature encoding module successfully isolates the polluted representations, while uncertainty awareness learning module further enhances structure contrast by filtering noise and aligning feature spaces. The results validate that both modules play complementary and essential roles in mitigating structure uncertainty and enhancing model robustness.

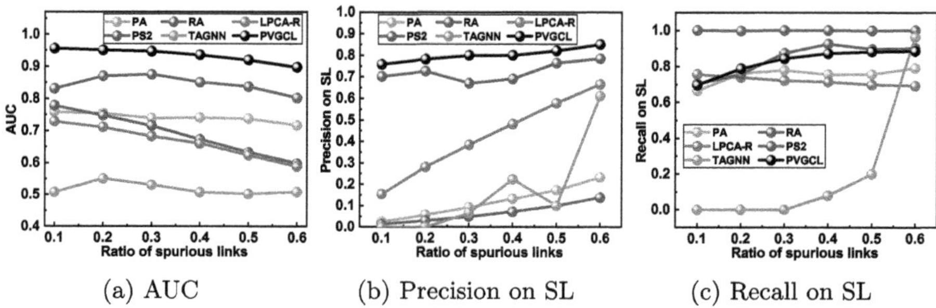

(a) AUC   (b) Precision on SL   (c) Recall on SL

Fig. 5. Performance comparison of different methods under varying spurious-link ratios on the PubMed dataset.

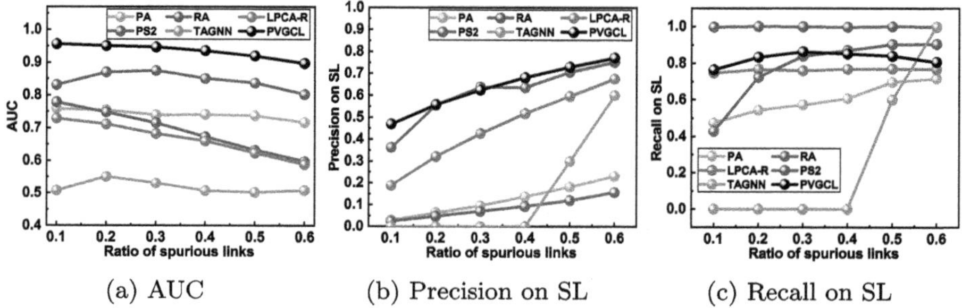

(a) AUC   (b) Precision on SL   (c) Recall on SL

Fig. 6. Performance comparison of different methods under varying spurious-link ratios on the Cora dataset.

Various Ratio of Spurious Links (RQ3). As illustrated in Figs. 5 and 6, we evaluate the robustness of various methods on the PubMed and Cora datasets under different ratios of injected spurious links (ranging from 0.1 to 0.6).

(i) Overall Trend: Across both datasets, the proposed PVGCL consistently achieves the highest AUC values and maintains stable performance as the proportion of spurious links increases, indicating strong robustness against structural noise. In contrast, traditional similarity-based baselines such as PA, RA, and LPCA-R show clear performance degradation with higher noise levels, while PS2 and TAGNN exhibit moderate stability but remain inferior to PVGCL.

(ii) Spurious Link Detection Capability: In terms of Precision on SL and Recall on SL, PVGCL exhibits a balanced and superior ability to accurately identify spurious links across all injection ratios. Specifically, PVGCL demonstrates steadily increasing precision without sacrificing recall, whereas other baselines often display unstable or overly biased behaviors—either achieving high recall with poor precision or vice versa. These results validate the effectiveness of PVGCL in distinguishing spurious connections, demonstrating its generalization capability across datasets and noise ratios.

32. Sun, X., Cheng, H., Li, J., Liu, B., Guan, J.: All in one: multi-task prompting for graph neural networks. In: Proceedings of the 29th ACM SIGKDD Conference on Knowledge Discovery and Data Mining, pp. 2120–2131 (2023)
33. Sun, X., et al.: Self-supervised hypergraph representation learning for sociological analysis. IEEE Trans. Knowl. Data Eng. **35**(11), 11860–11871 (2023)
34. Sun, X., et al.: Multi-level hyperedge distillation for social linking prediction on sparsely observed networks. In: Proceedings of the Web Conference 2021, pp. 2934–2945 (2021)
35. Sun, X., Zhang, J., Wu, X., Cheng, H., Xiong, Y., Li, J.: Graph prompt learning: a comprehensive survey and beyond. arXiv preprint arXiv:2311.16534 (2023)
36. Suresh, S., Li, P., Hao, C., Neville, J.: Adversarial graph augmentation to improve graph contrastive learning. Adv. Neural. Inf. Process. Syst. **34**, 15920–15933 (2021)
37. Tan, Q., et al.: S2gae: self-supervised graph autoencoders are generalizable learners with graph masking. In: Proceedings of the Sixteenth ACM International Conference on Web Search and Data Mining, pp. 787–795 (2023)
38. Tan, Q., et al.: Bring your own view: graph neural networks for link prediction with personalized subgraph selection. In: Proceedings of the Sixteenth ACM International Conference on Web Search and Data Mining, pp. 625–633 (2023)
39. Wang, H., Cui, Z., Liu, S., Ni, Q., Gong, Z.: Evaluating edge credibility in evolving noisy social networks. IEEE Trans. Knowl. Data Eng. **35**(11), 11342–11353 (2022)
40. Wang, H., Cui, Z., Yang, Y., Wang, B., Zhu, L., Zhang, W.: A network enhancement method to identify spurious drug-drug interactions. IEEE/ACM Trans. Comput. Biol. Bioinf. **21**(5), 1335–1347 (2024)
41. Wang, H., et al.: Resisting the edge-type disturbance for link prediction in heterogeneous networks. ACM Trans. Knowl. Discov. Data **18**(2), 1–24 (2023)
42. Wang, H., Qiao, C., Guo, X., Fang, L., Sha, Y., Gong, Z.: Identifying and evaluating anomalous structural change-based nodes in generalized dynamic social networks. ACM Trans. Web **15**(4), 1–22 (2021)
43. Wang, Q., Sun, X., Cheng, H.: Does graph prompt work? A data operation perspective with theoretical analysis. arXiv preprint arXiv:2410.01635 (2024)
44. Xu, J., et al.: Robust network enhancement from flawed networks. IEEE Trans. Knowl. Data Eng. **34**(7), 3507–3520 (2020)
45. Xu, X., Yu, Y., Li, B., Song, L., Liu, C., Gunter, C.: Characterizing malicious edges targeting on graph neural networks (2018)
46. Yacouby, R., Axman, D.: Probabilistic extension of precision, recall, and F1 score for more thorough evaluation of classification models. In: Proceedings of the First Workshop on Evaluation and Comparison of NLP Systems, pp. 79–91 (2020)
47. Yang, R., Ali, M.A., Wang, H., Chen, J., Wang, D.: Luster: link prediction utilizing shared-latent space representation in multi-layer networks. In: Proceedings of the ACM on Web Conference 2025, pp. 2476–2487 (2025)
48. Zeng, J., Xie, P.: Contrastive self-supervised learning for graph classification. In: Proceedings of the AAAI Conference on Artificial Intelligence, vol. 35, pp. 10824–10832 (2021)
49. Zeng, Z., et al.: Mitigating world biases: a multimodal multi-view debiasing framework for fake news video detection. In: Proceedings of the 32nd ACM International Conference on Multimedia, pp. 6492–6500 (2024)
50. Zeng, Z., et al.: Understand, refine and summarize: multi-view knowledge progressive enhancement learning for fake news video detection. In: Proceedings of the 33rd ACM International Conference on Multimedia, pp. 9216–9225 (2025)

51. Zeng, Z., et al.: IMOL: incomplete-modality-tolerant learning for multi-domain fake news video detection. In: Proceedings of the 63rd Annual Meeting of the Association for Computational Linguistics (Volume 1: Long Papers), pp. 30921–30933 (2025)
52. Zhang, G., Yuan, G., Cheng, D., Liu, L., Li, J., Zhang, S.: Mitigating propensity bias of large language models for recommender systems. ACM Trans. Inf. Syst. **43**(6), 1–26 (2025)
53. Zhang, G., Yuan, G., Cheng, D., Liu, L., Li, J., Zhang, S.: Towards fair graph representation learning by overcoming social homophily. ACM Trans. Intell. Syst. Technol. 1–25 (2025)
54. Zhang, G., Zhang, S., Yuan, G.: Bayesian graph local extrema convolution with long-tail strategy for misinformation detection. ACM Trans. Knowl. Discov. Data **18**(4), 1–21 (2024)
55. Zhang, X., Zhao, C., Wang, X., Yi, D.: Identifying missing and spurious interactions in directed networks. Int. J. Distrib. Sens. Netw. **11**(9), 507386 (2015)
56. Zhang, Z., Sun, S., Ma, G., Zhong, C.: Line graph contrastive learning for link prediction. Pattern Recogn. **140**, 109537 (2023)
57. Zhou, B., Li, R., Zheng, X., Wang, Y.G., Gao, J.: Graph denoising with framelet regularizers. IEEE Trans. Pattern Anal. Mach. Intell. **46**(12), 7606–7617 (2024)
58. Zhou, X., Shen, Z.: A tale of two graphs: freezing and denoising graph structures for multimodal recommendation. In: Proceedings of the 31st ACM International Conference on Multimedia, pp. 935–943 (2023)

Node Anomaly Detection via Multiscale Time-Frequency Fusion and Hidden Markov Generations in Complex Networks

Yifan Hong[1] and Jiao Luo[2(✉)]

[1] Data Space Research Institute, Hefei Comprehensive National Science Center, Hefei, China

[2] College of Informatics, Huazhong Agricultural University, Wuhan, Hubei, China
`luojj@webmail.hzau.edu.cn`

Abstract. Detecting anomalies within complex networks is essential for identifying malicious activities, system failures, and security vulnerabilities. However, existing anomaly detection methods face two key challenges: (1) Generative Adversarial Networks (GANs) used for anomaly synthesis often fail to capture the temporal evolution of node behaviors, leading to a lack of diversity in generated anomalies; and (2) Graph Neural Networks (GNNs) often suffer from information loss due to their inherent smoothing operations, which obscure high-frequency details critical for detecting subtle anomalies. To address these challenges, we propose HADNet, a novel framework incorporating three specialized modules. The Temporal Feature Evolution Extraction Module (TFEE) captures subtle temporal changes in node features, enabling improved differentiation between normal and anomalous behaviors. The Hidden Markov Anomaly Synthesis Module (HMAS) leverages hidden Markov models to generate diverse and temporally consistent anomaly samples, enhancing the robustness of training data beyond traditional GAN-based methods. The Multiscale Time-Frequency Fusion Prediction Module (MTFFP) integrates time-domain and frequency-domain information using discrete wavelet transforms, mitigating the loss of high-frequency details caused by GNN feature aggregation. Extensive experiments on Wikipedia, Reddit, and MOOC datasets demonstrate HADNet's superiority over state-of-the-art methods, achieving ROC-AUC improvements of 2.92%, 10.66%, and 7.72%, respectively.

Keywords: Anomalous node detection · Hidden markov anomaly synthesis · Graph neural networks · Generative adversarial networks · Dynamic networks

1 Introduction

IN the digital era, complex networks are fundamental to the operation of modern information systems, underpinning various applications such as network security,

H. Jung et al. (Eds.): DASFAA 2026, LNCS 16536, pp. 67–82, 2026.
https://doi.org/10.1007/978-981-92-0366-6_5

fraud detection, and infrastructure monitoring [18]. Anomaly detection in these networks plays a crucial role in identifying abnormal nodes that may represent malicious activities, system failures, or security vulnerabilities [9,16]. These anomalous nodes, which deviate significantly from the normal behavior of other nodes, threaten the integrity of the network and must be detected promptly to prevent disruptions, breaches, and enhance network optimization and fault diagnosis [8,19,21].

Existing approaches, particularly those employing Generative Adversarial Networks (GANs) for anomaly generation, often fall short in capturing the true dynamics of complex networks [3,31]. While GANs can generate plausible anomaly samples, they typically produce anomalies that resemble existing ones without accounting for the temporal evolution of node behaviors. This leads to a lack of diversity in generated anomalies, limiting their ability to represent the full spectrum of potential anomalies in dynamic environments. Furthermore, many graph-based anomaly detection techniques rely on Graph Neural Networks (GNNs) for feature extraction [6,15,24]. While GNNs are powerful for capturing graph topological patterns, their feature extraction process often involves aggregation and smoothing operations that can result in the loss of high-frequency information—critical for detecting subtle anomalies [13,25].

To address these challenges, two key issues must be resolved. First, generating diverse anomaly samples that reflect the evolving dynamics of node behavior requires methods that consider the temporal dependencies in network data, which traditional GAN-based approaches fail to fully capture [11,15]. Second, while GNNs extract useful features, the high-frequency information—important for detecting subtle and transient anomalies—is often lost due to the smoothing and aggregation processes inherent in GNN architectures [5,13].

To overcome these challenges, we propose a novel framework called Hidden Markov-enhanced Anomaly Detection Network (HADNet). The HADNet framework introduces three specialized modules: The Temporal Feature Evolution Extraction Module (TFEE) captures subtle temporal changes in node features, enabling better differentiation between normal and anomalous behaviors. The Hidden Markov Anomaly Synthesis Module (HMAS) utilizes Hidden Markov Models (HMMs) to generate diverse and temporally consistent anomaly samples based on historical node behaviors, enhancing anomaly diversity beyond GAN-based methods. The Multiscale Time-Frequency Fusion Prediction Module (MTFFP) combines time-domain and frequency-domain information using discrete wavelet transforms to mitigate the loss of high-frequency details caused by GNN feature aggregation.

1. We propose the *HFEE Module*, which effectively captures subtle temporal changes in node features, improving the ability to differentiate between normal and anomalous behaviors.
2. We design the *HMAS Module*, which leverages Hidden Markov Models to generate diverse and temporally consistent anomaly samples based on historical node behaviors.

3. We introduce the *MTFFP Module*, which integrates time-domain and frequency-domain information through discrete wavelet transforms to mitigate the loss of high-frequency details caused by GNN feature aggregation.
4. We conduct extensive experiments on multiple datasets to evaluate the effectiveness of HADNet, demonstrating its superior robustness and generalizability compared to state-of-the-art methods. Specifically, HADNet achieves ROC-AUC improvements of 2.92%, 10.66%, and 7.72% on the Wikipedia, Reddit, and Mooc datasets, respectively, highlighting its enhanced capability in dynamic anomaly detection.

2   Related Work

2.1   Hidden Markov Model-Based Approaches for Temporal Anomaly Detection

Hidden Markov Models (HMMs) have been widely used for modeling temporal dependencies in sequential data, making them suitable for anomaly detection in dynamic networks. HMM-based methods can capture the underlying state transitions of nodes or edges, enabling the identification of abnormal patterns that deviate from expected temporal behaviors. Early works applied HMMs to network traffic analysis and intrusion detection, leveraging their probabilistic modeling of normal and anomalous sequences [20,27].

Recent studies have extended HMMs to graph-based scenarios, where node or edge attributes evolve over time, and anomalies are detected by evaluating the likelihood of observed sequences under the learned HMM parameters [1,3,21]. Furthermore, some approaches integrate HMMs with deep learning models or generative frameworks to improve detection performance in high-dimensional or complex network structures, capturing both temporal dynamics and subtle deviations in node behavior [4,15,19]. These advances enable more accurate and robust temporal anomaly detection in dynamic and heterogeneous network environments.

2.2   Wavelet Transform-Based Multiscale Time-Frequency Feature Fusion Methods

The wavelet transform is a powerful tool for analyzing signals at multiple scales, providing both time and frequency localization. In the context of network anomaly detection, wavelet-based methods can extract multiscale features that capture both global trends and local irregularities in node or edge attributes [5]. By decomposing temporal signals into different frequency components, wavelet transforms help preserve high-frequency information that is often lost in traditional graph neural network (GNN) aggregation processes. Recent approaches combine wavelet-based feature extraction with GNNs or other machine learning models to enhance the detection of subtle and transient anomalies [13,17,22,25]. These methods demonstrate improved robustness and generalizability, especially in dynamic and heterogeneous network environments. Beyond dynamic graphs, robustness-oriented detection studies in other domains highlight general principles for improving reliability under distribution shifts and missing/biased observations, such as multi-view knowledge enhancement, incomplete-modality-tolerant learning, and multi-modal debiasing [28–30]. This perspective further motivates designing anomaly detectors that remain reliable under multi-source, evolving, and noisy observations.

Despite these advances, existing methods still face challenges in generating diverse, temporally consistent anomaly samples and in effectively integrating temporal evolution with multiscale feature analysis. To address these issues, we propose a novel framework that unifies temporal modeling and multiscale analysis for more robust anomaly detection in dynamic networks (Fig. 1).

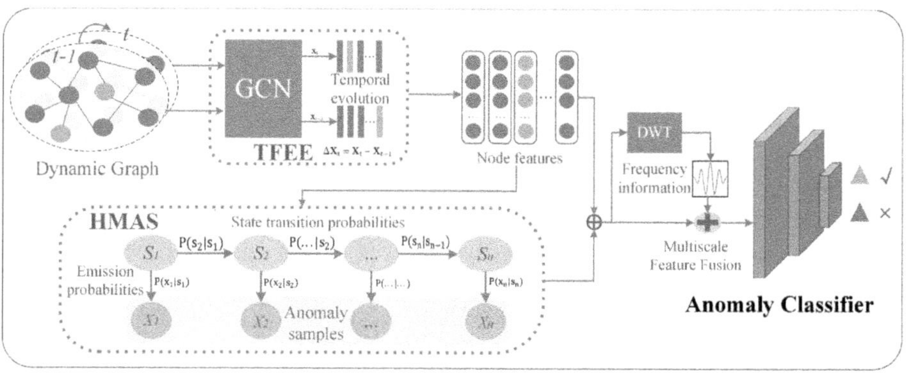

Fig. 1. The architecture of *HADNet*.

3 Methodology

Problem Definition. In dynamic networks, anomaly detection aims to identify nodes whose behavior deviates from normal temporal and structural patterns. Let $\mathcal{G} = \{G_1, G_2, \ldots, G_T\}$ denote a sequence of graph snapshots over T time steps, where each snapshot $G_t = (V_t, E_t, \mathbf{X}_t)$ contains a set of nodes V_t, edges E_t, and a node feature matrix $\mathbf{X}_t \in \mathbb{R}^{|V_t| \times d}$, with d being the feature dimension. The goal is to detect anomalous nodes in the latest snapshot G_T by analyzing the structural and temporal patterns of nodes across the historical snapshots G_1 to G_{T-1}. For each node $v \in V_T$, an anomaly score $\delta_v \in [0, 1]$ is computed, where $\delta_v = 0$ indicates normal behavior and $\delta_v = 1$ indicates an anomalous node. The anomaly score reflects the degree of deviation of a node's current features and temporal evolution from the expected patterns learned from historical data. The output is the set of anomaly scores for all nodes in G_T, i.e., $\mathcal{A}_T = \{(v, \delta_v) \mid v \in V_T, \delta_v \in [0, 1]\}$.

3.1 Temporal Feature Evolution Extraction Module (TFEE)

The goal of the TFEE module is to capture how node representations evolve across time by jointly modeling spatial correlations (through graph convolutions) and temporal variations (through feature differencing).

Graph-Based Structural Feature Encoding. At each time step t, the node features $\mathbf{X}_t$ are first processed by a K-layer graph convolutional network (GCN) to learn structure-aware latent representations. Let $\tilde{A}_t = A_t + I$ denote the adjacency matrix of G_t with self-loops added, and $\tilde{D}_t$ be the corresponding degree matrix. The normalized adjacency matrix is given by:

$$\hat{A}_t = \tilde{D}_t^{-\frac{1}{2}} \tilde{A}_t \tilde{D}_t^{-\frac{1}{2}}. \tag{1}$$

The GCN operation at the k-th layer is defined as:

$$\mathbf{H}_t^{(k)} = \sigma\left(\hat{A}_t\, \mathbf{H}_t^{(k-1)} W^{(k)}\right), \tag{2}$$

where $\mathbf{H}_t^{(0)} = \mathbf{X}_t$ is the input feature matrix, $\mathbf{H}_t^{(k)} \in \mathbb{R}^{|V_t| \times d_k}$ is the hidden representation at the k-th layer, $W^{(k)} \in \mathbb{R}^{d_{k-1} \times d_k}$ is the learnable weight matrix, and $\sigma(\cdot)$ denotes a nonlinear activation function such as ReLU. After K convolutional layers, we obtain the final structure-enhanced node representation $\mathbf{Z}_t = \mathbf{H}_t^{(K)}, \quad \mathbf{Z}_t \in \mathbb{R}^{|V_t| \times d_K}$.

Temporal Evolution Differencing. To capture temporal changes in node behaviors, we compute the difference between consecutive structure-enhanced representations:

$$\Delta \mathbf{Z}_t = \begin{cases} \mathbf{Z}_t - \mathbf{Z}_{t-1}, & t > 1, \\ \mathbf{Z}_1, & t = 1. \end{cases} \tag{3}$$

For $t > 1$, $\Delta \mathbf{Z}_t$ reflects the temporal evolution intensity of each node, highlighting short-term fluctuations or abrupt changes that may correspond to anomalies. At $t = 1$, since no previous snapshot exists, we set $\Delta \mathbf{Z}_1 = \mathbf{Z}_1$ as the initial baseline, which ensures scale consistency and preserves the temporal continuity of the sequence.

This differencing operation effectively removes static components while amplifying transient, potentially anomalous variations in node-level patterns. The resulting temporal evolution features $\{\Delta \mathbf{Z}_1, \Delta \mathbf{Z}_2, \ldots, \Delta \mathbf{Z}_T\}$ are then passed to the subsequent modules for anomaly synthesis and multiscale prediction.

3.2 Hidden Markov Anomaly Synthesis Module (HMAS)

The HMAS addresses the challenge of limited anomaly samples by leveraging the temporal dependencies inherent in sequential graph data. Based on the structure-enhanced and temporally evolved node representations obtained from the TFEE module, the HMAS focuses on learning the generative patterns of anomalous nodes and synthesizing additional samples to enrich the scarce anomaly set.

Model Training. Let $\mathcal{A}_t \subseteq V_t$ be the set of known or pseudo-labeled anomalous nodes at time t. The feature vector of anomalous node $v_i \in \mathcal{A}_t$ is given by

$$\mathbf{x}_i = \Delta\mathbf{Z}_t[v_i, :] \in \mathbb{R}^d \tag{4}$$

where $\Delta\mathbf{Z}_t[v_i, :]$ denotes the row of $\Delta\mathbf{Z}_t$ corresponding to node v_i. Aggregating all such vectors across all time steps yields the anomaly-related feature set $X_a = \{\mathbf{x}_1, \mathbf{x}_2, \ldots, \mathbf{x}_M\}, \quad \mathbf{x}_i \in \mathbb{R}^d$, where M is the total number of anomalous instances collected across all time steps. This construction ensures that each $\mathbf{x}_i$ directly represents the temporal evolution pattern of a specific anomalous node, providing the Hidden Markov Model (HMM) with a sequence of meaningful anomaly-related features to learn from.

The HMM is parameterized by $\theta = (\mathbf{A}, \mathbf{B}, \pi)$, where:

- **State transition probabilities** $\mathbf{A} = [a_{ij}]$, with

$$a_{ij} = P(s_{t+1} = j \mid s_t = i), \quad \sum_j a_{ij} = 1 \tag{5}$$

where s_t denotes the hidden state at time step t, i and j index the discrete hidden states, a_{ij} represents the probability of transitioning from state i at time t to state j at time $t + 1$, and the summation constraint ensures that the probabilities of all possible next states sum to 1.
- **Emission probabilities** $\mathbf{B} = [b_j(\mathbf{x})]$, with

$$b_j(\mathbf{x}) = P(\mathbf{x}_t \mid s_t = j) \tag{6}$$

where $\mathbf{x}_t \in \mathbb{R}^d$ is the observed feature vector at time t, s_t is the hidden state at time t, and $b_j(\mathbf{x})$ denotes the probability of observing $\mathbf{x}_t$ given that the hidden state is j.
- **Initial state distribution** $\pi = [\pi_i]$, with

$$\pi_i = P(s_1 = i), \quad \sum_i \pi_i = 1 \tag{7}$$

where s_1 is the hidden state at the first time step, π_i denotes the probability that the initial hidden state is i, and the summation constraint ensures that the probabilities over all initial states sum to 1.

The model parameters $\theta = (\mathbf{A}, \mathbf{B}, \pi)$ are estimated by maximizing the likelihood of the observed anomaly features:

$$\mathcal{L}(\theta) = \prod_{t=1}^{M} P(\mathbf{x}_t \mid \theta) \tag{8}$$

where $P(\cdot)$ is computed by marginalizing over all possible hidden state sequences. The Expectation-Maximization (EM) algorithm is used for iterative optimization, alternating between the E-step, which estimates the most probable hidden states given the current parameters, and the M-step, which updates θ to maximize $\mathcal{L}(\theta)$.

Synthetic Anomaly Generation. After training, the HMM can be used to generate synthetic anomaly samples that reflect the temporal evolution patterns learned from the real anomalies. Let N_{gen} denote the desired number of synthetic samples, and d the feature dimensionality. The generated anomaly feature matrix is denoted as:

$$\hat{X}_{\text{a}} = \{\hat{\mathbf{x}}_1, \hat{\mathbf{x}}_2, \ldots, \hat{\mathbf{x}}_{N_{\text{gen}}}\}, \quad \hat{\mathbf{x}}_i \in \mathbb{R}^d, \tag{9}$$

The generation procedure leverages the learned HMM parameters $\theta = (\mathbf{A}, \mathbf{B}, \pi)$ as follows: Sampling an initial hidden state $s_1 \sim \pi$; Iteratively sampling subsequent states $s_{t+1} \sim P(s_{t+1} \mid s_t) = a_{s_t, s_{t+1}}$; Sampling an observation from the emission probability $\hat{\mathbf{x}}_t \sim b_{s_t}(\mathbf{x})$.

This process produces a sequence of synthetic anomaly features that preserve the temporal coherence of real anomalies: consecutive observations are correlated through the hidden state transitions, while the emission distributions ensure that each generated feature remains consistent with the statistical properties of the original anomaly set.

3.3 Multiscale Time-Frequency Fusion Prediction Module (MTFFP)

The MTFFP enhances anomaly detection by capturing both global trends and local variations in node features using wavelet-based multiresolution analysis. MTFFP operates on the *augmented anomaly feature set* obtained from the HMAS module.

The augmented dataset is defined as:

$$X_{\text{aug}} = X_{\text{a}} \cup \hat{X}_{\text{a}} \in \mathbb{R}^{N_{\text{aug}} \times d}, \tag{10}$$

where $N_{\text{aug}} = M + N_{\text{gen}}$ is the total number of anomaly samples, and d is the feature dimension after TFEE processing.

Wavelet Decomposition. For each augmented node feature vector $\mathbf{x}_i \in X_{\text{aug}}$ and each feature dimension $f \in \{1, \ldots, d\}$, we treat the TFEE-enhanced feature vector as a temporal signal along the embedded evolution dimension, and apply a Discrete Wavelet Transform (DWT) to extract multiscale approximation and detail coefficients:

$$[\mathbf{A}_{i,f}^{(s)}, \mathbf{D}_{i,f}^{(s)}] = \text{DWT}^{(s)}(\mathbf{x}_i^{(f)}), \quad s = 1, \ldots, S \tag{11}$$

where $\mathbf{x}_i^{(f)} \in \mathbb{R}^T$ denotes the temporally evolved feature of dimension f for node i, S is the number of decomposition scales, $\mathbf{A}_{i,f}^{(s)}$ and $\mathbf{D}_{i,f}^{(s)}$ are the approximation and detail coefficients at scale s, respectively.

Multiscale Feature Fusion. The wavelet coefficients across all scales are fused to form a multiscale representation for each node:

$$\mathbf{H}_i = \text{Fusion}\left([\mathbf{A}_i^{(1)}, \mathbf{D}_i^{(1)}, \ldots, \mathbf{A}_i^{(S)}, \mathbf{D}_i^{(S)}]\right) \tag{12}$$

where $\mathbf{H}_i \in \mathbb{R}^{2\sum_{s=1}^{S} T_s}$ is the fused multiscale feature vector for node i, and T_s is the length of the coefficient vector at scale s.

Rationale: Different types of anomalies manifest at different temporal scales. High-frequency components ($\mathbf{D}_i^{(s)}$) capture short-term transient anomalies, such as sudden spikes or abrupt changes, while low-frequency components ($\mathbf{A}_i^{(s)}$) capture long-term trend deviations. Fusing these multiscale features ensures that the model can detect both transient and gradual anomalies. Using the augmented dataset X_{aug} allows the predictor to learn patterns from both real and synthetic anomalies, improving robustness and coverage of rare abnormal behaviors.

Anomaly Prediction. The fused multiscale representation $\mathbf{H}_i$ is mapped to an anomaly score:

$$\hat{y}_i = \sigma(\mathbf{W}_{\text{pred}}\mathbf{H}_i + \mathbf{b}_{\text{pred}}) \tag{13}$$

where $\hat{y}_i \in [0, 1]$ is the predicted anomaly score for node i, $\mathbf{W}_{\text{pred}}$ and $\mathbf{b}_{\text{pred}}$ are learnable parameters, and $\sigma(\cdot)$ is an sigmoid function.

4   Time Complexity Analysis

TFEE Complexity. The TFEE module applies a K-layer GCN to each snapshot. The per-layer GCN complexity at time t is $O(M_t d_{k-1} + N_t d_{k-1} d_k)$, accounting for message passing and linear transformations. Thus, the total complexity of TFEE over all T snapshots and K layers is:

$$O\left(\sum_{t=1}^{T}\sum_{k=1}^{K}\left(M_t d_{k-1} + N_t d_{k-1} d_k\right)\right). \tag{14}$$

Assuming $N_t \approx N$, $M_t \approx M$, and $d_k \approx F$ for all k, this reduces to:

$$O(TK(MF + NF^2)). \tag{15}$$

HMAS Complexity. The HMAS module involves training an HMM on the anomaly feature set X_{a} and generating synthetic anomalies. Let S_h be the number of hidden states in the HMM. The training (parameter estimation via the EM algorithm) has complexity $O(MS_h^2 T_s)$, dominated by the forward-backward computations. Sampling N_{gen} synthetic anomalies requires $O(N_{\text{gen}} S_h T_s)$, as each step involves state transition and emission sampling.

MTFFP Complexity. The MTFFP module applies a DWT to each feature of each augmented anomaly node and fuses multiscale coefficients. Let S be the number of wavelet decomposition scales, the complexity of a single DWT is $O(T_s)$, and there are $N_{\text{aug}} \cdot d \cdot S$ such transforms. Therefore, the total complexity of MTFFP is:

$$O(N_{\text{aug}} \, d \, S \, T_s). \tag{16}$$

Combining all three modules, the total time complexity of HADNet is:

$$O\Big(TK(MF + NF^2)\Big) + O(MS_h^2 T_s + N_{\text{gen}} S_h T_s) + O(N_{\text{aug}} dST_s), \tag{17}$$

5 Experiments

5.1 Datasets and Baselines

Our experiments employ three benchmark datasets: Wikipedia, Reddit, and Mooc [12]. Each dataset is divided into five temporal segments, where the last segment serves as the test set, and the first four are allocated for training and validation. To evaluate the performance of our anomaly detection model, we utilize two widely adopted metrics: Area Under the Receiver Operating Characteristic Curve (ROC-AUC) and Precision.

To evaluate the performance of our framework, five state-of-the-art comparison methods are introduced as follows:

- *TGAT* [26] utilizes the self-attention mechanism and introduces an innovative time-encoding technique based on Bochner's theorem from harmonic analysis.
- *GDN* [7] employs a limited number of labeled anomalies to ensure statistically significant distinctions between abnormal and normal nodes.
- *SAD* [23] is a comprehensive anomaly detection framework tailored for dynamic graphs. It integrates a time-equipped memory bank with a pseudo-label contrastive learning module, effectively harnessing large unlabeled samples to identify anomalies within graph streams.
- *TADDY* [14] formulates a node encoding that encapsulates both spatial and temporal knowledge. It utilizes a solitary transformer model to grasp the interlinked spatial-temporal information.
- *MAMF* [10] leverages Generative Adversarial Models (GANs) to augment the training with synthetic anomaly samples for learning anomaly patterns and combines meta-learning to combat concept drift.

5.2 Implementation Details

For the temporal feature extractor in HADNet, the node embedding dimension is set to $k = 128$. The wavelet-enhanced fusion predictor employs a four-layer perceptron after feature fusion, with an initial fully connected encoding layer of 512 dimensions, followed by hidden layers of 256, 128, and 64 dimensions. This progressive dimensionality reduction facilitates the extraction of high-order patterns while mitigating overfitting, ensuring generalization to unseen data. The models are trained on a server equipped with an NVIDIA RTX 4090 GPU (24GB VRAM) and 64 GB RAM, leveraging CUDA 12.2 and cuDNN 8.8 for GPU acceleration. The training uses a batch size of 128, the Adam optimizer with an initial learning rate of 1×10^{-3}, and a learning rate decay factor of 0.5 every 50 epochs. All models are trained for up to 100 epochs, with early stopping based on validation loss to prevent overfitting.

Table 1. Performance comparisons of different methods on all datasets in terms of $ROC - AUC$ and *Precision*. Boldface scores indicate the best results.

Method	Wikipedia		Reddit		Mooc	
	ROC-AUC	Precision	ROC-AUC	Precision	ROC-AUC	Precision
TGAT	83.23	1.84	67.06	3.23	66.88	6.23
GDN	85.12	6.78	67.02	0.75	66.21	3.86
SAD	86.77	1.67	68.77	0.16	69.44	3.29
TADDY	84.72	8.31	67.95	8.16	68.47	10.97
MAMF	91.21	89.36	75.64	78.42	71.35	56.59
HADNet	**94.13**	**90.11**	**86.30**	**81.58**	**79.07**	**82.32**

5.3   Performance Comparison

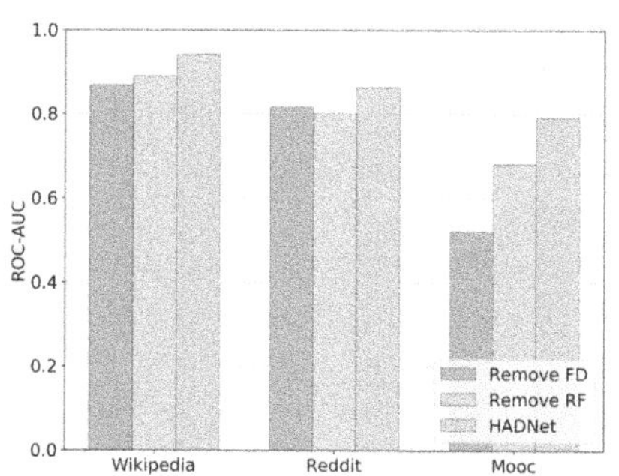
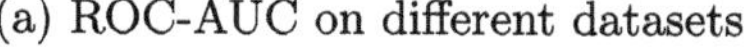

(a) ROC-AUC on different datasets

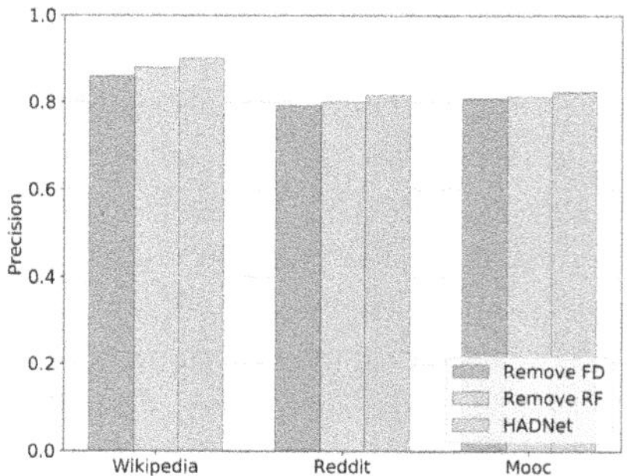

(b) Precision on different datasets

Fig. 2. Ablation study of HADNet with ROC-AUC and Precision metrics.

The performance of each method was evaluated using two primary metrics: Area Under the Curve (ROC-AUC) and Precision. To ensure robust results, each method was independently repeated 20 times for each dataset. As shown in Table 1, our framework, HADNet, demonstrated superior performance in terms of ROC-AUC and Precision across all datasets. Specifically, HADNet achieved ROC-AUC scores of 94.13%, 86.30%, and 79.07% on the Wikipedia, Reddit, and Mooc datasets, respectively.

Among baselines, TGAT showed competitive results but underperformed HADNet, indicating that self-attention and time encoding alone are insufficient for precise anomaly detection. GDN, which relies on a limited number of labeled anomalies, performed reasonably well but still lagged behind HADNet, high-lighting the potential of our approach to better utilize available data. SAD achieved promising results but was outstripped by HADNet, suggesting that our approach to feature evolution and multiscale analysis may be more effective. TADDY achieved reasonable performance through spatial-temporal modeling but failed to capture fine-grained temporal dependencies. MAMF achieved the second-highest ROC-AUC scores, yet still fell short overall, underscoring the advantage of HADNet's integrated multi-module design.

The experimental results clearly demonstrate the effectiveness of HADNet in detecting anomalies within complex networks. This suggests that our approach is not only capable of identifying anomalies with high accuracy but also of doing so with a high degree of confidence, which is crucial for practical applications where false positives can lead to significant consequences. The superior perfor-mance of HADNet validates the design choices and the innovative aspects of our framework, positioning it as a leading solution in the field of anomaly detection for complex networks.

5.4 Ablation Study

To evaluate the effectiveness of each key component in HADNet, we conduct a series of ablation experiments on three real-world datasets: Wikipedia, Reddit, and Mooc. Specifically, we consider the following variants:

- **w/o Frequency Domain (FD):** Removes the frequency domain analysis based on discrete wavelet transformation.
- **w/o Relative Features (RF):** Excludes relative temporal features that capture node state changes over time.
- **Full HADNet:** The complete model with all components included.

As shown in Fig. 2, removing either the frequency domain (FD) or relative features (RF) leads to a clear performance drop across all datasets. Among the two, the absence of frequency domain information causes a more substantial decline, especially in ROC-AUC scores, indicating its critical role in capturing hidden anomaly patterns. The relative features also contribute notably to model precision by helping distinguish subtle behavioral shifts.

The full HADNet model consistently achieves the best ROC-AUC and Pre-cision across datasets, validating the synergy between time-domain dynamics and frequency-domain representations. These results confirm the importance of integrating both types of features for robust and accurate anomaly detection.

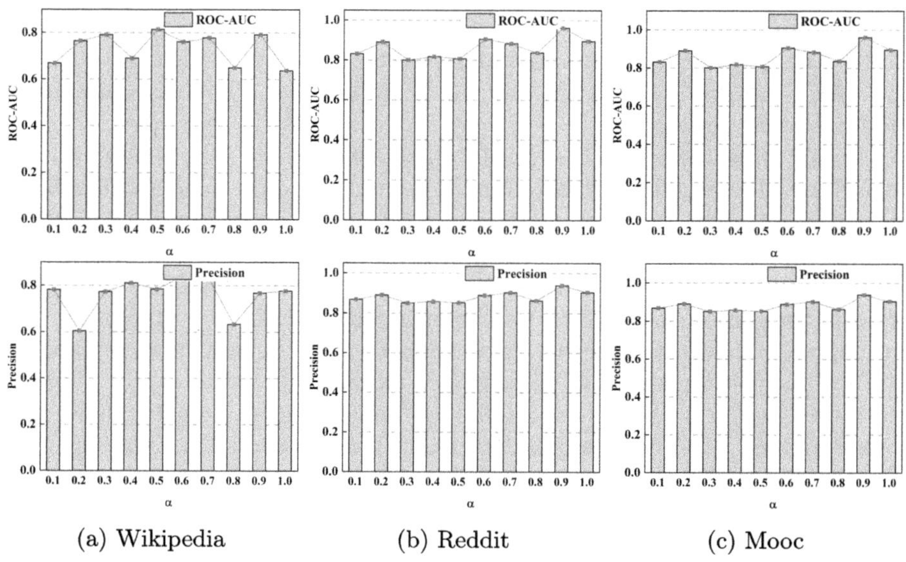

(a) Wikipedia　　　(b) Reddit　　　(c) Mooc

Fig. 3. ROC-AUC and Precision values of HADNet under different α values across the three datasets. Each column corresponds to a dataset, with the upper subfigure showing ROC-AUC and the lower showing Precision.

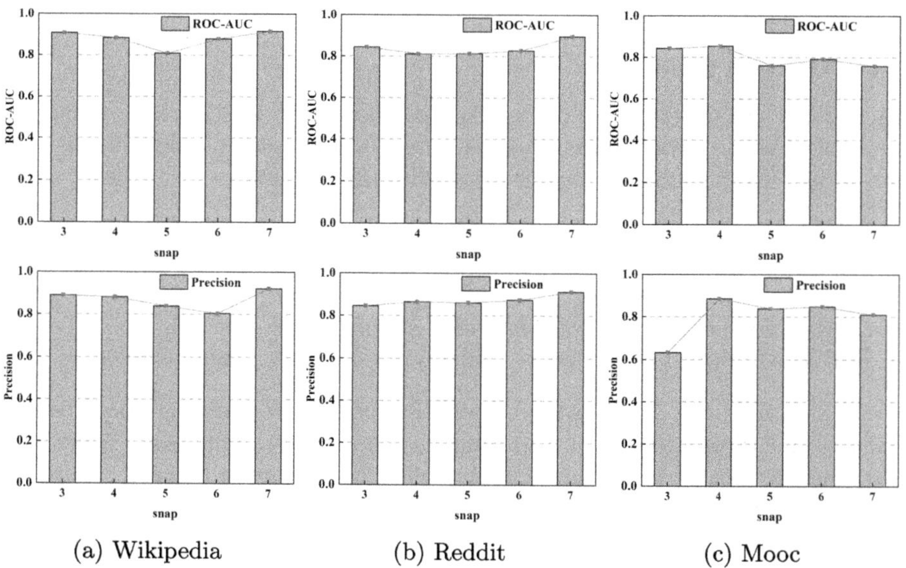

(a) Wikipedia　　　(b) Reddit　　　(c) Mooc

Fig. 4. ROC-AUC and Precision (bottom) values of HADNet under different *snap* values across the three datasets. Each column corresponds to a dataset, with the upper subfigure showing ROC-AUC and the lower showing Precision.

5.5   Parameter Analysis

To assess the robustness and adaptability of our proposed HADNet model, we conduct a parameter sensitivity analysis on two key parameters: the fusion coefficient α, which controls the contribution of frequency domain features, and the network slice granularity *snap*, which determines the temporal resolution of graph snapshots.

Analysis of Fusion Coefficient α. The fusion coefficient α regulates the weight assigned to high-frequency features extracted via discrete wavelet transformation. Figure 3 illustrates the model's ROC-AUC and Precision on three datasets under varying α values. On Wikipedia, both ROC-AUC and precision peak at $\alpha = 0.5$, indicating an optimal balance between frequency and original features. In contrast, the model on Reddit achieves maximal performance at $\alpha = 0.9$, showing that Reddit benefits more from frequency-aware information. For Mooc, peak performance is observed at $\alpha = 0.6$, after which performance slightly declines.

Analysis of Network Slice Granularity *snap*. The parameter *snap* controls the granularity of dynamic snapshots. Figure 4 reports the effect of different values of *snap* on ROC-AUC and Precision across datasets. On Wikipedia, performance improves with *snap* up to 7, after which it experiences a slight decline. In contrast, Reddit shows a consistent increase up to *snap* = 7, suggesting that it benefits from finer temporal resolution. For Mooc, performance remains relatively stable, indicating only moderate sensitivity to variations in *snap* values.

5.6   Visualization and Analysis of Feature Evolution

Figure 5 visualizes node feature distributions on Wikipedia, Reddit, and Mooc datasets before and after training via t-SNE [2]. Different colors denote distinct time snapshots, illustrating how the model adapts to temporal changes.

Before training, node features are clearly separated by time, indicating the model has yet to learn temporal generalization and is sensitive to time-specific fluctuations. After training, features from different snapshots intermingle, reflecting improved temporal generalization and reduced sensitivity to abrupt distribution shifts.

These visualizations confirm that HADNet's feature evolution extraction enables the model to capture evolving patterns over time, enhancing robustness against concept drift. This reduces false anomaly detections caused by short-term feature variations, especially in dynamic datasets like Wikipedia and Reddit. Overall, the integration of temporal features through HADNet's design leads to more stable and accurate anomaly detection in evolving networks.

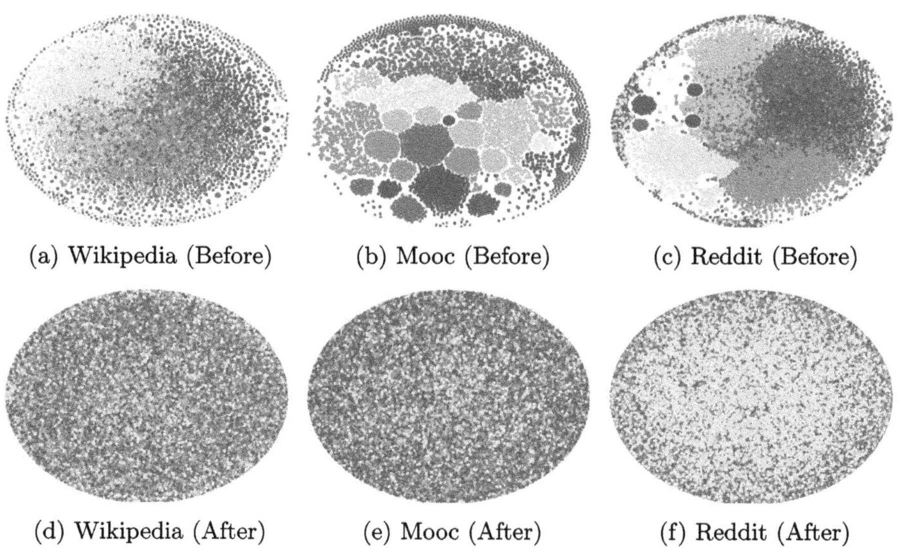

(a) Wikipedia (Before)   (b) Mooc (Before)   (c) Reddit (Before)

(d) Wikipedia (After)   (e) Mooc (After)   (f) Reddit (After)

Fig. 5. Node feature distributions of three datasets (Wikipedia, Mooc, and Reddit) visualized by t-SNE clustering. Different colors correspond to different time steps. The top row shows the feature spaces before training, and the bottom row shows the feature spaces after training.

6   Conclusion

This paper introduces HADNet, a novel model for anomaly detection in complex network environments. By integrating frequency-domain features via discrete wavelet transformation and leveraging temporal analysis, HADNet achieves superior performance on real-world datasets, including Wikipedia, Reddit, and Mooc. We examine two key parameters—the fusion coefficient α and network slice granularity *snap*—to assess their impact on performance. Extensive experiments using ROC-AUC and Precision metrics show that HADNet outperforms state-of-the-art methods in robustness to concept drift and generalization across diverse settings. High-frequency feature fusion enhances sensitivity to subtle anomalies, while temporal slicing effectively captures evolving dynamics.

References

1. Ahmed, M., Mahmood, A., Hu, J.: Network traffic analysis based on hidden Markov models. Comput. Secur. **61**, 1–14 (2016)
2. Arora, S., Hu, W., Kothari, P.K.: An analysis of the t-SNE algorithm for data visualization. In: Conference on Learning Theory, pp. 1455–1462. PMLR (2018)
3. Azzalini, D., et al.: HMMs for anomaly detection in autonomous robots. In: Proceedings of the 19th International Conference on Autonomous Agents and Multi-agent Systems (AAMAS 2020), pp. 105–113. IFAAMAS (2020)

4. Chen, L., et al.: Generative adversarial synthetic neighbors-based unsupervised anomaly detection. Sci. Rep. (2025)
5. Debnath, L.: Wavelet Transforms and Time-Frequency Signal Analysis. Springer (2012)
6. Ding, K., Li, J., Bhanushali, R., Liu, H.: Deep anomaly detection on attributed networks. In: SIAM International Conference on Data Mining (SDM), pp. 594–602 (2019)
7. Ding, K., Zhou, Q., Tong, H., Liu, H.: Few-shot network anomaly detection via cross-network meta-learning. In: Proceedings of the Web Conference 2021, pp. 2448–2456 (2021)
8. He, Y., et al.: Graph-enhanced anomaly detection framework in multivariate time series using graph attention and enhanced generative adversarial networks. Expert Syst. Appl. **271**, 126667 (2025)
9. Hong, Y., Ali, M.A., Wang, H., Chen, J., Wang, D.: Abnet: mitigating sample imbalance in anomaly detection within dynamic graphs. In: Proceedings of the Thirty-Fourth International Joint Conference on Artificial Intelligence, pp. 2910–2918 (2025)
10. Hong, Y., Shi, C., Chen, J., Wang, H., Wang, D.: Multitask asynchronous met-alearning for few-shot anomalous node detection in dynamic networks. IEEE Trans. Comput. Soc. Syst. 1–12 (2024). https://doi.org/10.1109/TCSS.2024.3442238
11. Jain, R., Abouzakhar, N.S.: Hidden Markov model based anomaly intrusion detection. In: 2012 International Conference for Internet Technology and Secured Transactions, pp. 528–533. IEEE (2012)
12. Kumar, S., Zhang, X., Leskovec, J.: Predicting dynamic embedding trajectory in temporal interaction networks. In: Proceedings of the 25th ACM SIGKDD International Conference on Knowledge Discovery & Data Mining, pp. 1269–1278 (2019)
13. Li, M., Shu, M., Lu, T.: Anomaly pattern detection in high-frequency trading using graph neural networks. J. Ind. Eng. Appl. Sci. **2**(6), 77–85 (2024)
14. Liu, Y., et al.: Anomaly detection in dynamic graphs via transformer. IEEE Trans. Knowl. Data Eng. (2021)
15. Ma, L., et al.: Generative adversarial message passing-based anomaly detection. J. Comput. Sci. Technol. (2025)
16. Ma, X., et al.: A comprehensive survey on graph anomaly detection with deep learning. IEEE Trans. Knowl. Data Eng. **35**(12), 12012–12038 (2021)
17. Narang, S.K., Gadde, A., Ortega, A.: Signal processing techniques for interpolation in graph structured data. In: 2013 IEEE International Conference on Acoustics, Speech and Signal Processing, pp. 5445–5449. IEEE (2013)
18. Qiao, H., Tong, H., An, B., King, I., Aggarwal, C., Pang, G.: Deep graph anomaly detection: a survey and new perspectives. IEEE Trans. Knowl. Data Eng. (2025)
19. Qiu, C., et al.: Raising the bar in graph-level anomaly detection. In: Proceedings of the Thirty-First International Joint Conference on Artificial Intelligence (IJCAI) (2022)
20. Rabiner, L.R.: A tutorial on hidden Markov models and selected applications in speech recognition. Proc. IEEE **77**(2), 257–286 (2002)
21. Ren, H., Ye, Z., Li, Z.: Anomaly detection based on a dynamic Markov model. Inf. Sci. **411**, 52–65 (2017)
22. Shuman, D.I., Narang, S.K., Frossard, P., Ortega, A., Vandergheynst, P.: The emerging field of signal processing on graphs: extending high-dimensional data analysis to networks and other irregular domains. IEEE Signal Process. Mag. **30**(3), 83–98 (2013)

23. Tian, S., et al.: SAD: semi-supervised anomaly detection on dynamic graphs. arXiv preprint arXiv:2305.13573 (2023)
24. Wu, Z., Pan, S., Chen, F., Long, G., Zhang, C., Philip, S.Y.: A comprehensive survey on graph neural networks. IEEE Trans. Neural Netw. Learn. Syst. **32**(1), 4–24 (2020)
25. Xu, B., Shen, H., Cao, Q., Qiu, Y., Cheng, X.: Graph wavelet neural network. arXiv preprint arXiv:1904.07785 (2019)
26. Xu, D., Ruan, C., Korpeoglu, E., Kumar, S., Achan, K.: Inductive representation learning on temporal graphs. arXiv preprint arXiv:2002.07962 (2020)
27. Ye, N., Emran, S.M., Chen, Q., Vilbert, S.: A Markov chain model for computer intrusion detection. IEEE Trans. Software Eng. **28**(7), 595–607 (2002)
28. Zeng, Z., et al.: Mitigating world biases: a multimodal multi-view debiasing framework for fake news video detection. In: Proceedings of the 32nd ACM International Conference on Multimedia, pp. 6492–6500 (2024)
29. Zeng, Z., et al.: Understand, refine and summarize: multi-view knowledge progressive enhancement learning for fake news video detection. In: Proceedings of the 33rd ACM International Conference on Multimedia, pp. 9216–9225 (2025)
30. Zeng, Z., et al.: IMOL: incomplete-modality-tolerant learning for multi-domain fake news video detection. In: Proceedings of the 63rd Annual Meeting of the Association for Computational Linguistics (Volume 1: Long Papers), pp. 30921–30933 (2025)
31. Zong, B., et al.: Deep autoencoding gaussian mixture model for unsupervised anomaly detection. In: International Conference on Learning Representations (ICLR) (2018)

Graph Clustering with Scalable Graph Filters and View-Specific Semantic Fusion

Wenxin Zhang[1], Xi Xuan[2], Renda Han[3], Desheng Wu[1], Cuicui Luo[1($\boxtimes$)], and Ljupco Kocarev[4]

[1] University of Chinese Academy of Sciences, Beijing, China
`luocuicui@ucas.ac.cn`
[2] City University of Hong Kong, Hong Kong SAR, China
[3] Hainan University, Haikou, China
[4] Ss Cyril and Methodius University in Skopje, Skopje, North Macedonia

Abstract. Multi-view graph clustering (MVGC) has made great progress in analyzing the interaction patterns of complex networks. Existing methods leverage different graph filters to obtain high- and low-pass signals and implement multi-view fusion. However, these filter-based methods face a scalability issue, which results in insufficient representation discrimination. Besides, they lack view-specific semantics in multi-view fusion, leading to poor information fidelity. To address these limitations, we propose a graph clustering framework with scalable graph filters and view-specific semantic fusion (SGSF-GC). SGSF-GC designs a Beta-based scalable graph filter and cohesion-based fusion mechanism to capture and integrate high- and low-frequency signals. Then SGSF-GC employs class activation mapping to capture semantics of view-specific representations for multi-view fusion. Finally, SGSF-GC conducts KL-based graph clustering. Extensive experiments on five public datasets with eleven baselines verify the utility and superiority of SGSF-GC.

Keywords: Graph clustering · Graph representation learning · Unsupervised learning

1 Introduction

Multi-view graph clustering (MVGC) has gained significant attention as a powerful framework for analyzing graph-structured data, leveraging complementary structural information across multiple perspectives. The rise and development of graph neural networks (GNNs) have notably advanced MVGC [4,5,31], showcasing enhanced performance. To adequately capture the interactive information on the graph topology, scholars develop many advanced models for MVGC. They focus on capturing the consistency of different views with auxiliary tasks to enhance the clustering performance [19], which explores the interactions between multi-view representations, driving advancements in graph clustering.

W. Zhang and X. Xuan—These authors contributed equally to this work.

H. Jung et al. (Eds.): DASFAA 2026, LNCS 16536, pp. 83–97, 2026.
https://doi.org/10.1007/978-981-92-0366-6_6

Due to the existence of heterophily, traditional GNNs fail to capture the discriminative features in each graph view, thereby resulting in over-smoothing problems. Recently, many studies manage to explore the potential of graph filters for clustering [24,32,43]. The prevailing scheme typically refers to the separation of graph signals into different frequencies using low-pass and high-pass filters and their subsequent cross-frequency fusion. However, although existing methods achieve significant performance, their capacity for capturing discriminative features is still highly limited. First, due to the high reliance on single low- and high-pass filters, **existing methods exhibit limited scalability, leading to insufficient representation learning**. Most methods, respectively, design an individual filter for different frequencies, lacking adaptive multi-scale filtering for specific frequency domains. This setting greatly limits the capability to capture frequency-domain-specific representations of different scales when encountering complex relational structures. Indeed, the low- and high-frequency domains contain multi-scale signals, which are crucial for learning fine-grained representations. Therefore, adaptively capturing and integrating frequency-domain-specific graph signals has great significance for graph clustering. Second, **most approaches exhibit poor fidelity and discriminability, overlooking the semantic correlation of view-specific representations**. This phenomenon can be attributed to simple and direct summation or concatenation when integrating view-specific representations [29]. According to [26], view-specific representations have different contributions due to discrepancies in semantic relevance, while these trivial fusion schemes pay less attention to corresponding semantic information. Inevitably, these view fusion schemes will directly lead to less fidelity and discriminability of the latent representations, resulting in distorted compounds. Therefore, extracting precise semantics to preserve the view-specific fidelity is crucial for the multi-view representation fusion in graph clustering.

To solve these problems, we introduce a novel graph clustering with scalable graph filters and view-specific semantic fusion (SGSF-GC). For the first limitation, SGSF-GC captures multi-scale graph signals in different frequency domains using a series of Beta wavelet graph filters and develops a cohesion-based fusion mechanism to integrate frequency-domain signals. In this way, SGSF-GC can derive comprehensive characteristics of filtering signals, enhancing the model's discrimination. For the second limitation, SGSF-GC deploys class activation mapping (CAM) to refine the semantics of view-specific representations for multi-view fusion. The contributions of SGSF-GC are as follows:

- We propose a novel SGSF-GC, which leverages CAM to refine critical semantic characteristics of view-specific representations and implement multi-view fusion without losing information fidelity about different graph views.
- We introduce a scalable Beta graph encoder and a cohesion-based fusion mechanism to generate high-quality fused graph signals, enhancing the scalability of filters to capture scalable graph signals and improving the discrimination of representations.
- Extensive comparative experiments are conducted on five public datasets with eleven baselines, demonstrating the effectiveness of SGSF-GC.

2   Related Work

Recent years have witnessed advancements in multi-view graph clustering research, which can be categorized into traditional GNN-based methods and graph filter-based methods. Traditional approaches aim to leverage traditional GNN to capture a comprehensive representation for nodes. Many effective methods are proposed. LHG-MVC [28] exploits the local high-order representation based on rotated tensor nuclear norm and combines consistent affinity features with their high-order features using adaptive weights. RJGC [7] introduces a $\ell 1$-norm-based unified graph matrix integrated with a similarity matrix to enhance the clustering robustness. BMGC [23] identifies a dominant view for graph clustering to alleviate the noise from different views. CEMR [36] employs diversified contrastive learning to achieve cross-view alignment while mitigating dimensional collapses. Some study [18] also addresses multi-view representation integration through consistency constraints. Graph filter-based methods aim to elaborate powerful graph filters to capture discriminative signals in different frequencies [3, 20–22]. AHGFC [29] designs a pair of high- and low-pass filters to capture the characteristics of the critical category. CGFMVC [43] establishes uniform multiple graph filters by integrating information from all views instead of relying on filtering each view. MvC-BDGF [32] develops a block diagonal graph filter and seamlessly combines graph filters with consensus to obtain optimal filters.

Although existing methods successfully deploy diverse graph filters to obtain signals with different frequencies, most methods use a single graph filter for each frequency domain. Ignoring multi-scale comprehensive frequency-domain representations leads to less scalability of graph filters when encountering complex graph topologies. Besides, existing methods lead to poor fidelity when implementing view-specific fusion due to the direct and simple fusion operators.

3   Methodology

3.1   Problem Definition

Consider a multi-view graph as $\mathcal{G} = \{\mathcal{V}, \mathcal{A}, \mathbf{X}\}$, where $\mathcal{V} = \{v_1, \cdots, v_N\}$ represents the node set and N is the number of nodes. $\mathbf{X} \in \mathbb{R}^{N \times d}$ denotes the node features, where d is the original feature dimension. $\mathcal{A} = \{\mathbf{A}^1, \cdots, \mathbf{A}^R\}$ is adjacency matrices and R is the number of relations. $\mathbf{A}^r \in \{0, 1\}^{N \times N}$ denotes the adjacency matrix under relation r, where $\mathbf{A}^r_{ij} = 1$ if an edge exists between node v_i and v_j, otherwise $\mathbf{A}^r_{ij} = 0$. The graph clustering problem aims to partition the node set $\mathcal{V}$ into m independent clusters $\mathcal{C} = \{C_1, \cdots, C_m\}$ and $\sum_i^m |C_i| = N$.

3.2   Overview

As illustrated in Fig. 1, SGSF-GC first employs a scalable Beta graph encoder to capture sufficient node features from different scales of frequencies and generates high- and low-pass signals using dynamical weights of different scales. Then SGSF-GC implements signal fusion based on the similarity between anchor nodes with their local representations in different frequency domains. Next, SGSF-GC obtains semantic embeddings of view-specific representations using CAM and achieves adaptive multi-view fusion based on attention, preserving critical view-specific semantics. Finally, it enhances category preference information using KL-based clustering regularization.

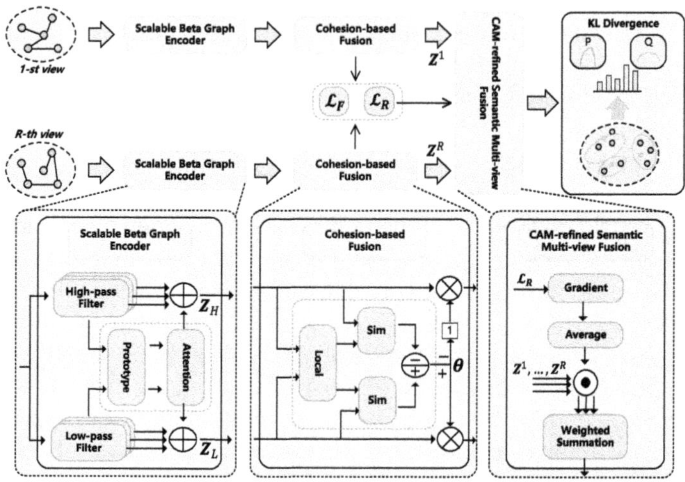

Fig. 1. The framework of SGSF-GC.

3.3   Scalable Beta Graph Encoder

Traditional filter-based approaches suffer from the scalability problem, significantly compromising the quality of representation learning and impairing the capture of critical characteristics of nodes. We introduce the Beta wavelet as graph filters for each subgraph under relation r, and for simplicity, we omit the superscript r. The process is defined as:

$$\mathbf{Z}_{(p,q)} = \mathcal{F}_{(p,q)}(\mathbf{X}, \mathbf{A}) = \beta_{(p,q)}(\hat{\mathbf{L}})\mathbf{X}, \quad (p + q = C), \tag{1}$$

where $\beta_{(p,q)}(\mathbf{L}) = \frac{\mathbf{L}^p(1-\mathbf{L})^q}{B_{(p,q)}}$, $B_{(p,q)} = \frac{p!q!}{(p+q+1)!}$, C are hyperparameters, and $\hat{\mathbf{L}} = \mathbf{I}_N - \mathbf{D}^{-\frac{1}{2}}\mathbf{A}\mathbf{D}^{-\frac{1}{2}}$ is normalized Laplacian matrix. To filter low- and high-pass graph signals, we define a constant C and generate a series of filters $\mathcal{F} = [\mathcal{F}_{(0,C)}, \mathcal{F}_{(1,C-1)}, \cdots, \mathcal{F}_{(C,0)}]$. By setting a tunable hyperparameter T, we split $\mathcal{F}$ into $\mathcal{F}_L$ and $\mathcal{F}_H$ to respectively capture low- and high-pass graph signals:

$$\mathcal{F}_L = [\mathcal{F}_{0,C}, \cdots, \mathcal{F}_{T,C-T}], \mathcal{F}_H = [\mathcal{F}_{T+1,C-T-1}, \cdots, \mathcal{F}_{C,0}]. \tag{2}$$

Since not all frequency bands show equal contributions, we introduce a prototype-based adaptive fusion. Specifically, we compute the prototypes of different frequency-specific domains:

$$\bar{\mathbf{Z}}_L = \frac{1}{T+1} \sum_{p=1}^{T} \mathbf{Z}_{(p,T-p)}, \quad \bar{\mathbf{Z}}_H = \frac{1}{C-T-1} \sum_{p=T+1}^{C} \mathbf{Z}_{(p,T-p)}, \tag{3}$$

where $\bar{\mathbf{Z}}_L \in \mathbb{R}^{N \times d'}$ and $\bar{\mathbf{Z}}_H \in \mathbb{R}^{N \times d'}$ respectively represent the prototype of low- and high-frequency domains and d' is the hidden dimension. Then we generate affinity weights according to the prototypes:

$$\kappa_{(p,T-p)} = \begin{cases} \dfrac{e^{\phi(\mathbf{Z}_{(p,T-p)}, \bar{\mathbf{Z}}_L)}}{\sum_{p=0}^{T} e^{\phi(\mathbf{Z}_{(p,T-p)}, \bar{\mathbf{Z}}_L)}}, & \text{if } 0 \leq p \leq T, \\ \dfrac{e^{\phi(\mathbf{Z}_{(p,T-p)}, \bar{\mathbf{Z}}_H)}}{\sum_{p=T+1}^{C} e^{\phi(\mathbf{Z}_{(p,T-p)}, \bar{\mathbf{Z}}_H)}}, & \text{if } T < p \leq C, \end{cases} \tag{4}$$

where $\phi(\cdot, \cdot)$ denotes cosine similarity. Finally, we integrate low- and high-pass graph signals respectively with their specific weights:

$$\mathbf{Z}_L = \sum_{p=0}^{T} \kappa_{(p,T-p)} \mathbf{Z}_{(p,T-p)}, \quad \mathbf{Z}_H = \sum_{p=T+1}^{C} \kappa_{(p,T-p)} \mathbf{Z}_{(p,T-p)}, \tag{5}$$

In this way, the model can capture more multi-scale graph signals in different frequency-specific domains. To adaptively learn the contributions of different frequency domains, we integrate high- and low-pass representations based on $\boldsymbol{\theta}^r$ to get latent representations $\mathbf{Z}^r$ under relation r:

$$\mathbf{Z}^r = (1 - \boldsymbol{\theta}^r) \cdot \mathbf{Z}_L^r + \boldsymbol{\theta}^r \cdot \mathbf{Z}_H^r. \tag{6}$$

The trade-off parameter $\boldsymbol{\theta}^r \in \mathbb{R}^{N \times 1}$ dynamically balances the weights of homophilic and heterophilic information within the graph structure. This adaptively learnable parameter effectively prevents signal distortion by evaluating the cohesion of node representations. The detailed resolution of $\boldsymbol{\theta}^r$ is illustrated in Sect. 3.4. Subsequently, we design a reconstruction loss and preserve view-specific topological characteristics through decoding $\mathbf{Z}^r$. The intuition is that the structural patterns of nodes under different relations exhibit discrepancies. We introduce scaled cosine error (SCE) [34] to evaluate reconstruction.

$$\hat{\mathbf{X}}^r = \text{MLP}(\mathbf{Z}^r),$$

$$\mathcal{L}_R = \frac{1}{R} \sum_{r=1}^{R} \text{SCE}((\mathbf{X}^r)^T, \hat{\mathbf{X}}^r), \tag{7}$$

where $\text{MLP}(\cdot)$ is a shared transform layer. Through view-specific reconstruction, the features imply the distinctive topological information under each relation, which helps to model the potential association between structural features and node category preferences.

3.4 Cohesion-Based Fusion Mechanism

The spectral characteristics of a graph are mainly comprised of homophilic and heterophilic compositions. Traditional graph signal fusion methods only use learnable parameters to integrate different compositions into the latent common space without considering cohesion, which is not conducive to generating discriminative features. To alleviate this problem, we propose a cohesion-based fusion mechanism to obtain $\boldsymbol{\theta}$. Inspired by the common consensus that homophilic components indicate the affinity between nodes, while heterophilic components suggest the discordance between nodes, the intuition of the cohesion-based fusion mechanism is that low-frequency node representation should exhibit strong affinity with local representation, while high-frequency node representation should represent significant inconsistency with local representation. Therefore, we first obtain the local representations of nodes, and then calculate the similarity between node representations and the local to evaluate the cohesion of nodes.

Specifically, we first generate local representations for low- and high-frequency representations of each node. For simplicity, we omit the superscript r:

$$\hat{\mathbf{Z}}_{\mathcal{P}}[i,:] = \frac{1}{|\mathcal{N}_{v_i}|} \sum_{v_j \in \mathcal{N}_{v_i}} \mathbf{Z}_{\mathcal{P}}[j,:] \tag{8}$$

where $\mathcal{N}_{v_i}$ is the neighbor set of node v_i, $\mathcal{P} = \{L, H\}$ and $\mathbf{Z}_{\mathcal{P}}[i,:] \in \mathbb{R}^{1 \times d'}$ denotes the i-th row of representation of $\mathbf{Z}_{\mathcal{P}}$, namely the representation of node v_i. After obtaining local representations $\hat{\mathbf{Z}}_L \in \mathbb{R}^{N \times d'}$ and $\hat{\mathbf{Z}}_H \in \mathbb{R}^{N \times d'}$, we calculate the trade-off parameter $\boldsymbol{\theta}$ according to the similarity between node representations and the local:

$$\boldsymbol{\theta} = \Phi[\phi(\hat{\mathbf{Z}}_H, \mathbf{Z}_H) - \phi(\hat{\mathbf{Z}}_L, \mathbf{Z}_L)], \tag{9}$$

where $\Phi(\cdot)$ is Softmax normalization function. Finally, for $\boldsymbol{\theta}^r$ under relation r, we define the entropy minimization loss function to polarize it:

$$\mathcal{L}_F = \frac{1}{R} \sum_{r=1}^{R} [-\frac{1}{N} \sum_{i=1}^{N} \boldsymbol{\theta}_i^r \log(\boldsymbol{\theta}_i^r)]. \tag{10}$$

The parameter $\boldsymbol{\theta}^r$ aims to generate node-wise attentions, effectively capturing the structural balance between node cohesion patterns within graph representations for each node. Different from traditional approaches that leverage labels to count edge properties, $\boldsymbol{\theta}^r$ serves as a viable alternative for quantifying network homophily characteristics, reducing computational cost and maintaining discriminative capacity.

3.5 CAM-Refined Semantic Multi-view Fusion

After obtaining $\mathbf{Z}^r$ under each view, we need to integrate the multi-view representations. However, traditional methods often leverage a concatenation function

[29,37] or learnable parameters [38,39] to derive the fused representations, losing vital semantic relevance information and causing poor fidelity. As a popular interpretable tool [13], CAM can capture valuable gradient information to better capture the semantic fusion and enhance view-specific fidelity and discriminability. To this end, we introduce a CAM-based semantic fusion, which leverages the topological semantic information to adjust view-specific representations.

Specifically, we compute the closed form gradient of $\mathcal{L}_R$ in Eq. 7:

$$\mathbf{S}^r = \frac{\partial \mathcal{L}_R}{\partial \mathbf{Z}^r} \tag{11}$$

where $\mathbf{S}^r \in \mathbb{R}^{N \times d'}$ denotes the gradient matrix of all dimensions over all nodes. Then we derive the importance vector $\mathbf{s}_p^r \in \mathbb{R}^{1 \times d'}$ of p-th dimension leveraging readout function on $\mathbf{S}^r$ along the node dimension:

$$\mathbf{s}_p^r = \sigma(\frac{1}{N} \sum_{i=1}^{N} \mathbf{S}^r[i,:]) \tag{12}$$

where $\sigma(\cdot)$ is ReLU activation function and $\mathbf{S}^r[i,:] \in \mathbb{R}^{1 \times d'}$ denote the representations of node v_i in $\mathbf{S}^r$. Inspired by CAM, We generate the refined representations based on the importance vector and view-specific representations:

$$\boldsymbol{\omega}^r = \text{NL}[\sigma(\mathbf{Z}^r (\mathbf{s}_p^r)^\top)] \tag{13}$$

where $\boldsymbol{\omega}^r \in \mathbb{R}^{N \times 1}$ records the importance scores of all nodes under relation r. Then we define a matrix $\boldsymbol{\Omega} = [\boldsymbol{\omega}^1, \cdots, \boldsymbol{\omega}^R] \in \mathbb{R}^{N \times R}$. Then we leverage a normalization function along the view dimension to regularize the view-specific importance scores, denoted as $\hat{\boldsymbol{\Omega}} = [\hat{\boldsymbol{\omega}}^1, \cdots, \hat{\boldsymbol{\omega}}^R]$. Finally, we generate the fused representation $\mathbf{Z}_F$ using multiplication with a broadcast mechanism:

$$\mathbf{Z}_F = \sum_{r=1}^{R} \mathbf{Z}_r \hat{\boldsymbol{\omega}}^r. \tag{14}$$

3.6   Graph Clustering

After obtaining the final embeddings of nodes, we introduce graph clustering to obtain the category preference of nodes. Specifically, we employ k-means to final node representations to obtain the soft cluster probability distributions:

$$\hat{\mathbf{y}} = \text{KMeans}(\mathbf{Z}_F), \tag{15}$$

where $\text{KMeans}(\cdot)$ denotes the k-means cluster algorithm. Then the soft distribution $\mathcal{Q}$ can be calculated:

$$q_{ij} = \frac{\eta_j (1 + ||\mathbf{Z}_F[i,:] - \mathbf{c}_j||^2)^{-1}}{\sum_{k=1}^{C} \eta_k (1 + ||\mathbf{Z}_F[i,:] - \mathbf{c}_k||^2)^{-1}}, \tag{16}$$

Algorithm 1. Algorithm of SGSF-GC.

Input: Graph $\mathcal{G} = (\mathbf{X}, \mathbf{A})$; Hyperparameters α, β and γ; Maximum accessibility neighbor order L_e; Batch size B; Epochs E; Learning rate lr; Temperature coefficient τ
Output: The anomaly scores s

1: Initialize the parameters of SGSF-GC.
2: **for** $e \in (1, 2, \cdots, E)$ **do**
3:     Obtain batch $\mathcal{B}$ according to B.
4:     **for** $b \in \mathcal{B}$ **do**
5:         **for** $r \in (1, 2, \cdots, R)$ **do**
6:             Generate multi-scale frequency-domain signals using Beta filters. $\rightarrow$ Eq. (3)

7:             Generate low-frequency and high-frequency representations. $\rightarrow$ Eq. (5)
8:             Calculate adaptive fusion parameter. $\rightarrow$ Eq. (9)
9:             Adaptively integrate high- and low-frequency representations. $\rightarrow$ Eq. (6)
10:           Calculate gradient $\mathbf{S}^r$ and importance vector $\boldsymbol{\omega}^r$, then integrate view-specific representations $\leftarrow$ Eq. (11)- Eq. (14)
11:         **end for**
12:         Calculate $\mathcal{L}_R$. $\rightarrow$ Eq. (7)
13:         Calculate $\mathcal{L}_F$. $\rightarrow$ Eq. (10)
14:         Implement clustering and calculate Calculate $\mathcal{L}_{KL}$. $\rightarrow$ Eq. (15)-Eq. (19)
15:         Calculate the final loss function. $\rightarrow$ Eq. (20)
16:         Update the model's parameters.
17:     **end for**
18: **end for**
19: **return** solution

---

where $\mathbf{c}_j \in \mathbb{R}^{1 \times d'}$ is the prototype center of cluster j, N_j is the number of nodes in the cluster $\mathcal{C}_j$, C is the number of clusters, q_{ij} indicates the distance between latent representation and the prototype center and η_j denotes the learnable parameter which measures the importance of cluster $\mathcal{C}_j$. However, directly optimizing the parameter is inevitably tough. To effectively and dynamically evaluate the quality of clusters, we calculate the average distance between the prototype center of cluster $\mathcal{C}_j$ and the nodes in cluster $\mathcal{C}_j$:

$$\eta_j = \text{softmax}(\frac{1}{N_j} \sum_{v_i \in \mathcal{C}_j} ||\mathbf{Z}_F[i, :] - \mathbf{c}_j||^2) \tag{17}$$

where $\text{softmax}(\cdot)$ is softmax normalization function. In this way, parameter $\eta_j \in [0, 1]$ satisfies the constraint $\sum_{m \in \mathcal{C}} \eta_m = 1$ and can be viewed as the learnable weights to reflect the quality of clusters. Then the target distribution $\mathcal{P}$ can be calculated as:

$$p_{ij} = \frac{(q_{ij} / \sum_{i=1}^{N} q_{ij})^2}{\sum_{k=1}^{C} q_{ik} / \sum_{i=1}^{N} q_{ik}}. \tag{18}$$

We introduce KL divergence to enhance the alignment between the target distribution $\mathcal{P}$ and the soft distribution $\mathcal{Q}$:

$$\mathcal{L}_{KL} = KL(\Omega(\mathcal{P})||\mathcal{Q}) + KL(\mathcal{P}||\Omega(\mathcal{Q})), \tag{19}$$

Table 1. Clustering performance on five datasets. The best and second-best observations are highlighted **in bold** and underlined, respectively. The results are in %.

Dataset	Metric	HAN 2019	VGAE 2016	DGI 2019	DLGR 2023	DMG 2023	MvAGC 2023	BTGF 2024	DOGC 2024	MCGF 2025	DSGC 2025	RJGC 2025	SGSF-GC
Amazon	NMI	40.37	1.63	5.32	27.67	12.18	42.16	38.53	48.73	49.16	<u>50.83</u>	49.12	**53.12**
	ARI	42.41	1.29	2.02	27.15	2.83	33.15	28.29	41.64	48.58	<u>49.16</u>	47.85	**52.93**
	ACC	74.37	31.94	37.62	61.23	38.87	64.39	66.03	73.89	<u>75.94</u>	75.24	74.26	**78.66**
	F1	74.33	27.25	28.59	62.15	34.41	62.10	66.12	72.46	<u>74.26</u>	73.29	72.89	**76.12**
ACM	NMI	68.64	49.10	63.64	73.28	75.61	61.80	75.80	75.20	77.51	<u>77.93</u>	76.42	**80.10**
	ARI	74.89	54.48	68.22	79.42	80.33	68.08	80.85	82.51	81.45	<u>83.16</u>	81.35	**84.32**
	ACC	90.88	82.28	88.16	92.71	93.02	88.06	92.22	92.90	93.52	<u>93.86</u>	92.15	**94.68**
	F1	90.85	82.31	88.29	92.70	93.06	88.35	91.31	92.33	93.24	<u>93.77</u>	92.28	**95.10**
DBLP	NMI	69.98	69.34	61.68	75.59	79.07	77.68	60.27	79.51	80.15	<u>80.29</u>	78.96	**82.62**
	ARI	76.41	74.13	56.53	81.68	83.84	82.67	65.34	<u>84.39</u>	83.18	84.03	82.85	**88.06**
	ACC	90.15	88.86	74.46	91.80	93.03	92.82	84.56	<u>93.41</u>	93.33	93.19	91.74	**94.72**
	F1	89.66	87.48	73.92	91.80	<u>93.03</u>	92.37	84.56	92.90	91.26	92.57	91.68	**93.41**
ACM2	NMI	64.35	45.07	57.79	59.88	63.41	78.19	64.83	68.52	71.02	<u>72.58</u>	70.89	**75.22**
	ARI	69.79	43.47	51.74	63.99	67.26	69.33	67.76	72.16	<u>73.68</u>	72.91	71.95	**76.49**
	ACC	89.43	73.58	81.14	86.53	87.96	82.16	88.53	89.14	89.08	<u>90.27</u>	88.92	**92.41**
	F1	89.55	71.01	82.61	86.53	87.73	80.47	88.87	<u>90.73</u>	90.23	90.72	89.18	**91.77**
Yelp	NMI	67.62	39.19	39.42	66.21	39.10	66.16	41.35	69.04	66.49	<u>70.86</u>	69.24	**73.26**
	ARI	<u>70.45</u>	42.57	42.62	68.47	42.61	71.15	35.64	70.16	65.93	68.15	68.92	**70.96**
	ACC	<u>90.82</u>	65.07	65.29	89.48	65.12	87.11	71.92	88.46	90.52	89.97	89.35	**91.88**
	F1	**91.63**	56.74	56.79	90.51	56.79	80.18	73.07	83.61	88.98	90.46	89.67	<u>91.45</u>

where $\Omega(\cdot)$ denotes the stop-gradient strategy, which stabilizes the training process and avoids model collapse. The total loss function is defined as:

$$\mathcal{L} = \mathcal{L}_{KL} + \alpha \mathcal{L}_R + \beta \mathcal{L}_F, \tag{20}$$

where α, β and γ are hyperparameters. The pseudo-code is in Algorithm 1.

4 Experiments

4.1 Experimental Setup

Datasets and Baselines. We utilize five real-world datasets, including three citation networks, ACM [1], ACM2 [2], and DBLP [41], and two review networks, Yelp [40] and Amazon [6]. We select eleven baselines: 1) Supervised multi-view graph approach HAN [27]; 2) Single-view graph clustering techniques VGAE [9] and DGI [25]; 3) Multi-view graph clustering frameworks DLGR [12], DMG [18], MvAGC [11], BTGF [19], DOGC [30], MCGF [8], DSGC [42], and RJGC [7]. **Settings and Evaluation.** For metrics and settings, four widely adopted clustering metrics are employed: Accuracy (ACC), Normalized Mutual Information (NMI) [10], F1 score [33], and Adjusted Rand Index (ARI) [16]. For the settings, the training epoch is set to 300. We use Adam optimizer [14,15,17,35]; the learning rate is set to 1e−2, and the weight decay is set to 5e−5. The C and T are respectively set to 4 and 2. The hidden dimension d' is set to 64 for ACM2 and 32 for other datasets. For Yelp and Amazon datasets, α and β are respectively

Table 2. The ablation study results on five datasets. The best and second-best observations are highlighted **in bold** and <u>underlined</u>, respectively. The results are in %.

Variants	Amazon			ACM			DBLP			ACM2			Yelp		
	NMI	ACC	F1	NMI	ACC	F1	NMI	ACC	F1	NMI	ACC	F1	NMI	ACC	F1
-KL	50.26	73.63	72.44	79.41	91.22	93.11	79.31	93.10	92.86	71.64	90.43	90.76	70.91	84.31	84.76
-R	46.13	62.77	69.99	70.16	85.44	83.36	72.16	88.66	87.16	70.13	89.63	86.28	61.49	79.64	76.42
-F	47.23	63.47	68.19	72.36	85.01	85.30	74.26	88.24	85.11	71.20	88.10	87.01	62.19	78.42	77.12
-KL/R	43.15	63.84	66.91	68.28	80.72	81.95	65.43	81.64	82.27	62.42	79.37	78.59	57.75	73.41	72.92
-KL/F	38.86	55.47	60.28	62.19	76.56	74.74	63.82	79.43	77.97	61.15	80.28	77.42	52.64	70.28	67.86
-R/F	41.28	56.91	60.42	65.63	75.85	77.17	64.95	80.42	77.75	61.59	80.15	79.28	52.93	71.28	70.84
Vanilla GNN	48.83	65.27	66.79	70.38	84.41	83.33	72.22	87.77	83.51	69.50	86.66	85.78	65.15	82.27	84.45
SGSF-GC	**53.12**	**78.66**	**76.12**	**80.10**	**94.68**	**95.10**	**80.62**	**94.72**	**93.41**	**75.22**	**92.41**	**91.77**	**73.26**	**90.88**	**91.45**

set to 1.0 and 0.8. For other datasets, α and β are set to 0.8 and 1.0. All experiments are conducted based on PyTorch in a Python 3.9.12 environment with a single NVIDIA A40 GPU, 40 GB of RAM, and a 2.60 GHz 6240 CPU.

4.2 Performance Comparison

To assess the performance of our model, comparative experiments with thirteen state-of-the-art baselines are conducted based on five public datasets. From the observations shown in Table 1, the performance of SGSF-GC surpasses all baselines on five datasets, which demonstrates the effectiveness of our proposed method. Notably, the average improvements of SGSF-GC on NMI, ARI, ACC, and F1 are, respectively, 2.37%, 2.38%, 1.61%, and 0.89%. The reason can be triple. First, SGSF-GC introduces a powerful graph encoder that employs multiscale graph filters to capture comprehensive different frequency-domain information. Second, SGSF-GC develops a distinct cohesion-based fusion mechanism to integrate high- and low-frequency signals, generating discriminative and high-quality representations. Third, SGSF-GC exploits the CAM-refined semantic multi-view fusion module, which enhances the integration of view-specific representations using valuable gradient information to adaptively learn the contributions of different views. However, SGSF-GC achieves suboptimal F1 on the Yelp dataset. The reason can be attributed to the supervised learning paradigm in HAN, which utilizes label information to capture critical features. Even so, SGSF-GC still holds competitiveness using an unsupervised paradigm compared with a supervised paradigm.

4.3 Ablation Study

Loss function. We conduct ablation studies on the loss function to verify the effectiveness of each component of SGSF-GC. We first introduce three variants in the single ablation study. "-KL" denotes the SGSF-GC training without $\mathcal{L}_{KL}$. "-R" denotes the SGSF-GC training without $\mathcal{L}_R$. "-F" denotes the SGSF-GC training without $\mathcal{L}_F$. According to Table 2, we can conclude that SGSF-GC outperforms all variants on three metrics, verifying the effectiveness of each component in SGSF-GC. Specifically, "-R" and "-F" have relatively worse performance than other variants, especially on Amazon and ACM datasets, suggesting poor ability to capture interaction modes. The performance of "-KL" indicates the utility of the rationality of distribution alignment based on the KL divergence. We further conduct joint ablation studies to validate the interactions of the loss function. As shown in Table 2, the performance of variants has significant degradation compared to the single-component ablation study, demonstrating the interactive influences of modules on the performance.

Graph Encoder. To demonstrate the importance of high-frequency signals, we then implement an ablation study on the graph encoder. We replace the hybrid graph encoder with only vanilla GNNs (low-pass filter) and ignore the adaptive fusion mechanism. The encoded features are directly fed into the reconstruction-guided multi-view fusion module. The observations are shown in Table 2. The performance of SGSF-GC with vanilla GNNs has a significant deterioration due to the inability to derive high-frequency information, validating the utility of hybrid graph filters. First, vanilla GNNs leverage traditional graph filters, which have relatively poor discrimination. Second, vanilla GNNs only capture low-frequency signals, ignoring high-frequency graph signals. Third, vanilla GNNs fail to capture multi-scale frequency-domain information, which leads to the inexactitude and insufficiency in frequency-domain characteristics.

4.4   Parameter Sensitivity

Loss Function. We first implement sensitivity experiments of a single hyper-parameter. As shown in Fig. 2, we have the following observations: (1) As the parameter value increases, the performance gradually improves and tends to be relatively stable. When the parameters are too low, the performance exhibits significant degradation. This is because small values of the parameters result in the fundamental deactivation of corresponding functions, thereby decreasing the

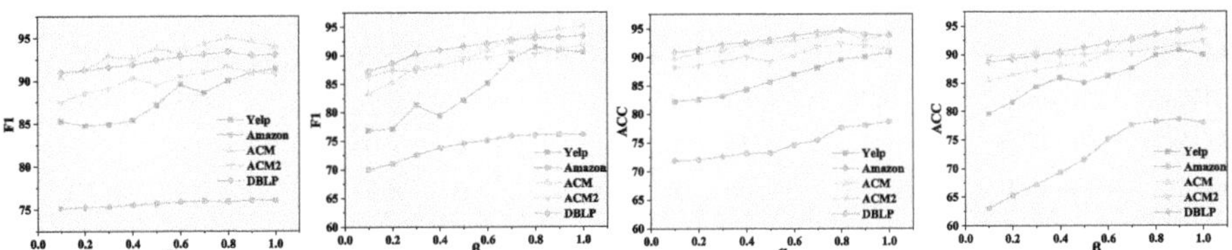

Fig. 2. The single-hyperparameter sensitivity on five datasets.

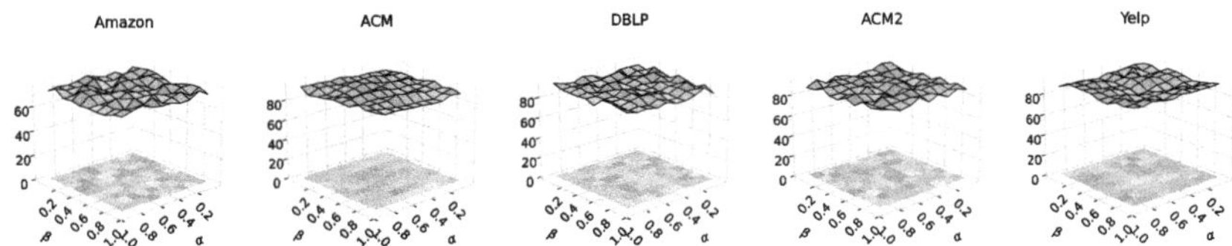

Fig. 3. The interactive parameter sensitivity of α and β on F1 metric.

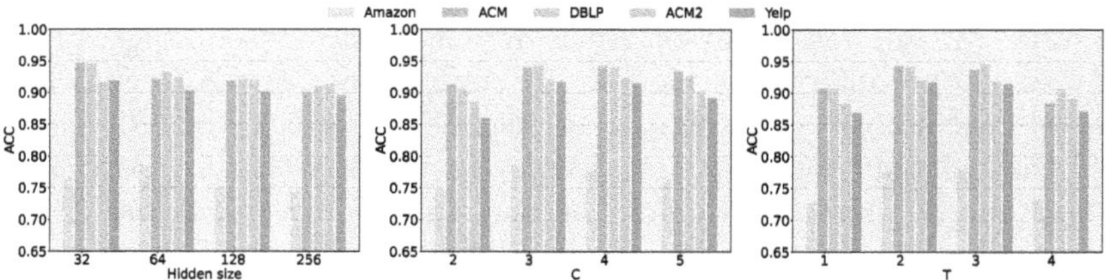

Fig. 4. The parameter sensitivity of hidden size, C, and T on five datasets.

ability of the model to extract essential attributes; (2) Hyperparameters have different influences on the performance. For example, on the Amazon dataset, α has less impact on the performance metrics than β, while on the DBLP dataset, α and β have similar impact on the performance. (3) The performance of the same parameter varies across different datasets. For example, according to the F1-score, Yelp is more sensitive to the change of α compared to other datasets.

We then validate the interactive influence between hyperparameters in the loss function, namely α and β. As shown in Fig. 3, the performance indicates relative stability and showcases the best configurations for datasets. Specifically, the performance on ACM is more stable than the performance on Amazon, according to color changes. This can be attributed to the discrepancy in the data properties of different datasets. However, through parameter sensitivity experiments, the model can select corresponding parameter ranges based on different datasets, thereby enhancing the robustness of the model.

Table 3. The efficiency experiment on five datasets compared with eleven baselines. The best and runner-up are highlighted **in bold** and <u>underlined</u>, respectively.

Dataset	HAN	VGAE	DGI	MGDCR	DLGR	DMG	MvAGC	BTGF	DOGC	MCGF	DSGC	SGSF-GC
Amazon	1.55	1.25	1.03	1.17	1.04	0.89	0.99	0.89	0.92	<u>0.62</u>	0.73	**0.53**
ACM	1.39	1.16	0.97	1.03	0.98	0.82	0.86	0.81	0.70	**0.51**	0.75	<u>0.53</u>
DBLP	1.34	1.22	1.10	0.99	0.89	0.93	0.87	0.82	<u>0.73</u>	0.75	0.87	**0.61**
ACM2	1.89	1.56	1.45	1.37	1.43	1.25	1.09	1.11	1.03	**0.87**	0.97	<u>0.90</u>
Yelp	1.93	1.69	1.46	1.25	1.09	1.18	1.10	0.91	0.97	<u>0.79</u>	0.93	**0.71**

Modules. We validate the influence of hidden size d', the constant C, and the tunable hyperparameter T. The observations of the ACC are respectively shown in Fig. 4. We can see that SGSF-GC exhibits relatively good stability to hidden size compared with C and T.

From the observations, we can see that the performance has significant improvement when C is set to 3 or 4, indicating that capturing multi-scale signals is vital to obtain critical characteristics. However, when C is set to 5, it brings redundant or even adverse signals that cause impure representations. We change the tunable hyperparameter T from 1 to 4 with a step size of 1 and set

C to a fixed value of 5. The model outperforms when T is set to 2 or 3 because the model can capture comprehensive high- and low-frequency graph signals. However, setting T to 1 or 4 leads to a significant discrepancy in the quality of the extracted characteristics.

4.5 Efficiency

To evaluate the computational efficiency of SGSF-GC, we measure its runtime per epoch compared with baseline methods, as presented in Table 3. Traditional GNN-based methods have relatively higher complexity because of their heavy parameter composition in the models, such as HAN, which needs to compute attentions of each node with its neighbors. Recent graph clustering methods show competitive efficiency compared with traditional methods. MCGF shows competitive efficiency because of its lightweight filter design and relatively small training parameter scale. SGSF-GC achieves superior efficiency over most baselines, showing relatively low complexity. This can be attributed to the lightweight computation of the Beta convolutional kernel and efficient computation.

5 Conclusion

In this study, we propose a novel graph clustering with scalable graph filters and view-specific semantic fusion, SGSF-GC. SGSF-GC leverages scalable Beta wavelet graph filters to capture multi-scale frequency-domain signals and introduces a cohesion-based fusion mechanism to adaptively integrate the representations from different frequency domains. SGSF-GC employs CAM-based semantic multi-view fusion to preserve the semantic fidelity in different view-specific representations. Finally, SGSF-GC implements graph clustering based on the KL-divergence to model the critical characteristics of different categories. Extensive experiments demonstrate the effectiveness and superiority of SGSF-GC compared with eleven baselines. In the future, we will explore more efficient methods to enhance the scalability and fidelity of graph clustering algorithms.

References

1. Fan, S., Wang, X., Shi, C., Lu, E., Lin, K., Wang, B.: One2multi graph autoencoder for multi-view graph clustering. In: The Web Conference 2020, pp. 3070–3076 (2020)
2. Fu, X., Zhang, J., Meng, Z., King, I.: MAGNN: metapath aggregated graph neural network for heterogeneous graph embedding. In: The Web Conference 2020, pp. 2331–2341 (2020)
3. Han, R., et al.: Federated graph-level clustering network with adaptive knowledge compensation. Neural Netw., 108212 (2025)
4. Han, R., et al.: Dual boost-driven graph-level clustering network. In: International Joint Conference on Neural Networks, pp. 1–8 (2025)
5. Han, R., et al.: Multi-relation graph-kernel strengthen network for graph-level clustering. In: International Joint Conference on Neural Networks, pp. 1–8 (2025)

6. He, R., McAuley, J.J.: Ups and downs: modeling the visual evolution of fashion trends with one-class collaborative filtering. In: Proceedings of the 25th International Conference on World Wide Web, pp. 507–517 (2016)
7. He, Y., Yusof, U.K.: Robust joint graph learning for multi-view clustering. IEEE Trans. Big Data **11**(2), 722–734 (2025)
8. Jiang, H., Du, L.: Enhanced multi-view clustering with multiple linear graph filtering. In: Proceedings of the 2025 International Conference on Multimedia Retrieval, pp. 597–606 (2025)
9. Kipf, T.N., Welling, M.: Variational graph auto-encoders. CoRR **abs/1611.07308** (2016)
10. Liang, K., et al.: A survey of knowledge graph reasoning on graph types: static, dynamic, and multi-modal. IEEE Trans. Pattern Anal. Mach. Intell. (2024)
11. Lin, Z., Kang, Z., Zhang, L., Tian, L.: Multi-view attributed graph clustering. IEEE Trans. Knowl. Data Eng. **35**(2), 1872–1880 (2023)
12. Ling, Y., et al.: Dual label-guided graph refinement for multi-view graph clustering. In: 37th AAAI Conference on Artificial Intelligence, pp. 8791–8798 (2023)
13. Liu, J., Chen, S., Cai, M., Shao, H., Gui, W.: Semi-heterogeneous graph-perception network with gradient-weighted class activation mapping for class-incremental industrial fault recognition and root cause diagnosis. IEEE Trans. Neural Networks Learn. Syst. **36**(9), 16825–16839 (2025)
14. Liu, M., Liang, K., Wang, S., Hu, X., Zhou, S., Liu, X.: Deep temporal graph clustering: a comprehensive benchmark and datasets. IEEE Trans. Pattern Anal. Mach. Intell. (2025)
15. Liu, M., et al.: Multiview temporal graph clustering. IEEE Trans. Neural Netw. Learn. Syst., 1–14 (2025)
16. Liu, M., et al.: Self-supervised temporal graph learning with temporal and structural intensity alignment. IEEE Trans. Neural Netw. Learn. Syst. (2024)
17. Liu, M., et al.: Deep temporal graph clustering. In: The 12th International Conference on Learning Representations (2024)
18. Mo, Y., Lei, Y., Shen, J., Shi, X., Shen, H.T., Zhu, X.: Disentangled multiplex graph representation learning. In: International Conference on Machine Learning, vol. 202, pp. 24983–25005 (2023)
19. Qian, X., Li, B., Kang, Z.: Upper bounding Barlow twins: a novel filter for multi-relational clustering. In: Thirty-Eighth AAAI Conference on Artificial Intelligence, pp. 14660–14668 (2024)
20. Qiu, X., et al.: TFB: towards comprehensive and fair benchmarking of time series forecasting methods. In: Proc. VLDB Endow, pp. 2363–2377 (2024)
21. Qiu, X., et al.: DBLoss: decomposition-based loss function for time series forecasting. In: NeurIPS (2025)
22. Qiu, X., Wu, X., Lin, Y., Guo, C., Hu, J., Yang, B.: DUET: dual clustering enhanced multivariate time series forecasting. In: SIGKDD, pp. 1185–1196 (2025)
23. Shen, Z., He, H., Kang, Z.: Balanced multi-relational graph clustering. In: Proceedings of the 32nd ACM International Conference on Multimedia, pp. 4120–4128 (2024)
24. Shen, Z., Kang, Z.: When heterophily meets heterogeneous graphs: latent graphs guided unsupervised representation learning. IEEE Trans. Neural Networks Learn. Syst. **36**(6), 10283–10296 (2025)
25. Velickovic, P., Fedus, W., Hamilton, W.L., Liò, P., Bengio, Y., Hjelm, R.D.: Deep graph infomax. In: 7th International Conference on Learning Representations (2019)

26. Wang, W., Tran, D., Feiszli, M.: What makes training multi-modal classification networks hard? In: Proceedings of the IEEE/CVF Conference on Computer Vision and Pattern Recognition, pp. 12695–12705 (2020)
27. Wang, X., et al.: Heterogeneous graph attention network. In: The World Wide Web Conference, pp. 2022–2032 (2019)
28. Wang, Z., Lin, Q., Ma, Y., Ma, X.: Local high-order graph learning for multi-view clustering. IEEE Trans. Big Data **11**(2), 761–773 (2025)
29. Wen, Z., et al.: Homophily-related: adaptive hybrid graph filter for multi-view graph clustering. In: The 38th AAAI Conference on Artificial Intelligence, pp. 15841–15849 (2024)
30. Wen, Z., et al.: Dual-optimized adaptive graph reconstruction for multi-view graph clustering. In: Proceedings of the 32nd ACM International Conference on Multimedia, pp. 1819–1828 (2024)
31. Xia, W., Wang, S., Yang, M., Gao, Q., Han, J., Gao, X.: Multi-view graph embedding clustering network: joint self-supervision and block diagonal representation. Neural Netw. **145**, 1–9 (2022)
32. Xin, H., Wu, D., Lu, J., Wang, R., Nie, F., Li, X.: Multiview clustering via block diagonal graph filtering. IEEE Trans. Neural Networks Learn. Syst. **36**(8), 14269–14282 (2025)
33. Xuan, X., et al.: Fake-mamba: real-time speech deepfake detection using bidirectional mamba as self-attention's alternative. In: Proceedings of the IEEE ASRU (2025)
34. Xuan, X., et al.: Multilingual source tracing of speech deepfakes: a first benchmark. In: 5th Symposium on Security and Privacy in Speech Communication, pp. 27–34 (2025). https://doi.org/10.21437/SPSC.2025-5
35. Xuan, X., et al.: WaveSP-Net: learnable wavelet-domain sparse prompt tuning for speech deepfake detection. In: IEEE International Conference on Acoustics, Speech and Signal Processing. IEEE (2026)
36. Yuan, R., Tang, Y., Wu, Y., Zhang, W.: Clustering enhanced multiplex graph contrastive representation learning. IEEE Trans. Neural Networks Learn. Syst. **36**(1), 1341–1355 (2025)
37. Zhang, W., Luo, C.: GE-GNN: gated edge-augmented graph neural network for fraud detection. IEEE Trans. Big Data **11**(4), 1664–1676 (2025)
38. Zhang, W., Luo, C.: MCLASt: multi-hierarchy contrastive learning graph anomaly detection with structure-awareness. Neurocomputing, 132480 (2025)
39. Zhang, W., et al.: Robust rumor detection against noise. Neurocomputing, 132741 (2026)
40. Zhang, W., et al.: Dual-channel heterophilic message passing for graph fraud detection. In: International Joint Conference on Neural Networks, pp. 1–8 (2025)
41. Zhao, J., Wang, X., Shi, C., Liu, Z., Ye, Y.: Network schema preserving heterogeneous information network embedding. In: Bessiere, C. (ed.) Proceedings of the Twenty-Ninth International Joint Conference on Artificial Intelligence, pp. 1366–1372 (2020)
42. Zhao, P., Li, X., Zhang, Z., Wang, M., Zhu, X., Liao, L.: Robust deep signed graph clustering via weak balance theory. In: Proceedings of the ACM on Web Conference 2025, pp. 3819–3830 (2025)
43. Zou, J., Chen, Y., Zhou, P., Wen, C., Du, L., Qian, Y.: Consensus graph filter learning for multiple graph clustering. In: 2025 IEEE International Conference on Acoustics, Speech and Signal Processing, pp. 1–5 (2025)

DGTC: Dynamic Graph Transformer for Graph-Level Classification

Zhe Wang[1,2], Jiawei Chen[1,2,3(✉)], Sheng Zhou[1], Canghong Jin[4], Chun Chen[1,2], and Can Wang[2,3]

[1] College of Computer Science and Technology, Zhejiang University, Hangzhou, China
`{zhewangcs,zhousheng_zju,chenc}@zju.edu.cn`
[2] State Key Laboratory of Blockchain and Data Security, Zhejiang University, Hangzhou, China
`{sleepyhunt,wcan}@zju.edu.cn`
[3] Hangzhou High-Tech Zone (Binjiang) Institute of Blockchain and Data Security, Hangzhou, China
[4] Hangzhou City University, Hangzhou, China
`jinch@hzcu.edu.cn`

Abstract. Graph-level tasks on dynamic graphs are prevalent in real-world applications, yet methods for this problem remain limited. Existing models exhibit poor parallel processing capabilities and struggle to capture high-order subgraph evolution patterns. To bridge this gap, we propose the Dynamic Graph Transformer for Graph-level Classification (DGTC), which integrates a global self-attention mechanism to adaptively aggregate node information across the entire graph. This architecture eliminates the need for sequential processing of temporal edges and enables highly parallel computation. To comprehensively model the spatio-temporal evolution of dynamic graphs, we introduce (i) dynamic degree encoding to capture local structural variation; and (ii) dynamic distance encoding to track changes in higher-order subgraph structure via time-varying pairwise distances. Extensive experiments show that DGTC delivers an average performance gain of 14.28% and up to 35 × speedup over ten competitive baselines. Code is available at https:// github.com/wangz3066/DGTC.

Keywords: Dynamic Graph Learning · Graph-level Task · Transformer

1 Introduction

Dynamic graphs provide a natural abstraction for many real-world dynamic systems in which entities and their interactions evolve over time [37]. An increasing number of studies aim to learn effect low-dimensional representations of dynamic graphs that can support downstream tasks such as future link prediction. The dominating tools for dynamic graph representation learning are Dynamic graph

H. Jung et al. (Eds.): DASFAA 2026, LNCS 16536, pp. 98–113, 2026.
https://doi.org/10.1007/978-981-92-0366-6_7

neural networks (DGNNs), which rely on temporal message passing to aggregate information from historically interacting neighbors and to incorporate temporal sequence information for node representation learning [10,21,28,32,39]. These techniques have enabled impactful applications in financial fraud detection [7,11], recommender systems [6,40], and urban traffic forecasting [15,23].

Beyond node- and edge-centric problems, many practical settings require graph-level reasoning on dynamic graphs [9]. For instance, in financial fraud detection, illegal activities often occur within organized groups whose memberships and transactional relationships between groups evolve continuously over time. Identifying such criminal networks is a prototypical graph-level classification task. The crux of such graph-level tasks is to capture the evolution of node groups that exhibit salient structural features in dynamic graphs, such as the temporal changes in subgraph patterns including cycles, cliques, communities, and clusters [17,19,34]. However, most prior work focuses on learning node- or edge-level representations by aggregating information from local time-ordered neighborhoods, leaving crucial graph-level signals underexplored.

To the best of our knowledge, TP-GNN [16] is the only recent approach explicitly targeting dynamic graph-level classification. TP-GNN updates node states through single-hop temporal propagation and subsequently aggregates these node states to produce a graph-level representation. Despite this progress, two limitations remain: (i) Sequential chronological updates of all node/edge representations induce strict temporal dependencies that preclude parallel processing and degrade efficiency; and (ii) TP-GNN relies solely on message-passing mechanisms, which exhibit limited expressive power for capturing higher-order time-evolving subgraph structures (e.g., cycles, cliques) that are critical for dynamic graph classification tasks.

To address these limitations, we propose Dynamic Graph Transformer for Dynamic Graph Classification (DGTC), which employs global self-attention to adaptively capture node dependencies across the entire graph. This design overcomes the expressive bottlenecks of message passing and removes the requirement for strictly sequential temporal updates, enabling efficient parallelization. To further enhance modeling of the spatio-temporal evolution of dynamic graphs, we introduce three positional encodings: The dynamic degree encoding captures local structural variation via changes in node degrees, while the dynamic distance encoding tracks higher-order subgraph structural shifts through time-varying pairwise distances. A temporal encoding is also integrated to capture the temporal signals present in dynamic graphs. Overall, The main contributions of this work are as follows: (1) We propose comprehensive positional encodings to model the evolution of higher-order structures. (2) Equipped with comprehensive positional encodings, we propose DGTC which leverages global self-attention to learn graph-level representations. (3) Extensive experiments demonstrate that DGTC achieves an average improvement of 14.28% and provides up to 35 × speedup over baselines.

2   Related Works

2.1   Dynamic Graph Learning

Dynamic graphs are networks whose structures or node attributes evolve over time. Depending on whether the interacted timestamps are discrete or continuous, they can be categorized into Discrete-Time Dynamic Graphs (DTDGs) and Continuous-Time Dynamic Graphs (CTDGs). Early studies predominantly focused on learning representations for DTDGs [18,22,27,38]. However, DTDGs neglect the temporal ordering of events within each snapshot, leading to the loss of fine-grained temporal information. In contrast, recent research increasingly targets CTDGs, a more general and expressive form of dynamic graphs. The dominant paradigm in this setting is Message-PassingâĂŞbased Dynamic Graph Neural Networks (MP-DGNs), where information from historically interacting neighbors is recursively aggregated to update the representations of target nodes [2,29,30,32]. Some works further integrate memory mechanisms to preserve long-term interaction histories [14,21,25]. Moreover, some approaches utilize temporal random walks to extract temporal motifs, which serve as informative patterns for representation learning [10,28].

Most existing dynamic graph studies predominantly address node-level or edge-level tasks, focusing on the representations of individual nodes or edges. Recently, TP-GNN [16] was proposed to address the dynamic graph classification problem. Different from existing works, the proposed DGTC is a Transformer-based architecture targeting at dynamic graph classification problem. Moreover, unlike existing methods that require sequential processing of temporal edges, DGTC feeds all nodes of the dynamic graph into the transformer in a single step, yielding higher computational efficiency.

2.2   Graph Transformer

Several studies have investigated the use of Transformers for representation learning on both dynamic and static graphs. To enable the vanilla Transformer to incorporate graph structural information, these approaches embed representative graph-specific features into the positional encoding—such as eigenvectors of the graph Laplacian [3,13], node degrees [8,36], or the diagonal of a random-walk matrix [4,20]. For CTDGs, existing approaches typically generate temporal node embeddings with graph-specific positional encodings, such as common-neighbor features [39] or spatialâĂŞtemporal distance metrics [29]. The proposed DGTC differs fundamentally from these prior works: unlike the above methods, which focus on node- or edge-level tasks and compute attention scores only among nodes within enclosing subgraphs of target edges, DGTC targets the dynamic graph classification problem and computes attention scores between any pair of nodes across the entire dynamic graph, enabling comprehensive modeling of high-order structural dependencies.

3 Preliminaries

3.1 Dynamic Graph Classification

In this work, we primarily focus on Continuous-Time Dynamic Graphs (referred to as Dynamic Graphs in the remainder of this work) due to their generality. A *Dynamic Graph (DG)* is defined as $\mathcal{DG} = (\mathcal{V}, \mathcal{E})$, where $\mathcal{V} = \{1, 2, ..., N\}$ is the node set. $\mathcal{E} = \{(u_i, v_i, t_i)\}_{i=1}^{M}$ with $t_i \leq t_{i+1}$ is a sequence of (undirected) interactions, where (u_i, v_i, t_i) represents that node u_i and node v_i have an interaction event at time t_i. The raw node feature matrix of $\mathcal{DG}$ is denoted $\boldsymbol{X}_{raw} \in \mathbb{R}^{|\mathcal{V}| \times d_N}$. Given N dynamic graphs $\{\mathcal{DG}_i\}_{i=1}^{N}$, each belongs to a ground-truth label $c \in [1, C]$ where C is the class number. The *Dynamic Graph Classification* problem aims to learn a parameterized predictor $f : \mathcal{DG}_i \rightarrow [0, 1]^C$ which predicts the label of $\mathcal{DG}_i$.

The dynamic graph classification tasks are prevalent in real-world applications. For instance, in financial transaction networks, it is essential to identify criminal organizations whose membership and transactional relationships evolve over time. Despite its significance, dynamic graph classification problems poses unique challenges. Firstly, it requires to capture complex, higher-order temporal patterns and structural changes at the graph level. Secondly, Since the number of dynamic graphs can be large, it is necessary to process them in batches to ensure the efficiency of the model. However, most existing methods require sequential processing of temporal edges and cannot handle multiple dynamic graphs in parallel.

3.2 Message-Passing Based Dynamic Graph Neural Networks (MP-DGNs)

Message-Passing based Dynamic Graph neural Networks (MP-DGNs) have been widely explored in recent literature and have achieved state-of-the-art performance. Most existing MP-DGNs are tailored for node-level or edge-level tasks. In the training phase, MP-DGNs processes each edge of the dynamic graph sequentially based on temporal order, producing node embeddings at each specific time step. To compute the representation of node u at time t, MP-DGNs recursively sample its historical neighbors and integrate both the neighbors' historical information and temporal difference encodings. The resulting aggregated representation is subsequently used to update the embedding of the central node. Formally, the *historical neighbor* of node u at time t is defined as $\mathcal{N}(u, t) = \{(w, t')|t' < t, (u, w, t') \in \mathcal{E} \vee (w, u, t') \in \mathcal{E}\}$. The 0-th layer embedding of node u is the node feature $\boldsymbol{h}_t^{(0)}(u) = \boldsymbol{X}_{raw}(u)$. The l-th layer ($l > 0$) embedding is computed as:

$$\tilde{\boldsymbol{h}}_t^{(l)}(u) = \text{AGG}(\{\!\!\{([\boldsymbol{h}_{t'}^{(l-1)}(w)||\sigma(t - t')]|(w, t') \in \mathcal{N}(u, t)\}\!\!\})$$

$$\boldsymbol{h}_t^{(l)}(u) = \text{UPDATE}(\boldsymbol{h}_t^{(l-1)}(u), \tilde{\boldsymbol{h}}_t^{(l)}(u)) \tag{1}$$

where $\sigma : \mathbb{R}^+ \rightarrow \mathbb{R}^{d_T}$ projects the time interval to a vector and $||$ denotes the concatenation. $\{\!\!\{\cdot\}\!\!\}$ denotes the multiset. Some methods also leverage memory mechanisms to record the long-term historical interactions of each node [14,21].

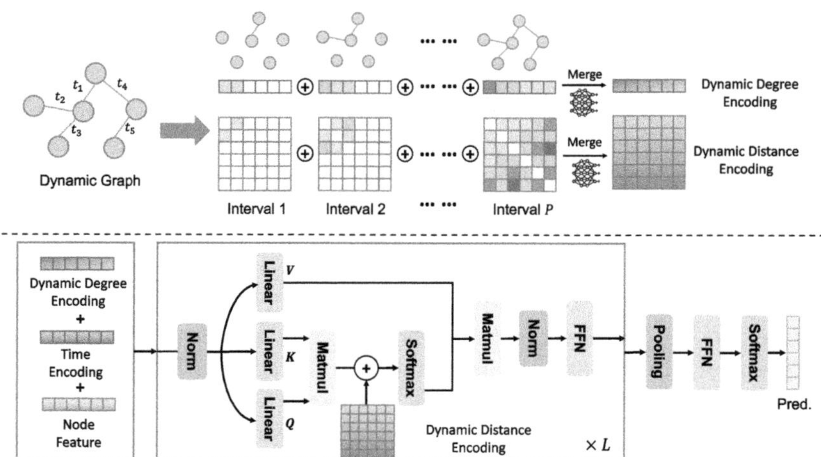

Fig. 1. Framework of the proposed DGTC model. DGTC partitions the dynamic graph into time intervals, computing node degrees and shortest path distances in each to obtain dynamic degree and distance encodings. The degree encoding is used as additional node features, while the distance encoding serves as a bias in self-attention.

4   Methods

In this section, we introduce the proposed **Dynamic Graph Transformer for Graph-Level Classification (DGTC)** model. The general framework of the proposed DGTC is presented in Fig. 1. We employ a transformer-based model and input all nodes in the graph as tokens simultaneously, thereby avoiding the inefficiency caused by sequentially processing temporal edges. Given a dynamic graph, we firstly divide it into P time intervals. To capture the evolving pattern of high-order subgraphs, we compute the degree encoding (capturing the local subgraphs information) and the distance encoding (capturing the high-order subgraphs information) within each time interval, then aggregating both encodings within these intervals to capture their evolving patterns. The interaction timestamps of nodes are also encoded as time encodings. Note that the computation of comprehensive encoding can be performed during the pre-processing stage, thereby introducing no additional computational overhead during training. Next, to enable the Transformer to effectively perceive the spatio-temporal structural changes of dynamic graphs, dynamic degree encodings and time encodings are concatenated with raw node features, which are input into the Transformer model. Dynamic distance encodings are leveraged as the bias term of self-attention matrix. We will elaborate on each module.

4.1   Comprehensive Positional Encodings

Dynamic Degree Encoding. Changes in local structures are crucial for understanding and discriminating dynamic graphs, as they often correspond to significant events such as anomalies. In DGTC, the evolution of local subgraphs is captured through the temporal dynamics of node degrees. Specifically, given a dynamic graph $\mathcal{DG}$ with time range $T = [t_{min}, t_{max}]$, we insert $P-1$ quantiles to T, thus obtaining P time intervals $[t_{min}, t_i](i = 1, 2..., P, t_P = t_{max})$. The edges within these intervals formulates P graphs $\mathcal{G}_1, \mathcal{G}_2, ..., \mathcal{G}_P$. We compute the degree of node u in $\mathcal{G}_i$, denoted as $deg_i(u)$. Then, $deg_i(u)$ is mapped into a learnable embedding. Embeddings from all time intervals are concatenated to construct the dynamic degree encoding:

$$\boldsymbol{x}_{deg}(u) = [\boldsymbol{z}_{deg_1(u)}||\boldsymbol{z}_{deg_2(u)}||...||\boldsymbol{z}_{deg_P(u)}] \in \mathbb{R}^{d \cdot P} \tag{2}$$

where $\boldsymbol{z}$ denotes a d-dimensional learnable embedding. Dynamic degree encoding effectively captures the temporal dynamics of node degrees, thereby representing the evolving patterns of local subgraph structures around each node.

Dynamic Distance Encoding. In addition to local structures, high-order subgraph structures can reflect the macro-level evolutionary characteristics of dynamic graphs. We use the changes of shortest path distance to represent the evolution of high-order subgraph structures. Specifically, based on the obtained graphs $\mathcal{G}_1, \mathcal{G}_2, ..., \mathcal{G}_P$ in corresponding time intervals (illustrated in Sect. 4.1), the shortest path length between nodes u and v in $\mathcal{G}_i$ is denoted as $spd_i(u, v)$. Then, $spd_i(u, v)$ is mapped into a learnable embedding. Embeddings from all time intervals are concatenated to construct the dynamic distance encoding:

$$\boldsymbol{x}_{dis}(u, v) = [\boldsymbol{c}_{spd_1(u,v)}||\boldsymbol{c}_{spd_2(u,v)}||...||\boldsymbol{c}_{spd_P(u,v)}] \in \mathbb{R}^{d \cdot P} \tag{3}$$

where $\boldsymbol{c}$ denote a d-dimensional learnable embedding. The dynamic distance encoding will be incorporated into the self-attention mechanism as an additional bias.

Time Encoding. Temporal signals can reflect the temporal and evolutionary patterns of node interactions, thereby helping dynamic graph models learning. However, the temporal distributions may vary significantly across different dynamic graphs, and directly using the raw timestamps as features may lead to unstable training. To this end, given a dynamic graph $\mathcal{DG}$, we sort all timestamps in $\mathcal{DG}$ in ascending order and convert each timestamp into its corresponding *rank*. Next, the interaction timestamps of node u, $\boldsymbol{t}(u) = [t_1, t_2, ..., t_m]$, is transformed to the rankings: $rank(\boldsymbol{t}(u)) = [rank(t_1), rank(t_2), ..., rank(t_m)]$. As the lengths of interaction time series differ among nodes, we retain the most recent S timestamps for node u. For sequences shorter than S, padding is applied to ensure a uniform length of S. Then, $rank(\boldsymbol{t}(u))$ is embeded to obtain the time encoding:

$$\boldsymbol{x}_{time}(u) = [\boldsymbol{w}_{rank(t_1)}||\boldsymbol{w}_{rank(t_2)}, ...] \in \mathbb{R}^{d \cdot S} \tag{4}$$

where $\boldsymbol{w}$ denotes a learnable d-dimensional embedding.

4.2   Network Architecture

The aforementioned comprehensive encoding effectively captures the spatio-temporal variation patterns of the dynamic graph. To enable the model to leverage these patterns, we augment the raw node features with time encoding and dynamic degree encoding, which are then used as the input tokens for the Transformer model.

$$\boldsymbol{x}_u = f_1(\boldsymbol{x}_{raw}(u)) + f_2(\boldsymbol{x}_{deg}(u)) + f_3(\boldsymbol{x}_{time}(u)) \in \mathbb{R}^{d_m} \tag{5}$$

where f_1, f_2, f_3 are Multi-Layer Perceptions (MLPs) that project corresponding features into a d_m-dimensional space. $\boldsymbol{x}_{raw}$ denotes the raw node feature. Moreover, we incorporate the dynamic distance encoding into the self-attention matrix, enabling each node to adaptively attend to other nodes based on the graph structure information. The self-attention matrix in DGTC is computed as:

$$\tilde{\boldsymbol{A}}_{u,v} = \frac{(\boldsymbol{h}_u \boldsymbol{W}_q)(\boldsymbol{h}_v \boldsymbol{W}_k)^T}{\sqrt{d_k}} + f_4(\boldsymbol{x}_{dis}(u, v)) \tag{6}$$

where f_4 is a MLP that projects into a scalar. $\boldsymbol{W}_q, \boldsymbol{W}_k \in \mathbb{R}^{d_m \times d_k}$ are the projecting matrices. $\boldsymbol{h}_u$ and $\boldsymbol{h}_v$ are the node embeddings.

DGTC is built upon the Transformer architecture. For easier optimization, we adopt the PreNorm architecture [31] where the Layer Normalization (LN) is applied before self-attention and Feed-Forward Networks (FFN). Formally, let $\boldsymbol{H}^{(0)} = \boldsymbol{X} = [\boldsymbol{x}_u]_{u \in \mathcal{V}} \in \mathbb{R}^{|\mathcal{V}| \times d_m}$ be the input of DGTC, the l-th ($1 \leq l \leq L$) layer of DGTC is calculated as:

$$\boldsymbol{H}'^{(l)} = \mathrm{MHSA}(\mathrm{LN}(\boldsymbol{H}^{(l-1)})) + \boldsymbol{H}^{(l-1)}$$
$$\boldsymbol{H}^{(l)} = \mathrm{FFN}(\mathrm{LN}(\boldsymbol{H}'^{(l)})) + \boldsymbol{H}'^{(l)} \tag{7}$$

The output of L-th DGTC layer $\boldsymbol{H}^{(L)} \mathbb{R}^{|\mathcal{V}| \times d_m}$ is the learned node embeddings matrix. We then apply mean-pooling on the node embeddings to obtain the graph-level representation of $\mathcal{DG}$:

$$h_{\mathcal{DG}} = \frac{1}{|\mathcal{V}|} \sum_{i=1}^{|\mathcal{V}|} \boldsymbol{H}^{(L)}[i, :] \tag{8}$$

Subsequently, the dynamic graph representation $h_{\mathcal{DG}}$ is fed into a MLP to produce the unnormalized class scores (logits) $\boldsymbol{z} \in \mathbb{R}^C$. These logits are then converted into a probability distribution $\boldsymbol{p} \in \mathbb{R}^C$ through the Softmax function. DGTC is trained using the cross-entropy loss:

$$\mathcal{L} = \sum_{i=1}^{C} -\boldsymbol{y}_i \log \boldsymbol{p}_i \tag{9}$$

where $\boldsymbol{y}$ denotes the ground truth label.

4.3 Complexity Analysis

The computational cost of the DGTC stems from two components: the computation of comprehensive positional encodings and the network's forward propagation. In the comprehensive positional encoding stage, calculating the dynamic degree encodings, dynamic distance encodings, and time encodings for each dynamic graph costs $O(P|\mathcal{E}|)$, $O(P|\mathcal{V}|(|\mathcal{V}| + |\mathcal{E}|))$ and $O(P|\mathcal{E}|)$, respectively. These computations involve only the calculation of graph-level statistical information on the CPU, and can be pre-computed and stored prior to training to accelerate the overall training process. During the network's forward propagation, the time complexities of forwarding f_1, f_2 and f_3 are $O(|\mathcal{V}|(d_m d_N + d_m^2))$, $O(|\mathcal{V}|(Pdd_m + d_m^2))$, $O(|\mathcal{V}|(Sdd_m + d_m^2))$, respectively. The time cost of forwarding f_4 is $O(|\mathcal{V}|^2(Pdd_m + d_m^2))$. The computation cost of MHSA is $O(|\mathcal{V}|^2 d_m + |\mathcal{V}|d_m^2)$. Therefore, the overall computation cost of DGTC is $O(|\mathcal{V}|^2(Pdd_m + d_m^2) + |\mathcal{V}|(d_m d_N + Sdd_m))$. DGTC supports batch processing of multiple dynamic graphs, with training relying exclusively on tensor operations. This design enables full GPU-based parallel computation, thereby significantly enhancing runtime efficiency. In contrast, most existing DyGNN architectures require sequential traversal of all temporal edges within each dynamic graph, resulting in an additional complexity of $O(B|\mathcal{E}|)$ that is inherently resistant to parallelization, where B denotes the mini-batch size. By circumventing this constraint through its tensor-based batch processing design, DGTC attains markedly higher runtime efficiency. We will compare the running efficiency of DGTC and baselines in Sect. 5.4.

4.4 Comparison with Existing Works

Comparison with MP-DGNs. When compared to MP-DGNs, the distinctions of DGTC and MP-DGNs lies in following aspects. First, when computing graph-level representations of a dynamic graph, DGTC processes all nodes jointly in a single Transformer forward pass, thereby producing representations for all nodes simultaneously This process can be highly parallelized on GPUs, thereby granting DGTC superior computational efficiency. In contrast, existing MP-DGNs update node embeddings by traversing temporal edges sequentially. The neighbor sampling step in this process involves CPU-bound operations that are inherently difficult to parallelize. Second, the Transformer architecture in DGTC enables each node to adaptively attend to and aggregate information from any other node in the graph. This design allows high-order subgraphs information to be captured. Most existing MP-DGNs, however, only consider local subgraph structures—typically restricted to historical neighbors within two hops.

Comparison with static Graph Transformer The key distinction between DGTC and static graph Transformers (e.g., Graphormer [36] and GraphGPS [20] lies in DGTC 's comprehensive positional encodings which can capture the evolving structural patterns of dynamic graphs. Specifically, DGTC leverages the dynamic degree encoding and dynamic distance encoding to characterize the evolutions of local subgraphs and high-order subgraphs. In addition, DGTC incorporates time encoding to embed temporal signal information. The positional encodings employed in static-graph Transformers are intrinsically tailored to the structural properties of static graphs, and thus fail to represent the evolving topological patterns inherent in dynamic graphs, which in turn results in degraded performance when applied to dynamic graphs.

5   Experiments

In this section, we conduct extensive experiments to evaluate the proposed DGTC, including dynamic graph classification experiments, ablation studies, efficiency evaluation and parameter sensitivity analysis.

Table 1. Statistics of datasets.

Datasets	Forum-java	HDFS	Gowalla	FourSquare	Brightkite
Graph Number	172,443	130,344	105,862	347,848	44,693
Avg. # Node	27	12	72	61	46
Avg. # Edge	30	31	117	135	188
# Node Feature	3	3	3	3	3

5.1   Experimental Settings

Datasets. The dataset configurations process closely follows [16]. We adopt five datasets for dynamic graph classification experiments: (1) **Forum-java** [16]. The Forum-java dataset originates from logs collected from an open-source Java-based forum platform. In each network, nodes correspond to individual log events. Edges represent the sequential order in which these events take place. (2) **HDFS** [33]. The HDFS dataset is an openly accessible log dataset maintained within the HDFS distributed file system. The log entries are processed into 575,061 dynamic session networks. Among these, 16,838 networks have been precisely annotated as "negative" by domain experts. Each log event's level, originating module, and thread ID are embedded into the node features through a label coding process. (3) **Brightkite** [1], **Gowalla** [1] and **FourSquare** [35]. These three public datasets captures user movement trajectories. The original datasets are split into a series of dynamic networks determined by user IDs. Within each network, nodes denote the geographic locations where check-ins occur, while edges illustrate transitions in a user's position. The longitude, latitude, and country ID associated with each check-in location are retained as the original node features.

The HDFS, Brightkite, Gowalla and FourSquare are public available datasets which lacks negative samples. Following [16], we adopt two strategies to generate the negative samples for these datasets, namely context-dependent negative sampling strategy and edge-order random shuffling. The statistics of pre-processed datasets are presented in Table 1.

Baselines. We select 10 baseline models for comparison as follows. **Static GNNs**: GCN [12], GAT [26] and GraphSAGE [5]; **Static Graph Transformer**: GraphGPS [20] and Graphormer [36]; **MP-DGNs**: TGAT [32], Graph-Mixer [2] and FreeDyG [24]; **Dynamic Graph Transformer**: DyGFormer [39]. **Dynamic Graph Classification Model**: TP-GNN [16].

Evaluation Protocols. Our evaluation protocols closely follow [16]. We randomly assign 70% of the dynamic graphs to the training set, and the remaining graphs are used as the test set. After training the model on the training set, we evaluate it on the test set. The model's performance is measured using the F1 score and Area Under the ROC curve (AUC).

Implementation Details

Baseline Configurations. For all models, we train for 10 epochs with a learning rate set to 0.001. For all static graph models, we disregard the timestamps in the dynamic graphs and treat them as static graphs. The output embedding dimension is set to 32, with 2 layers in the architecture. In GraphSAGE, we employ the mean aggregator. For GraphGPS, we adopt RWPE for node-level positional encoding. For both GraphGPS and Graphormer, the number of attention heads is set to 2. We use mean-pooling to aggregate all node representations in a graph into a single graph-level representation. For dynamic graph models, the model processes nodes or edges sequentially according to temporal order, after which the node- or edge-level representations are aggregated via mean-pooling to obtain the dynamic graph representation. All dynamic graph models are implemented based on DyGLib[1]. The architecture consists of 2 layers, and the output dimension of node or edge representations is set to 32. In GraphMixer and TGAT, the number of historical neighbors sampled is set to 20. For TP-GNN [16], the default hyperparameters provided in its official implementation are used. The edge aggregation function is set to mean.

DGTC Configurations. We train DGTC for 10 epochs with a hidden dimension of $d_m = d = 32$, which is also used for the output graph representation dimension. The number of attention heads is set as 2, and the model contains $L = 2$ Transformer layers. The number of splited time intervals P is set as 5. The timestamps number S in Time Encoding is set as $S = 10$. All experiments are conducted on an Ubuntu 24.04 Server with eight NVIDIA GeForce RTX 5090 GPUs.

5.2   Dynamic Graph Classification

The F1 scores and AUC values of dynamic graph classification are presented in Table 2. From Table 2, we observe that DGTC achieves the best performance on all five datasets on both metrics. Specifically, the average performance

---

[1] https://github.com/yule-BUAA/DyGLib.

Table 2. F1 scores and AUC values of dynamic graph classification experiments (all values multiplied by 100). The best performing model under each metric is marked in **bold**.

Model	Forum-java		HDFS	
	F1	AUC	F1	AUC
GCN	84.41 ± 2.24	66.67 ± 8.46	82.49 ± 0.00	50.00 ± 0.00
GAT	80.58 ± 0.00	50.00 ± 0.00	82.49 ± 0.00	50.00 ± 0.00
GraphSAGE	97.58 ± 0.20	95.10 ± 0.19	84.86 ± 0.19	58.03 ± 0.62
GraphGPS	67.19 ± 2.44	59.05 ± 6.37	57.26 ± 2.46	48.79 ± 1.31
Graphormer	97.27 ± 0.05	94.88 ± 0.11	92.67 ± 0.22	81.80 ± 0.20
TGAT	89.06 ± 0.00	84.55 ± 0.00	84.93 ± 0.00	58.23 ± 0.00
GraphMixer	89.05 ± 0.00	78.98 ± 0.00	84.37 ± 0.00	56.47 ± 0.00
DyGFormer	93.50 ± 0.00	90.05 ± 0.00	84.71 ± 0.00	57.70 ± 0.00
FreeDyG	79.96 ± 0.00	49.90 ± 0.00	84.52 ± 0.00	56.94 ± 0.00
TP-GNN	98.73 ± 0.16	98.20 ± 0.12	97.02 ± 0.32	95.24 ± 0.70
DGTC (ours)	$\mathbf{99.62 \pm 0.02}$	$\mathbf{99.45 \pm 0.01}$	$\mathbf{98.62 \pm 0.02}$	$\mathbf{98.52 \pm 0.05}$
impr. (%)	+0.90%	+1.27%	+1.62%	+3.33%

Model	Gowalla		Brightkite		FourSquare	
	F1	AUC	F1	AUC	F1	AUC
GCN	83.18 ± 0.00	50.00 ± 0.00	82.14 ± 0.00	50.00 ± 0.00	84.47 ± 0.00	50.00 ± 0.00
GAT	83.18 ± 0.00	50.00 ± 0.00	82.14 ± 0.00	50.00 ± 0.00	83.56 ± 0.00	50.00 ± 0.00
GraphSAGE	85.85 ± 0.21	61.83 ± 1.06	82.08 ± 0.05	50.03 ± 0.07	87.53 ± 0.28	70.25 ± 0.57
GraphGPS	61.80 ± 2.44	69.41 ± 4.99	71.52 ± 3.40	69.73 ± 2.29	59.67 ± 4.17	67.59 ± 3.10
Graphormer	84.42 ± 0.06	70.81 ± 0.19	82.75 ± 0.07	68.24 ± 0.37	84.86 ± 0.03	74.05 ± 0.07
TGAT	84.60 ± 0.00	55.00 ± 0.00	83.11 ± 0.00	53.25 ± 0.00	83.26 ± 0.00	53.77 ± 0.00
GraphMixer	83.77 ± 0.00	52.12 ± 0.00	82.71 ± 0.00	51.91 ± 0.00	83.11 ± 0.00	53.24 ± 0.00
DyGFormer	83.83 ± 0.00	52.31 ± 0.00	82.70 ± 0.00	51.88 ± 0.00	82.98 ± 0.00	56.04 ± 0.00
FreeDyG	83.76 ± 0.00	52.06 ± 0.00	82.68 ± 0.00	51.80 ± 0.00	83.11 ± 0.00	53.24 ± 0.00
TP-GNN	89.10 ± 0.24	70.54 ± 0.10	84.66 ± 2.21	61.15 ± 6.33	81.56 ± 0.04	73.99 ± 0.06
DGTC (ours)	$\mathbf{89.99 \pm 0.12}$	$\mathbf{88.66 \pm 0.12}$	$\mathbf{86.61 \pm 0.07}$	$\mathbf{83.83 \pm 0.46}$	$\mathbf{89.44 \pm 0.04}$	$\mathbf{89.89 \pm 0.02}$
impr. (%)	+1.00%	+25.21%	+2.30%	+20.22%	+2.18%	+21.39%

improvement of DGTC in terms of F1 and AUC is 2.24% and 14.28%, respectively. These results demonstrate the superiority of DGTC on dynamic graph classification experiments over existing models. Moreover, we have following observations: Firstly, compared with static graph transformers (i.e., GraphGPS and Graphormer), DGTC exhibits substantial performance improvements. This demonstrates that the comprehensive positional encoding in DGTC can effectively capture the spatio-temporal variation patterns of dynamic graph structures, making it more suitable for dynamic graph classification tasks. Secondly, DGTC also achieves significant performance gains over existing dynamic graph models. This is because most current dynamic graph models neglect learning the variation of global patterns of dynamic graphs. In contrast, DGTC employs global self-attention to effectively capturing the overall variation patterns of dynamic graphs.

Table 3. Ablation studies results of DGTC. AUC values are presented (all values multiplied by 100).

	HDFS	Gowalla	Brightkite	FourSquare
Full model	98.52 ± 0.05	88.66 ± 0.12	83.83 ± 0.46	89.89 ± 0.02
w/o dist	98.47 ± 0.03	87.18 ± 0.09	81.69 ± 0.40	88.57 ± 0.00
w/o time	97.84 ± 0.01	87.24 ± 0.05	75.47 ± 2.46	88.58 ± 0.01
w/o deg	98.44 ± 0.01	66.66 ± 0.18	69.38 ± 3.86	87.77 ± 0.01

5.3 Ablation Studies

In this section, we conduct ablation studies to verify the effectiveness of each component in DGTC . The experimental results are presented in Table 3. In the table, "w/o deg", "w/o dist", and "w/o time" indicate that the dynamic degree encoding, dynamic distance encoding, and time encoding are removed from DGTC , respectively. It can be observed that the dynamic degree encoding has the most significant impact on the performance of DGTC . When removing the dynamic degree encoding, the performance of DGTC drops by 22% and 14.15% on *Gowalla* and *Brightkite*, respectively. Both the dynamic distance encoding and time encoding also have notable effects on model performance. Specifically, on *Brightkite*, removing the dynamic distance encoding and time encoding results in performance reductions of 2.14% and 8.36%, respectively. These findings demonstrate that both the dynamic distance encoding and time encoding play important roles in enhancing the performance of DGTC . Overall, all three positional encodings contribute effectively to the model's performance improvement.

5.4 Efficiency Analysis

In this chapter, we evaluate the runtime efficiency of DGTC . Specifically, we compare the time required for DGTC and other dynamic graph models to complete one training epoch (i.e., processing all dynamic graphs) on the HDFS and Gowalla datasets. The experimental results are presented in Fig. 2. As shown, DGTC demonstrates substantially higher efficiency than all competing dynamic graph models. On HDFS, DGTC (28 s) is 35× faster than the second fastest model, GraphMixer (985 s), and 69× faster than TP-GNN (1,940 s). On Gowalla, DGTC (134 s) outperforms TP-GNN (4,124 s) by a margin of 30×. This efficiency gain stems from the fact that, unlike other models, DGTC does not require iteration over all temporal edges in the dynamic graph. Instead, it processes all nodes in a single pass through the Transformer, enabling a high degree of parallelism.

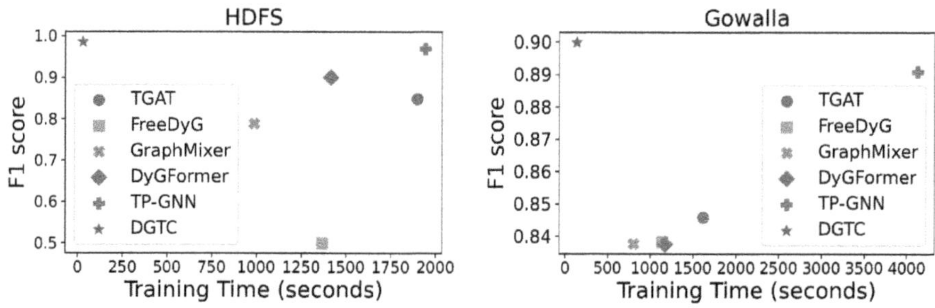

Fig. 2. Running time per epoch (seconds) and F1 scores of dynamic graph models on HDFS (left) and Gowalla (right).

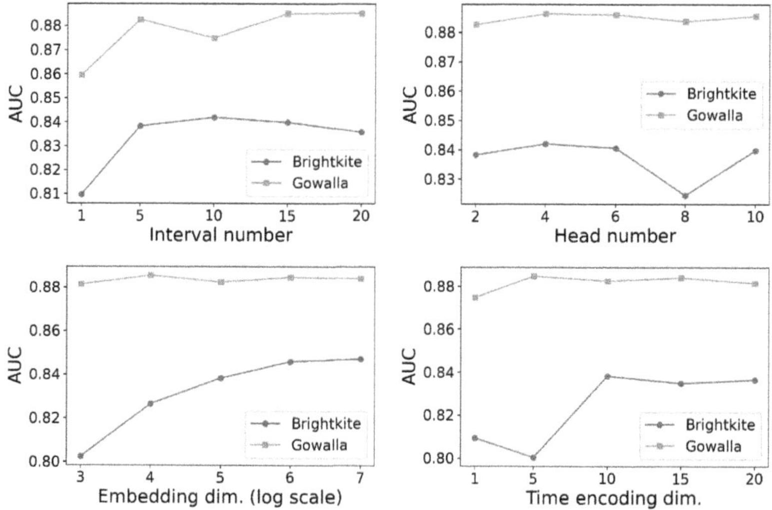

Fig. 3. Parameter sensitivity analysis of DGTC. The AUC values on Brightkite and Gowalla datasets are presented. **Top left**: Time interval number P; **Top right**: Attention head number; **Bottom left**: Embedding dimension d_m; **Bottom right**: Time encoding dimension S.

5.5   Parameter Sensitivity Analysis

In this section, we investigate the sensitivity of DGTC to several key hyperparameters—namely, the number of temporal intervals P, the number of attention heads, the embedding dimension d_m, and the time encoding dimension S—using the Brightkite and Gowalla datasets. The results are shown in Fig. 3, from which we draw the following observations. First, setting the number of temporal intervals to $P = 1$ leads to a substantial drop in performance, as the structural variation patterns of the dynamic graph cannot be effectively captured through positional encodings alone when time interval partitioning is absent. This finding validates the importance of adopting temporal interval partitioning. Second,

DGTC exhibits low sensitivity to the number of attention heads, with optimal performance attained at 4 heads. Third, variations in the embedding dimension d_m and the time encoding dimension S cause performance fluctuations in the Brightkite dataset, whereas performance remains relatively stable on the Gowalla dataset.

6 Conclusions

Existing studies on dynamic graph classification face challenges in parallelism and modeling high-order subgraph structures. To address these issues, we propose DGTC, a Transformer-based architecture that leverages global self-attention to achieve high parallel computational efficiency. To comprehensively capture the spatio-temporal evolution patterns of dynamic graphs, we adopt dynamic degree encoding and dynamic distance encoding to represent changes in local structures and high-order subgraphs, respectively. Extensive experiments on dynamic graph classification tasks demonstrate that DGTC achieves higher accuracy and efficiency than existing methods.

Acknowledgments. This work is supported by the Zhejiang Province "JianBingLingYan+X" Research and Development Plan (2025C02020)

References

1. Cho, E., Myers, S.A., Leskovec, J.: Friendship and mobility: user movement in location-based social networks. In: Proceedings of the 17th ACM SIGKDD International Conference on Knowledge Discovery and Data Mining, pp. 1082–1090 (2011)
2. Cong, W., et al.: Do we really need complicated model architectures for temporal networks? In: International Conference on Learning Representations (2023)
3. Dwivedi, V.P., Bresson, X.: A generalization of transformer networks to graphs. arXiv preprint arXiv:2012.09699 (2020)
4. Dwivedi, V.P., Luu, A.T., Laurent, T., Bengio, Y., Bresson, X.: Graph neural networks with learnable structural and positional representations. In: International Conference on Learning Representations (2021)
5. Hamilton, W., Ying, Z., Leskovec, J.: Inductive representation learning on large graphs. Adv. Neural Inf. Process. Syst. **30** (2017)
6. Huang, L., Ma, Y., Liu, Y., Du Danny, B., Wang, S., Li, D.: Position-enhanced and time-aware graph convolutional network for sequential recommendations. ACM Trans. Inf. Syst. **41**(1), 1–32 (2023)
7. Huang, X., et al.: DGraph: a large-scale financial dataset for graph anomaly detection. Adv. Neural. Inf. Process. Syst. **35**, 22765–22777 (2022)
8. Hussain, M.S., Zaki, M.J., Subramanian, D.: Global self-attention as a replacement for graph convolution. In: Proceedings of the 28th ACM SIGKDD Conference on Knowledge Discovery and Data Mining, pp. 655–665 (2022)
9. Jansen, B.J.: Search log analysis: what it is, what's been done, how to do it. Libr. Inf. Sci. Res. **28**(3), 407–432 (2006)

10. Jin, M., Li, Y.F., Pan, S.: Neural temporal walks: Motif-aware representation learning on continuous-time dynamic graphs. Adv. Neural. Inf. Process. Syst. **35**, 19874–19886 (2022)
11. Khodabandehlou, S., Golpayegani, A.H.: FiFrauD: unsupervised financial fraud detection in dynamic graph streams. ACM Trans. Knowl. Discov. Data **18**(5), 1–29 (2024)
12. Kipf, T.N., Welling, M.: Semi-supervised classification with graph convolutional networks. In: International Conference on Learning Representations (2017)
13. Kreuzer, D., Beaini, D., Hamilton, W., Létourneau, V., Tossou, P.: Rethinking graph transformers with spectral attention. Adv. Neural. Inf. Process. Syst. **34**, 21618–21629 (2021)
14. Kumar, S., Zhang, X., Leskovec, J.: Predicting dynamic embedding trajectory in temporal interaction networks. In: Proceedings of the 25th ACM SIGKDD International Conference on Knowledge Discovery & Data Mining, pp. 1269–1278 (2019)
15. Li, Y., Yu, R., Shahabi, C., Liu, Y.: Diffusion convolutional recurrent neural network: Data-driven traffic forecasting. In: International Conference on Learning Representations (2018)
16. Liu, J., Liu, J., Zhao, K., Tang, Y., Chen, W.: TP-GNN: continuous dynamic graph neural network for graph classification. In: 2024 IEEE 40th International Conference on Data Engineering (ICDE), pp. 2848–2861. IEEE (2024)
17. Niepert, M., Ahmed, M., Kutzkov, K.: Learning convolutional neural networks for graphs. In: International Conference on Machine Learning, pp. 2014–2023. PMLR (2016)
18. Pareja, A., et al.: EvolveGCN: evolving graph convolutional networks for dynamic graphs. In: Proceedings of the AAAI Conference on Artificial Intelligence, vol. 34, pp. 5363–5370 (2020)
19. Peng, H., Li, J., Gong, Q., Ning, Y., Wang, S., He, L.: Motif-matching based subgraph-level attentional convolutional network for graph classification. In: Proceedings of the AAAI Conference on Artificial Intelligence, vol. 34, pp. 5387–5394 (2020)
20. Rampášek, L., Galkin, M., Dwivedi, V.P., Luu, A.T., Wolf, G., Beaini, D.: Recipe for a general, powerful, scalable graph transformer. Adv. Neural. Inf. Process. Syst. **35**, 14501–14515 (2022)
21. Rossi, E., Chamberlain, B., Frasca, F., Eynard, D., Monti, F., Bronstein, M.: Temporal graph networks for deep learning on dynamic graphs. arXiv preprint arXiv:2006.10637 (2020)
22. Sankar, A., Wu, Y., Gou, L., Zhang, W., Yang, H.: DySat: deep neural representation learning on dynamic graphs via self-attention networks. In: Proceedings of the 13th International Conference on Web Search and Data Mining, pp. 519–527 (2020)
23. Shao, Z., et al.: Decoupled dynamic spatial-temporal graph neural network for traffic forecasting. Proc. VLDB Endow. **15**(11), 2733–2746 (2022)
24. Tian, Y., Qi, Y., Guo, F.: FreeDyG: frequency enhanced continuous-time dynamic graph model for link prediction. In: International Conference on Learning Representations (2024)
25. Trivedi, R., Farajtabar, M., Biswal, P., Zha, H.: DyRep: learning representations over dynamic graphs. In: International Conference on Learning Representations (2019)
26. Veličković, P., Cucurull, G., Casanova, A., Romero, A., Lio, P., Bengio, Y.: Graph attention networks. In: International Conference on Learning Representations (2017)

27. Wang, X., et al.: APAN: asynchronous propagation attention network for real-time temporal graph embedding. In: Proceedings of the 2021 International Conference on Management of Data, pp. 2628–2638 (2021)
28. Wang, Y., Chang, Y.Y., Liu, Y., Leskovec, J., Li, P.: Inductive representation learning in temporal networks via causal anonymous walks. In: International Conference on Learning Representations (2021)
29. Wang, Z., et al.: Dynamic graph transformer with correlated spatial-temporal positional encoding. In: Proceedings of the Eighteenth ACM International Conference on Web Search and Data Mining, pp. 60–69 (2025)
30. Wen, Z., Fang, Y.: Trend: temporal event and node dynamics for graph representation learning. In: Proceedings of the ACM Web Conference 2022, pp. 1159–1169 (2022)
31. Xiong, R., et al.: On layer normalization in the transformer architecture. In: International Conference on Machine Learning, pp. 10524–10533. PMLR (2020)
32. Xu, D., Ruan, C., Korpeoglu, E., Kumar, S., Achan, K.: Inductive representation learning on temporal graphs. In: International Conference on Learning Representations (2020)
33. Xu, W., Huang, L., Fox, A., Patterson, D., Jordan, M.: Largescale system problem detection by mining console logs. In: Proceedings of SOSP, vol. 9, pp. 1–17 (2009)
34. Yang, C., Liu, M., Zheng, V.W., Han, J.: Node, motif and subgraph: leveraging network functional blocks through structural convolution. In: 2018 IEEE/ACM International Conference on Advances in Social Networks Analysis and Mining (ASONAM), pp. 47–52. IEEE (2018)
35. Yang, D., Zhang, D., Qu, B.: Participatory cultural mapping based on collective behavior data in location-based social networks. ACM Trans. Intell. Syst. Technol. (TIST) **7**(3), 1–23 (2016)
36. Ying, C.: Do transformers really perform badly for graph representation? Adv. Neural. Inf. Process. Syst. **34**, 28877–28888 (2021)
37. You, J., Du, T., Leskovec, J.: Roland: graph learning framework for dynamic graphs. In: Proceedings of the 28th ACM SIGKDD Conference on Knowledge Discovery and Data Mining, pp. 2358–2366 (2022)
38. Yu, B., Yin, H., Zhu, Z.: Spatio-temporal graph convolutional networks: a deep learning framework for traffic forecasting. In: Proceedings of the Twenty-Seventh International Joint Conference on Artificial Intelligence (2017)
39. Yu, L., Sun, L., Du, B., Lv, W.: Towards better dynamic graph learning: new architecture and unified library. Adv. Neural. Inf. Process. Syst. **36**, 67686–67700 (2023)
40. Zhang, S., Chen, L., Wang, C., Li, S., Xiong, H.: Temporal graph contrastive learning for sequential recommendation. In: Proceedings of the AAAI Conference on Artificial Intelligence, vol. 38, pp. 9359–9367 (2024)

MACA: Multimodal Aspect-Based Sentiment Analysis with Dynamic Adaptive Attention, Contrastive Alignment, and LLM Augmentation

Xuelin Xu[1,2] and Jun Lu[1,2(✉)]

[1] College of Computer Science and Technology, Heilongjiang University, Harbin 150080, China
[2] Jiaxiang Industrial Technology Research Institute, Heilongjiang University, Jining 272400, Shandong, China
`lujun111_lily@sina.com`

Abstract. Multimodal Aspect-Based Sentiment Analysis (MABSA) is an important task. Its goal is to predict the emotional polarity. Existing methods mainly realize the association between aspects and multimodal features through question and answer modeling, but there are still limitations in the dynamics of modal fusion, the accuracy of aspect-modal alignment, and the use of external knowledge. Therefore, this article proposes MACA (Multimodal Aspect-Based Sentiment Analysis with Dynamic Adaptive Gating, Contrastive Alignment, and LLM Augmentation): A unified framework for MABSA. In order to solve the problem of modal difference and aspect evidence mismatch at the sample level, a method combining dynamic adaptive gating (DAG), aspect-modal contrast alignment (ACA) and LLM prior dual path integration is proposed. The MACA model maintains consistent modal preferences. It adaptively allocates text/image/description/prior contributions through learnable gating, and uses contrastive learning to reinforce consistency between aspect and evidence region. The decision-making layer introduces temperature calibration and confidence modulation, amplifying their effects when the prior is reliable and automatically contracting when the prior is uncertain. On Twitter-15/17, MACA outperforms the strong baseline in multiple tasks. Ablation shows that the three modules reinforce each other, significantly improving the accuracy and robustness of the model.

Keywords: MABSA · Contrastive Alignment · Dynamic adaptive gating · Prior-Driven Adaptive Fusion

1 Introduction

MABSA aims at combining multimodal information such as text, image and auxiliary description to make a fine-grained prediction of emotional tendencies in specific aspects. Compared to traditional unimodal sentiment analysis, MABSA

H. Jung et al. (Eds.): DASFAA 2026, LNCS 16536, pp. 114–130, 2026.
https://doi.org/10.1007/978-981-92-0366-6_8

Aspect	NFL	Cowboys	Elliott
Sentiment	NEU	NEU	POS

GPT-4: A Dallas Cowboys player, wearing a helmet inscribed with "ELLIOTT" and "COWBOYS", blue-and-white gloves, and a jersey with an AT&T logo, is catching a football.

Fig. 1. MABSA task. This article uses GLM-4.5 to construct prior knowledge

can utilize the complementarity of visual and linguistic context to achieve more accurate predictions. As shown in Fig. 1, MATE is committed to recognizing aspect terms from text image pairs, MASC classifies the sentiment polarity of aspect words, and MABSA jointly extracts aspect terms and predicts their emotions, thus it can be regarded as a fusion solution of MATE and MASC.

In recent years, question and answer modeling frameworks have gradually attracted attention in MABSA. Typical model is DEQA [3], which has achieved good results by designing question and answer templates (such as "What is the sentential of?") to associate aspect terms with multimodal features. However, DEQA class models still have three limitations: 1. Modal fusion lacks dynamics. 2. Aspect modal alignment is insufficient. 3. External knowledge utilization is limited.

This article proposes MACA to address the aforementioned issues, with the following specific design motivations:

First, for the modal fusion static problem, MACA adaptively assigns the weights of text, image and description information based on the modal correlation scoring function to improve the modal adaptability of different samples.

Secondly, in order to strengthen the correspondence between aspect and modal, MACA introduces comparative learning strategy and constructs positive and negative sample pairs to enhance the alignment between aspect words and modal features.

Finally, with the help of large language model, MACA introduces the aspect-emotion pairs generated by LLM as an auxiliary modal and a decision priori dual path, maintaining consistent modal and semantic preferences in the two stages, supplementing the existing modal signal gap, and improving the accuracy and robustness of the model.

The main contributions of this paper are as follows:

1. Propose a unified Multimodal Aspect-Based Sentiment Analysis framework, MACA, and integrate multiple technologies to solve the shortcomings of existing methods in modal interaction, aspect alignment, and external knowledge utilization. Large scale experiments are conducted on multiple datasets,

and the results show that MACA significantly outperforms existing methods on all indicators.

2. Propose aspect modal contrast alignment strategy (ACA) and dynamic adaptive gating (DAG) technology to optimize multimodal evidence utilization and matching accuracy, and systematically solve the problem of dynamic difference in modal contribution and feature mismatch in multi-aspect scenes.

3. LLM is proposed as an auxiliary modal to introduce the whole link process, and the precise injection and adaptive adjustment of external knowledge are realized through a priori dual path integration strategy, so as to improve the robustness of emotion discrimination.

2   Related Work

MABSA combines text and images for fine-grained emotional understanding, and is usually divided into two sub tasks: Multimodal Aspect Term Extraction (MATE) and Multimodal Aspect-oriented Sentiment Classification (MASC): the former identifies aspect terms in sentences, and the latter predicts emotional polarity based on given aspects. This paradigm has been systemized recently, and the impact of cross modal evidence on emotional judgment is intuitively reflected through examples, such as the complementarity of text and image information.

Pipeline Framework and Joint Framework: systematic carding shows that most early implementations of MABSA are pipeline solutions: pair MATE model with MASC model, to obtain complete solutions,such as "UMT+TomBERT", "OCSGA+TomBERT", etc. Then the research turns to the joint and multi task framework, such as JML's [4] end-to-end joint learning, VLP-MABSA's [5] task specific visual - language pre-training, CMMT's [13] cross modal multi task transformer. More recent work further emphasizes "aspect oriented" alignment and prompt/energy based expert modules, such as AoM [16], DQPSA [8], TCMT [17], EKMG [7], showing a gradual refinement trend around the "aspect - evidence - discrimination" chain.

Q&A **Modeling and Templating Tips:** Recently, the *Q&A* paradigm formalizes MABSA into multi instance QA: complete aspect extraction, verification and emotion classification through three types of natural language queries, and resolve multi instance ambiguity in the same aspect with special tags. This idea improves the interpretability and controllability with linguistic task prompts, and provides a structured interface for unified processing of two sub-tasks (MATE/MASC).

MACA inherits the DEQA question and answer framework and two-stage structure, and realizes key improvements in three aspects: dynamic adaptive gating, aspect modal contrast alignment, and LLM prior dual path. It systematically solves the problem of its modal fusion static and insufficient use of external semantic prior.

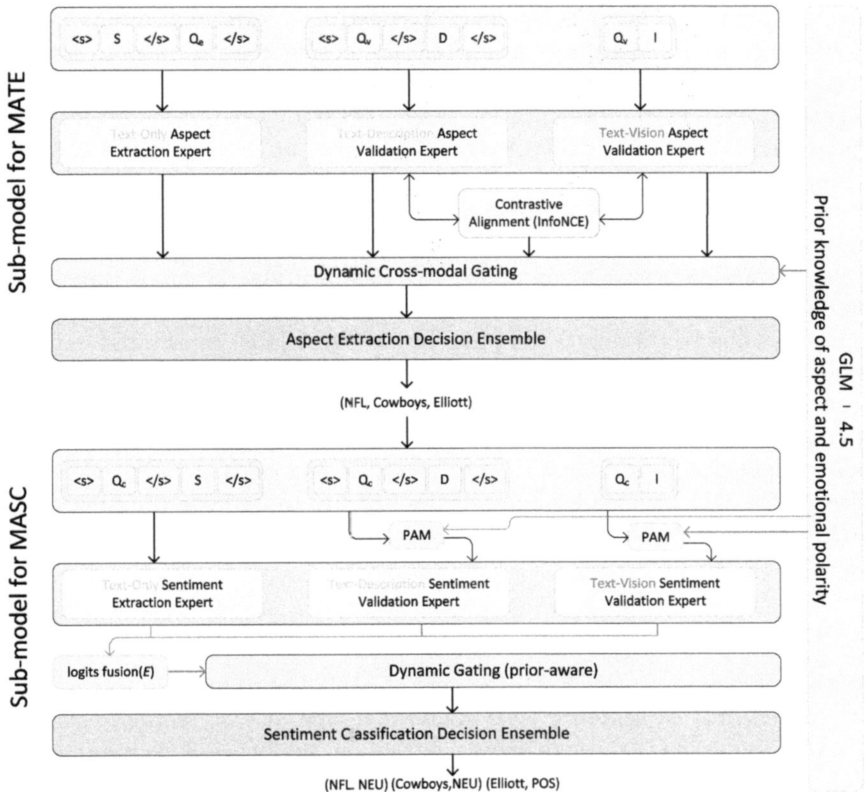

Fig. 2. Framework of MACA. Contains sub modules for MATE and MASC, $\langle s \rangle$ is the beginning token, and $\langle /s \rangle$ is the segment token.

3   Method

3.1   Main Framework of MACA

The MABSA task aims to recognize $\{(a_1, s_1), (a_2, s_2), \ldots, (a_i, s_i), \ldots\}$ (a_i is an aspect item, $s_i \in \{pos, neu, neg\}$) in tweets containing image I and sentence S. The input includes text T, image I, and image description D. As shown in Fig. 2, MACA first executes the MATE submodule: constructs the input according to the "aspect extraction query - aspect validation query", combines sequence annotation and conditional random field (CRF) to generate candidate aspects. In the verification phase, three experts (text, text+description and text+vision) are introduced to provide evidence scoring, and dynamic adaptive gating is added between expert output and decision integration to adaptively allocate $\{T, I, D\}$ contributions. At the same time, the "aspect modal" alignment is strengthened through comparative learning to reduce multi aspect scene confusion.

After obtaining the candidate aspect set, MACA enters the MASC stage: it still uses the same three experts as the MATE construction. In the MASC stage, these three experts independently construct conditional inputs, and the classification head and gating parameters are not shared. In order to introduce external knowledge, prior absorption is achieved through dual path LLM prior integration, that is, the presentation layer uses prior perception gating (PAM) to modulate the internal modal weights of experts at the sample level, and the decision-making layer implements the mechanism of temperature calibration and learnable fusion for expert logits. The training adopts segmented optimization: first minimize the MATE extraction target and fix its parameters, and then fix the extraction result to minimize the classification target of three MASC experts. To reduce the distribution gap, the sentiment classification stage performs annealing based mixed sampling on the "gold label aspect and extraction prediction" to gradually adapt the classifier to the true inference distribution. In order to alleviate the inconsistency of cross stage modal preferences, the gating consistency regularization is added without returning to MATE, and the modal weight distribution in the same aspect is aligned. On the whole, MACA makes use of evidence quality and priori reliability to improve the robustness of MABSA through dynamic gating, contrast alignment and LLM priori dual path integration.

3.2   Contrastive Alignment and Dynamic Adaptive Gating

Given the image text pair (S, I), the goal of MATE is to identify the aspect set $\mathcal{A} = \{a_1, \ldots, a_n\}$ in sentence level context. In the MATE section, this article first extracts query Q_e through aspect extraction, performs sequence annotation on $[Q_e; S]$ to obtain candidate aspects, further verify the candidate aspect binary classification of the text-description and text-visual expert channels using aspect validation query Q_v.

Subsequently, in the text extraction stage, DeBERTa is used to encode the input data and generate token level scores $X \in \mathbb{R}^{L \times |\mathcal{Y}|}$ ($\mathcal{Y} = \{B, I, O\}$). Linear chain CRF is used to model inter label dependencies:

$$\mathcal{L}_{\mathrm{CRF}} = -\log \frac{\exp(\sum_{i=1}^{L} (A_{y_{i-1}, y_i} + X_{i, y_i}))}{\sum_{y' \in \mathcal{Y}^L} \exp(\sum_{i=1}^{L} (A_{y'_{i-1}, y'_i} + X_{i, y'_i}))} \tag{1}$$

Among them, L represents the number of tokens in the text sequence. $\mathcal{Y}$ represents the set of labels for sequence annotation, A_{y_{i-1}, y_i} represents the labels from the $i-1$th token. y' represents any label sequence of length L. $\mathcal{Y}^L$ represents the set of all label sequences of length L.

Based on the output of the CRF model mentioned above, the candidate aspect set $\mathcal{A}^{(0)}$ can be obtained through Viterbi decoding. In the candidate verification stage, candidate a is annotated as </target> and inserted into the text to construct $[Q_v; S_{\langle \mathrm{target} \rangle a}; D]$, which is then fed into the text encoder and the first </target> hidden vector h_{tgt} is taken. Text – Description expert judges the validity of candidates through the following Eq. (2):

$$\mathcal{L}_{ad} = -\sum_{c \in \{0,1\}} y_c \log(\mathrm{softmax}(W_{ad} h_{\mathrm{tgt}} + b_{ad}))_c \tag{2}$$

where $\mathcal{L}_{ad}$ represents text-description expert classification loss. C represents the category label of candidate.

In addition to the text-description channel, the model also designs a text-visual expert channel to use visual modal information. This channel first inputs $[Q_v; S_{\langle\text{target}\rangle a}]$ into the vector t of the CLIP text encoder, and image I is input into the patch sequence V of the CLIP visual encoder. Through multi head cross attention, the alignment representation g is obtained, and the text side is affined to obtain the transformed text representation t', as shown in Eq. (3) (4):

$$g = \text{CA}(t, V) \tag{3}$$

$$t' = \text{Dropout}(\text{ReLU}(W_t t + b_t), 0.5) \tag{4}$$

Then, MFB fusion is used to fuse t' and g, and cross modal fusion feature f is obtained. Based on f, classification loss $\mathcal{L}_{av}$ of text-visual expert is defined:

$$f = MFB(t', g) \tag{5}$$

$$\mathcal{L}_{av} = - \sum_{c \in \{0,1\}} y_c \log\left(\text{softmax}(W_{av} f + b_{av})\right)_c . \tag{6}$$

where, Y_c represents the effective true label of candidate aspects.

Although the above two expert channels can achieve candidate verification, in order to overcome the limitation of DEQA static fusion, MACA introduces four modal adaptive fusion and dynamic expert mixing: linearize the "aspect emotion pair" generated by LLM into a special token sequence $\mathcal{R}_L$, and pool it into h_L via text encoder. The description sequence is pooled into $\bar{H}_D$ and the alignment is expressed as g. On this basis, the four modal features are respectively linearly mapped to obtain the unified dimension modal feature vector:

$$z_T = W_T h_{\text{tgt}}, \quad z_I = W_I g, \quad z_D = W_D \bar{H}_D, \quad z_L = W_L h_L \tag{7}$$

In order to achieve the dynamic weight allocation at the sample level, the shared gating network is designed to generate the adaptive weight α_m of each modal, which is normalized through Softmax. The final modal fusion indicates that h_{fuse} is obtained from the weighted sum of modal features. At the same time, in order to avoid too centralized weight distribution, the model introduces the entropy regularization loss $\mathcal{L}_{\text{gate}}$. The specific equation is as follows:

$$\alpha_m = \frac{\exp\left(\mathbf{w}^\top \tanh(U_m z_m)\right)}{\sum_{m' \in \{T,I,D,L\}} \exp\left(\mathbf{w}^\top \tanh(U_{m'} z_{m'})\right)}, \quad m \in \{T, I, D, L\} \tag{8}$$

$$h_{\text{fuse}} = \sum_{m \in \{T,I,D,L\}} \alpha_m z_m, \quad \mathcal{L}_{\text{gate}} = -\beta \sum_m \alpha_m \log \alpha_m \tag{9}$$

Among them, β represents the entropy regularization coefficient.

In the dynamic expert mixing stage, the model first refers to the linear head logits of text-description expert as s_{ad}, and the linear head logits of text-visual expert as s_{av}. Subsequently, by driving the gating mechanism, adaptive weighted integration of two expert channels is achieved, and the loss function $\mathcal{L}_{\text{mix}}$ is denoted as:

$$\mathcal{L}_{\text{mix}} = - \sum_{c \in \{0,1\}} y_c \log(\text{softmax}(\pi_{ad}\, s_{ad} + \pi_{av}\, s_{av} + W_f h_{\text{fuse}} + b_f))_c \qquad (10)$$

Among them, π_{ad} and π_{av} correspond to the weight coefficients of text-description experts and text-visual expert, respectively. This expert mixture mechanism uses h_{fuse} to adaptively select trusted experts to improve robustness under modal loss and noise conditions.

Although the dynamic expert mixing mechanism mentioned above improves robustness, it is still prone to aspect-evidence mismatch in scenarios where multiple aspects coexist. To suppress this problem, this paper introduces a contrastive alignment mechanism: the aspect anchor point h_a is represented by the interval average annotated as </target>, and the positive sample r^+ that is most relevant to a is selected from the visual patch and description fragment, while the remaining regions or opposite aspects form a negative sample set $\{r^-\}$. The InfoNCE objective is used to strengthen the alignment of aspect and positive sample representations and suppress mismatches. As shown in Eq. (11):

$$\mathcal{L}_{\text{con}} = - \log \frac{\exp(\text{sim}(h_a, r^+)/\tau)}{\exp(\text{sim}(h_a, r^+)/\tau) + \sum_{r^-} \exp(\text{sim}(h_a, r^-)/\tau)} \qquad (11)$$

Among them, $\mathcal{L}_{\text{con}}$ represents aspect-modal contrast alignment loss (InfoNCE loss). h_a represents aspect anchor feature, and τ represents temperature parameter.

Overall, the optimization objective of MATE is:

$$\mathcal{L}_{\text{MATE}} = \mathcal{L}_{\text{CRF}} + \mathcal{L}_{ad} + \mathcal{L}_{av} + \mathcal{L}_{\text{mix}} + \lambda_1 \mathcal{L}_{\text{con}} + \lambda_2 \mathcal{L}_{\text{gate}} \qquad (12)$$

Among them, λ_1, λ_2 is optimized by validation set, and the final aspect set $\hat{A} = \{a : (P_a)_{\text{true}} \geq \theta\}$ is selected through threshold θ.

The core difference between MACA and DEQA in MATE is that four modal adaptive gating is incorporated into LLM prior. Dynamic expert mixing based on h_{fuse} replaces static addition. Introducing comparative alignment during the verification phase can alleviate various mismatches and improve accuracy and robustness.

3.3   Prior-Driven Adaptive Fusion

After obtaining the candidate aspect set $\hat{A}$, MASC aims to determine the sentiment polarity y_a corresponding to each $a \in \hat{A}$. Consistent with the MATE part, the four modals mapping z_m and basic weight distribution need to be obtained first.

Subsequently, the representation layer fusion adopts the sample level gating idea of MATE and further introduces prior aware gating modulation (PAM): firstly, the basic weight distribution α_m is generated through a scoring function consistent with MATE, and then the logarithmic domain correction is performed on the gating through bilinear matching score κ_m to obtain the recalibrated weight $\tilde{\alpha}_m$, as shown in Eq. (13).

$$\tilde{\alpha}_m = \frac{\alpha_m \exp(\rho\,\kappa_m)}{\sum_{m'} \alpha_{m'} \exp(\rho\,\kappa_{m'})}, \qquad \rho \geq 0 \tag{13}$$

where, κ_m is the bilinear matching score of the m-th modal feature z_m and LLM prior feature z_L. κ_m is used to measure the correlation between z_m and z_L. R_m is the weight matrix of bilinear matching. ρ is a correction intensity hyperparameter, and the larger its value, the stronger the correct effect.

This design can enlarge the corresponding modal weight when the prior is consistent with the evidence, and reduce its impact when the conflict occurs, so as to avoid treating L as the static fourth modal. Then obtain the fusion vector h_{cls} based on $\tilde{\alpha}$, and use h_{cls} as the basis to obtain the model personal emotion logits s_{mod} generated by the linear classification head, and the sentiment posterior distribution p_{mod} is obtained through Softmax.

To suppress modulation instability in the early stages of training, gate offset constraint is added:

$$\mathcal{L}_{\mathrm{pam}} = \left\| \tilde{\boldsymbol{\alpha}} - \boldsymbol{\alpha} \right\|_2^2 \tag{14}$$

The prior calibration of the decision layer is the second path: let LLM output three types of logarithmic probabilities $\ell_{\mathrm{llm}} \in \mathbb{R}^3$, which are temperature calibrated and mapped to the label space, and then learnable fused with the model logits:

$$p_{\mathrm{llm}} = \mathrm{softmax}\left(\frac{\ell_{\mathrm{llm}}}{T_{\mathrm{llm}}}\right), s_{\mathrm{llm}} = W_{\mathrm{llm}}\ell_{\mathrm{llm}} + b_{\mathrm{llm}}, p = \mathrm{softmax}(s_{\mathrm{mod}} + \gamma\, s_{\mathrm{llm}}) \tag{15}$$

Among them, T_{llm} is the temperature parameter, and p_{llm} is the posterior distribution of LLM logarithmic probabilities obtained after temperature calibration. s_{llm} is the logits of LLM logarithmic probability mapping to the model label space, and γ is the fusion coefficient of LLM prior. P is the final posterior distribution of emotions obtained by fusing the model logits and LLM mapping logits.

To suppress unreliable prior interference, the confidence level c_{llm} is defined using prior distribution entropy, and the fusion coefficient can be obtained by Eq. (16).

$$\gamma = \gamma_0 \cdot \sigma\big(w_\gamma^\top [h_{\mathrm{cls}}; h_L]\big) \cdot c_{\mathrm{llm}} \tag{16}$$

Thus, the fine-grained control of prior injection intensity is achieved.

The training objectives is set layer by layer according to the above reasoning chain. Firstly, the final posterior p is supervised by standard cross entropy:

$$\mathcal{L}_{\mathrm{cls}} = -\sum_{y \in \mathcal{Y}} [y = y^*] \log p(y) \tag{17}$$

Secondly, this paper uses the confidence weighted KL divergence to constraint model consistency between the posterior and prior distributions, and then to transform external knowledge into statistical regularity.

$$\mathcal{L}_{\text{cons}} = c_{\text{llm}} \cdot \text{KL}(p_{\text{llm}} \parallel p) \tag{18}$$

Furthermore, entropy regularization is applied to the gate control distribution at the classification stage to avoid single modal dominance and improve the fusion robustness:

$$\mathcal{L}_{\text{gate}}^{\text{MASC}} = -\beta \sum_{m \in \{T,I,D,L\}} \tilde{\alpha}_m \log \tilde{\alpha}_m \tag{19}$$

Subsequently, to stabilize the gating offset caused by PAM and prevent prior from being "over driven" in the representation layer, $\mathcal{L}_{\text{pam}}$ is added.

For expert $E \in \mathcal{E} = \{E_T, E_{TD}, E_{TV}\}$, the PAM loss is:

$$\mathcal{L}_{\text{pam}}^{(E)} = \frac{1}{Z_E} \sum_i \sum_{a \in \tilde{\mathcal{A}}_i} \sum_{m \in \mathcal{M}_E} \left(1 - c_{\text{llm},i,a}^{(E)}\right) \cdot \left(\tilde{\alpha}_{i,a,m}^{(E)} - \alpha_{i,a,m}^{(E)}\right)^2 \tag{20}$$

Among which $\alpha_{i,a,m}^{(E)}$ is the modal weight before PAM module. $\tilde{\alpha}_{i,a,m}^{(E)}$ fuses the prior modal weight. $c_{\text{llm},i,a}^{(E)} \in [0,1]$ is prior confidence level, positively correlated with the prior consistency of aspect a. Z_E is a normalization constant to ensure the average loss to "sample-aspect-modal".

Finally, Weighted sum of losses from different experts to obtain the total PAM loss:

$$\mathcal{L}_{\text{pam}} = \sum_{E \in \{E_T, E_{TD}, E_{TV}\}} \eta_E \cdot \mathcal{L}_{\text{pam}}^{(E)} \tag{21}$$

Thus, the overall goal of the emotion classification stage can be comprehensively obtained

$$\mathcal{L}_{\text{MASC}} = \mathcal{L}_{\text{cls}} + \lambda_{\text{cons}}\mathcal{L}_{\text{cons}} + \lambda_{\text{reg}}\mathcal{L}_{\text{gate}}^{\text{MASC}} + \lambda_{\text{pam}}\mathcal{L}_{\text{pam}} \tag{22}$$

Among them, $\lambda_{\text{cons}}, \lambda_{\text{reg}}, \lambda_{\text{pam}} > 0$ are determined by the validation set. In actual training, experts E_T, E_{TD}, and E_{TV} are trained independently to minimize the objective function that is isomorphic to the above (parameters and hyperparameters are not shared with each other), and the gradient is strictly not returned to MATE. L is only used as a prior modulation/calibration signal for each expert branch.

For the MABSA task, this paper adopts a two-stage training strategy: first independently optimize the MATE module, and then fix its parameters to train the MASC module. A scheduled mixing mechanism during training is introduced to gradually mix the real aspect annotation sets and the aspect sets predicted by MATE, allowing the classifier to smoothly transition to the real distribution. The prior generated by the large model only takes effect in the MASC stage and does not return gradients. At the same time, the consistency of modal attention is ensured through the regular gate alignment, and the soft start strategy of "data first, then a priori" is adopted to prioritize the data driven boundary discrimination, and then gradually integrate external knowledge to achieve efficient decoupling and performance optimization of the training process.

4   Experiment

4.1   Settings

Dataset and Metrics: Similar to existing work, this article uses Twitter2015 and Twitter2017 as benchmark datasets for testing. At the same time, this article selects Micro-F1 score (F1), Precision (P), and Recall (R) as evaluation indicators on MABSA and MATE tasks, and uses Accuracy (Acc) and F1 for evaluation on MASC tasks.

Implementation Details: This paper uses NVIDIA RTX 4090 (48G) GPU to train the model. The overall model parameters are consistent with DEQA. GLM-4.5 is used to generate prior knowledge.

4.2   Main Results

MABSA: As shown in Table 1, after transferring the above two returns to the unified MABSA indicator, it can be seen that MACA increases by 1.9% and 1.2% respectively compared to the second best model on both datasets, R increases by 1% and 1.1% respectively, and F1 increases by 1.54% and 1.2% respectively. Compared with the single modal upper limit of text, such as F1 score of BART is only 63.9, the advantage is more significant, indicating the synergistic effectiveness of dynamic gating+contrastive alignment+prior dual path in real open domain tweets. It should be pointed out that this article still retains the training splitting strategy consistent with DEQA (MATE joint, MASC expert independent), so the improvement can be attributed to the internal mechanism of the module rather than changes in the training protocol.

MATE: It can be seen from Table 2 that the F1 of MACA reaches 89.29 (P = 88.50, R = 90.10) and 95.00 (P = 94.40, R = 95.60) on MATE task, respectively 1.29% and 0.60% higher than 88.00/94.40 of DEQA; Compared with the existing optimal methods, such as DQPSA 87.70/94.30, it also has advantages. More importantly, the R growth of MACA on Twitter 2015 is 0.60%, which forms a complementary improvement with the P growth on Twitter 2017, in line with the design motivation of the comparative aspect-modal alignment and dynamic adaptive gating in 3.1 section. Alignment constraints reduce "false extraction", sample level gating suppresses noise modal interference, and achieves P/R balanced gain.

MASC: It can also be seen from Table 1 that the Acc and F1 of MACA are 84.10/80.00 and 77.25/76.30 respectively on MASC task, compared with 82.10/77.60 and 75.80/75.10 of DEQA, the Acc and F1 are increased by 2.00%/1.45% and 2.40%/1.20% respectively. Compared with AoM, VLP-MABSA, DQPSA and other methods, MACA is in the lead or flat and superior in the two data sets. Only on Twitter 2017, the Acc/F1 of MACA is 0.05%/0.3% lower than that of the optimal model, because there are a lot of statements in this data set that are likely to cause confusion of large models. Combined with the

Table 1. Compare of different model in MABSA.* denotes the results are from VLP-MABSA.

Methods	twitter2015			twitter2017		
	P	R	F1	P	R	F1
Text-based methods						
SPAN*	53.7	53.9	53.8	59.6	61.7	60.6
D-GCN*	58.3	58.8	59.4	64.2	64.1	64.1
BART*	62.9	65	63.9	65.2	65.6	65.4
Multimodal methods						
JML [4]*	65.0	63.2	64.1	66.5	65.5	66.0
VLP-MABSA [5]*	65.1	68.3	66.6	66.9	69.2	68.0
AOM [16]	67.9	69.3	68.6	68.4	71.0	69.7
M2DF [15]	67.0	68.3	67.6	67.9	68.8	68.3
Atlantis [10]	65.6	69.2	67.3	68.6	70.3	69.4
MCPL-VLP [14]	67.2	69.2	68.2	69.0	69.4	69.2
ADAR [1]	71.6	71.5	71.2	<u>71.6</u>	71.0	71.4
RNG [6]	67.8	69.5	68.6	69.5	71.0	70.2
EKMG [7]	68.1	70.0	69.0	69.0	70.3	69.7
TCMT [17]	69.3	70.4	69.8	70.2	71.5	70.8
DQPSA [8]	<u>71.7</u>	72.0	71.9	71.1	70.2	70.6
DEQA [3]	71.4	<u>73.9</u>	<u>72.7</u>	71.4	<u>72.4</u>	<u>71.9</u>
MACA(Ours)	**73.60**	**74.90**	**74.24**	**72.80**	**73.50**	**73.15**

analysis in 3.2, the performance improvement comes from: i) the prior perception gating (PAM) can maintain the stability of discrimination when it conducts the sample level recalibrates the T, I, D weights at the presentation layer, even the visual-evidence is thin or the description noise is large. ii) The fusion of temperature calibration and confidence modulation at the decision-making level will increase the logarithm probability of correct class when the prior is reliable, and automatically shrink the impact when the prior is uncertain, thus obtaining a higher macro average F1.

In addition, this paper compares MACA on MASC tasks with the multimodal big language model. The results are shown in Table 3. It can be observed that the MACA proposed in this paper is significantly superior to all multimodal big language models, including GPT-4V, Claude3-V, Gemini-V, etc.

To sum up, MACA improves fusion quality through adaptive cross modal gating, reduces mismatches through comparative aspect modal alignment, and adopts a dual path integration strategy with LLM prior, ultimately enhancing discriminative robustness. The model has demonstrated good performance on MABSA, MASC, and MATE, validating the general effectiveness of the proposed method in MABSA.

Table 2. Comparison of different models in MATE and MASC tasks. * denotes existing baseline models; †denotes the use of Micro-F1 as the evaluation metric; "-" denotes no available data.

Methods	MATE Task						MASC Task			
	Twitter2015			Twitter2017			Twitter2015		Twitter2017	
	P	R	F1	P	R	F1	ACC	F1	ACC	F1
JML [4]	83.6	81.2	82.4	92.0	90.7	91.4	78.7	-	72.7	-
VLP-MABSA* [5]	83.6	87.9	85.7	90.8	92.6	91.7	78.6	73.8	73.8	71.8
AOM [16]	84.6	87.9	86.2	91.8	92.8	92.3	80.2	75.9	76.4	75.0
M2DF [15]	85.2	87.4	86.3	91.5	93.2	92.4	78.9	74.8	74.3	73.0
MGFN-SD [12]	-	-	-	-	-	-	79.36	74.81	72.77	72.07
MCPL-VLP [14]	84.8	87.4	86.1	91.9	92.4	92.2	79.3	74.9	75.1	74.0
A2II [2]	-	-	-	-	-	-	79.46	75.16	74.39	72.35
MPFIT [11]	-	-	-	-	-	-	77.53	73.53	70.35	68.84
DPFN [9]	-	-	-	-	-	-	79.9	76.0	73.9	72.6
ADAR [1]	86.5	88.0	87.9	93.0	93.9	93.8	81.3	77.1	77.2	**76.6**
EKMG [7]	84.6	89.4	87.0	92.5	93.7	93.1	80.3	76.5	75.7	74.3
TCMT [17]	85.8	89.4	87.6	93.5	94.1	93.8	81.4	76.7	**77.3**	75.8
DQPSA [8]	<u>88.3</u>	87.1	87.7	**95.1**	93.5	94.3	81.1	81.1†	75.0	75.0†
DEQA [3]	86.6	<u>89.5</u>	<u>88.0</u>	93.8	<u>95.1</u>	<u>94.4</u>	<u>82.1</u>	<u>77.6</u>	75.8	75.1
MACA (Ours)	**88.5**	**90.1**	**89.29**	<u>94.4</u>	**95.6**	**95.0**	**84.10**	**80.0**	<u>77.25</u>	<u>76.30</u>

4.3 Ablation Study

In order to verify the effectiveness and synergy of DAG, ACA and LLM in MACA, this section adopts the step by step removal (w/o) scheme (the experimental settings remain unchanged), and the results are shown in Table 4. Compared with baseline DEQA, the complete MACA achieves stable improvement in three tasks: the F1 value on Twitter 2015/2017 dataset in MABSA task is increased by 1.54% and 1.25% respectively. Acc/F1 on two data sets in MASC task are increased by 2.00%/2.40% and 1.45%/1.20% respectively. The F1 values on the two data sets in MATE task is increased by 1.29% and 0.60% respectively. The benefit in the MATE task on the Twitter 2017 data set is small, and the difficulty of cross modal alignment is increased due to more metaphors/fuzzy expressions. The single module removal experiment reveals the core role of each component: the removal of DAG mainly led to the reduction of recall rate (R), for example, the R values on two data sets in MABSA is decreased by 1.00% and 0.60% respectively, and the R values on two data sets in MATE is decreased by 0.90% and 0.60% respectively. It confirms that DAG can retain weak evidence and suppress noise through sample level gating. Removing ACA will damage accuracy (P). For example, the P values in MABSA on MATE-Twitter2017 data sets are reduced by 1.20% and 0.90% respectively, indicating that it can alleviate

Table 3. Results of different LLM for the MASC task. The evaluation metric is Accuracy. All results of multimodal LLM are from DEQA.

Large Models	Twitter2015	Twitter2017
GPT-4V	53.9	<u>60.2</u>
Claude3-V	38.5	54.5
Gemini-V	54.5	59.3
LLaVA-v1.6-13B	58.7	56.1
Fuyu-8B	58.8	50.8
Qwen-VL-Chat	<u>65.5</u>	59.7
MACA (Ours)	**84.1**	**77.25**

Table 4. Ablation results for **MABSA**, **MATE**, and **MASC** on Twitter2015/2017.

Methods	MABSA						MATE						MASC			
	Twitter2015			Twitter2017			Twitter2015			Twitter2017			Twitter2015		Twitter2017	
	P	R	F1	P	R	F1	P	R	F1	P	R	F1	Acc	F1	Acc	F1
Full MACA (Ours)	**73.6**	**74.9**	**74.24**	**72.8**	**73.5**	**73.15**	**88.5**	**90.1**	**89.29**	**94.4**	**95.6**	**95.0**	**84.1**	**80.0**	**77.25**	**76.3**
MACA w/o DAG	73.0	73.9	73.45	72.2	72.9	72.55	87.8	89.2	88.49	94.0	95.0	94.50	82.9	78.7	76.6	75.8
MACA w/o ACA	72.4	74.3	73.34	71.9	73.2	72.54	87.1	89.8	88.43	93.5	95.2	94.34	82.6	78.2	76.3	75.6
MACA w/o LLM	73.1	74.3	73.70	72.7	73.1	72.90	87.9	89.9	88.89	94.1	95.3	94.70	83.2	78.8	76.9	75.9
MACA w/o DAG&ACA	71.9	73.1	72.50	71.1	72.0	71.55	86.2	88.1	87.14	93.1	94.5	93.79	82.3	77.9	76.0	75.4
MACA w/o DAG&LLM	71.8	73.2	72.49	71.8	71.9	71.85	86.5	88.0	87.24	93.3	94.7	93.99	82.0	77.7	75.9	75.2
MACA w/o ACA&LLM	71.6	73.8	72.68	71.0	72.6	71.79	86.0	88.4	87.18	92.9	94.8	93.84	82.2	77.8	75.85	75.35
Baseline (DEQA)	71.4	73.9	72.7	71.4	72.4	71.9	86.6	89.5	88.0	93.8	95.1	94.4	82.10	77.60	75.80	75.10

cross modal mismatch detection. The removal of LLM causes a mild degradation of performance (most sensitive to MASC task), and the F1 value of each sub-model is lower than the complete model. For example, the F1 values on two datasets in MABSA are reduced by 0.54% and 0.25% respectively, and the F1 values on two datasets in MASC are reduced by 1.20% and 0.40% respectively. It confirms the gains in emotional discrimination.

The dual module removal experiment further reflects the component synergy: after removing any two modules (w/o DAG&ACA, w/o DAG&LLM, w/o ACA&LLM) at the same time, the model performance significantly declines and approaches the baseline DEQA. For example, in w/o DAG&ACA, the F1 value on MABSA-Twitter2017 dataset decreases by 2.69% (close to the DEQA level), the F1 value in MATE task decreases by 1.21%, and the F1 value in MASC task decreases by 0.90%. The sub-model w/o DAG&LLM or the sub-model w/o ACA&LLM also shows a similar trend. For example, the F1 values on the MABSA-Twitter2015 dataset are decrease by 1.75% and 1.56% respectively, indicating that the loss of DAG or ACA will aggravate the performance shrinkage in the absence of LLM prior calibration. To sum up, DAG and ACA jointly determine the quality of "evidence acquisition and matching", LLM provides the steady-state gain of the discrimination layer, and the three together constitute the core advantages of MACA.

In addition, this paper also focuses on entropy regularization parameter β. It is analyzed that β acts on the validation gating in MATE and the dynamic gating in MASC at the same time, so it can synchronously affect the two subtasks in phased training, and the impact on the indicators of the three tasks is more intuitive.

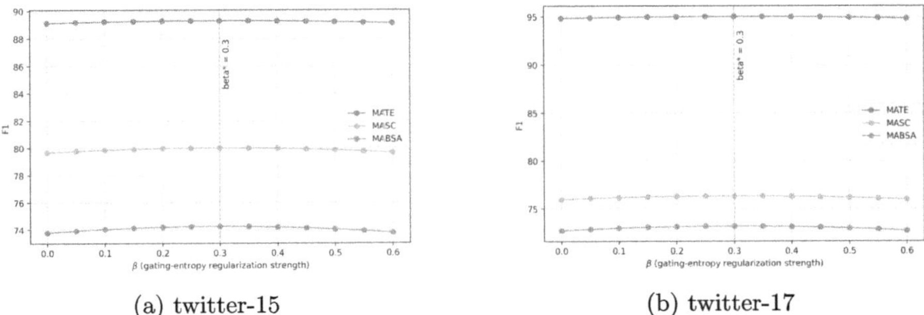

(a) twitter-15   (b) twitter-17

Fig. 3. The impact of β on MATE, MASC and MABSA.

As can be seen from Fig. 3, MATE and MASC have the same consistent response to β on Twitter 2015 and 2017, the optimal interval is [0.2, 0.4], and the peak value is at $\beta = 0.3$. The MASC curve is significantly steeper, indicating that the classification stage is more dependent on the modal weight that is stable without over polarization. MATE curve is relatively flat. The curve of MABSA is approximately proportional to the quality of the two stages. Too small β results in weight collapse and sensitivity to noise modal, and too large β makes the gating approximately uniform and weakens the sample level adaptation. Considering robustness and peak performance comprehensively, $\beta = 0.3$ is taken as the gating strength index in this paper.

4.4   Case Study

In order to further illustrate the effectiveness of the model proposed in this paper, this section gives prediction examples of different models. The comparison methods are VLP-MABSA, DEQA and the MACA method proposed in this paper, as shown in Fig. 4.

As can be seen from Fig. 4, in the first example, VLP-MABSA and DEQA generates the wrong emotional polarity. In the second example, VLP-MABSA ignores the generation of the aspect Haiti, and incorrectly predicts the emotional polarity of ADSS Global to be neutral, while DEQA incorrectly matches the emotional polarity of Haiti to be positive. In the last example, VLP-MABSA and DEQA fails to correctly identify the emotional polarity of Grabovo. After

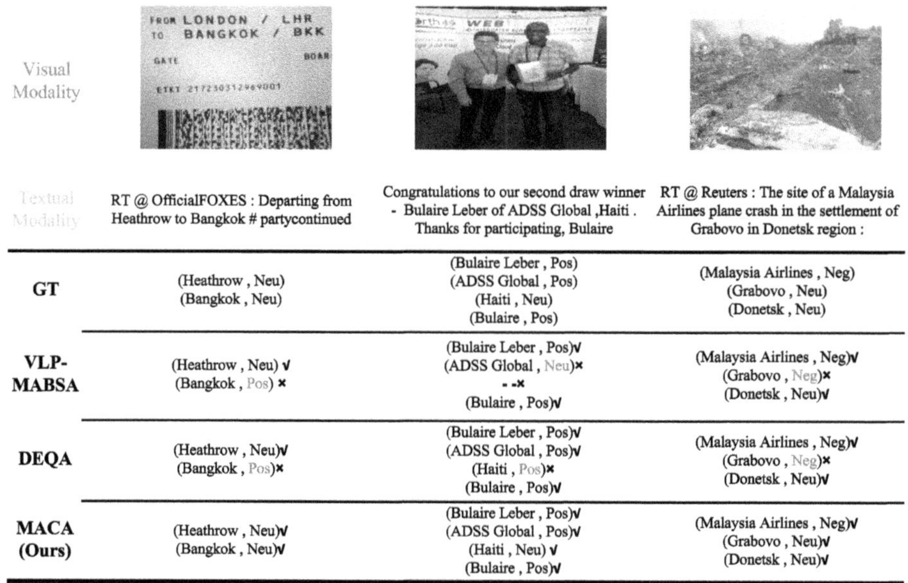

Visual Modality			
Textual Modality	RT @ OfficialFOXES : Departing from Heathrow to Bangkok # partycontinued	Congratulations to our second draw winner - Bulaire Leber of ADSS Global ,Haiti . Thanks for participating, Bulaire	RT @ Reuters : The site of a Malaysia Airlines plane crash in the settlement of Grabovo in Donetsk region :
GT	(Heathrow , Neu) (Bangkok , Neu)	(Bulaire Leber , Pos) (ADSS Global , Pos) (Haiti , Neu) (Bulaire , Pos)	(Malaysia Airlines , Neg) (Grabovo , Neu) (Donetsk , Neu)
VLP-MABSA	(Heathrow , Neu) ✔ (Bangkok , Pos) ✘	(Bulaire Leber , Pos)✔ (ADSS Global , Neu)✘ - -✘ (Bulaire , Pos)✔	(Malaysia Airlines , Neg)✔ (Grabovo , Neg)✘ (Donetsk , Neu)✔
DEQA	(Heathrow , Neu)✔ (Bangkok , Pos)✘	(Bulaire Leber , Pos)✔ (ADSS Global , Pos)✔ (Haiti , Pos)✘ (Bulaire , Pos)✔	(Malaysia Airlines , Neg)✔ (Grabovo , Neg)✘ (Donetsk , Neu)✔
MACA (Ours)	(Heathrow , Neu)✔ (Bangkok , Neu)✔	(Bulaire Leber , Pos)✔ (ADSS Global , Pos)✔ (Haiti , Neu) ✔ (Bulaire , Pos)✔	(Malaysia Airlines , Neg)✔ (Grabovo , Neu)✔ (Donetsk , Neu)✔

Fig. 4. Case study

full information utilization, the model MACA correctly extracts all aspects and correctly predicts the emotional polarity, which reflects the advantages of this model in the MABSA task.

5   Conclusions

This article proposes MACA, which inherits the two-stage paradigm of DEQA question answering and implements three key technologies. Firstly, dynamic adaptive gating to adaptively allocate text/visual/descriptive contributions at the sample level. Secondly, aspect-modal comparison align explicitly constrains the consistency between anchor points and evidence regions. Thirdly, inject the "aspect-sentiment pairs" generated by the large model as auxiliary modalities and decision level priors through dual paths. This article uses joint objective optimization of MATE and MASC, and mitigates the error propagation caused by static fusion and external prior mismatch through consistency regularization of gating consistency and confidence weighting. It improves robustness and interpretability in noise, modal loss, and long tail scenarios. The limitations are mainly reflected in the prior domain shift and additional computational overhead. Future work will focus on prior retrieval enhancement and adaptive calibration, lightweight training through parameter sharing and distillation, cross task transfer under open vocabulary settings, and systematic robustness evaluation covering adversarial perturbations and modal loss.

References

1. Chen, Z., Zhu, Z., Xu, W., Zhang, Y., Wu, X., Zheng, Y.: *Aspects Are Anchors: towards multimodal aspect-based sentiment analysis via aspect-driven alignment and refinement*. In: Proceedings of the 32nd ACM International Conference on Multimedia, pp. 2292–2300 (2024)
2. Feng, J., Lin, M., Shang, L., Gao, X.: Autonomous aspect-image instruction a2II: Q-former guided multimodal sentiment classification. In: Proceedings of the 2024 Joint International Conference on Computational Linguistics, Language Resources and Evaluation (LREC-COLING 2024), pp. 1996–2005. IJCAI Proceedings (2024)
3. Han, Z., Hu, M., Bai, Y., Wang, X., Luo, B.: DEQA: descriptions enhanced question-answering framework for multimodal aspect-based sentiment analysis (2025)
4. Ju, X., et al.: Joint multi-modal aspect-sentiment analysis with auxiliary cross-modal relation detection. In: Proceedings of the 2021 Conference on Empirical Methods in Natural Language Processing, pp. 4395–4405 (2021)
5. Ling, Y., Yu, J., Xia, R.: Vision-language pre-training for multimodal aspect-based sentiment analysis. In: Proceedings of the 60th Annual Meeting of the Association for Computational Linguistics (Volume 1: Long Papers), pp. 2149–2159 (2022)
6. Liu, Y., et al.: RNG: reducing multi-level noise and multi-grained semantic gap for joint multimodal aspect-sentiment analysis. In: 2024 IEEE International Conference on Multimedia and Expo (ICME), pp. 1–6 (2024)
7. Liu, Z., Lin, J., Chen, Y., Dong, Y.: Multimodal aspect-based sentiment analysis with external knowledge and multi-granularity image-text features. Neural Process. Lett. **57**(2), 25 (2025)
8. Peng, T., Li, Z., Wang, P., Zhang, L., Zhao, H.: A novel energy based model mechanism for multi-modal aspect-based sentiment analysis. Proc. AAAI Conf. Artif. Intell. **38**(17), 18869–18878 (2024)
9. Wang, D., Tian, C., Liang, X., Zhao, L., He, L., Wang, Q.: Dual-perspective fusion network for aspect-based multimodal sentiment analysis. IEEE Trans. Multimedia **26**, 4028–4038 (2024)
10. Xiao, L., Wu, X., Xu, J., Li, W., Jin, C., He, L.: Atlantis: aesthetic-oriented multiple granularities fusion network for joint multimodal aspect-based sentiment analysis. Inf. Fusion **106**, 102304 (2024)
11. Yang, D., Li, X., Li, Z., Zhou, C., Wang, X., Chen, F.: Prompt fusion interaction transformer for aspect-based multimodal sentiment analysis. In: 2024 IEEE International Conference on Multimedia and Expo (ICME), pp. 1–6 (2024)
12. Yang, J., Xiao, Y., Du, X.: Multi-grained fusion network with self-distillation for aspect-based multimodal sentiment analysis. Knowl.-Based Syst. **293**, 111724 (2024)
13. Yang, L., Na, J.C., Yu, J.: Cross-modal multitask transformer for end-to-end multimodal aspect-based sentiment analysis. Inf. Process. Manag. **59**(5), 103038 (2022)
14. Zhang, J., Qu, J., Liu, J., Wang, Z.: MCPL: multi-model co-guided progressive learning for multimodal aspect-based sentiment analysis. Knowl.-Based Syst. **301**, 112331 (2024)
15. Zhao, F., Li, C., Wu, Z., Ouyang, Y., Zhang, J., Dai, X.: M2DF: multi-grained multi-curriculum denoising framework for multimodal aspect-based sentiment analysis. In: Proceedings of the 2023 Conference on Empirical Methods in Natural Language Processing, pp. 9057–9070 (2023)

16. Zhou, R., Guo, W., Liu, X., Yu, S., Zhang, Y., Yuan, X.: AoM: detecting aspect-oriented information for multimodal aspect-based sentiment analysis. In: Findings of the Association for Computational Linguistics: ACL 2023, pp. 8184–8196 (2023)
17. Zou, W., et al.: TCMT: target-oriented cross modal transformer for multimodal aspect-based sentiment analysis. Expert Syst. Appl. **264**, 125818 (2025)

DynSpectral: A Multi-channel Temporal Spectral GNN with Frequency Decomposition for Dynamic Graphs

Runguo Tao[1], Tianpeng Li[3(✉)], Minglai Shao[3], Wenjun Wang[1,2], Xuan Guo[1], and Yueheng Sun[1]

[1] College of Intelligence and Computing, Tianjin University, Tianjin 300072, China
[2] Yazhou Bay Innovation Institute, Hainan Tropical Ocean University, Sanya 572022, China
[3] School of New Media and Communication, Tianjin University, Tianjin 300072, China
ltpnimeia@tju.edu.cn

Abstract. Spectral graph neural networks (GNNs) can leverage frequency domain properties to effectively analyze graph-structured data. However, existing dynamic GNNs typically model temporal graph evolution as a holistic process, overlooking the inherent diversity of spectral frequencies within dynamic graphs: low-frequency components that capture stable, gradually evolving structures, and high-frequency components that reflect abrupt, localized changes. To address this gap, we propose DynSpectral, a multi-channel temporal spectral GNN that explicitly decomposes and models these co-existing frequency signals. Our approach introduces a novel frequency-aware graph decomposition strategy. For each temporal snapshot, we construct three complementary views: (1) a Union Graph that aggregates edges across consecutive snapshots to preserve low-frequency structural stability; (2) a Difference Graph that isolates newly appearing and disappearing edges to capture high-frequency dynamics; and (3) the Original Graph as a full-spectrum reference. Each view is then processed by a specialized temporal encoder tailored to its unique evolutionary characteristics, effectively modeling both smooth, long-term trends and abrupt, short-term changes. To ensure coherent and robust representations, we employ a contrastive learning framework that aligns these multi-frequency views while preserving their distinct information. Extensive experiments on dynamic link prediction benchmarks validate the efficacy of our frequency decomposition approach, demonstrating its capability to provide a more comprehensive modeling of temporal graph dynamics while achieving promising and competitive performance.

Keywords: Dynamic network embedding · Spectral Graph Neural Network · dynamic evolution

1   Introduction

Dynamic graphs are fundamental for modeling evolving systems like social networks and financial transactions [23]. Learning effective representations from these graphs is crucial for tasks ranging from fraud detection to traffic forecasting [2,8]. A core challenge lies in capturing the diverse nature of graph evolution, which typically involves a mixture of stable, gradual structural changes and abrupt, localized events.

While recent Graph Transformers like TGAT [21] and DyGFormer [22] have achieved impressive performance, they predominantly treat graph evolution as a holistic process. This "one-size-fits-all" paradigm overlooks the inherent spectral diversity within dynamic graphs: the coexistence of low-frequency components (representing stable, long-term structures) and high-frequency components (reflecting transient, short-term dynamics). Neglecting these distinct spectral characteristics limits the model's ability to disentangle complex evolutionary patterns.

To address this critical gap, we introduce **DynSpectral**, a novel Synergistic Spectral-Temporal Learning framework built upon a multi-channel architecture. The core of our approach is a frequency-aware graph decomposition strategy that factorizes the graph's evolution at each snapshot into three complementary views: (1) a **Union Graph**, aggregating historical edges to preserve low-frequency structural stability; (2) a **Difference Graph**, isolating new and dissolved edges to capture high-frequency changes; and (3) the **Original Graph**, serving as a full-spectrum reference. Each view is then processed by a specialized temporal encoder tailored to its unique evolutionary characteristics, enabling a more comprehensive and nuanced understanding of the graph dynamics.

Our primary contributions across theory, methodology, and empirical validation: 1) We propose a novel spectral-frequency decomposition for dynamic graphs and formally prove that our decomposed views exhibit a consistent ordering of total variation (Dirichlet energy). This result establishes a crucial theoretical bridge between the graph's structural evolution and its spectral properties. 2) We design **DynSpectral**, a novel multi-channel framework with specialized temporal encoders and a contrastive learning objective to synergistically model these decomposed views. 3) Extensive experiments on five benchmark datasets validate the superiority of our approach. DynSpectral achieves remarkable performance gains of over 13% points on both the MOOC and Bitcoin datasets. Collectively, these contributions empirically validate our core thesis that decomposing dynamic graphs based on their spectral properties leads to richer and more effective representations.

2   Related Work

Our work is situated at the confluence of dynamic graph learning and spectral graph theory. We review key literature from both areas to contextualize our contributions. Research in dynamic graph representation learning has produced a diverse array of architectures. Early works, such as the continuous-time models **JODIE** [6] and **DyRep** [17], leverage recurrence to model temporal dynamics. More recently, powerful Transformer-based frameworks like **TGAT** [21] and **DyGFormer** [22] have emerged to capture long-range dependencies through sophisticated attention mechanisms. In parallel, a line of research advocating for simpler, more efficient models has produced architectures such as **GraphMixer** [1] and **LightDyG** [11]. Despite their architectural diversity, a common thread unites these approaches: they typically model graph evolution as a **holistic process**. Such a uniform treatment lacks the necessary granularity to differentiate between stable structural trends and transient events. Our work directly addresses this limitation by proposing a multi-channel architecture to explicitly disentangle these dynamics. Spectral GNNs, grounded in Graph Signal Processing (GSP) [10,16], analyze graphs in the frequency domain, where the graph Laplacian's spectrum defines notions of signal smoothness via the Dirichlet energy [15]. Modern spectral GNNs have advanced beyond fixed filters, employing learnable polynomial bases [4,7,19] that adapt to graph properties. Our encoder is inspired by UniBasis [5], which adjusts its filters based on graph homophily. However, these powerful methods are fundamentally designed for static graphs. Their application to dynamic settings is hindered by the prohibitive computational cost of repeatedly diagonalizing the Laplacian and the challenge of aligning inconsistent spectral bases across time. Our framework circumvents this by factorizing the graph evolution into spectrally distinct and more stable views. The concepts of multi-view learning [20] and frequency decomposition are also relevant. For instance, some GNNs for static graphs perform contrastive learning between low-pass and high-pass filtered views to improve representations [24], while other GSP works analyze signals evolving over dynamic graphs [9]. A critical gap, however, remains. In prior work, views are typically pre-defined, or frequency decomposition is applied to node features on a static structure, rather than to the **graph's structural evolution itself**. To our knowledge, our work is the first to bridge this gap by **dynamically constructing frequency-aware views** (Union and Difference) from the graph's own evolution history. This enables a specialized and more comprehensive modeling of its co-existing low-frequency stability and high-frequency volatility.

3   Spectral Decomposition Theory for Dynamic Graphs

To bridge the structural evolution of dynamic graphs with spectral graph signal processing (GSP), we formalize the frequency characteristics of our multi-view decomposition. Existing spectral GNNs, such as those based on UniBasis [5], primarily address static heterophily through polynomial bases, but overlook temporal dynamics. In dynamic settings, graph signals evolve across snapshots, necessitating a theoretical foundation that associates union views with low-frequency stability (smooth, gradual structures) and difference views with high-frequency changes (abrupt, localized variations). Drawing from GSP principles [10,15,16], where total variation (Dirichlet energy) quantifies signal smoothness—low energy indicating smooth, low-frequency signals and high energy indicating oscillatory, high-frequency ones—we establish an ordering of Dirichlet energies across views. This serves as the foundational bridge, enabling subsequent integration with graph homophily for multi-view basis construction and temporal modeling in FAT.

To facilitate the spectral analysis of dynamic graph evolution, we first formalize the multi-view decomposition used throughout this work. At each time snapshot t, we consider three graph views over the same node set V.

Definition 1 (Multi-View Graph Decomposition and Laplacians). *Let $\mathcal{G} = \{G_t = (V, E_t)\}_{t=1}^{T}$ be a dynamic graph sequence over a fixed node set V. At each snapshot $t \in [T]$, we define three graph views:*

- ***Original Graph:*** *The graph at the current snapshot.*

$$G_t^{orig} = (V, E_t)$$

- ***Union Graph:*** *The aggregation of all historical edges up to the current snapshot.*

$$G_t^{union} = \left(V, \bigcup_{i=1}^{t} E_i \right)$$

- ***Difference Graph:*** *The graph of newly appeared edges at the current snapshot.*

$$G_t^{diff} = (V, \ E_t \setminus E_{t-1}), \quad \text{where } E_0 := \emptyset.$$

Each view induces a corresponding non-normalized Laplacian matrix:

$$L_t^{orig} = L(G_t^{orig}),$$
$$L_t^{union} = L(G_t^{union}),$$
$$L_t^{diff} = L(G_t^{diff}),$$

where $L(G) = D_G - A_G$ is the standard unnormalized graph Laplacian. For $t = 1$, we have $G_1^{diff} = G_1^{orig}$ by convention.

Theorem 1 (Total Variation Ordering of Dynamic Graph Views). *For any graph signal $f \in \mathbb{R}^{|V|}$, the total variation (Dirichlet energy) of f on these graphs satisfies:*

$$E(f, G_t^{diff}) \leq E(f, G_t^{orig}) \leq E(f, G_t^{union}),$$

where $E(f, G) = f^\top L_G f = \sum_{(u,v) \in E(G)} (f(u) - f(v))^2$ is the Dirichlet energy of signal f on graph G.

The detailed proof is provided in Appendix A. This theorem establishes that the difference graph—defined as the subgraph containing only edges newly appearing at time t—imposes the weakest smoothness constraints on any signal f, resulting in the lowest Dirichlet energy. In graph signal processing (GSP) terms [10,16], this does *not* imply that the difference graph itself possesses a high-frequency spectrum. Rather, the signal supported on the difference graph, when analyzed in the Laplacian eigenspace of the *original* or *union* graph at time t, exhibits high spectral energy—i.e., it behaves as a high-frequency component relative to the stable structural backbone. Conversely, the union graph, by aggregating historical edges into a denser topology, enforces stronger smoothness and yields the highest Dirichlet energy for any given signal f, aligning with low-frequency, long-term trends. The original graph serves as an intermediate full-spectrum reference. This spectral duality motivates our multi-view decomposition: union for low-pass filtering (gradual trends), difference for high-pass, and original for balanced modeling.

Building upon the spectral energy ordering established in Theorem 1, we now articulate the theoretical motivation for our Frequency-Adaptive Transformer (FAT). The core intuition is that the temporal modeling strategy should be aligned with the inherent evolutionary pattern of each spectral view. Specifically, the *Union Graph* view, which aggregates historical edges, tends to preserve stable, long-term structural trends. Signals on this view are expected to be smooth over time, suggesting that a temporal encoder should be biased towards capturing long-range dependencies. Conversely, the *Difference Graph* view, which isolates newly appearing or disappearing edges, captures abrupt, short-term changes. Its signals are expected to exhibit high temporal variability, necessitating a temporal encoder that is sensitive to local, immediate dynamics. To operationalize this intuition, we design two key mechanisms in FAT:

1. **Frequency-Adaptive Decay Bias**: We introduce a learnable decay bias $-\lambda_{\text{view}} \cdot |t - \tau|^{\beta}$ into the attention scores. We initialize λ_{union} to a small value (weak decay) to encourage the Union channel to attend to distant historical states, while initializing λ_{diff} to a larger value (strong decay) to focus the Difference channel on recent states.
2. **Dynamic Adaptive Mapping (DAM)**: We employ a dynamic nonlinearity $\tanh(\alpha_t^{\text{view}} \cdot x)$, where the strength α_t^{view} is a function of the dynamic homophily h_t^{view} and its temporal derivative Δh_t^{view}. This allows the model to adaptively amplify its response to sudden structural shifts (high $|\Delta h_t^{\text{view}}|$).

By conditioning the temporal modeling on these spectral and homophily characteristics, FAT provides a principled way to handle the co-existing long-term stability and short-term volatility in dynamic graphs.

4   Method

Our proposed framework, DynSpectral, is designed to learn comprehensive temporal representations from dynamic graphs through a novel frequency decomposition paradigm. The core idea is to factorize the complex graph evolution into distinct spectral-frequency channels, each capturing a unique type of structural dynamics. As illustrated in Fig. 1, this is realized through a multi-stage architecture that first constructs these frequency-aware views, then processes them via a parallel dual-encoder, and finally fuses their representations for the downstream task under a contrastive learning objective. In the following sections, we will detail each of these components.

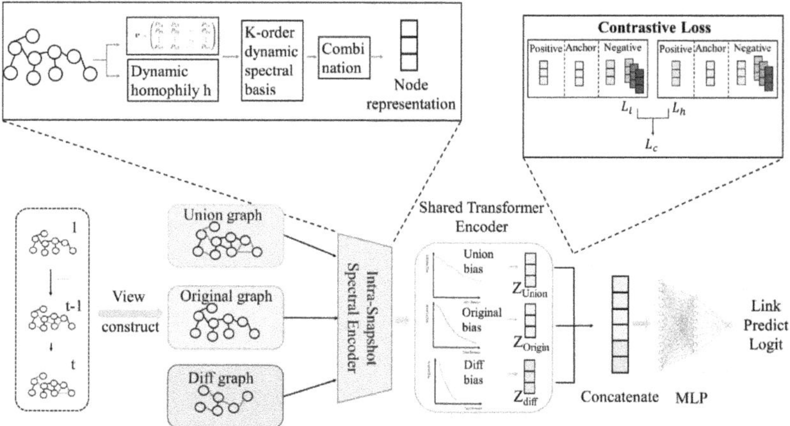

Fig. 1. The overall architecture of the DynSpectral framework

4.1   Intra-Snapshot Spectral Encoder

The Intra-Snapshot Spectral Encoder transforms each graph snapshot at time t into a set of distinct spectral representations. This is achieved through two sequential steps: first constructing our multi-frequency views, and then applying dynamic spectral filtering to each view.

Multi-frequency View Construction. To effectively disentangle the complex dynamics of the evolving graph, we first decompose each graph snapshot G_t into three semantically distinct views. As theoretically motivated in Sect. 3, this decomposition establishes a crucial bridge between the graph's structural evolution and its spectral-frequency properties.

- The **Union Graph** ($G_t^{\text{union}} = \bigcup_{i=1}^{t} G_i$) acts as a *low-pass filter*. By aggregating historical structures, it reinforces stable communities and smooths out transient changes, thereby preserving the low-frequency signals of the network's backbone.
- The **Difference Graph** ($G_t^{\text{diff}} = G_t \Delta G_{t-1}$) acts as a *high-pass filter*. It isolates the structural novelties—newly formed edges—which correspond to the high-frequency signals representing changes and irregularities in the graph.
- The **Original Graph** ($G_t^{\text{orig}} = G_t$) serves as a *full-spectrum* reference, containing a mixture of both low and high-frequency components.

These three views provide a factorized perspective on the graph's state at time t, enabling us to model their different evolutionary patterns separately.

Dynamic Spectral Filtering. With the three views constructed, we then apply a spectral filtering process to generate node representations for each. A key challenge in dynamic graphs is that their spectral properties, such as homophily, are often time-varying. A fixed spectral filter is therefore suboptimal for capturing these evolving characteristics. To address this, our encoder leverages the core principles of UniBasis [5], a universal polynomial basis adaptable to varying graph homophily. We extend this concept to the dynamic setting. Specifically, for each view (e.g., G_t^{view} where view $\in$ {union, diff, orig}), we first compute its **time-smoothed homophily ratio** $\hat{h}_t^{\text{view}}$. The raw homophily h_t^{view} is estimated on the training subgraph of G_t^{view} using the standard edge homophily metric. To stabilize temporal variations, we apply exponential smoothing: $\hat{h}_t^{\text{view}} = (1 - \lambda_h) \cdot h_t^{\text{view}} + \lambda_h \cdot \hat{h}_{t-1}^{\text{view}}$, where $\lambda_h \in [0, 1)$ is a hyperparameter. This smoothed ratio $\hat{h}_t^{\text{view}}$ then governs the construction of a customized, K-order spectral basis for that specific time step and view. The filtering process transforms the initial node features X_t into a single-step spectral representation Z_t^{view}:

$$Z_t^{\text{view}} = \text{SpecFilter}_{\hat{h}_t^{\text{view}}}(G_t^{\text{view}}, X_t, K) \tag{1}$$

where $\text{SpecFilter}_{\hat{h}_t^{\text{view}}}$ denotes our adapted polynomial filtering operation parameterized by the time-smoothed homophily $\hat{h}_t^{\text{view}}$. The variation in $\hat{h}_t^{\text{view}}$ across time reflects the changing structural properties of the graph, which in turn leads to an adapted spectral basis. This produces a sequence of spectral representations $\{Z_1^{\text{view}}, Z_2^{\text{view}}, \dots, Z_t^{\text{view}}\}$ that encodes the evolution of the graph's spectral characteristics. This dynamic sequence serves as a rich, informative input—a critical touchpoint—for the subsequent Inter-Snapshot Temporal Encoder, providing a strong rationale for modeling its temporal dependencies.

4.2 Inter-Snapshot Temporal Encoder

Building upon the spectral energy ordering established in Theorem 1, we articulate the core intuition that motivates our Frequency-Adaptive Transformer (FAT). The Union Graph view, which aggregates historical edges, tends to preserve stable, long-term structural trends and yields smooth temporal signals. In contrast, the Difference Graph view, which isolates newly appearing edges, captures abrupt structural shifts and exhibits high temporal variability. The Original Graph serves as a full-spectrum reference that balances both dynamics. This spectral duality—grounded in graph signal processing principles and the frequency-adaptive nature of the UniBasis filter [5]—guides our design of FAT with two key mechanisms: (1) a frequency-adaptive decay bias in attention to differentiate long-range versus short-term temporal focus, and (2) a dynamic adaptive mapping (DAM) that modulates nonlinearity strength based on dynamic homophily and its temporal derivative.

To model these distinct evolutionary patterns in a unified framework, we process the sequences of spectral node embeddings $\{Z_1^{\text{view}}, \dots, Z_t^{\text{view}}\}$—produced by the Intra-Snapshot Spectral Encoder—for each view using a shared Transformer-based architecture. Unlike prior dynamic GNNs that treat graph evolution holistically (e.g., TGN [13], DyRep [17]) or employ channel-specific RNNs with limited scalability, our approach leverages the parallelism and global receptive field of the Transformer encoder [18] while injecting view-specific spectral biases. Temporal positional encodings are added to preserve snapshot order, and the final representation H_t^{view} is taken from the last time step of the encoder output. Crucially, the standard attention mechanism is augmented with two frequency-adaptive components described below.

Frequency-Adaptive Decay Bias. To align temporal attention with the spectral characteristics of each view, we introduce a learnable decay bias into the attention scores:

$$\text{bias}_{i,j} = -\lambda_{\text{view}} \cdot |i - j|^{\beta},$$

where $\lambda_{\text{view}} > 0$ and $\beta > 0$ are learnable parameters. A causal mask ensures temporal consistency by setting future positions to $-\infty$. The full attention score is computed as:

$$\text{score}_{i,j} = \frac{\mathbf{Q}_i \mathbf{K}_j^{\top}}{\sqrt{d_k}} + \text{mask}_{i,j} + \text{bias}_{i,j}.$$

This bias encourages the Union channel to attend to distant time steps (reflecting low-frequency, long-term stability) and the Difference channel to focus on recent steps (reflecting high-frequency, short-term volatility). The parameters λ_{view} and β are shared across nodes but view-specific, enabling each channel to learn its optimal temporal receptive field.

Dynamic Adaptive Mapping (DAM). To further adapt the model's nonlinearity to the evolving structural properties of each spectral view, we introduce *Dynamic Adaptive Mapping* (DAM), which modulates the activation strength based on dynamic homophily characteristics. Specifically, after the Transformer encoder, we apply a view-specific, time-adaptive nonlinearity:

$$\mathbf{H}_t^{\text{view}} = \tanh\left(\alpha_t^{\text{view}} \cdot \mathbf{Z}_t^{\text{view}}\right),$$

where $\mathbf{Z}_t^{\text{view}}$ denotes the encoder output at time t, and $\alpha_t^{\text{view}} > 0$ is a scalar adaptivity coefficient computed as:

$$\alpha_t^{\text{view}} = \text{Softplus}\left(\mathbf{w}_{\text{view}}^{\top} \begin{bmatrix} \tilde{h}_t^{\text{view}} \\ \widetilde{\Delta h}_t^{\text{view}} \end{bmatrix} + b_{\text{view}}\right) + 0.5.$$

Here, h_t^{view} is the homophily ratio of view at time t, and we define:

$$\tilde{h}_t^{\text{view}} = \frac{h_t^{\text{view}} - \mu_{\text{global}}}{\sigma_{\text{global}}}, \quad \Delta h_t^{\text{view}} = h_t^{\text{view}} - h_{t-1}^{\text{view}}, \quad \widetilde{\Delta h}_t^{\text{view}} = \tanh\left(\frac{\Delta h_t^{\text{view}}}{\sigma_{\Delta h}}\right),$$

where μ_{global} and σ_{global} are precomputed statistics of homophily ratios over the training set, and $\sigma_{\Delta h}$ is a scaling factor for homophily changes. The tanh squashing on Δh ensures robustness to extreme fluctuations, while the Softplus activation combined with the $+0.5$ offset guarantees $\alpha_t^{\text{view}} > 0.5$, preserving sufficient nonlinearity and stable gradient flow.

Each view employs its own lightweight MLP (parameterized by $\mathbf{w}_{\text{view}}, b_{\text{view}}$), allowing the model to learn distinct response strategies: the Difference view naturally emphasizes $\widetilde{\Delta h}_t^{\text{view}}$ to amplify sensitivity to abrupt structural shifts, while the Union view prioritizes $\tilde{h}_t^{\text{view}}$ for stable long-term modeling. This extends graph-adaptive activation principles to dynamic heterogeneous graphs, enabling the model to self-adjust its expressivity in response to temporal structural evolution.

Spectral-Consistent Contrastive Regularization. To encourage semantic coherence across the three spectral views while preserving their distinct frequency characteristics, we introduce a contrastive regularization objective during training. This objective enforces alignment between the full-spectrum *Original* view and the two specialized views (*Union* and *Difference*), treating the Original representation as an anchor that balances long-term stability and short-term volatility.

Specifically, let $\mathbf{h}_t^{\text{orig}}$, $\mathbf{h}_t^{\text{union}}$, and $\mathbf{h}_t^{\text{diff}} \in \mathbb{R}^d$ denote the final node representations from the three channels at time t. Since the channels may produce embeddings of different dimensions (e.g., in ablation variants), we first project $\mathbf{h}_t^{\text{union}}$ and $\mathbf{h}_t^{\text{diff}}$ into the embedding space of the Original channel via learnable linear mappings. The contrastive loss is then formulated as a symmetric InfoNCE objective:

$$\mathcal{L}_{\text{CL}} = \frac{1}{2}\left(\mathcal{L}_{\text{orig-union}} + \mathcal{L}_{\text{orig-diff}}\right),$$

where each term $\mathcal{L}_{\text{orig-view}}$ (view $\in \{\text{union}, \text{diff}\}$) is computed as:

$$\mathcal{L}_{\text{orig-view}} = -\log \frac{\exp\left(\text{sim}(\mathbf{h}_t^{\text{orig}}, \mathbf{h}_t^{\text{view}})/\tau\right)}{\exp\left(\text{sim}(\mathbf{h}_t^{\text{orig}}, \mathbf{h}_t^{\text{view}})/\tau\right) + \sum_{\mathbf{h}^- \in \mathcal{N}} \exp\left(\text{sim}(\mathbf{h}_t^{\text{orig}}, \mathbf{h}^-)/\tau\right)}.$$

Here, $\text{sim}(\cdot, \cdot)$ denotes cosine similarity, $\tau > 0$ is a temperature hyperparameter, and $\mathcal{N}$ is a set of negative samples constructed as follows:

– **Cross-view negatives**: the representation from the *other* specialized view (e.g., $\mathbf{h}_t^{\text{diff}}$ when contrasting orig-union);
– **Structural negatives**: representations of neighboring nodes in the current Original graph, sampled to capture local structural context and prevent trivial alignment.

This contrastive scheme is activated only when all three spectral channels are enabled, and the total training loss is defined as $\mathcal{L} = \mathcal{L}_{\text{task}} + \lambda_{\text{cl}}\mathcal{L}_{\text{CL}}$, where

$\mathcal{L}_{\text{task}}$ is the primary link prediction loss and λ_{cl} is a balancing coefficient. The contrastive module is discarded during inference, ensuring no computational overhead at test time.

5   Experiment

To empirically validate the effectiveness and architectural design of our proposed **DynSpectral** framework, we conduct a series of comprehensive experiments. While we begin by establishing our model's competitive performance against state-of-the-art baselines, the primary goal of our evaluation is to move beyond simple performance metrics. We aim to rigorously dissect the contributions of our core architectural innovations. Specifically, we will analyze the necessity and functional specialization of each spectral channel (Union, Difference, and Original) through targeted ablation studies. Furthermore, we will investigate the internal behavior of our frequency-adaptive mechanisms to provide clear, empirical evidence that they operate as theoretically intended, thus validating the architectural soundness and interpretability of our approach.

5.1   Experiment Setup

Datasets. We evaluate our model on five widely-used public benchmark datasets to demonstrate its generalizability. These datasets cover a diverse range of domains, including the **MOOC** [6] education network, the **UCI** [12] and **fb-forum** [14] social networks, the highly dynamic **Bitcoin** [3] transaction network, and the **MovieLens** [14] user-item recommendation network. For each dataset, the raw temporal event data is processed by aggregating interactions into a sequence of discrete, non-overlapping graph snapshots. Key statistics for these datasets are summarized in Table 1.

Table 1. Statistics of the benchmark datasets used for evaluation.

Dataset	Nodes	Edges (Dynamic)	Snapshots	Domain
MOOC	7,144	411,749	50	Education
UCI	1,899	59,835	95	Social Network
Bitcoin	5,881	35,592	137	Transaction
movielens	10,415	857,014	78	Recommendation
fb-forum	899	33,720	120	Social Network

Baselines. We compare the performance of DynSpectral against a comprehensive suite of state-of-the-art methods for dynamic graph representation learning. The selected baselines include: JODIE [6], DyRep [17], TGAT [21], GraphMixer [1], DyGFormer [22], and LightDyG [11]. These models represent a diverse range of architectural approaches, from recurrent and temporal point process-based methods to those leveraging Transformers. The detail of the baselines are introduced in related work.

Evaluation Protocol and Metrics. The primary task for evaluation is **dynamic link prediction**. We adopt a rigorous temporal evaluation protocol known as rolling prediction. For each snapshot at time t in the validation or test set, the model is tasked with predicting the links that will form at the subsequent snapshot $t + 1$, using all historical graph information up to and including time t. This setup closely mimics real-world scenarios where models must forecast future interactions based on past observations.

To measure performance, we use two standard metrics: AUC (Area Under the ROC Curve): This metric assesses the model's ability to correctly rank positive instances over negative ones AP (Average Precision): This metric summarizes the precision-recall curve and is particularly sensitive to the ranking of positive instances.

Implementation Details. Our framework is implemented in PyTorch and all experiments were conducted on a single NVIDIA 3090 GPU. Models were trained using the Adam optimizer with a learning rate of 0.005 and a weight decay of 0.0005. We trained for a maximum of 200 epochs and employed an early stopping strategy with a patience of 50 epochs, based on the AUC performance on the validation set.

For the core components of DynSpectral, the polynomial order of the Spectral Basis filter, K, was set to 10 across all experiments. Crucially, to bootstrap the learning of our frequency-adaptive mechanisms in alignment with our theoretical motivation, the initial values for the learnable decay parameter λ in the Frequency-Adaptive Transformer (FAT) were set based on each channel's spectral role: $\lambda_{\text{union}} = 0.1$ (encouraging focus on long-term trends), $\lambda_{\text{diff}} = 0.5$ (focusing on recent changes), and $\lambda_{\text{orig}} = 0.3$. Our code will be made publicly available to ensure full reproducibility.

5.2   Experiment Result

Overall Performance. As presented in Table 2, DynSpec establishes a new state-of-the-art on the majority of benchmarks with enhanced stability. This advantage stems from its unique spectral decomposition capability. On the MOOC dataset, the model leverages the **Union channel** to aggregate stable historical communities (low-frequency signals), achieving a remarkable performance leap of over 13% points. Conversely, on the highly dynamic Bitcoin network, it relies on the **Difference channel** to precisely capture transient transactional bursts (high-frequency signals), securing a similarly pronounced lead. This dual success on structurally disparate networks provides strong evidence for our core thesis that explicit spectral disentanglement is more effective than holistic modeling.

The model's performance also reveals its architectural adaptability. On the fb-forum dataset, for instance, the global attention-based DyGFormer performs

Table 2. Performance

Metrics	Model	MOOC	UCI	Bitcoin	movielens	fb-forum
AUC	JODIE	$84.00_{\pm0.73}$	$88.07_{\pm0.89}$	$86.00_{\pm0.68}$	$83.20_{\pm0.91}$	$67.58_{\pm1.05}$
	DyRep	$84.82_{\pm0.57}$	$68.20_{\pm1.18}$	$80.17_{\pm0.94}$	$81.05_{\pm0.82}$	$68.92_{\pm1.14}$
	TGAT	$87.17_{\pm0.48}$	$77.90_{\pm1.03}$	$78.33_{\pm1.09}$	$84.85_{\pm0.76}$	$72.58_{\pm0.88}$
	GraphMixer	$83.59_{\pm0.79}$	$91.63_{\pm0.41}$	$94.43_{\pm0.29}$	$87.91_{\pm0.54}$	$74.71_{\pm0.97}$
	DyGFormer	$85.81_{\pm0.63}$	$94.73_{\pm0.35}$	$84.69_{\pm0.86}$	$85.75_{\pm0.67}$	$\mathbf{93.49}_{\pm0.38}$
	LightDyG	$82.19_{\pm0.84}$	$93.48_{\pm0.47}$	$90.62_{\pm0.59}$	$86.43_{\pm0.71}$	$91.25_{\pm0.52}$
	DynSpectral	$\mathbf{99.09}_{\pm0.18}$	$\mathbf{95.47}_{\pm0.33}$	$\mathbf{97.74}_{\pm0.23}$	$\mathbf{92.80}_{\pm0.44}$	$92.17_{\pm0.49}$
AP	JODIE	$79.41_{\pm0.62}$	$83.54_{\pm0.77}$	$84.47_{\pm0.81}$	$81.56_{\pm0.88}$	$63.87_{\pm1.12}$
	DyRep	$81.57_{\pm0.93}$	$64.72_{\pm1.20}$	$78.28_{\pm1.07}$	$79.45_{\pm0.96}$	$65.24_{\pm1.16}$
	TGAT	$85.83_{\pm0.69}$	$79.06_{\pm0.98}$	$80.41_{\pm1.02}$	$84.99_{\pm0.74}$	$71.83_{\pm0.91}$
	GraphMixer	$82.36_{\pm0.75}$	$93.07_{\pm0.56}$	$94.88_{\pm0.31}$	$87.95_{\pm0.64}$	$76.15_{\pm0.85}$
	DyGFormer	$85.77_{\pm0.58}$	$\mathbf{95.98}_{\pm0.37}$	$87.86_{\pm0.79}$	$88.16_{\pm0.61}$	$\mathbf{93.85}_{\pm0.42}$
	LightDyG	$83.24_{\pm0.82}$	$94.81_{\pm0.46}$	$91.46_{\pm0.53}$	$87.58_{\pm0.66}$	$92.28_{\pm0.48}$
	DynSpectral	$\mathbf{98.52}_{\pm0.21}$	$95.83_{\pm0.28}$	$\mathbf{97.85}_{\pm0.25}$	$\mathbf{93.88}_{\pm0.39}$	$92.74_{\pm0.43}$

slightly better. We hypothesize that while our frequency decomposition excels at modeling localized structural evolution, certain global dynamics like viral cascades may benefit more from DyGFormer's mechanism. These observations collectively point to a key finding: effective dynamic graph modeling requires *adaptive spectral bandwidth*. DynSpec provides this adaptability through its multi-channel architecture, enabling it to remain highly competitive across diverse dynamic environments.

5.3   Ablation Study

To isolate and verify the contribution of each component within our proposed multi-channel architecture, we conducted a series of ablation studies, with the results presented in Table 3. The ablation results presented in Table 3 validate the

Table 3. Ablation Study

Configuration	MOOC		UCI		Bitcoin		movielens		fb-forum	
	AUC	AP	AUC	AP	AUC	AP	AUC	AP	AUC	AP
Full Model	**99.09**	**98.52**	**95.47**	**95.83**	**97.74**	**97.85**	**92.80**	**93.88**	**92.17**	**92.74**
w/o Channel Origin	98.15	97.68	94.82	95.21	96.89	97.03	91.54	92.76	90.83	91.45
w/o Channel Diff	98.47	97.91	94.95	95.38	96.52	96.71	91.12	92.38	89.67	90.52
w/o Channel Union	98.23	97.73	93.88	94.62	97.21	97.46	90.78	91.95	91.29	91.88
Only Channel Origin	96.84	96.12	92.76	93.54	95.38	95.69	89.45	90.73	88.51	89.28

architectural necessity of our multi-channel framework. The consistent performance degradation across all 'w/o' variants confirms the complementary, rather than redundant, roles of the decomposed spectral views. The functional specialization of these channels becomes evident when correlating performance impacts with dataset characteristics. For instance, the model's reliance on the Difference channel is most pronounced on networks characterized by transient structural changes, such as Bitcoin and fb-forum, where its removal incurs the most substantial performance penalty. This result empirically supports its hypothesized function as a high-pass filter, crucial for modeling abrupt network dynamics. Conversely, the necessity of the Union channel is highlighted on datasets defined by long-term, cumulative patterns (e.g., UCI and movielens), where its absence leads to the greatest loss in predictive power, thereby validating its role in capturing low-frequency structural stability. Ultimately, the performance margin between our Full Model and the 'Only Origin' baseline validates the fundamental premise that our spectral decomposition strategy offers a more comprehensive modeling paradigm than a conventional, single-view approach to dynamic graphs.

5.4   Analysis of Frequency-Adaptive Mechanisms

To verify that DynSpectral's performance gains are rooted in its proposed adaptive mechanisms, we investigate the model's internal behavior on the Bitcoin dataset. The results, presented in Table 4 and Table 5, provide strong evidence that the model learns the intended frequency-specific specializations. The efficacy of the Dynamic Adaptive Mapping (DAM) is detailed in Table 4. The activation strength α in the **Difference** channel exhibits the highest correlation (0.791) with structural changes ($|\Delta h|$) and the greatest sensitivity (2.19×). This empirically confirms that the model learns to apply stronger non-linearity to amplify its

Table 4. DAM Adaptivity Analysis: Correlation between α and $|\Delta h|$ on Bitcoin Dataset

Channel	Corr(α, $|\Delta h|$)	Mean α ($|\Delta h| < 0.05$)	Mean α ($|\Delta h| > 0.15$)	Sensitivity
Union	0.528	0.69	0.93	1.35×
Original	0.615	**0.66**	1.14	1.73×
Difference	**0.791**	0.68	**1.49**	**2.19×**

Table 5. Decay Parameter Evolution During Training on Bitcoin Dataset

Channel	λ (Init)	λ (Final)	Change
Union	0.10	0.082	−18.0%
Original	0.30	0.355	+18.3%
Difference	0.50	0.723	+44.6%

response to high-frequency events, consistent with the channel's role as a high-pass filter. Furthermore, Table 5 shows that the model learns distinct temporal receptive fields by amplifying the inductive biases set in our initialization. The decay parameter λ for the **Union** channel decreases, lengthening its temporal focus to capture stable patterns, while λ for the **Difference** channel increases significantly ($+44.6\%$), narrowing its focus to recent, transient dynamics. Together, these results provide internal validation that our frequency-adaptive components enable the model to learn specialized, interpretable roles for each spectral channel.

6   Conclusion

In this paper, we addressed a fundamental limitation in existing dynamic graph neural networks: their tendency to model temporal evolution as a holistic process, thereby overlooking the coexistence of stable and abrupt structural changes. We proposed DynSpectral, a novel multi-channel framework that tackles this challenge by introducing a frequency-aware graph decomposition strategy. By explicitly disentangling the graph's evolution into low-frequency (Union Graph) and high-frequency (Difference Graph) components and processing them with specialized temporal encoders, our model achieves a more comprehensive and nuanced understanding of temporal dynamics.

Extensive experiments confirmed the efficacy of our approach. DynSpectral not only established new state-of-the-art performance on multiple benchmarks but also demonstrated remarkable adaptability, yielding significant gains on datasets with disparate dynamic characteristics like MOOC and Bitcoin. These results empirically validate our core thesis that decomposing dynamic graphs based on their spectral properties leads to richer and more effective representations. We also acknowledge that the unbounded accumulation of historical edges in the Union Graph may potentially lead to over-smoothing in extremely long sequences. Future work could mitigate this via adaptive sliding window mechanisms or graph sparsification to maintain spectral distinctiveness. Our work advocates for a shift towards more principled, decomposed architectures in dynamic graph learning and opens up new avenues for future research into adaptive, multi-scale temporal modeling.

Acknowledgments. The research was supported by the National Natural Science Foundation of China, Grant No. 62472305 and Hainan Province Science and Technology Special Fund, Grant No: ZDYF2024SHFZ051.

Disclosure of Interests. The authors have no competing interests to declare that are relevant to the content of this article.

A   Proof of Theorem 1

We begin with the edge set relations from the multi-frequency views. By definition:

- $E_t^{\text{union}} = E_t \cup \left(\bigcup_{i=1}^{t-1} E_i\right)$ (includes current and historical edges),
- $E_t^{\text{add}} = E_t \setminus E_{t-1}$ (newly appearing edges only),
- $E_t^{\text{orig}} = E_t$ (current edges).

To proceed, introduce auxiliary edge sets:

- $E^{\text{hist}} = \bigcup_{i=1}^{t-1} E_i \setminus E_t$ (historical edges not in current graph),
- $E^{\text{persist}} = E_t \cap \left(\bigcup_{i=1}^{t-1} E_i\right)$ (persistent edges from history to current).

The Laplacian matrices satisfy:

- $L_t^{\text{union}} = L_t^{\text{orig}} + L^{\text{hist}}$, where L^{hist} is the Laplacian for E^{hist} (positive semi-definite, as it corresponds to a subgraph),
- $L_t^{\text{orig}} = L_t^{\text{add}} + L^{\text{persist}}$, where L^{persist} is the Laplacian for E^{persist} (positive semi-definite).

For any signal $f \in \mathbb{R}^{|V|}$, the Dirichlet energy is $E(f, G) = f^T L_G f$. Since Laplacians of subgraphs are positive semi-definite, adding such a matrix non-decreases the quadratic form.

1. **Upper bound:** $E(f, G_t^{\mathbf{orig}}) \leq E(f, G_t^{\mathbf{union}})$. Since $L^{\text{hist}} \succeq 0$,

$$f^T L_t^{\text{union}} f = f^T (L_t^{\text{orig}} + L^{\text{hist}}) f = f^T L_t^{\text{orig}} f + f^T L^{\text{hist}} f \geq f^T L_t^{\text{orig}} f,$$

as $f^T L^{\text{hist}} f \geq 0$ by positive semi-definiteness. Thus, $E(f, G_t^{\text{orig}}) \leq E(f, G_t^{\text{union}})$.

2. **Lower bound:** $E(f, G_t^{\mathbf{add}}) \leq E(f, G_t^{\mathbf{orig}})$. Similarly, since $L^{\text{persist}} \succeq 0$,

$$f^T L_t^{\text{orig}} f = f^T (L_t^{\text{add}} + L^{\text{persist}}) f = f^T L_t^{\text{add}} f + f^T L^{\text{persist}} f \geq f^T L_t^{\text{add}} f,$$

as $f^T L^{\text{persist}} f \geq 0$ by positive semi-definiteness. Thus, $E(f, G_t^{\text{add}}) \leq E(f, G_t^{\text{orig}})$.

Combining these inequalities yields $E(f, G_t^{\text{add}}) \leq E(f, G_t^{\text{orig}}) \leq E(f, G_t^{\text{union}})$. This holds for arbitrary f, relying solely on the positive semi-definiteness of subgraph Laplacians, independent of specific spectral structure.

References

1. Cong, W., et al.: Do we really need complicated model architectures for temporal networks? arXiv preprint arXiv:2302.11636 (2023)
2. Fan, J., Weng, W., Chen, Q., Wu, H., Wu, J.: PDG2Seq: periodic dynamic graph to sequence model for traffic flow prediction. Neural Netw. **183**, 106941 (2025)
3. Gao, C., Zhu, J., Zhang, F., Wang, Z., Li, X.: A novel representation learning for dynamic graphs based on graph convolutional networks. IEEE Trans. Cybern. **53**(6), 3599–3612 (2022)
4. Guo, Y., Wei, Z.: Graph neural networks with learnable and optimal polynomial bases. In: International Conference on Machine Learning, pp. 12077–12097. PMLR (2023)

5. Huang, K., Wang, Y.G., Li, M., et al.: How universal polynomial bases enhance spectral graph neural networks: heterophily, over-smoothing, and over-squashing. arXiv preprint arXiv:2405.12474 (2024)
6. Kumar, S., Zhang, X., Leskovec, J.: Predicting dynamic embedding trajectory in temporal interaction networks. In: Proceedings of the 25th ACM SIGKDD International Conference on Knowledge Discovery & Data Mining, pp. 1269–1278 (2019)
7. Li, G., Yang, J., Liang, S., Luo, D.: Polynomial selection in spectral graph neural networks: an error-sum of function slices approach. In: Proceedings of the ACM on Web Conference 2025, pp. 3276–3287 (2025)
8. Liu, A., Zhang, Y.: Spatial-temporal dynamic graph convolutional network with interactive learning for traffic forecasting. IEEE Trans. Intell. Transp. Syst. **25**(7), 7645–7660 (2024)
9. Nazari, M., Korshøj, A.R., ur Rehman, N.: Multiscale dynamic graph signal analysis. Sig. Process. **222**, 109519 (2024)
10. Ortega, A., Frossard, P., Kovačević, J., Moura, J.M., Vandergheynst, P.: Graph signal processing: overview, challenges, and applications. Proc. IEEE **106**(5), 808–828 (2018)
11. Pan, Z., Gao, C., Cai, F., Chen, H., Li, Y.: Light dynamic graph learning on temporal networks. ACM Trans. Inf. Syst. (2025)
12. Poursafaei, F., Huang, S., Pelrine, K., Rabbany, R.: Towards better evaluation for dynamic link prediction. In: Advances in Neural Information Processing Systems, vol. 35, pp. 32928–32941 (2022)
13. Rossi, E., Chamberlain, B., Frasca, F., Eynard, D., Monti, F., Bronstein, M.: Temporal graph networks for deep learning on dynamic graphs. arXiv preprint arXiv:2006.10637 (2020)
14. Rossi, R., Ahmed, N.: The network data repository with interactive graph analytics and visualization. In: Proceedings of the AAAI Conference on Artificial Intelligence, vol. 29 (2015)
15. Sandryhaila, A., Moura, J.M.: Discrete signal processing on graphs: frequency analysis. IEEE Trans. Signal Process. **62**(12), 3042–3054 (2014)
16. Shuman, D.I., Narang, S.K., Frossard, P., Ortega, A., Vandergheynst, P.: The emerging field of signal processing on graphs: extending high-dimensional data analysis to networks and other irregular domains. IEEE Signal Process. Mag. **30**(3), 83–98 (2013)
17. Trivedi, R., Farajtabar, M., Biswal, P., Zha, H.: DyRep: learning representations over dynamic graphs. In: International Conference on Learning Representations (2019)
18. Vaswani, A., et al.: Attention is all you need. In: Advances in Neural Information Processing Systems, vol. 30 (2017)
19. Wang, X., Zhang, M.: How powerful are spectral graph neural networks. In: International Conference on Machine Learning, pp. 23341–23362. PMLR (2022)
20. Xiao, S., Li, J., Lu, J., Huang, S., Zeng, B., Wang, S.: Graph neural networks for multi-view learning: a taxonomic review. Artif. Intell. Rev. **57**(12), 341 (2024)
21. Xu, D., Ruan, C., Korpeoglu, E., Kumar, S., Achan, K.: Inductive representation learning on temporal graphs. arXiv preprint arXiv:2002.07962 (2020)
22. Yu, L., Sun, L., Du, B., Lv, W.: Towards better dynamic graph learning: new architecture and unified library. In: Advances in Neural Information Processing Systems, vol. 36, pp. 67686–67700 (2023)

23. Zheng, Y., Yi, L., Wei, Z.: A survey of dynamic graph neural networks. Front. Comp. Sci. **19**(6), 196323 (2025)
24. Zou, Z., Jiang, Y., Shen, L., Liu, J., Liu, X.: LOHA: direct graph spectral contrastive learning between low-pass and high-pass views. In: Proceedings of the AAAI Conference on Artificial Intelligence, vol. 39, pp. 13492–13500 (2025)

Not All Neighbors are Temporally Relevant: An Adaptive Neighborhood Aggregation Framework for Dynamic Graph Learning

Bingce Wang, Weiping Li[✉], Tong Mo, Xiang Yuan, Xu Chu,
and Liwen Zhang

School of Software and Microelectronics, Peking University, Beijing, China
{wangbingce,xiangyuan,chuxu,zhanglw0529}@stu.pku.edu.cn,
{wpli,motong}@ss.pku.edu.cn

Abstract. Discrete-Time Dynamic Graphs (DTDGs) are represented as a sequence of static graphs, i.e., graph snapshots, in which nodes and edges may change from snapshot to snapshot. Most current Graph Neural Networks (GNNs) for DTDGs follow a paradigm where the sequence encoder and the graph encoder are combined within a fixed receptive field. Although this paradigm can capture the temporal evolution of subgraphs, it may also include temporally irrelevant neighborhoods because of the unchanged receptive field. The method of determining a temporally correlated neighborhood in DTDGs remains unexplored. To fill the gap, we innovatively propose **GraphANA**, an **A**daptive **N**eighborhood **A**ggregation framework for dynamic graph embedding. **For the first time**, we transfer the prevailing concept of "one node one receptive field" to the DTDGs. Specifically, GraphANA first defines multi-hop neighborhood propagation within a shortest-path kernel function to derive neighborhood representations for each snapshot. Then, GraphANA employs a hop-aware sequence encoder to generate time-series tokens that capture the temporal properties of each neighborhood's feature. Finally, a dual-stage adaptive neighborhood aggregation is conducted to eliminate irrelevant neighborhoods and to aggregate each time-series token from the remaining neighborhoods with temporal relevance scores. Extensive experiments demonstrate that GraphANA achieves competitive performance on downstream tasks.

Keywords: Graph Neural Networks · Discrete-Time Dynamic Graphs · Graph Representation Learning · Adaptive Neighborhood Aggregation

1 Introduction

Graph data with complex attributes and relations is used across multiple fields, including user analysis in social media, product recommendation on e-commerce

H. Jung et al. (Eds.): DASFAA 2026, LNCS 16536, pp. 148–164, 2026.
https://doi.org/10.1007/978-981-92-0366-6_10

platforms, and integration with large language models [1]. A dynamic graph is a type of graph in which both the graph structure and node attributes change over time. Existing methods for learning dynamic graph representations are mainly categorized into Continuous-Time Dynamic Graphs (CTDGs) and Discrete-Time Dynamic Graphs (DTDGs) [19]. The former represents the dynamic graph as a sequence of graph events. Graph events are typically defined by a triple comprising the node associated with the graph event, the event type, and timestamp information. The latter divides the timestamps in the dynamic graph into multiple time intervals, representing the dynamic graph as a sequence of graph snapshots. In this work, we mainly discuss the discrete-time methods.

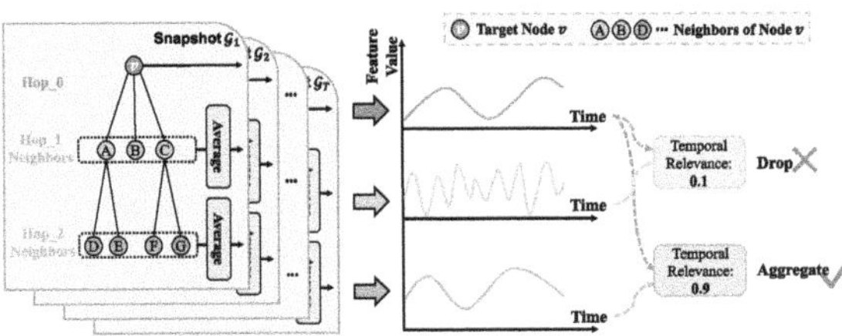

Fig. 1. An illustration of the concept "one node one receptive field" in DTDGs.

Recently, a novel paradigm, the Snowflake Hypothesis [15], rooted in the concept of "one node, one receptive field" has gained significant attention for its efficacy in addressing the over-smoothing and over-fitting issues in GNNs. This hypothesis posits that for an L-layer GNN, every node in the graph harbors an optimal receptive field width denoted as l ($1 \leq l \leq L$). Intuitively, the snowflake hypothesis demonstrates even greater vitality and significance in the context of dynamic graphs. The dynamic nature of DTDGs introduces continuous changes in both structure and semantics, amplifying variation in information density across nodes and, consequently, accentuating the heterogeneity of information distribution. This heterogeneity challenges the use of fixed receptive field designs in dynamic graph learning.

Building on this, we define the Snowflake Hypothesis on DTDGs as the principle that each node should dynamically select the most temporally relevant neighborhood for aggregation, rather than aggregating all neighbors based solely on proximity. We further illustrate this issue in Fig. 1. It shows node v and its neighbor nodes at varying hop distances, with the 0-hop neighborhood corresponding to node v itself. Assuming each node is associated with a single feature, we compute the average feature values of nodes within the 1-hop and 2-hop neighborhoods. This allows us to observe how the aggregated feature representations of different neighborhoods evolve over time. As observed, the feature values from the 1-hop neighborhood exhibit no temporal correlation with the target node v. Aggregating information from the 1-hop neighborhood in this case would introduce substantial noise, negatively affecting the model's performance. In contrast, the 2-hop neighborhood shows strong behavioral synchronization with the target node. Therefore, during dynamic graph learning for node v, it is more appropriate to aggregate information from the 2-hop neighborhood, even though it is farther away from the target node compared to the 1-hop neighborhood.

To address the challenge mentioned above, we propose **GraphANA**, an **A**daptive **N**eighborhood **A**ggregation framework for dynamic graph embedding. **For the first time**, we transfer the prevailing concept of "one node one receptive field" to the dynamic graph. The core idea behind this approach is to model the temporal correlation between neighborhood features and eliminate irrelevant information, rather than capturing the temporal evolution of the entire subgraph. This enables more accurate representations for each node in the dynamic graph. Specifically, GraphANA first defines the multi-hop neighborhood of a target node based on the shortest-path kernel function. Then, GraphANA employs a hop-aware sequence encoder to tokenize the time series of each neighborhood's feature across different snapshots. Within a dual-stage adaptive aggregation block, GraphANA selects the most temporally relevant neighborhoods utilizing the neighborhood attention mechanism and aggregates their temporal tokens using a feature-level Transformer, thereby generating node representations that reduce neighborhood noise.

Our contributions are highlighted as follows:

- To the best of our knowledge, **we are the first** to incorporate the prevailing concept of "one node one receptive field" into dynamic graphs. To achieve this target, we propose an adaptive neighborhood aggregation framework, namely GraphANA, which primarily consists of multi-hop information propagation, a hop-aware sequence encoder, and a dual-stage adaptive neighborhood aggregation block.
- GraphANA can precisely capture temporal properties and accurately measure temporal correlations across neighborhoods. GraphANA enables each node to adaptively select the most temporally relevant neighborhood for aggregation, while filtering out irrelevant noise, rather than aggregating all neighbors based solely on proximity.
- Finally, we conduct comprehensive experiments on several real-world datasets. The experimental results demonstrate that our model achieves remarkable improvements in the dynamic node classification task.

2   Related Work

For modeling discrete-time dynamic graphs, the most straightforward approach is to process each snapshot with a separate graph neural network and pass the GNN output to the time-series component. For example, RgCNN [9] combines spatial-domain graph convolution with Long Short-Term Memory (LSTM) networks. HTGN [17] applies hyperbolic manifolds to dynamic graphs, utilizing Poincaré discs for graph embedding to achieve good embedding results.

DySAT [13] introduces a stacked architecture made up entirely of self-attention blocks. SpikeNet [6] utilizes spiking neural networks instead of RNNs to capture the temporal and structural patterns of temporal graphs efficiently.

Other approaches combine GNNs and RNNs within the same encoder layer to simultaneously model both temporal and spatial information. EvolveGCN [11] integrates RNNs into GCNs and uses RNN components to update the weights of GCNs efficiently. ROLAND [18] utilizes a meta-learning framework that forwards the embedding and loss learned from a snapshot to the following snapshot. WinGNN [21] establishes representation associations between two neighboring snapshots and removes all time-specific encoders to improve generalization and parameter efficiency. STAA [2] employs a spatiotemporal activity-aware random walk strategy on both current and previous snapshots, allowing the model to acquire spatial and temporal awareness simultaneously.

The above approaches effectively model dynamic graphs with different characteristics and achieve excellent results. However, they fail to consider temporal relevance among neighborhoods during the aggregation process, which may introduce temporally irrelevant neighbors, leading to increased memory consumption and a higher risk of overfitting.

3   Preliminary and Motivation Example

3.1   Discrete-Time Dynamic Graph

In this paper, given the availability of data, we focus on the discrete-time dynamic graphs, which are defined as a series of snapshots $\{\mathcal{G}^1, \mathcal{G}^2 \ldots, \mathcal{G}^T\}$, where T is the total number of snapshots. A snapshot at time t, $\mathcal{G}^t = (\mathcal{V}^t, \mathcal{E}^t)$, is a snapshot with a node set $\mathcal{V}^t$ and an edge set $\mathcal{E}^t \subseteq \mathcal{V}^t \times \mathcal{V}^t$ of the graph. Notably, the graph's connecting edges may appear or disappear over time. In each snapshot, nodes can be paired with evolving features $\mathbf{X}^t = \{\mathbf{x}_v^t \mid \forall v \in \mathcal{V}\} \in \mathbb{R}^{|\mathcal{V}| \times N}$, where N is the feature dimension and is commonly stable over time. The entire node features with time dimension can be denoted as $\mathcal{X} = \{\mathbf{X}^1, \mathbf{X}^2, \ldots, \mathbf{X}^T\} \in \mathbb{R}^{T \times |\mathcal{V}| \times N}$. We aim to learn node embeddings $\mathbf{Z}$ for all nodes in the temporal graph $\mathcal{G}$.

3.2   Motivation Example

To further explain the concept of "one node one receptive field" in DTDGs, we conducted experiments on two real-world datasets and present the results in Fig. 2. We compute the average Pearson Correlation coefficient between each target node and its multi-hop neighborhoods. Let the initial feature of node v at snapshot $\mathcal{G}_1$ be denoted as $\mathbf{x}_v \in \mathbb{R}^{N \times 1}$, and its 1-hop neighborhood as $\mathcal{N}_v^l$. We obtain the representation of each neighborhood using a single-layer message-passing operation: $m_{v,1}^l = Mean\left(\mathbf{x}_u | u \in \mathcal{N}_v^l\right)$. Treating the variation of each feature across different snapshots as a time series, we can represent all feature time series as a matrix $\mathbf{M}_v^l = \{m_{v,1}^l, m_{v,2}^l, \ldots, m_{v,T}^l\} \in \mathbb{R}^{N \times T}$. Based on this,

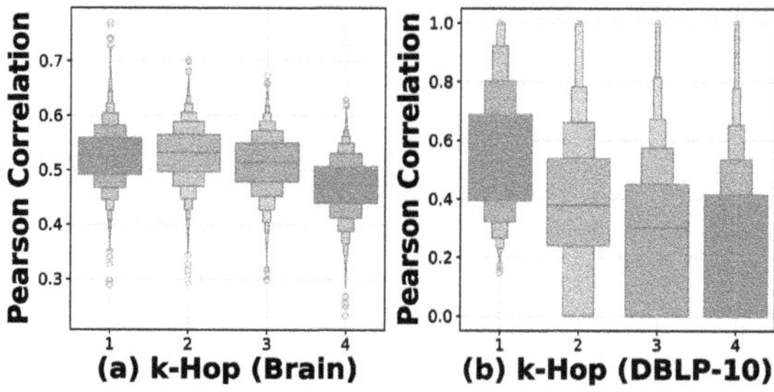

Fig. 2. The distribution of Pearson Correlation Coefficients between the target node and its neighborhoods.

the average Pearson Correlation coefficient between node v and its l-hop neighborhood can be computed using:

$$\mathcal{P}\left(\mathbf{X}_v, \mathbf{M}_v^l\right) = \frac{1}{N} \sum_n \left| \rho\left(\mathbf{X}_v[n,:], \mathbf{M}_v^l[n,:]\right) \right|$$

$$\rho(x, y) = \frac{\sum_t (x_t - \bar{x})(y_t - \bar{y})}{\sqrt{\sum_t (x_t - \bar{x})^2} \sqrt{\sum_t (y_t - \bar{y})^2}} \tag{1}$$

The value of $\mathcal{P}$ lies in the range $[0, 1]$, with values closer to 1 indicating stronger temporal correlation, and values closer to 0 indicating weaker correlation.

The distribution of Pearson Correlation Coefficients between the target node and its neighborhoods is shown in Fig. 2(a) for the Brain dataset and Fig. 2(b) for the DBLP1 dataset. The distributions reveal significant variation in the correlation levels among nodes. Even within the 1-hop neighborhood, some nodes exhibit strong temporal correlations, while others show very weak correlations. Overall, in the DBLP-10 dataset, neighborhood correlation tends to decrease as the hop distance increases. In contrast, in the Brain dataset, the correlation remains relatively stable across different hop levels. These observations highlight the importance of enabling the model to adaptively select appropriate neighborhoods for each node, either across different datasets or among different nodes within the same dataset. This adaptivity is crucial for avoiding the introduction of irrelevant or noisy information into the representation learning process.

4   Method

The overall framework of GraphANA is shown in Fig. 3. GraphANA consists of three main components: **(a) Multi-hop Information Propagation.** We first define multi-hop neighborhoods using a shortest path kernel function. Based on this, we propagate messages from neighbors by averaging their feature vectors at different hop distances to derive the neighborhood representations under each snapshot. The message from the L-hop neighborhood under snapshot $\mathcal{G}_1$ is denoted as $\mathbf{m}_{v,1}^l \in \mathbb{R}^{N \times 1}$. Besides, the representation of the L-hop neighborhood

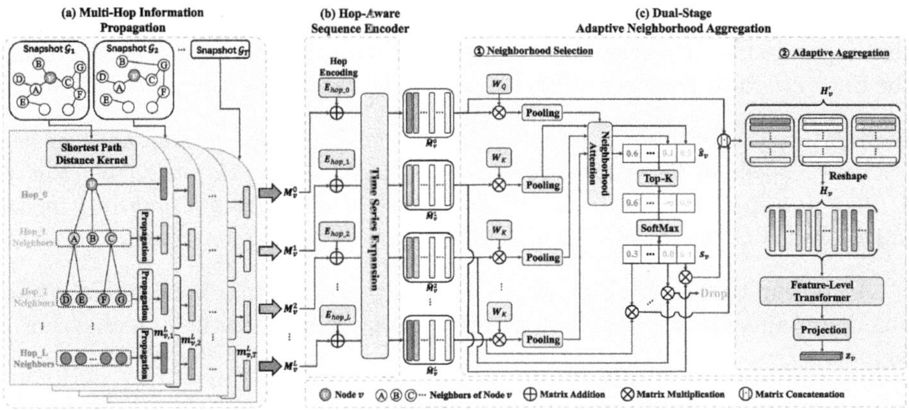

Fig. 3. An illustration of the proposed GraphANA. For clarity, we show the pipeline of GraphANA working on a single node v.

across all snapshots is denoted as matrix $\mathbf{M}_v^L \in \mathbb{R}^{N \times T}$, while the 0-hop neighborhood indicates the target node v itself. **(b) Hop-aware Sequence Encoder.** We design a set of learnable hop encodings $\mathbf{E}_{hop}$ for each hop. The neighborhood representations are added to their corresponding hop encodings and then passed through a time series expansion function, i.e., a Multi-Layer Perceptron (MLP), to expand the temporal dimension T to a higher-dimensional representation $\widehat{\mathbf{M}}_v^L \in \mathbb{R}^{N \times D}$. This process yields time-series tokens corresponding to the temporal evolution of each feature. **(c) Dual-Stage Adaptive Neighborhood Aggregation.** The first stage is neighborhood selection, where the model computes neighborhood attention scores $\hat{\mathbf{s}}_v \in \mathbb{R}^L$ between the target node and tokens from different neighborhoods. Only the top-k attention scores are retained and normalized, while the rest are set to zero. These normalized scores $\mathbf{s}_v$ are then multiplied by the corresponding neighborhood tokens, and the resulting tokens are concatenated with the target node's representation to form $\mathbf{H}_v' \in \mathbb{R}^{N \times (K+1) \times D}$. The figure illustrates the case when $K = 2$. The second stage is adaptive aggregation, where each feature's time-series token is treated as an individual element. All feature tokens from different neighborhoods are considered as a sequence $\mathbf{H}_v \in \mathbb{R}^{N(K+1) \times D}$ whose length is denoted as $N(K+1)$, and are fed into a feature-level Transformer. This enables each feature of the target node to aggregate temporal information from all features within the selected neighborhoods. Finally, the output of the Transformer is passed through a projection function, implemented using a Multi-Layer Perceptron (MLP), to perform dimensionality reduction and obtain the final node representation $\mathbf{z}_v \in \mathbb{R}^N$. Each component of GraphANA is described in detail below.

4.1   Multi-hop Neighborhood Propagation

Currently, most existing GNNs are designed based on the 1-hop message passing framework. The 1-hop message-passing framework can be directly generalized to

l hop message-passing as it shares the same message mechanism. The difference is that separate message aggregation functions can be used for each hop. We utilize the Shortest Path Distance (SPD) kernel [3] to define the L-hop neighborhood:

Definition 1. *For node v in the graph $\mathcal{G}$, the L-th hop neighborhood of v is the set of nodes in the graph whose shortest path distance to node v is equal to L. In this paper, we denote $\mathcal{N}_v^{l,spd}$ as the set of nodes that belong to l-th hop neighborhood in the graph $\mathcal{G}$, and $\mathcal{N}_v^{0,spd} = \{v\}$ is the node v itself.*

According to the definition above, the multi-hop neighborhood message that a node obtained by the graph structure learning module in the t-th snapshot is as follows:

$$\mathbf{m}_{v,t}^l = \frac{1}{\phi(\cdot)} \sum_{u \in \mathcal{N}_v^{l,spd}} \mathbf{x}_{u,t} \in \mathbb{R}^N \tag{2}$$

where $\mathbf{m}_{v,t}^l$ is the node representation output by the message-passing process, $\mathbf{x}_{u,t} \in \mathbb{R}^N$ is the original input feature representation of node u, $\phi(\cdot)$ is the normalization function, and $\mathcal{N}_v^{l,spd}$ is l-th neighborhood of node v under the Shortest Path Distance kernel. Based on the above equation, the representation of the different hop neighborhoods is preserved for the subsequent sequence encoder.

4.2   Hop-Aware Sequence Encoder

We propose a hop-aware sequence encoder to generate the time series token for each feature separately. The propagation process above treats each hop's information independently. Hop encoding is necessary to make the model aware of the multi-hop structural information. Given a target node v, the multi-hop neighborhoods message with hop encoding can be formulated as follows:

$$\mathbf{m}_{v,t}^l = \sum_{u \in \mathcal{N}_v^{l,spd}} \frac{\mathbf{x}_{u,t}}{|\mathcal{N}_v^{l,spd}|} + \mathbf{E}_{hop}[l,:] \in \mathbb{R}^N \tag{3}$$

where $\mathbf{E}_{hop}$ denotes learnable embeddings that stored as a $L \times N$ matrix. The index number l represents the l-th hop neighborhood, which will be assigned a learnable hop embedding $\mathbf{E}_{hop}[l,:]$ (i.e., the l-th row of matrix $\mathbf{E}_{hop}$) to indicate its hop value. By stacking the messages from each snapshot, we can denote the l-th hop neighborhood representation in a matrix form $\mathbf{M}_v^l$. The sequence encoder with a specific input feature $n \in N$ can then be formulated as follows:

$$\widehat{\mathbf{M}}_v^l[n,:] = \text{Embedding}\left(\mathbf{M}_v^l[n,:]\right) \in \mathbb{R}^D,$$
$$\widehat{\mathbf{M}}_v^l \in \mathbb{R}^{N \times D}, \ \mathbf{M}_v^l \in \mathbb{R}^{N \times T} \tag{4}$$

where $\widehat{\mathbf{M}}$ contains embedded tokens in dimension D for entire N features. Embedding function $\mathbb{R}^T \mapsto \mathbb{R}^D$ is implemented by a multi-layer perceptron (MLP) and encodes each feature's time-series identically. It projects time series from a low-dimensional to a high-dimensional token space to model the predicted temporal evolution. The neurons in the MLP are instructed to depict the intrinsic properties of any time series, such as amplitude and periodicity, which has been proven competent by previous work [7].

4.3   Dual-Stage Adaptive Neighborhood Aggregation

Neighborhood Selection. We define the Snowflake Hypothesis on DTDGs as the principle that each node should dynamically select the most temporally relevant neighborhood for aggregation, rather than aggregating all neighbors based solely on proximity. Therefore, we propose an adaptive neighborhood selection module based on the neighborhood attention mechanism to select messages from different neighborhoods. The attention score for each neighborhood is calculated by measuring the average similarity between its temporal tokens and the corresponding target nodes. The neighborhood attention mechanism can be formulated as follows:

$$
s_v^l = \begin{cases} \mathrm{Softmax}\left(\hat{s}_v^l\right), & \hat{s}_v^l \in \mathrm{Top}\left(\hat{\mathbf{s}}_v, \Phi\right), \\ 0, & \text{otherwise}, \end{cases}
$$

$$
\hat{s}_v^l = \frac{1}{N} \sum_{n \in N} \mathbf{W}_q \cdot \left(\widehat{\mathbf{M}}_v^0[n,:] \,\|\, \widehat{\mathbf{M}}_v^l[n,:]\right)
$$

$$(5)$$

where $\mathbf{W}_q \in \mathbb{R}^{2D}$ is a learnable weight vector, $\hat{s}_v^l$ denotes the average score among N features, $\hat{\mathbf{s}}_v = \left[\hat{s}_v^1, \hat{s}_v^2, \ldots, \hat{s}_v^L\right]$ stores attention scores from all neighborhoods, and Φ in function Top is a parameter that limits the number of chosen neighborhoods. Different attention scores determine the relevance between multi-hop neighborhoods and their corresponding target nodes. In subsequent computations, we discard neighborhoods with zero relevance scores. In particular, we treat the time-series token of the node v itself as a 0-hop neighborhood and combine it with the remaining neighborhoods.

$$
\mathbf{H}_v = \widehat{\mathbf{M}}_v^0 \,\|\, \widehat{\mathbf{H}}_v \in \mathbb{R}^{\Phi N \times D}
$$

$$
\widehat{\mathbf{H}}_v = \overset{s_v^l > 0}{\|} s_v^l \cdot \widehat{\mathbf{M}}_v^l
$$

$$(6)$$

Adaptive Aggregation. We argue that the same feature may exhibit different, or even opposite, temporal trends across different neighborhoods. In traditional GNNs, the message aggregation process tends to smooth feature tokens across neighborhoods, which may offset these potentially conflicting trends. To fill this gap, we employ a feature-level Transformer encoder in the Adaptive Aggregation process. To simplify, we abbreviate $\mathbf{H}_v[n,:]$ as $\mathbf{h}_n$. We stack multiple Transformer layers and update the feature token in the k-th layer:

$$
\begin{aligned}
\mathbf{h}_n^0 &= \mathbf{h}_n, \\
\mathbf{h}_n^{k+1} &= \text{Concat}\left(\text{head}_1, \ldots, \text{head}_M\right)\mathbf{O}^k, \\
\text{head}_m &= \text{Attention}\left(\mathbf{Q}^{m,k}\mathbf{h}_n^k, \mathbf{K}^{m,k}\mathbf{h}_j^k, \mathbf{V}^{m,k}\mathbf{h}_j^k\right) \\
&= \sum_{j \in N} w_{nj}\left(\mathbf{V}^{m,k}\mathbf{h}_j^k\right) \\
w_{nj} &= \text{softmax}_j\left(\frac{\mathbf{Q}^{m,k}\mathbf{h}_n^k \cdot \mathbf{K}^{m,k}\mathbf{h}_j^k}{\sqrt{D}}\right)
\end{aligned}
\tag{7}
$$

where $\mathbf{Q}^{m,k}$, $\mathbf{K}^{m,k}$, $\mathbf{V}^{m,k}$ are the learnable weights of the m-th attention head, $\mathbf{h}_n^k$ is the n-th feature token of layer k and $\mathbf{O}^k$ is a projection to match the dimensions between adjacent layers. The output of the self-attention module is then passed through a feed-forward network layer, followed by a residual connection and layer normalization. Since the snapshot order is implicitly stored in the feed-forward network's neuron permutation, the position embedding in the vanilla Transformer is no longer needed here. Finally, the output of the K-th layer feature-level Transformer $\mathbf{H}_v^K \in \mathbb{R}^{\Phi N \times D}$ is computed as follows:

$$
\begin{aligned}
\widehat{\mathbf{H}}^K &= \text{LayerNorm}\left(\mathbf{H}^{K-1} + \text{Self-Atten}\left(\mathbf{H}^{K-1}\right)\right), \\
\mathbf{H}^K &= \text{LayerNorm}\left(\widehat{\mathbf{H}}^K + \text{Feed-Forward}\left(\widehat{\mathbf{H}}^K\right)\right).
\end{aligned}
\tag{8}
$$

Although GraphANA employs a Transformer in the adaptive aggregation stage, where the self-attention mechanism can also be viewed as selecting more important neighborhood features, we still apply the neighborhood selection before the Transformer. This design is motivated by two main considerations. **(i) Soft attention mechanism.** Conventional attention mechanisms assume all neighbors are reliable and aggregate them with soft (i.e., positive) values. Vast malicious neighbors can accumulate smaller but positive attention values and ultimately dominate the receptive field of GNNs, thereby misleading classification results. Based on this fact, the power of assigning zero attention values to unreliable neighborhoods is significant. **(ii) High computation cost.** Although the self-attention mechanism can identify relevance scores between any feature pairs, it is still very time-consuming and resource-intensive with $(L \times N)^2$ feature pairs. Therefore, we adopt a simple yet effective neighborhood selection mechanism to identify more informative neighborhoods, which reduces the time complexity from $\mathcal{O}((LN)^2)$ to $\mathcal{O}((\Phi N)^2)$, where $\Phi < L$.

4.4 Projection for Final Representation

The final output of node v in the feature-level Transformer is $\mathbf{H}_v^K$. We extract the token vectors corresponding to the target node $\mathbf{H}_v^K[:N,:] \in \mathbb{R}^{N \times D}$ and then project them into the final representation. We can simply adopt a projection layer: $\mathbb{R}^D \to \mathbb{R}^1$ after the Transformer encoder.

$$\mathbf{z}_v = \text{Projection}\left(\mathbf{H}_v^K[: N, :]\right) \in \mathbb{R}^N \qquad (9)$$

The final node representation $\mathbf{z}_v$ of node v can be utilized for subsequent downstream tasks.

5   Experiment

In this section, we evaluate our model to demonstrate its effectiveness in node classification tasks for dynamic graphs.

5.1   Experimental Setting

Datasets. Table 1 summarizes four public datasets used in this work [16]. Brain is a network of brain tissues where a node represents a tiny cube of brain tissue, and two nodes are connected if they show a similar degree of activation. DBLP-3 and DBLP-10 [5] are extracted from the DBLP, and they are co-authorship networks where nodes represent authors. Reddit is a post network where a node is a post, and two posts are connected if they share similar keywords.

Table 1. Summary of datasets.

Datasets	Nodes	Edges	Features	Snapshots	Classes
DBLP-3	4,257	23,540	100	10	3
Brain	5,000	1,955,488	20	12	10
Reddit	8,291	264,050	20	10	4
DBLP-10	28,085	236,894	128	27	10

Baseline Models. We compare GraphANA to the following baselines: (i) static graph embedding methods: DeepWalk [12], Node2Vec [4]; (ii) unsupervised temporal graph embedding methods: HTNE [22], M^2DNE [8], and DynamicTriad [20]; (iii) discrete-time temporal graph neural networks: MPNN [10], STAR [16], tNodeEmbed [14], EvolveGCN [11], ROLAND [18], SpikeNet [6], and GraphSSM [5]. For baselines that are originally designed for static graphs, we accumulate historical information (edges) in the graph snapshot sequence and represent the static graph structure at the last snapshot.

Training Details The hyperparameters in GraphANA are tuned based on the validation set's performance. We utilize ADAM with an initial learning rate in {0.001, 0.0003, 0.005, 0.008, 0.01}. The multi-hop neighborhoods range is searched from {2,3,4,5}, and the topk is chosen from {1,2}. We set the dimension of hidden embedding in {32, 64, 128, 256}, the number of Transformer blocks in {2,3,4}, the number of multi-heads in {2,4}, and the dropout rate in {0.2,0.4,0.6,0.8}.

5.2 Overall Performance

In our experiments, we consider node classification to evaluate representation quality. Following [16], we split all nodes into train/validation/test sets by the 81%/9%/10% ratio for the DBLP-3, Brain, and Reddit datasets. For the DBLP-10 dataset, we follow the experimental settings as suggested in [5], where 80% of the nodes are randomly selected as the training set, and the remaining nodes are used as the test set. For the baseline model, we report the results as presented in [5]. For the GraphANA, we conduct five independent runs and report the average Macro-F1 and Micro-F1 scores with standard deviation.

Table 2. Node classification results on the node classification task, where the best results in each row are highlighted in **bold**.

Methods	DBLP-3		Brain		Reddit		DBLP-10	
	Micro-F1	Macro-F1	Micro-F1	Macro-F1	Micro-F1	Macro-F1	Micro-F1	Macro-F1
DeepWalk [12]	47.53±0.4	47.21±0.2	51.45±0.6	51.03±0.8	26.82±0.6	26.75±0.4	66.38	67.12
Node2Vec [4]	48.79±0.3	48.42±0.4	53.51±0.5	52.95±0.6	25.47±0.6	25.44±0.5	67.31	66.93
HTNE [22]	48.98±0.2	48.74±0.3	22.31±0.8	22.12±0.5	26.96±0.5	26.80±0.7	68.79	68.36
M^2DNE [8]	49.12±0.5	48.87±0.4	23.79±0.5	23.54±0.6	25.79±0.6	25.61±0.4	69.71	69.75
DynamicTriad [20]	48.78±0.5	48.63±0.6	21.71±0.7	21.94±0.7	28.76±0.5	28.51±0.5	66.95	66.42
MPNN [10]	81.78±0.6	81.46±1.2	90.97±1.4	91.01±1.5	40.85±1.3	40.64±1.2	67.74±0.3	65.05±0.5
STAR [16]	84.74±1.0	84.20±1.2	92.08±1.3	92.23±1.3	43.42±2.3	43.43±2.4	72.98±1.5	72.03±1.2
tNodeEmbed [14]	84.51±1.2	83.57±1.1	92.35±0.8	92.30±1.0	42.11±1.8	42.06±1.3	74.19±1.8	74.23±2.2
EvolveGCN [11]	84.01±1.5	83.12±1.5	92.20±1.3	92.00±1.0	41.24±1.3	41.11±1.5	71.32±0.5	71.20±0.7
SpikeNet [6]	83.92±1.5	83.04±1.1	92.00±1.2	91.97±1.2	40.42±2.0	40.20±2.1	74.86±0.5	74.65±0.5
ROLAND [18]	84.21±1.4	84.06±1.5	92.14±1.2	91.85±1.1	44.22±2.2	44.25±1.9	75.01±1.1	74.98±1.0
GraphSSM [5]	85.26±0.9	85.00±1.3	93.52±1.0	93.54±0.9	49.21±0.5	49.05±0.7	76.80±0.3	76.00±0.4
GraphANA	**87.58 ± 0.8**	**87.86 ± 0.5**	**95.32 ± 0.6**	**95.47 ± 0.5**	**49.33 ± 0.9**	**49.96 ± 0.8**	**80.11 ± 0.4**	**79.69 ± 0.3**

The node classification performance of all methods is presented in Table 2. Compared with the state of the art, GraphANA achieves average improvements of 2.3% and 3.1% in Micro-F1 and Macro-F1 scores, respectively, among four datasets. Based on the standard deviation across five training rounds, GraphANA demonstrates more stable performance than other graph neural network models in most cases. Notably, our model achieves the most significant improvement on the largest dataset, i.e., DBLP-10, with Micro-F1 and Macro-F1 scores increasing by 4.3% and 4.9%, respectively, compared to the best baseline model. As shown in Table 2, DBLP-10 contains a significantly larger number of snapshots than the other datasets. This suggests that existing baseline models struggle to effectively extract dynamic changes in graph structure over extended time spans. Since our model separately computes the temporal evolution of different neighbors and propagates this change information to the target node, the dynamic information of the graph structure is more comprehensively preserved. These results demonstrate that GraphANA can retain dynamic graph information more effectively and learn node representations that are more beneficial for downstream tasks.

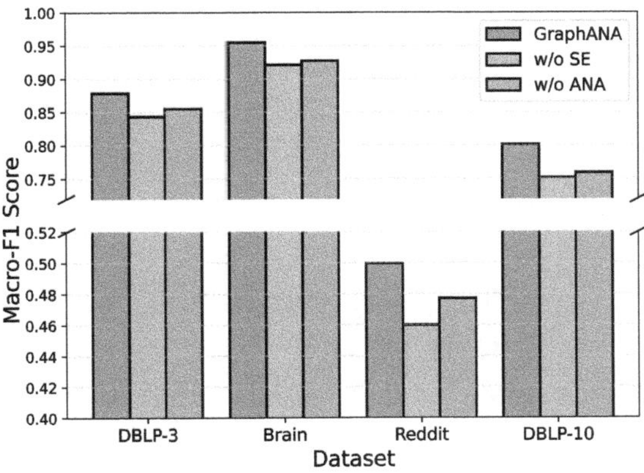

Fig. 4. Ablation study to demonstrate the effectiveness of each component in GraphANA.

5.3   Ablation Study

To answer how each component contributes to GraphANA's performance, we replace parts of the entire framework for an ablation study. The results are reported in Fig. 4. $-w/o$SE refers to the variant where the hop-aware sequence encoder is removed. In this setting, the raw feature sequences over snapshots are directly used as input to the subsequent modules. As shown in the figure, removing the sequence encoder leads to a performance drop across all datasets. This demonstrates that the proposed sequence encoder is more effective in capturing the sequential properties of node features. $-w/o$ANA denotes the variant in which the dual-stage adaptive neighborhood aggregation module is replaced with simple mean aggregation. To ensure a fair comparison of different neighborhood aggregation strategies, the target node temporal tokens obtained via mean aggregation are still fed into the Transformer for self-attention-based updates. However, this replacement results in a significant performance degradation. The reason is that mean aggregation fails to filter out irrelevant or noisy information from the neighborhood, thereby impairing the model's decision-making capability. In contrast, the proposed adaptive neighborhood aggregation mechanism dynamically selects the scope of aggregation. It assigns feature-specific weights based on the temporal correlation between nodes and their neighborhoods, enabling more accurate aggregation of informative neighborhood signals.

5.4   Selection of Multi-hop Neighborhoods

To evaluate the effectiveness of the neighborhood selection, we compare the model's performance under different neighborhood combinations in Table 3. **Bold** indicates the best performance and <u>underline</u> indicates the runner-up. The GraphANA-fixed method in Table 3 replaces neighborhood selection with a fixed neighborhood combination, i.e., each node chooses the same neighborhood.

Q^k denotes the set of k-hop neighbor nodes obtained through the shortest path kernel-based sampling, ensuring that different neighborhoods of a certain node do not overlap.

Table 3. The node classification results on four datasets under different neighborhood combinations.

Methods	Neighborhood			DBLP-3	Brain	Reddit	DBLP-10
	Q^1	Q^2	Q^3				
GraphANA -fixed	✓			86.96	94.92	47.60	79.18
		✓		86.00	94.22	47.81	78.07
			✓	86.41	93.89	47.19	77.12
	✓	✓		86.13	94.82	47.18	79.16
	✓		✓	85.67	94.83	47.57	79.15
		✓	✓	86.45	94.32	48.09	77.45
	✓	✓	✓	86.27	93.65	47.32	79.51
GraphANA	✓	✓	✓	**87.86**	**95.47**	**49.28**	**79.78**

From Table 3, it can be observed that different neighborhood combinations yield significant differences in the final node classification results. Additionally, more distant neighbors do not necessarily lead to a performance decline in graph neural networks. For instance, in the DBLP-3 dataset, the results from the 3-hop neighborhood outperform those from the 2-hop neighborhood. Besides, incorporating multiple neighborhoods does not always yield better results than using a single neighborhood. In some cases, the resulting performance is close to the average obtained by using different neighborhoods separately. This is partly due to the fact that each node has its own optimal neighborhood selection. While introducing more neighborhoods may improve classification accuracy for some nodes, it inevitably introduces additional noise for others. Our method, through neighborhood gating, can adaptively learn the optimal neighborhood for each node within the multi-hop range, maximizing retention of beneficial information while filtering out irrelevant information, thereby enhancing node classification performance.

5.5   Analysis of Feature Correlations

To validate whether the feature-level Transformer effectively captures temporal correlations among features, we visualize the learned self-attention weights as heatmaps. We first compute the Pearson Correlation Coefficients between the features of target nodes and their neighborhoods. We present results as heatmaps, where yellow indicates stronger positive correlations and blue indicates stronger negative correlations. Next, we retrieve the self-attention weights in the feature-level Transformer and present them as heatmaps in the same way.

Figure 5(a) illustrates a case in the Reddit dataset, and Fig. 5(b) depicts a case in the DBLP3 dataset. For each case, the first heatmap displays the correlation coefficients between features, while the second heatmap shows the learned attention weights. The numerical labels on the axes correspond to different features.

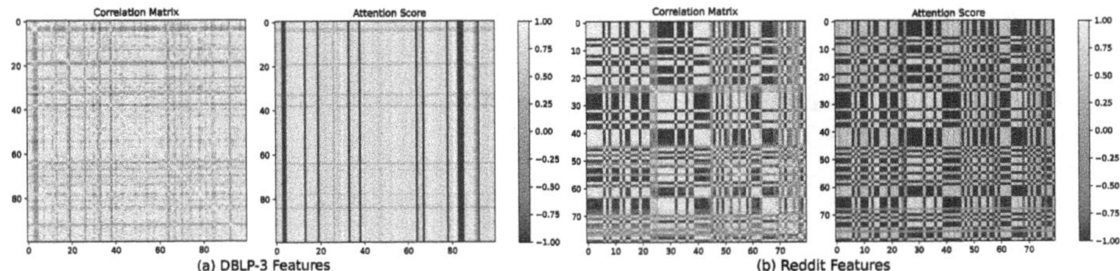

Fig. 5. Heatmaps of the Pearson Correlation Coefficients and the learned self-attention weights. (Color figure onine)

It can be observed that the attention weights learned by GraphANA are similar to the original correlation coefficients between features. This indicates that the feature-level Transformer can learn interpretable attention weights, facilitating the aggregation of dynamically changing information across features and neighborhoods. Furthermore, the case visualization from the Reddit dataset demonstrates that even among 4-hop neighbors (the coordinate range is [80,100]), certain features exhibit high correlations with the target node's features (the coordinate range is [0,20]). Compared to traditional graph neural networks, the feature-wise Transformer can directly aggregate information from distant neighbors, thereby avoiding the information loss that typically occurs due to multi-layer message passing.

5.6 Sensitivity Analysis

To assess the robustness of our model, we evaluate the impact of specific parameters and factors on the performance of our GraphANA, and the results are illustrated by Fig. 6.

The Dimension of Hidden Embedding. Our experiments reveal that overly large dimensions will lead to overfitting, particularly when the dimension of the node's initial feature is small (e.g., Brain and Reddit), where test accuracy drops by 5–10% compared to optimal settings. Small dimensions lead to slight underfitting, with test accuracy dropping by 2–3% compared to optimal settings. This suggests insufficient representation power for complex graph structures.

Dropout Rate. Our findings show that the low dropout rate severely facilitates performance (10–12% improvement), particularly on sparser graphs where

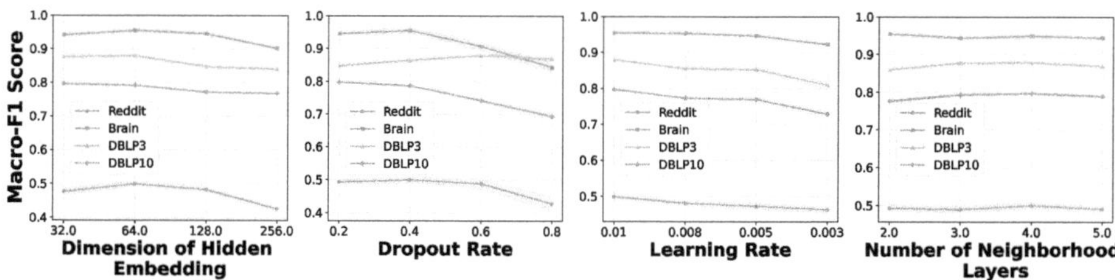

Fig. 6. The effect of parameters on the Macro-F1 score is examined concerning four critical hyperparameters.

message-passing relies heavily on retained node features. Interestingly, DBLP3, with its sparse connections, tolerates slightly higher dropout (0.6) before performance declines, likely due to the richness of initial node features. This suggests that dropout should be carefully tuned based on graph density, as overly aggressive regularization harms information propagation.

Initial Learning Rate. On the DBLP-3 and DBLP-10 datasets, a moderate learning rate facilitates the convergence of the model to better solutions. In contrast, on the Reddit and Brain datasets, the model performance remains relatively stable even with lower learning rates, indicating a lower sensitivity to this hyperparameter. Overall, the choice of learning rate has a significant impact on model performance, and different datasets exhibit varying degrees of sensitivity to it.

The Number of Neighborhood Layers. As the number of layers increases, thanks to the neighborhood selection module, GraphANA does not suffer from performance degradation due to over-smoothing. This demonstrates the critical importance of our proposed neighborhood selection module, which selectively aggregates the most temporally relevant information from multi-hop neighborhoods while filtering out irrelevant noise.

These findings emphasize the need for careful hyperparameter tuning that accounts for both task requirements and intrinsic graph properties.

6   Conclusion

In this paper, we propose an effective discrete-time dynamic graph learning method termed GraphANA. For the first time, we transfer the prevailing concept of "one node one receptive field" to the dynamic graph. Building on this, GraphANA can model temporal correlations between neighborhood features and eliminate irrelevant information, rather than aggregating all neighbors based solely on proximity. Extensive experiments demonstrate that GraphANA achieves competitive performance on downstream tasks.

Acknowledgment. This work is supported by the National Key R&D Program of China (2023YFC3304900).

Disclosure of Interest. The authors have no competing interests to declare that are relevant to the content of this article.

References

1. Chu, X., Xue, H., Tan, Z., Wang, B., Mo, T., Li, W.: Graphsos: graph sampling and order selection to help LLMs understand graphs better. ArXiv (2025)
2. Chu, X., et al.: Adaptive spatiotemporal augmentation for improving dynamic graph learning. In: ICASSP, pp. 1–5 (2025)
3. Feng, J., Chen, Y., Li, F., Sarkar, A., Zhang, M.: How powerful are k-hop message passing graph neural networks. In: Proceedings of the 36th International Conference on Neural Information Processing Systems (2022)
4. Grover, A., Leskovec, J.: node2vec: scalable feature learning for networks. In: KDD, pp. 855–864 (2016)
5. Li, J., Wu, R., Jin, X., Ma, B., Chen, L., Zheng, Z.: State space models on temporal graphs: a first-principles study. In: Proceedings of the 38th International Conference on Neural Information Processing Systems (2024)
6. Li, J., et al.: Scaling up dynamic graph representation learning via spiking neural networks. In: AAAI, pp. 8588–8596 (2023)
7. Liu, Y., et al.: itransformer: inverted transformers are effective for time series forecasting. ArXiv (2023)
8. Lu, Y., Wang, X., Shi, C., Yu, P.S., Ye, Y.: Temporal network embedding with micro- and macro-dynamics. In: CIKM, pp. 469–478 (2019)
9. Narayan, A., O'N Roe, P.H.: Learning graph dynamics using deep neural networks. IFAC-PapersOnLine **51**(2), 433–438 (2018)
10. Panagopoulos, G., Nikolentzos, G., Vazirgiannis, M.: Transfer graph neural networks for pandemic forecasting. In: Proceedings of the AAAI Conference on Artificial Intelligence, vol. 35, no. 6, pp. 4838–4845 (2021)
11. Pareja, A., et al.: EvolveGCN: evolving graph convolutional networks for dynamic graphs. In: Proceedings of the Thirty-Fourth AAAI Conference on Artificial Intelligence (2020)
12. Perozzi, B., Al-Rfou, R., Skiena, S.: Deepwalk: online learning of social representations. In: KDD, pp. 701–710 (2014)
13. Sankar, A., Wu, Y., Gou, L., Zhang, W., Yang, H.: Dysat: deep neural representation learning on dynamic graphs via self-attention networks. In: WSDM, pp. 519–527 (2020)
14. Singer, U., Guy, I., Radinsky, K.: Node embedding over temporal graphs. In: IJCAI, pp. 4605–4612 (2019)
15. Wang, K., et al.: The heterophilic snowflake hypothesis: training and empowering GNNs for heterophilic graphs. In: KDD, pp. 3164–3175 (2024)
16. Xu, D., Cheng, W., Luo, D., Liu, X., Zhang, X.: Spatio-temporal attentive RNN for node classification in temporal attributed graphs. In: IJCAI, pp. 3947–3953 (2019)
17. Yang, M., Zhou, M., Kalander, M., Huang, Z., King, I.: Discrete-time temporal network embedding via implicit hierarchical learning in hyperbolic space. In: KDD, pp. 1975–1985 (2021)

18. You, J., Du, T., Leskovec, J.: Roland: graph learning framework for dynamic graphs. In: Proceedings of the 28th ACM SIGKDD Conference on Knowledge Discovery and Data Mining, pp. 2358–2366 (2022)
19. Zheng, Y., Yi, L., Wei, Z.: A survey of dynamic graph neural networks. In: Frontiers of Computer Science, vol. 19, p. 196323 (2024)
20. Zhou, L., Yang, Y., Ren, X., Wu, F., Zhuang, Y.: Dynamic network embedding by modeling triadic closure process. In: AAAI (2018)
21. Zhu, Y., et al.: Wingnn: dynamic graph neural networks with random gradient aggregation window. In: KDD, pp. 3650–3662 (2023)
22. Zuo, Y., Liu, G., Lin, H., Guo, J., Hu, X., Wu, J.: Embedding temporal network via neighborhood formation. In: KDD, pp. 2857–2866 (2018)

Homophilic and Heterophilic-Aware Multi-view Graph Clustering

Yeqin Zhou[1,2] and Lirong Qiu[1(✉)]

[1] Beijing University of Posts and Telecommunications, Beijing, China
`{zhouyeqin,qiulirong}@bupt.edu.cn`
[2] Shandong Police College, Jinan, China

Abstract. Learning on homophilic and heterophilic graphs has gradually gained attention. Compared to homophilic graphs, where neighborhood information is reliable, nodes in heterophilic graphs may face neighborhoods that are dissimilar to their own, containing inappropriate information. However, existing multi-view graph clustering methods address this issue in heterophilic graphs solely by masking or weighting structural information. Such solutions make it harder for these nodes to access valid information–their overall information content becomes impoverished. While reducing the aggregation of inappropriate information, it also discards much suitable information. To address this challenge, we propose a Cluster-Guided Contrastive Encoding method called H2MGC, which unifies the modeling of homophilic and heterophilic graphs and aims to enhance data by generating positive samples for nodes using prior knowledge. In particular, building upon contrastive learning with adaptive adjustment of topological weights, we introduce cluster center-guided data augmentation. By leveraging information entropy to measure the latent relationship between cluster centers and nodes, we construct positive samples for each node, thereby achieving unified data augmentation for both homophilic and heterophilic graphs. Extensive experiments demonstrate that H2MGC achieves state-of-the-art performance across diverse homophilic regimes while exhibiting stable convergence.

Keywords: Graph clustering · Homophilic and heterophilic graph · Multi-View graph

1 Introduction

Graph-structured data naturally arises in diverse real-world scenarios, such as social networks [2], citation networks [24], and biological systems [16]. Mining and learning from such graph data enables researchers to efficiently and accurately fulfill practical application requirements, such as prediction, reasoning, and detection. Some graph data structures often manifest in diverse forms. For instance, within citation networks, collaborative relationships between authors, the fields of different authors, and citation relationships between papers can all

H. Jung et al. (Eds.): DASFAA 2026, LNCS 16536, pp. 165–181, 2026.
https://doi.org/10.1007/978-981-92-0366-6_11

be constructed into distinct graph structures. These structures represent multiple views of the data. By analyzing and integrating these various views [14,25], one can more comprehensively extract the rich, latent information within graph data [12,27]. This has led to the emergence of multi-view graph clustering.

Multi-view graph clustering aims to partition nodes into meaningful groups by leveraging structural and semantic information from different perspectives, making it a fundamental task in unsupervised learning [3,34]. Although existing multi-view graph clustering methods have achieved success on homophilic graph datasets, insufficient attention has been paid to heterophilic graphs.

Compared to homophilic graphs, where neighborhood information is reliable, nodes in heterophilic graphs may face neighborhoods that are dissimilar to their own, containing inappropriate information. However, existing multi-view graph clustering methods address this issue in heterophilic graphs solely by masking or weighting structural information. Such solutions make it harder for these nodes to access valid information–their overall information content becomes impoverished. While reducing the aggregation of inappropriate information, it also discards much suitable information. Moreover, existing methods often neglect the potential of cluster structure itself as a rich source of supervision signals. Cluster centers represent high-confidence prototypes of class distributions, yet most approaches fail to explicitly leverage them for guiding representation learning and enriching node representations.

To address these challenges, we propose Cluster-Guided Contrastive Encoding (H2MGC), a comprehensive framework that unifies the modeling of homophilic and heterophilic graphs by enhancing data through generating positive samples using prior knowledge. Rather than simply removing noisy structural information, the key innovation lies in three synergistic components: (a) We introduce contrastive learning with denoising guidance that estimates graph-specific homophilic weights and adaptively balances feature similarity with original topology, enabling robust learning across both homophilic and heterophilic graphs. (b) We consider cluster center-guided data augmentation that constructs adaptive positive samples based on soft assignment uncertainty measured by information entropy. This mechanism progressively aligns node embeddings with cluster prototypes while preserving individual characteristics, providing direct supervision signals that complement structure-aware learning and achieve unified augmentation for both graph types. (c) We employ reconstruction mechanisms to regularize the learned embeddings, ensuring they preserve both feature and structural information. Experiments on multiple real-world homophilic and heterophilic graph datasets demonstrate the effectiveness of H2MGC. The main contributions can be summarized as follows:

(1) **Problem.** We point out that existing multi-view graph clustering methods merely mask and weight the topological structure when processing heterophilic graph data, overlooking that such operations make it difficult for nodes in heterophilic graphs to obtain sufficient information.

(2) **Algorithm.** We propose a novel framework that addresses the heterophilic challenge in multi-view graph clustering through adaptive homophilic weight estimation and structure-aware contrastive learning.

(3) **Experiment.** Extensive experiments on multiple datasets demonstrate that H2MGC achieves state-of-the-art performance on both homophilic and heterophilic graphs.

2   Related Work

Deep graph clustering is an important task in the field of graph neural networks, which aims to divide the nodes in a graph into different clusters. In this process, deep learning techniques are needed to learn the corresponding embedding vector for each node. Graph clustering can help real-world application scenarios such as biomedical analysis, interest recommendation, smart cities, etc. In the field of graph clustering, researchers are focusing on different areas. For example, some work focuses on graph clustering with an unknown number of clusters [20], some work focuses on fuzzy graph clustering [23], some work focuses on temporal graph clustering [19], and some work uses graph structures to solve clustering problems [8]. Among them, multi-view graph clustering is gradually gaining attention.

Multi-view graph clustering (MVGC), that is, performing node clustering on graph-structured data with multiple views. In real-world scenarios, graph structures are not singular; they can be composed in a variety of ways, that is, from multiple perspectives. Consequently, researchers are increasingly interested in this type of complex graph data that better reflects real-world scenarios, and have proposed numerous research methods.

Such methods typically employ graph encoders to learn latent node representations that capture underlying semantic information, thereby facilitating the division of nodes into distinct clusters. A variety of GNN-based MVGC approaches have been proposed. Pioneering this integration, O2MAC [9] utilized a graph encoder-decoder architecture to map multi-view graphs into a shared low-dimensional space. Subsequent work, such as [7], introduced a two-pathway encoder to acquire view-consistent information. Concurrently, other methods have focused on self-supervised and contrastive learning paradigms. For instance, [10] developed a self-supervised framework to learn both node- and graph-level representations, while [21] leveraged contrastive learning to identify common geometric and semantic structures, enhancing the formation of a consensus graph. More recently, [33] explored cluster structures by training a graph convolutional encoder to learn a self-expression coefficient matrix.

However, a significant limitation shared by these approaches is their high sensitivity to graph homophily. Consequently, their performance tends to degrade substantially when applied to graphs characterized by strong heterophily.

Homophilic Graph and Heterophilic Graph. The distinction between homophilic and heterophilic graphs lies in their connectivity patterns. A homophilic graph is a graph where connected nodes are likely to have the same class label or similar attributes. The connections are predominantly intra-class. A heterophilic graph is one where connected nodes are likely to have different class labels or dissimilar attributes. The connections are predominantly inter-class.

The vast majority of existing MVGC methods are built upon standard GNN architectures, which implicitly operate under the homophily assumption. There is a significant research gap in the development of MVGC frameworks specifically designed to handle heterophilic graphs. The performance of current methods often degrades substantially in such scenarios, as the feature aggregation process can corrupt a node's representation by mixing it with features from dissimilar neighbors. To date, the challenge of performing effective multi-view clustering on heterophilic graphs remains a largely underexplored area.

3   Method

3.1   Problem Definition

One multi-view graph can be regarded as graph data composed of the same set of nodes and multiple sets of topological structures. Let $\mathcal{G} = \{\mathcal{G}^1, \mathcal{G}^2, \cdots, \mathcal{G}^m\}$ denote the view set, where each view $\mathcal{G}^m$ is represented by an adjacency matrix $\mathbf{A}^m \in \{0,1\}^{N \times N}$ and a shared feature matrix $\mathbf{X} \in \mathbb{R}^{N \times d}$. The goal of unsupervised clustering is to partition nodes into K clusters without labels. We seek view-specific embeddings $\mathbf{h}^m \in \mathbb{R}^{N \times d}$ and a consensus embedding $\mathbf{H} \in \mathbb{R}^{N \times d}$ such that the induced clustering structure is compact within clusters and well separated across clusters.

3.2   Overall Framework

Figure 1 introduces the overall framework of H2MGC. Such a framework can be divided into three parts: (1) heterophilic-aware contrastive learning part, which estimates homophilic weights from consensus embeddings to construct joint adjacency matrices that balance feature similarity and original topology; (2) Cluster-guided augmentation part, which leverages cluster centers as supervision signals to enhance representations; (3) Reconstruction regularization part, which utilizes consensus embeddings to refine adjacency matrices across views.

3.3   Heterophilic-Aware Contrastive Learning

Traditional GNN-based methods fail in heterophilic settings due to the neighbor similarity assumption. To solve this problem, we consider contrastive learning with denoising guidance that adaptively adjusts graph structures based on estimated homophilic degrees, dynamically balancing feature similarity and original topology. The m-th view's similarity matrix between node embeddings can be calculated as $\mathbf{S}^m = \cos(\mathbf{Z}^m, \mathbf{Z}^{m\top})$.

Furthermore, we introduce an adaptive encoding strategy that estimates the graph's homophilic degree: higher values favor similarity information while lower values emphasize topology. Since ground-truth labels are unavailable, we infer

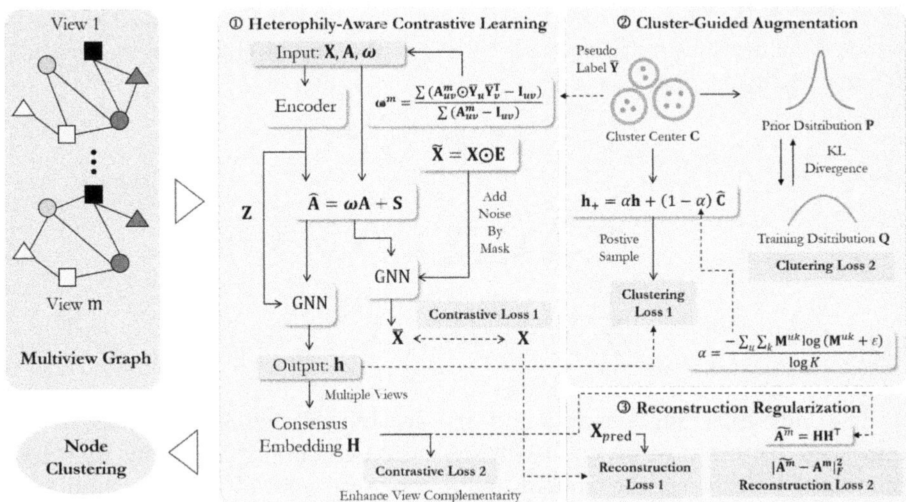

Fig. 1. The illustration of H2MGC graph clustering framework, which addresses multi-view graph clustering through three interconnected components following a principled learning pipeline.

pseudo-labels [1] from the consensus embedding $\mathbf{H}$ and combine them with adjacency information to approximate the homophilic weight ω^m as

$$\omega^m = \frac{\sum_{u,v} \left(\mathbf{A}^m_{u,v} \odot \overline{\mathbf{Y}}_u \overline{\mathbf{Y}}^\top_v - \mathbf{I}_{u,v}\right)}{\sum_{u,v} \left(\mathbf{A}^m_{u,v} - \mathbf{I}_{u,v}\right)}, \tag{1}$$

where ω^m reflects the homophilic degree from the previous iteration, $\odot$ denotes the Hadamard product, and $\overline{\mathbf{Y}} \in \{0,1\}^{N \times K}$ represents one-hot encoded pseudo-labels from clustering $\mathbf{H}$, which preserves cross-view consistency. When $\omega \approx 1$ (homophilic), we emphasize the original structure; when $\omega \approx 0$ (heterophilic), we rely more on the learned similarity matrix to avoid aggregating noisy information from dissimilar neighbors. The joint adjacency matrix is formulated as

$$\widehat{\mathbf{A}}^m = \mathbf{S}^m + \omega^m \mathbf{A}^m, \tag{2}$$

where $\widehat{\mathbf{A}}^m$ integrates feature similarity ($\mathbf{S}^m$), original structure ($\mathbf{A}^m$), and adaptive adjustment via ω^m from Eq. 1.

Building on the adaptive adjacency, we denoise node features with a parameter-free GCN inspired by DAE [30] and GraphMAE [11]. Original feature matrix $\mathbf{X} \in \mathbb{R}^{n \times d}$ is randomly masked by $\mathbf{E} \sim \mathrm{Bern}(1 - p_e)$, giving corrupted input $\widetilde{\mathbf{X}} = \mathbf{X} \odot \mathbf{E}$. GCN can be used to reconstruct clean embeddings that preserve both homophilic and heterophilic signals, which recovers features via $\overline{\mathbf{X}}^m = \mathrm{GCN}(\widehat{\mathbf{A}}^m, \widetilde{\mathbf{X}})$, that is,

$$\mathbf{H}^{(l+1)} = \sigma(\mathbf{D}^{-\frac{1}{2}} \widehat{\mathbf{A}} \mathbf{D}^{-\frac{1}{2}} \mathbf{H}^{(l)}) + \widetilde{\mathbf{X}}, \tag{3}$$

where $\mathbf{H}^{(0)} = \widetilde{\mathbf{X}}$, $\mathbf{H}^{(l)}$ denotes the l-th layer output, and $\mathbf{D}$ is the degree matrix. Residual connections mitigate gradient issues. We enforce that recovered features $\overline{\mathbf{X}}^m$ match the original $\mathbf{X}$ via

$$
\begin{aligned}
\mathcal{L}_{\mathrm{Con}_1} &= \sum_m \frac{1}{N^2} \sum \left(\overline{\mathbf{X}}^m - \mathbf{X} \right)^2 \\
&= \sum_m \frac{1}{N^2} \left(\sum_u \sum_v \mathbb{I}_{uv}^1 \left(\overline{\mathbf{X}}_{uv}^m - \widetilde{\mathbf{X}}_{uv} \right)^2 \right. \\
&\quad \left. + \sum_u \sum_v \mathbb{I}_{uv}^0 \left(\left(\overline{\mathbf{X}}_{uv}^m - \widetilde{\mathbf{X}}_{uv} \right) - \mathbf{X}_{uv} \right)^2 \right).
\end{aligned}
\tag{4}
$$

The indicator functions $\mathbb{I}_{uv}^0$ and $\mathbb{I}_{uv}^1$ are binary switches for masked and unmasked features, respectively (detailed derivation in [26]). The first term minimizes discrepancy for masked nodes, while the second preserves unmasked information. This establishes an adaptive mechanism where unpreserved features serve as positive samples and preserved feature variations as negative samples. To enable cross-view fusion, we apply parameter-free GCNs for dimensionality reduction, yielding compact embeddings $\mathbf{h}^m = \mathrm{GCN}(\widehat{\mathbf{A}}^m, \mathbf{Z}^m)$ aligned with node embeddings $\mathbf{Z}^m$.

To enhance discriminative power, we refine the joint embedding $\mathbf{H}$ with contrastive learning in the fine-tuning phase ($t \geq T_1$). We adopt a variant of NT-Xent loss [6], i.e., for node u, its N_{knn} nearest neighbors serve as positive samples, while cluster centers from different classes (C_k, $k \neq c_u$) serve as negative samples after cluster centers stabilize [3]. Thus their contrastive loss can be calculated as

$$
\mathcal{L}_{\mathrm{Con}_2} = -\mathcal{A}\,\mathbb{I}_{[t \geq T_1]} \sum_{u=1}^{N} \log \left(\frac{\sum_{n=1}^{N_{\mathrm{knn}}} e^{S(h_u, p_{u,n})}}{\sum_{n=1}^{N_{\mathrm{knn}}} e^{S(h_u, p_{u,n})} + \sum_{k=1}^{K} \mathbb{I}_{[k \neq c_u]} e^{S(h_u, C_k)}} \right),
\tag{5}
$$

where $\mathbf{h}_u$ is the feature of node u in $\mathbf{H}$, $\mathbf{p}_{u,n}$ is its n-th nearest neighbor from $P(\mathbf{h}_u)$ (the set of N_{knn} nearest neighbors, fixed at 20), $S(\cdot, \cdot)$ denotes cosine similarity, and $\mathcal{A} = 1/(N(K-1))$ is the loss scaling factor for normalization.

By integrating both denoising and contrastive learning mechanisms, we formulate the comprehensive contrastive loss as $\mathcal{L}_{\mathrm{Con}} = \mathcal{L}_{\mathrm{Con}_1} + \mathcal{L}_{\mathrm{Con}_2}$, where the first component focuses on feature reconstruction consistency and the second component emphasizes cluster-aware embedding learning.

3.4 Cluster-Guided Data Augmentation

Having established adaptive graph structures that handle heterophilic graphs, we now leverage cluster-guided data augmentation to enhance representation quality through prototype-driven learning. By constructing adaptive positive samples based on soft cluster assignment uncertainty, we enable each view's embedding to progressively align with cluster prototypes while preserving node-specific characteristics. This cluster-guided mechanism provides direct supervision signals that complement the structure-aware learning from the previous stage. For each view $m \in \{1, \ldots, V\}$, let $\mathbf{h} \in \mathbb{R}^{N \times d}$ denote node embeddings, and $\mathbf{C} \in \mathbb{R}^{K \times d}$ represent the cluster center embeddings. We calculate similarities using ℓ_2-normalized embeddings and centers as $\mathbf{M} = \frac{\mathbf{h}}{\|\mathbf{h}\|_2} \cdot \frac{\mathbf{C}^\top}{\|\mathbf{C}\|_2}$.

The resulting $\mathbf{M} \in \mathbb{R}^{N \times K}$ serves as the soft assignment weight matrix. The cluster center embedding based on soft assignment weights can be formulated as $\widehat{\mathbf{C}} = \mathrm{softmax}\left(\frac{\mathbf{M}}{\tau}\right) \cdot \mathbf{C}$, where τ is the temperature parameter, and the softmax function is applied to compute the weights. We calculate and normalize the assignment entropy as the adaptive interpolation coefficient, that is,

$$\alpha = \frac{-\sum_{u=1}^{N} \sum_{k=1}^{K} \mathbf{M}^{u,k} \log(\mathbf{M}^{u,k} + \varepsilon)}{\log K}, \tag{6}$$

where $\mathbf{M}^{u,k}$ is the (u,k)-th element of the soft assignment matrix, and ε is a very small value to avoid negative infinity in the logarithm. We then clamp the adaptive interpolation coefficient α to the range $[0.1, 0.9]$, setting values below 0.1 to 0.1 and above 0.9 to 0.9. Higher entropy yields a larger α (uncertain boundary nodes retain more of their current embedding), whereas lower entropy yields a smaller α (confident nodes are more strongly guided by cluster information). The positive sample can be obtained as $\mathbf{h}_+ = \alpha \mathbf{h} + (1 - \alpha)\widehat{\mathbf{C}}$.

Here $\mathbf{h}_+ \in \mathbb{R}^{N \times d}$ is generated by combining the current view embedding and cluster center through the adaptive interpolation value. We then encourage each view's embedding to match its adaptive positive using a view-weighted mean-squared reconstruction objective, that is,

$$\mathcal{L}_{\mathrm{Clu}_1} = \sum_{m=1}^{V} \omega_h^m \left\|\mathbf{h}^m - \mathbf{h}_+^m\right\|_2^2, \tag{7}$$

where ω_h^m is the contribution weight of view m, which is calculated from Eq. 15. This loss is evaluated separately for each view and then pooled, so the multi-view signal is weighted on-the-fly by the node-wise uncertainty delivered from the clusters.

To keep the partitions consistent across views, we also force every view-specific embedding to respect the global attribute distribution; in this way the clustering structure remains coherent without any additional trainable parameters. Assuming the autoencoder has a total of L layers, where ℓ represents a specific layer number, the updates of the encoder and decoder can be uniformly represented as

$$\mathbf{H}_t^{(\ell)} = \phi\left(\mathbf{W}_t^{(\ell)} \mathbf{H}_t^{(\ell-1)} + \mathbf{b}_t^{(\ell)}\right), \quad t \in \{e, d\}, \tag{8}$$

where ϕ is the activation function of the fully connected layer (e.g., ReLU or Sigmoid), $\mathbf{W}_t^{(\ell)}$ and $\mathbf{b}_t^{(\ell)}$ denote the weight matrix and bias of the ℓ-th layer in the encoder ($t = e$) or decoder ($t = d$), respectively. Additionally, we use $H_e^{(0)}$ to represent the original data $\mathbf{X}$, and $\mathbf{H}_d^{(0)} = \mathbf{H}_e^{(L)}$ to represent the encoded features passed to the decoder.

During the initial embedding learning phase, we calculate distribution consistency losses to ensure each view's clustering structure aligns with the prior knowledge from attribute features. The cluster centers can be initialized by the classic K-means algorithm. Furthermore, we construct a prior distribution $\mathbf{P}$ from the original attribute features processed through the autoencoder, and a training distribution $\mathbf{Q}^m$ for each view's current embeddings. Both distributions are calculated based on the Student's t-distribution:

$$p_{uk} = \frac{(1 + \|\mathbf{Z}_u^{AE} - \mathbf{C}_k^{AE}\|^2/\alpha)^{-\frac{\alpha+1}{2}}}{\sum_{k'=1}^{K}(1 + \|\mathbf{Z}_u^{AE} - \mathbf{C}_{k'}^{AE}\|^2/\alpha)^{-\frac{\alpha+1}{2}}}, \tag{9}$$

$$q_{uk}^m = \frac{(1 + \|\mathbf{Z}_u^m - \mathbf{C}_k^m\|^2/\alpha)^{-\frac{\alpha+1}{2}}}{\sum_{k'=1}^{K}(1 + \|\mathbf{Z}_u^m - \mathbf{C}_{k'}^m\|^2/\alpha)^{-\frac{\alpha+1}{2}}}, \tag{10}$$

where $\mathbf{Z}_u^{AE}$ denotes the u-th node embedding from the attribute feature autoencoder, $\mathbf{C}_k^{AE}$ is its corresponding k-th cluster center. Similarly, $\mathbf{Z}_u^m$ is the u-th node embedding of the m-th view, and $\mathbf{C}_k^m$ is its k-th cluster center. The parameter α is the degrees of freedom in the Student's t-distribution; we lock it to 1 for speed. Here, $\mathbf{P} = \{p_{uk}\}$ serves as the prior distribution derived from pure attribute features, providing a stable reference for clustering structure, while $\mathbf{Q}^m = \{q_{uk}^m\}$ represents the training distribution of the m-th view's embeddings during optimization. To enhance the discriminability of the prior distribution, we apply a sharpening operation as

$$\bar{p}_{uk} = \frac{p_{uk}^2/\sum_{u=1}^{N} p_{uk}}{\sum_{k'=1}^{K}(p_{uk'}^2/\sum_{u=1}^{N} p_{uk'})}. \tag{11}$$

To ensure that each view's training distribution remains consistent with the prior distribution, we introduce the distribution consistency loss as

$$\mathcal{L}_{\text{Clu}_2} = \mathbb{I}_{[t<T_1]} \sum_{m=1}^{V} \omega_h^m \cdot \text{KL}(\bar{\mathbf{P}}\|\mathbf{Q}^m) = \mathbb{I}_{[t<T_1]} \sum_{m=1}^{V} \omega_h^m \sum_u \sum_k \bar{p}_{uk} \log \frac{\bar{p}_{uk}}{q_{uk}^m}, \tag{12}$$

where V is the number of views, ω_h^m is the weight of the m-th view calculated from Eq. 15, and $\text{KL}(\cdot\|\cdot)$ is the KL divergence that measures the discrepancy between the sharpened prior distribution $\bar{\mathbf{P}}$ and each view's training distribution $\mathbf{Q}^m$. The indicator $\mathbb{I}[\cdot]$ equals 1 if the condition in brackets holds and 0 otherwise. Here, t is the current training epoch, and T_1 is a predefined hyperparameter; thus, this loss term is activated only during the initial phase when $t < T_1$, encouraging each view's embeddings to maintain consistency with the clustering structure derived from pure attribute features. Furthermore, the cluster loss can be summarized as $\mathcal{L}_{\text{Clu}} = \mathcal{L}_{\text{Clu}_1} + \mathcal{L}_{\text{Clu}_2}$.

3.5   Feature and Structure Reconstruction

In order to supplement heterophilic-aware contrastive learning and cluster-guided enhancement, we adopted a reconstruction mechanism to ensure that the learned embeddings retain feature and structural information. This reconstruction process acts as regularization to maintain data fidelity while the model learns discriminative representations. An autoencoder is introduced to extract and refine node features as

$$\mathbf{X}^m_{\text{pred}} = g^m(\sigma(\mathbf{Z}^m; \varphi^m)), \quad \mathbf{Z}^m = f^m(\sigma(\mathbf{X}; \theta^m)), \tag{13}$$

where $\mathbf{Z}^m$ is the learned node embeddings of the m-th view, θ^m and φ^m are the unshared learnable parameters of the encoder and decoder respectively, $\mathbf{X}$ is the original node features, and $\sigma(\cdot)$ is the activation function. To ensure the quality of feature reconstruction, the feature reconstruction loss can be defined as

$$\mathcal{L}_{\text{Rec}_1} = \sum_m l(g^m(\mathbf{X}^m_{\text{pred}}; \mathbf{X})). \tag{14}$$

For structural reconstruction, we construct a consensus embedding by integrating node-guided joint embeddings $\mathbf{h}^m$ from all views [13,32], leveraging the complementary information across different graph perspectives. Since the informational value of each view differs, we introduce a weighting mechanism that evaluates the quality of each view and adjusts its contribution accordingly. Essentially, a view whose embedding exhibits higher similarity to the current consensus embedding $\mathbf{H}$ is regarded as more informative and is assigned a larger weight, whereas views with lower similarity receive smaller weights. The consensus embedding is derived as $\mathbf{H} = \sum_{m=1}^{V} \omega_h^m \mathbf{h}^m$, where ω_h^m denotes the weight assigned to the node-guided joint embedding of the m-th view, determined as follows. In our implementation, these weights are initially set to 1×10^{-12}. Furthermore, the joint embedding contribution ratio for the m-th view can be calculated as

$$\omega_h^m = \left(\frac{eva^m}{\max(eva^1, eva^2, \ldots, eva^V)} \right)^\rho, \tag{15}$$

Such weight ω_h^m is obtained by an evaluation function that quantifies the agreement between the consensus embedding $\mathbf{H}$ and the m-th view embedding $\mathbf{h}^m$, where $eva^m = evaluation(\mathbf{h}^m, \mathbf{H})$. The evaluation function can be implemented using similarity metrics such as cosine similarity or other similarity measures. The hyperparameter ρ modulates view weights, enhancing or diminishing each view's contribution. By default, ρ is set to 3. Finally, the consensus embedding $\mathbf{H}$ undergoes K-means clustering to ascertain the clustering outcomes.

To refine noisy adjacency matrices across views, we reconstruct them using the consensus embedding, thereby enhancing structural consistency. Based on the fused embeddings, a simplified yet effective method is used to reconstruct the adjacency matrices for each view as $\widetilde{\mathbf{A}}^m = \mathbf{H}\mathbf{H}^\top$, where $\widetilde{\mathbf{A}}^m \in \mathbb{R}^{N \times N}$ is the reconstructed adjacency matrix for the m-th view. Thus, the adjacency matrix reconstruction loss can be calculated as

$$\mathcal{L}_{\text{Rec}_2} = \sum_{m=1}^{V} \omega^m \cdot \|\widetilde{\mathbf{A}}^m - \mathbf{A}^m\|_F^2, \tag{16}$$

where $\| \cdot \|_F$ is the Frobenius norm, and ω^m is the homophilic weight from Eq. 1. The total reconstruction loss combines both feature and structure reconstruction: $\mathcal{L}_{\text{Rec}} = \mathcal{L}_{\text{Rec}_1} + \mathcal{L}_{\text{Rec}_2}$.

Finally, the total loss function can be formulated as

$$\mathcal{L} = \lambda_1 \mathcal{L}_{\text{Rec}} + \lambda_2 \mathcal{L}_{\text{Clu}} + \lambda_3 \mathcal{L}_{\text{Con}}. \tag{17}$$

4 Experiment

4.1 Datasets

This study employs five multi-view graph datasets. The homophilic graph datasets include: **ACM, DBLP**, and **IMDB** [9]. The heterophilic graph datasets comprise: **Chameleon** [28] and **Wisconsin** [22]. The homophilic graph datasets are as follows: ACM, derived from the ACM database, consists of two graphs: the co-authorship network and the co-citation network; DBLP, sourced from the DBLP database, encompasses three graphs: co-authorship, co-conference, and co-term; IMDB, originating from the IMDB dataset, represents a movie network that includes graphs of both co-actors and co-directors. The heterophilic graph datasets are described as follows: Chameleon is a subset of the Wikipedia network; Wisconsin is a webpage graph from WebKB.

4.2 Baselines

We consider multiple SOTA methods for comparison. **VGAE** [15] is a classical single-view graph clustering method. **O2MAC** [9] is a method that learns from both node features and graphs. **MvAGC** [17] and **MCGC** [21] are two approaches based on graph filters to learn a consensus graph for clustering. **DualGR** [18] utilizes soft labels and pseudo-labels to guide the graph refinement and fusion process for clustering. **CMGEC** [31] aims to encode more complementary information from multiple graphs and provide a more comprehensive data description. **BMGC** [29] addresses the view imbalance issue in multirelational graphs by proposing unsupervised dominant view mining and dual signals guided embedding learning. **SMVC** [4] is a structural deep multi-view clustering method that integrates top-level abstraction with underlying details to jointly optimize cluster assignments and feature embeddings. **VGMGC** [5] introduces a variational graph generator to infer a reliable consensus graph from multiple graphs, aiming to better utilize both view-specific and view-common information for multi-view graph clustering. **NGCE** [26] integrates homogeneous and heterogeneous information, and strengthens cross-view contrastive learning capabilities.

4.3 Experimental Settings

For all comparison methods, we keep their default hyperparameters. For our method, we set the embedding configurations as follows: ACM/DBLP/IMDB use hidden $= 512$, latent $= 128$, and order $= 2/2/3$; Chameleon uses hidden $= 256$, latent $= 64$, and order $= 2$; Wisconsin uses hidden $= 128$, latent $= 32$, and order $= 3$.

Our experiments include the main clustering task, parameter sensitivity analysis, ablation study, and convergence analysis. All experiments were conducted on an Intel i9-12900K CPU with 32 GB memory and an NVIDIA RTX 3090 GPU with 24 GB memory.

Table 1. The results of clustering on heterophilic graph datasets. The best and runner-up results are highlighted in bold and underlined, respectively.

Methods	Chameleon				Wisconsin			
	NMI	ARI	ACC	F1	NMI	ARI	ACC	F1
VGAE (2016)	15.1	12.4	35.4	29.6	10.5	13.7	49.3	34.1
DAEGC (2019)	9.1	5.6	32.2	31.2	10.6	3.4	32.7	28.3
AGE (2020)	8.6	7.6	32.4	32.4	9.3	1.3	31.1	31.1
O2MAC (2020)	12.3	8.9	33.5	28.6	11.0	8.9	40.0	27.9
MvAGC (2020)	10.8	3.3	29.2	24.3	8.1	4.8	47.7	20.6
AGCN (2021)	6.7	6.1	32.5	20.4	6.4	6.8	49.8	24.9
MCGC (2021)	9.5	5.9	30.0	19.1	12.9	5.9	51.8	30.7
DCRN (2022)	8.7	5.7	30.9	21.9	10.8	16.0	50.2	34.1
DualGR (2023)	19.5	16.0	41.1	37.7	34.1	28.8	56.4	47.1
BMGC (2024)	9.4	5.9	30.8	30.7	34.0	24.3	51.5	40.8
VGMGC (2025)	22.4	13.4	40.1	<u>39.5</u>	41.6	34.8	56.6	<u>49.6</u>
NGCE (2025)	<u>22.6</u>	**19.3**	<u>42.2</u>	38.4	<u>46.8</u>	<u>46.4</u>	<u>73.7</u>	47.2
H2MGC (Ours)	**22.9**	<u>17.1</u>	**46.3**	**44.9**	**56.2**	**60.5**	**80.1**	**54.4**

4.4   Node Clustering Performance

Tables 1 and 2 report the clustering results on several homophilic and heterophilic graph datasets, from which we draw the following observations:

(1) H2MGC outperforms other methods on heterophilic graph datasets, which are typically challenging for most models. Our model performs better than the baselines in most evaluation metrics. Especially on Wisconsin, our proposed method achieves significant performance improvements.

(2) In addition, results on homophilic graph datasets demonstrate that H2MGC is highly competitive with state-of-the-art models, showing significant improvements across all evaluative metrics. These datasets are commonly used to train comparative methods and are more universally meaningful.

Table 2. The results of clustering on homophilic datasets. The best and runner-up results are highlighted in bold and underlined, respectively.

Methods	IMDB				ACM				DBLP			
	NMI	ARI	ACC	F1	NMI	ARI	ACC	F1	NMI	ARI	ACC	F1
VGAE (2016)	0.4	0.9	44.2	35.7	49.1	54.4	82.2	82.3	69.3	74.1	88.6	87.4
DAEGC (2019)	0.6	1.0	37.9	35.3	63.8	70.1	89.0	88.9	30.8	33.4	66.5	65.6
AGE (2020)	4.4	4.6	43.2	42.2	73.5	78.9	92.4	92.4	45.0	47.6	75.3	74.6
O2MAC (2020)	0.3	0.2	40.2	35.4	69.2	73.9	90.4	90.5	72.9	77.8	90.7	90.1
MvAGC (2020)	1.3	-1.8	48.5	28.2	67.4	72.1	89.8	89.9	77.2	82.8	92.8	92.3
AGCN (2021)	0.3	1.4	54.5	31.1	68.4	74.2	90.6	90.6	39.7	42.5	73.3	72.8
MCGC (2021)	5.2	10.3	<u>58.3</u>	38.8	71.3	76.3	91.5	91.6	**83.0**	77.5	93.0	92.5
DCRN (2022)	0.2	0.1	53.4	25.5	71.6	77.6	91.9	91.9	49.0	53.6	79.7	79.3
DualGR (2023)	6.2	12.5	52.0	44.7	73.2	79.4	92.7	92.7	75.5	81.7	92.4	91.8
CMGEC (2023)	5.1	4.7	48.4	**51.0**	69.1	72.3	90.9	90.7	72.4	78.6	91.0	90.4
BMGC (2024)	5.5	4.9	44.1	40.7	78.4	83.3	94.1	94.2	<u>80.1</u>	**85.4**	**94.0**	**93.6**
SMVC (2024)	<u>8.0</u>	7.2	41.3	37.2	72.4	78.0	92.3	92.0	76.1	81.6	92.4	92.0
VGMGC (2025)	0.8	3.2	52.6	32.8	76.3	81.9	93.6	93.6	78.3	83.7	93.2	92.7
NGCE (2025)	5.6	<u>12.7</u>	54.6	43.4	<u>80.5</u>	<u>85.0</u>	<u>94.7</u>	<u>94.8</u>	79.1	84.0	93.3	92.8
H2MGC (Ours)	**9.2**	**17.0**	**60.3**	<u>45.2</u>	**80.9**	**85.5**	**94.9**	**95.0**	79.9	<u>85.1</u>	<u>93.8</u>	<u>93.4</u>

(3) Note that compared to the ACM and DBLP datasets, the IMDB dataset presents greater challenges, as its nodes are more difficult to clearly partition, as evidenced by the low NMI and ARI scores achieved by many existing methods. Under these conditions, our method significantly improves NMI and ARI performance, demonstrating that it not only unifies the modeling of homophilic and heterophilic graphs but also delves deeper into the latent structures within graph data to extract more comprehensive information.

(4) Overall, our method H2MGC achieves advantages in most metrics, and has a clear advantage in heterophilic graphs, which have received little attention from researchers. Furthermore, experimental results show that methods focusing on multi-view graph clustering are more competitive than general graph clustering methods, demonstrating the value of focusing on multi-view graph data.

4.5 Parameter Sensitivity Study

In this part, we conduct two different parameter sensitivity experiments: loss weights $\lambda_1, \lambda_2, \lambda_3$ and graph filter order.

(1) Figures 2, 3, and 4 report the sensitivity analysis of the loss weights λ_1, λ_2, and λ_3 on homophilic datasets ACM and DBLP, and heterophilic dataset Wisconsin. It can be seen from the figure that, as the values of λ_1, λ_2, and λ_3 vary, the four metrics ACC, NMI, ARI, and F1 remain relatively stable. This implies

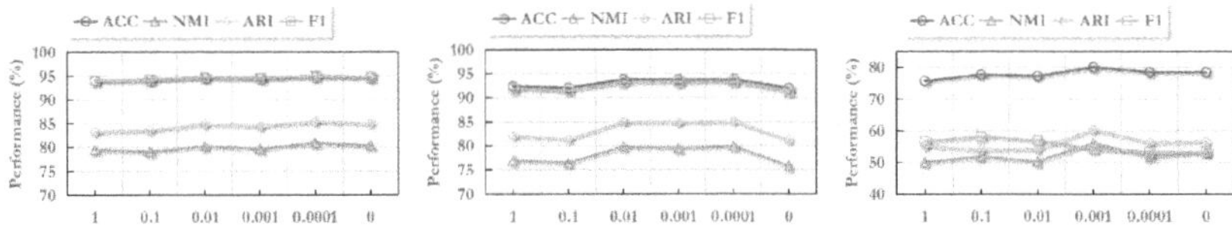

Fig. 2. Parameter analysis of λ_1 on ACM, DBLP, and Wisconsin datasets.

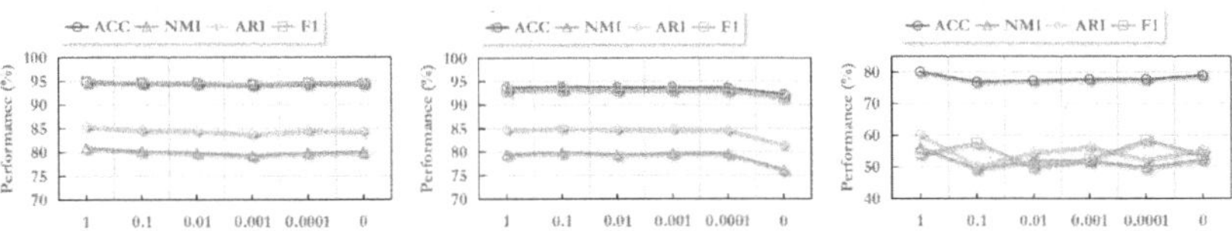

Fig. 3. Parameter analysis of λ_2 on ACM, DBLP, and Wisconsin datasets.

that despite considering losses at different levels, they are complementary and do not exhibit crowding-out effects or negative impacts on loss values. Therefore, adjusting the weighting of these losses in any manner will not significantly affect the performance.

Note that while these losses do not influence each other, their values differ across different datasets. We attribute this to variations in data distribution, which in turn necessitate different focal points for the model. However, since weight adjustments do not significantly impact performance, setting universal default values for different datasets remains acceptable.

(2) Figure 5 presents the sensitivity analysis of the hyperparameter order on homophilic datasets (DBLP, IMDB) and the heterophilic dataset (Chameleon). From a spatial perspective, order controls the effective depth of the graph filter: higher order aggregates information from more distant nodes, whereas lower order restricts aggregation to local neighborhoods. The curves show that performance remains relatively stable for small orders (approximately within 10) but begins to decline as the order increases; notably, a pronounced drop appears when the order is around 13âĂŞ18. This suggests over-smoothing or noise propagation at large receptive fields; therefore, we recommend using a moderate order (e.g., below 10) to balance information range and robustness.

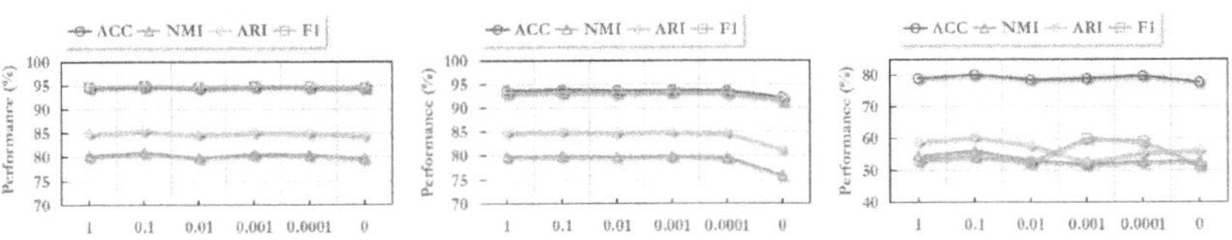

Fig. 4. Parameter analysis of λ_3 on ACM, DBLP, and Wisconsin datasets.

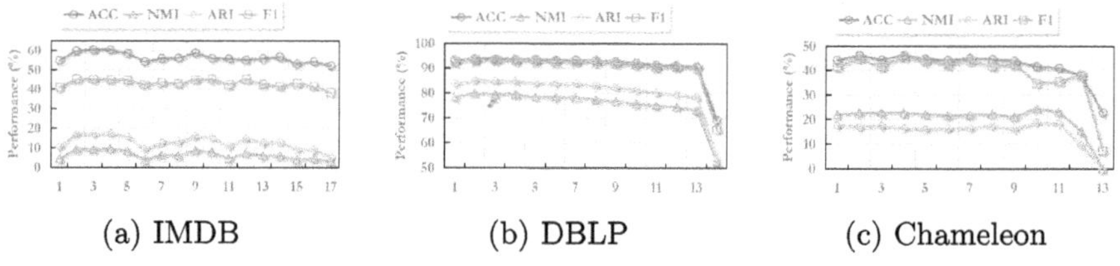

(a) IMDB   (b) DBLP   (c) Chameleon

Fig. 5. Parameter analysis of order on IMDB, DBLP, and Chameleon datasets.

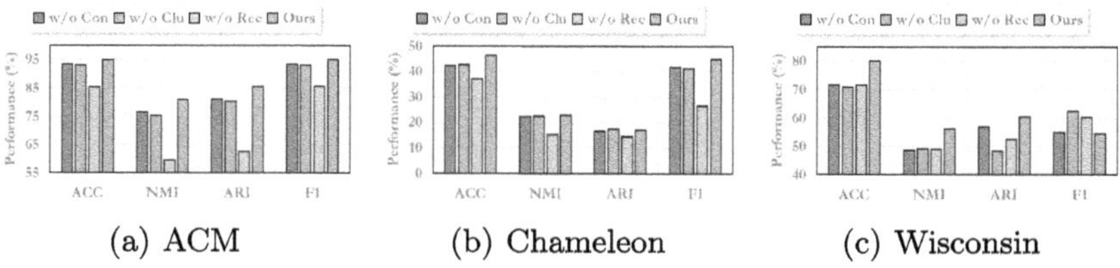

(a) ACM   (b) Chameleon   (c) Wisconsin

Fig. 6. Ablation study of different modules.

4.6 Ablation Study

Our model is founded on the concept of guiding graph encoding and consensus embedding with clustering information. As shown in Fig. 6, we map each loss to the numbered modules in the figure: $\mathcal{L}_{Attr}$ corresponds to Module 1 (feature alignment), $\mathcal{L}_{Adj}$ corresponds to Module 2 (structure reconstruction), and $\mathcal{L}_{Clu}$ corresponds to Module 3 (cluster-guided positives). We conducted an ablation study by starting from NGCE and incrementally adding these components.

Across ACM, DBLP, Chameleon, and Wisconsin, each module consistently improves performance: Module 1 mainly enhances cross-view consistency, Module 2 strengthens structural quality, and Module 3 further boosts embedding compactness via prototype guidance. The full model achieves the best results on all datasets, as depicted in Fig. 6.

4.7 Convergence Analysis

As shown in Fig. 7, we report the training dynamics on the homophilic dataset IMDB and the heterophilic dataset Wisconsin. The blue curve denotes the training loss (right axis) and the yellow curve denotes NMI (left axis), with the horizontal axis indicating epochs. The loss consistently decreases and quickly settles into a stable region, while NMI steadily increases and plateaus after a moderate number of epochs, indicating progressive improvement of embedding quality and cluster structure.

Compared with IMDB, Wisconsin exhibits slightly larger fluctuations and a slower stabilization, yet both cases converge reliably. These results demonstrate a stable and well-behaved optimization process across homophilic and

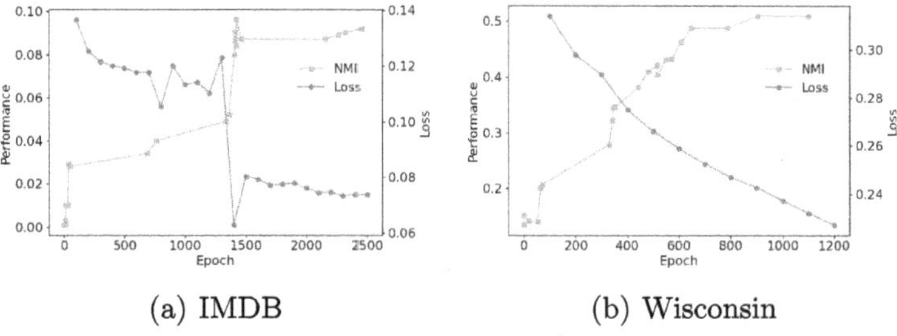

(a) IMDB   (b) Wisconsin

Fig. 7. Convergence of H2MGC model on IMDB and Wisconsin datasets.

heterophilic settings, validating the effectiveness and robustness of the proposed joint encoding and cluster-guided mechanisms.

5 Conclusion

We propose H2MGC, a cluster-guided contrastive encoding method for multi-view graph clustering. By combining adaptive joint adjacency construction, node-guided denoising, uncertainty-aware cluster-positive augmentation, two-stage feature alignment, and structure reconstruction, H2MGC delivers robust performance on both homophilic and heterophilic graphs. Experiments on five datasets show consistent accuracy gains and stable convergence.

Advancing Multi-View Graph Clustering (MVGC) to effectively operate on heterophilic graphs represents a critical and promising frontier. This endeavor necessitates a paradigm shift away from the prevailing homophily assumption that underpins most current GNN-based methods. The ultimate objective is to develop versatile MVGC frameworks that are agnostic to the underlying graph homophily, unlocking their applicability to a much broader and more complex range of real-world problems.

Acknowledgments. This paper is supported by Beijing Natural Science Foundation (funding number L257023).

References

1. Arazo, E., Ortego, D., Albert, P., OConnor, N.E., McGuinness, K.: Pseudo-labeling and confirmation bias in deep semi-supervised learning. In: IJCNN (2020)
2. Backstrom, L., Huttenlocher, D., Kleinberg, J., Lan, X.: Group formation in large social networks: membership, growth, and evolution. In: KDD (2006)
3. Chao, G., Jiang, Y., Chu, D.: Incomplete contrastive multi-view clustering with high-confidence guiding. In: AAAI (2024)
4. Chen, B., et al.: Structural deep multi-view clustering with integrated abstraction and detail. Neural Netw. (2024)

5. Chen, J., et al.: Variational graph generator for multiview graph clustering. IEEE TNNLS (2025)
6. Chen, M., Wei, Z., Huang, Z., Ding, B., Li, Y.: Simple and deep graph convolutional networks. In: ICML. PMLR (2020)
7. Cheng, J., Wang, Q., Tao, Z., Xie, D., Gao, Q.: Multi-view attribute graph convolution networks for clustering. In: IJCAI (2021)
8. Dong, Z., et al.: Enhanced then progressive fusion with view graph for multi-view clustering. In: CVPR (2025)
9. Fan, S., Wang, X., Shi, C., Lu, E., Lin, K., Wang, B.: One2Multi graph autoencoder for multi-view graph clustering. In: The Web Conference (2020)
10. Hassani, K., Khasahmadi, A.H.: Contrastive multi-view representation learning on graphs. In: ICML (2020)
11. Hou, Z., Liu, X., Cen, Y., Dong, Y., Yang, H., Wang, C., Tang, J.: GraphMAE: self-supervised masked graph autoencoders. In: KDD (2022)
12. Huang, Z., Ren, Y., Pu, X., He, L.: Non-linear fusion for self-paced multi-view clustering. In: ACM MM (2021)
13. Jia, X., Jing, X.Y., Zhu, X., Cai, Z., Hu, C.H.: Co-embedding: a semi-supervised multi-view representation learning approach. Neural Comput. Appl. (2022)
14. Ke, J., et al.: Integrating vision-language semantic graphs in multi-view clustering. In: IJCAI (2024)
15. Kipf, T.N., Welling, M.: Variational graph auto-encoders. arXiv preprint arXiv:1611.07308 (2016)
16. Krogan, N.J., et al.: Global landscape of protein complexes in the yeast saccharomyces cerevisiae. Nature (2006)
17. Lin, Z., Kang, Z.: Graph filter-based multi-view attributed graph clustering. In: IJCAI (2021)
18. Ling, Y., et al.: Dual label-guided graph refinement for multi-view graph clustering. In: AAAI (2023)
19. Liu, M., et al.: Deep temporal graph clustering. In: ICLR (2024)
20. Liu, Y., et al.: Reinforcement graph clustering with unknown cluster number. In: ACM MM (2023)
21. Pan, E., Kang, Z.: Multi-view contrastive graph clustering. In: NeurIPS (2021)
22. Pei, H., Wei, B., Chang, K.C.C., Lei, Y., Yang, B.: Geom-GCN: geometric graph convolutional networks. arXiv preprint arXiv:2002.05287 (2020)
23. Peng, Y., Zhu, X., Nie, F., Kong, W., Ge, Y.: Fuzzy graph clustering. Inf. Sci. (2021)
24. Perozzi, B., Akoglu, L., Iglesias Sanchez, P., Muller, E.: Focused clustering and outlier detection in large attributed graphs. In: KDD (2014)
25. Ren, Y., et al.: A novel federated multi-view clustering method for unaligned and incomplete data fusion. Inf. Fus. (2024)
26. Ren, Y., et al.: Multi-view graph clustering via node-guided contrastive encoding. In: ICML (2025)
27. Ren, Y., et al.: Dynamic weighted graph fusion for deep multi-view clustering. In: IJCAI (2024)
28. Rozemberczki, B., Allen, C., Sarkar, R.: Multi-scale attributed node embedding. J. Complex Netw. (2021)
29. Shen, Z., He, H., Kang, Z.: Balanced multi-relational graph clustering. In: ACM MM (2024)
30. Vincent, P., Larochelle, H., Bengio, Y., Manzagol, P.A.: Extracting and composing robust features with denoising autoencoders. In: ICML (2008)

31. Wang, Y., Chang, D., Fu, Z., Zhao, Y.: Consistent multiple graph embedding for multi-view clustering. IEEE TMM (2021)
32. Wu, F., et al.: Semi-supervised multi-view individual and sharable feature learning for webpage classification. In: The Web Conference (2019)
33. Xia, W., Wang, S., Yang, M., Gao, Q., Han, J., Gao, X.: Multi-view graph embedding clustering network: joint self-supervision and block diagonal representation. Neural Netw. (2022)
34. Xu, J., et al.: Deep incomplete multi-view clustering via mining cluster complementarity. In: AAAI (2022)

Untying the Knots of Heterophily in Hypergraphs via Mixed-Curvature Manifolds

Lizhi Liu[✉][iD]

China UnionPay, Shanghai, China
liulizhi1996@gmail.com

Abstract. Hypergraph neural networks (HNNs) have recently emerged as a promising paradigm for modeling higher-order relations, yet their study under heterophilic settings remains highly limited. Empirical evidence shows that classical message-passing–based HNNs suffer severe performance degradation on hypergraphs with low homophily ratios. We attribute this primarily to heterophily mixing, where semantic signals from neighbors of different classes become entangled, thereby eroding the discriminative power of node representations. To address this challenge, we propose HyperUnmix, a novel method that disentangles heterophily mixing through a mixed-curvature manifold. Guided by the intuition that nodes of different classes exhibit distinct distributional characteristics, we model the representation space as a Cartesian product of multiple hyperbolic submanifolds, each aligned with a specific class. By constraining information flow to propagate mainly within the submanifold corresponding to its class, HyperUnmix effectively alleviates mixing during aggregation. Extensive experiments on both heterophilic and homophilic hypergraph benchmarks demonstrate that our model establishes new state-of-the-art performance, providing fresh insights into heterophilic hypergraph learning. The source code is available at https://github.com/liulizhi1996/HyperUnmix.

Keywords: Hypergraph neural network · Heterophilic graph learning · Mixed-curvature product manifold

1 Introduction

Hypergraphs provide a natural framework for modeling higher-order relations, where a single hyperedge can simultaneously capture interactions among multiple entities. Such structures appear in diverse domains, from co-purchasing groups in e-commerce [29] to functional modules in biological systems [17], making hypergraph-based learning increasingly essential. Hypergraph neural networks (HNNs) have thus emerged as a powerful extension of message-passing architectures originally introduced by graph neural networks (GNNs) [21], enabling effective representation learning over these complex patterns.

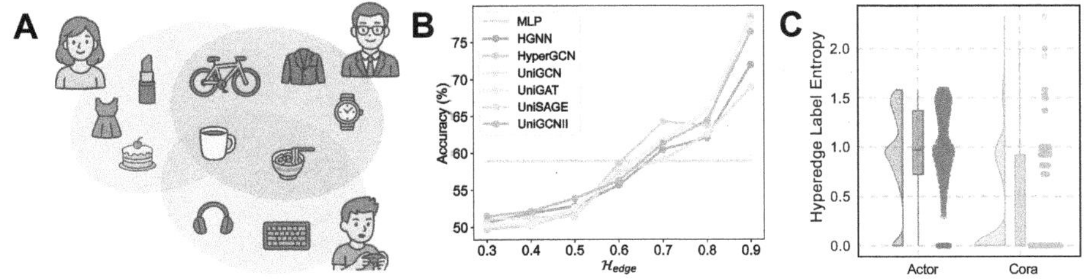

Fig. 1. (A) Toy hypergraph of co-purchase relationships in an e-commerce platform, illustrating heterophily. (B) Performance of MLP and HNNs on synthetic hypergraphs with varying homophily rates. (C) Hyperedge label entropy distributions in Actor (heterophilic) vs. Cora (homophilic).

Most existing HNNs, however, are designed under the homophily assumption, where nodes connected by a hyperedge are likely to share the same labels. In practice, this assumption often fails. As illustrated in Fig. 1A with a toy example from an e-commerce platform, the purchasing history of a gaming enthusiast may include not only electronic devices but also items such as coffee and instant food, which are favored by other consumer groups. This phenomenon–where a single hyperedge connects nodes belonging to diverse classes–is referred to as *heterophily*, and it poses severe challenges to message-passing architectures [15]. Empirical evidence in Fig. 1B shows that the performance of several classical HNNs deteriorates sharply under high heterophily, falling below that of graph-agnostic models such as the Multilayer Perceptron (MLP).

In this work, we argue that this counterintuitive performance drop is primarily caused by *heterophily mixing* during message passing: semantic signals from neighbors of different classes become entangled, eroding the discriminative power of node representations. This leads to "feature over-smoothing," where representations from different classes become indistinguishable in feature space. Figure 1C highlights this phenomenon by comparing hyperedge label entropy distributions: in the homophilic dataset Cora, 65.7% of hyperedges have zero entropy, meaning all incident nodes share the same label, whereas in the heterophilic dataset Actor, only 14.7% of hyperedges exhibit zero entropy. Prior studies [35] further proved that heterophily is even more prevalent in hypergraphs than in simple graphs, since it is unrealistic to expect all entities within a large hyperedge to share a common label. With deeper hypergraph convolutions, harmful neighbors increasingly introduce irrelevant information, intensifying the mixing effect. Although recent years have witnessed substantial advances in heterophilic learning for simple graphs [27], research on heterophilic hypergraphs remains scarce. Existing attempts [9,20,24] offer partial remedies but lack explicit mechanisms for disentangling heterophilic signals, leaving significant room for improvement.

To address this gap, we propose HyperUnmix, a novel HNN that explicitly disentangles heterophily mixing through a mixed-curvature manifold. The core

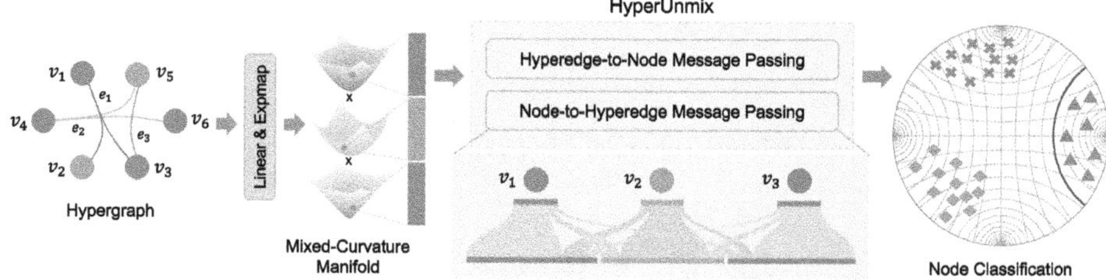

Fig. 2. Illustration of the core design of HyperUnmix. The initial features are projected onto a mixed-curvature manifold, where each submanifold corresponds to a specific class. Message passing is performed bidirectionally between nodes and hyperedges, while the information flow is guided to flow predominantly toward the submanifold associated with its class, thereby disentangling heterophily mixing.

intuition is straightforward: *heterophily mixing can be alleviated if messages are constrained to propagate along class-specific semantic channels.* Guided by the assumption that *nodes of different classes exhibit distinct distributional characteristics*, we model the representation space as a Cartesian product of multiple Riemannian submanifolds, each associated with a particular class. Motivated by recent findings [25] that hyperbolic geometry excels at embedding hierarchical structures, we adopt hyperbolic submanifolds for class-aligned message flow. As illustrated in Fig. 2, HyperUnmix constrains information from neighboring nodes to primarily flow into the submanifold of the target that corresponds to its associated class, thereby preserving semantic purity and discriminability. Since node labels are not available during training, we further design a self-attention mechanism to assign soft labels that dynamically guide class-specific routing.

Our contributions are twofold:

(1) We present a principled solution to disentangle heterophily mixing in hypergraph neural networks. By modeling node representations on a Cartesian product of class-aligned submanifolds, HyperUnmix separates message flows into per-class channels, preventing harmful signal entanglement.
(2) Extensive experiments on real-world homophilic and heterophilic hypergraph benchmarks demonstrate the superiority of HyperUnmix over state-of-the-art HNNs. Further analysis shows its ability to mitigate over-smoothing, suggesting that it retains discriminability by avoiding heterophily mixing.

2   Preliminaries

2.1   Notation

Consider a hypergraph $\mathcal{G} = (\mathcal{V}, \mathcal{E})$, where the node set $\mathcal{V} = \{v_i\}_{i=1}^N$ consists of N nodes, and the hyperedge set $\mathcal{E} = \{e_j\}_{j=1}^M$ contains M hyperedges. Unlike ordinary graphs, each hyperedge $e_j \subseteq \mathcal{V}$ in a hypergraph can join any number of nodes. Both nodes and hyperedges are associated with feature matrices $\mathbf{X}_n \in \mathbb{R}^{N \times f_n}$ and $\mathbf{X}_e \in \mathbb{R}^{M \times f_e}$, respectively, where the i-th row $\mathbf{X}_n[i,:]$ corresponds to node v_i and the j-th row $\mathbf{X}_e[j,:]$ corresponds to hyperedge e_j. In this work, we focus on the semi-supervised node classification task. Specifically, a subset of nodes $\mathcal{V}' \subset \mathcal{V}$ has known labels $y_i \in \{1, 2, \cdots, C\}$, and the goal is to predict the labels of the remaining unlabeled nodes $\mathcal{V} \setminus \mathcal{V}'$.

2.2 Hypergraph Homophily Measures

Analogous to ordinary graphs, two connected nodes in a hypergraph are considered *homophilic* if they share the same label, and *heterophilic* otherwise. The homophily level of a hypergraph can generally be quantified from both hyperedge and node perspectives [24]:

$$
\begin{aligned}
\mathcal{H}_{edge} &= \frac{1}{|\mathcal{E}|} \sum_{e_j \in \mathcal{E}} \frac{|\{(u,v) \in e_j \mid \mathbb{I}(y_u = y_v)\}|}{\binom{|e_j|}{2}}, \\
\mathcal{H}_{node} &= \frac{1}{|\mathcal{V}|} \sum_{v_i \in \mathcal{V}} \frac{1}{|\mathcal{N}_{\mathcal{E}}(v_i)|} \sum_{e_j \in \mathcal{N}_{\mathcal{E}}(v_i)} \frac{|\{(u,v) \in e_j \mid \mathbb{I}(y_u = y_v)\}|}{|e_j|},
\end{aligned}
\tag{1}
$$

where $\mathcal{N}_{\mathcal{E}}(v_i) = \{e_j \in \mathcal{E} \mid v_i \in e_j\}$ denotes the set of hyperedges containing node v_i. $\mathcal{H}_{edge}$ measures the proportion of node pairs within a hyperedge that belong to the same class, while $\mathcal{H}_{node}$ reflects the distribution of nodes belonging to the same class across the hyperedges associated with each node. Typically, a hypergraph is considered heterophilic if $\mathcal{H}. < 0.5$, and homophilic otherwise.

2.3 Hyperbolic Geometry

Hyperbolic geometry refers to a class of Riemannian manifolds characterized by constant negative curvature. Among its equivalent models, the Poincaré ball model is one of the most widely adopted due to its mathematical elegance and practical utility. Formally, the Poincaré ball with curvature $\kappa < 0$ is defined as an open n-dimensional ball $\mathbb{B}_\kappa^n = \{\mathbf{x} \in \mathbb{R}^n \mid -\kappa\|\mathbf{x}\|_2^2 < 1\}$, equipped with the Riemannian metric $g_\mathbf{x}^\kappa = (\lambda_\mathbf{x}^\kappa)^2 \mathbf{I}$, where the conformal factor at point $\mathbf{x}$ is given by $\lambda_\mathbf{x}^\kappa = \frac{2}{1+\kappa\|\mathbf{x}\|_2^2}$.

For any $\mathbf{x} \in \mathbb{B}_\kappa^n$, the tangent space $\mathcal{T}_\mathbf{x}\mathbb{B}_\kappa^n$, which is isomorphic to $\mathbb{R}^n$, provides a first-order local approximation. The exponential and logarithmic maps serve as the fundamental bridges between the hyperbolic space and its Euclidean tangent space. Specifically, for $\mathbf{v} \neq \mathbf{0}$ and $\mathbf{y} \neq \mathbf{x}$:

$$
\begin{aligned}
\exp_\mathbf{x}^\kappa(\mathbf{v}) &= \mathbf{x} \oplus_\kappa \left(\tanh \left(\sqrt{|\kappa|} \frac{\lambda_\mathbf{x}^\kappa \|\mathbf{v}\|_2}{2} \right) \frac{\mathbf{v}}{\sqrt{|\kappa|}\|\mathbf{v}\|_2} \right), \\
\log_\mathbf{x}^\kappa(\mathbf{y}) &= \frac{2}{\sqrt{|\kappa|}\lambda_\mathbf{x}^\kappa} \tanh^{-1} \left(\sqrt{|\kappa|} \|-\mathbf{x} \oplus_\kappa \mathbf{y}\|_2 \right) \frac{-\mathbf{x} \oplus_\kappa \mathbf{y}}{\|-\mathbf{x} \oplus_\kappa \mathbf{y}\|_2}.
\end{aligned}
\tag{2}
$$

Here, $\oplus_\kappa$ denotes the Möbius addition for any $\mathbf{x}, \mathbf{y} \in \mathbb{B}_\kappa^n$:

$$
\mathbf{x} \oplus_\kappa \mathbf{y} = \frac{(1 - 2\kappa\langle\mathbf{x}, \mathbf{y}\rangle - \kappa\|\mathbf{y}\|_2^2)\mathbf{x} + (1 + \kappa\|\mathbf{x}\|_2^2)\mathbf{y}}{1 - 2\kappa\langle\mathbf{x}, \mathbf{y}\rangle + \kappa^2\|\mathbf{x}\|_2^2\|\mathbf{y}\|_2^2}.
\tag{3}
$$

Moreover, within the Poincaré ball, the weighted midpoint of a set of points $\{\mathbf{x}_i \in \mathbb{B}^n_\kappa\}_{i=1}^N$ with real weights $\{\alpha_i \in \mathbb{R}\}_{i=1}^N$ can be formulated via the Möbius gyromidpoint as

$$\bar{\mathbf{x}} = \bigodot_{i=1}^N [\mathbf{x}_i, \alpha_i]_\kappa = \frac{1}{2} \otimes_\kappa \left(\frac{\sum_{i=1}^N \alpha_i \lambda^\kappa_{\mathbf{x}_i} \cdot \mathbf{x}_i}{\sum_{i=1}^N \alpha_i (\lambda^\kappa_{\mathbf{x}_i} - 1)} \right). \tag{4}$$

Here, $\otimes_\kappa$ denotes Möbius scalar-vector multiplication in the Poincaré ball, defined for $r \in \mathbb{R}$ and $\mathbf{x} \in \mathbb{B}^n_\kappa \setminus \{\mathbf{0}\}$ as:

$$r \otimes_\kappa \mathbf{x} = \frac{1}{\sqrt{|\kappa|}} \tanh\left(r \tanh^{-1}\left(\sqrt{|\kappa|} \|\mathbf{x}\|_2 \right) \right) \frac{\mathbf{x}}{\|\mathbf{x}\|_2}. \tag{5}$$

2.4   Mixed-Curvature Manifold

A mixed-curvature manifold is defined as the Cartesian product of multiple component manifolds, each endowed with distinct curvatures: $\mathbb{P} = \prod_{i=1}^m \mathbb{B}^{n^{[i]}}_{\kappa^{[i]}}$. The overall curvature is collectively represented by the set of individual curvatures $\mathfrak{K} = \{\kappa^{[1]}, \cdots, \kappa^{[m]}\}$. A point in this product space is decomposable into its components, i.e., any $\mathbf{x} \in \mathbb{P}$ can be expressed as $\mathbf{x} = (\mathbf{x}^{[1]}, \cdots, \mathbf{x}^{[m]})$ where $\mathbf{x}^{[i]} \in \mathbb{B}^{n^{[i]}}_{\kappa^{[i]}}$. Similarly, a tangent vector $\mathbf{v} \in \mathcal{T}_\mathbf{x}\mathbb{P}$ decomposes into $\mathbf{v} = (\mathbf{v}^{[1]}, \cdots, \mathbf{v}^{[m]})$ with each component lying in the corresponding tangent space $\mathbf{v}^{[i]} \in \mathcal{T}_{\mathbf{x}^{[i]}}\mathbb{B}^{n^{[i]}}_{\kappa^{[i]}}$. Operations such as exponential and logarithmic maps naturally extend to product manifolds by applying them componentwise to each factorized embedding, followed by concatenation. The only exception is the geodesic distance, which is defined as the root-sum-square of the distances across individual manifolds: $D_\mathfrak{K} = \sqrt{\sum_{i=1}^m d^2_{\kappa^{[i]}}}$, where $d_{\kappa^{[i]}}$ denotes the distance within the i-th submanifold.

3   Methodology

We posit that one of the principal reasons why classical message-passing frameworks underperform in heterophilic scenarios lies in what we call *heterophily mixing*: during aggregation, semantic information from dissimilar neighbors is indiscriminately fused, thereby diluting the distinctive characteristics of the central target and degrading its discriminability. To overcome this challenge, we introduce HyperUnmix, a novel framework that leverages mixed-curvature manifolds to disentangle heterophilic information. The core motivation is straightforward yet effective: *nodes of different classes should naturally reside on distinct low-dimensional manifolds*. By aligning submanifolds with classes and constraining information flow primarily within class-consistent submanifolds, HyperUnmix effectively suppresses interference from heterophilic neighbors. Below we elaborate on the complete pipeline.

3.1 Projecting Features Onto the Mixed-Curvature Manifold

Both raw node and hyperedge features are initially embedded in Euclidean space. To exploit the representational power of curved geometries, we first project them into a mixed-curvature product manifold. For a hypergraph with C classes, we construct the manifold as $\mathbb{P} = \prod_{i=1}^{C} \mathbb{B}_{\kappa^{[i]}}^{d/C}$, where each $\kappa^{[i]} \in \mathbb{R}$ is a learnable curvature parameter associated with class i. For simplicity, each submanifold is assigned the same dimensionality d/C. Accordingly, we transform the initial node and hyperedge features through a linear projection followed by the exponential map:

$$\mathbf{H}_n^{(0)} = \mathrm{Exp}_{\mathbf{0}}^{\mathfrak{K}}(\mathbf{X}_n \mathbf{W}_n^{(0)} + \mathbf{b}_n^{(0)}), \quad \mathbf{H}_e^{(0)} = \mathrm{Exp}_{\mathbf{0}}^{\mathfrak{K}}(\mathbf{X}_e \mathbf{W}_e^{(0)} + \mathbf{b}_e^{(0)}), \qquad (6)$$

with $\mathbf{W}_n^{(0)} \in \mathbb{R}^{f_n \times d}, \mathbf{W}_e^{(0)} \in \mathbb{R}^{f_e \times d}$ as weight matrices, and $\mathbf{b}_n^{(0)}, \mathbf{b}_e^{(0)} \in \mathbb{R}^d$ as biases. The exponential map on the product manifold is given by $\mathrm{Exp}_{\mathbf{0}}^{\mathfrak{K}}(\mathbf{x}) = (\exp_{\mathbf{0}}^{\kappa^{[1]}}(\mathbf{x}^{[1]}), \cdots, \exp_{\mathbf{0}}^{\kappa^{[C]}}(\mathbf{x}^{[C]}))$. Thus, each projected feature decomposes into C chunks, each aligned with a class-specific submanifold.

3.2 Disentangling Heterophily in Message Passing

The essence of HyperUnmix lies in class-aware aggregation: messages are restricted to propagate within class-consistent submanifolds, thereby minimizing cross-manifold contamination. A key difficulty, however, is that node classes are unknown a priori. To address this, we employ an attention mechanism that assigns soft labels, modulating the intensity of flows into each submanifold channel. In the hypergraph setting, message passing occurs from the nodes to hyperedges ($\mathcal{V} \rightarrow \mathcal{E}$) and vice versa ($\mathcal{E} \rightarrow \mathcal{V}$) simultaneously. We first describe the node-to-hyperedge case at layer l.

Step 1: Assigning Attention Scores. Given the previous-layer embeddings $\mathbf{H}_n^{(l-1)} \in \mathbb{P}^N$ and $\mathbf{H}_e^{(l-1)} \in \mathbb{P}^M$, we first project them into a unified space:

$$\mathbf{Z}_{n,n2e}^{(l)} = \mathrm{Log}_{\mathbf{0}}^{\mathfrak{K}}(\mathbf{H}_n^{(l-1)})\mathbf{W}_{n,n2e}^{(l)}, \quad \mathbf{Z}_{e,n2e}^{(l)} = \mathrm{Log}_{\mathbf{0}}^{\mathfrak{K}}(\mathbf{H}_e^{(l-1)})\mathbf{W}_{e,n2e}^{(l)}, \qquad (7)$$

where $\mathbf{W}_{n,n2e}^{(l)}, \mathbf{W}_{e,n2e}^{(l)} \in \mathbb{R}^{d \times (d/C)}$. The logarithmic map on the product manifold is defined componentwise: $\mathrm{Log}_{\mathbf{0}}^{\mathfrak{K}}(\mathbf{x}) = (\log_{\mathbf{0}}^{\kappa^{[1]}}(\mathbf{x}^{[1]}), \cdots, \log_{\mathbf{0}}^{\kappa^{[C]}}(\mathbf{x}^{[C]}))$. For each node v_j incident to hyperedge e_i, we then compute attention scores as

$$\mathsf{S}_{n2e}^{(l)}[i,j,:] = \mathrm{Softmax}\left(\frac{\sigma(\mathbf{Z}_{e,n2e}^{(l)}[i,:] + \mathbf{Z}_{n,n2e}^{(l)}[j,:])\mathbf{W}_{s,n2e}^{(l)}}{\sqrt{C}}\right) \in \mathbb{R}^C, \qquad (8)$$

with $\mathbf{W}_{s,n2e}^{(l)} \in \mathbb{R}^{(d/C) \times C}$ and activation function $\sigma(\cdot)$ chosen as LeakyReLU. Intuitively, $\mathsf{S}_{n2e}^{(l)}[i,j,c]$ reflects the extent to which node v_j transmits information to the c-th submanifold of hyperedge e_i. Ideally, the attention concentrates mass on the submanifold corresponding to the true class of v_j, while suppressing others to near zero.

Step 2: Class-Aware Message Aggregation. Guided by these attention scores, node messages are selectively projected into class-specific channels. For node v_j contributing to hyperedge e_i, the message sent to the c-th submanifold is:

$$\mathsf{M}_{n2e}^{(l)}[i,j,c,:] = \mathsf{S}_{n2e}^{(l)}[i,j,c] \otimes_{\kappa^{[c]}} \exp_{\mathbf{0}}^{\kappa^{[c]}}\left(\mathbf{Z}_{n,n2e}^{(l)}[j,:]\right). \tag{9}$$

Messages from all incident nodes are then aggregated via the Möbius gyromidpoint:

$$\mathsf{P}_e^{(l)}[i,c,:] = \bigodot_{v_j \in e_i} \left[\mathsf{M}_{n2e}^{(l)}[i,j,c,:], 1\right]_{\kappa^{[c]}} = \frac{1}{2} \otimes_{\kappa^{[c]}} \left(\frac{\sum_{v_j \in e_i} \lambda_{\mathsf{M}_{n2e}^{(l)}[i,j,c,:]}^{\kappa^{[c]}} \mathsf{M}_{n2e}^{(l)}[i,j,c,:]}{\sum_{v_j \in e_i}\left(\lambda_{\mathsf{M}_{n2e}^{(l)}[i,j,c,:]}^{\kappa^{[c]}} - 1\right)}\right). \tag{10}$$

Finally, the outputs across C channels are concatenated componentwise to form the updated hyperedge embedding:

$$\mathbf{P}_e^{(l)}[i,:] = \left(\mathsf{P}_e^{(l)}[i,1,:], \cdots, \mathsf{P}_e^{(l)}[i,C,:]\right). \tag{11}$$

Step 3: Residual Connections. To preserve the intrinsic properties of hyperedges, we incorporate residual connections [18]. Unlike the Euclidean case where a simple addition suffices, here we adopt the Möbius gyromidpoint to compute a weighted average between the initial feature and the aggregated representation:

$$\mathbf{H}_e^{(l)}[i,:] = \frac{1}{2} \otimes_{\mathfrak{K}} \left(\frac{\alpha\lambda_{\mathbf{P}_e^{(l)}[i,:]}^{\mathfrak{K}}\mathbf{P}_e^{(l)}[i,:] + (1-\alpha)\lambda_{\mathbf{H}_e^{(0)}[i,:]}^{\mathfrak{K}}\mathbf{H}_e^{(0)}[i,:]}{\alpha\lambda_{\mathbf{P}_e^{(l)}[i,:]}^{\mathfrak{K}} + (1-\alpha)\lambda_{\mathbf{H}_e^{(0)}[i,:]}^{\mathfrak{K}} - 1}\right), \tag{12}$$

where $\alpha \in (0,1)$ is a scaling factor.

The same procedure applies symmetrically to hyperedge-to-node propagation, ultimately producing refined node embeddings.

3.3 Downstream Task: Node Classification

After passing through L stacked HyperUnmix layers, the final node embeddings $\mathbf{H}_n^{(L)}$ can be utilized for various downstream tasks. Here we focus on multi-class node classification to demonstrate HyperUnmix's efficacy in mitigating heterophily mixing.

As discussed earlier, multinomial logistic regression (MLR) [7,14] on a product manifold aggregates classification results from each submanifold. We begin with the single hyperbolic manifold. The hyperplane $\tilde{H}_{\mathbf{a},\mathbf{p}}^{\kappa}$, defined by point $\mathbf{p} \in \mathbb{B}_{\kappa}^d$ and tangent vector $\mathbf{a} \in \mathcal{T}_{\mathbf{p}}\mathbb{B}_{\kappa}^d \setminus \{\mathbf{0}\}$, is the set of geodesics orthogonal to $\mathbf{a}$ and passing through $\mathbf{p}$:

$$\tilde{H}_{\mathbf{a},\mathbf{p}}^{\kappa} = \{\mathbf{x} \in \mathbb{B}_{\kappa}^d \mid \langle -\mathbf{p} \oplus_{\kappa} \mathbf{x}, \mathbf{a}\rangle = 0\}. \tag{13}$$

The distance from $\mathbf{x} \in \mathbb{B}_{\kappa}^d$ to this hyperplane is

$$d_{\kappa}(\mathbf{x}, \tilde{H}_{\mathbf{a},\mathbf{p}}^{\kappa}) = \frac{1}{\sqrt{|\kappa|}} \sinh^{-1}\left(\frac{2\sqrt{|\kappa|}\langle -\mathbf{p} \oplus_{\kappa} \mathbf{x}, \mathbf{a}\rangle}{(1 + \kappa\|-\mathbf{p} \oplus_{\kappa} \mathbf{x}\|_2^2)\|\mathbf{a}\|_2}\right). \tag{14}$$

Thus, for a C-class problem, the probability of $\mathbf{x}$ belonging to class k becomes

$$P(y = k \mid \mathbf{x}) = \mathrm{Softmax}\big(\mathrm{logits}_{\mathbb{B}_\kappa^d}(\mathbf{x}, k)\big),$$

$$\mathrm{logits}_{\mathbb{B}_\kappa^d}(\mathbf{x}, k) = \frac{\lambda_{\mathbf{p}_k}^\kappa \|\mathbf{a}_k\|_2}{\sqrt{|\kappa|}} \sinh^{-1}\left(\frac{2\sqrt{|\kappa|}\langle -\mathbf{p}_k \oplus_\kappa \mathbf{x}, \mathbf{a}_k\rangle}{(1 + \kappa\|-\mathbf{p}_k \oplus_\kappa \mathbf{x}\|_2^2)\|\mathbf{a}_k\|_2} \right), \tag{15}$$

where $\mathbf{p}_k \in \mathbb{B}_\kappa^d$ and $\mathbf{a}_k \in \mathcal{T}_{\mathbf{p}_k}\mathbb{B}_\kappa^d \setminus \{\mathbf{0}\}$ are class-specific parameters.

Extending to the mixed-curvature case, logits are aggregated across submanifolds via an ℓ_2-norm, signed by the direction of the summed inner products:

$$P(y = k \mid \mathbf{x}) = \mathrm{Softmax}\big(\mathrm{logits}_{\mathbb{P}}(\mathbf{x}, k)\big),$$

$$\mathrm{logits}_{\mathbb{P}}(\mathbf{x}, k) = \sqrt{\sum_{\mathbb{B}_{\kappa^{[i]}}^{d/C} \in \mathbb{P}} \mathrm{logits}^2_{\mathbb{B}_{\kappa^{[i]}}^{d/C}}(\mathbf{x}^{[i]}, k)} \cdot \mathrm{sgn}\left(\sum_{\mathbb{B}_{\kappa^{[i]}}^{d/C} \in \mathbb{P}} \langle -\mathbf{p}_k^{[i]} \oplus_{\kappa^{[i]}} \mathbf{x}^{[i]}, \mathbf{a}_k^{[i]}\rangle \right). \tag{16}$$

Here, $\mathbf{p}_k^{[i]}$ and $\mathbf{a}_k^{[i]}$ are the parameters for class k in the i-th submanifold. The final node prediction is then obtained as $\hat{y}_i = \arg\max_k P(y = k \mid \mathbf{H}_n^{(L)}[i, :])$.

Table 1. Statistics of heterophilic and homophilic hypergraph datasets.

Category	Dataset	Nodes	Hyperedges	Features	Classes	$\mathcal{H}_{node}$	$\mathcal{H}_{edge}$
Heterophilic	Actor	16,255	10,164	50	3	0.4815	0.4675
	Twitch-gamers	16,812	2,627	7	2	0.4893	0.4857
	Pokec	14,998	2,406	65	2	0.4952	0.4529
	Senate	282	315	100	2	0.4793	0.4642
	House	1,290	340	100	2	0.5049	0.4851
Homophilic	Cora	2,708	1,579	1,433	7	0.6399	0.7462
	Citeseer	3,312	1,079	3,703	6	0.5771	0.6814
	Pubmed	19,717	7,963	500	3	0.5499	0.7765
	Cora-CA	2,708	1,072	1,433	7	0.7279	0.7797
	DBLP-CA	43,413	22,535	1,425	6	0.8557	0.8656

4   Experiments

4.1   Experimental Settings

Datasets. We evaluate HyperUnmix on ten real-world hypergraph datasets spanning diverse domains and structural characteristics. Specifically, five datasets are categorized as heterophilic, including Actor, Twitch-gamers, Pokec [24], Senate, and House [13], while five datasets are homophilic, including Cora, Citeseer, Pubmed, Cora-CA, and DBLP-CA [40]. The detailed statistics of all datasets are summarized in Table 1.

Competing Methods. To comprehensively assess the effectiveness of Hyper-Unmix, we compare it against a broad range of representative baselines. These include: (1) a simple non-graph baseline, MLP; (2) spectral-based HNNs, namely HGNN [12], HyperGCN [40], HyperND [31], and TF-HNN [34]; (3) spatial-based HNNs, where we select the most competitive variant of UniGNN [19], i.e., UniGCNII, along with AllDeepSets, AllSetTransformer [6], HyperGT [26], Hypergraph-MLP [33], and KHGNN [39]; (4) tensor-based HNNs, including EHNN [20] and T-HyperGNN [36]; and (5) heterophily-oriented HNNs, namely SheafHyperGNN [9], ED-HNN [37], and HyperUFG [24].

Evaluation Protocols. For fair comparison and consistent with prior studies [12,24], we adopt different data partitioning strategies depending on the benchmark. For the recently established heterophilic datasets (Actor, Twitch, and Pokec), the data is split into 40%/20%/40% for training, validation, and testing, respectively. For all remaining datasets, we follow a 50%/25%/25% partition. We report the average classification accuracy and standard deviation over ten independent random splits.

Implementation Details. HyperUnmix is trained using cross-entropy loss optimized with the Riemannian Adam optimizer [3]. Hyperparameters are tuned via Optuna [1], with the following search ranges: hidden channel dimension $\{32, 64, 128, 256, 512\}$, number of layers $\{2, 3, 4\}$, balancing factor of residual connections $\alpha \in [0.1, 0.9]$, dropout rate $(0, 1)$, learning rate $[10^{-4}, 10^{-2}]$, and weight decay $[10^{-5}, 10^{-3}]$. Training is conducted for up to 200 epochs, and the test performance corresponding to the best validation result is reported. Baseline results are reproduced using the officially released implementations, with hyperparameters aligned to those reported in their original papers.

4.2 Results on Benchmark Datasets

Table 2 reports the classification accuracy on heterophilic hypergraph benchmarks. HyperUnmix achieves the best performance across all datasets and consistently ranks above all competing methods. Notably, on the particularly challenging Twitch-gamers dataset, HyperUnmix is the only model that surpasses MLP. This observation highlights both the inherent difficulty of heterophilic scenarios and the distinctive advantage of our approach in handling heterophilic hypergraphs. While certain heterophily-oriented HNNs outperform conventional HNNs on selected datasets, none achieve the same level of consistent superiority as HyperUnmix, underscoring the effectiveness of our proposed heterophily-mixing disentanglement mechanism.

Beyond excelling in heterophilic settings, HyperUnmix also demonstrates strong performance in homophilic scenarios. As shown in Table 3, HyperUnmix consistently ranks among the top-performing methods on 4 out of 5 datasets, and on Pubmed it achieves accuracy on par with the best models. It indicates that

Table 2. Evaluation results of node classification on heterophilic hypergraphs: average accuracy (%) ± standard deviation. The best-performing method is boldfaced, and the runner-up is underlined.

Method	Actor	Twitch-gamers	Pokec	Senate	House	*Avg. Rank*
MLP	85.45 ± 1.21	$\underline{52.77} \pm 1.81$	56.92 ± 2.46	52.25 ± 5.17	51.86 ± 2.34	8.6
HGNN	74.47 ± 0.32	51.88 ± 0.26	49.82 ± 0.27	48.59 ± 4.52	61.39 ± 2.96	13
HyperGCN	68.67 ± 4.38	51.32 ± 1.02	52.43 ± 3.68	42.45 ± 3.67	48.32 ± 2.93	14.6
HyperND	83.19 ± 0.92	51.44 ± 0.67	55.94 ± 0.45	52.82 ± 3.20	51.70 ± 3.37	11.2
TF-HNN	85.96 ± 0.41	52.39 ± 0.60	57.65 ± 1.08	58.31 ± 7.59	67.55 ± 1.33	5.8
UniGCNII	80.48 ± 1.13	50.84 ± 0.76	54.25 ± 2.70	49.30 ± 4.25	67.25 ± 2.57	12.4
AllDeepSets	82.00 ± 2.33	50.72 ± 0.96	51.11 ± 1.04	48.17 ± 5.67	67.82 ± 2.40	13
AllSetTransformer	83.39 ± 1.73	50.45 ± 0.76	58.40 ± 0.42	51.83 ± 5.22	69.33 ± 2.20	10
HyperGT	62.43 ± 0.55	52.10 ± 0.40	57.55 ± 0.54	48.14 ± 5.15	46.93 ± 2.60	12.8
Hypergraph-MLP	84.72 ± 0.54	51.95 ± 0.53	59.63 ± 0.28	68.00 ± 4.20	71.24 ± 2.16	5.2
KHGNN	62.31 ± 0.42	50.91 ± 0.40	52.32 ± 0.62	58.14 ± 3.38	55.40 ± 2.25	13
EHNN	86.21 ± 0.49	52.14 ± 0.76	58.23 ± 1.07	53.80 ± 7.26	65.45 ± 2.21	6.8
T-HyperGNN	85.32 ± 0.48	51.82 ± 0.38	58.82 ± 0.49	56.06 ± 3.92	64.58 ± 2.42	8
SheafHyperGNN	80.09 ± 2.45	51.03 ± 0.76	55.34 ± 4.39	$\underline{68.73} \pm 4.68$	$\underline{73.84} \pm 2.30$	8.2
ED-HNN	85.77 ± 0.46	50.86 ± 0.88	59.11 ± 0.57	64.79 ± 5.14	72.45 ± 2.28	6.4
HyperUFG	$\underline{89.32} \pm 0.75$	52.35 ± 0.04	$\underline{62.30} \pm 0.12$	67.61 ± 7.00	72.82 ± 2.22	3
HyperUnmix	$\mathbf{90.57} \pm 0.38$	$\mathbf{53.94} \pm 0.41$	$\mathbf{62.98} \pm 0.57$	$\mathbf{70.29} \pm 6.05$	$\mathbf{73.91} \pm 2.03$	1

the convolutional design of HyperUnmix, though tailored to heterophily, generalizes well to homophilic cases. Interestingly, heterophily-specific HNNs also perform competitively on homophilic hypergraphs, suggesting that their design principles may benefit both regimes. In contrast, while some traditional methods achieve reasonable performance on homophilic datasets, their performance deteriorates significantly in heterophilic settings, highlighting their limited applicability. Furthermore, accuracy levels on homophilic hypergraphs are generally higher, with narrower gaps among different methods. This trend suggests that tasks on homophilic hypergraphs are relatively less challenging, as most effective methods can achieve strong performance, whereas heterophilic tasks demand more sophisticated model designs.

4.3   Results on Synthetic Heterophilic Hypergraph Datasets

To further investigate the ability of HyperUnmix to address heterophily mixing, we conduct experiments on synthetic datasets with controlled hyperedge- and node-level homophily. Specifically, we generate hypergraphs using the Hypergraph Stochastic Block Model (HSBM) [24], varying homophily ratios between 0.3 and 0.9. Each dataset contains 10,000 nodes from two classes, with hyperedges of size three. We adopt the 50%/25%/25% train/validation/test splits and repeat experiments ten times. The results, illustrated in Fig. 3, demonstrate that HyperUnmix consistently outperforms all baselines across different homophily

Table 3. Evaluation results of node classification on homophilic hypergraphs: average accuracy (%) ± standard deviation. The best-performing method is boldfaced, and the runner-up is underlined.

Method	Cora	Citeseer	Pubmed	Cora-CA	DBLP-CA	*Avg. Rank*
MLP	$75.16_{\pm1.41}$	$71.71_{\pm1.01}$	$87.20_{\pm0.34}$	$75.17_{\pm1.41}$	$84.37_{\pm0.33}$	14.4
HGNN	$79.39_{\pm1.36}$	$72.45_{\pm1.16}$	$86.44_{\pm0.44}$	$82.64_{\pm1.65}$	$91.03_{\pm0.20}$	10.2
HyperGCN	$78.45_{\pm1.26}$	$71.28_{\pm0.82}$	$82.84_{\pm8.67}$	$79.48_{\pm2.08}$	$89.38_{\pm0.25}$	14
HyperND	$79.20_{\pm1.14}$	$72.62_{\pm1.49}$	$86.68_{\pm0.43}$	$80.62_{\pm1.32}$	$90.35_{\pm0.26}$	11
TF-HNN	$79.47_{\pm1.31}$	$73.00_{\pm1.27}$	$87.90_{\pm0.37}$	$84.19_{\pm0.89}$	$91.38_{\pm0.24}$	7.2
UniGCNII	$78.81_{\pm1.05}$	$73.05_{\pm2.21}$	$88.25_{\pm0.40}$	$83.60_{\pm1.14}$	$91.69_{\pm0.19}$	7.2
AllDeepSets	$76.88_{\pm1.80}$	$70.83_{\pm1.63}$	$88.75_{\pm0.33}$	$81.97_{\pm1.50}$	$91.27_{\pm0.27}$	10
AllSetTransformer	$78.58_{\pm1.47}$	$73.08_{\pm1.20}$	$88.72_{\pm0.37}$	$83.63_{\pm1.47}$	$91.53_{\pm0.23}$	7.2
HyperGT	$75.16_{\pm1.32}$	$72.33_{\pm0.91}$	$80.31_{\pm0.70}$	$80.19_{\pm2.17}$	$90.20_{\pm0.24}$	14
Hypergraph-MLP	$79.80_{\pm1.82}$	$73.90_{\pm1.57}$	$87.89_{\pm0.55}$	$78.57_{\pm1.12}$	$90.29_{\pm0.26}$	9.4
KHGNN	$80.67_{\pm0.76}$	$\underline{74.80}_{\pm1.10}$	$88.47_{\pm0.47}$	$84.25_{\pm0.74}$	$87.34_{\pm0.92}$	6.2
EHNN	$76.51_{\pm1.52}$	$68.67_{\pm0.75}$	$87.12_{\pm0.31}$	$81.68_{\pm0.81}$	$90.47_{\pm0.43}$	12.8
T-HyperGNN	$74.20_{\pm1.37}$	$71.21_{\pm0.87}$	$86.28_{\pm0.62}$	$75.01_{\pm1.44}$	$85.44_{\pm0.14}$	16
SheafHyperGNN	$81.30_{\pm1.70}$	$74.71_{\pm1.23}$	$87.68_{\pm0.60}$	$\underline{85.52}_{\pm1.28}$	$91.59_{\pm0.24}$	4.8
ED-HNN	$80.31_{\pm1.35}$	$73.70_{\pm1.38}$	$\mathbf{89.03}_{\pm0.53}$	$83.97_{\pm1.55}$	$\underline{91.90}_{\pm0.19}$	4
HyperUFG	$\underline{81.51}_{\pm0.99}$	$74.72_{\pm2.10}$	$88.73_{\pm0.42}$	$85.18_{\pm0.69}$	$91.67_{\pm0.31}$	3.2
HyperUnmix	$\mathbf{84.34}_{\pm1.55}$	$\mathbf{75.96}_{\pm1.19}$	$\underline{88.93}_{\pm0.38}$	$\mathbf{87.00}_{\pm1.25}$	$\mathbf{92.56}_{\pm0.45}$	1.2

levels. Although recent methods such as SheafHyperGNN and ED-HNN integrate heterophily-aware designs and can outperform MLP even under low homophily ratios, their performance gains plateau as homophily increases, whereas Hyper-Unmix continues to improve at a faster rate. Overall, these results demonstrate the adaptability and effectiveness of HyperUnmix in handling diverse neighborhood contexts.

4.4   Analysis of Over-Smoothing

Vanilla GNNs suffer from over-smoothing, where predictive performance deteriorates rapidly as model depth increases [5]. This issue persists in hypergraph learning. Figure 4 illustrates accuracy trends with respect to the number of layers. HyperUnmix exhibits remarkably flat accuracy curves on both homophilic and heterophilic hypergraphs, showing little degradation as depth increases, thereby providing strong evidence of its ability to mitigate over-smoothing. While UniGCNII incorporates strategies to alleviate over-smoothing and performs reasonably well on homophilic graphs, it struggles on heterophilic ones. Other advanced methods such as SheafHyperGNN and ED-HNN achieve competitive results with shallow architectures but degrade significantly at larger depths, resembling the behavior of traditional approaches.

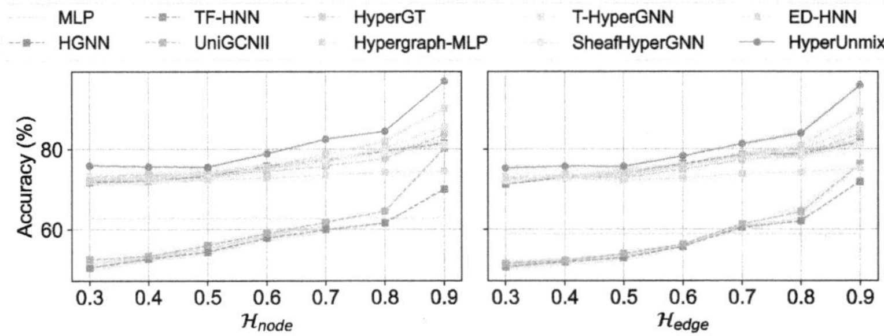

Fig. 3. Performance on synthetic hypergraphs with different homophily ratios.

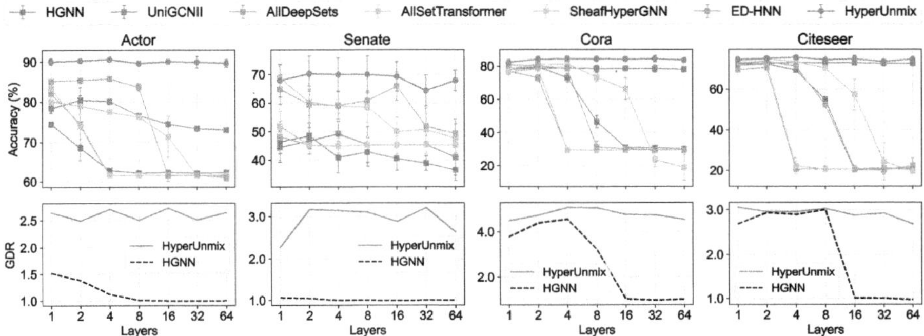

Fig. 4. Accuracy and Group Distance Ratio (GDR) w.r.t. the number of layers.

To quantify this effect, we further employ the group distance ratio R_g [43], defined as the ratio between inter-group and intra-group distances of node representations. Lower values indicate more severe over-smoothing. Figure 4 compares HyperUnmix with the baseline HGNN. While HGNN's R_g drops sharply with increasing depth, HyperUnmix consistently maintains high R_g values across both homophilic and heterophilic datasets, providing further evidence of its robustness against over-smoothing.

4.5 Robustness Analysis

Robustness is a critical dimension for evaluating the resilience of a model to real-world perturbations. To this end, we conduct systematic analyses under three types of perturbations: (1) structural perturbations, where we randomly add or remove a proportion of node–hyperedge incidences; (2) feature perturbations, where Gaussian noise is injected or random feature masking is applied; and (3) label perturbations, where a fraction of training labels are corrupted or removed. As shown in Fig. 5, HyperUnmix demonstrates strong robustness across all perturbation types, maintaining high accuracy even under severe noise conditions. Although models such as ED-HNN exhibit relatively stable degradation under

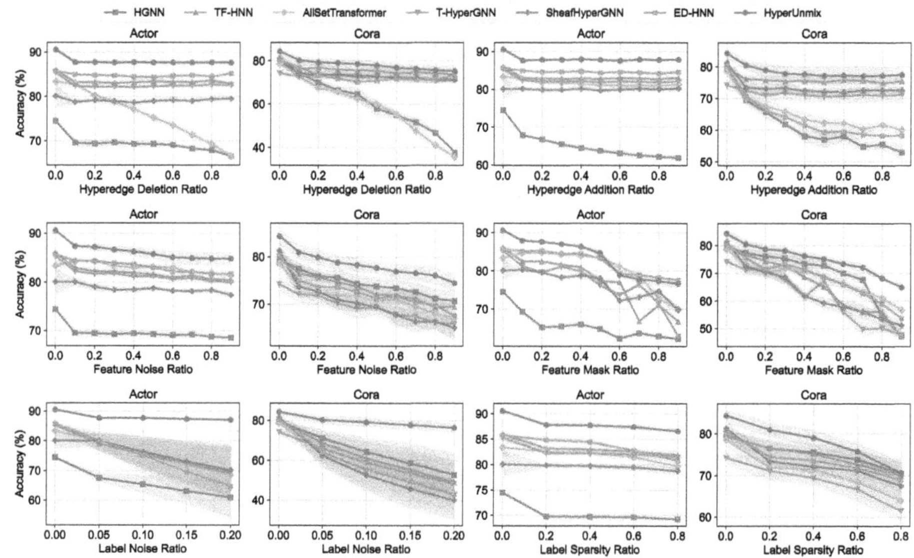

Fig. 5. Robustness to structure-, feature-, and supervision-level perturbations on representative heterophilic and homophilic hypergraphs.

specific perturbations like feature masking, their overall robustness remains inferior to HyperUnmix. These findings highlight the reliability of HyperUnmix in resisting both random noise and adversarial attacks in practical applications.

4.6   Ablation Study on Hyperbolic Manifolds

HyperUnmix models the representation space as a mixed-curvature manifold, where each submanifold corresponds to a specific class. In this work, we adopt hyperbolic geometry to characterize these submanifolds. A natural question arises: is a simple Euclidean space sufficient? To investigate this, we conduct an ablation study by replacing all hyperbolic operations with their trivial Euclidean counterparts. As reported in Tab. 4, incorporating hyperbolic geometry yields gains ranging from 0.28% to 3.96%, demonstrating the rationality and effectiveness of our design. This finding further aligns with recent studies [25] suggesting that hyperbolic spaces are particularly well-suited for representing graph-structured data with inherent hierarchical properties.

5   Related Work

5.1   Hypergraph Neural Networks

Hypergraph neural networks (HNNs) have emerged as a powerful paradigm for modeling higher-order interactions beyond simple pairwise relations, enabling richer representation learning compared to conventional graph neural networks

Table 4. Performance comparison of HyperUnmix and its Euclidean variant.

	Actor	Twitch-gamers	Pokec	Senate	House
HyperUnmix	90.57 ± 0.38	53.94 ± 0.41	62.98 ± 0.57	70.29 ± 6.05	73.91 ± 2.03
w/ Euclidean	88.76 ± 1.15	52.77 ± 0.77	59.02 ± 0.58	68.00 ± 3.95	71.55 ± 2.97
	Cora	Citeseer	Pubmed	Cora-CA	DBLP-CA
HyperUnmix	84.34 ± 1.55	75.96 ± 1.19	88.93 ± 0.38	87.00 ± 1.25	92.56 ± 0.45
w/ Euclidean	82.88 ± 1.60	73.55 ± 1.00	88.65 ± 0.33	84.83 ± 1.33	92.12 ± 0.29

(GNNs) [21]. Existing approaches can be broadly categorized into spectral-based and spatial-based methods. Early spectral approaches extend classical spectral convolution to hypergraphs by designing novel operators tailored to hypergraph structures [2,12,31], or by reducing hyperedge relationships to pairwise connections so that standard GNN architectures can be applied [40–42], albeit at the expense of structural fidelity. More recent advances [9,23,24,38] have introduced diverse techniques such as energy-based regularization, sheaf theory, and framelet transforms to enhance the exploitation of topological properties. While these methods offer strong theoretical foundations, their spectral formulations often entail relatively high computational overheads and lack the flexibility to filter out "bad neighbors" from intricate context.

Conversely, spatial methods implement localized message-passing schemes [15] directly on hypergraph incidence structures. Some generalize graph convolution by alternating vertex-to-hyperedge and hyperedge-to-vertex aggregation [8,19], while others integrate attention mechanisms to refine information flow [6]. More recent architectures [10,26,32,39] attempt to overcome locality limitations by incorporating long-range dependencies and inducing global topology-aware inductive biases. Beyond message passing, alternative perspectives leverage functional approximators such as the Kolmogorov–Arnold network [11] or adopt structure-aware supervision strategies [33] to innovate in capturing and exploiting structural features. Additionally, tensor-based models [20,36] encode hypergraph structures as higher-order tensors, thereby enabling more sophisticated interaction modeling and enhanced expressive power.

Despite these advances, the majority of existing HNNs implicitly assume homophilic settings where information from neighbors is beneficial. In heterophilic contexts, however, indiscriminate message fusion leads to heterophily mixing, severely degrading performance. This limitation highlights the necessity of developing new methodologies explicitly specialized to heterophilic hypergraphs.

5.2 Heterophilic Graph Learning

Heterophilic graph learning has emerged as a major challenge in GNNs. Conventional models typically assume homophily and rely on aggregating information from similar neighbors, but this assumption breaks down in heterophilic settings, leading to significant performance degradation [27]. To address this issue, recent studies have explored several directions. Spectral perspectives interpret GNNs as low-pass filters that overlook high-frequency signals, which capture neighborhood dissimilarity and have proven beneficial for heterophily; corresponding improvements introduce high-pass filtering through adaptive mixing [28] or signed message passing [4]. From the spatial perspective, some approaches enlarge the receptive field beyond local neighborhoods to incorporate long-range dependencies and mitigate the interference of dissimilar neighbors [30,44]. Inspired by the success of attention mechanisms, Transformer-based architectures have also been employed to flexibly regulate information propagation beyond hardwired interactions [22]. Another line of research focuses on graph rewiring [16], attempting to prune harmful or task-irrelevant edges and construct new connections, though often at the cost of losing latent structural cues.

Compared with simple graphs, research on heterophily in hypergraphs remains limited, despite evidence that it is even more prevalent in such settings [35]. Early attempts include incorporating learnable diffusion operators to adaptively control the influence of different classes of neighbors [37], or enriching hypergraph structure with additional geometric representations such as cellular sheaves, which have been shown to enhance performance under heterophily [9]. More recently, framelet-based hypergraph model HyperUFG [24] has combined low-pass and high-pass filters within a spectral framework, achieving state-of-the-art results. Building on these developments, our work introduces a mixed-curvature manifold formulation that enables class-specific message routing, offering a novel direction for advancing heterophilic hypergraph learning.

6   Conclusion

In this work, we introduce HyperUnmix, a novel hypergraph neural network that disentangles heterophily mixing through mixed-curvature product manifolds. By establishing a principled correspondence between node classes and geometric subspaces, and constraining message passing to predominantly occur within same-class submanifolds, HyperUnmix effectively suppresses interference from heterophilic neighbors. Extensive empirical evaluations on a wide range of real-world heterophilic and homophilic datasets establish new state-of-the-art performance and demonstrate the superiority of HyperUnmix, particularly in its resilience against over-smoothing and external perturbations. In the future, we plan to extend this methodology to more complex networked structures with higher-order interactions, such as simplicial complexes, and to further explore more efficient mechanisms for addressing heterophily mixing.

References

1. Akiba, T., Sano, S., Yanase, T., Ohta, T., Koyama, M.: Optuna: a next-generation hyperparameter optimization framework. In: KDD, pp. 2623–2631 (2019)

2. Bai, S., Zhang, F., Torr, P.H.S.: Hypergraph convolution and hypergraph attention. Pattern Recognit. **110**, 107637 (2021)
3. Bécigneul, G., Ganea, O.: Riemannian adaptive optimization methods. In: ICLR (2019)
4. Bo, D., Wang, X., Shi, C., Shen, H.: Beyond low-frequency information in graph convolutional networks. In: AAAI, pp. 3950–3957 (2021)
5. Chen, M., Wei, Z., Huang, Z., Ding, B., Li, Y.: Simple and deep graph convolutional networks. In: Proceedings of Machine Learning Research (ICML), vol. 119, pp. 1725–1735 (2020)
6. Chien, E., Pan, C., Peng, J., Milenkovic, O.: You are AllSet: a multiset function framework for hypergraph neural networks. In: ICLR (2022)
7. Chlenski, P., Chu, Q., Khan, R.R., Du, K., Moretti, A.K., Pe'er, I.: Mixed-curvature decision trees and random forests. In: ICML (2025)
8. Dong, Y., Sawin, W., Bengio, Y.: HNHN: hypergraph networks with hyperedge neurons. In: ICML Graph Representation Learning and Beyond Workshop (2020)
9. Duta, I., Cassarà, G., Silvestri, F., Lió, P.: Sheaf hypergraph networks. In: NeurIPS (2023)
10. Fang, X., Huan, C., Wang, B., Ma, S., Zhang, H., Zhao, C.: HyperSF: a hypergraph representation learning method based on structural fusion. In: ICASSP, pp. 1–5 (2025)
11. Fang, X., Wang, B., Huan, C., Ma, S., Zhang, H., Zhao, C.: HyperKAN: hypergraph representation learning with Kolmogorov-Arnold networks. In: ICASSP, pp. 1–5 (2025)
12. Feng, Y., You, H., Zhang, Z., Ji, R., Gao, Y.: Hypergraph neural networks. In: AAAI, pp. 3558–3565 (2019)
13. Fowler, J.H.: Legislative cosponsorship networks in the US house and senate. Soc. Netw. **28**(4), 454–465 (2006)
14. Ganea, O., Bécigneul, G., Hofmann, T.: Hyperbolic neural networks. In: NeurIPS, pp. 5350–5360 (2018)
15. Gilmer, J., Schoenholz, S.S., Riley, P.F., Vinyals, O., Dahl, G.E.: Neural message passing for quantum chemistry. In: Proceedings of Machine Learning Research (ICML), vol. 70, pp. 1263–1272 (2017)
16. Gong, S., Zhou, J., Xie, C., Xuan, Q.: Neighborhood homophily-based graph convolutional network. In: CIKM, pp. 3908–3912 (2023)
17. Gu, S., et al.: Functional hypergraph uncovers novel covariant structures over neurodevelopment. Hum. Brain Mapp. **38**(8), 3823–3835 (2017)
18. He, K., Zhang, X., Ren, S., Sun, J.: Deep residual learning for image recognition. In: CVPR, pp. 770–778 (2016)
19. Huang, J., Yang, J.: UniGNN: a unified framework for graph and hypergraph neural networks. In: IJCAI, pp. 2563–2569 (2021)
20. Kim, J., Oh, S., Cho, S., Hong, S.: Equivariant hypergraph neural networks. In: Avidan, S., Brostow, G., Cissé, M., Farinella, G.M., Hassner, T. (eds.) ECCV. LNCS, vol. 13681, pp. 86–103. Springer, Cham (2022). https://doi.org/10.1007/978-3-031-19803-8_6
21. Kipf, T.N., Welling, M.: Semi-supervised classification with graph convolutional networks. In: ICLR (2017)
22. Kong, K., Chen, J., Kirchenbauer, J., Ni, R., Bruss, C.B., Goldstein, T.: GOAT: a global transformer on large-scale graphs. In: Proceedings of Machine Learning Research (ICML), vol. 202, pp. 17375–17390 (2023)
23. Li, M., et al.: Deep hypergraph neural networks with tight framelets. In: AAAI, pp. 18385–18392 (2025)

24. Li, M., et al.: When hypergraph meets heterophily: new benchmark datasets and baseline. In: AAAI, pp. 18377–18384 (2025)
25. Liu, Q., Nickel, M., Kiela, D.: Hyperbolic graph neural networks. In: NeurIPS, pp. 8228–8239 (2019)
26. Liu, Z., Tang, B., Ye, Z., Dong, X., Chen, S., Wang, Y.: Hypergraph transformer for semi-supervised classification. In: ICASSP, pp. 7515–7519 (2024)
27. Luan, S., et al.: The heterophilic graph learning handbook: Benchmarks, models, theoretical analysis, applications and challenges. arXiv arXiv:2407.09618 (2024)
28. Luan, S., et al.: Revisiting heterophily for graph neural networks. In: NeurIPS (2022)
29. Mao, M., Lu, J., Han, J., Zhang, G.: Multiobjective e-commerce recommendations based on hypergraph ranking. Inf. Sci. **471**, 269–287 (2019)
30. Pei, H., Wei, B., Chang, K.C., Lei, Y., Yang, B.: Geom-GCN: geometric graph convolutional networks. In: ICLR (2020)
31. Prokopchik, K., Benson, A.R., Tudisco, F.: Nonlinear feature diffusion on hypergraphs. In: Proceedings of Machine Learning Research (ICML), vol. 162, pp. 17945–17958 (2022)
32. Saxena, S., Ghatak, S., Kolla, R., Mukherjee, D., Chakraborty, T.: DPHGNN: a dual perspective hypergraph neural networks. In: KDD, pp. 2548–2559 (2024)
33. Tang, B., Chen, S., Dong, X.: Hypergraph-MLP: learning on hypergraphs without message passing. In: ICASSP, pp. 13476–13480 (2024)
34. Tang, B., Liu, Z., Jiang, K., Chen, S., Dong, X.: Training-free message passing for learning on hypergraphs. In: ICLR (2025)
35. Veldt, N., Benson, A.R., Kleinberg, J.: Combinatorial characterizations and impossibilities for higher-order homophily. Sci. Adv. **9**(1), eabq3200 (2023)
36. Wang, F., Pena-Pena, K., Qian, W., Arce, G.R.: T-HyperGNNs: hypergraph neural networks via tensor representations. IEEE Trans. Neural Netw. Learn. Syst. **36**(3), 5044–5058 (2025)
37. Wang, P., Yang, S., Liu, Y., Wang, Z., Li, P.: Equivariant hypergraph diffusion neural operators. In: ICLR (2023)
38. Wang, Y., Gan, Q., Qiu, X., Huang, X., Wipf, D.: From hypergraph energy functions to hypergraph neural networks. In: Proceedings of Machine Learning Research (ICML), vol. 202, pp. 35605–35623 (2023)
39. Xie, L., Gao, S., Liu, J., Yin, M., Jin, T.: K-hop hypergraph neural network: a comprehensive aggregation approach. In: AAAI, pp. 21679–21687 (2025)
40. Yadati, N., Nimishakavi, M., Yadav, P., Nitin, V., Louis, A., Talukdar, P.P.: HyperGCN: a new method for training graph convolutional networks on hypergraphs. In: NeurIPS, pp. 1509–1520 (2019)
41. Yan, Y., Chen, Y., Wang, S., Wu, H., Cai, R.: Hypergraph joint representation learning for hypervertices and hyperedges via cross expansion. In: AAAI, pp. 9232–9240 (2024)
42. Yang, C., Wang, R., Yao, S., Abdelzaher, T.F.: Semi-supervised hypergraph node classification on hypergraph line expansion. In: CIKM, pp. 2352–2361 (2022)
43. Zhou, K., Huang, X., Li, Y., Zha, D., Chen, R., Hu, X.: Towards deeper graph neural networks with differentiable group normalization. In: NeurIPS (2020)
44. Zhu, J., Yan, Y., Zhao, L., Heimann, M., Akoglu, L., Koutra, D.: Beyond homophily in graph neural networks: current limitations and effective designs. In: NeurIPS (2020)

Towards Real-Time Maintenance of HNSW for ANN Search on Edge Devices

Wenbo Zhao[1], Yuheng Chang[1,2(✉)], Hui Li[1,3(✉)], and Jiangtao Cui[1]

[1] School of Computer Science and Technology, Xi'an 710126, China
{wbzhao,hli,cuijt}@xidian.edu.cn, changyuheng2000@163.com
[2] Alibaba Cloud Computing Co. Ltd., Hangzhou, China
[3] Shanghai Yunxi Technology, Shanghai, China

Abstract. Edge intelligence requires fast approximate nearest neighbor (ANN) search on mobile devices with tight memory and power budgets. Under frequent updates, the mainstream HNSW index suffers from node isolation, where nodes lose incoming edges. This leads to fragmented graph connectivity, degraded recall, and increased latency. Such issues are exacerbated on resource-limited edge devices, where rapid data changes render traditional index reconstruction impractical. This paper presents a lightweight, real-time maintenance framework for HNSW that dynamically detects and repairs isolated nodes. By analyzing the root causes of connectivity loss, we classify isolated nodes into direct and indirect types, each with a tailored maintenance strategy. Indirect isolations are mitigated by integrating auxiliary edges to the main index, while direct isolations are managed via a compact, extra index. During queries, both indexes are searched concurrently to ensure complete coverage with minimal overhead. Experimental results show that the proposed approach effectively eliminates isolated nodes and stabilizes recall, with at most 10% latency overhead and less than 2% additional memory footprint, enabling practical on-device vector search under dynamic data updates.

Keywords: On-device ANN search · HNSW · Dynamic updates · Isolated nodes

1 Introduction

Advances in mobile chips and model compression [1,2] have migrated AI services from cloud to edge devices for enhanced privacy and personalized services. To bridge the gap between static parametric knowledge and evolving local contexts, on-device models increasingly rely on vector search for retrieving streaming local data [3]. In such environments, local data is not static but dynamically changing over time. For instance, a meeting assistant [4] indexes dialogue segments in

---

W. Zhao and Y. Chang–Contributed equally to this work.

H. Jung et al. (Eds.): DASFAA 2026, LNCS 16536, pp. 199–215, 2026.
https://doi.org/10.1007/978-981-92-0366-6_13

real time and must physically remove invalidated vectors as later utterances often revise previous contexts. Similarly, in incremental visual mapping [5], the system extracts keyframe vectors from visual streams, continuously replacing redundant or low-quality frames with updated representations.

To support high-quality vector retrieval, HNSW indexes have emerged as the mainstream choice due to their near-optimal search latency and recall [6, 7]. Nevertheless, a fundamental mismatch exists between HNSW's static design and the fluid nature of on-device data. HNSW lacks the inherent flexibility to handle the frequent data turnover typical of edge applications. Unlike server-side deployments that can tolerate index expansion with elastic resources, on-device applications must operate within a strictly capped memory envelope. This hard constraint necessitates physical slot reuse: the system must actively reclaim and overwrite the memory slots of stale vectors as new data arrives.

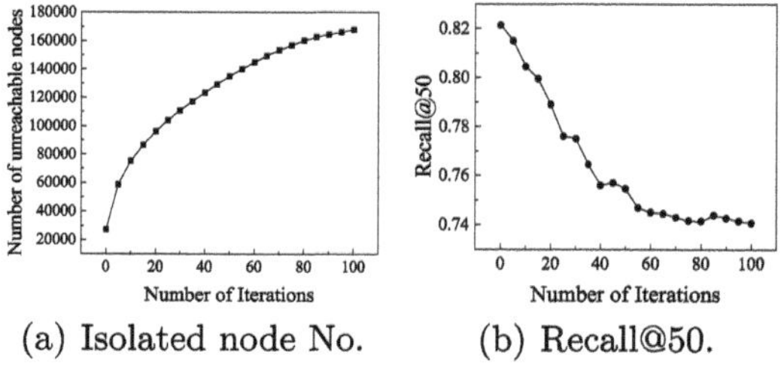

Fig. 1. HNSW update statistics on ImageNet.

Such mandatory memory reuse, however, may interfere with HNSW's marked-deletion mechanism [6]. Since HNSW does not explicitly track in-edges, overwriting a node and modifying its associated neighbors can sever critical routing paths, creating *isolated source nodes* (nodes without incoming edges). The issue threatens graph integrity, as shown in Fig. 1, where iterative updates on the ImageNet dataset (e.g., replacing 1% of nodes per iteration) lead to a cumulative increase in hard-to-reach nodes, causing a sharp drop in recall.

We address the issue of topological decay within a fully in-memory on-device setting. This is motivated by the fact that user-scale vector stores are inherently capped by application memory budgets rather than aiming for billion-scale growth [3,5,8]. Given this finite scale, storage capacity is no longer the primary concern; instead, the challenge shifts to avoiding the unpredictable latency and power drain incurred by frequent batch rebuilds or flash-based indexing. Thus, developing a lightweight and online maintenance strategy that preserves search quality within a fixed memory envelope is a critical yet unresolved challenge.

Prior works have attempted to mitigate this issue, but fail to align with the stringent constraints of real-time, in-memory, on-device search. Disk-based indices rely on periodic merges and external storage [9–12], incurring prohibitive overhead and power drain on mobile devices. Within the in-memory domain, existing solutions typically suffer from either high latency or performance fluctuations. Specifically, bath-rebuild approaches require periodic global scans to repair the graph [13,14], which becomes very costly when executed frequently. Other real-time maintenance methods [15,16] involve expensive searches in locating in-neighbors for local repairs during node updates.

In this paper, we propose **HNSW-DM** (Dynamic Maintenance), a lightweight algorithm designed for robust HNSW updates. By distinguishing the root cause of isolated source nodes, our approach employs two complementary strategies: maintaining a compact extra index and incorporating auxiliary edges within the main index to restore reachability. To ensure memory efficiency, nodes are added to the extra index only when local topological repair is not applicable. Furthermore, a joint search mechanism is conducted concurrently over both indices to enhance recall robustness. Experimental results demonstrate that, with only 2% additional index memory usage and no loss in recall, our approach effectively minimizes unreachable nodes, offering an efficient solution for real-time, on-device vector search.

2 Related Works

Overview of Graph-based ANNS. Traditional ANN methods like IVF [17] and LSH [18] reduce search costs through clustering or hashing but often suffer from recall loss or high memory overhead in high-dimensional spaces. In contrast, graph-based indexes like HNSW [6], NSG [19], and DiskANN [9] use data connectivity and greedy search to achieve high accuracy with low latency. Variants such as DPG [20], NSSG [21], and HCNNG [22] further improve efficiency. These methods perform well in static settings but struggle with dynamic updates.

Maintenance of Disk-Based HNSW. To support large-scale vector retrieval, DiskANN [9] offloads lower graph layers to disk but cannot efficiently handle deletions. FreshDiskANN [10] introduces lazy deletion and batch merging, though recall still fluctuates with the merging cycle. IP-DiskANN [11] performs global re-search during deletions for in-neighbor reconstruction and adds replacement edges for instant repair. Topology-Aware Update [12] focuses on small-batch updates with localized repair to minimize redundant I/O. However, these methods rely on batch reconstructions [10], global computations [11], or complex graph structures [12], thus unsuitable for in-memory or on-device environments.

Maintenance of In-memory HNSW. To address dynamic update in HNSW, MINT [13] detects isolated nodes via batch scanning and stores them in an auxiliary index, but triggers an $O(N \log N)$ global rebuild when isolated nodes exceed a threshold, causing recall degradation between reconstructions. PRO-HNSW [14] performs periodic graph-wide scans and BFS-style repairs (e.g., finding disconnected parts and reconnecting nodes), which are prohibitively expensive for high-rate on-device updates. IPGM [15] explicitly maintains in-neighbors and repairs links during deletions and updates, but incurs high memory overhead.

CleANN [16] avoids in-neighbor storage and global scans by dynamically detecting and repairing tombstone links during queries, but frequent repairs lead to unpredictable latency spikes. While these methods [15,16] address general graph maintenance, they lack HNSW-specific structural optimizations and remain too computationally intensive for on-device deployment.

3 Problem Statement

Preliminaries and Notation. This work studies online maintenance of HNSW under dynamic vector updates on client-side devices. User interactions on devices, such as taking photos or refreshing feeds, produce frequent, small updates, leading to continuous insertions, deletions, or replacements of vector embeddings. We adopt a *mark-deletion* strategy, where deleted nodes are marked invalid but retained for reuse by new insertions. While this reduces rebuilding overhead, it can create *isolated source nodes* with outgoing but no incoming edges, weakening graph connectivity and lowering recall. Given limited resources, we aim for a lightweight and online repair mechanism that avoids frequent global rebuilds.

3.1 Definitions

We consider a set of n vectors $X = \{x_1, x_2, \cdots, x_n\}$, where each $x_i \in \mathbb{R}^d$ denotes a d-dimensional vector. For any $x_i, x_j \in X$, their dissimilarity is measured by a distance function $\mathsf{dist}(x_i, x_j)$ (e.g., Euclidean distance or cosine similarity).

ANN Search and Recall@k. Given a query vector $x_q \in \mathbb{R}^d$, an ANN search returns a set $R \subset X$ of k approximate nearest neighbors of x_q under $\mathsf{dist}(x_i, x_j)$. Let $\mathcal{G}$ denote the ground-truth top-k neighbors of x_q. Since computing $\mathcal{G}$ exactly is infeasible for large-scale datasets, ANN methods provide efficient approximations that are close to but not necessarily optimal. The search accuracy is measured by *Recall@k*, defined as $\frac{|R \cap \mathcal{G}|}{k}$.

HNSW Index and Its Updates. An HNSW index is a multi-layer directed graph with L layers, represented as $G = (G_1, G_2, \cdots, G_L)$, where each layer $G_l = (V_l, E_l)$ is a proximity graph. Here, V_l is the node set of the l−th layer, and edges are recorded in $E_l \subset V_l \times V_l$, with $V_l \subset V_{l-1}$ for all $l > 0$.

For a node $p \in V_l$, $N_{\text{out}}^l(p)$ and $N_{\text{in}}^l(p)$ represent its outgoing and incoming neighbor sets in G_l, respectively. We consider dynamic updates with insertions and mark-deletion, where a deleted node is marked invalid and its memory may be reused for later insertions.

Isolated Source Nodes. These are nodes with no incoming edges in any layer of the HNSW graph, i.e., $\mathcal{I}(G) = \bigcup_{\forall l \in [1,L]} \{v \mid \exists v \in V_l \text{ and } N_{\text{in}}^l(v) = \emptyset\}$. For brevity, we refer to them as isolated nodes hereafter.

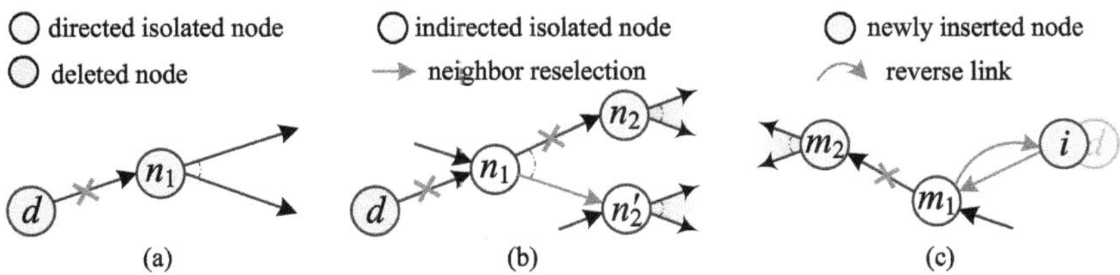

Fig. 2. Formation of isolated nodes caused by node deletion and neighbor reselection.

3.2 Causes of Isolated Source Nodes

Isolated nodes primarily result from bounded neighbor lists and local, asymmetric pruning during updates. When the neighbor list of a node overflows, heuristic selection removes some links. However, the removed nodes are not necessarily reconnected elsewhere, so their inbound references can gradually disappear. Based on whether a node is an original neighbor of the recycled (deleted) node, isolated nodes can be categorized into two types.

Direct Isolated Node. It is a node whose only incoming edge originates from a deleted node, causing its in-degree to drop to zero after deletion. Formally, in layer l with $G_l = (V_l, E_l)$, deleting a node $d \in V_l$ isolates any one-hop neighbor $n_1 \in N_{out}^l(d)$ satisfying $deg_{in}(n_1) = 1$ and $(d, n_1) \in E_l$ (Fig. 2(a)).

Indirect Isolated Node. It is a node whose only incoming edge is removed by *local neighbor reselection* or *reverse-link replacement*, even though its in-neighbor still exists in the index. Formally, in layer l with $G_l = (V_l, E_l)$, each node $p \in V_l$ maintains a bounded outgoing neighbor set $N_{out}^l(p)$ with maximum degree M. After neighbor reselection, p's outgoing neighbor set becomes $N_{out}^{\prime l}(p)$. If there exists a node $v \in N_{out}^l(p) \setminus N_{out}^{\prime l}(p)$, such that $deg_{in}(v) = 1$ and $(p, v) \in E_l$, then removing (p, v) eliminates v's only incoming edge, making v an indirect isolated node. Figure 2(b) illustrates *deletion-triggered reselection*: after a deletion, n_1 may drop (n_1, n_2), isolating n_2 if n_1 was n_2's only in-neighbor. Figure 2(c) shows *insertion-triggered updates*: when inserting a node i, its neighbor m_1 may add a reverse edge to i while replacing (m_1, m_2). If that edge is m_2's only incoming link, m_2 becomes indirectly isolated.

4 System Overview

Core Design Insight. We propose a hybrid maintenance framework with tailored strategies for direct and indirect isolated nodes. Since the in-neighbors of direct isolated nodes have been deleted, repairing them in the main index would require a full-graph search to find new incoming neighbors, which is prohibitively expensive. We therefore store them in an auxiliary HNSW index, keeping them retrievable without rewiring the main index.

In contrast, indirect isolated nodes retain their original in-neighbors. We therefore repair them locally by restoring a small number of incoming edges. To

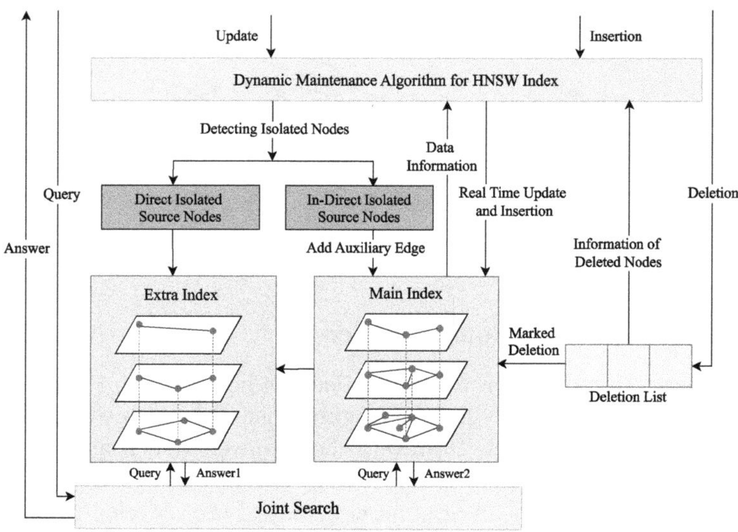

Fig. 3. System architecture of dynamic maintenance and search for HNSW-DM.

bound memory overhead, each node reserves a small extra edge budget under a global budget M_e. If the original in-neighbor p has no free reserved slot to restore (p, v), we offload v to the auxiliary index. This design preserves connectivity while ensuring bounded memory overhead and efficient updates.

System Architecture. Figure 3 shows the HNSW-DM (Dynamic Maintenance) system, designed to address the recall degradation during updates. It consists of four core modules. *The main index* stores all active nodes and supports dynamic insertions and deletions. Deletions are handled lazily by queuing deleted nodes instead of immediate removal. During insertions, the system updates neighbor connections and triggers maintenance when isolated nodes are detected. *The extra index* stores direct isolated nodes and those indirect ones that cannot be repaired locally. It operates independently but participates in the joint search with the main index to ensure consistent recall.

...	Number of Neighbors	Neighbor List	Data Vector	Label	In-Degree	Extra Nbr List	...

(a) Memory layout of level-0 nodes.

...	Number of Neighbors	Neighbor List	In-Degree	Extra Nbr List	...

(b) Memory layout of upper layer nodes.

Fig. 4. Memory layout of nodes in the main index.

After each update, *the maintenance module* applies a hybrid repair strategy based on the types of isolated nodes and updates the main index. During queries, *the joint search module* searches both indexes in parallel, and merges results to

---

Algorithm 1: UpdateInDeg: In-Degree maintenance

Input: Node p whose neighbors are updated; layer l of the HNSW index where the update occurs; *flag* indicating whether p is newly inserted (true) or undergoing neighbor reselection (false), updated neighbor set *newNbs*.

Output: Set S of isolated source nodes.

1 $S \leftarrow \emptyset$
2 **for** each node $i \in newNbs$ **do**
3 $\quad$ i.inDegree++

4 **if** ! *flag* **then**
$\quad$ // case of re-selecting neighbors
5 $\quad$ **for** each old neighbor $j \in N^l_{\text{out}}(p) \bigcup EN^l_{\text{out}}(p)$ **do**
6 $\quad\quad$ j.inDegree-
7 $\quad\quad$ **if** j.inDegree $= 0$ **then**
8 $\quad\quad\quad$ $S \leftarrow S \bigcup \{j\}$

9 **return** S

---

return the final k neighbors. A key benefit is that hard-to-repair items are routed to the extra index, leaving only bounded local modifications to the main index.

Generally, HNSW-DM detects isolated nodes in real time and applies local maintenance. Details of detection and maintenance of isolated nodes, as well as the dual-index search algorithm, are provided in Sect. 5.

5 The Real-Time Maintenance of HNSW

5.1 Detecting Isolated Nodes in Real-Time

As defined in Sect. 3.1, an isolated source node has zero in-degree in any layer of the HNSW index. To detect such nodes, the system records and tracks the in-degree of each node in real-time.

Data Structure of Nodes in the Main Index. The in-degree information is stored directly in the node structure by adding an extra field. This method reduces addressing overhead compared to storing all in-degrees in a separate memory area, allowing direct access to the in-degree information and improving system response time. In HNSW, upper-layer nodes store only adjacency information, while level-0 nodes store both vector and label data, as shown in Fig. 4. The Label field serves as a unique identifier for each node, linking it to its corresponding data stored externally (e.g., in a database or file system). The extra neighbor list deals with the indirect isolated nodes, which will be discussed in Sect. 5.2.

When a node's neighbor set is updated, the system synchronously adjusts the in-degree of the affected nodes to maintain consistency. Algorithm 1 details this process: all new neighbors have their in-degree incremented, while in the case of neighbor reselection (indicated by *flag* $=$ false), the in-degree of old neighbors is decremented accordingly (step 5). $EN^l_{\text{out}}(p)$ denotes the extra neighbor set of node p. Nodes whose in-degree drops to zero are immediately recorded as isolated source nodes.

5.2   Maintaining Reachability of Isolated Source Nodes

We next describe how the system restores reachability for isolated source nodes once they are detected. Different maintenance strategies are applied to direct and indirect isolated nodes, respectively, as presented in the following sections.

Maintaining the Direct Isolated Nodes. In Fig. 2(a), we have shown that direct isolated nodes occur when a node loses all incident edges after deletion, rendering it unreachable. Reconnecting such nodes by searching for nearby vertices is a potential solution, but this requires a full-graph search, which is prohibitively expensive during updates.

To preserve the reachability of direct isolated nodes without affecting the main index, the system maintains an extra HNSW index $eIdx$. The overall maintenance process is outlined in Algorithm 2. Once an isolated node is detected, it is asynchronously inserted into $eIdx$ and remains searchable independent of ongoing updates. To prevent redundant insertions, $eIdxSet$ tracks nodes already added to $eIdx$, as a node may be isolated in multiple layers. Insertion occurs only if the node is not in $eIdxSet$, and it runs in a separate thread, ensuring the main update process proceeds without blocking and minimizing latency.

Data Structure of Nodes in the Extra Index. The extra HNSW differs from the main index in the layer-0 nodes layout. As shown in Fig. 5(a), layer-0 nodes omit vector storage to save memory. When vector data are needed, the system retrieves the node's internal id from the main index via its external label. The vector data for nodes in the extra index is accessed by first looking up the internal id using the label, and then retrieving the vector from the main index. This design requires only two additional memory accesses while significantly reducing memory consumption by eliminating redundant storage.

During query processing, both the main and extra indexes are searched jointly to ensure full reachability of isolated nodes. Details would be given in Sect. 5.3.

Maintaining the Indirect Isolated Nodes. Indirect isolated nodes are mainly caused by neighbor reselection, with their original in-neighbors still existing. Thus, connectivity can be restored by adding missing edges. To do this, each node reserves a small amount of additional edge space, as shown in Fig. 4.

Algorithm 2: DirectMaintain: Maintain direct isolated source nodes

> **Input**: A direct node p to be maintained; extra HNSW index $eIdx$ containing all the direct isolated nodes; set $eIdxSet$ storing node identifiers in $eIdx$;
> **Output**: The updated $eIdx$ and $eIdxSet$ with node p newly added;

```
1  spawn thread
2  |  Lock
3  |  |  if p.internalId ∉ eIdxSet then
4  |  |  |  eIdx.addNode(p) ;        // initialize a new thread to asynchronously
                                        insert p into the extra index
5  |  |  |  eIdxSet ← eIdxSet ∪ {p.internalId}
```

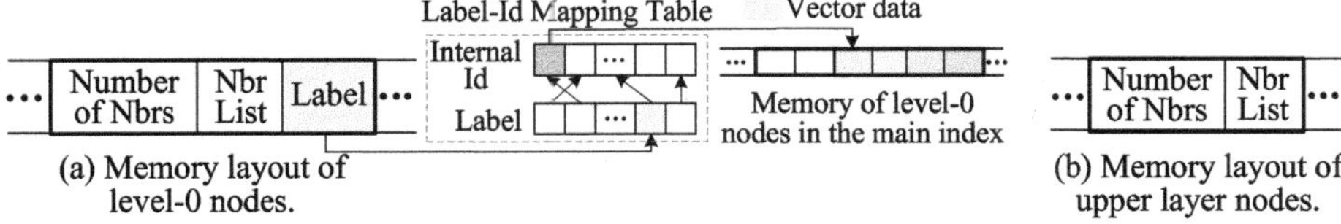

(a) Memory layout of level-0 nodes.

(b) Memory layout of upper layer nodes.

Fig. 5. Memory layout of nodes in the extra index.

Algorithm 3: IndirectMaintain: Maintain indirect isolated nodes

Input: Indirect isolated node q; former in-neighbor p of q; layer l of the main index where q resides; extra index $eIdx$ for maintaining direct isolated nodes; set $eIdxSet$ storing node IDs in $eIdx$.

Output: Updated main and extra indexes.

1   $q_l \leftarrow \text{getNode}(q, l)$
2   $p_l \leftarrow \text{getNode}(p, l)$
3   **if** $q_l.\texttt{internalId} \notin eIdxSet$ **then**
4   **if** $|N_{\text{out}}^l(p)| < M$ **then**
5       $N_{\text{out}}^l(p) \leftarrow N_{\text{out}}^l(p) \cup \{q_l\}$
6   **else if** $|EN_{\text{out}}^l(p)| < M_e$ **then**
7       $EN_{\text{out}}^l(p) \leftarrow EN_{\text{out}}^l(p) \cup \{q_l\}$
8   **else**
9       $\text{DirectMaintain}(eIdx, q, eIdxSet)$

Algorithm 3 describes the maintenance of indirect isolated nodes. Suppose an indirect isolated node q is detected in layer l, and its original in-neighbor in this layer is p. We use p_l and q_l to denote node instances of p and q in layer l. Node p restores reachability by attempting to add q_l to its neighbor list $N_{\text{out}}^l(p)$, or extra neighbor list $EN_{\text{out}}^l(p)$, thereby re-establishing the missing link (p_l, q_l). To bound memory overhead and preserve cache locality, each $EN_{\text{out}}^l(p)$ is capped at M_e entries. Once $|EN_{\text{out}}^l(p)| = M_e$, no more extra links can be added from p at layer l, and q is instead offloaded to the extra index. In practice, this fallback is rare. For million-scale datasets, $M_e \in \{1, 2\}$ is sufficient to accommodate almost all indirect isolated nodes with only a few MB of extra memory, yielding a practical trade-off between memory efficiency and update cost.

5.3 HNSW-DM Algorithm

So far, we have presented the real-time detection of isolated nodes and the corresponding type-aware maintenance strategies, which form the core components of HNSW-DM. By maintaining reachability during updates and including the extra index in search, HNSW-DM mitigates accuracy degradation under continuous updates. Next, we describe the update and searching processes of HNSW-DM.

Node Reuse Algorithm. The HNSW index update includes three operations: insertion, deletion, and replacement. For deletion, HNSW-DM retains the original mark-deletion strategy. Since both insertion and replacement essentially

---

Algorithm 4: NodeReuse: Reusing deleted nodes upon insertions

Input: Reused node d whose memory slot is reassigned; node i reusing d's space; main index $mIdx$; extra index $eIdx$; set $eIdxSet$ storing node IDs in $eIdx$.

Output: Updated main and extra indexes.

1   $L_{max} \leftarrow \text{getMaxLayer}(mIdx, d)$
2   **for** each layer $l \in [0, L_{max}]$ **do**
3     $C \leftarrow N_{out}^l(d)$ // C: candidate set for neighbor reselection
4   **for** each node $v \in N_{out}^l(d)$ **do**
5       $C \leftarrow C \bigcup N_{out}^l(v)$
6   **for** each node $p \in N_{out}^l(d)$ **do**
7       $newNbs \leftarrow \text{getNbrHeristic}(p, C) \bigcup EN_{out}^l(p)$
8       $S \leftarrow \text{UpdateInDeg}(p, l, false, newNbs)$
9   **if** $S \ != \emptyset$ **then**
10   **for** each node u *in* S **do**
11         $\text{IndirectMaintain}(u, p, l, eIdx, eIdxSet)$
12       $N_{out}^l(p) \leftarrow newNbs$
13     $d.vector \leftarrow i.vector$ // replace node d with node i inserted
14     $newNbs \leftarrow \text{getMNN}(d, l)$
15     $S \leftarrow \text{UpdateInDeg}(d, l, false, newNbs)$
16   **if** $S! = \emptyset$ **then**
17   **for** each node $v \in S$ **do**
18        $\text{DirectMaintain}(v, eIdx, eIdxSet)$
19     $N_{out}^l(d) \leftarrow newNbs$
20     $EN_{out}^l(d) \leftarrow \emptyset$
21   **for** each node $p \in newNbs$ **do**
22   **if** $d \notin N_{out}^l(p)$ **then**
23   **if** $|N_{out}^l(p)| = M$ **then**
24         $C \leftarrow N_{out}^l(p) \bigcup \{d\}$
25         $newNbs \leftarrow \text{getNbrHeristic}(p, C) \bigcup EN_{out}^l(p)$
26         $S \leftarrow \text{UpdateInDeg}(p, l, false, newNbs)$
27   **if** $S \ != \emptyset$ **then**
28   **for** each node v *in* S **do**
29           $\text{IndirectMaintain}(v, p, l, eIdx, eIdxSet)$
30         $N_{out}^l(p) \leftarrow newNbs$
31   **else**
32         $N_{out}^l(p) \leftarrow N_{out}^l(p) \bigcup \{d\}$
33         $\text{UpdateInDeg}(p, l, true, \{d\})$

---

reuse the memory location of nodes, HNSW-DM improves the existing node reuse algorithm (see Algorithm 4). For a reused node d, node reuse operations are performed at each layer where d exists. During the processing of each layer, HNSW-DM sequentially executes the neighbor reselection (steps 3–12), node replacement (steps 13–19), and backward edge construction (steps 20–32), with each step dynamically maintaining potential isolated nodes.

First, each one-hop neighbor of the reused node d undergoes neighbor reselection. Since this may alter nodes' in-degrees and create indirect isolated nodes, the system triggers in-degree maintenance (steps 7–8) and handles any resulting isolation (steps 9–11). To ensure the degree invariance for extra neighbors during this process, they are explicitly merged into the $NewNbs$ set. Next, node d is physically overwritten with the new node's vector and label, followed by the selection of its new neighbors (steps 13–14). The in-degree maintenance algorithm is invoked again, and if any former neighbor's in-degree drops to zero, the system initiates direct isolated node maintenance (steps 16–18). Finally, reverse edges are established from each new neighbor p to the inserted node. If p's out-degree is below M, the new node is added directly (steps 31–33). Otherwise, M neighbors are chosen via heuristic reselection (steps 23–25). If the new node is excluded, reverse edge addition fails. The process concludes with a final round of in-degree and indirect isolation maintenance (steps 26–29).

For insertions, Algorithm 4 is called with the new node i and the deleted node d being reused. For updates, the same algorithm applies, where d is the old node and i is the new one built from updated data. This ensures that all isolated nodes arising during insertions or updates can be safely maintained.

Joint Search Algorithm. HNSW-DM maintains direct isolated nodes in the extra index and repairs connections to indirect isolated nodes with extra edges in the main index. Next, we introduce the dual search algorithm of HNSW-DM, which incorporates both extra edges and the extra index into the search range.

(1) Search Strategy Considering Extra Edges. During graph search on an HNSW index, the neighbors of the current node are traversed. In the original search algorithm, only the standard neighbor list is considered. However, after dynamically maintaining indirect isolated nodes, some neighbors are stored separately in the extra edge list. Therefore, the search needs to retrieve both the original and extra neighbor lists when expanding a node. In our proposed main index search algorithm, the node's neighbor list and the extra edge list are combined as the current node's neighbor set, with the rest of the procedure remaining the same as the original search algorithm. Due to space limitations, we have omitted the pseudocode description of the algorithm.

(2) Search Strategy with Extra Index. HNSW-DM searches the main index and extra index separately to obtain two sets of k nearest neighbors, merges the results, and selects the final top-k candidates from the combined set.

The search process is shown in Algorithm 5. The main and extra indexes are searched concurrently: the main thread searches the main index with extra edges (steps 8–11), while an auxiliary thread handles the extra index (steps 3–7). After the main thread completes, it obtains the temporary result R_m and waits for the auxiliary thread to finish and return R_e. Since the extra index contains fewer nodes, its search typically finishes by the time the main index search is complete, allowing the main thread to retrieve results without delay. The two result sets are then merged, and the top-k nearest neighbors are selected as the final output (step 13). The concurrent search guarantees access to both types of isolated nodes efficiently.

To keep the system simple, we do not migrate nodes from the extra index back to the main index. Direct isolations are inherently rare: a node n becomes

---

Algorithm 5: dualSearchKNN: Joint search over two indexes

Input: Target node $target$; number of nearest neighbors k; main HNSW index $mIdx$; extra HNSW index $eIdx$.

Output: k nearest neighbors of $target$.

1 $L_m \leftarrow$ getMaxLayer($mIdx$)
2 $L_e \leftarrow$ getMaxLayer($eIdx$)
3 **spawn thread**
4      $ep \leftarrow$ getRandomEntry($eIdx, L_e$)// ep: entry point for searching layer L_e
5   **for** $l = L_e$ **to** 1 **do**
6          $ep \leftarrow$ searchKNNLayer($ep, target, 1, l, eIdx$) // search in the extra index
7      $R_e \leftarrow$ searchKNNLayer($ep, target, k, 0, eIdx$) // get k nearest neighbors in the bottom layer of $eIdx$

8 $ep \leftarrow$ getRandomEntry($mIdx, L_m$)
9 **for** $l = L_m$ **to** 1 **do**
10     $ep \leftarrow$ searchKNNLayer($ep, target, 1, l, mIdx$) // search in the main index

11 $R_m \leftarrow$ SearchKNNLayer($ep, target, k, 0, mIdx$)
12 **thread**.join() // wait for the thread to finish and get result
13 $Knns \leftarrow$ top-k elements from $R_e \bigcup R_m$
14 **return** $Knns$

---

directly isolated only when its unique predecessor d is replaced, whereas indirect isolation can be triggered by replacing any predecessor of d. This large gap in trigger frequency keeps the extra index compact. Moreover, extra neighbor slots are reclaimed on deletions, and most indirect isolations are fixed by extra edges, with only rare failures falling back to the extra index. This slows the growth of the extra index. Finally, on-device workloads are typically time-bounded, so isolated nodes do not accumulate unboundedly. If the extra index ever exceeds a preset threshold, a global HNSW rebuild can be triggered, which is expected to be extremely infrequent and has a negligible impact.

6 Evaluation

6.1 Experimental Setup

Environment and Basic Settings. All experiments were implemented in C++ on a lightweight desktop environment simulating on-device hardware constraints. The platform runs Ubuntu Linux with a dual-core CPU and 16 GB of RAM. We cap the memory at 4 GB to reflect typical mobile environment limitations. We evaluate different methods on standard ANN benchmarks summarized in Table 1. On-device queries require top-50 search using Euclidean distance as the metric. To avoid localized topological fragmentation, insertions and deletions are uniformly distributed across indexed nodes. Query and update operations are executed sequentially to ensure a consistent evaluation.

Comparison Methods. We compare three methods: **HNSW-Original** [24], the **MINT algorithm** [13], and the proposed **HNSW-DM** algorithm. HNSW-Original follows the original HNSW update procedure and does not maintain isolated nodes generated during the update process. We use the official C++ implementation from the GitHub repository [24] for index construction, updates,

Table 1. Statistics of datasets used in experiments.

Datasets	Size	Item	Dimension	M	M_e	ef_c	ef_s	Metric
ImageNet [23]	1.2 GB	2,340,373	150	16	1	200	1200	L_2
Gist1M [17]	3.6 GB	1,000,000	960	8	1	200	400	L_2
Sift1M [17]	501 MB	1,000,000	128	4	1	100	200	L_2

and evaluation. MINT [13] is a non-real-time HNSW maintenance algorithm. It preserves reachability by setting a threshold τ for the node update rate. Once exceeded, a full-graph traversal locates unreachable nodes, which are then inserted into an auxiliary HNSW index and queried alongside the main index. HNSW-DM supports real-time dynamic maintenance of the HNSW index by detecting isolated nodes through in-degree tracking and maintaining reachability via extra edges in the main index and an auxiliary HNSW index.

Parameters. For each dataset, the following parameters were configured (see Table 1): the maximum number of outgoing neighbors per node is M, and $2M$ for level 0 nodes. $ef_{construction}$ (ef_c) controls the size of the candidate list during index construction. ef_{search} (ef_s) determines the size of the candidate list during search. For HNSW-DM, the main index allows up to $M_e = 1$ extra edges per node. The update threshold τ is set to 4% for MINT.

Testing Procedure. Each experiment runs for 100 iterations. In each iteration, 1% of the nodes are deleted and then reinserted, forming one update cycle. Every five iterations, we record the update latency, count the unreachable nodes, compute the recall rate of the current graph index and then plot the results.

Metrics. We consider the following experimental metrics. **Number of Unreachable Nodes:** used as a direct measure of graph reachability and maintenance quality, indicating nodes cannot be accessed via routing from the entry point during the search. **Recall@50:** measures search accuracy by the proportion of true top-50 nearest neighbors found within the top-50 returned results. Ground truth is obtained via exhaustive search, with exact nearest neighbors computed through full pairwise distance calculation. **Update Latency:** evaluates update efficiency by measuring the average latency introduced by different algorithms. **Index Memory Usage:** accesses the impact on memory inflation, as maintaining real-time reachability in the graph index requires adding extra data fields to the original node structures, thus increasing memory usage.

6.2   Number of Unreachable Nodes

During evaluation, the number of unreachable nodes is calculated as the graph size minus the number of reachable nodes. As shown in Fig. 6, even without updates, a small number of unreachable nodes remain due to isolated islands, i.e., subgraphs with nonzero node in-degrees but no global connections. As updates proceed, the number of unreachable nodes steadily increases with HNSW-Original, indicating its failure to maintain connectivity. MINT, with periodic

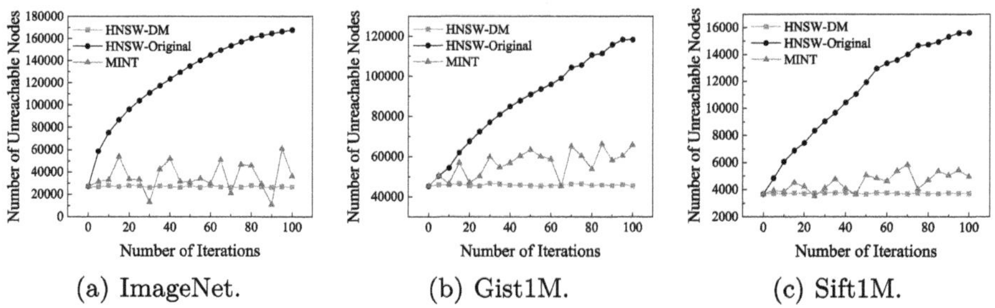

Fig. 6. The number of unreachable nodes changes with the update process.

reconstructions of the auxiliary index, exhibits oscillations as maintenance is not real-time but depends on the update frequency. After reconstruction, fewer unreachable nodes are found, but continuous updates before the next reconstruction lead to more unreachable nodes. In contrast, the proposed HNSW-DM performs real-time maintenance, keeping isolated nodes at zero and ensuring a stable and minimal unreachable node count throughout the updates.

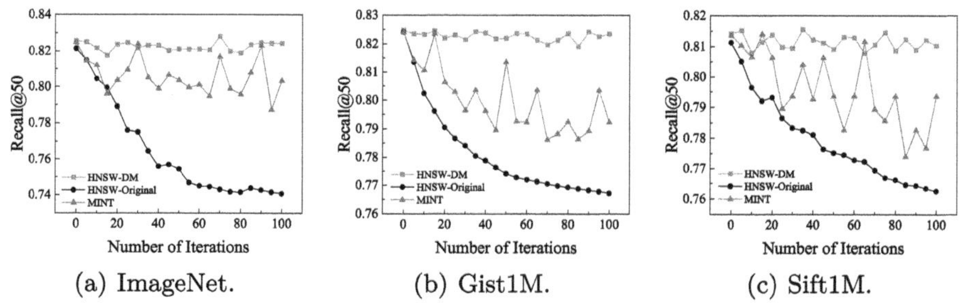

Fig. 7. Comparison of recall@50 rates across methods on each dataset.

6.3   Recall@50

Before any updates, all methods achieve about 82% recall due to intentionally constrained memory settings that simulate edge-side hardware, where parameters such as M and $ef_{construction}$ are kept low. This does not affect the relative comparison among algorithms. Figure 7 shows that HNSW-DM consistently maintains high and stable recall throughout updates, effectively preserving reachability. In contrast, HNSW-Original experiences a clear decline (8% drop on ImageNet and 6% on Gist1M and Sift1M), whereas MINT fluctuates with its periodic reconstructions and performs consistently worse.

6.4   Update Latency

We evaluate the update efficiency by measuring the average time taken to update a single node in each iteration. Because HNSW-DM performs additional operations for reachability maintenance, its update latency is slightly higher than that of HNSW-Original (see Fig. 8). Across all datasets, our method adds less than 10% delay, which is moderate and acceptable given the improved connectivity and recall stability it provides. We did not test MINT's update latency, as its periodic maintenance only affects updates during maintenance periods, with latency similar to HNSW-Original otherwise.

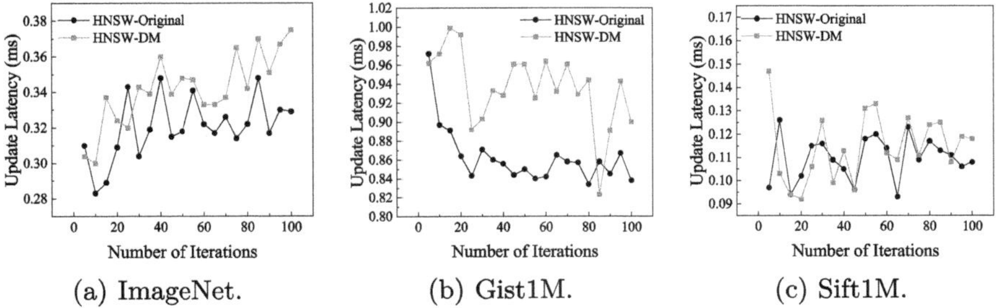

(a) ImageNet.   (b) Gist1M.   (c) Sift1M.

Fig. 8. Comparison of update latency across methods on each dataset.

6.5 Index Memory Usage

After 100 iterations, we measure index size by serializing it to disk and checking the file size. Figure 9 shows the memory performance of the proposed method. Across all datasets, indirect isolated nodes consistently make up about 90% of isolated nodes, while direct ones account for roughly 10% (see Fig. 9(a)). This indicates that only a small fraction of nodes need to be stored in the auxiliary index, keeping memory overhead low. Indirect isolated nodes are handled through lightweight extra edges, with negligible cost. Figure 9(b) compares memory footprints of HNSW-Original and HNSW-DM. The memory expansion rate is 1.3% on ImageNet, 0.25% on Gist, and 1.46% on Sift, confirming that the overhead of our strategy remains minimal.

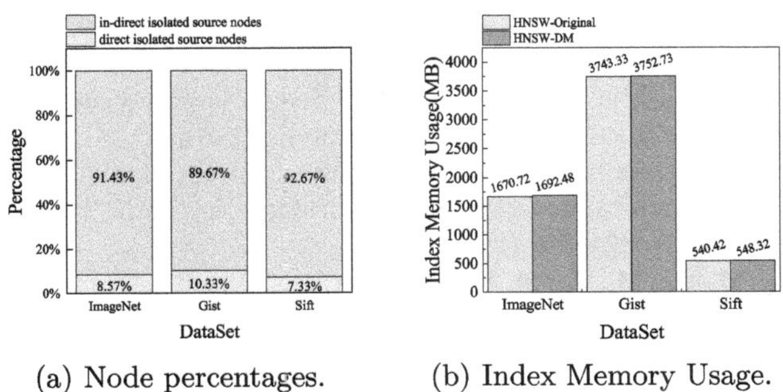

(a) Node percentages.   (b) Index Memory Usage.

Fig. 9. Memory performance of the proposed method.

7   Conclusion

This paper addresses the issue of isolated source nodes arising from graph updates in HNSW, which degrade recall over time. We propose a real-time maintenance approach that detects isolated nodes by tracking in-degrees and restores their reachability through adding auxiliary edges in the main index and constructing an auxiliary index. To minimize structural redundancy, our method further enables dynamic conversion between these two strategies. In addition, a dual-index search strategy is designed to ensure all nodes remain accessible during queries. Experimental results demonstrate that our method effectively suppresses the growth of isolated nodes and stabilizes recall, while introducing only a modest 10% latency overhead and a slight increase in index memory usage within 2%. Overall, the proposed approach provides a practical and efficient solution for maintaining dynamic HNSW indexes under constrained memory environments.

Acknowledgments. This work is supported by the Jing-Jin-Ji Regional Integrated Environmental Improvement-National Science and Technology Major Project of Ministry of Ecology and Environment of China (No. 2025ZD1200600), and the National Natural Science Foundation of China (No. 62372352, 62272369).

Disclosure of Interests. The authors have no competing interests to declare that are relevant to the content of this article.

References

1. Han, S., Mao, H., Dally, W.J.: Deep compression: compressing deep neural networks with pruning, trained quantization and Huffman coding. In: Proceedings of ICLR '16 (2016)
2. Howard, A., Sandler, M., Chu, G., Chen, L.-C., Chen, B., et al.: Searching for MobileNetV3. In: Proceedings of IEEE/CVF ICCV '19, pp. 1314–1324 (2019)
3. Fan, W., Ding, Y., Ning, L., Wang, S., et al.: A survey on rag meeting LLMs: towards retrieval-augmented large language models. In: Proceedings of ACM SIGKDD '24, pp. 6491–6501 (2024)
4. Park, T., Lee, G., Kim. , M.-S.: MobileRAG: a fast, memory-efficient, and energy-efficient method for on-device RAG. arXiv preprint arXiv:2507.01079v1 (2025)
5. Chen, K., Xiao, J., Liu, J., Tong, Q., et al.: Semantic visual simultaneous localization and mapping: a survey. IEEE Trans. Intell. Transp. Syst. (2025)
6. Malkov, Y.A., Yashunin, D.A.: Efficient and robust approximate nearest neighbor search using hierarchical navigable small world graphs. IEEE Trans. Pattern Anal. Mach. Intell. **42**(4), 824–836 (2018)
7. Johnson, J., Douze, M., Jégou, H.: Billion-scale similarity search with GPUs. IEEE Trans. Big Data **7**(3), 535–547 (2019)
8. Yao, J., Zhang, S., Yao, Y., Wang, F., Ma, J., et al.: Edge-Cloud Polarization and Collaboration: A Comprehensive Survey for AI. IEEE Trans. Knowl. Data Eng. **35**(7), 6866–6886 (2023)

- **State-of-the-Art Performance.** Experiments on five diverse real-world hypergraph datasets verify the effectiveness and superiority of RAHC-HPP. Our model consistently outperforms eight baseline methods, delivering gains of 2.6% to 45.4% over the SOTA model across datasets from diverse domains.

2 Related Work

2.1 Hypergraph Neural Networks

Driven by the success of GNNs [14,23], HGNNs typically expand hypergraphs into standard graphs to leverage the powerful message passing algorithms. Early HGNNs use clique expansion, passing messages between nodes in the expansion graph. This paradigm includes spectral methods like HGNN [9], GCN-based refinements like HyperGCN [28], and attention-based models like HCHA [2]. However, clique expansion suffers from edge explosion and information loss, as different hypergraphs can expand into identical graphs.

To address these limitations, subsequent studies have adopted a two-stage message-passing framework on star-expanded hypergraphs, aggregating information first at hyperedges then propagating back to nodes. HNHN [8] pioneered this approach, which theoretically generalizes clique expansion. Building on this, UniGCN [11] unified graph and hypergraph frameworks, HAT [13] integrated self-attention, and AllSet [6], inspired by Set Transformer [16], treated message passing as a permutation-invariant operation on sets.

Despite their utility, star expansions introduce intermediate hyperedge nodes that dilute message flow and weaken direct interactions among original nodes. To counter this, WHATsNet [7] injected node-centrality rankings within each hyperedge as positional encodings, while ID-HAN [5] leveraged edge-dependent node labels to better measure the importance of hyperedges. Cross Expansion [29] simultaneously retains the dense connectivity of clique expansion and the explicit hyperedge-node structure of star expansion. While this method mitigates signal weakening, it inevitably increases computational complexity by introducing additional connections between nodes. The trade-off between signal preservation and computational efficiency remains a significant challenge in designing hypergraph neural networks.

2.2 Post-processing Models

Post-processing modules have demonstrated effectiveness across numerous domains. Leveraging sequential relationships, LSTM-CRF [12] and BERT-CRF [18] position CRF [15] after LSTM or BERT layers, modeling optimal label sequences through transition potentials. ConvCRF [20] adds CRF after convolutional neural networks to refine segmentation boundaries, exploiting relative positional relationships in the image. For graph-structured data, GNN-CRF [19] utilizes CRF modules to model implicit connections between pairwise edges. However, these methods are built upon binary relationships, making them difficult to apply directly to complex constraints between sets. While hyperedges can be transformed into cliques, their computational complexity increases exponentially with |clique| size, limiting practical applications for hypergraph structures.

3   Preliminaries

3.1   Hypergraphs

A hypergraph $\mathcal{G} = (\mathcal{V}, \mathcal{E})$ consists of a vertex set $\mathcal{V}$, and a hyperedge set $\mathcal{E}$. Each hyperedge $e \in \mathcal{E}$ is a non-empty subset of $\mathcal{V}$. Incidence relation $\mathcal{R}_{v,e}$ holds if and only if $v \in e$. In our representation learning setting, for hyperedge $e \in \mathcal{E}$, let $\mathbf{V}_e = \{\mathbf{h}_v \mid v \in e\}$ be the set of embeddings of nodes contained in e. $\mathbf{R}_e = \{\mathbf{r}_{v,e} \mid v \in e\}$ contains the relation embeddings between hyperedge e and each node v in e. Analogously, for node $v \in \mathcal{V}$, let $\mathbf{E}_v = \{\mathbf{h}_e \mid v \in e\}$ be the set of embeddings of hyperedges containing node v. $\mathbf{R}_v = \{\mathbf{r}_{v,e} \mid v \in e\}$ is the set of embeddings of relation between v and e, but arranged from the node's perspective.

3.2   Problem Formulation

Classical node classification assigns each node a single global label, yet nodes may play different roles across hyperedges. The EDNC task recovers these context-specific roles.

Problem 1 (Edge-Dependent Node Classification [7]). Given (a) a hypergraph $\mathcal{G} = (\mathcal{V}, \mathcal{E})$, (b) a set of edge-dependent node labels in $\mathcal{E}' \in \mathcal{E}$ (i.e., $y_{v,e}, \forall \mathcal{R}_{v,e}, e \in \mathcal{E}'$), and optionally (c) a node feature matrix X, the problem is to correctly predict the unknown edge-dependent node labels in $\mathcal{E} \setminus \mathcal{E}'$ (i.e., $y_{v,e}, \forall \mathcal{R}_{v,e}, e \in \mathcal{E} \setminus \mathcal{E}'$).

However, by treating each node-hyperedge pair as an independent prediction, EDNC fails to capture the coordinated label assignments required for hyperedge consistency. This motivates a more stringent, hyperedge-centric problem formulation. We thus introduce HC-NC, which is formally defined as follows.

Problem 2 (Hyperedge-Consistent Node Classification). Given (a) a hypergraph $\mathcal{G} = (\mathcal{V}, \mathcal{E})$, (b) a set of edge-dependent node labels in $\mathcal{E}' \in \mathcal{E}$ (i.e., $y_{v,e}, \forall \mathcal{R}_{v,e}, e \in \mathcal{E}'$), and optionally (c) a node feature matrix X, the problem is to correctly predict all unknown edge-dependent node labels for each hyperedge. Prediction for hyperedge $e \in \mathcal{E} \setminus \mathcal{E}'$ is considered correct if and only if all edge-dependent node labels $y_{v,e}$ for all vertices $v \in e$ are predicted correctly.

4   Methodology

In this section, we propose RAHC-HPP, which integrates clique-aware message passing with label-specific post-processing for HC-NC. As shown in Fig. 2, RAHC augments star expansion with clique-aware message passing, while HPP enhances hyperedge-level label consistency via a post-processing network. Training employs a dual loss combining the HC-NC supervised objective with a contrastive alignment between star and clique-aware representations. We then present the full model and compare its design with existing hypergraph learning methods.

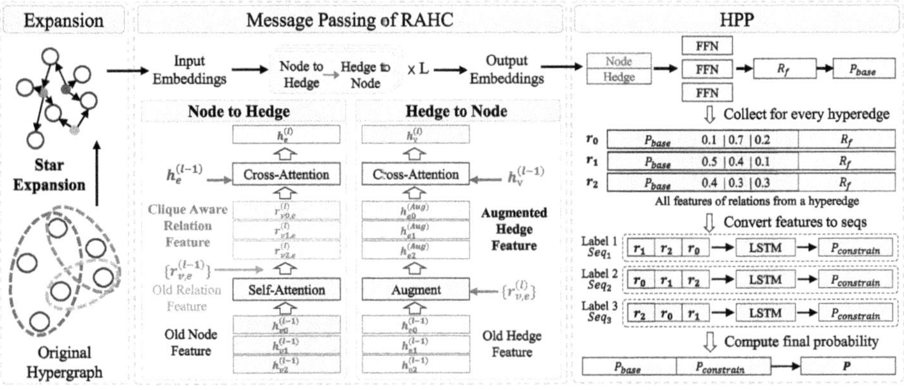

Fig. 2. The overall architecture of RAHC-HPP. We first apply star-expansion to the hypergraph, then the RAHC network enhances two-step message passing on the expanded graph with clique-based relation features. Finally, the HPP network produces hyperedge-consistent, edge-dependent node label predictions. In the HPP network, we show a 3-node hyperedge example, we calculate base probabilities, rank them to form label-specific sequences, and process these sequences into final probability outputs.

4.1   Relation-Augmented Hypergraph Convolutional Network

The RAHC network is designed to capture direct node interaction information that is typically lost in traditional star expansion models for hypergraphs. Unlike cross-expansion, which incurs a higher computational cost by aggregating information from both hyperedges and all neighboring nodes, RAHC explicitly preserves and leverages clique-aware messages within the star expansion framework, restricting relational modeling to nodes inside each hyperedge. It operates in two stages: (1) Node-to-Hyperedge Aggregation: Nodes within each hyperedge interact to generate clique-aware representations, which are then aggregated into hyperedge embeddings. (2) Hyperedge-to-Node Aggregation: The clique-aware messages are propagated back to the nodes, enriched with hyperedge information to refine node representations.

Formally, RAHC takes the node embeddings $\{\mathbf{h}_v\}^{(l-1)}$, the hyperedge embeddings $\{\mathbf{h}_e\}^{(l-1)}$, and the node-hyperedge relationship embedding $\{\mathbf{r}_{v,e}\}^{(l-1)}$ at layer l as input. Through the aforementioned two-phase aggregation, it outputs updated embeddings $\{\mathbf{h}_v\}^{(l)}$, $\{\mathbf{h}_e\}^{(l)}$, and $\{\mathbf{r}_{v,e}\}^{(l)}$. Unless otherwise stated, superscripts (l) and $(l-1)$ denote the current and previous layers, respectively.

Node-to-Hyperedge Aggregation. To address the limitations of star expansion, RAHC simulates clique-aware information through the star-expanded architecture. Specifically, we approximate explicit node–node interactions from clique expansion using a standard Multi-Head Attention block (MHA) [22]. For a set of queries, keys, and values, MHA operates by running H attention "heads" in parallel. In each head, it computes attention weights using scaled dot-product attention between queries and keys to create a weighted sum of the values. Finally, the outputs of all H heads are concatenated and linearly projected to produce the final output. By modeling the nodes in each hyperedge as a clique through self-attention, the model effectively captures each node's contribution to the hyperedge and its interactions with other nodes within it. Based on these interactions, RAHC updates its relation embeddings with a residual connection as follows:

$$\mathbf{R}'_e = \mathrm{MHA}(\mathbf{V}^{(l-1)}_{e\boxplus}, \mathbf{V}^{(l-1)}_{e\boxplus}, \mathbf{V}^{(l-1)}_{e\boxplus}),$$

$$\mathbf{R}^{(l)}_e = \mathrm{Layer\ Norm}(\mathbf{R}^{(l-1)}_e + \mathrm{FFN}(\mathbf{R}'_e))),$$

where $\boxplus$ denotes the node-centrality ranking positional encoding from WHAT-sNet [7], and FFN denotes a two-layer feed-forward network with ReLU activation function. RAHC retains these representations and applies them in the next message aggregation stage.

We then aggregate this information into hyperedge embeddings through cross-attention and FFN modules:

$$\mathbf{h}^{(l)}_e = \mathbf{h}^{(l-1)}_e + \mathrm{FFN}(\mathrm{MHA}(\mathbf{h}^{(l-1)}_e, \mathbf{R}^{(l)}_e, \mathbf{R}^{(l)}_e)).$$

Hyperedge-to-Node Aggregation. The relation representation of direct node interactions $\{r_{v,e}\}$ is injected into hyperedge features, yielding an augmented hyperedge representation E^{Aug}_v. This enhancement enables nodes to aggregate messages from incident hyperedges and neighboring node interactions jointly.

$$\mathbf{E}^{\mathrm{Aug}}_v = \mathrm{FFN}(\{\mathbf{h}^{(l)}_e \oplus \mathbf{r}_{v,e} \mid v \in e, \mathbf{h}_e \in \mathbf{E}_v, \mathbf{r}_{v,e} \in \mathbf{R}_v\}),$$

where $\oplus$ denotes the concatenation operation. Nodes participate in hyperedges with varying intensities, causing each hyperedge to contribute heterogeneous information to its member nodes. To address this, we incorporate a self-attention module into the hyperedge-to-node aggregation. This computes attention scores to adaptively weight each hyperedge message, producing a refined intermediate representation E'_v.

$$\mathbf{E}'_v = \mathrm{MHA}(\mathbf{E}^{Aug}_{v\boxplus}, \mathbf{E}^{Aug}_{v\boxplus}, \mathbf{E}^{Aug}_{v\boxplus}),$$

where $\boxplus$ denotes the node-centrality ranking positional encoding from WHAT-sNet. Finally, RAHC aggregates vertex messages using the intermediate features.

$$\mathbf{h}^{(l)}_v = \mathrm{MHA}(\mathbf{h}_{v^{(l-1)}}, \mathbf{E}'_v, \mathbf{E}'_v).$$

4.2 Hyperedge Post-processing Network

To capture node constraints within hyperedges, we propose HPP, which learns intra-hyperedge constraints to optimize predictions. As shown in Fig. 2, we first perform label-specific feature transformations to produce base probabilities. Inspired by Set2Seq [24], we construct sequences to generate constraint-aware probabilities and incorporate them into the final predictions.

Base Probability. To exploit the rich hypergraph information captured by RAHC, we compute the basis probabilities based on its output embeddings. Given hyperedge features $\{\mathbf{h}_e\}$ and node features $\{\mathbf{h}_v\}$, we first concatenate them to get the star-expansion based relation $\mathbf{R}_s$:

$$\mathbf{R}_s = \{\mathbf{r}_{i,e}^{(s)} = \mathbf{h}_v \oplus \mathbf{h}_e \mid v \in e\}.$$

Then we process them using independent feed-forward networks for each label category:

$$\mathbf{R}_e^f = [\mathrm{FFN}_1(\mathbf{R}_s) \oplus \cdots \oplus \mathrm{FFN}_n(\mathbf{R}_s)],$$

where $\mathcal{L}$ represents possible labels, and $n = |\mathcal{L}|$. These transformed features generate a probability distribution:

$$\mathrm{P}_{\mathrm{base}} = \mathrm{Softmax}(\mathbf{R}_e^f).$$

Sequence-Based Contextual Dependency Modeling. Different from traditional methods that predict node labels independently, HPP explicitly leverages the correlation among node labels. We hypothesize that nodes with higher confidence in a certain label should contribute more significantly to label predictions of related nodes within the same hyperedge. Therefore, we construct ordered sequences for each label. Specifically, given a hyperedge e, we define a permutation function π_i that sorts nodes in descending order of their base probabilities for label i:

$$\pi_i : \{1, \ldots |e|\} \to \{1, \ldots |e|\}, i \in (1, ..., n),$$

$$\mathrm{P}_{\mathrm{base}, \pi_i(1)}[i] \geq \cdots \geq \mathrm{P}_{\mathrm{base}, \pi_i(|e|)}[I].$$

Based on this ordering, we construct a sequence for label i:

$$\mathbf{Seq}_i = [\mathrm{P}_{base, \tau_i(j)}[i] \oplus \mathbf{R}_{e, \pi_i(j)}^f[i]], j = 1^{|e|}.$$

This sequence structure explicitly prioritizes nodes with higher initial confidence, thereby enhancing their influence on contextual modeling. We input each Seq_i into an LSTM [10], which processes the sequence through gating mechanisms to capture contextual dependencies among nodes within the hyperedge:

$$\mathbf{S}_i = \mathrm{LSTM}_i(\mathbf{Seq}_i), i \in (1, ..., n),$$

where $\mathbf{S}_i \in \mathbb{R}^{|e| \times 1}$ denotes the label-specific scores. We map these scores back to the original order via inverse permutation, and concatenate all score vectors and transform them into a constraint-aware probability distribution over labels:

$$\mathrm{P}_{\mathrm{constrain}} = \mathrm{Softmax}([\hat{\mathbf{S}}_1 \oplus \cdots \oplus \hat{\mathbf{S}}_n]), \text{ where } \hat{\mathbf{S}}_i[\pi_i(j)^{-1}] = \mathbf{S}_i[j], \text{ for } j = 1, \ldots, |e|.$$

Constraint-Aware Probability Integration. The final probability distribution is obtained through the element-wise summation of the initial probability and hyperedge-level constraints:

$$\mathrm{P} = \frac{1}{2}\left(\mathrm{P}_{\mathrm{base}} + \mathrm{P}_{\mathrm{constrain}}\right).$$

4.3 Dual-Objective Loss Function

Our model is optimized with a dual-objective loss function combining supervised and contrastive learning, which encourages the model to simultaneously achieve accurate label prediction and learn view-invariant representations from the hypergraph's structure.

For the label prediction task, we compute the cross-entropy loss between the final probability distribution P produced by HPP and the ground-truth labels Y. We can define the supervised loss $\mathcal{L}_{\text{pred}}$:

$$\mathcal{L}_{\text{pred}} = -\frac{1}{M} \sum_{i=1}^{M} \sum_{c=1}^{C} y_{ic} \log(\mathrm{P}_{ic}),$$

where M denotes the number of node-hedge relations, C is the number of classes, y_{ic} is the ground-truth label, and P_{ic} is the predicted probability that relation i belongs to class c.

The star expansion generates relation vectors $\mathbf{R}_s$ by concatenating node and hyperedge features. Meanwhile, the clique-like relation embeddings $\mathbf{R}_c = \{r_{v,e}^{(L)}\}$ are obtained by applying self-attention among nodes within each hyperedge. To ensure consistency between these complementary representations, we implement a contrastive learning loss inspired by the SimSiam architecture [4]. Both $\mathbf{R}_s$ and $\mathbf{R}_c$ are processed through a shared feed-forward network $\mathbf{D}$ with ReLU as activation. The contrastive loss is then defined as:

$$\mathcal{L}_{\text{contrastive}} = \|\mathbf{R}_s - \text{stopgrad}(\mathbf{R}_c')\|_2 + \|\mathbf{R}_c - \text{stopgrad}(\mathbf{R}_s')\|_2,$$

$$\text{where } \mathbf{R}_s' = \mathbf{D}(\mathbf{R}_s), \ \mathbf{R}_c' = \mathbf{D}(\mathbf{R}_c), \ \mathbf{D}(\mathbf{x}) = \text{ReLU}(\mathbf{W}(\mathbf{x})) + \mathbf{b},$$

where $\mathbf{W}$ is a learnable weight matrix, $\mathbf{b}$ denotes the bias vector. The $\text{stopgrad}(\cdot)$ prevents gradient back propagation through the respective terms, creating an asymmetric learning dynamic that prevents representation collapse while promoting semantic alignment between the two views.

The Final loss is calculated as follows:

$$\mathcal{L}_{\text{total}} = \mathcal{L}_{\text{pred}} + \lambda \mathcal{L}_{\text{contrastive}}.$$

where λ is a hyperparameter that balances the contribution of the contrastive term. This optimization strategy enables our model to leverage both supervised signals from labeled data and information captured by different hypergraph structures.

4.4 Our Final Model

Our final model integrates RAHC and HPP into a comprehensive end-to-end learning framework. The pipeline begins by converting the input hypergraph $\mathcal{G}$ into a bipartite graph via star expansion. Node embeddings $\mathbf{h}_v^{(0)} \in \mathbb{R}^{d_v}$ are initialized using second-order random walks [3], while hyperedge embeddings $\mathbf{h}_e^{(0)}$ and node-hyperedge relation embeddings $\mathbf{r}_{v,e}^{(0)}$ are initially set to zero.

The hypergraph is then passed through L RAHC layers. At each layer l, RAHC leverages a clique-information-enhanced star expansion mechanism for message passing. We obtain the final node embeddings $\mathbf{h}_v^{(L)}$, hyperedge embeddings $\mathbf{h}_e^{(L)}$ and relation embeddings under clique-like interaction $\mathbf{R}_c = \{\mathbf{r}_{v,e}^{(c,L)}\}$. We concatenate the final node and hyperedge embeddings $\mathbf{h}_v^{(L)}$ and $\mathbf{h}_e^{(L)}$ to derive the relation representations $\mathbf{R}_s$ under star expansion. These representations are subsequently fed into the HPP module to produce the final label probability distribution.

The model's optimization is guided by a dual-objective loss function, which combines a prediction loss $\mathcal{L}_{\mathrm{pred}}$ based on HPP output and a contrastive loss from relation representations derived under both star expansion and clique-aware interaction. This dual objective promotes accurate label prediction and robust, view-invariant relational representation learning.

4.5 Model Design Principles and Comparison to Existing Methods

A known limitation of star expansion is reduced direct node-to-node interaction within hyperedges. Cross-Expansion [29] addresses this by introducing additional clique-like connections into the graph structure. In contrast, RAHC mitigates this limitation without adding edges to the expansion graph, instead systematically preserving intermediate clique-like interactions from the star expansion's message passing to enhance attenuated information.

We also observe that both WHATsNet [7] and ID-HAN [5], though not explicitly stated, implicitly enhance star expansion through clique-like interactions via node-centrality rankings in hyperedge and edge-dependent labels, respectively. Our method builds on this principle but goes further by explicitly modeling clique-like interactions during message aggregation and aligning them with representation from star expansion through $\mathcal{L}_{\mathrm{contrastive}}$, offering a more effective solution than relying solely on static structural or label-based cues.

5   Experiments

In this section, we will evaluate the proposed method and answer the following questions:

- How effectively can RAHC-HPP address the HC-NC task across diverse hypergraph datasets, and how does it compare to state-of-the-art methods?
- What is the contribution of each component in RAHC-HPP? Is the HPP module compatible with other models?
- How well does RAHC-HPP generalize to the EDNC problem? Is it necessary to propose the HC-NC problem after the EDNC task?
- How does RAHC-HPP capture edge-consistent node labels in Real-world hypergraphs?

We describe the datasets, baseline methods, and experimental setup in Sect. 5.1, and answer the above questions in Sect. 5.2 to 5.5.

Table 1. Summary of Datasets Used in Experiments.

Dataset	Num of Nodes	Num of Hyperedges	Max of Node Deg.	Max of Hyperedge Size	Num of Class 0	Num of Class 1	Num of Class 2	Avg. Entropy
Coauth-DBLP	108,484	91,266	236	36	91,266	138,479	91,266	0.13
Email-Eu	986	209,508	8,659	59	209,508	332,334	-	0.48
Stack-Biology	15,490	26,823	1,318	12	26,290	18,400	11,523	0.10
Stack-Physics	80,936	200,811	6,332	48	194,575	201,121	84,113	0.12
BOM	13,705	25,505	4794	3	25,505	25,505	25,505	0.12

5.1 Experiment Setup

Datasets. We evaluate on five real-world hypergraphs–DBLP[1], Eu [17], Biology[2], Physics (see Footnote 2), and BOM. The first four are adapted for EDNC by WHATsNet [7], while BOM is from our production data. Each dataset provides structure and edge-dependent node labels: Co-author (DBLP) uses publications as hyperedges with authors as nodes and labels {first, last, other}; Email (Eu) uses emails with participants and labels {sender, recipient}; StackOverflow (Biology/Physics) uses Q/A threads with contributors and labels {questioners, accepted answerers, other answerers}. In BOM, Hyperedges represent material substitution, and nodes are computer series or components. Edge-dependent node labels denote {series, main part, substitute part}. Table 1 summarizes statistics and mean label entropy. All mean entropies are non-zero, indicating that node labels vary across hyperedges.

Baselines. We evaluated 8 hypergraph neural networks as baselines for the HC-NC task: 2 clique-based methods (HGNN, HCHA); 4 star-based approaches (HNHN, HAT, UniGCNII, AllSetTransformer); Cross-Expansion, a hybrid framework with a GCN module; and WHATsNet, enhancing star-expansion with node-centrality positional encodings. We exclude ID-HAN because edge-dependent node labels are input to the model, and HyperGCN because it cannot generate the required explicit hyperedge embeddings. For most competitors not originally designed for this task, we adapted them by concatenating hyperedge and node embeddings, then processing them through a two-layer FFN for final class logits.

Implementation Details. We randomly split hyperedges 60%/20%/20% for training, validation, and testing, respectively. Both our RAHC model and all baselines are implemented in the Deep Graph Library (DGL) [25] and optimized end-to-end with $\text{AdamW}(\beta_1 = 0.9,\ \beta_2 = 0.999,\ \text{weight decay} = 1 \times 10^{-2})$. Baselines are performed a grid search over learning rates $\{5 \times 10^{-3},\ 1 \times 10^{-2},\ 3 \times 10^{-2},\ 5 \times 10^{-2},\ 1 \times 10^{-1}\}$ and model depths of one and two layers. Our RAHC uses two convolutional layers with an embedding dimension of 64, and the λ in $\mathcal{L}_{\text{total}}$ is set to 1. We conducted 5 independent runs for each set of experiments.

---

[1] https://dblp.uni-trier.de/xml/.
[2] https://archive.org/download/stackexchange.

5.2 Hyperedge-Consistent Node Classification

Table 2. Experimental Results of Hyperedge-Consistent Node Classification. For each dataset, the top three results are shown in **bold**, and the best one is also **underlined**

	Coauth-DBLP	Email-Eu	Stack-Biology	Stack-Physics	BOM
HGNN	0.127 ± 0.007	0.353 ± 0.002	0.471 ± 0.002	0.474 ± 0.001	0.744 ± 0.002
HCHA	0.153 ± 0.000	0.033 ± 0.003	0.384 ± 0.005	0.413 ± 0.003	0.508 ± 0.003
HNHN	0.138 ± 0.002	0.359 ± 0.005	0.436 ± 0.005	0.455 ± 0.002	0.570 ± 0.003
HAT	0.092 ± 0.001	0.368 ± 0.007	0.439 ± 0.004	0.447 ± 0.006	0.642 ± 0.009
UniGCNII	0.118 ± 0.002	0.159 ± 0.008	0.398 ± 0.001	0.396 ± 0.001	0.532 ± 0.004
AST	0.010 ± 0.002	0.401 ± 0.007	0.457 ± 0.004	0.494 ± 0.006	0.229 ± 0.009
WHATsNet	$\mathbf{0.364 \pm 0.004}$	$\mathbf{0.442 \pm 0.005}$	$\mathbf{0.521 \pm 0.006}$	$\mathbf{0.558 \pm 0.003}$	$\mathbf{0.818 \pm 0.002}$
CrossExpansion	$\mathbf{0.268 \pm 0.002}$	$\mathbf{0.470 \pm 0.003}$	$\mathbf{0.502 \pm 0.007}$	$\mathbf{0.513 \pm 0.007}$	$\mathbf{0.792 \pm 0.002}$
RAHC-HPP (ours)	$\mathbf{0.390 \pm 0.001}$	$\mathbf{0.896 \pm 0.001}$	$\mathbf{0.566 \pm 0.004}$	$\mathbf{0.567 \pm 0.002}$	$\mathbf{0.877 \pm 0.001}$

We evaluated the performance of eight different methods for the HC-NC task across five datasets in four scenarios. As the evaluation metric for the HC-NC task involves jointly predicting multiple nodes with underlying constraints, accuracy is used to measure the model's capability in capturing the structure within each hyperedge. Table 2 presents the results of all models on HC-NC, where higher scores indicate better performance.

Based on the experimental results, we draw the following conclusions: (a) There is no clear superiority between clique-based and star-based expansion networks on the HC-NC. For example, HGNN achieves better accuracy on Stack-Biology and BOM, while AST exhibits better predictive capability on Stack-Physics. (b) Cross expansion and relation-enhanced star expansion methods consistently outperform clique-based and star-based expansion methods. Cross expansion preserves the message from both expansion types, achieving excellent results even with simple single-layer GCN aggregation. WHATsNet, based on node centrality ordering within each hyperedge, effectively embeds clique structure information to enhance star expansion, yielding the second-best results across all datasets. **The result also confirms the observation** of the cross-expansion paper that pure star expansion attenuates the direct interaction signal between nodes. (c) RAHC-HPP consistently surpasses state-of-the-art methods, achieving up to 45.44% improvement over the second-best method on Email-Eu. These results validate our architecture's proficiency in capturing higher-order node interactions within hyperedges through robustly coherent outputs.

5.3 Analysis of RAHC-HPP

We conduct a comprehensive analysis focusing on two key aspects: (1) the effectiveness of individual components within RAHC-HPP, and (2) the compatibility of the HPP module with other hypergraph neural networks. We first conduct an ablation study to assess each component's contribution, then demonstrate HPP's plug-and-play capability through integration into existing models.

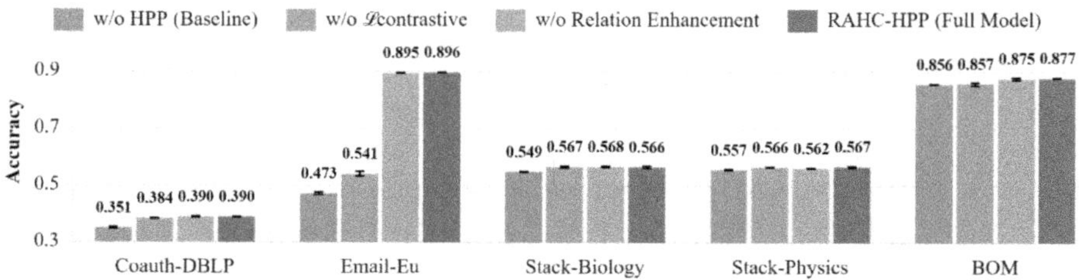

Fig. 3. Ablation results of RAHC-HPP.

Table 3. Results of Cross-Model Integration with HPP. Accuracy ($\pm$ std), with change vs. original model in parentheses: $\uparrow$ improvement, $\downarrow$ drop.

Dataset	WHATsNet-HPP	Cross-HPP	RAHC-HPP
Coauth-DBLP	0.367 ± 0.007 ($\uparrow$ 0.0033)	0.287 ± 0.004 ($\uparrow$ 0.0187)	0.390 ± 0.001($-$)
Email-Eu	0.774 ± 0.014 ($\uparrow$ 0.3321)	0.712 ± 0.060 ($\uparrow$ 0.2417)	0.896 ± 0.001($-$)
Stack-Biology	0.535 ± 0.014 ($\uparrow$ 0.0143)	0.538 ± 0.002 ($\uparrow$ 0.0364)	0.566 ± 0.004($-$)
Stack-Physics	0.563 ± 0.002 ($\uparrow$ 0.0049)	0.532 ± 0.013 ($\uparrow$ 0.0185)	0.567 ± 0.002($-$)
BOM	0.854 ± 0.005 ($\uparrow$ 0.0360)	0.778 ± 0.007 ($\uparrow$ 0.0360)	0.877 ± 0.001($-$)

Ablation Study. We evaluate each component's contribution by designing three variants: (a) RAHC-HPP w/o HPP: Replace the HPP module with a two-layer FFN for direct prediction. (b) RAHC-HPP w/o $\mathcal{L}_{contrastive}$: Optimize only using the prediction loss $\mathcal{L}_{pred}$. (c) RAHC-HPP w/o Relation Enhancement: Remove augmentation during hyperedge-to-node aggregation.

Figure 3 reports accuracy on HC–NC. **Importance of HPP.** RAHC-HPP consistently outperforms the variant without HPP, which is the weakest across all datasets. The effect is most pronounced on EmailEu (42.32% drop), indicating that HPP effectively refines within-hyperedge label distributions for coherent predictions in large hyperedges. **Importance of $\mathcal{L}_{contrastive}$.** The contrastive loss generally improves performance by providing complementary structural views. The exception is stack-Biology with low node-label entropy, which suggests that with highly discriminative labels, star expansion captures sufficient global context, limiting the marginal benefit of contrastive learning. **Importance of Relation Enhancement.** Modeling direct node–node interactions improves most datasets, as standard star expansion attenuates such signals. Consistent with the above, it provides a marginal benefit when labels are easily separable in stack-Biology.

Cross-Model Integration. To evaluate HPP's generalization capability, we integrate it with two existing models (WHATsNet and Cross-Expansion) on HC-NC benchmarks. The resulting architectures, WHATsNet-HPP and Cross-HPP, feed node and hyperedge embeddings from base models into HPP for final predictions.

Table 3 reports accuracy on HC–NC with average performance changes compared to RAHC-HPP. **Generalizability of HPP.** HPP consistently boosts HC–NC performance across diverse backbones. Cross-HPP achieves 24.17% improvement on Email-Eu, while WHATsNet-HPP attains up to 33.21% gain, demonstrating HPP's plug-and-play capability and its effectiveness in modeling label coordination within hyperedges. **Compatibility of HPP and RAHC-HPP.** RAHC-HPP consistently outperforms both WHATsNet-HPP and Cross-HPP across all datasets, with improvements ranging from 0.3% (Stack-Physics) to 18.4% (Email-Eu), confirming the architectural compatibility between HPP and RAHC and establishing RAHC-HPP as the optimal configuration.

5.4 Hyperedge-Consistent Node Classification vs. Edge-Dependent Node Classification

To assess performance on edge-dependent node classification (ED-NC), we evaluate the three best-performing models from HC-NC: WHATsNet, Cross-Expansion, and RAHC-HPP. Results are shown in Table 4 using both macro-F1 and micro-F1 metrics, with best results in **bold**. Note that these models were optimized for HC-NC and not retrained for ED-NC; thus, WHATsNet shows minor deviations from its original paper (WHATsNet-Ori).

The results reveal three key findings: (a) RAHC-HPP achieves the best overall performance across nearly all ED-NC tasks, with only marginal gaps on DBLP (macro-F1) and Stack-Biology (3.3% micro-F1 behind WHATsNet), demonstrating strong generalization. (b) Models performing well on HC-NC generally perform well on ED-NC, indicating aligned modeling capabilities. (c) Despite this alignment, the tasks present distinct challenges. For instance, WHATsNet outperforms RAHC-HPP on ED-NC (DBLP: +0.1% macro-F1; Stack-Biology: +3.3% micro-F1) but lags on HC-NC (−2.6% and −4.5% respectively), justifying HC-NC as an independent research problem.

5.5 Case Study

As shown by three DBLP hyperedges in Table 5 under the HC-NC criterion, RAHC-HPP displays three typical behaviors: (E1) an easy case where RAHC alone already yields perfectly consistent labels; (E2) a case where HPP corrects two inconsistencies produced by RAHC by coordinating labels within the

Table 4. Results of Edge-Dependent Node Classification.

Dataset	Metric	WHATsNet-Ori	WHATsNet	Cross Expansion	RAHC-HPP (ours)
Coauth-DBLP	MicroF1	0.605 ± 0.002	$\mathbf{0.603 \pm 0.001}$	0.544 ± 0.004	$\mathbf{0.603 \pm 0.001}$
	MacroF1	0.595 ± 0.002	$\mathbf{0.593 \pm 0.001}$	0.545 ± 0.004	0.592 ± 0.001
Email-Eu	MicroF1	0.671 ± 0.000	0.667 ± 0.001	0.651 ± 0.000	$\mathbf{0.951 \pm 0.001}$
	MacroF1	0.646 ± 0.003	0.620 ± 0.013	0.641 ± 0.000	$\mathbf{0.947 \pm 0.001}$
Stack-Biology	MicroF1	0.742 ± 0.003	$\mathbf{0.775 \pm 0.001}$	0.632 ± 0.004	0.737 ± 0.001
	MacroF1	0.686 ± 0.004	0.653 ± 0.008	0.630 ± 0.003	$\mathbf{0.680 \pm 0.002}$
Stack-Physics	MicroF1	0.770 ± 0.003	0.720 ± 0.001	0.678 ± 0.004	$\mathbf{0.765 \pm 0.001}$
	MacroF1	0.707 ± 0.004	0.702 ± 0.004	0.669 ± 0.005	$\mathbf{0.706 \pm 0.003}$
BOM	MicroF1	-	0.918 ± 0.000	0.894 ± 0.002	$\mathbf{0.922 \pm 0.001}$
	MacroF1	-	0.918 ± 0.000	0.895 ± 0.002	$\mathbf{0.922 \pm 0.001}$

Table 5. Results of Three Hyperedges in DBLP. For each author, the notation is (True Label, RAHC Prediction, RAHC-HPP Prediction). Labels are F (First), L (Last), or O (Other). Red bold characters highlight predictions that differed from the true label.

E1: Co-Regularized Ensemble for Feature Selection.

Han Y. (FFF), Yang Y. (OOO), Zhou X. (LLL)

E2: Thinking of Images as What They Are: Compound Matrix Regression for Image Classification. Ma Z. (FOF), Yang Y. (OFO), Nie F. (OOO), Sebe N. (LLL)

E3: Complex Event Detection via Event Oriented Dictionary Learning. Yan Y. (FOF), Yang Y. (OOO), Shen H. (OLO), Meng D. (OOO), Liu G. (OFO), Hauptmann A. G. (OOL), Sebe N. (LOO)

---

hyperedge; and (E3) a challenging case in which, although HPP recovers the correct first author (RAHC had misassigned it) and fixes several nodes, the final assignment remains partially incorrect. Overall, these examples illustrate a common pattern: RAHC's message passing effectively captures node–hyperedge interactions but can induce within-hyperedge inconsistencies, while HPP leverages label-wise constraints to improve internal consistency. However, when base predictions conflict substantially with the ground truth, the scope of post-hoc correction is limited.

6 Conclusion

This work introduces Hyperedge-Consistent Node Classification (HC-NC), a novel problem setting that shifts the paradigm from instance-level accuracy to hyperedge-level validity. To address this, we propose RAHC-HPP, which leverages group interaction information and a learnable post-processing module to enforce hyperedge label consistency. Extensive experiments demonstrate that our method consistently outperforms state-of-the-art baselines on real-world HC-NC datasets. A promising future direction is joint hyperedge prediction and HC-NC, enabling simultaneous inference of node labels and latent group structures, which is particularly valuable for applications like chemical formula prediction.

Acknowledgments. The research is partially supported by the National Natural Science Foundation of China under Grant Nos. 92567301, 62132018, and 62231015, the Plans for Major Provincial Science & Technology Projects (No. 202523o09050019), Hefei Key Technology Research and Development Project (No. 2024SZD005).

References

1. Antelmi, A., Cordasco, G., Polato, M., Scarano, V., Spagnuolo, C., Yang, D.: A survey on hypergraph representation learning. ACM Comput. Surv. **56**(1), 1–38 (2023)
2. Bai, S., Zhang, F., Torr, P.H.: Hypergraph convolution and hypergraph attention. Pattern Recogn. **110**, 107637 (2021)
3. Carletti, T., Battiston, F., Cencetti, G., Fanelli, D.: Random walks on hypergraphs. Phys. Rev. E **101**(2), 022308 (2020)
4. Chen, X., He, K.: Exploring simple siamese representation learning. In: Proceedings of the IEEE/CVF Conference on Computer Vision and Pattern Recognition, pp. 15750–15758 (2021)
5. Chen, Y., Wang, X., Chen, C.: Hyperedge importance estimation via identity-aware hypergraph attention network. In: Proceedings of the 33rd ACM International Conference on Information and Knowledge Management, pp. 334–343 (2024)
6. Chien, E., Pan, C., Peng, J., Milenkovic, O.: You are allset: a multiset learning framework for hypergraph neural networks. In: 10th International Conference on Learning Representations, ICLR 2022 (2022)
7. Choe, M., Kim, S., Yoo, J., Shin, K.: Classification of edge-dependent labels of nodes in hypergraphs. In: Proceedings of the 29th ACM SIGKDD Conference on Knowledge Discovery and Data Mining, pp. 298–309 (2023)
8. Dong, Y., Sawin, W., Bengio, Y.: HNHN: hypergraph networks with hyperedge neurons. arXiv preprint arXiv:2006.12278 (2020)
9. Feng, Y., You, H., Zhang, Z., Ji, R., Gao, Y.: Hypergraph neural networks. In: Proceedings of the AAAI Conference on Artificial Intelligence, vol. 33, pp. 3558–3565 (2019)
10. Graves, A., Graves, A.: Long short-term memory. In: Supervised Sequence Labelling with Recurrent Neural Networks, pp. 37–45 (2012)
11. Huang, J., Yang, J.: UniGNN: a unified framework for graph and hypergraph neural networks. In: Proceedings of the Thirtieth International Joint Conference on Artificial Intelligence. International Joint Conferences on Artificial Intelligence Organization (2021)
12. Huang, Z., Xu, W., Yu, K.: Bidirectional LSTM-CRF models for sequence tagging. arXiv preprint arXiv:1508.01991 (2015)
13. Hwang, H., Lee, S., Shin, K.: HyFER: a framework for making hypergraph learning easy, scalable and benchmarkable. In: WWW Workshop on Graph Learning Benchmarks (2021)

14. Kipf, T.N., Welling, M.: Semi-supervised classification with graph convolutional networks. In: International Conference on Learning Representations (2017)
15. Lafferty, J., McCallum, A., Pereira, F., et al.: Conditional random fields: probabilistic models for segmenting and labeling sequence data. In: ICML, vol. 1, p. 3. Williamstown, MA (2001)
16. Lee, J., Lee, Y., Kim, J., Kosiorek, A., Choi, S., Teh, Y.W.: Set transformer: a framework for attention-based permutation-invariant neural networks. In: International Conference on Machine Learning, pp. 3744–3753. PMLR (2019)
17. Paranjape, A., Benson, A.R., Leskovec, J.: Motifs in temporal networks. In: Proceedings of the Tenth ACM International Conference on Web Search and Data Mining, pp. 601–610 (2017)
18. Souza, F., Nogueira, R., Lotufo, R.: Portuguese named entity recognition using BERT-CRF. arXiv preprint arXiv:1909.10649 (2019)
19. Tang, S., Chen, D., Bai, L., Liu, K., Ge, Y., Ouyang, W.: Mutual CRF-GNN for few-shot learning. In: Proceedings of the IEEE/CVF Conference on Computer Vision and Pattern Recognition, pp. 2329–2339 (2021)
20. Teichmann, M.T., Cipolla, R.: Convolutional CRFs for semantic segmentation. arXiv preprint arXiv:1805.04777 (2018)
21. Tian, C., et al.: HyperMixer: specializable hypergraph channel mixing for long-term multivariate time series forecasting. In: Proceedings of the AAAI Conference on Artificial Intelligence, vol. 39, pp. 20885–20893 (2025)
22. Vaswani, A., et al.: Attention is all you need. Adv. Neural Inf. Process. Syst. **30** (2017)
23. Veličković, P., Cucurull, G., Casanova, A., Romero, A., Liò, P., Bengio, Y.: Graph attention networks. In: International Conference on Learning Representations (2018)
24. Vinyals, O., Bengio, S., Kudlur, M.: Order matters: sequence to sequence for sets. In: ICLR (Poster) (2016)
25. Wang, M.Y.: Deep graph library: towards efficient and scalable deep learning on graphs. In: ICLR Workshop on Representation Learning on Graphs and Manifolds (2019)
26. Wu, T., Qu, J., Wang, D., Cui, Z., Liu, G., Zhao, P.: Contrasting transformer and hypergraph network for cooperative sequential recommendation. In: Onizuka, M., et al. (eds.) Database Systems for Advanced Applications, DASFAA 2024. LNCS, vol. 14852, pp. 83–98. Springer, Singapore (2025). https://doi.org/10.1007/978-981-97-5555-4_6
27. Xuan, P., Wu, S., Cui, H., Li, P., Nakaguchi, T., Zhang, T.: Interactive multi-hypergraph inferring and channel-enhanced and attribute-enhanced learning for drug-related side effect prediction. Comput. Biol. Med. **184**, 109321 (2025)
28. Yadati, N., Nimishakavi, M., Yadav, P., Nitin, V., Louis, A., Talukdar, P.: Hyper-GCN: a new method for training graph convolutional networks on hypergraphs. Adv. Neural Inf. Process. Syst. **32** (2019)
29. Yan, Y., Chen, Y., Wang, S., Wu, H., Cai, R.: Hypergraph joint representation learning for hypervertices and hyperedges via cross expansion. In: Proceedings of the AAAI Conference on Artificial Intelligence, vol. 38, pp. 9232–9240 (2024)

HyReaL: Clustering Attributed Graph via Hyper-complex Space Representation Learning

Junyang Chen[1], Yang Lu[2], Mengke Li[3], Cuie Yang[4], Yiqun Zhang[5,6(✉)], and Yiu-ming Cheung[6]

[1] Shenzhen International Graduate School, Tsinghua University, Shenzhen, China
chenjuny25@mails.tsinghua.edu.cn

[2] School of Informatics, Xiamen University, Xiamen, China
luyang@xmu.edu.cn

[3] College of Computer Science and Software Engineering, Shenzhen University, Shenzhen, China
mengkeli@szu.edu.cn

[4] State Key Laboratory of Synthetical Automation for Process Industries, Northeastern University, Shenyang, China
yangcuie@mail.neu.edu.cn

[5] School of Computer Science and Technology, Guangdong University of Technology, Guangzhou, China
yqzhang@gdut.edu.cn

[6] Department of Computer Science, Hong Kong Baptist University, Hong Kong, Hong Kong S.A.R., China
ymc@comp.hkbu.edu.hk

Abstract. In modern database systems, a large portion of data naturally exists in the form of attributed graphs, such as knowledge graphs, social networks, and recommender systems. Although common, these data types remain underexplored, particularly in learning expressive representations that support effective clustering. However, Graph Convolutional Networks (GCNs) often suffer from the Over-Smoothing (OS) effect, which homogenizes node embeddings, while existing OS solutions mainly focus on topology information rather than attribute learning, which is inconsistent with the objective of attributed graph clustering. To address this limitation, we propose **Hy**per-complex space **Re**presentation Le**a**rning (**HyReaL**), a generalized framework that introduces hyper-complex (quaternion) feature transformation to enhance attribute representation. The HyReaL bridges arbitrary-dimensional attributes to quaternion algebra and connects the learned embeddings to a generalized clustering objective without being restricted to a specific number of clusters k. By strengthening attribute coupling and reducing the need for deep graph convolution layers, HyReaL alleviates the OS problem and produces more discrim-

---

Supplementary Information The online version contains supplementary material available at https://doi.org/10.1007/978-981-92-0366-6_15.

inative node representations. Extensive experiments, including significance tests and ablation studies, demonstrate that HyReaL achieves superior and scalable performance for attributed graph clustering in modern database systems. The source code is here https://github.com/Juny-Chen/HyReaL.git.

Keywords: Attributed Graph Clustering · Graph Representation Learning · Quaternion Neural Networks

1   Introduction

Learning clustered distributions of data in an unsupervised setting is a fundamental analytic process in artificial intelligence and database systems [7]. Graph contains richer relational information among data objects (also called nodes interchangeably), making them increasingly important in applications such as knowledge graphs [9], social networks [13], recommender systems [30], and anomaly detection [42]. Clustering complex data represented as graphs has therefore attracted considerable attention [39–41], where effective graph representation [16,43,46] is critical to clustering accuracy.

Some recent works [3,16,20] further leverage the attribute values of graph nodes, which reflect inherent similarity relationships, to achieve more information-comprehensive clustering. Despite these advances, many database systems still underexplore attributed graph data, particularly in learning expressive node representations that can support accurate clustering.

To perform attributed graph clustering, conventional representation learning approaches [22,23] usually adopt multiple kernel functions for node embedding. However, this type of approach involves the non-trivial selection of kernels. Under such circumstances, end-to-end deep graph representation learning based on Graph Convolution Network (GCN) [12] provides an effective way to enhance the performance of attributed graph clustering. GCN and its variants [2,29,37] can simultaneously learn the embedding of graph structure and attribute values to achieve a more informative representation.

To explore the cluster distribution of data from a global perspective, relationships among nodes that span far in the graph are also critical. Although stacking more graph convolutional layers may theoretically help extract long-span information, the high overlap of over-hopped nodes will tend to homogenize the node representations, which is known as the widely discussed *Over-Smoothing* (OS) effect [14]. Existing solutions for mitigating such effect can be categorized into training tricks [24,45,49], dynamic hopping strategies [6,25], residual connections [4,31], and more powerful representation enhancement paradigms, e.g., contrastive learning [17,34,35].

Most existing OS solutions focus more on relieving the smoothing of nodes in terms of topological information for serving supervised learning tasks. However, from the perspective of unsupervised clustering, overemphasizing the graph

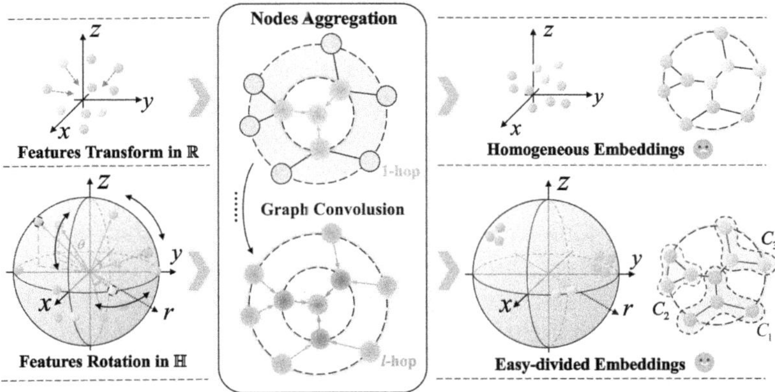

Fig. 1. Vanilla graph encoders (upper) vs. quaternion graph encoders (lower). After the node information aggregation through several hops, nodes represented in real-value space $\mathbb{R}$ by vanilla graph encoders tend to be homogeneous due to the "over-smoothing" (OS) and "over-dominating" (OD) effects. By contrast, the four views of data are flexibly rotated in a hyper-complex space $\mathbb{H}$ by the quaternion graph encoders to facilitate representation learning with a higher degree of learning freedom.

topology and the distinction of nodes may hamper the learning of compact clusters. That is, the similarity of nodes' embeddings can be easily *dominated* by the topology while attribute information tends to be overlooked as the graph convolution operation is guided by the topology. As a result, such an *Over-Dominating* (OD) effect will improperly drive the learning process away from the objective of clustering. Although a wide and shallow network can emphasize the coupling of attributes without incurring OS, "shallow" restricts the acquisition of abstract-level representations, and "wide" limits the scalability and efficiency of the model. In summary, conventional OS and OD solutions are contradictory, and may also introduce side effects to representation learning. Therefore, simultaneously coping with the OS and OD effects is the key to obtaining clustering-friendly representations on attributed graphs.

This paper therefore performs graph convolution in a four-axis hyper-complex space to achieve representation learning with adequate attribute interaction (counter OD) and shallow model architecture (counter OS). More detailed justifications for adopting hyper-complex space are provided in the following. Feature rotation in the hyper-complex space $\mathbb{H}$ is usually described by the Hamilton product of two quaternions [21]. Figure 1 illustrates that the feature quaternion can be transformed (through the parameter quaternion of a model) to simulate feature interaction. Since the intractable Gimbal lock problem in real-valued space $\mathbb{R}$ is circumvented by the quaternion transformation, feature interaction with four times the Degrees of Freedom (DoF) under the same parameter scale can thus be achieved. This can efficiently facilitate coupling learning of input features without incurring too many extra parameters. The high DoF of quaternion transformation enhances the attributes' contribution to the representation learning, thereby considerably mitigating the OD effect, whilst the strong fitting capability brought by the quaternion operation can relieve the need to stack more GCN layers, naturally circumventing the OS issue.

As for the model design, the framework of a graph auto-encoder [11] is adopted to perform **Hy**per-complex space **Re**present**a**tion **L**earning (**HyReaL**). We project input attributes into the hyper-complex space $\mathbb{H}$ with four views corresponding to the four parts of quaternion to facilitate the efficient feature quaternion transformation through the learnable parameter quaternion of the model. Such a four-view feature amplification also helps enhance the attributes' contribution for relieving the OD effect. Through learning the parameters of the Four-View Projection (FVP) and the Quaternion Graph Encoders (QGE) with a loss composed of graph reconstruction loss and a clustering objective, the proposed HyReaL model can output clustering-friendly embeddings to fit different clustering tasks. Considering that an optimal k is usually unavailable in real clustering scenarios, we integrate a general clustering objective to the loss without requiring a pre-specified k. This allows the model to train more general representations and, more importantly, makes the model feasible in most cases where the 'true' k is unknown in advance. Theoretical analysis and extensive experiments on various real benchmark graph datasets have illustrated the superiority of HyReaL. The main contributions are summarized in four-fold:

- A generalized representation learning framework is proposed for attributed graph clustering, which attempts to: 1) introduce quaternion to unsupervised learning by bridging arbitrary dimensional attributes to the four-part quaternion algebra, and 2) connect the representation learning to clustering without requiring k as prior knowledge.
- The OD effect is first discovered and explored in the context of graph clustering. This effect reasonably explains that the performance bottleneck encountered by current deep graph clustering is due to the disregard of the attribute information, and also provides potential guidance to subsequent graph clustering research.
- Quaternion is introduced to jointly address the OS and OD effects. For OD, FVP amplifies feature learning while QGE enhances attribute coupling. For OS, the '4×DoF' in quaternion transformation strengthen model capacity, reducing the need for deep GCN layers.
- HyReaL achieves high clustering accuracy and flexibility by learning generalized representations adaptable to different ks, avoiding retraining for varying cluster granularities.

2   Related Work

2.1   Deep Attributed Graph Clustering

Deep attributed graph clustering aims to partition nodes with both topological links and attribute values into compact groups. Some studies focus on the heterogeneity of attribute values [3,44,47], while most work assumes that attribute values are pure numerical. Early numerical clustering methods such as GAE and VGAE [11] combine graph convolution with autoencoders for representation learning. Subsequent methods enhance this framework with attention mechanisms [29], adversarial regularization [20], or joint optimization objectives, as in EGAE [37]. DFCN [27] integrates autoencoder and graph autoencoder features via a fusion module, and employs triplet self-supervision for robust clustering. Meanwhile, contrastive learning has emerged as a powerful paradigm. CCGC [34] leverages clustering as a proxy task to enhance representation discrimination. CONVERT [35] introduces a reversible perturbâĂŞrecover mechanism to preserve semantic consistency across augmented views. More recently, GLAC [32] jointly exploits local and global topologies via dual GCN pipelines with adaptive contrastive learning; SCGDN [18] proposes an augmentation-free framework using attentional diffusion; and MAGI [17] bridges contrastive learning and modularity maximization by defining community-aware contrastive signals.

2.2 Quaternion Representation Learning

A quaternion is a four-dimensional extension of complex numbers with a complete algebraic foundation. Since the Hamilton product [36] efficiently facilitates the interaction between the four parts of quaternions through the quaternion vector rotation upon the three imaginary axes, the quaternion operator is considered promising to enhance the representation learning ability [21], especially for the data with natural relations among its feature tuples, e.g., the three channels of colored image [48] and the 3D sound signals [5]. The work [38] is considered the first attempt to introduce the powerful quaternion for knowledge graph embedding. Most existing quaternion usage is in supervised scenarios, where the input data are usually with inherent tuples or multiple interdependent feature components corresponding to the three imaginary parts of the quaternion, making it especially suitable for quaternion representation learning. However, how to inherit the merits of the quaternion for unsupervised learning tasks has been relatively unexplored.

3   Proposed Method

We first provide the preliminaries, and then introduce the proposed graph clustering method HyReaL. The overview of HyReaL is demonstrated in Fig. 2.

3.1   Preliminaries

A quaternion is denoted as $Q = r + x\mathbf{i} + y\mathbf{j} + z\mathbf{k}$ where r is the real part and x, y, z are the imaginary parts. The Hamilton product between two quaternions $Q_1 = r_1 + x_1\mathbf{i} + y_1\mathbf{j} + z_1\mathbf{k}$ and $Q_2 = r_2 + x_2\mathbf{i} + y_2\mathbf{j} + z_2\mathbf{k}$ can be denoted as $Q_1 \otimes Q_2$, which follows the laws of association and distribution, but does not follow the law of commutativity. Benefiting from the orthogonality among imaginary axes, the essence of the product is to rotate Q_1 according to Q_2 in the

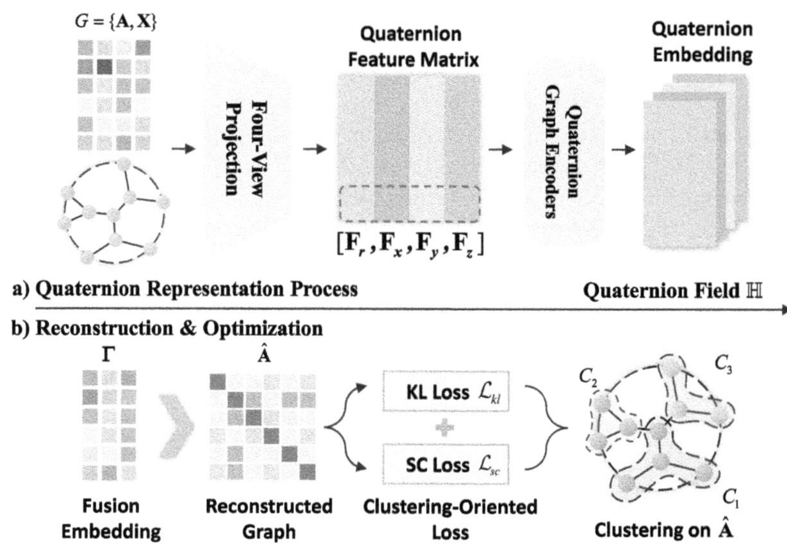

Fig. 2. Overview of HyReaL. Given attributed graph $G = \{\mathbf{A}, \mathbf{X}\}$, the attributes $\mathbf{X}$ are first projected into four views to form a feature quaternion $\mathbf{F} = \mathbf{F}_r + \mathbf{F}_x \mathbf{i} + \mathbf{F}_y \mathbf{j} + \mathbf{F}_z \mathbf{k}$. Then $\mathbf{F}$ is encoded by a quaternion graph convolution module to capture the local graph structure. The graph clustering-friendly embedding $\boldsymbol{\Gamma}$ learned according to the joint graph reconstruction Kullback-Leibler (KL) loss $\mathcal{L}_{kl}$ and the graph clustering loss $\mathcal{L}_{sc}$ is finally obtained for clustering.

hyper-complex space $\mathbb{H}$ spanned by the three imaginary axes. This is considered a critical characteristic expected by representation learning, especially for the learning of naturally coupled feature components.

Given an undirected attributed graph $G = \{\mathbf{A}, \mathbf{X}\}$, where $\mathbf{X} \in \mathbb{R}^{n \times d}$ is the node feature matrix and $\mathbf{A} \in \mathbb{R}^{n \times n}$ is a non-negative adjacency matrix ($\mathbf{A}_{ij} \geq 0$), with n and d denoting the number of nodes and feature dimensions, respectively. The node features can be written as $\mathbf{X} = [\mathbf{x}1, \cdots, \mathbf{x}n]^{\top}$, which are to be clustered into k disjoint subgraphs $\{G_1, \ldots, G_k\}$. The degree matrix $\mathbf{D}$ is diagonal with entries $\mathbf{D}_{ii} = \sum_j \mathbf{A}_{ij}$, representing node connectivity. To facilitate representation learning, we use the symmetrically normalized adjacency matrix $\tilde{\mathbf{A}} = \mathbf{D}^{-1/2}(\mathbf{A} + \mathbf{I})\mathbf{D}^{-1/2}$, where $\mathbf{I}$ adds self-loops and $\mathbf{D}$ corresponds to $\mathbf{A} + \mathbf{I}$. The self-loop preserves self-information, and normalization avoids bias toward high-degree nodes. The non-negativity of $\mathbf{A}$ ensures numerical stability of $\tilde{\mathbf{A}}$.

3.2   Generalized Quaternion Representation Learning

FVP: Four-View Projection. Unlike most existing hyper-complex space representation learning scenarios that the datasets are with tuple feature components (e.g., RGB images), attributed graph data are with different numbers of attributes and various graph structures. To leverage quaternion in attributed graph representation learning, we design a learnable projection mechanism to project the attributes $\mathbf{X}$ into four views. Such a mechanism acts to lift the tuple restriction of input features in quaternion and also leverages the Hamilton product for efficient coupling learning of the attributes.

Specifically, four independent initial MLPs are utilized to project the nodes represented by the attributes $\mathbf{X} = [\mathbf{x}_1, \mathbf{x}_2, \cdots, \mathbf{x}_n]^\top$ into four views $\mathbf{F}_r$, $\mathbf{F}_x$, $\mathbf{F}_y$, and $\mathbf{F}_z$, which can be written as

$$\mathbf{F}_\rhd = \mathcal{F}_\rhd(\mathbf{X}) = \mathbf{W}_\rhd^L \mathbf{X} + \mathbf{B}_\rhd^L, \quad \rhd \in \{r, x, y, z\}, \tag{1}$$

where $\mathcal{F}_\rhd(\cdot)$ is a shared MLP with learnable parameters $\mathbf{W}^L{}_\rhd$ and $\mathbf{B}_\rhd^L$. It encodes each node into a latent feature of dimension $\hat{d}$, forming a quaternion embedding $\mathbf{F} \in \mathbb{H}^{n \times (4 \times \hat{d})}$ for n nodes:

$$\mathbf{F} = \mathbf{F}_r + \mathbf{F}_x \mathbf{i} + \mathbf{F}_y \mathbf{j} + \mathbf{F}_z \mathbf{k}, \tag{2}$$

where $\mathbf{F}_r, \mathbf{F}_x, \mathbf{F}_y, \mathbf{F}_z \in \mathbb{R}^{n \times \hat{d}}$ are the real and imaginary components of the quaternion representation, respectively.

To project the quaternion embeddings into a new latent space, we introduce a learnable quaternion weight matrix $\mathbf{W}^Q \in \mathbb{H}^{4\hat{d} \times 4\tilde{d}}$, where $\tilde{d}$ is the target dimensionality per quaternion component. The transformation is performed via the Hamilton product between $\mathbf{F}$ and $\mathbf{W}^Q$:

$$\mathbf{F} \otimes \mathbf{W}^Q = \begin{bmatrix} \mathbf{F}_r \\ \mathbf{F}_x \\ \mathbf{F}_y \\ \mathbf{F}_z \end{bmatrix}^\top \begin{bmatrix} \mathbf{W}_r^Q & \mathbf{W}_x^Q & \mathbf{W}_y^Q & \mathbf{W}_z^Q \\ -\mathbf{W}_x^Q & \mathbf{W}_r^Q & -\mathbf{W}_z^Q & \mathbf{W}_y^Q \\ -\mathbf{W}_y^Q & \mathbf{W}_z^Q & \mathbf{W}_r^Q & -\mathbf{W}_x^Q \\ -\mathbf{W}_z^Q & -\mathbf{W}_y^Q & \mathbf{W}_x^Q & \mathbf{W}_r^Q \end{bmatrix}, \tag{3}$$

where $\otimes$ indicates the Hamilton product. The above $\mathcal{F}_\rhd(\cdot)$ and $\otimes$ processes can convert an arbitrary-dimensional attribute set into the quaternion field, and associate different parts of the feature quaternion through their shared weights in the weight quaternion $\mathbf{W}^Q$. By tuning the weights in $\mathbf{W}^Q$, features in $\mathbf{F}$ can be efficiently transformed with a higher DoF to capture feature couplings.

Remark 1. **Quaternion transformation facilitates a Higher Degree of Freedom (DoF).** According to Eq. (3), learnable parameters of $\mathbf{W}^Q$, i.e., $\{\mathbf{W}_r^Q, \mathbf{W}_x^Q, \mathbf{W}_y^Q, \mathbf{W}_z^Q\}$, yield 16 pairs of feature transformation to determine arbitrary rotation of $\mathbf{F}$ in hyper-complex space $\mathbb{H}$, while in real-value space $\mathbb{R}$, realizing the same transformation requires four times of parameters. Therefore, the DoF of quaternion feature transformation is four times the transformation in real-value space.

Remark 2. **Benefits of high DoF in coping with OD and OS effects.** A higher DoF can improve the learning efficiency of the model, i.e., using smaller parameter scale to leverage stronger learning capability. Specifically in our model, there are two key benefits: 1) the model can tolerate a wider network structure for FVP to amplify the input attributes and thus offset the OD effect, and 2) ensuring learning capability with a wider structure can relieve OS effect caused by staking too many encoding layers.

QGE: Quaternion Graph Encoders. To further make a fusion of the feature quaternion $\mathbf{F}$ with the normalized Laplacian $\tilde{\mathbf{A}}$, $\mathbf{F}$ is feed forward to a quaternion graph convolutional module composed of stacked encoders, where the operation of the l-th encoder can be written as

$$H_l = \varphi_l(\tilde{\mathbf{A}} \cdot H_{l-1} \otimes \mathbf{W}_l^Q) \tag{4}$$

where the operation priority of $\otimes$ is higher than the matrix product.

Here, $\varphi_l(\cdot)$ is the activation function that ensures the non-negative of H_l. Each encoder aggregates l-hop quaternion features via $\tilde{\mathbf{A}}$ to produce a higher-level representation H_l. The output of the m-layer quaternion graph encoder is fused into a single embedding matrix $\mathbf{\Gamma}$ via

$$\mathbf{\Gamma} = \mathrm{Re}(H_m) \circledast \mathrm{Im}(H_m), \tag{5}$$

where $\mathrm{Re}(\cdot)$ and $\mathrm{Im}(\cdot)$ extract the real and imaginary parts of H_m, and $\circledast$ denotes averaging over quaternion components. The reconstructed adjacency matrix $\hat{\mathbf{A}} \in \mathbb{R}^{n \times n}$ is then obtained by

$$\hat{\mathbf{A}} = \mathbf{\Gamma} \cdot \mathbf{\Gamma}^\top, \tag{6}$$

serving as a decoder to preserve the graph topology.

3.3   Clustering-Oriented Optimization

From a global perspective, the FVP and QGE modules jointly emphasize attribute semantics in the node representations $\mathbf{\Gamma}$, while graph reconstruction encourages structural consistency. To further promote sparsity in the learned graph for clustering purposes, we define the overall loss as a combination of reconstruction, spectral clustering, and regularization terms:

$$\mathcal{L} = \mathcal{L}_{kl} + \alpha \mathcal{L}_{reg} + \beta \mathcal{L}_{sc}, \tag{7}$$

where α and β are balancing hyperparameters. The reconstruction loss $\mathcal{L}_{kl}$ encourages $\mathbf{\Gamma}$ to reflect the structure encoded in the original graph adjacency $\tilde{\mathbf{A}}$:

$$\mathcal{L}_{kl} = \frac{1}{n^2} \sum_{i=1}^{n} \sum_{j=1}^{n} \tilde{\mathbf{A}}_{ij} \log \frac{1}{\hat{\mathbf{A}}_{ij}}, \tag{8}$$

where $\hat{\mathbf{A}}$ is the reconstructed adjacency from Eq. (6). Minimizing $\mathcal{L}_{kl}$ aligns topology-aware and attribute-aware structures via $\mathbf{\Gamma}$. The regularization term $\mathcal{L}_{reg}$ prevents overfitting by controlling model complexity.

To encourage clustering-friendly embeddings, the spectral loss is defined as:

$$\mathcal{L}_{sc} = \mathrm{Tr}(\mathbf{\Gamma}^\top \mathbf{L}_{\hat{\mathbf{A}}} \mathbf{\Gamma}), \tag{9}$$

where $\mathbf{L}_{\hat{\mathbf{A}}} = \mathbf{D}_{\hat{\mathbf{A}}} - \hat{\mathbf{A}}$ is the unnormalized Laplacian constructed from the learned adjacency $\hat{\mathbf{A}}$. This term encourages $\mathbf{\Gamma}$ to capture high intra-cluster similarity (larger $\hat{\mathbf{A}}_{ij}$) and reduced inter-cluster connections (smaller $\mathbf{D}_{\hat{\mathbf{A}}}$), consistent with the spectral clustering objective.

After training, we obtain the final adjacency $\hat{\mathbf{A}}$, from which the Laplacian $\mathbf{L}$ is formed. Spectral clustering solves the relaxed objective:

$$\arg\min_{\mathbf{H}} \mathrm{Tr}(\mathbf{H}^{\top}\mathbf{L}\mathbf{H}) \quad \text{s.t.} \quad \mathbf{H}^{\top}\mathbf{H} = \mathbf{I}, \tag{10}$$

where $\mathbf{H} \in \mathbb{R}^{n \times k}$ is the cluster indicator matrix. The optimal solution is given by the k smallest eigenvectors of $\mathbf{L}$, forming a matrix $\mathbf{E}$, whose rows serve as node embeddings for K-Means clustering, referring to [10,28]. Only the number of clusters k needs to be specified.

Finally, the overall time complexity of HyReaL is $\mathcal{O}(T[nd\hat{d} + n^2\tilde{d}^2])$, where T is the number of training iterations. The dominant costs come from quaternion feature projection and stacked quaternion graph convolution operations, with lower-order terms ignored.

4   Experiment

4.1   Experimental Settings

Datasets. Experiments are conducted on ten real benchmark attributed graph datasets, including CORA [26], CITESEER [26], DBLP [1], ACM [1], WIKI [33], FILM [16], and the four, i.e., CORNELL, WISC, UAT, and AMAP, from [15]. The CORA and DBLP datasets are the citation network. ACM and DBLP datasets are paper citation relationships. WIKI and FILM datasets are the relationships of Wikipedia links and films, respectively. CORNELL, WISC, UAT, and AMAP datasets are American university website links. Detailed dataset statistics are sorted in the Table 1.

Training Process. All experiments are conducted in PyTorch 1.8.0 on an NVIDIA A5000 GPU with 64GB RAM. The model is first warmed up for 10 epochs using only the KL loss $\mathcal{L}_{kl}$ and regularization loss $\mathcal{L}_{reg}$. Following recent graph clustering works [17,18,27,29,32,34,35,37], each result is the average performance with standard deviation from ten runs. The model is trained for 50 epochs, with four iterations per epoch, and clustering is performed after each epoch. The best performance across the 50 epochs is reported.

Experiment Settings. We compare eleven clustering methods, including K-Means [8] and ten deep graph clustering approaches, which fall into two categories: (1) Representation learning-based methods: GAE [11], EGAE [37], and DFCN [27]. (2) Contrastive learning-based methods: CONVERT [35], CCGC [34], SCDGN [18], MAGI [17], and GLAC [32].

All baselines are evaluated under their original settings. K-Means and Spectral Clustering are repeated 10 times and averaged, operating on node features and graph topology, respectively. GAE and VGAE are trained for 200 unsupervised epochs followed by K-Means clustering, while advanced methods are reproduced using official implementations with default hyper-parameters. Detailed hyper-parameter settings are provided in the released code and the extended version on arXiv.

Table 1. Statistics of the ten graph datasets. n is the number of nodes, d is the attribute dimension, and k is the number of clusters.

No.	Dataset	n	d	k	No.	Dataset	n	d	k
1	ACM	3025	1870	3	6	CORNELL	183	1703	5
2	WIKI	2405	4973	17	7	CORA	2708	1433	7
3	CITESEER	3327	3703	6	8	WISC	251	1703	5
4	DBLP	4057	334	4	9	UAT	1190	239	4
5	FILM	7600	932	5	10	AMAP	7650	745	8

Validity Metrics. Six evaluation metrics are used including three external metrics, Clustering Accuracy (ACC), Normalized Mutual Information (NMI), and Average Rand Index (ARI), which evaluate performance based on data labels, with values in $[0, 1]$, $[0, 1]$, and $[-1, 1]$, respectively. Three internal metrics, i.e., Silhouette Coefficient (SC), Davies-Bouldin Index (DBI), and Calinski-Harabasz Index (CHI), do not rely on labels, but indicate the separability and/or compactness of obtained clusters, are with values ranges $[-1, 1]$, $[0, +\infty)$, and $[0, +\infty)$, respectively. Except for DBI, higher values indicate a better performance.

4.2   Quantitative Results

Four quantitative experiments are conducted to evaluate HyReaL: (1) clustering performance based on external metrics; (2) internal metrics under varying k to assess embedding separability and generality; (3) execution time comparison for efficiency; and (4) ablation study to verify the effectiveness of key modules.

Clustering Performance Evaluation. Table 2 reports the clustering performance of all the compared methods by using k provided by the data labels. Overall, compared to recent contrastive learning methods, HyReaL shows competitive performance, clearly outperforming traditional embedding models while remaining comparable to state-of-the-art contrastive approaches. Notably, several contrastive methods (e.g., CONVERT, CCGC, MAGI and GLAC) benefit from explicitly mining positive and negative sample pairs, allowing them to extract richer relational cues despite simpler base encoders. However, the fact that HyReaL achieves similar or better performance without contrastive

Table 2. Clustering performance compared with existing methods. The best and second-best results on each dataset are marked in **boldface** and underline, respectively. "AR" in the last row reports the average performance rank of each method.

Dataset	Metric	K-Means Classical	GAE Classical	CCGC AAAI'23	DFCN AAAI'21	EGAE TNNLS'22	CONVERT MM'23	SCDGN MM'23	MAGI KDD'24	GLAC TAI'25	HyReaL (ours)
ACM	ACC	36.78±0.01	44.22±4.11	86.94±1.37	86.04±2.18	85.54±3.62	80.53±2.91	81.64±0.37	89.37±0.51	66.94±12.16	**90.37±0.41**
	NMI	0.82±0.01	14.67±4.53	58.20±3.29	59.56±4.51	56.09±8.26	47.45±4.35	51.63±0.46	66.00±1.07	34.18±12.69	**67.50±1.17**
	ARI	0.24±0.01	3.66±2.39	64.94±3.29	63.94±4.79	62.10±8.21	51.30±6.03	54.02±0.37	71.22±0.51	34.03±12.16	**73.57±1.06**
WIKI	ACC	25.81±0.89	33.11±2.08	44.47±3.66	43.10±3.67	47.49±1.13	51.41±1.15	44.66±0.20	52.61±1.50	23.88±1.66	**52.95±0.88**
	NMI	22.69±1.21	31.62±1.51	44.13±2.65	38.33±2.91	43.33±1.99	48.46±0.62	44.02±0.55	**51.93±0.68**	14.86±2.75	49.26±1.32
	ARI	2.54±0.32	5.61±0.89	24.44±3.24	17.17±3.75	28.99±1.58	28.39±1.33	24.33±0.20	**34.61±1.50**	6.52±1.66	33.77±0.87
CITESEER	ACC	26.10±1.33	32.93±3.01	54.37±2.96	42.37±2.05	58.71±3.68	62.14±1.53	59.26±0.40	66.52±0.70	38.47±3.96	**66.57±1.14**
	NMI	6.92±1.36	20.11±2.63	27.54±2.85	23.90±1.83	33.15±2.99	34.68±1.78	34.70±0.16	**41.24±0.88**	19.24±5.00	40.36±1.21
	ARI	0.31±1.93	4.64±2.01	25.11±3.66	19.19±2.43	31.46±4.61	34.69±1.88	29.58±0.40	40.98±0.70	16.64±3.96	**41.43±1.94**
DBLP	ACC	32.74±0.06	46.10±1.43	54.97±6.88	38.91±0.04	53.64±1.46	54.52±2.37	45.42±0.02	71.57±1.73	36.62±3.17	**72.46±2.24**
	NMI	2.98±0.01	19.71±1.83	22.61±5.44	8.11±0.04	18.19±1.07	22.33±1.93	13.63±0.05	**43.40±0.87**	6.02±2.90	39.12±2.27
	ARI	15.31±1.87	5.78±0.87	17.70±5.12	6.63±0.02	15.07±2.02	17.81±1.17	8.56±0.02	**43.30±1.73**	5.45±3.17	41.24±2.55
FILM	ACC	24.21±0.01	25.64±0.02	24.31±±1.32	25.91±1.64	22.79±0.25	**27.43±0.23**	26.32±0.01	23.62±0.18	26.47±0.18	26.81±0.65
	NMI	0.01±0.00	0.09±0.01	0.22±0.39	0.28±0.03	0.21±0.08	0.79±0.07	1.04±0.00	0.26±0.02	0.50±0.17	**1.47±0.22**
	ARI	0.00±0.01	0.13±0.01	0.31±0.51	0.27±±0.04	0.17±0.08	1.34±0.17	0.35±0.01	0.18±0.18	0.84±0.18	**1.78±0.26**
CORNELL	ACC	42.40±0.65	38.03±1.09	36.55±2.65	39.72±1.90	39.23±0.53	41.86±2.98	45.52±0.60	40.05±1.27	**55.63±0.41**	38.25±1.84
	NMI	2.71±0.17	5.35±0.36	3.19±0.56	3.25±0.37	6.49±0.73	**9.80±2.68**	4.57±0.56	9.20±1.45	6.32±1.39	8.86±2.51
	ARI	−2.14±0.10	2.11±0.48	0.85±1.18	−1.10±1.23	3.17±0.91	6.19±3.25	5.08±0.60	**7.67±1.27**	1.79±0.41	5.48±1.32
CORA	ACC	31.14±3.76	49.47±5.76	66.72±3.04	45.94±5.80	72.11±1.35	66.34±1.80	74.30±0.37	67.78±2.89	41.11±4.77	**75.82±1.51**
	NMI	6.67±5.28	40.86±4.81	48.96±2.62	36.46±3.44	52.89±1.17	46.84±1.68	56.06±0.39	50.56±1.23	23.71±4.32	**59.02±1.20**
	ARI	7.83±1.69	22.49±7.27	42.80±2.69	23.95±5.52	48.49±2.16	40.13±1.67	51.75±0.37	44.45±2.89	18.38±4.77	**55.64±3.06**
WISC	ACC	42.03±2.04	42.11±1.73	37.17±2.58	40.95±5.44	37.01±2.01	**47.61±1.91**	45.78±1.03	36.97±2.64	47.05±1.39	44.46±1.58
	NMI	6.25±1.13	8.09±0.63	5.35±3.26	7.01±0.58	11.02±1.16	9.70±3.35	11.09±1.90	11.52±1.18	5.23±1.15	**16.31±1.66**
	ARI	−3.02±1.68	2.85±0.54	2.02±1.63	4.45±0.99	6.00±1.12	4.76±2.69	5.39±1.03	6.19±2.64	1.94±1.39	**9.92±1.38**
UAT	ACC	32.69±0.12	44.55±0.07	41.85±1.63	39.33±4.72	53.10±0.79	**55.18±1.34**	44.99±2.01	43.71±0.33	46.59±2.59	53.84±0.27
	NMI	20.63±0.63	18.61±9.44	15.86±2.38	13.95±1.67	21.80±0.85	**27.31±1.18**	12.64±1.32	19.18±0.49	17.04±2.13	24.15±1.16
	ARI	6.42±0.44	11.61±5.90	10.33±2.82	7.28±3.16	20.77±0.64	19.46±1.90	11.27±2.01	16.45±0.33	13.73±2.59	**22.54±0.91**
AMAP	ACC	22.66±0.31	60.47±0.87	N/A	58.51±3.96	76.37±1.32	66.28±1.86	**77.52±0.02**	68.99±0.82	42.13±5.07	76.02±0.94
	NMI	2.37±0.09	58.01±0.56	N/A	55.95±1.13	65.44±1.61	52.57±0.79	**70.26±0.69**	59.95±0.59	27.49±5.71	66.47±1.50
	ARI	0.37±0.03	33.31±1.02	N/A	41.76±1.58	57.51±1.74	42.89±1.49	**61.21±0.02**	47.88±0.82	19.51±5.07	57.73±1.50
-	AR	9.1	7.0	6.0	6.8	4.8	3.6	4.3	3.2	6.9	1.7

supervision underscores the representational strength of our quaternion design. This validates its ability to model complex structural and semantic information in a compact and clustering-friendly form. Moreover, HyReaL consistently surpasses conventional representation learning baselines (e.g., GAE, EGAE, and SCDGN) across all metrics, highlighting its effectiveness in solving the OS and OD problems. A promising direction for future work is to integrate contrastive mechanisms into the quaternion framework, enabling more effective mining of intra-graph positive and negative relations.

Separability and Universality Studies. The clustering separability is assessed using internal metrics. To evaluate the effectiveness of HyReaL's generalized loss, its performance under varying k is compared with EGAE, CCGC, CONVERT, SCDGN, MAGI, and GLAC, the strongest baselines in Table 2. As shown in Fig. 3, HyReaL consistently outperforms all counterparts across different k values and metrics, which significantly proves the outstanding separability and universality of the embeddings learned by the HyReaL. More specific observations are as follows: 1) As k increases, HyReaL's performance slightly degrades, as internal metrics emphasize inter-cluster separation and intra-cluster compactness. With larger k, clusters become smaller and more compact, making metric

Table 3. Ablation study of the key HyReaL modules.

Dataset	Baseline			HyReaL w/o FVP			HyReaL w/o QGE			HyReaL w/o β			**HyReaL** (ours)		
	ACC	NMI	ARI	ACC	NMI	ARI	ACC	NMI	ARI	ACC	NMI	ARI	ACC	NMI	ARI
ACM	89.34	64.54	70.92	89.90	65.80	72.27	84.53	56.38	60.85	90.44	67.48	73.70	**90.53**	**67.67**	**73.93**
WIKI	51.60	49.15	32.43	51.57	47.91	31.99	51.64	48.66	32.45	**53.16**	**50.13**	34.24	52.59	50.01	**34.35**
CITESEER	65.73	40.24	40.72	66.21	40.44	41.28	66.57	40.38	41.35	67.31	41.08	42.07	**67.41**	**41.14**	**42.17**
DBLP	67.89	35.20	35.48	71.41	38.15	39.73	67.26	36.59	35.71	69.93	37.80	38.15	**72.45**	**39.58**	**40.96**
FILM	26.95	1.12	1.76	27.41	1.28	1.97	**27.70**	**1.51**	**2.01**	27.42	1.49	1.86	26.84	1.39	1.74
CORNELL	35.85	6.87	3.69	36.99	6.52	4.51	36.61	6.52	3.93	**38.14**	**9.37**	4.99	37.65	8.14	**5.40**
CORA	72.73	55.84	**51.44**	73.12	55.13	50.61	72.11	55.35	49.68	70.89	53.25	48.05	**73.28**	**56.22**	51.34
WISC	40.92	13.54	8.03	43.03	16.09	9.37	**44.74**	**17.21**	**10.45**	43.63	16.40	9.78	44.14	16.59	10.38
UAT	53.68	23.69	21.87	53.71	23.77	22.36	**54.37**	23.26	22.19	53.47	23.68	22.11	53.82	**24.03**	**22.38**
AMAP	73.23	61.13	53.81	74.69	63.65	55.53	74.98	64.34	56.37	77.00	69.01	59.85	**77.00**	**69.01**	**59.85**
Average Rank	4.20			3.33			3.17			2.60			**1.67**		

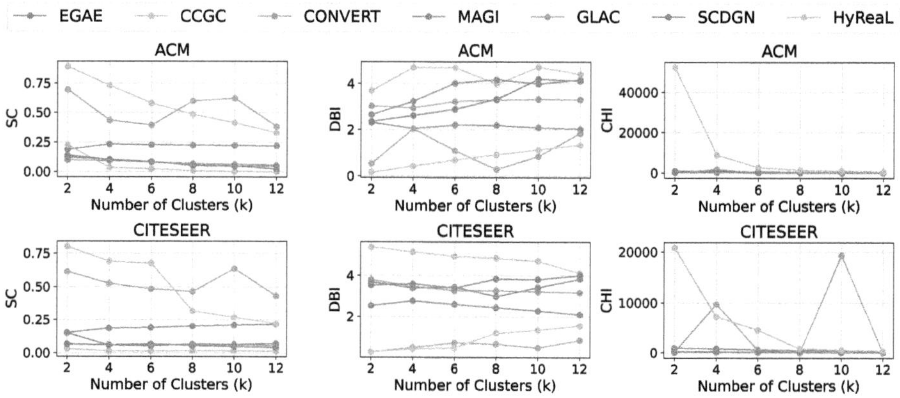

Fig. 3. Clustering performance w.r.t. internal metrics under different ks.

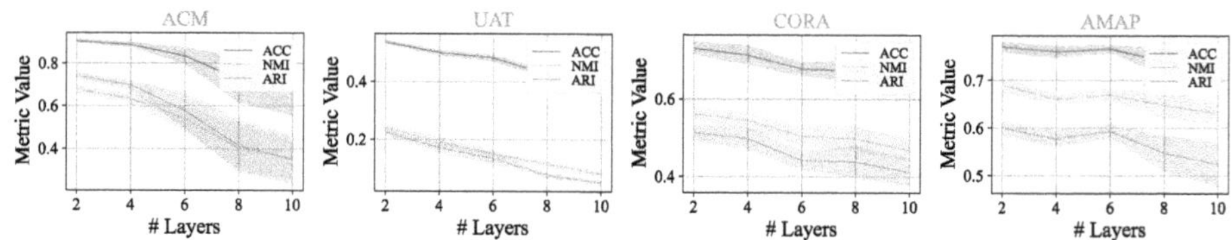

Fig. 4. The OS effect examined under different numbers of QGE layers.

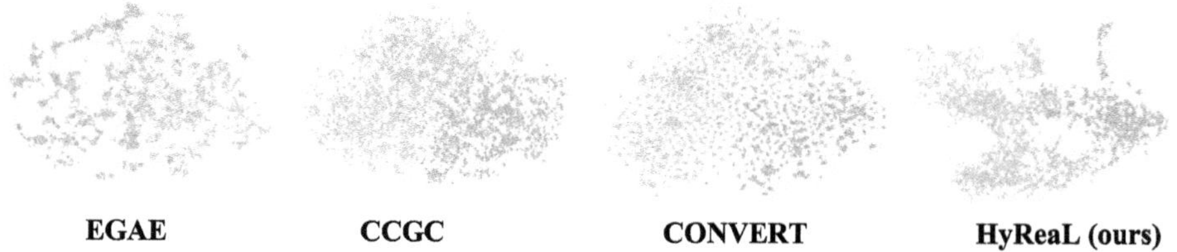

Fig. 5. t-SNE visualization of ACM dataset represented by the four advanced methods.

values converge and less sensitive to performance differences. 2) The k values in Fig. 3 include the ground-truth cluster numbers listed in Table 1. At these 'true' k values, HyReaL achieves superior performance, demonstrating the higher separability of its representations.

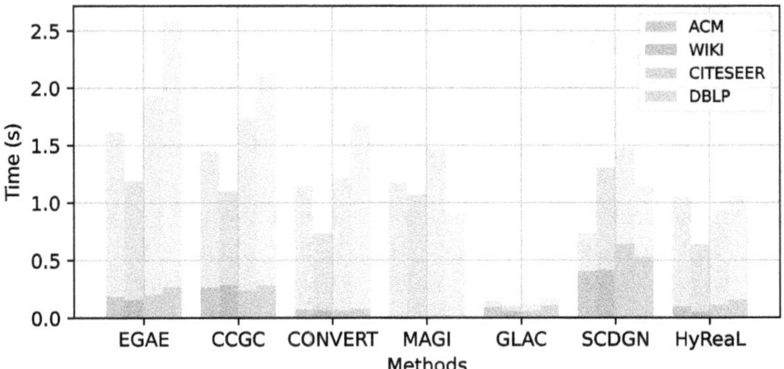

Fig. 6. Average execution time on different k values. Different colors indicate the average execution time on different datasets. Deep and shallow colors indicate the execution time of model training and clustering.

Effectiveness of QGE on Mitigating OS Effect. To evaluate the effectiveness of stacked QGE layers in mitigating the over-smoothing (OS) effect, we conduct a depth ablation study by varying the number of QGE layers from 2 to 12, with a step size of 2. The two-layer model reduces the input from 512 to 256 and then to the final 128-dimensional output. For deeper variants, additional QGE layers with fixed 256-dimensional input and output are inserted between the initial and final layers to form hierarchical representations.

Figure 4 presents results, showing a consistent decline in all metrics across datasets as layer depth increases. These trends confirm that deeper graph convolution exacerbates the OS effect, resulting in homogeneous representations and degraded clustering performance. Furthermore, the results of ACM and AMAP become more unstable, reflecting that over-smoothing causes the randomness of the representation. In summary, the above observations support that high DoF quaternion representation learning alleviates the OS problem and contributes to stronger feature interaction representation.

Efficiency Evaluation. Corresponding to the four datasets and six advanced methods in the previous experiment as shown in Fig. 3, we also compare their execution time with the proposed HyReaL averaged on the six different k values. Execution time comparison is shown in Fig. 6. HyReaL demonstrates lower runtime than five baselines, as it learns general representations applicable to multiple

k values without retraining. In contrast, other methods require re-training for each k, resulting in higher computational overhead. While GLAC achieves the lowest runtime by encoding k directly into its network output, it lacks flexibility for dynamic clustering and generalization across unseen cluster configurations.

Ablation Studies. Table 3 compares HyReaL with four variants: (1) Baseline using one MLP and two GCN layers; (2) HyReaL w/o FVP, replacing FVP with a linear layer; (3) HyReaL w/o QGE, replacing QGE with two standard GCN encoders; and (4) HyReaL w/o β, removing the spectral clustering loss $\mathcal{L}_{sc}$. The comparisons validate the effectiveness of the FVP module and clustering-oriented loss, the necessity of QGE, and the interplay between FVP and QGE. Furthermore, the performance gains confirm that FVP and QGE jointly mitigate the OD effect by preserving and extracting discriminative attribute information.

Four observations are provided below: 1) HyReaL outperforms the baseline in 29 out of 30 cases, demonstrating the effectiveness of the proposed FVP and QGE modules. 2) HyReaL achieves better performance than HyReaL w/o FVP in 27 out of 30 cases, indicating that FVP is essential for learning meaningful feature quaternions for QGE. 3) HyReaL surpasses HyReaL w/o QGE in 22 out of 30 cases, showing that QGE effectively aggregates node information and mitigates the OD effect. 4) HyReaL outperforms HyReaL w/o β in 23 out of 30 cases, validating the contribution of the cluster-oriented loss.

4.3 Qualitative Results

To illustrate the effectiveness of the representations obtained by HyReaL from the perspective of clustering, we visualize the 2D embedding distributions of EGAE, CCGC, CONVERT, and HyReaL on the ACM dataset using the unsupervised t-SNE [19], as shown in Fig. 5, with colors that indicate clusters of ground truth. Compared to EGAE, the contrastive methods CCGC and CONVERT, and the proposed HyReaL yield more separable clusters. This is attributed to the use of contrastive augmentation in CCGC and CONVERT, and quaternion rotation in HyReaL, which enhances representation learning. HyReaL further improves cluster separability by structurally rotating multi-view attributes, better preserving global node distributions. In contrast, EGAE may be overly influenced by topology, lacking mechanisms to preserve attribute semantics.

5 Concluding Remarks

In modern database systems, a large portion of data naturally exists in the form of attributed graphs, whose efficient clustering requires expressive node representations that preserve both structural and attribute information. This paper proposes HyReaL to leverage the efficient Hamilton product of quaternions, jointly addressing the OS and OD issues that bottleneck clustering performance. The generalized design of HyReaL integrates learnable FVP and QGE for clustering-oriented representation learning. FVP bridges arbitrary-dimensional attributes to the four-part quaternion operations in QGE, and they jointly enhance 1) model learning capacity and 2) attribute information preservation. A generalized clustering objective further guides the model to learn high-DoF, universal representations without being restricted to a fixed number of clusters. Consequently, HyReaL produces more discriminative and clustering-consistent node embeddings across different ks, offering strong applicability for real-world data analysis. Extensive experiments validate its superiority. Although HyReaL proves effective, it is not exempt from limitations: The generality and efficiency of HyReaL are established on static data. Future research can focus on improving it for the adaptation to streaming and time-series data with concept drift.

Acknowledgements. This work was supported in part by the National Natural Science Foundation of China (NSFC) under grants: 62476063, 62376233, and 62306181, the NSFC Outstanding Youth Science Fund Project (Overseas) under grant: HWYQ-2023-03, the General Research Fund of RGC under grant: 12202924, and the Guangdong and Hong Kong Universities "1+1+1" Joint Research Collaboration Scheme with grant: 2025A0505000004.

References

1. Bo, D., Wang, X., Shi, C., Zhu, M., Lu, E., Cui, P.: Structural deep clustering network. In: Proceedings of the 29th Web Conference, April 2020, pp. 1400–1410 (2020)
2. Bowman, S.R., Vilnis, L., Vinyals, O., Dai, A.M., Jozefowicz, R., Bengio, S.: Generating sentences from a continuous space. arXiv preprint arXiv:1511.06349 (2015)
3. Chen, J., Ji, Y., Zou, R., Zhang, Y., Cheung, Y.: QGRL: quaternion graph representation learning for heterogeneous feature data clustering. In: Proceedings of the 30th ACM SIGKDD Conference on Knowledge Discovery and Data Mining, pp. 297–306 (2024)
4. Chen, M., Wei, Z., Huang, Z., Ding, B., Li, Y.: Simple and deep graph convolutional networks. In: Proceedings of the 37th International Conference on Machine Learning, vol. 119, pp. 1725–1735 (2020)
5. Comminiello, D., Lella, M., Scardapane, S., Uncini, A.: Quaternion convolutional neural networks for detection and localization of 3D sound events. In: Proceedings of the 44th International Conference on Acoustics, Speech, and Signal Processing, pp. 8533–8537 (2019)
6. Eliasof, M., Haber, E., Treister, E.: PDE-GCN: novel architectures for graph neural networks motivated by partial differential equations. In: Proceedings of Advances in Neural Information Processing Systems, pp. 3836–3849 (2021)
7. Ester, M., Kriegel, H., Xu, X.: A database interface for clustering in large spatial databases. In: Proceedings of the First International Conference on Knowledge Discovery and Data Mining, pp. 94–99 (1995)

8. Hamerly, G., Elkan, C.: Learning the k in k-means. In: Proceedings of the 16th International Conference on Neural Information Processing Systems, pp. 281–288 (2003)
9. Hogan, A., et al.: Knowledge graphs. ACM Comput. Surv. **54**(4), 1–37 (2021)
10. Ikotun, A.M., Ezugwu, A.E., Abualigah, L., Abuhaija, B., Heming, J.: K-means clustering algorithms: a comprehensive review, variants analysis, and advances in the era of Big Data. Inf. Sci. **622**, 178–210 (2023)
11. Kipf, T.N., Welling, M.: Variational graph auto-encoders. arXiv preprint arXiv:1611.07308 (2016)
12. Kipf, T.N., Welling, M.: Semi-supervised classification with graph convolutional networks. In: Proceedings of the 5th International Conference on Learning Representations (2017)
13. Kumar, S., Mallik, A., Khetarpal, A., Panda, B.S.: Influence maximization in social networks using graph embedding and graph neural network. Inf. Sci. **607**, 1617–1636 (2022)
14. Li, Q., Han, Z., Wu, X.: Deeper insights into graph convolutional networks for semi-supervised learning. In: Proceedings of the 32nd AAAI Conference on Artificial Intelligence, pp. 3538–3545 (2018)
15. Liu, Y., Xia, J., Zhou, S., et al.: A survey of deep graph clustering: taxonomy, challenge, and application. arXiv preprint arXiv:2211.12875 (2022)
16. Liu, Y.: A survey of deep graph clustering: taxonomy, challenge, and application. arXiv preprint arXiv:2211.12875 (2022)
17. Liu, Y., et al.: Revisiting modularity maximization for graph clustering: a contrastive learning perspective. In: Proceedings of the 30th SIGKDD Conference on Knowledge Discovery and Data Mining, pp. 1968–1979 (2024)
18. Ma, Y., Zhan, K.: Self-contrastive graph diffusion network. In: Proceedings of the 31st ACM International Conference on Multimedia, pp. 3857–3865 (2023)
19. Van der Maaten, L., Hinton, G.: Visualizing data using t-SNE. J. Mach. Learn. Res. **9**(11) (2008)
20. Pan, S., Hu, R., Long, G., Jiang, J., Yao, L., Zhang, C.: Adversarially regularized graph autoencoder for graph embedding. In: Proceedings of the 27th International Joint Conference on Artificial Intelligence, pp. 2609–2615 (2018)
21. Parcollet, T., Morchid, M., Linarès, G.: A survey of quaternion neural networks. Artif. Intell. Rev. **53**, 2957–2982 (2020)
22. Ren, Z., Sun, Q., Wei, D.: Multiple kernel clustering with kernel k-means coupled graph tensor learning. In: Proceedings of the 35th AAAI Conference on Artificial Intelligence, February 2021, vol. 35, pp. 9411–9418 (2011)
23. Ren, Z., Yang, S.X., Sun, Q., Wang, T.: Consensus affinity graph learning for multiple kernel clustering. IEEE Trans. Cybern. **51**(6), 3273–3284 (2020)
24. Rong, Y., Huang, W., Xu, T., Huang, J.: DropEdge: towards deep graph convolutional networks on node classification. In: Proceedings of 8th International Conference on Learning Representations (2020)
25. Rusch, T.K., Chamberlain, B., Rowbottom, J., Mishra, S., Bronstein, M.M.: Graph-coupled oscillator networks. In: Proceedings of International Conference on Machine Learning, vol. 162, pp. 18888–18909 (2022)
26. Sen, P., Namata, G., Bilgic, M., Getoor, L., Gallagher, B., Eliassi-Rad, T.: Collective classification in network data. AI Mag. **29**(3), 93–106 (2008)
27. Tu, W., et al.: Deep fusion clustering network. In: Proceedings of the 35th AAAI Conference on Artificial Intelligence, pp. 9978–9987 (2021)
28. Von Luxburg, U.: A tutorial on spectral clustering. Stat. Comput. **17**, 395–416 (2007)

29. Wang, C., Pan, S., Hu, R., Long, G., Jiang, J., Zhang, C.: Attributed graph clustering: a deep attentional embedding approach. In: Kraus, S. (ed.) Proceedings of the 28th International Joint Conference on Artificial Intelligence, pp. 3670–3676 (2019)
30. Wu, S., Sun, F., Zhang, W., Xie, X., Cui, B.: Graph neural networks in recommender systems: a survey. ACM Comput. Surv. **55**(5), 97:1–97:37 (2023)
31. Xu, K., Li, C., Tian, Y., Sonobe, T., Kawarabayashi, K., Jegelka, S.: Representation learning on graphs with jumping knowledge networks. In: Proceedings of the 35th International Conference on Machine Learning, vol. 80, pp. 5449–5458 (2018)
32. Xu, Y., Huang, D., Wang, C., Lai, J.: GLAC-GCN: global and local topology-aware contrastive graph clustering network. IEEE Tans. Artif. Intell. **6**(6), 1448–1459 (2025)
33. Yang, C., Liu, Z., Zhao, D., Sun, M., Chang, E.Y.: Network representation learning with rich text information. In: Proceedings of the 24th International Joint Conference on Artificial Intelligence, pp. 2111–2117 (2015)
34. Yang, X., et al.: Cluster-guided contrastive graph clustering network. In: Proceedings of the 37th AAAI Conference on Artificial Intelligence, pp. 10834–10842 (2023)
35. Yang, X., et al.: CONVERT: contrastive graph clustering with reliable augmentation. In: Proceedings of the 31st ACM International Conference on Multimedia, pp. 319–327 (2023)
36. Zhang, F.: Quaternions and matrices of quaternions. Linear Algebra Appl. **251**, 21–57 (1997)
37. Zhang, H., Li, P., Zhang, R., Li, X.: Embedding graph auto-encoder for graph clustering. IEEE Trans. Neural Netw. Learn. Syst. (2022)
38. Zhang, S., Tay, Y., Yao, L., Liu, Q.: Quaternion knowledge graph embeddings. In: Processing of the 32nd Conference on Neural Information Processing Systems, pp. 2731–2741 (2019)
39. Zhang, Y., Cheung, Y.: Graph-based dissimilarity measurement for cluster analysis of any-type-attributed data. IEEE Trans. Neural Netw. Learn. Syst. **34**(9), 6530–6544 (2022)
40. Zhang, Y., Cheung, Y., Zeng, A.: Het2Hom: representation of heterogeneous attributes into homogeneous concept spaces for categorical-and-numerical-attribute data clustering. In: Proceedings of the 34th International Joint Conference on Artificial Intelligence, pp. 3758–3765 (2022)
41. Zhang, Y., et al.: Learning self-growth maps for fast and accurate imbalanced streaming data clustering. IEEE Trans. Neural Netw. Learn. Syst. **36**(9), 16049–16061 (2025)
42. Zhang, Y., Tan, Z., Luo, X., Liu, Y.: Hierarchical reference sets for robust unsupervised detection of scattered and clustered outliers. IEEE IoT J. (2025)
43. Zhang, Y., Zhao, M., Chen, Y., Lu, Y., Cheung, Y.: Learning unified distance metric for heterogeneous attribute data clustering. Exp. Syst. Appl. **273**, 126738 (2025)
44. Zhang, Y., Zhao, M., Jia, H., Li, M., Lu, Y., Cheung, Y.: Categorical data clustering via value order estimated distance metric learning. Proc. ACM Manage. Data **3**(6), 1–24 (2025)
45. Zhao, L., Akoglu, L.: PairNorm: tackling oversmoothing in GNNs. In: Proceedings of 8th International Conference on Learning Representations (2020)
46. Zhao, M., Feng, S., Zhang, Y., Li, M., Lu, Y., Cheung, Y.: Learning order forest for qualitative-attribute data clustering. In: Proceedings of the 27th European Conference on Artificial Intelligence, pp. 1943–1950 (2024)

47. Zhao, M., et al.: Break the tie: learning cluster-customized category relationships for categorical data clustering. In: Proceedings of the 40th AAAI Conference on Artificial Intelligence (2026)
48. Zheng, Z., Huang, G., Yuan, X., Pun, C., Liu, H., Ling, W.: Quaternion-valued correlation learning for few-shot semantic segmentation. IEEE Trans. Circ. Syst. Video Technol. **33**(5), 2102–2115 (2023)
49. Zhou, K., et al.: Dirichlet energy constrained learning for deep graph neural networks. In: Proceedings of Advances in Neural Information Processing Systems, pp. 21834–21846 (2021)

Mitigating Generic Token Dominance in Cross-Domain Foundation Model for Text-Attributed Graphs

Heng Zhang[1], Haochen You[2], Zijian Zhang[3], Lubin Gan[4], Hao Zhang[5], Wenjun Huang[6], and Jin Huang[1](✉)

[1] South China Normal University, Guangzhou, China
{2024025450,huangjin}@m.scnu.edu.cn
[2] Columbia University, New York, USA
hy2354@columbia.edu
[3] University of Pennsylvania, Philadelphia, USA
zz_harry@umich.edu
[4] University of Science and Technology of China, Hefei, China
ganlubin@mail.ustc.edu.cn
[5] University of Chinese Academy of Sciences, Beijing, China
zhang_hao1999@yeah.net
[6] Sun Yat-sen University, Guangzhou, China
huangwj98@mail2.sysu.edu.cn

Abstract. Graph Foundation Models have emerged as powerful tools for transferring knowledge across diverse graph domains and tasks. However, we identify a critical limitation in current graph-text contrastive learning approaches: generic token dominance. Our analysis reveals that over 70% of attention weights concentrate on non-discriminative generic terms like "paper" and "research", creating a semantic bottleneck that impairs cross-domain transfer capabilities. This phenomenon arises from optimization shortcuts in contrastive learning, where models favor high-frequency generic tokens over sparse discriminative features. To address this challenge, we propose **GraphRefiner**, a training-free calibration framework that enhances discriminative feature utilization without requiring model retraining. GraphRefiner employs token decomposition to identify and suppress generic terms, adaptive semantic modulation to preserve essential context, and discriminative similarity reasoning for fine-grained matching. Extensive experiments demonstrate that GraphRefiner achieves an average accuracy gain of 8.3% on cross-domain zero-shot tasks and 6.7% on few-shot transfer scenarios across standard benchmarks. Our findings highlight the importance of addressing optimization shortcuts in foundation models and provide a practical solution applicable to existing pretrained models.

Keywords: Graph Foundation Models · Contrastive Learning · Text Encoding · Cross-Domain Transfer · Text-Attributed Graphs

H. Jung et al. (Eds.): DASFAA 2026, LNCS 16536, pp. 251–265, 2026.
https://doi.org/10.1007/978-981-92-0366-6_16

1   Introduction

Graph Foundation Models [21] represent a pivotal shift in graph learning, focusing on transferring representations across varied domains and tasks. By employing large-scale pretraining techniques [13,31], these models reduce reliance on labeled data and adapt effectively with limited task-specific annotations. Through integrating graph structures with textual descriptions [15,34], they capture both relational patterns and semantic details efficiently. Their transferability has been demonstrated in academic networks, e-commerce platforms, and social media analysis without extensive retraining [21].

Recent years have witnessed substantial progress through graph-text contrastive learning frameworks [15]. These approaches establish semantic alignment between graph structures and textual descriptions in unified embedding spaces. Methods like ConGraT [1] and OFA [16] employ batch-wise contrastive objectives to train graph neural networks and language models jointly. Current methods fall into three paradigms: language models as feature enhancers for enriching node attributes [4,37], as direct predictors converting graphs into natural language sequences [8,26], and as semantic aligners establishing cross-modal correspondences [1,34]. These approaches demonstrate strong performance on standard benchmarks. However, current methods predominantly focus on global alignment between graph and text representations. This coarse-grained strategy overlooks fine-grained semantic details essential for distinguishing similar instances. Additionally, many approaches exhibit limited cross-domain zero-shot capabilities due to overfitting source domain patterns during pretraining (Fig. 1).

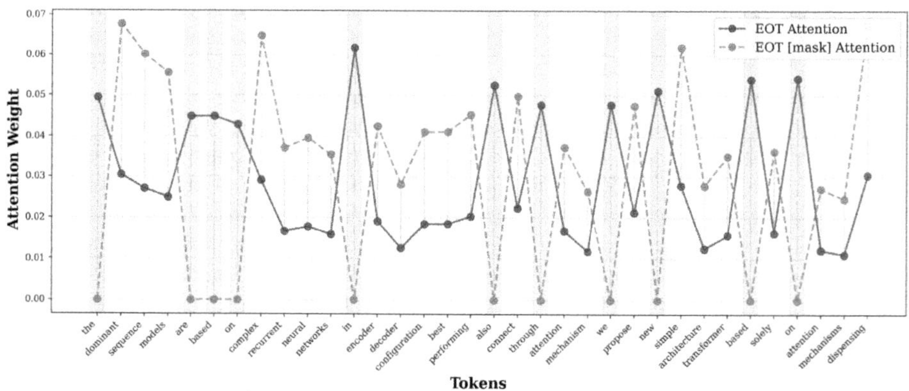

Fig. 1. Comparison of [EOT] token attention weights. When generic tokens are masked, remaining discriminative tokens receive substantially higher attention weights, demonstrating the suppression effect of generic token dominance.

Despite these advancements, our investigation reveals a critical limitation termed "generic token dominance". Through attention weight visualization across multiple graph-text datasets, we observe that approximately 70% of total attention concentrates on generic descriptive terms such as "paper", "research", "method", and "network", while discriminative concepts receive significantly less focus. These generic tokens include domain-generic terms appearing frequently across different graph types, structural description words providing universal topology characterizations, and template phrases from LLM generation patterns. To validate this observation, we designed a masking experiment replacing high-attention generic tokens with mask symbols and found that attention weights on remaining discriminative tokens increase by an average of 42% across all tested datasets, indicating that generic tokens actively suppress discriminative information representation. This dominance arises because high-frequency co-occurrence makes such associations statistically easier to establish, creating a semantic bottleneck where models fail to capture nuanced differences between graph instances and reducing cross-domain transfer performance.

The root cause lies in the optimization dynamics of contrastive learning objectives. These models maximize similarity between positive graph-text pairs while minimizing similarity between negative pairs, creating implicit bias toward features providing the easiest loss reduction. Generic terms appear frequently across training samples and offer consistent gradient signals. Nearly all academic graph summaries contain words like "paper" and "research", allowing the model to achieve satisfactory contrastive loss by emphasizing these ubiquitous tokens and learning corresponding coarse-grained patterns in graph encoding. In contrast, discriminative concepts occur sporadically and result in sparse gradient updates, as specific method names or unique attributes appear only in topically focused summaries, demanding fine-grained semantic understanding for alignment. LLMs further exacerbate this by producing template-based expressions with high-frequency vocabulary, creating a feedback loop where models increasingly rely on generic terms for alignment. This manifests as diminished sensitivity to fine-grained semantic differences, particularly in cross-domain zero-shot scenarios where generic term distributions shift dramatically between domains.

To address this challenge, we propose GraphRefiner, a training-free calibration framework that enhances discriminative feature utilization in Graph Foundation Models through a dual-component mechanism. First, we implement token decomposition that categorizes summary tokens into generic and discriminative subspaces based on attention weights and term frequency analysis, enabling precise identification of dominant generic terms that suppress discriminative information. Second, we introduce adaptive semantic modulation that dynamically adjusts generic token influence based on discriminative density, so that summaries rich in discriminative content experience stronger suppression of generic terms. Finally, we develop discriminative similarity reasoning with specialized tokens for fine-grained matching during inference, operating through two-stage retrieval that selects initial candidates using original representations and then re-ranks them using enhanced discriminative signals. Our hypothesis is that suppressing generic token dominance compels text encoders to rely more on discriminative tokens for representation construction. As the entire framework requires no additional training, GraphRefiner is directly applicable to existing pretrained Graph Foundation Models.

The main contributions of this paper can be summarized as follows:

- We identify that generic token dominance in text encoders causes existing Graph Foundation Models to exhibit significant performance degradation in cross-domain zero-shot transfer scenarios. Our analysis reveals that over 70% of attention weights concentrate on non-discriminative generic terms, creating a semantic bottleneck.
- We propose GraphRefiner, a training-free calibration framework that employs adaptive token modulation and discriminative similarity reasoning to suppress generic tokens and enhance discriminative feature utilization. This approach directly addresses the optimization shortcut problem without requiring costly model retraining.
- Extensive experiments demonstrate that GraphRefiner achieves consistent improvements across multiple benchmarks, with an average accuracy gain of 8.3% on cross-domain zero-shot tasks and 6.7% on few-shot transfer scenarios across standard graph datasets.

2   Related Works

2.1   Text-Attributed Graph Learning

Text-attributed graphs have emerged as a powerful representation for capturing both structural relationships and semantic information in networked data. Early approaches focus on combining graph neural networks with text encoders to learn joint representations. Methods like GraphSAINT [32] and GraphZoom [5] develop sampling strategies to handle large-scale graphs with textual attributes. Recent advances introduce more sophisticated architectures for processing text-attributed graphs. TAPE [11] proposes a unified framework for encoding both graph topology and node text descriptions through dual-branch networks. GIANT [4] leverages neighborhood text aggregation to enhance node representations in web-scale graphs. ConGraT [1] establishes cross-modal contrastive learning between graph structures and textual summaries to enable better semantic alignment. These methods demonstrate strong performance on standard benchmarks like citation networks and product graphs.

2.2   LLMs for Graphs

Large language models have demonstrated remarkable capabilities in understanding and generating natural language descriptions of graph structures. GPT-based methods [9] convert graph topology into sequential text representations to leverage the reasoning capabilities of pretrained language models. Instruct-GLM [30] fine-tunes large language models on graph-specific instructions to perform node classification and link prediction tasks. GraphGPT [22] integrates graph neural networks with language model architectures to enable better text generation. OFA [16] proposes a one-for-all framework for unifying multiple graph learning tasks through natural language interfaces. These approaches benefit from the vast knowledge encoded in large-scale language models and achieve impressive zero-shot performance on various graph tasks. LLM-based graph learning methods also introduce new paradigms for graph reasoning. GraphText [34] transforms graph structures into descriptive narratives to enable complex query answering through language understanding. NLGraph [28] uses natural language as an intermediate representation for transferring knowledge between different graph domains. LLaGA [3] combines large language models with graph adapters to enhance cross-modal alignment between linguistic and structural information.

3   Preliminary

3.1   Graph Foundation Models

Graph Foundation Models aim to learn transferable representations from large-scale source graph data and adapt to diverse downstream tasks through zero-shot or few-shot transfer. Consider a text-attributed graph $\mathcal{G} = (\mathcal{V}, \mathcal{E}, \mathcal{X}, \mathcal{T})$ where $\mathcal{V}$ denotes the node set with $|\mathcal{V}| = N$ nodes, $\mathcal{E}$ represents the edge set, $\mathcal{X} \in \mathbb{R}^{N \times d}$ contains node features derived from text encodings, and $\mathcal{T} = \{T_1, T_2, \ldots, T_N\}$ represents the raw text associated with each node. For each node $v_i \in \mathcal{V}$, its text description T_i provides semantic information complementing the structural patterns captured by the graph topology. To enable scalable learning on large graphs, a sampling function $\Gamma(\cdot)$ extracts ego-graphs $\mathcal{I} = \{G_1, G_2, \ldots, G_N\}$ where each G_i represents the subgraph centered at node v_i.

The pretraining objective of Graph Foundation Models can be formulated as minimizing a pretext loss over sampled subgraphs from source domains. Given source graph data $\mathcal{G}^s$ and the set of sampled subgraphs $\mathcal{I}^s$, the optimal pretrained model f_{θ^*} is obtained through

$$f_{\theta^*} = \arg \min_{f_\theta} \mathbb{E}_{G_i^s \in \mathcal{I}^s} \mathcal{L}_{\text{pretext}}(f_\theta; G_i^s), \tag{1}$$

where f_θ represents a graph neural network parameterized by θ and $\mathcal{L}_{\text{pretext}}$ denotes the pretext task loss such as contrastive learning objectives. After pre-training, the model can be directly applied to target graph data $\mathcal{G}^t$ for downstream tasks including node classification and link prediction (Fig. 2).

3.2   Graph-Text Contrastive Learning

Graph-text contrastive learning establishes semantic alignment between graph structures and textual descriptions through cross-modal representation learning. The framework employs dual encoders to project graphs and texts into a shared embedding space. Given a subgraph G_i and its corresponding text summary S_i, a graph encoder g_ϕ processes the graph structure to obtain a global graph

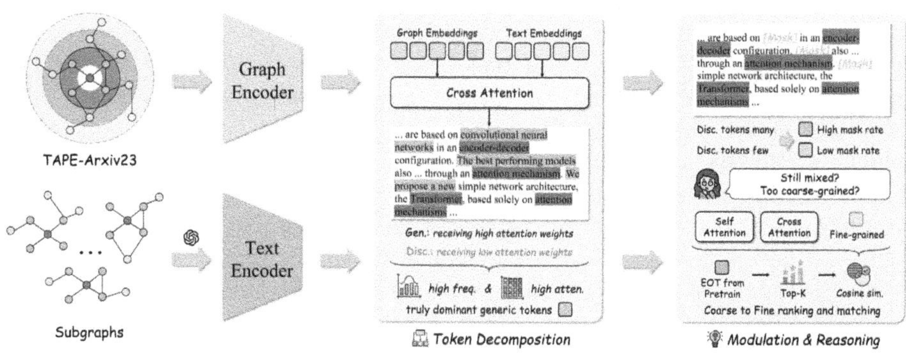

Fig. 2. GraphRefiner framework for mitigating generic token dominance. The method decomposes tokens based on attention weights and frequency, applies adaptive modulation to suppress generic tokens while preserving discriminative features, and employs two-stage retrieval for fine-grained graph-text matching. The entire process is training-free and applicable to existing pretrained models.

representation $\mathbf{v}_{\text{cls}} = g_\phi(G_i)$ where the [CLS] token aggregates information from all node embeddings through self-attention mechanisms. Similarly, a text encoder h_ψ maps the summary into a text representation $\mathbf{t}_{\text{eot}} = h_\psi(S_i)$ where the [EOT] token serves as the global text semantic representation.

The contrastive learning objective aims to maximize the similarity between matched graph-text pairs and minimize the similarity between unmatched pairs. For a batch of B graph-text pairs, the loss function is formulated as

$$\mathcal{L}_{\text{contrast}} = -\frac{1}{B} \sum_{i=1}^{B} \log \frac{\exp(\text{sim}(\mathbf{v}_{\text{cls}}^i, \mathbf{t}_{\text{eot}}^i)/\tau)}{\sum_{j=1}^{B} \exp(\text{sim}(\mathbf{v}_{\text{cls}}^i, \mathbf{t}_{\text{eot}}^j)/\tau)}, \qquad (2)$$

where $\text{sim}(\cdot, \cdot)$ denotes cosine similarity, τ represents the temperature parameter controlling the sharpness of the distribution, and the numerator captures positive pair similarity while the denominator includes similarities with all negative pairs in the batch. Through this optimization process, the model learns to align semantically related graph-text pairs in the embedding space. However, the model may exploit optimization shortcuts by relying on easily alignable generic tokens rather than learning discriminative semantic correspondences.

4 Methodology

4.1 Token Decomposition

The first stage identifies generic tokens that dominate semantic aggregation in text encoders. Given a text summary S encoded by a pretrained text encoder with L layers, we analyze the attention weights between the [EOT] token and individual subword tokens across all encoding layers. Let $\mathbf{A}^l \in \mathbb{R}^{1 \times n}$ denote the attention weights at layer l where n represents the number of tokens in the summary. To compute the aggregated importance of each token, we introduce layer-wise importance weights $\gamma_l = \|\mathbf{h}^l_{\text{eot}}\|_2$ capturing the magnitude of the [EOT] token representation at layer l. The rationale stems from the observation that feature magnitude correlates with semantic importance in transformer architectures.

The aggregated attention score for the i-th token is computed as

$$\alpha_i = \frac{\sum_{l=1}^{L} \gamma_l A^l_i}{\sum_{l=1}^{L} \gamma_l}, \tag{3}$$

where A^l_i represents the attention weight assigned to token i at layer l. This formulation provides a comprehensive measure of how much each token contributes to the global semantic representation across the entire encoding process. To distinguish generic tokens from discriminative ones, we establish an adaptive threshold τ_t based on the distribution of aggregated attention scores. Tokens with $\alpha_i \geq \tau_t$ are categorized as generic attributes $\mathcal{T}_{\text{gen}}$ while tokens with $\alpha_i < \tau_t$ form the discriminative attribute set $\mathcal{T}_{\text{disc}}$.

However, attention weights alone may incorrectly classify genuinely important high-frequency terms as generic. We refine this classification through term frequency analysis across the pretraining corpus. For each token $t_j \in \mathcal{T}_{\text{gen}}$, we compute its corpus-level frequency $\text{freq}(w_j)$ where w_j denotes the corresponding word. Tokens satisfying both high attention and high frequency conditions constitute the truly dominant generic token set

$$\mathcal{T}_{\text{dom}} = \{t_j \in \mathcal{T}_{\text{gen}} \mid \text{freq}(w_j) > \tau_f\}, \tag{4}$$

where τ_f serves as the frequency threshold. This dual-criterion approach ensures we identify tokens that are both statistically overrepresented and semantically generic. The remaining tokens in $\mathcal{T}_{\text{gen}} \setminus \mathcal{T}_{\text{dom}}$ and all tokens in $\mathcal{T}_{\text{disc}}$ are preserved as they carry essential semantic information despite varying attention patterns.

4.2   Adaptive Semantic Modulation

After identifying dominant generic tokens, the second stage dynamically modulates their representations to suppress their overwhelming influence on semantic aggregation. A naive approach of uniformly masking or removing generic tokens risks discarding essential contextual information. Instead, we propose adaptive modulation that adjusts suppression strength based on the discriminative content density within each summary.

We quantify discriminative density as the ratio of discriminative tokens to total tokens

$$\rho = \frac{|\mathcal{T}_{\text{disc}}|}{|\mathcal{T}_{\text{gen}}| + |\mathcal{T}_{\text{disc}}|}, \tag{5}$$

where $|\cdot|$ denotes set cardinality. This metric captures how much discriminative information the summary contains relative to generic descriptions. Summaries with high ρ values are rich in specific concepts and can tolerate aggressive suppression of generic terms. Conversely, summaries with low ρ values rely more on generic terms for semantic coherence and require gentler modulation to avoid information loss.

The modulated text representation is constructed through weighted aggregation

$$t_a = \sum_{j \in \mathcal{T}_{\text{dom}}} (1 - \rho) \cdot t_j + \sum_{k \in \mathcal{T}_{\text{disc}}} t_k, \tag{6}$$

where t_j and t_k represent the embeddings of generic and discriminative tokens respectively. The modulation coefficient $(1 - \rho)$ implements adaptive suppression where high discriminative density leads to stronger suppression of generic tokens. When ρ approaches 1, indicating abundant discriminative content, the coefficient approaches 0 and generic tokens are heavily suppressed. When ρ approaches 0, indicating sparse discriminative content, the coefficient approaches 1 and generic tokens retain most of their original influence. This adaptive mechanism prevents over-suppression in cases where generic terms provide necessary semantic scaffolding while effectively reducing their dominance when sufficient discriminative information exists.

4.3 Discriminative Similarity Reasoning

The final stage enhances fine-grained semantic matching through discriminative similarity reasoning during retrieval. While the modulated representation t_a reduces generic token dominance, the original [EOT] token still encodes global semantics that may obscure nuanced discriminative features. We introduce a complementary token specifically designed to capture discriminative information for improved matching precision.

We construct this discriminative token through multi-stage attention aggregation. First, we concatenate a learnable query token r with the modulated representation t_a and discriminative token embeddings $\{t_k\}_{k \in \mathcal{T}_{\text{disc}}}$. A self-attention layer processes this concatenated sequence to produce an initial discriminative representation

$$t_r = \text{SelfAttn}([r; t_a; \{t_k\}_{k \in \mathcal{T}_{\text{disc}}}]), \tag{7}$$

where $[\cdot; \cdot]$ denotes concatenation and $\text{SelfAttn}(\cdot)$ represents the self-attention operation that allows the query token r to aggregate information from both modulated and discriminative tokens. Subsequently, a cross-attention layer refines this representation by attending specifically to the discriminative content

$$\hat{t}_r = \text{CrossAttn}(t_r, [t_a; \{t_k\}_{k \in \mathcal{T}_{\text{disc}}}]), \tag{8}$$

where t_r serves as the query and the concatenation of modulated and discriminative tokens provides keys and values. This two-stage attention mechanism ensures $\hat{t}_r$ primarily encodes discriminative semantic information essential for distinguishing similar instances.

The retrieval process employs a two-stage pipeline to balance efficiency and precision. In the coarse ranking stage, we compute standard similarity between the original [EOT] token and graph representations to select the top-K candidates forming a candidate set $\mathcal{C}_K$. This stage leverages global semantic alignment for high recall. In the fine ranking stage, we compute discriminative similarity for each candidate graph $G^{(i)} \in \mathcal{C}_K$ as

$$\text{sim}_{\text{disc}}(\hat{\mathbf{t}}_r, \mathbf{v}_{\text{cls}}^{(i)}) = \frac{\langle \hat{\mathbf{t}}_r, \mathbf{v}_{\text{cls}}^{(i)} \rangle}{\|\hat{\mathbf{t}}_r\| \cdot \|\mathbf{v}_{\text{cls}}^{(i)}\|}, \tag{9}$$

where $\mathbf{v}_{\text{cls}}^{(i)}$ represents the graph representation of the i-th candidate and $\langle \cdot, \cdot \rangle$ denotes inner product. The final ranking score combines both global and discriminative similarities through a weighted fusion

$$\text{score}^{(i)} = \lambda \cdot \text{sim}(\mathbf{t}_{\text{eot}}, \mathbf{v}_{\text{cls}}^{(i)}) + (1 - \lambda) \cdot \text{sim}_{\text{disc}}(\hat{\mathbf{t}}_r, \mathbf{v}_{\text{cls}}^{(i)}), \tag{10}$$

where $\lambda \in [0, 1]$ is a hyperparameter controlling the balance between global alignment and discriminative matching. This two-stage design ensures computational efficiency by applying expensive discriminative reasoning only to a small candidate set while maintaining high retrieval accuracy through fine-grained semantic matching.

5   Experiments

5.1   Datasets

We conduct experiments on seven text-attributed graph datasets spanning academic and e-commerce domains. For the academic domain, we utilize Cora [20], CiteSeer [20], Pubmed [20], ArXiv [27], and WikiCS [18] to evaluate cross-domain transferability. These datasets contain citation networks where nodes represent papers with textual abstracts and edges denote citation relationships. For the e-commerce domain, we employ Amazon product co-purchase networks including Children, History, Photo [17], and Ele-Photo [10], as well as Instagram [14] for social network analysis. These datasets capture diverse graph structures and semantic contexts from both domains.

5.2   Baselines

We compare GraphRefiner against diverse baseline methods. Large language models including Qwen2-7B/72B [29] and LLaMA3.1-8B/70B [6] to demonstrate the capabilities of general-purpose language models on graph tasks. We further evaluate against graph foundation models including GraphGPT [22], LLaGA [3], OFA [16], GraphEdit [8], TEA-GLM [37], ZeroG [19], GraphCLIP [36], GraphInsight [2], GraphTranslator [33], and UniGLM [7]. We also include graph self-supervised learning methods such as DGI [25], GRACE [35], BGRL [23], Graph-MAE [12]. These methods represent diverse paradigms in graph learning and provide comprehensive comparison across different technical approaches.

Table 1. Zero-shot inference results for node classification across various target datasets. Boldface indicates the best performance.

Methods	Cora	CiteSeer	WikiCS	Instagram	Ele-Photo
Qwen2-7B	61.44 ± 1.29	53.57 ± 0.86	58.72 ± 0.25	39.13 ± 0.78	45.55 ± 0.12
Qwen2-72B	62.18 ± 0.98	60.97 ± 0.87	60.91 ± 0.08	47.70 ± 0.31	52.41 ± 0.39
LLaMA3.1-8B	57.75 ± 1.21	53.54 ± 1.71	58.32 ± 0.21	39.37 ± 1.14	34.38 ± 0.25
LLaMA3.1-70B	65.72 ± 1.24	62.79 ± 1.24	62.82 ± 0.04	43.68 ± 0.52	51.26 ± 0.53
GraphGPT	23.25 ± 1.45	18.04 ± 1.45	6.30 ± 0.26	45.12 ± 1.16	7.62 ± 0.22
LLaGA	21.44 ± 0.65	16.07 ± 1.15	2.65 ± 0.72	41.12 ± 0.94	6.50 ± 0.53
OFA	37.25 ± 1.38	29.64 ± 0.19	45.52 ± 1.06	32.71 ± 0.16	33.03 ± 0.64
ZeroG	62.32 ± 1.91	52.55 ± 1.23	54.93 ± 0.06	48.97 ± 0.78	45.12 ± 0.65
GraphClip	67.31 ± 1.76	63.13 ± 1.13	70.19 ± 0.10	64.05 ± 0.34	53.40 ± 0.64
GraphInsight	64.19 ± 2.34	65.87 ± 0.89	66.78 ± 0.27	66.64 ± 1.58	50.73 ± 1.12
GraphTranslator	69.85 ± 0.94	60.18 ± 1.87	73.50 ± 0.35	61.38 ± 0.76	55.92 ± 0.38
TEA-GLM	70.62 ± 1.53	66.29 ± 0.67	67.38 ± 0.52	67.30 ± 0.91	56.07 ± 0.29
DGI	24.03 ± 1.40	18.71 ± 1.22	18.86 ± 0.25	61.42 ± 1.12	13.96 ± 0.17
GRACE	13.69 ± 1.27	22.88 ± 1.49	16.07 ± 0.32	62.23 ± 0.93	10.16 ± 0.13
BGRL	20.80 ± 1.06	26.50 ± 1.22	18.35 ± 0.22	61.45 ± 0.82	5.21 ± 0.22
GraphMAE	23.25 ± 1.07	20.75 ± 0.88	12.14 ± 0.20	62.39 ± 0.84	12.53 ± 0.08
GraphRefiner	$\mathbf{72.41 \pm 1.12}$	$\mathbf{66.53 \pm 0.43}$	$\mathbf{74.13 \pm 0.19}$	$\mathbf{68.94 \pm 1.01}$	$\mathbf{57.61 \pm 0.39}$

5.3　Implementation Details

We implement GraphRefiner on top of pretrained graph foundation models without modifying their parameters. The text encoder uses 12 transformer layers with 768 hidden dimensions, while the graph encoder follows a 3-layer Graph Attention Network [24] with 256-dimensional embeddings. For token decomposition, we set attention threshold τ_t to the 70th percentile and frequency threshold τ_f to 0.01. The discriminative token uses single-layer self-attention and cross-attention with 8 heads. During two-stage retrieval, we select top-100 candidates and set fusion weight $\lambda = 0.6$. All experiments run on NVIDIA A100 GPUs with 80GB memory using cosine similarity and temperature $\tau = 0.07$. For few-shot learning, we fine-tune only a classifier head with 0.001 learning rate for 100 epochs.

5.4　Main Result

Table 1 presents the zero-shot node classification performance across five datasets. GraphRefiner consistently achieves the best accuracy, with 72.41% on Cora and 74.13% on WikiCS, surpassing the second-best baseline by 1.79% and 2.23% respectively. By decomposing text representations into generic and discriminative components, our method addresses the semantic bottleneck through adaptive modulation that dynamically adjusts suppression strength without

Table 2. Link prediction AUC Score (cross-task) under different shot setting. The best results are in bold.

Dataset K-shot	GraphGPT	GraphEdit	LLaGA	TEA-GLM	**GraphRefiner**
Cora 0-shot	$50.20_{\pm0.68}$	$58.20_{\pm0.26}$	$58.80_{\pm0.81}$	$55.30_{\pm0.39}$	$\mathbf{61.74_{\pm0.52}}$
1-shot	$50.74_{\pm0.12}$	$59.14_{\pm0.65}$	$\mathbf{62.01_{\pm0.43}}$	$56.00_{\pm0.27}$	$60.38_{\pm0.71}$
3-shot	$52.21_{\pm0.83}$	$60.01_{\pm0.91}$	$59.51_{\pm0.82}$	$57.10_{\pm0.89}$	$\mathbf{63.47_{\pm0.94}}$
5-shot	$51.49_{\pm0.34}$	$60.46_{\pm0.38}$	$61.24_{\pm0.55}$	$57.57_{\pm0.49}$	$\mathbf{64.18_{\pm0.38}}$
Pubmed 0-shot	$49.90_{\pm0.98}$	$42.80_{\pm0.43}$	$56.10_{\pm0.18}$	$67.80_{\pm0.97}$	$\mathbf{72.38_{\pm0.81}}$
1-shot	$50.39_{\pm0.37}$	$44.47_{\pm0.33}$	$56.91_{\pm0.85}$	$69.08_{\pm0.35}$	$\mathbf{71.05_{\pm0.59}}$
3-shot	$50.95_{\pm0.42}$	$43.44_{\pm0.20}$	$57.96_{\pm0.96}$	$\mathbf{70.52_{\pm0.63}}$	$69.84_{\pm0.77}$
5-shot	$52.00_{\pm0.18}$	$45.01_{\pm0.71}$	$58.97_{\pm0.64}$	$69.59_{\pm0.32}$	$\mathbf{77.31_{\pm1.05}}$
Children 0-shot	$45.80_{\pm0.33}$	$38.80_{\pm0.64}$	$43.40_{\pm0.97}$	$57.40_{\pm0.27}$	$\mathbf{63.79_{\pm0.56}}$
1-shot	$46.60_{\pm0.44}$	$39.47_{\pm0.15}$	$44.17_{\pm0.53}$	$57.88_{\pm0.91}$	$\mathbf{63.15_{\pm0.48}}$
3-shot	$47.18_{\pm0.37}$	$41.23_{\pm0.96}$	$44.86_{\pm1.02}$	$\mathbf{61.46_{\pm0.69}}$	$61.42_{\pm0.83}$
5-shot	$47.99_{\pm0.46}$	$40.12_{\pm0.42}$	$45.81_{\pm0.51}$	$58.88_{\pm0.03}$	$\mathbf{68.53_{\pm0.67}}$
History 0-shot	$28.20_{\pm0.50}$	$42.20_{\pm0.78}$	$48.10_{\pm0.68}$	$53.10_{\pm0.64}$	$\mathbf{67.23_{\pm1.12}}$
1-shot	$29.15_{\pm0.22}$	$42.78_{\pm0.55}$	$50.07_{\pm0.21}$	$53.78_{\pm0.07}$	$\mathbf{65.17_{\pm0.93}}$
3-shot	$30.14_{\pm0.30}$	$43.58_{\pm0.58}$	$49.07_{\pm0.30}$	$55.26_{\pm1.03}$	$\mathbf{64.92_{\pm0.41}}$
5-shot	$31.12_{\pm0.63}$	$44.04_{\pm0.06}$	$50.96_{\pm1.01}$	$54.43_{\pm0.73}$	$\mathbf{66.81_{\pm0.28}}$

requiring model retraining. On e-commerce datasets, GraphRefiner achieves 68.94% on Instagram and 57.61% on Ele-Photo with improvements of 1.64% and 1.54%. Table 2 shows consistent gains across shot settings, particularly 77.31% AUC on Pubmed in 5-shot scenarios. These results confirm that suppressing generic token dominance enables better utilization of discriminative features for robust cross-domain transfer.

5.5 Model Analysis

Table 3 demonstrates GraphRefiner's generalization across multi-dataset transfer settings. Training on Arxiv and Cora then testing on PubMed achieves 49.8% accuracy with 2.9% improvement, while transferring from Arxiv and PubMed to Cora yields 70.3% with 4.6% gain. Our training-free calibration framework directly addresses optimization shortcuts by identifying and suppressing high-frequency generic tokens, compelling text encoders to rely on discriminative semantic information. This proves especially effective when domains exhibit different generic term distributions, where conventional methods suffer from semantic misalignment. The consistent gains across all four transfer configurations val-

Table 3. Generalization performance of GraphRefiner on zero-shot node classification across multi-dataset transfer settings.

Model	(Arxiv+Cora) → PubMed	(Arxiv+Cora) → Arxiv	(Arxiv+PubMed) → Cora	(Arxiv+PubMed) → Arxiv
LLaGA	42.1%	48.2%	16.4%	46.7%
TEA-GLM	38.6%	50.3%	21.9%	54.7%
GraphAny	41.2%	47.8%	24.6%	62.1%
GraphCLIP	46.9%	55.7%	19.3%	58.4%
GraphInsight	44.7%	53.8%	23.5%	63.2%
GraphTranslator	45.2%	54.6%	24.8%	65.7%
GraphRefiner	49.8%(2.9% ↑)	58.5%(2.8% ↑)	26.4%(1.6% ↑)	70.3%(4.6% ↑)

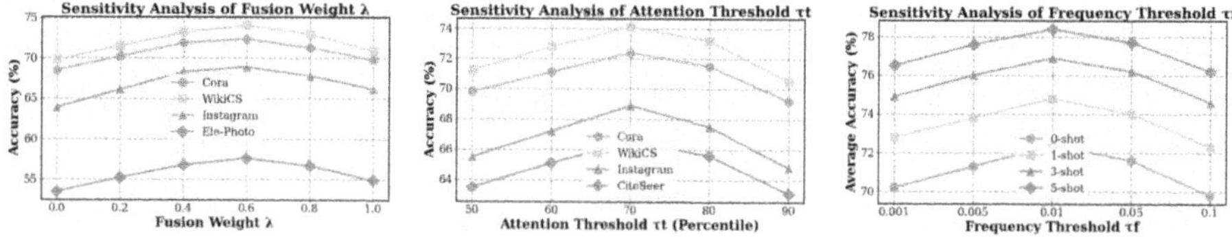

Fig. 3. Hyperparameter sensitivity analysis. (Left) Fusion weight $\lambda = 0.6$ optimally balances global alignment and discriminative matching across datasets. (Middle) Attention threshold at the 70th percentile achieves an optimal balance between generic token suppression and discriminative feature preservation. (Right) Frequency threshold $\tau_f = 0.01$ effectively identifies generic tokens while preserving domain-specific terms.

idate that generic token dominance is a fundamental bottleneck in cross-modal contrastive learning.

5.6 Hyper-parameter Analysis

Figure 3 illustrates the sensitivity of three key hyperparameters. The fusion weight achieves optimal performance at 0.6 where Cora reaches 72.41% and WikiCS attains 74.13%, with deviations causing drops up to 2.6%. The attention threshold performs best at the 70th percentile, confirming that moderate suppression preserves context while reducing generic token influence. Our dual-criterion approach combines attention weights with frequency statistics to target truly dominant terms. The frequency threshold shows 0.01 as optimal, with deviations to 0.001 or 0.1 causing 2.6% accuracy drops in zero-shot scenarios. These behaviors align with our analysis that contrastive learning exploits high-frequency tokens as shortcuts.

5.7 Ablation Study

Figure 4 quantifies each component's contribution through systematic removal. Token decomposition is most critical, with removal causing drops of 4.2% on

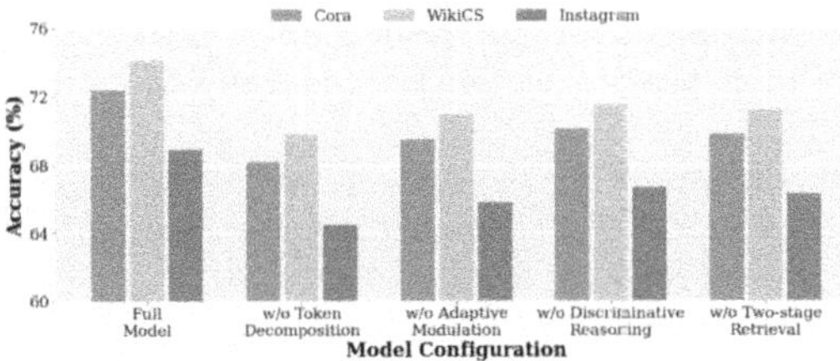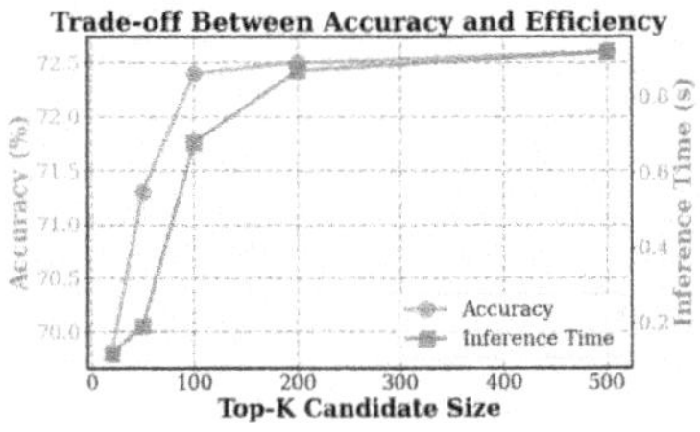

Fig. 4. Trade-off between retrieval accuracy and inference time under varying Top-K sizes, and ablation results across datasets. K = 100 achieves near-maximum accuracy with 69.6% less inference time, showing an efficient balance. Removing token decomposition causes the largest drop, confirming its critical role, while all components contribute positively to overall performance.

Cora, 4.3% on WikiCS, and 4.4% on Instagram. Adaptive modulation contributes 2.9% to 3.2%, discriminative reasoning provides 2.2% to 2.6%, and two-stage retrieval adds 2.6% to 2.9%. The right panel shows candidate size 100 achieves 72.41% accuracy with 0.28 s inference time versus 0.92 s for size 500. These results validate that effective calibration requires coordinated operation of all components rather than simple masking.

6   Conclusion

We address generic token dominance in graph foundation models, where over 70% of attention concentrates on non-discriminative terms, impairing cross-domain transfer. We propose GraphRefiner, a training-free calibration framework using token decomposition, adaptive semantic modulation, and discriminative similarity reasoning. Experiments show 8.3% and 6.7% accuracy gains on zero-shot and few-shot tasks respectively, providing a practical solution applicable to pretrained models without retraining.

Acknowledgments. This work was supported by the Natural Science Foundation of Guangdong Province, China. "Research on Key Theories and Technologies for Nano-learning".

References

1. Brannon, W., et al.: ConGraT: self-supervised contrastive pretraining for joint graph and text embeddings. In: Proceedings of the 2023 SIAM International Conference on Data Mining, pp. 172–180 (2023)
2. Cao, Y., Han, S., Gao, Z., Ding, Z., Xie, X., Zhou, S.K.: GraphInsight: unlocking insights in large language models for graph structure understanding. arXiv preprint arXiv:2409.03258 (2024)
3. Chen, R., Zhao, T., Jaiswal, A., Shah, N., Wang, Z.: LLaGA: large language and graph assistant. arXiv preprint arXiv:2402.08170 (2024)

4. Chien, E., Pan, W.C., Peng, J., Li, C.S., Milenković, O.: Node feature extraction by self-supervised multi-scale neighborhood prediction. In: International Conference on Learning Representations (2022)
5. Deng, C., Zhao, Z., Wang, Y., Zhang, Z., Feng, Z.: GraphZoom: a multi-level spectral approach for accurate and scalable graph embedding. In: International Conference on Learning Representations (2020)
6. Dubey, A., et al.: The Llama 3 herd of models. arXiv preprint arXiv:2407.21783 (2024)
7. Fang, Y., Fan, D., Ding, S., Liu, N., Tan, Q.: UniGLM: training one unified language model for text-attributed graph embedding. arXiv preprint arXiv:2406.12052 (2024)
8. Fatemi, B., Halcrow, J., Perozzi, B.: Talk like a graph: encoding graphs for large language models. In: The Twelfth International Conference on Learning Representations (2024)
9. Guo, J., Du, L., Liu, H., Zhou, M., He, X., Han, S.: GPT4Graph: can large language models understand graph structured data? an empirical evaluation and benchmarking. arXiv preprint arXiv:2305.15066 (2023)
10. He, R., McAuley, J.: Ups and downs: modeling the visual evolution of fashion trends with one-class collaborative filtering. In: Proceedings of the 25th International Conference on World Wide Web, pp. 507–517 (2016)
11. He, X., Bresson, X., Laurent, T., Hooi, B.: Harnessing explanations: LLM-to-LM interpreter for enhanced text-attributed graph representation learning. arXiv preprint arXiv:2305.19523 (2023)
12. Hou, Z., Liu, X., Dong, Y., Wang, C., Tang, J.: GraphMAE: self-supervised masked graph autoencoders. In: Proceedings of the 28th ACM SIGKDD Conference on Knowledge Discovery and Data Mining, pp. 594–604 (2022)
13. Hu, Z., Dong, Y., Wang, K., Chang, K.W., Sun, Y.: GPT-GNN: generative pre-training of graph neural networks. In: Proceedings of the 26th ACM SIGKDD International Conference on Knowledge Discovery & Data Mining, pp. 1857–1867 (2020)
14. Hu, Z., Dong, Y., Wang, K., Sun, Y.: Heterogeneous graph transformer. In: Proceedings of The Web Conference 2020, pp. 2704–2710 (2020)
15. Jin, B., Liu, G., Han, C., Jiang, M., Ji, H., Han, J.: Large language models on graphs: a comprehensive survey. arXiv preprint arXiv:2312.02783 (2023)
16. Liu, H., et al.: One for all: towards training one graph model for all classification tasks. In: The Twelfth International Conference on Learning Representations (2024)
17. McAuley, J., Targett, C., Shi, Q., Van Den Hengel, A.: Image-based recommendations on styles and substitutes. In: Proceedings of the 38th International ACM SIGIR Conference on Research and Development in Information Retrieval, pp. 43–52 (2015)
18. Mernyei, P., Cangea, C.: Wiki-CS: a Wikipedia-based benchmark for graph neural networks. In: ICML Workshop on Graph Representation Learning (2020)
19. Pushp, P.K., Srivastava, M.M.: Train once, test anywhere: zero-shot learning for text classification. In: Proceedings of the 29th ACM SIGKDD Conference on Knowledge Discovery and Data Mining, pp. 1837–1849 (2017)
20. Sen, P., Namata, G., Bilgic, M., Getoor, L., Galligher, B., Eliassi-Rad, T.: Collective classification in network data. AI Mag. **29**, 93–106 (2008)
21. Sun, X., Cheng, H., Li, J., Liu, B., Guan, J.: All in one: multi-task prompting for graph neural networks. In: Proceedings of the 29th ACM SIGKDD Conference on Knowledge Discovery and Data Mining, pp. 2120–2131 (2023)

22. Tang, J., et al.: GraphGPT: graph instruction tuning for large language models. In: Proceedings of the 47th International ACM SIGIR Conference on Research and Development in Information Retrieval, pp. 491–500 (2024)
23. Thakoor, S., Tallec, C., Azar, M.G., Munos, R., Veličković, P., Valko, M.: Bootstrapped representation learning on graphs. In: ICLR Workshop on Geometrical and Topological Representation Learning (2022)
24. Veličković, P., Cucurull, G., Casanova, A., Romero, A., Liò, P., Bengio, Y.: Graph attention networks. In: International Conference on Learning Representations (2018)
25. Veličković, P., Fedus, W., Hamilton, W.L., Liò, P., Bengio, Y., Hjelm, R.D.: Deep graph infomax. In: International Conference on Learning Representations (2019)
26. Wang, H., Feng, S., He, T., Tan, Z., Han, X., Tsvetkov, Y.: Can language models solve graph problems in natural language? In: Advances in Neural Information Processing Systems, vol. 36 (2024)
27. Wang, K., Shen, Z., Huang, C., Wu, C.H., Dong, Y., Kanakia, A.: Microsoft academic graph: when experts are not enough. Quantit. Sci. Stud. 1, 396–413 (2020)
28. Wang, R., et al.: Natural language is all a graph needs. arXiv preprint arXiv:2308.07134 (2023)
29. Yang, A., et al.: Qwen2 technical report. arXiv preprint arXiv:2407.10671 (2024)
30. Ye, R., Zhang, C., Wang, R., Xu, S., Zhang, Y.: Language is all a graph needs. arXiv preprint arXiv:2308.07134 (2024)
31. You, Y., Chen, T., Sui, Y., Chen, T., Wang, Z., Shen, Y.: Graph contrastive learning with augmentations. In: Advances in Neural Information Processing Systems, vol. 33, pp. 5812–5823 (2020)
32. Zeng, H., Zhou, H., Srivastava, A., Kannan, R., Prasanna, V.: GraphSAINT: graph sampling based inductive learning method. In: International Conference on Learning Representations (2020)
33. Zhang, M., Shi, M., Zhu, P., Liu, X., Ye, C., Liang, J.: GraphTranslator: aligning graph model to large language model for open-ended tasks. arXiv preprint arXiv:2402.07197 (2024)
34. Zhao, J., et al.: Learning on large-scale text-attributed graphs via variational inference. arXiv preprint arXiv:2210.14709 (2023)
35. Zhu, Y., Xu, Y., Yu, F., Liu, Q., Wu, S., Wang, L.: Deep graph contrastive representation learning. In: ICML Workshop on Graph Representation Learning and Beyond (2020)
36. Zhu, Y., et al.: GraphCLIP: enhancing transferability in graph foundation models for text-attributed graphs. arXiv preprint arXiv:2410.10329 (2024)
37. Zhu, Y., Wang, Y., Shi, H., Tang, S.: Efficient tuning and inference for large language models on textual graphs. arXiv preprint arXiv:2401.15569 (2024)

Disentanglement-Based Contrastive Learning and Optimization for User Identity Linkage

Yue Yang[1], Yichao Zhang[1(✉)], Jihong Guan[1(✉)], Shuigeng Zhou[2,3],
and Wengen Li[1]

[1] School of Computer Science and Technology, Tongji University, Shanghai
200092, China
`{yuey_95,yichaozhang,jhguan,lwengen}@tongji.edu.cn`
[2] Shanghai Key Laboratory of Intelligent Information Processing, Fudan University,
Shanghai 200433, China
`sgzhou@fudan.edu.cn`
[3] School of Computer Science, Fudan University, Shanghai 200433, China

Abstract. The proliferation of social networks has led users to engage across multiple platforms, creating a critical need to link their identities for comprehensive user profiling and personalized recommendations. Existing approaches face semantic discrepancies caused by platform-specific functions, interference from structurally similar but non-aligned nodes, many-to-one matching in similarity-based methods, and high computational complexity in Optimal Transport (OT)-based approaches. To overcome these issues, we propose a novel Disentanglement-Based Contrastive Learning and Optimization for user identity linkage. DCLO first disentangles user representations into network-specific and network-shared features while preserving semantic integrity with reconstruction constraints. A cross-network fusion layer integrates shared features of aligned users while aggregating neighborhood and network-specific features. DCLO employs cross-network and intra-network contrastive learning to enhance alignment and discrimination. Furthermore we design a unified node consistency transport cost that jointly captures semantic and structural consistency to optimize alignment inference. Extensive experiments on three real-world social network datasets demonstrate that DCLO consistently outperforms ten state-of-the-art baselines, achieving superior effectiveness, robustness, and generalizability.

Keywords: User Identity Linkage (UIL) · Representation Disentanglement · Online Social networks · Optimal Transport

1 Introduction

User identity linkage (UIL) refers to the process of identifying and connecting different accounts that belong to the same individual across multiple online

---

Y. Yang and Y. Zhang—are contribute equally to this work.

H. Jung et al. (Eds.): DASFAA 2026, LNCS 16536, pp. 266–282, 2026.
https://doi.org/10.1007/978-981-92-0366-6_17

social platforms. With the increasing popularity of online social platforms, users often maintain accounts on multiple platforms or services to meet different social needs, such as social platforms (Facebook, Twitter, Weibo), as well as e-commerce platforms (Taobao, Amazon), among others. As accounts exist independently across platforms, the need for user identity linkage has grown significantly in recent years due to its importance in social network analysis. Linking user identities accurately has significant downstream applications, such as personalized recommendations [13,14,36], user profiling [7,24], and privacy protection [18,27], among others.

Previous UIL studies have leveraged available information such as user demographics [5,19] and behavioral information [38] shared by users. However, aligned user attributes across platforms are often inconsistent, incomplete, or even false information, making it difficult to obtain sufficient semantic information for accurate alignment. Researchers have suggested that social network structure offers a more reliable and robust feature. The most direct method is to measure structural similarity between aligned users [6,9]. Recent studies project networks into a unified low-dimensional latent space [39] and minimize distances between aligned representations to infer anchor links [25]. However, anchor users may not share similar topologies, as neighboring nodes within the same network often have more similar structures. Therefore, aligning users with topological structures faces significant challenges.

When aligning users, traditional alignment methods typically use cosine similarity and select the most similar user, which can result in many-to-one matches. Optimal Transport (OT) jointly considers all alignment relationships to achieve a globally optimal and more stable matching. Recent studies have applied OT to user alignment with notable success [2,20,29], modeling networks as resource distributions and defining transport costs based on node consistency [30]. However, they require computing multiple transport cost distances separately, increasing computational overhead and complicating optimization.

In a word, existing approaches for UIL are currently facing four significant challenges: (1) Due to the functional differences across social platforms, users may maintain distinct friend circles and interaction patterns. Even if two accounts belong to the same real-world user, their neighboring nodes are not always similar, resulting in inevitable semantic discrepancies between aligned users. (2) Adjacent users within the same network are generally more similar, which may interfere with alignment process. Therefore, fully exploiting the common features between aligned nodes while distinguishing them from similar but non-aligned node features remains a challenging problem. (3) Again, traditional alignment methods rely on cosine similarity and greedily assign each user to its most similar counterpart, which may lead to a many-to-one matching and impractical alignment. (4) Recent OT-based methods [2,30] improve alignment by constructing node-level and link-level consistency and compute the costs using WD and GWD. However, computing multiple transport costs increases complexity, and GWD's reliance on structural distances makes it sensitive to noisy or missing links, leading to unstable results.

To address the challenges above, we propose a Disentanglement-based Contrastive Learning and Optimization for User Identity Linkage (DCLO). To address challenge (1), we first design a representation disentanglement component that projects node representations into a network-specific and a network-shared space, introducing a reconstruction loss to preserve the semantic completeness. Then, we propose a cross-network fusion layer that fuses the shared features of aligned users across networks and combines them with their specific features. To address challenge (2), we first aggregate neighborhood information for networks and reconstructed networks and employ a cross-network contrastive loss to align user representations, along with intra-network contrastive loss to distinguish similar but non-aligned nodes. To address challenges (3) and (4), we model alignment task as an optimal transport problem. We propose a unified node consistency transport cost function capturing both node and structural consistency. By seeking an optimal transport plan to minimize the transport cost distance, DCLO achieves more robust, semantically consistent alignment while reducing computational complexity and improving efficiency.

In summary, our contributions are fourfold:

- We propose a Representation Disentanglement module to extract fine-grained shared features between aligned users and integrate them via a cross-network fusion layer.
- We introduce contrastive learning to align user representations across networks and views within the source and target networks, including original and reconstructed networks.
- We define a unified node consistency transport cost matrix incorporating user features and structural features to align all the target nodes simultaneously.
- We extensively evaluate our model on three pairs of real social network datasets. Experimental results demonstrate that our model outperforms 10 state-of-the-art models in terms of generalizability and robustness.

2   Related Work

We present a brief survey on related work of UIL based on network structure, including embedding-based methods and OT-based methods.

Embedding-based methods usually learn node representations to compare node similarities under the assumption of node alignment consistency [3,8, 23,26]. With the vigorous development of graph neural networks (GNNs), researchers embeded nodes from different networks into a unified low-dimensional representation space according to known aligned user pairs [6,34,39]. To alleviate the over-smoothing problem caused by excessive message aggregation, researchers focus on capturing richer and more fine-grained semantic representations [11,15,33,35]. Seo et al. constructed node features based on similarity between node and aligned node, incorporating multi-aspect structural characteristic [17]. The alignment process may be affected by inconsistencies in user behavior across networks, making even aligned user pairs be identified as false negatives. Some researchers proposed hard negative sampling strategy to distinguish aligned users from their similar nodes [31].

Optimal Transport (OT) is widely applied to cross-graph tasks, such as graph matching and cross-domain recommendation. Some researchers applied OT to UIL tasks [20, 29]. They typically treat graphs as distributions and align them by minimizing the transport cost distance from one distribution to another. Zeng et al. designed the regularization terms to define the transport cost from node, link, and neighborhood consistency between aligned nodes [30]. Chen et al. designed an ensemble learning module to combine OT and node embedding [2]. Yu et al. constructed transport cost matrices with learnable node embeddings and adopted the FGW distance and multi-level ranking loss as optimization objective function [28]. However, while GWD and FGW capture link-level consistency, they are sensitive to structural noise and computationally expensive due to their dependence on the number of link pairs.

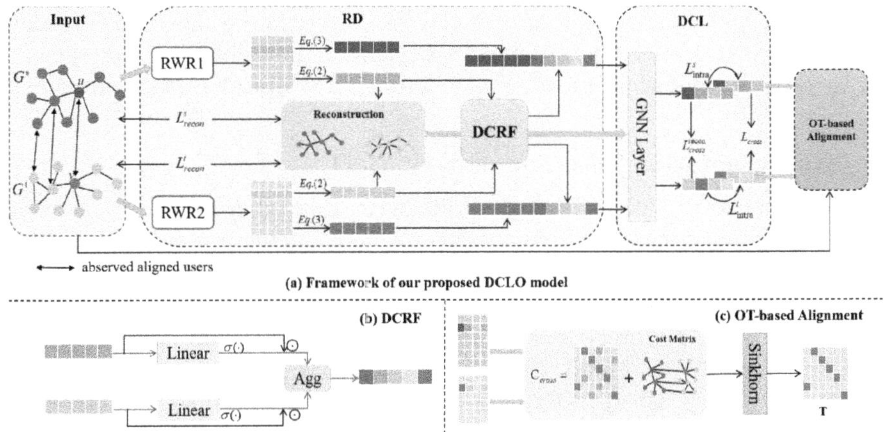

Fig. 1. The Framework of our DCLO model.

3   Problem Formulation

Before introducing our model, we present the formal definition of the UIL problem. Suppose there are two social networks, the source network $G^s = \{V^s, A^s\}$ and the target network $G^t = \{V^t, A^t\}$.

The users for each network are abstracted as nodes and represented by $V^s = \{v_1^s, v_2^s, \cdots, v_{|V^s|}^s\}$ and $V^t = \{v_1^t, v_2^t, \cdots, v_{|V^t|}^t\}$, respectively. $A^s \in \mathbb{R}^{|V^s| \times |V^s|}$ and $A^t \in \mathbb{R}^{|V^t| \times |V^t|}$ denote the adjacency matrix of the two networks, respectively. The entry A_{ij} is 1 if there is a link between node v_i and v_j; otherwise, $A_{ij} = 0$.

Given an observed aligned user pairs seed set $S = \{(u, v)|u \in V^s, v \in V^t\}$, UIL tasks aim to optimize and output an alignment matrix $T \in \mathbb{R}^{|V^s| \times |V^t|}$

between nodes from different networks, where the entry T_{ij} denotes the likelihood that user $v_i \in V^s$ and $v_j \in V^t$ are the same individual.

4   Methodology

Figure 1(a) illustrates the framework of the DCLO, which comprises three components: (1) Representation Disentanglement (RD) first extracts structural information as initial node representations. Then, it maps node features into a network-shared space that captures users' consistent features across different networks, and a network-specific space that preserves network-specific user features. After that, we fuse the shared features of aligned nodes for complementarity, as shown in Fig. 1(b). We introduce a reconstruction loss to preserve semantic completeness. (2) Disentanglement-based Contrastive Learning (DCL) first learns node local features and constructs contrastive loss to preserve aligned node consistency and distinguish semantically similar nodes effectively. (3) OT-based Alignment constructs a transport cost matrix based on node embedding and topology to match identities with prediction probability T, as shown in Fig. 1(c). In the following, we provide a detailed description of each component.

4.1   Representation Disentanglement (RD)

Node Representation. Random Walk with Restart (RWR) [21], a widely adopted proximity measure, extends RW by introducing a fixed probability of returning to the starting node. Considering the time complexity, we employ RWR to construct relationships between each node and observed aligned seed users.

Then, the RWR scores for source and target networks are derived from an RWR-based iterative process [17,26,28]: $r_i^s = \alpha W^s r_i^s + (1 - \alpha)e_i^s$ and $r_i^t = \alpha W^t r_i^t + (1 - \alpha)e_i^t$, where $\alpha \in (0, 1)$ is the restart probability, $e_i^s \in \mathbb{R}^{|V^s|}$ and $e_i^t \in \mathbb{R}^{|V^t|}$ is the one-hot vector with 1 at the i-th position for source network and target network. $W^s = ((D^s)^{-1}A^s)^\top$ and $W^t = ((D^t)^{-1}A^t)^\top$ is the row-normalized adjacency matrix, where D denotes the diagonal degree matrix and $(\cdot)^\top$ denotes the transpose operator. Finally, node representation matrices are constructed by concatenating the RWR scores and referred to as X^s and X^t.

Representation Disentanglement Layer. The node representations contain rich structural and local contextual information. However, due to differences in user behavior patterns and relationships across networks, directly aligning user representations is typically impeded by network-specific information, leading to a poor matching performance. To improve the robustness and accuracy of alignment, we propose to disentangle node representations into shared features, capturing consistent semantic information between aligned users, and specific features, preserving network-specific features.

Specifically, we first map node representations into a shared feature space for each node via a Multi Layer Perceptron (MLP), respectively. The shared features $\hat{X}_{sh}^{s} \in \mathbb{R}^{|V^s| \times d}$ and $\hat{X}_{sh}^{t} \in \mathbb{R}^{|V^t| \times d}$ are derived from

$$\hat{X}_{sh}^{s} = \text{MLP}_{sh}(X^s) \quad and \quad \hat{X}_{sh}^{t} = \text{MLP}_{sh}(X^t), \tag{1}$$

where MLP_{sh} comprises two hidden layers with activation function $ReLU()$. Then the network-specific features

$$\hat{X}_{sp}^{s} = \text{MLP}_{sp_s}(X^s) \quad and \quad \hat{X}_{sp}^{t} = \text{MLP}_{sp_t}(X^t), \tag{2}$$

where MLP_{sp_s} and MLP_{sp_t} likewise comprise two hidden layers with $ReLU()$.

To ensure that the disentangled shared features retain the local structural relationships of the original network, we introduce a structure reconstruction constraint, which reconstructs adjacency matrices based on the similarity between nodes for each network. The reconstructed adjacency matrices are $\hat{A}^s = \hat{X}_{sh}^{s} \cdot \left(\hat{X}_{sh}^{s}\right)^{\top}$ and $\hat{A}^t = \hat{X}_{sh}^{t} \cdot \left(\hat{X}_{sh}^{t}\right)^{\top}$, where $(\cdot)^{\top}$ denotes the transpose operator. The reconstruction optimization objective for each network is

$$\mathcal{L}_{\text{recon}}^{*} = \frac{1}{|V^*|^2} \sum_{i=1}^{|V^*|} \sum_{j=1}^{|V^*|} \left(\hat{A}_{ij}^{*} - A_{ij}^{*}\right)^2, \tag{3}$$

where $*$ can take s or t denoting the source or target network, respectively. Combining them, the final reconstruction loss is $\mathcal{L}_{\text{recon}} = \mathcal{L}_{\text{recon}}^{s} + \mathcal{L}_{\text{recon}}^{t}$.

Disentanglement-Based Cross-Graph Representation Fusion (DCRF). The node shared features extracted by the representation disentanglement layer effectively capture the shared semantics of aligned users across networks. Integrating these features into the final node representations not only preserves the intrinsic attributes of individual users but also incorporates complementary information from their aligned counterparts in other networks, thereby enabling collaborative cross-domain feature expression. This fusion of shared features further mitigates the interference from semantic discrepancies between aligned nodes, resulting in more consistent and stable representations. Consequently, the expressiveness and robustness of user representations for identity alignment tasks are significantly improved.

Specifically, we first apply a linear transformation followed by a sigmoid activation to the shared features of aligned users, generating feature-wise weights. These weights are then applied element-wise to the original feature vectors, enabling the model to emphasize alignment-relevant information. Finally, we obtain a more robust node representation across networks by fusing the aligned users' shared features. Therefore, for an aligned node pair $(v_i, v_j) \in S$, the fused node representation z_i is

$$z_i = \hat{x}_{i,sh}^{s} \odot \sigma(\mathbf{W_s}\hat{x}_{i,sh}^{s} + b_s) + \hat{x}_{j,sh}^{t} \odot \sigma(\mathbf{W_t}\hat{x}_{j,sh}^{t} + b_t), \tag{4}$$

where $\sigma(\cdot)$ is sigmoid activation function and $\odot$ is element-wise multiplication. W_s and W_t are the learnable weight matrices. b_s and b_t are the bias vectors.

4.2 Disentanglement-Based Contrastive Learning (DCL)

Graph Encoder Layer. After obtaining fused node representations that capture consistent semantic information across networks, we adopt Graph Attention Network (GAT) [22] to aggregate node representations in neighborhoods.

Specifically, we first concatenate shared features with the network-specific features to obtain a complete node representation. The complete node representations are $Z^s = [Z\|\hat{X}_{sp}^s]$ and $Z^t = [Z\|\hat{X}_{sp}^t]$, where $[\cdot\|\cdot]$ denotes concatenation operation. Then we feed Z^s and Z^t into GAT. The attention score α_{ij} between node i and its neighbor j is

$$\alpha_{ij} = \frac{\exp\left(\text{LeakyReLU}\left(\mathbf{a}^\top[\mathbf{W}z_i \| \mathbf{W}z_j]\right)\right)}{\sum_{k\in\mathcal{N}(i)} \exp\left(\text{LeakyReLU}\left(\mathbf{a}^\top[\mathbf{W}z_i \| \mathbf{W}z_k]\right)\right)}, \tag{5}$$

where $\mathcal{N}(i)$ denotes i's neighbors. $\exp(\cdot)$ denotes the exponential function, and LeakyReLU$(\cdot)$ is the activation function. $\mathbf{W}$ and $\mathbf{a}$ are a learnable weight matrix and a learnable weight vector, respectively. Each node representation is updated by aggregating the features of its neighbors by the attention scores. The aggregated node representation $\hat{z}_i \in \mathbb{R}^{d'}$ of node i is

$$\hat{z}_i = \sigma\left(\sum_{j\in\mathcal{N}(i)} \alpha_{ij}\mathbf{W}_\mathbf{h}z_j\right), \tag{6}$$

where $\mathbf{W}_\mathbf{h}$ is a learnable weight matrix. $\sigma(\cdot)$ is $ReLU(\cdot)$ activation function.

To enhance the robustness and generalization of the model, we further adopt GAT to reconstruct networks using the reconstructed adjacency matrices $\hat{A}^s$ and $\hat{A}^t$ to strengthen local features between nodes. Similarly, for the reconstructed graph, we likewise feed Z^s and Z^t into GAT and employ Eqs. (5) and (6) to derive $\hat{z}_i^{recon,s}$ and $\hat{z}_i^{recon,t}$.

Cross-graph Contrastive Loss. The user alignment across networks aims to enhance the consistency of representations between aligned users while effectively distinguishing non-aligned users. To achieve this, we employ a cross-graph contrastive loss as the optimization objective for user alignment. Specifically, we employ InfoNCE [16] to minimize the distance between aligned nodes while maximizing the distance between nodes and their negative samples, aiming at improving the quality of the node representations.

Formally, the constructed cross-graph contrastive loss

$$\mathcal{L}_{\text{cross}}^o = \sum_{(i,j)\in\mathcal{S}} -\log\left(\frac{\phi\left(\hat{z}_i^s \cdot \hat{z}_j^t\right)}{\phi\left(\hat{z}_i^s \cdot \hat{z}_j^t\right) + \sum_{j'\in\text{Neg}(i)} \phi\left(\hat{z}_i^s \cdot \hat{z}_{j'}^t\right)}\right), \tag{7}$$

where $\phi(\cdot) = e^{(\cdot)/\tau}$ measures similarity function and τ is temperature coefficient. Neg(i) denotes the set of negative samples set for node i.

Next, we leverage reconstructed node representations as auxiliary signals for optimization. The corresponding cross-graph contrastive loss $\mathcal{L}_{\mathrm{cross}}^{\mathrm{recon}}$ is defined as Eq. (7) to further align reconstructed node embeddings. The overall cross-graph contrastive loss $\mathcal{L}_{\mathrm{cross}}$ is formulated as a combination of the original and reconstructed objectives $\mathcal{L}_{\mathrm{cross}} = \mathcal{L}_{\mathrm{cross}}^{\mathrm{o}} + \lambda\mathcal{L}_{\mathrm{cross}}^{\mathrm{recon}}$, where λ is a hyperparameter that balances the weights of the reconstructed cross-graph loss and the cross-graph contrastive loss.

Intra-Graph Contrastive Loss. To prevent the oversmoothing problem of node representations, we adopt an intra-graph contrastive loss between the original graph and the reconstructed graph to bring the representations of the same node closer. Again, we utilize InfoNCE to calculate the intra-graph contrastive loss, which is formulated as

$$\mathcal{L}_{\mathrm{intra}}^{*} = \sum_{i \in \mathcal{V}^{*}} -\log\left(\frac{\phi\left(\hat{z}_i^{*} \cdot \hat{z}_i^{*,recon}\right)}{\phi\left(\hat{z}_i^{*} \cdot \hat{z}_i^{*,recon}\right) + \sum\limits_{j' \in \mathrm{Neg}(i)} \phi\left(\hat{z}_i^{*} \cdot \hat{z}_{j'}^{*,recon}\right)}\right), \qquad (8)$$

where $*$ can take s or t denoting the source or target network, respectively. The intra-contrastive loss is defined as $\mathcal{L}_{\mathrm{intra}} = \mathcal{L}_{\mathrm{intra}}^{s} + \mathcal{L}_{\mathrm{intra}}^{t}$.

The final optimization objective function is $\mathcal{L} = \mathcal{L}_{\mathrm{cross}} + \beta\mathcal{L}_{\mathrm{intra}} + \gamma\mathcal{L}_{\mathrm{recon}}$ where β and γ are hyperparameters that balance the contributions of intra-graph contrastive loss and reconstruction loss, respectively.

4.3 OT-Based Alignment

After learning the node embeddings, we optimize user identity linkage via an OT-based method. We define a transport cost C_{cost} that captures both node consistency and local structural consistency of aligned users. Specifically, the cost $C_{cost} \in \mathbb{R}^{|V^s| \times |V^t|}$ is

$$C_{cost} = \|\hat{\mathbf{Z}}^{\mathbf{s}} - \hat{\mathbf{Z}}^{\mathbf{t}}\|_2 + exp(-P) \qquad (9)$$

where $\|\cdot\|_2$ denotes the Euclidean distance and $P = A^s S A^t$ denotes the first-order proximity between aligned users. To align user identities across networks, we minimize Wasserstein Distance (WD) to solve an optimal transport plan $\mathbf{T}$. The objective function is $\mathcal{L}_{\mathrm{OT}} = \sum_{i \in V^s} \sum_{j \in V^t} C_{cost}(i, j) \cdot \mathbf{T}_{\mathbf{ij}}$, which enables simultaneously considering node-level similarity and structural consistency in the optimal transport process, and efficiently computes the cost via the Sinkhorn [4] algorithm, achieving globally optimal user alignment.

4.4 Time Complexity Analysis

The time complexity during training primarily depends on three key components: RWR-based node representation, structure reconstruction within the representation disentanglement layer, and the GAT Layer. First, the time complexity of

RWR-based node representation is $\mathcal{O}((|E^s| + |E^t|) \cdot I_r)$, where I_r is the number of iterations. The time complexity of the representation disentanglement layer is $(|V^s|^2 + |V^t|^2)d$. For the GAT layer, the complexity is $\mathcal{O}((|E^s| + |E^t|) \cdot d)$. Therefore, the total time complexity during training can be approximated as $\mathcal{O}((|E^s| + |E^t| + |V^s|^2 + |V^t|) \cdot d + |E^s| + |E^t|) \cdot I_r)$. At the inference stage, the time complexity is $\mathcal{O}(|V^s| \cdot |V^t| \cdot (I + d'))$, where I denotes the number of iterations for OT-based optimization process. Thus, the running time is determined by the number of nodes and links in the source and target networks.

We adopt the Adaptive Moment Estimation (Adam) optimizer [12] to train the model, utilize the learning rate schedule to accelerate convergence, and leverage dropout and batch normalization to avoid overfitting and gradient extinction. Please refer to Sect. 5.1 for other detailed settings.

Table 1. Statistics of the two datasets used in this paper.

Dataset	Foursquare	Twitter	Facebook	Twitter	Douban	Weibo
#users	5312	5120	2458	2458	3154	3154
#links	76970	164920	40298	95034	301074	241736
#aligned users	3138		2458		3154	

5 Experiments

5.1 Experimental Settings

Datasets. In our experiments, we implement our model and baselines on three pairs of datasets from real-world social platforms to evaluate the models' performance, including **Foursquare-Twitter** [32], **Facebook-Twitter** [1], **Douban-Weibo** [1]. Table 1 summarizes the statistical information of the datasets.

Compared Methods. We compare our method with ten state-of-the-art models for user identity linkage in two categories, which are embedding-based methods (**DeepLink** [37], **DANA** [10], **BRIGHT** [26], **NeXtAlign** [33], **CCNE** [31], **CINA** [11], **ASSISTANT** [17]) and OT-based methods (**PARROT** [30], **HOT** [29], **JOENA** [28]).

Hyper-Parameter Settings and Performance Metrics. We implement our model on the PyTorch platform. The learning rate is initially set to 0.001, and the decay rate is 0.1. The mini-batch size is set to 256. The dropout rate is 0.5. We train the model with 2000 epochs and take the average of 5 experiments for the experimental results. For the parameters of our model, the restart probability α is set to 0.15. The disentangled representation dimension d is set to 200. The dimension d' of $\hat{Z}$ is set to 128, which are optimized with the grid search. The number of negative samples is set to 10, and the temperature coefficient is set to 0.1. We train all models with 20% of the aligned users. We run our model and the baselines on a GPU server equipped with Intel(R) Xeon(R) Gold 6226R CPU @ 2.90 GHz and GeForce RTX 4090 GPU.

We adopt two widely used metrics, $Hit@K$ and $MRR(Mean Reciprocal Rank)$ to evaluate the alignment performance of our model and baselines.

$$\text{Hit@K} = \frac{1}{|\mathcal{T}_{test}|} \sum_{(i,\cdot)\in\mathcal{T}_{test}} \mathbb{I}\{\text{rank}_i \leq K\}, \text{MRR} = \frac{1}{|\mathcal{T}_{test}|} \sum_{(i,\cdot)\in\mathcal{T}_{test}} \frac{1}{\text{rank}_i},$$

$$(10)$$

where $\mathbb{I}\{\cdot\}$ is the indicator function, which returns 1 if the aligned user of user v_i appears in top-K of the candidate list, $|\mathcal{T}_{test}|$ is the number of tested user pairs and $rank_{v_i}$ is the rank of correctly aligned user. We report the experimental results under the settings of $K = 1$, $K = 10$, and $K = 30$.

5.2 Performance Comparison

To demonstrate the effectiveness and advantage of our model, we compare it with ten existing methods on three pairs of real social network datasets. Results are presented in Table 2. we can see that (1) all methods achieve their best performance on the denser Douban–Weibo dataset, while sparser datasets increase the difficulty of alignment. (2) OT-based methods such as PARROT and JOENA outperform embedding-based methods, particularly in terms of Hits@1, suggesting the advantage of OT in identifying more precisely alignment. (3) For the sparse Foursquare–Twitter dataset, embedding-based methods perform better performance on Hits@30 metric, indicating that embedding-based methods can capture latent similarity between aligned nodes. (4)Our method, DCLO, consistently outperforms all baseline approaches across all metrics, improving MRR by 11.82%, 8.91%, and 3.41%, respectively, demonstrating its superior robustness and accuracy, particularly on relatively sparse datasets.

Table 2. Performance comparison with existing user identity linkage methods.

	Foursquare-Twitter				Facebook-Twitter				Douban-Weibo			
	Hits@1	Hits@10	Hits@30	MRR	Hits@1	Hits@10	Hits@30	MRR	Hits@1	Hits@10	Hits@30	MRR
DeepLink	0.002	0.1012	0.2048	0.0274	0.0071	0.1933	0.4069	0.0514	0.0361	0.2449	0.4437	0.073
DANA	0.0319	0.1197	0.1935	0.0621	0.0493	0.2341	0.4161	0.1128	0.1089	0.3766	0.582	0.1962
BRIGHT	0.1331	0.2726	0.3336	0.1796	0.15	0.3173	0.4301	0.1751	0.1377	0.2776	0.3412	0.1843
NeXtAlign	0.055	0.1581	0.2322	0.0908	0.0646	0.2883	0.4713	0.127	0.0963	0.254	0.367	0.1439
CCNE	0.1043	0.3413	0.4934	0.183	0.1622	0.4942	0.7158	0.2664	0.3011	0.6359	0.7821	0.4115
CINA	0.0681	0.2911	0.456	0.1403	0.0946	0.3803	0.6243	0.187	0.2575	0.5967	0.7552	0.3707
ASSISTANT	0.2647	0.4429	0.5694	0.3202	0.2761	0.5165	0.7006	0.352	0.4271	0.6244	0.771	0.493
PARROT	0.2808	0.4747	0.5639	0.348	0.3503	0.6528	0.7865	0.451	0.7061	0.8714	0.9265	0.767
HOT	0.0096	0.0266	0.0491	0.0163	0.1618	0.4336	0.5446	0.248	0.2419	0.3914	0.44	0.2933
JOENA	0.2103	0.3639	0.4396	0.2624	0.4418	0.6739	0.7516	0.5186	0.5967	0.8451	0.9188	0.6888
DCLO	**0.3939**	**0.6023**	**0.7097**	**0.4642**	**0.5137**	**0.7999**	**0.8981**	**0.6077**	**0.7448**	**0.9061**	**0.9416**	**0.8031**

5.3   Ablation Study

To better understand the contribution of each module in our model, we conduct an ablation study on three real datasets. We evaluate the following four variants of our framework to isolate the impact of individual modules and verify their contribution to the overall performance.

- **w/o-RDL:** removing the representation disentanglement layer and taking node representation from RWR as the input of the cross-network fusion layer.
- **w/o-DCRF:** removing the disentanglement-based cross-network representation fusion layer and feeding the disentangled node representations into the graph encoder layer.
- **w/o-Recon:** removing the reconstruction loss term.
- **w/o-Intra-CL:** removing the intra-network contrastive loss term.
- **w/o-AO:** removing the alignment optimization module and inferring node alignment based on similarity between node representations across networks.

Table 3 presents the performance of our model and its variants, from which we can see that (1) our proposed method, DCLO, outperforms all other variants on three datasets, demonstrating the contribution of each component. (2) w/o-DCRF causes a more significant drop in MRR and Hits@1, demonstrating its function in precise matching. However, in terms of Hits@30, w/o DCRF performs better than w/o-RDL, suggesting that the RDL may help the model capture underlying similar features. (3) w/o-Recon results in a slight performance slide, indicating that preserving structural information between nodes plays a supportive role in identifying aligned users. On the other hand, w/o-Intra-CL leads to a more significant performance drop, highlighting the importance of promoting the discriminative power of node representations and modeling local structural patterns for alignment. (4) w/o-AO has a relatively significant impact across all datasets, particularly on Hits@1. This result demonstrates the contribution of global optimization during alignment.

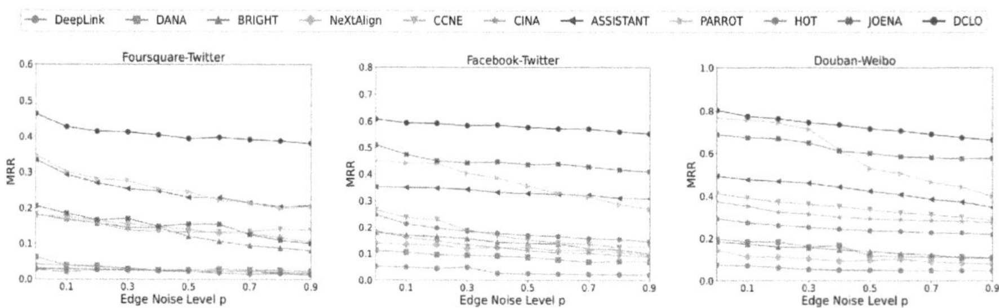

Fig. 2. Performance results with link noise levels p.

Table 3. Performance comparison with variants of our model.

Dataset	Metric	w/o-RDL	w/o-DCRF	w/o-Recon	w/o-Intra-CL	w/o-AO	DCLO
Foursquare|Twitter	Hits@1	0.3771	0.3626	0.3833	0.3568	0.3232	**0.3939**
	Hits@10	0.5711	0.5633	0.5860	0.5430	0.5629	**0.6023**
	Hits@30	0.6619	0.6746	0.6908	0.6364	0.6826	**0.7097**
	MRR	0.4431	0.4316	0.4518	0.4184	0.4014	**0.4642**
Facebook|Twitter	Hits@1	0.4634	0.4751	0.4990	0.4382	0.3836	**0.5137**
	Hits@10	0.7407	0.7664	0.7789	0.7214	0.714	**0.7999**
	Hits@30	0.8508	0.8729	0.8879	0.8348	0.8472	**0.8981**
	MRR	0.5553	0.5697	0.5939	0.5288	0.4876	**0.6077**
Douban|Weibo	Hits@1	0.7145	0.6985	0.7135	0.6999	0.5077	**0.7448**
	Hits@10	0.8782	0.8782	0.8863	0.8726	0.767	**0.9061**
	Hits@30	0.9221	0.9279	0.9297	0.9212	0.8641	**0.9416**
	MRR	0.7724	0.7626	0.7740	0.7606	0.5962	**0.8031**

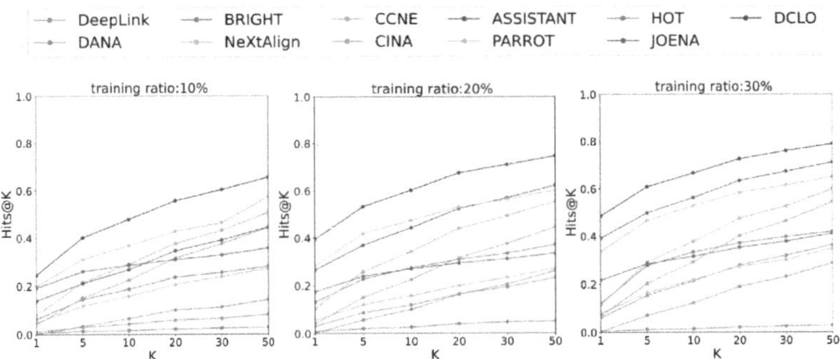

Fig. 3. Performance results on Foursquare-Twitter under different training ratio.

5.4 Robustness Evaluation

To verify the robustness of our model, we perform a series of experiments by introducing link noise and adjusting the training ratios.

Link Noise. We evaluate all models under link noise by randomly adding links at noise levels p ranging from 0.1 to 0.9 and the performance is reported in Fig. 2. As noise increases, all models show a consistent performance degradation. OT-based methods such as PARROT show significant performance degradation due to their reliance on the consistency of known aligned nodes, which becomes unreliable under excessive noise. Embedding-based methods are relatively more robust in noisy settings. Nevertheless, DCLO consistently outperforms all baselines across datasets and noise levels, demonstrating superior robustness and alignment capability under structural perturbations.

Training Ratios. Figure 3 shows the Hit@K results of the tested models with different training ratios ranging from 0.1 to 0.3. Compared to other baseline methods, our proposed method consistently outperforms them under the setting of all training ratios across all datasets, particularly in terms of Hit@K (K=1 and K=5). Notably, our method exhibits a more significant performance advantage when the training set is limited to just 10% of aligned data, demonstrating its strong generalization capability and robustness under sparse data conditions.

5.5   Time-Cost Comparison

To demonstrate the efficiency of our model, we compare its inference time with the existing OT-based methods on both datasets; the results are shown in Fig. 4. One can observe that PARROT is more sensitive to the scale of the networks. JOENA significantly reduces inference time while maintaining competitive performance. Our method further compresses the inference time across all three datasets, achieving lower time costs while improving alignment performance.

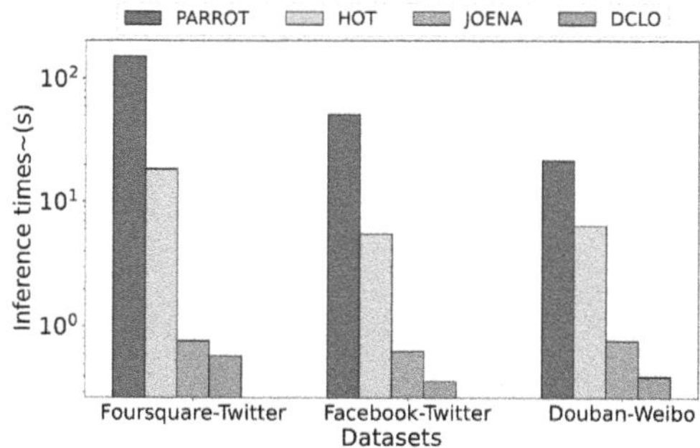

Fig. 4. The inference time of different methods.

5.6   Parameter Sensitivity Analysis

We evaluate the sensitivity of our method to its hyperparameters by varying the reconstructed cross-graph loss coefficient λ, intra-contrastive learning coefficient β, and reconstruction loss coefficient γ on three datasets. We report the performance of our model in terms of MRR under different hyperparameters.

Impact of Coefficient λ. The coefficient λ controls the weight of the recontrusted cross-network contrastive learning loss. The results are shown in Fig. 5(a). One can observe that the performance has a significant improvement after adding the cross-graph reconstruction loss, especially on the Douban-Weibo dataset. The performance peaks at $\lambda = 0.6$ and then gradually decays.

Impact of Coefficient β. The coefficient β controls the weight of the intra-network contrastive learning loss. As shown in Fig. 5(b), we can see that the

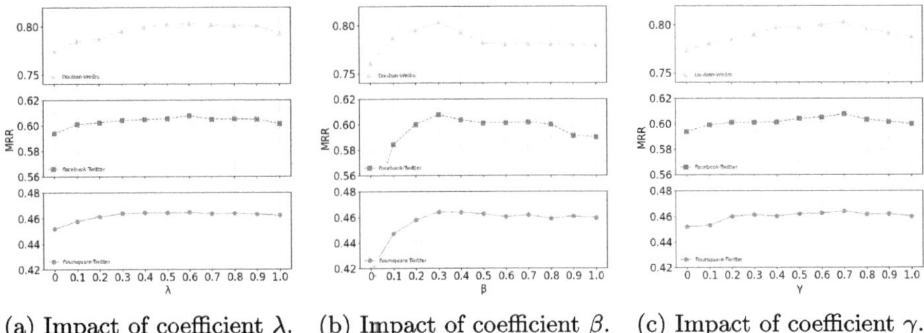

(a) Impact of coefficient λ.   (b) Impact of coefficient β.   (c) Impact of coefficient γ.

Fig. 5. Performance comparison under different parameter settings.

model's performance on the relatively sparse Foursquare-Twitter dataset fluctuates slightly. In contrast, on the relatively dense Douban-Twitter dataset, the performance of our model gradually grows with β, reaching its peak at $\beta = 0.3$, indicating that appropriately strengthening intra-network contrastive learning can effectively distinguish semantically similar node representations. Further increasing β may push semantically similar nodes excessively apart in the embedding space, destroying their semantic similarity and raising the risk of mismatches during cross-network alignment.

Impact of Coefficient γ. The coefficient γ controls the weight of the reconstruction loss in the overall objective function. The results are shown in Fig. 5(c) One can observe that the performance gradually grows with the reconstruction loss coefficient γ and achieves its peak at $\gamma = 0.7$. This observation indicates that the model requires a certain degree of reconstruction constraint to effectively capture and preserve the structural information of networks, which is beneficial to node representation learning.

6   Conclusion

This paper proposes a novel Disentanglement-Based Contrastive Learning framework for user identity linkage. Our model employs a representation disentanglement module to project node embeddings into network-shared and network-specific feature spaces, simultaneously fusing shared features of aligned users. We then apply contrastive losses from both cross-network and intra-network perspectives to preserve alignment consistency while enhancing feature discrimination. To further refine alignment, we introduce a unified transportation cost optimization. Experimental results on three real-world datasets demonstrate that our approach outperforms ten state-of-the-art baselines, maintaining robustness against link noise and varying training ratios. Ablation studies and inference time analyses confirm the framework's effectiveness. Future work could explore: (1) More efficient optimization strategies to enhance scalability on large-scale networks; (2) Identification of non-aligned users (those without cross-network counterparts) to streamline training and improve alignment accuracy.

Acknowledgement. This work was supported in part by National Natural Science Foundation of China (No. 62576248 and No. 62372326) and Major Program of National Social Science Fund (24&ZD185).

References

1. Cao, X., Yu, Y.: Asnets: a benchmark dataset of aligned social networks for cross-platform user modeling. In: Proceedings of the 25th ACM International on Conference on Information and Knowledge Management, pp. 1881–1884. ACM (2016)
2. Chen, S., et al.: Combining optimal transport and embedding-based approaches for more expressiveness in unsupervised graph alignment. arXiv preprint arXiv:2406.13216 (2024)
3. Chu, X., Fan, X., Yao, D., Zhu, Z., Huang, J., Bi, J.: Cross-network embedding for multi-network alignment. In: The World Wide Web Conference. pp. 273–284. ACM (2019)
4. Cuturi, M.: Sinkhorn distances: lightspeed computation of optimal transport. In: Advances in Neural Information Processing Systems, vol. 26 (2013)
5. Diao, M., Zhang, Z., Su, S., Gao, S., Cao, H.: Upon: user profile transferring across networks. In: Proceedings of the 29th ACM International Conference on Information & Knowledge Management, pp. 265–274. ACM (2020)
6. Du, X., Yan, J., Zhang, R., Zha, H.: Cross-network skip-gram embedding for joint network alignment and link prediction. IEEE Trans. Knowl. Data Eng. **34**(3), 1080–1095 (2020)
7. Fan, W., et al.: Adversarial attacks for black-box recommender systems via copying transferable cross-domain user profiles. IEEE Trans. Knowl. Data Eng. **35**(12), 12415–12429 (2023)
8. Guo, X., Liu, Y., Gong, D., Liu, F.: Dual graph convolutional networks for social network alignment. IEEE Trans. Big Data **11**(02), 684–695 (2025)
9. Heimann, M., Shen, H., Safavi, T., Koutra, D.: Regal: representation learning-based graph alignment. In: Proceedings of the 27th ACM International Conference on Information and Knowledge Management, pp. 117–126. ACM (2018)
10. Hong, H., Li, X., Pan, Y., Tsang, I.W.: Domain-adversarial network alignment. IEEE Trans. Knowl. Data Eng. **34**(7), 3211–3224 (2020)
11. Jiao, P., Liu, Y., Wang, Y., Zhang, G.: Cina: curvature-based integrated network alignment with hypergraph. In: 2024 IEEE 40th International Conference on Data Engineering (ICDE), pp. 2709–2722. IEEE (2024)
12. Kingma, D.P., Ba, J.: Adam: A Method for Stochastic optimization. In: 3rd International Conference on Learning Representations, ICLR 2015, San Diego, CA, USA, 7–9 May 2015, Conference Track Proceedings (2015)
13. Li, P., Brost, B., Tuzhilin, A.: Adversarial learning for cross domain recommendations. ACM Trans. Intell. Syst. Technol. **14**(1), 1–25 (2023)
14. Liu, J., Sun, L., Nie, W., Jing, P., Su, Y.: Graph disentangled contrastive learning with personalized transfer for cross-domain recommendation. In: Proceedings of the AAAI Conference on Artificial Intelligence, pp. 8769–8777 (2024)

15. Long, M., Chen, S., Du, X., Wang, J.: Deguil: degree-aware graph neural networks for long-tailed user identity linkage. In: Joint European Conference on Machine Learning and Knowledge Discovery in Databases, pp. 122–138. Springer (2023)
16. Oord, A.V.D., Li, Y., Vinyals, O.: Representation learning with contrastive predictive coding. arXiv preprint arXiv:1807.03748 (2018)
17. Seo, D.H., Lim, J.H., Shin, W.Y., Kim, S.W.: Leveraging trustworthy node attributes for effective network alignment. In: Proceedings of the 33rd ACM International Conference on Information and Knowledge Management, pp. 2004–2013 (2024)
18. Shao, J., Wang, Y., Shi, B., Gao, H., Shen, H., Cheng, X.: Adversarial for social privacy: a poisoning strategy to degrade user identity linkage. arXiv preprint arXiv:2209.00269 (2022)
19. Solanki, P., Harwood, A., et al.: User identification across social networking sites using user profiles and posting patterns. In: 2021 International Joint Conference on Neural Networks (IJCNN), pp. 1–8. IEEE (2021)
20. Tang, J., Zhang, W., Li, J., Zhao, K., Tsung, F., Li, J.: Robust attributed graph alignment via joint structure learning and optimal transport. In: 2023 IEEE 39th International Conference on Data Engineering (ICDE), pp. 1638–1651. IEEE (2023)
21. Tong, H., Faloutsos, C., Pan, J.Y.: Fast random walk with restart and its applications. In: Sixth International Conference on Data Mining (ICDM 2006), pp. 613–622. IEEE (2006)
22. Veličković, P., Cucurull, G., Casanova, A., Romero, A., Lio, P., Bengio, Y.: Graph attention networks. arXiv preprint arXiv:1710.10903 (2017)
23. Wang, C., Jiang, P., Zhang, X., Wang, P., Qin, T., Guan, X.: Gtcalign: global topology consistency-based graph alignment. IEEE Trans. Knowl. Data Eng. **36**(5), 2009–2025 (2023)
24. Wang, M., Wang, W., Chen, W., Zhao, L.: EEUPL: towards effective and efficient user profile linkage across multiple social platforms. World Wide Web **24**(5), 1731–1748 (2021). https://doi.org/10.1007/s11280-021-00882-7
25. Xiong, H., Yan, J., Pan, L.: Contrastive multi-view multiplex network embedding with applications to robust network alignment. In: Proceedings of the 27th ACM SIGKDD Conference on Knowledge Discovery & Data Mining, pp. 1913–1923 (2021)
26. Yan, Y., Zhang, S., Tong, H.: Bright: a bridging algorithm for network alignment. In: WWW '21: The Web Conference 2021, Virtual Event/Ljubljana, Slovenia, 19–23 April 2021, pp. 3907–3917. ACM/IW3C2 (2021)
27. Yao, X., Zhang, R., Zhang, Y.: Differential privacy-preserving user linkage across online social networks. In: 2021 IEEE/ACM 29th International Symposium on Quality of Service (IWQOS), pp. 1–10. ACM (2021)
28. Yu, Q., Zeng, Z., Yan, Y., Ying, L., Srikant, R., Tong, H.: Joint optimal transport and embedding for network alignment. In: Proceedings of the ACM on Web Conference 2025, pp. 2064–2075 (2025)
29. Zeng, Z., Du, B., Zhang, S., Xia, Y., Liu, Z., Tong, H.: Hierarchical multi-marginal optimal transport for network alignment. In: Proceedings of the AAAI Conference on Artificial Intelligence, vol. 38, pp. 16660–16668 (2024)
30. Zeng, Z., Zhang, S., Xia, Y., Tong, H.: Parrot: position-aware regularized optimal transport for network alignment. In: Proceedings of the ACM Web Conference 2023, pp. 372–382 (2023)

31. Zhang, H.F., Ren, G., Ding, X., Zhou, L., Zhang, X.: Collaborative cross-network embedding framework for network alignment. IEEE Trans. Netw. Sci. Eng. **11**(3), 2989–3001 (2024)
32. Zhang, J., Kong, X., Yu, P.S.: Transferring heterogeneous links across location-based social networks. In: Proceedings of the 7th ACM International Conference on Web Search and Data Mining, pp. 303–312. ACM (2014)
33. Zhang, S., Tong, H., Jin, L., Xia, Y., Guo, Y.: Balancing consistency and disparity in network alignment. In: Proceedings of the 27th ACM SIGKDD Conference on Knowledge Discovery & Data Mining, pp. 2212–2222. ACM (2021)
34. Zhang, S., Tong, H., Xia, Y., Xiong, L., Xu, J.: Nettrans: neural cross-network transformation. In: Proceedings of the 26th ACM SIGKDD International Conference on Knowledge Discovery & Data Mining, pp. 986–996. ACM (2020)
35. Zhang, Y., Tang, J., Yang, Z., Pei, J., Yu, P.S.: COSNET: connecting heterogeneous social networks with local and global consistency. In: Proceedings of the 21th ACM SIGKDD International Conference on Knowledge Discovery and Data Mining, pp. 1485–1494. ACM (2015)
36. Zhao, Q., Wang, Q.: Contrastive learning for extracting transferable user profiles in cross-domain recommendation system. In: 2024 International Joint Conference on Neural Networks (IJCNN), pp. 1–8. IEEE, IEEE (2024)
37. Zhou, F., Liu, L., Zhang, K., Trajcevski, G., Wu, J., Zhong, T.: Deeplink: a deep learning approach for user identity linkage. In: IEEE INFOCOM 2018-IEEE Conference on Computer Communications, pp. 1313–1321. IEEE, IEEE (2018)
38. Zhou, J.: Mane: a multi-cascade adversarial network embedding model for anchor link prediction. In: International Conference on Database Systems for Advanced Applications, pp. 168–184. Springer (2024)
39. Zhou, J., Fan, J.: TransLink: user identity linkage across heterogeneous social networks via translating embeddings. In: IEEE INFOCOM 2019 - IEEE Conference on Computer Communications, pp. 2116–2124. IEEE (2019)

Collaborative Pattern Mining in Activity Graphs

Beilei Ling[1], Ziyu Guan[1(✉)], Wei Zhao[1], Yiheng Lu[1], Meng Yan[1], Cai Xu[1], Weigang Lu[2], and Beizeng Ling[3]

[1] School of Computer Science and Technology, Xidian University, Xi'an, China
{lingbeilei,mengyan}@stu.xidian.edu.cn,
{zyguan,luyiheng,cxu}@xidian.edu.cn, ywzhao@mail.xidian.edu.cn
[2] Department of Civil and Environmental Engineering, The Hong Kong University of Science and Technology, Hong Kong, SAR, China
weianglu314@outlook.com
[3] RealAi, Beijing, China
beizengling82@gmail.com

Abstract. Existing graph pattern mining paradigms focus on repeated substructures or event orders and therefore overlook collaborative behavior, where correlated activities occur within short time intervals on structurally cohesive neighborhoods. We formalize this as Collaborative Pattern Mining (CPM), defining collaboration by joint temporal and structural proximity. We propose the support measure Collaborative Minimum Image Support (CMIS), defined as the minimum image count across participating labels, which is anti-monotone under the joint constraints and enables effective pruning. Building on this, we design piCPM, combining CMIS-based pruning with a partition-based index (PBI) constructed via diameter-constrained label propagation (DLPA) and augmented with precomputed multi-hop distance labels for fast graph queries. Across four real-world and synthetic datasets, piCPM uncovers meaningful dependencies among topology, attributes, and time, and it consistently outperforms strong baselines in accuracy and scalability, achieving up to three orders of magnitude speedups. These results establish CPM as a practical and scalable approach.

Keywords: Activity Graph · Collaborative Pattern · Anti-monotonicity · Temporal Proximity · Graph Indexing

1 Introduction

Graph pattern mining has become increasingly important with the rapid growth of large-scale networks across cybersecurity, e-commerce, and online social platforms, in which nodes record timestamped activities such as intrusions and transactions. In such networks, collaboration is characterized by temporally concentrated, correlated activity within structurally compact neighborhoods. Methods that do not encode these semantics often partition a genuine collaboration into

H. Jung et al. (Eds.): DASFAA 2026, LNCS 16536, pp. 283–299, 2026.
https://doi.org/10.1007/978-981-92-0366-6_18

disconnected fragments or merge unrelated events into spurious co-occurrences, resulting in incomplete and inaccurate modeling of collective behavior.

Consider intrusion alert networks, where events like `portsweep`, `smurf`, and `pod` often occur within short temporal intervals on structurally proximate nodes. Although each event is distinct, their nearly synchronous occurrence indicates coordinated multi-stage behavior. Prevailing paradigms, including frequent subgraph mining, sequential pattern mining, and dense subgraph mining, primarily target repeated substructures or sequences and treat temporal and structural constraints separately. This inductive bias limits their ability to detect collaborative patterns in which all required activities co-occur within a short interval on a compact neighborhood of nodes.

Recent efforts relax exact matching by mining co-occurrence under structural proximity [12,14]. Structural methods such as aFP [16] discover itemsets over connected nodes without modeling temporal proximity, which aggregates events that are far apart in time and fails to capture collaborative behavior. Temporal association frameworks such as GTAR [19] enforce temporal proximity but encode topology through predefined templates, which prevents enforcing structural proximity among arbitrary participating nodes. More recent systems emphasize scalability on large attributed graphs [28] or durability under temporal locality [1], but they typically optimize only one dimension, structural or temporal, leaving their conjunction underexplored.

To address these limitations, we propose **Collaborative Pattern Mining (CPM)**, which jointly enforces temporal and structural proximity in activity graphs without predefined templates and enables scalable detection of time and topology constrained collaborations, including rapid multi-stage attacks. Defining collaboration as the conjunction of temporal and structural proximity creates scalability challenges since the search space exhibits a combinatorial explosion with pattern size and the combined constraints, while verification is expensive due to repeated proximity checks over numerous candidates.

We introduce **piCPM (a partition-based index for CPM)**, an efficient framework for CPM. It encodes collaboration through joint temporal and structural proximity under anti-monotone constraints and prunes infeasible supersets early, which reduces the search space. The framework incorporates a **partition-based index (PBI)** built with a **diameter-constrained label propagation algorithm (DLPA)**, so intra-partition proximity holds by construction and cross-partition candidates are pruned using precomputed distance labels and landmark nodes, improving scalability on large graphs.

To quantify support, piCPM propose **Collaborative Minimum Image Support (CMIS)**, which counts co-occurring label sets under joint proximity and takes the minimum image count across labels to reflect collaboration semantics, avoid overcounting, and enable linear-time computation. Extensive experiments on four real and synthetic datasets confirm that piCPM effectively uncovers meaningful topology–attribute–time dependencies and outperforms baselines in accuracy and scalability, demonstrating its practical value in network security and social analysis.

This paper makes three contributions:

- We formalize CPM as collaboration activities defined by the conjunction of structural and temporal proximity, distinguishing it from frequent, sequential, and dense subgraph mining.
- We propose piCPM, which combines CMIS-based pruning with a partition-based index (PBI) that leverages precomputed multi-hop neighborhoods, achieving nearly linear complexity in graph size.
- We demonstrate on real and synthetic datasets that piCPM uncovers high quality collaborative patterns and achieves speedups of up to three orders of magnitude over strong baselines.

2   Related Work

This section reviews related work across four domains and contrasts them with CPM, emphasizing their limitations on activity graphs.

Frequent Subgraph Mining (FSM). FSM discovers subgraphs in a single large graph that exceed a threshold. To avoid double counting, anti-monotone supports such as MIS [10], MNI [5], and their occurrence-hypergraph variants MI and MVC [18] were proposed. Methods like GraMi [9] and SIGRAM [17] implement these supports via constraint evaluation and grow-and-store enumeration. However, they measure structural frequency on static graphs and cannot capture joint temporal–structural proximity or permutation-invariant co-occurrences. Adding temporal constraints afterward modifies the counting semantics and weakens anti-monotone pruning, whereas CPM defines support directly under joint temporal–structural proximity, preserving pruning guarantees.

Correlated Graph Pattern Mining (CGPM). Recent studies relax the rigid constraints of FSM to reduce isomorphism costs. Proximity patterns combine graph connectivity with frequent itemsets [16], and correlated subgraphs identify closely connected pattern pairs using the Replica structure with a best-first search for efficient mining [24]. Related work includes cohesive itemsets in attributed graphs [7], attribute covariation [23], neighborhood patterns linking topology and labels [11], and structural correlations between dense subgraphs and attributes [27]. Some methods further mine frequent label sets to reveal correlations [24]. However, these methods capture static topological proximity.

Dynamic Graph Pattern Mining (DGPM). Dynamic graphs have inspired various approaches for mining evolving patterns [4,20]. Sampling-based methods identify frequent k-vertex subgraphs, such as TipTAP [20] and MANIACS [25], while others track vertex co-evolution through attribute trends and temporal regularities [8,13,21,26]. These studies mainly emphasize topological and frequency evolution, whereas CPM captures collaborative dependencies under joint temporal-structural proximity.

Graph Querying and Indexing (GQI). Exact distance queries employ pruned BFS labeling [3], while reachability is accelerated via landmark and path labeling [30] and their label-constrained variants [29]. HCL introduces a tree-structured highway scheme balancing index size and latency [15], and other studies target k-hop reachability [6,22]. These methods provide efficient querying on large, mostly static graphs. Whereas prior work primarily emphasizes fast distance and reachability queries, CPM focuses on mining time-sensitive patterns under joint temporal and structural proximity.

3   Preliminaries

In this section, we introduce activity graph notation and an example, then define Collaborative Patterns with temporal and structural constraints, together with the notions of embedding and occurrence.

3.1   Activity Graph

An activity graph $G = (V, E, L)$ is an attributed graph with node set V, edge set E, and label set L. Each node $v \in V$ generates an activity sequence $\mathcal{A}_v = \langle a_v[1], \ldots, a_v[n_v] \rangle$ where $a_v[i] = (t_v[i], l_v[i])$ and $l_v[i] \in L$. We treat labels and attributes as the same concept and record them as timestamped activities $a_v[i]$. For brevity, we write $a_v[i].t = t_v[i]$ and $a_v[i].l = l_v[i]$. We analyze activity on a fixed topology, with dynamics arising from timestamped labels.

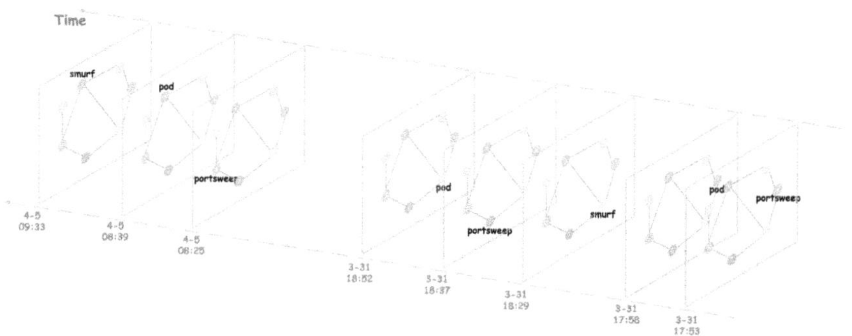

Fig. 1. An activity graph from the DARPA 2000 Intrusion Alert Network, where nodes represent IP addresses, edges denote host interactions, and activities capture timestamped alerts (e.g., node 5: **portsweep** at 17:53, **pod** at 17:58 on 3–31).

Example 1. Figure 1 illustrates an activity graph from the DARPA 2000 intrusion benchmark, where nodes represent IP addresses, edges denote host interactions, and activities are timestamped alerts (e.g., **portsweep**, **pod**). For node 5, the activity sequence $\mathcal{A}_5 = \langle (\text{3-31 } 17\text{:}53, \textbf{portsweep}), (\text{3-31 } 17\text{:}58, \textbf{pod}) \rangle$ shows two alerts within five minutes, a typical reconnaissance-to-attack transition. piCPM captures such temporal–structural proximities to uncover collaborative multi-step intrusions (Fig. 2).

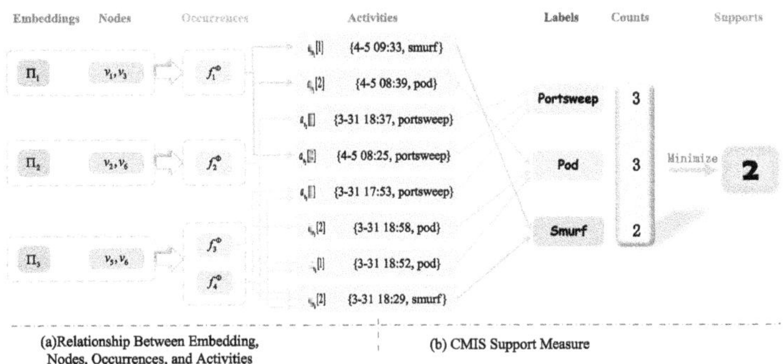

Fig. 2. (a) Relationship between embeddings, nodes, occurrences, and activities. (b) CMIS support measure for the Collaborative Pattern **portsweep, pod, smurf** in the DARPA 2000 Intrusion Alert Network.

3.2 Collaborative Patterns

CPM targets to discover activity sets that co-occur within proximate structural and temporal. Joint proximity is crucial because interaction strength decreases rapidly with graph distance, and events separated by long time intervals are unlikely to belong to the same collaborative process. We therefore impose two constraints: a temporal window $\mathcal{T}$ and a structural bound δ.

Definition 1. (Embedding). *Given an activity graph G and a candidate $I \subseteq L$, an embedding of I in G is a surjective map $\phi : I \to \Pi$ with $\Pi \subseteq V$ such that for every $l \in I$ there exists an index i with $a_{\phi(l)}[i].l = l$. Let $\Pi = \phi(I) = \{\phi(l) \mid l \in I\}$ denote the embedding node set of I.*

Definition 2. (Occurrence). *Let ϕ be an embedding with $\Pi = \phi(I)$, and let $\mathcal{A}_\Pi = \bigcup_{v \in \Pi} \mathcal{A}_v$. An occurrence is a map $f^\phi : I \to \mathcal{A}_\Pi$ such that $f^\phi(l) \in \mathcal{A}_{\phi(l)}$ and $f^\phi(l).l = l$ for all $l \in I$.*

Definition 3. (Temporal Constraint). *For each occurrence of a Collaborative Pattern I, the duration must not exceed a predefined temporal window $\mathcal{T}$.*

Definition 4. (Structural Constraint). *Let $G = (V, E, L)$ be an activity graph and let $I \subseteq L$ be a candidate label set. For any embedding node set $\Pi \subseteq V$ of I in G, the structural constraint requires $\mathrm{ediam}(\Pi) = \max\{ d_G(u, v) \mid u, v \in \Pi \} \leq \delta$, where $d_G(u, v)$ is the shortest path hop distance in G and $\delta > 0$.*

Diameter is chosen as the structural measure because it is anti-monotone under node expansion, allowing efficient pruning of candidates that exceed the structural constraint. It also offers a balanced trade-off between expressiveness and computational cost, as hop-based diameter can be verified through incremental reachability checks.

Definition 5. (Collaborative Pattern). *Given an activity graph G and a label set I, I is a Collaborative Pattern if the number of its occurrences satisfying both temporal and structural constraints exceeds a predefined threshold τ.*

4 Collaborative Pattern Mining

This section introduces Collaborative Minimum Image Support (CMIS) as the support measure for CPM, proves that CMIS and the structural and temporal constraints are anti-monotone, then presents the CoPM framework.

4.1 Support Computation

We propose a novel support measure called Collaborative Minimum Image Support (CMIS), an adaptation of the Minimum Image-based Support (MNI) by Bringmann and Nijssen [5], tailored for Collaborative Patterns. In CPM, an *image* of a label is a distinct activity selected for that label by a valid occurrence that satisfies the structural and temporal constraints. Identifying images per label is the basis for computing CMIS.

Definition 6 (Collaborative Minimum Image Support (CMIS)). *Given an activity graph G and a candidate collaborative pattern I, let $\{f_1^\phi, f_2^\phi, \ldots, f_m^\phi\}$ be the set of valid occurrences of I satisfying the structural and temporal constraints. The CMIS of I in G is*

$$\sigma_{\mathrm{CMIS}}(I, G) = \min_{l \in I} \left| \{ f_i^\phi(l) \mid i = 1, \ldots, m \} \right|. \tag{1}$$

It measures the smallest number of distinct activities assigned to any label of I.

A candidate I is regarded as a collaborative pattern if $\sigma_{\mathrm{CMIS}}(I, G) \geq \tau$. This measure is anti-monotone (Sect. 4.2), enabling efficient pruning of the search space. Unlike frequent subgraph mining, which counts subgraph embeddings, CPM evaluates constraint-satisfying occurrences, where a single embedding may correspond to multiple timestamped occurrences. Although this finer granularity enlarges the apparent search space, CMIS mitigates overestimation by aggregating distinct images per label across occurrences. In practice, it can be computed in linear time with respect to the number of activities examined for a candidate.

4.2   Anti-monotone Property

The anti-monotone property enables efficient search in CPM. In this section, we show that the structural and temporal constraints, as well as the CMIS measure, all satisfy this property.

Lemma 1. *The structural constraint satisfies the anti-monotone property.*

Proof. Let $\mathrm{ediam}(\Pi) = \max_{u,v \in \Pi} d_G(u,v)$ and let $\Pi' \subseteq \Pi$. Since $\{(u,v) : u,v \in \Pi'\} \subseteq \{(u,v) : u,v \in \Pi\}$, we have $\mathrm{ediam}(\Pi') \leq \mathrm{ediam}(\Pi)$. Hence if $\mathrm{ediam}(\Pi') > \delta$, then $\mathrm{ediam}(\Pi) > \delta$. Therefore, the constraint $\mathrm{ediam}(\Pi) \leq \delta$ is anti-monotone with respect to node set inclusion. Moreover, since any embedding for a superset restricts to an embedding for its subset, the constraint is also anti-monotone under pattern extension.

Lemma 2. *The temporal constraint satisfies the anti-monotone property.*

Proof. The proof is the same as above.

Theorem 1. *The CMIS support in CPM satisfies the anti-monotone property.*

Proof. Let $Occ(I,G)$ be the valid occurrences of I in G, define $c_l(I) = \left| \{ f^\phi(l) \mid f^\phi \in Occ(I,G) \} \right|$ and $\sigma_{\mathrm{CMIS}}(I,G) = \min_{l \in I} c_l(I)$. For $I' \subseteq I$, Lemmas 1 and 2 imply that restricting any $f^\phi \in Occ(I,G)$ to I' produces a valid element of $Occ(I',G)$, hence for every $l' \in I'$ we have $\{ f^\phi(l') \mid f^\phi \in Occ(I,G) \} \subseteq \{ f^\phi(l') \mid f^\phi \in Occ(I',G) \}$ and therefore $c_{l'}(I) \leq c_{l'}(I')$. It follows that $\sigma_{\mathrm{CMIS}}(I,G) \leq \min_{l' \in I'} c_{l'}(I) \leq \min_{l' \in I'} c_{l'}(I') = \sigma_{\mathrm{CMIS}}(I',G)$, which proves anti-monotonicity.

4.3   Algorithm Framework

We propose **CoPM**, a pattern-growth framework for collaborative pattern mining (Algorithm 1). CoPM expands patterns under a total order $\prec$ on labels, defined by descending global frequency, breaking ties lexicographically. Each candidate I is extended only with labels l satisfying $\max_{\prec}(I) \prec l$, thus avoiding duplicates and ensuring completeness. Starting from frequent single-label patterns, CoPM recursively explores larger candidates. For each candidate, it verifies the structural and temporal constraints, computes CMIS support, and prunes the branch if $\sigma_{\mathrm{CMIS}}(I,G) < \tau$.

4.4   Significance Evaluation

To assess whether a pattern CMIS support exceeds random expectation, we conduct a randomization test that preserves the graph topology. All activity tuples $\langle timestamp, label \rangle$ are pooled and uniformly redistributed to generate K randomized graphs $P_1, \ldots, P_K$, preserving node activity counts as well as label and time distributions. For each pattern I, let $p = \sigma_{\mathrm{CMIS}}(I,G)$ and $q = \frac{1}{K} \sum_{k=1}^{K} \sigma_{\mathrm{CMIS}}(I, P_k)$. We compute the G-test statistic as

$$G(I) = 2[p \ln(p/q) - (p - q)], \tag{2}$$

where a larger $G(I)$ value indicates a stronger deviation from randomness.

Algorithm 1. CollaborativePatternMining (CoPM)

1: **Input:** $G = (V, E, L)$, structural constraint δ, window $\mathcal{T}$, frequency threshold τ
2: **Output:** Collaborative patterns
3: Filter infrequent labels; sort $\rightarrow fL$; $\mathcal{W} \leftarrow \{\{l\} \mid l \in fL\}$; $\mathcal{R} \leftarrow \emptyset$
4: **while** $\mathcal{W} \neq \emptyset$ **do**
5: $\quad$ $I \leftarrow \mathrm{pop}(\mathcal{W})$; **for** $l \in I$: $S_l \leftarrow \emptyset$
6: $\quad$ **for all** $\phi : I \rightarrow \Pi \subseteq V$ **with** $\mathrm{ediam}(\Pi) \leq \delta$ **do**
7: $\quad\quad$ **for all** $f^\phi : I \rightarrow \mathcal{A}_\Pi$ **with** $\mathrm{duration}(f^\phi) \leq \mathcal{T}$ **do**
8: $\quad\quad\quad$ **for** $l \in I$: $S_l \leftarrow S_l \cup \{f^\phi(l)\}$
9: $\quad\quad$ **end for**
10: $\quad$ **end for**
11: $\quad$ **if** $\min_{l \in I} |S_l| \geq \tau$ **then**
12: $\quad\quad$ $\mathcal{R} \leftarrow \mathcal{R} \cup \{I\}$; $\alpha \leftarrow \max_\prec(I)$; **for** $l \in fL$ **with** $\alpha \prec l$: push $I \cup \{l\}$ into $\mathcal{W}$
13: $\quad$ **end if**
14: **end while**
15: **return** $\mathcal{R}$

---

5   PiCPM: Collaborative Pattern Mining Partition Index

The primary bottleneck of CPM is structural constraint verification: checking the diameter constraint δ for each candidate embedding requires many h-hop reachability queries. Conventional reachability queries lack precision for exact h-hop verification, whereas shortest-path queries incur redundant computation. We therefore propose **piCPM**, which integrates a diameter-constrained *partition index* to accelerate repeated CPM searches, particularly for small h and localized structures. DLPA generates compact partitions that satisfy intra-partition constraints, and bridge-node landmarks provide tight cross-partition constraints, invoking exact verification only when necessary. Section 5.1 introduces DLPA, and Sect. 5.2 describes the partition-based index for h-hop reachability.

5.1   Graph Partitioning with Diameter Constraint

To ensure that each partition satisfies the structural constraint required by CPM, we adopt a **Diameter-Constrained Label Propagation Algorithm (DLPA)**. DLPA extends label propagation by enforcing that every partition formed has a hop-diameter not exceeding δ. At each iteration, a node adopts the majority label among its neighbors only if joining the target partition keeps its diameter within δ, thereby maintaining local cohesion and preventing oversized or inconsistent partitions.

DLPA Procedure. (1) Initialization: $C_v \leftarrow v$. (2) Update: in random order, let v tentatively adopt the majority neighbor label c_{maj}; accept if and only if $\max_{x,y \in \mathcal{C}_{c_{\mathrm{maj}}} \cup \{v\}} d_G(x, y) \leq \delta$, otherwise keep C_v. (3) Terminate when no labels change.

$\quad$ Each partition maintains a few representative nodes and cached hop distances. If v is within $\lfloor \delta/2 \rfloor$ hops of a preselected node u, then any w with

$d_G(w, u) \leq \lfloor \delta/2 \rfloor$ satisfies $d_G(v, w) \leq \delta$ by the triangle inequality. Pairs not certified by this test are verified via the h-hop reachability test. Consequently, all nodes within a DLPA partition inherently satisfy the diameter constraint δ, substantially reducing the number of explicit structural checks in CPM (Fig. 3).

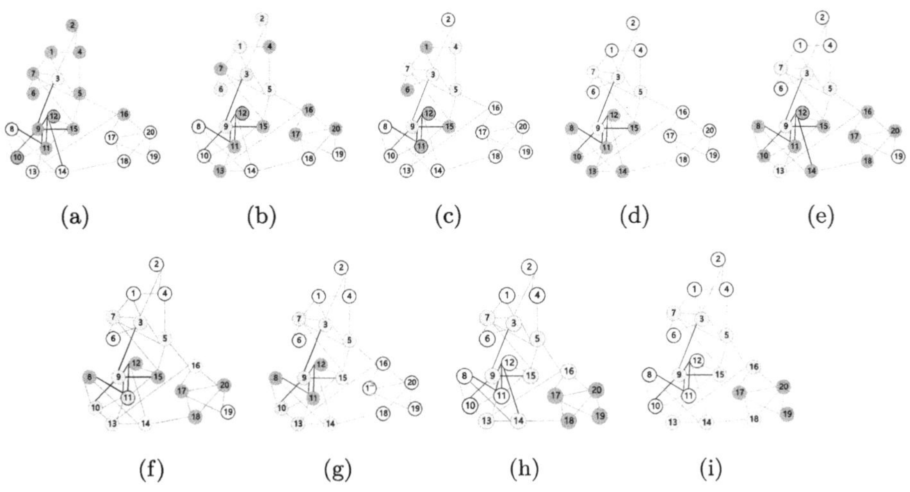

Fig. 3. Examples of graph partition index based on the pruned 2-hop BFS. Yellow nodes denote the roots; blue nodes denote those visited and labeled in this BFS round; red nodes denote those visited and pruned in this round; gray nodes indicate the roots that have already been used. (Color figure online)

5.2   Partition-Based Index for h-Hop Reachability

To accelerate structural constraint verification in CPM, we design a **Partition-Based Index (PBI)** supporting repeated h-hop reachability queries. Given $G = (V, E, L)$, we partition V as $C = \{C_1, \ldots, C_k\}$ via DLPA and precompute cross-partition distances via *bridge nodes*, denoted by V_b. For any nodes s, t, v, the hop distances satisfy the triangle inequalities [3]:

$$|d_G(s, v) - d_G(v, t)| \leq d_G(s, t) \leq d_G(s, v) + d_G(v, t). \tag{3}$$

For each node $u \in V$, the index label $L(u)$ stores pairs $(v, d_G(u, v))$ for $v \in V_b$. The distance between nodes s and t is queried as

$$\text{Query}(s, t, L) = \min\{ d_G(s, v) + d_G(t, v) \mid (v, d_G(s, v)) \in L(s),\ (v, d_G(t, v)) \in L(t) \}. \tag{4}$$

If $L(s)$ and $L(t)$ share no bridge node, then $\text{Query}(s, t; L) = \infty$. The index (Algorithm 2) scans bridge nodes in decreasing degree and, for each root v_b, runs a pruned h-hop BFS. A label $(v_b, d_G(u, v_b))$ is added for u only if it strictly improves the current estimate, i.e., $\mathrm{Query}(v_b, u; L) > d_G(u, v_b)$; otherwise u is pruned without further expansion. This dominance condition ensures label minimality, while the degree order exposes useful paths early and strengthens pruning. Setting $h = \delta$ to match the embedding-diameter constraint, we determine h-hop reachability as follows: if u and v belong to the same partition, $d_G(u, v) \leq h$ by construction; otherwise, the pair is reachable if $(v, d_G(u, v)) \in L(u)$, $(u, d_G(v, u)) \in L(v)$, or $\mathrm{Query}(u, v; L) \leq h$. Under the PBI construction, any true cross-partition pair with $d_G(u, v) \leq h$ always has at least one shared bridge node s, ensuring $\mathrm{Query}(u, v; L) = d_G(u, v)$.

6   Experimental Results

We evaluate CoPM and piCPM on four real and synthetic datasets, reporting top-k patterns and analyzing parameter sensitivity, efficiency, and scalability. Ablation studies further isolate the effects of CMIS pruning, structural and temporal constraints, and the PBI index. Experiments are conducted using Python 3.10 on Ubuntu 24.04.1 with 16 Intel Xeon 2.1 GHz cores and 125 GB RAM. Code and datasets are available at https://anonymous.4open.science/r/CPM-79F0/.

6.1   Experimental Settings

In this section, we describe the datasets, baseline methods, and implementation details used in our experiments.

- **Intrusion Alert Network** (DARPA 2000) is an activity graph containing 94 nodes, 201 edges, and 64 alert labels such as `portsweep` and `smurf`. It is used to detect multi-step intrusions that are both topology-aware and time-sensitive.

Algorithm 2. Partition-Based Index (PBI)

1: **Input:** Activity graph $G = (V, E, L)$, hop limit h
2: **Output:** Index L
3: Identify bridge nodes V_b (degree-sorted); **for** $u \in V$: $L(u) \leftarrow \emptyset$
4: **for all** $v_b \in V_b$ **do**
5:     **for** $u \in V$: $\mathrm{dist}(u) \leftarrow \infty$; $\mathrm{dist}(v_b) \leftarrow 0$; $Q \leftarrow [v_b]$
6:     **while** $Q \neq \emptyset$ **do**
7:         $u \leftarrow \mathrm{dequeue}(Q)$
8:         **if** $\mathrm{Query}(v_b, u, L) > \mathrm{dist}(u)$ **then**
9:             $L(u) \leftarrow L(u) \cup \{(v_b, \mathrm{dist}(u))\}$
10:            **for all** neighbor w of u **with** $\mathrm{dist}(w) = \infty$ **do**
11:                $\mathrm{dist}(w) \leftarrow \mathrm{dist}(u) + 1$; **if** $\mathrm{dist}(w) \leq h$ **then** enqueue w into Q
12:            **end for**
13:        **end if**
14:    **end while**
15: **end for**
16: **return** L

- **Last.fm** is a social music network with 1,892 users, 12,717 friendships, and 17,632 artists. We extract $\langle$timestamp, artist$\rangle$ tuples for the top three favorite artists of each user to capture temporally synchronized preferences among friends.
- **Delicious** is a social bookmarking network with 1,867 users, 7,668 edges, and 53,388 tags. We extract the top three tuples for each user to identify coordinated tagging behaviors across the network.
- **IMDB** is an online movie database where we construct a user co-rating network with 2,112 users, 81,492 edges, and 13,222 tags. We extract the top three tuples for each film to identify users who show temporally coordinated rating interests.

Baselines. We compare **piCPM** with several representative baselines under standard or recommended parameter settings. Apriori [2] treats each node label set as a transaction with a support threshold of $\sigma = 2$. GRAMI [9] mines frequent subgraphs through constraint satisfaction, with minimum support 2 and maximum subgraph size 5. MaNIACS [25] performs sampling-based approximate mining under the MNI measure, using a sampling rate of 0.1 and a confidence level of 0.95. aFP [16] applies propagation-based approximate mining on connected nodes with parameters $\alpha = 1$, $\epsilon = 0.12$, and $d = 2$. CSM-E [24] employs a best-first search guided by frequency and topological proximity, with a depth limit of 5 and a proximity threshold of 0.8. GTAR [19] extends association rules to temporal graphs by applying graph pattern queries within 60 min and a support threshold of 2. CoPM, introduced in Sect. 4.3, is the CPM framework without index optimization, using the same structural constraint δ, temporal window $\mathcal{T}$, and support threshold τ as **piCPM**.

Experimental Details. We set piCPM hyperparameters based on dataset characteristics and domain knowledge. For the Intrusion Alert Network, $\mathcal{T} = 10$ min and $\delta = h = 2$ align with its average path length of about 2.1. For Last.fm, Delicious, and IMDB, $\mathcal{T} = 60$ min and $\delta = h = 3$ match their sparser structures with average path lengths between 3.2 and 3.8. Support thresholds are $\tau = 4$ for Intrusion, $\tau = 50$ for Last.fm and Delicious, and $\tau = 40$ for IMDB, balancing pattern coverage and efficiency. Baselines use standard parameters: Apriori $\sigma = 2$, aFP $\alpha = 1, \epsilon = 0.12, d = 2$, and CoPM adopts the same δ, $\mathcal{T}$, and τ as **piCPM**.

6.2 Effectiveness Evaluation

In this section, we present the top 5 Collaborative Patterns, proximity patterns, and frequent itemsets for real-world datasets using piCPM, aFP, and Apriori.

Table 1 shows that piCPM uncovers meaningful Collaborative Patterns missed by conventional methods. The top pattern `portsweep, nmap, backdoor` (G-score 3.68) reveals a typical backdoor implantation sequence, reconnaissance (`portsweep`), target refinement (`nmap`), and persistent access (`backdoor`), all within 30 min across 2-hop neighbors, reflecting fast-paced intrusions like worm

Table 1. Top-5 Collaborative Patterns, Proximity Patterns, and Frequent Itemsets (Intrusion Alert Network)

Num	Proximity Patterns [16]	Frequent Itemsets [2]	Collaborative Patterns	G-scores
1	portsweep, ipsweep, mscan	ipsweep, mscan	portsweep, nmap,backdoor	**3.68**
2	smurf, ipsweep, mscan	portsweep, ipsweep	portsweep, pod, smurf	**3.29**
3	portsweep, smurf, ipsweep	portsweep, mscan	portsweep, syslogd, ps	**2.95**
4	portsweep, smurf, mscan	portsweep, queso	dosnuke, crashiis, yaga	**2.42**
5	ipsweep, neptune, queso	neptune, queso	portsweep, netcat, **secret**	**2.08**

propagation (e.g., WannaCry) often observed in security datasets (e.g., DARPA 2000). Baselines lose this time-sensitive and topology-aware pattern: aFP, prioritizing only spatial proximity, returns temporally ambiguous patterns (e.g., portsweep, ipsweep, mscan), while Apriori ignores the network context (e.g., ipsweep, mscan). By integrating temporal and structural constraints, piCPM identifies actionable attack chains and significantly outperforms baselines in pattern significance (average G-score 2.88 vs. 1.5 for aFP), indicating clear potential for enhancing real-time threat detection.

Table 2. Evaluation of Discovered Pattern Significance (G-test)

Dataset	aFP	GTAR	GRAMI	Apriori	CSM-E	MaNIACS	piCPM
Intrusion Alert Network	$1.53_{\pm 0.38}$	$2.51_{\pm 0.69}$	$1.82_{\pm 0.51}$	$1.24_{\pm 0.45}$	$1.77_{\pm 0.42}$	$1.53_{\pm 0.50}$	$\mathbf{2.88_{\pm 0.13}}$
Last.fm	$0.95_{\pm 0.11}$	$1.81_{\pm 0.42}$	$1.00_{\pm 0.20}$	$0.78_{\pm 0.07}$	$0.93_{\pm 0.16}$	$0.88_{\pm 0.43}$	$\mathbf{2.32_{\pm 0.31}}$
Delicious	$0.84_{\pm 0.18}$	$0.92_{\pm 0.46}$	$0.93_{\pm 0.06}$	$0.62_{\pm 0.04}$	$0.85_{\pm 0.09}$	$0.72_{\pm 0.36}$	$\mathbf{1.13_{\pm 0.28}}$
IMDB	$0.93_{\pm 0.21}$	$1.21_{\pm 0.44}$	$1.19_{\pm 0.47}$	$0.74_{\pm 0.13}$	$1.07_{\pm 0.48}$	$0.93_{\pm 0.35}$	$\mathbf{1.38_{\pm 0.22}}$

Table 2 evaluates the quality and significance of discovered patterns, and piCPM achieves the highest G-test scores across all datasets, further validating this coherent integration of temporal and structural constraints and confirming statistically significant collaborative activities.

6.3   Parameter Sensitivity

We study how key parameters influence the number of k-itemsets by varying one at a time while holding the others fixed, and Fig. 4 summarizes the effects of the support threshold, temporal window, and diameter.

Support threshold τ. On Last.fm we vary τ from 2 to 20 and observe that the number of k-itemsets decreases monotonically as τ increases, with larger k dropping more rapidly. At moderate τ values, itemsets with $k \geq 4$ become rare, indicating that longer patterns require lower support to appear.

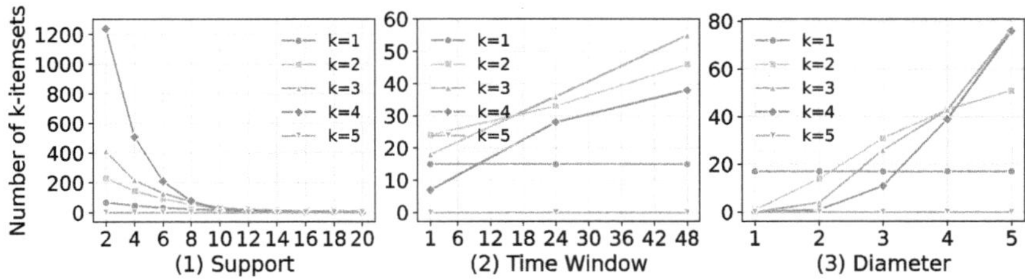

Fig. 4. Influence of support thresholds, temporal windows, and diameters on the number of k-itemsets.

Temporal window $\mathcal{T}$. We vary $\mathcal{T}$ from 1 to 48 h and find that counts increase for $k = 2$–4. Larger temporal windows admit more temporally correlated co-occurrences and therefore expand the candidate set for intermediate k.

Diameter δ. We vary δ from 1 to 5 and observe that counts rise for $k = 2$–4, while $k = 1$ remains roughly constant and $k = 5$ stays at zero. The growth under changing δ is stronger than under varying $\mathcal{T}$, indicating that diameter has a greater influence on pattern discovery on Last.fm.

6.4   Efficiency and Scalability

Table 3 reports running times on four real-world datasets under the settings in Sect. 6.1. Experimental results confirm that piCPM achieves substantial speedups compared to baseline methods, notably demonstrating orders-of-magnitude improvement over the unindexed CoPM approach. This efficiency gains arise from two primary factors: effective anti-monotone pruning strategies that significantly reduce the candidate search space, and the Partition-Based Index (PBI) which significantly accelerates the frequent h-hop queries essential for structural constraint verification.

Table 3. Average Running Time (minutes) of Graph Pattern Mining Algorithms on Four Real-World Datasets

Algorithm	Intrusion Alert Network	Last.fm	Delicious	IMDB
piCPM	$0.0012_{\pm 0.00006}$	$10.96_{\pm 0.55}$	$42.98_{\pm 2.15}$	$0.43_{\pm 0.022}$
aFP [16]	$0.81_{\pm 0.04}$	$134.96_{\pm 6.75}$	$387.62_{\pm 19.38}$	$69.03_{\pm 3.45}$
CSM-E [24]	$1.48_{\pm 0.074}$	$380.46_{\pm 19.02}$	$225.53_{\pm 11.28}$	$302.50_{\pm 15.13}$
GRAMI [9]	$1.68_{\pm 0.084}$	$105.83_{\pm 5.29}$	$63.92_{\pm 3.20}$	$679.17_{\pm 33.96}$
GTAR [19]	$1.47_{\pm 0.073}$	$43.4_{\pm 2.17}$	$387.62_{\pm 19.38}$	$69.03_{\pm 3.45}$
MaNIACS [25]	$0.096_{\pm 0.0048}$	$98.25_{\pm 4.91}$	$13.42_{\pm 0.671}$	$142.08_{\pm 7.10}$
CoPM	$4.80_{\pm 0.24}$	$892.21_{\pm 44.61}$	$1690.65_{\pm 84.53}$	$278.04_{\pm 13.90}$

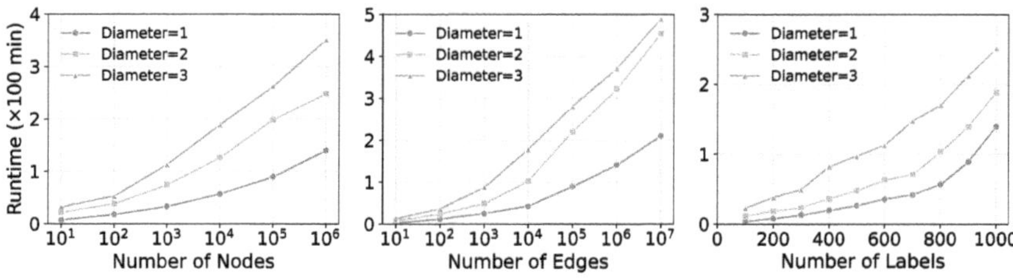

Fig. 5. Influence of varying numbers of nodes, edges, and labels on the **running time** of piCPM on synthetic data.

Scalability tests on synthetic data (Figs. 5–6) show that piCPM scales reasonably as nodes and edges increase, with runtime and memory growing near-linearly or slightly super-linearly, and performance being more sensitive to the number of unique labels. Overall, piCPM provides an efficient and scalable solution for mining Collaborative Patterns in large activity graphs.

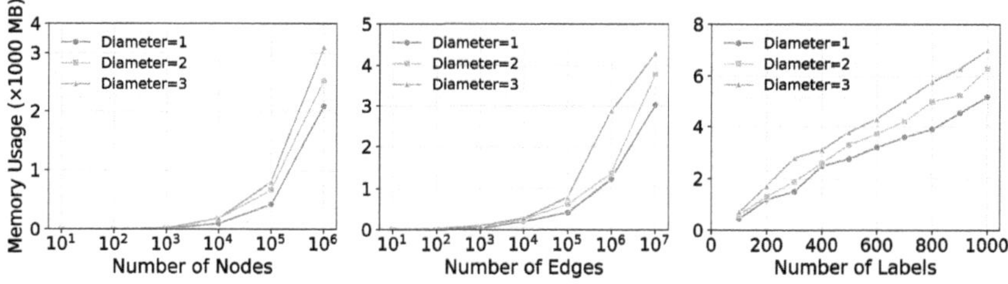

Fig. 6. Influence of varying numbers of nodes, edges, and labels on the **memory consumption** of piCPM on synthetic data.

6.5 Ablation Study

We conduct ablations to quantify the contribution of each component in **piCPM** by comparing the complete framework against four variants. We report the average G-test score and the runtime under the same hyperparameters as in Sect. 6.1.

- **piCPM w/o CMIS**: replaces CMIS with MIS, the traditional support measure in frequent subgraph mining, to assess the contribution of CMIS.
- **piCPM w/o SC**: remove the structural constraint to evaluate the contribution of structural proximity.
- **piCPM w/o TC**: remove the time-window constraint to evaluate the contribution of temporal proximity.
- **piCPM w/o PBI**: remove the partition-based index to evaluate the contribution of PBI.

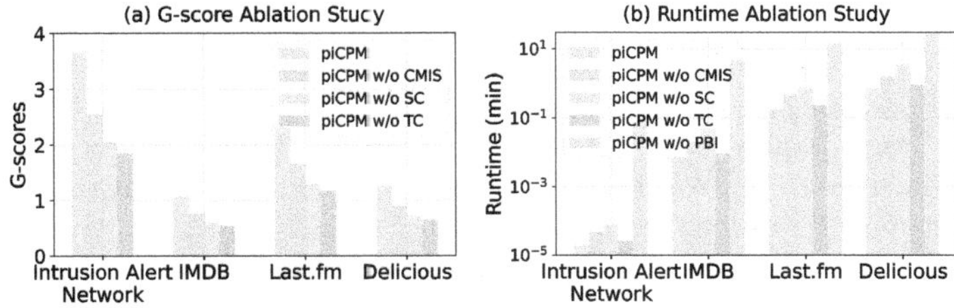

Fig. 7. Ablation study of **piCPM**, comparing the complete framework with ablated variants in terms of G-test significance and runtime.

As shown in Fig. 7, removing any component consistently lowers G-test significance and tends to increase runtime, indicating that the components of **piCPM** act synergistically. Replacing CMIS with MIS reduces the average G-test, suggesting overcounting and weaker pruning under the joint temporal and structural criterion. Removing either the structural constraint or the temporal window leads to larger declines, with the most pronounced drop observed on the Intrusion Alert Network without the temporal constraint, where collaboration arises from tightly synchronized events. On the social datasets, removing the structural constraint still causes a noticeable decrease in significance, underscoring the importance of compact neighborhoods in capturing collaborative behavior.

7   Conclusion

We introduced Collaborative Pattern Mining (CPM) for activity graphs to discover label sets that co-occur over nodes and activities that are proximate in structure and time. We formalized embeddings and occurrences, defined structural and temporal proximity, proved their anti-monotone properties, and introduced the CMIS support measure. We presented CoPM and piCPM, which uses the partition-based index (PBI) to accelerate h-hop reachability checks and reduce unnecessary embeddings. Experiments on four real-world and synthetic graphs show that piCPM efficiently finds statistically significant patterns that integrate topology, attributes, and timestamps. Future work will extend CPM to dynamic graphs and to richer patterns with higher order temporal and structural dependencies.

Acknowledgments. This work was supported in part by the National Natural Science Foundation of China under Grants 62425605, 62133012, and 62303366, in part by the Key Research and Development Program of Shaanxi under Grants 2022ZDLGY01-10, 2024CY2-GJHX-15, and 2025CY-YBXM-041, and in part by the Xidian University Specially Funded Project for Interdisciplinary Exploration under Grant TZJHF202506.

References

1. Agarwal, P.K., Hu, X., Sintos, S., Yang, J.: On reporting durable patterns in temporal proximity graphs. Proc. ACM Manag. Data **2**(2), 1–26 (2024)
2. Agrawal, R., Imieliński, T., Swami, A.: Mining association rules between sets of items in large databases. In: SIGMOD, pp. 69–84 (1993)
3. Akiba, T., Iwata, Y., Yoshida, Y.: Fast exact shortest-path distance queries on large networks by pruned landmark labeling. In: SIGMOD, pp. 349–360 (2013)
4. Aslay, Ç., Nasir, M.A.U., De Francisci Morales, G., Gionis, A.: Mining frequent patterns in evolving graphs. In: CIKM, pp. 923–932 (2018)
5. Bringmann, B., Nijssen, S.: What is frequent in a single graph? In: Washio, T., Suzuki, E., Ting, K.M., Inokuchi, A. (eds.) PAKDD 2008. LNCS (LNAI), vol. 5012, pp. 858–863. Springer, Heidelberg (2008). https://doi.org/10.1007/978-3-540-68125-0_84
6. Cai, Y., Zheng, W.: ESTI: efficient k-hop reachability querying over large general directed graphs. In: Jensen, C.S., et al. (eds.) DASFAA 2021. LNCS, vol. 12680, pp. 71–89. Springer, Cham (2021). https://doi.org/10.1007/978-3-030-73216-5_6
7. Cule, B., Goethals, B., Hendrickx, T.: Mining interesting itemsets in graph datasets. In: Pei, J., Tseng, V.S., Cao, L., Motoda, H., Xu, G. (eds.) PAKDD 2013. LNCS (LNAI), vol. 7818, pp. 237–248. Springer, Heidelberg (2013). https://doi.org/10.1007/978-3-642-37453-1_20
8. Desmier, E., Plantevit, M., Robardet, Boulicaut, J.F.: Cohesive co-evolution patterns in dynamic attributed graphs. In: International Conference on Discovery Science, pp. 110–124 (2012)
9. Elseidy, M., Abdelhamid, E., Skiadopoulos, S., Kalnis, P.: GRAMI: frequent subgraph and pattern mining in a single large graph. In: VLDB, pp. 517–528 (2014)
10. Fiedler, M., Borgelt, C.: Support computation for mining frequent subgraphs in a single graph. In: MLG, pp. 25–30 (2007)
11. Han, J., Wen, J.R.: Mining frequent neighborhood patterns in a large labeled graph. In: CIKM, pp. 259–268 (2013)
12. Hendrickx, T., Cule, B., Meysman, P., Naulaerts, S., Laukens, K., Goethals, B.: Mining association rules in graphs based on frequent cohesive itemsets. In: Advances in Knowledge Discovery and Data Mining: 19th Pacific-Asia Conference, PAKDD. pp. 637–648 (2015)
13. Ho, N., Pedersen, T.B., Vu, M., et al.: Efficient and distributed temporal pattern mining. In: 2021 IEEE International Conference on Big Data (Big Data), pp. 335–343. IEEE (2021)
14. Jazayeri, A., Yang, C.C.: Frequent subgraph mining algorithms in static and temporal graph-transaction settings: a survey. IEEE Trans. Big Data **8**(6), 1443–1462 (2021)
15. Jin, R., Ruan, N., Xiang, Y., Lee, V.: A highway-centric labeling approach for answering distance queries on large sparse graphs. In: SIGMOD, pp. 445–456 (2012)
16. Khan, A., Yan, X., Wu, K.L.: Towards proximity pattern mining in large graphs. In: SIGMOD. pp. 867–878 (2010)
17. Kuramochi, M., Karypis, G.: Finding frequent patterns in a large sparse graph. Data Min. Knowl. Disc. **11**(3), 243–271 (2005)
18. Meng, J., Tu, Y.c.: Flexible and feasible support measures for mining frequent patterns in large labeled graphs. In: SIGMOD, pp. 391–402 (2017)
19. Namaki, M.H., Wu, Y., Song, Q., Lin, P., Ge, T.: Discovering graph temporal association rules. In: Proceedings of the 2017 ACM on Conference on Information and Knowledge Management, pp. 1697–1706 (2017)

20. Nasir, A., Aslay, C., Morales, G.D.F., Riondato, M.: Tiptap: approximate mining of frequent k-subgraph patterns in evolving graphs. ACM Trans. Knowl. Discov. Data **15**(3), 1–35 (2021)
21. Ng, R.T., Lakshmanan, L.V., Han, J., Pang, A.: Mining persistent activity in continually evolving networks. In: Proceedings of the 26th ACM SIGKDD International Conference on Knowledge Discovery & Data Mining, pp. 934–944 (2020)
22. Peng, Y., Lin, X., Zhang, Y., Zhang, W., Qin, L.: Answering reachability and k-reach queries on large graphs with label constraints. VLDB J. 1–27 (2022)
23. Prado, A., Plantevit, M., Robardet, C., Boulicaut, J.-F.: Mining graph topological patterns: finding covariations among vertex descriptors. IEEE TKDE **25**(9), 2090–2104 (2013)
24. Prateek, A., Khan, A., Goyal, A., Ranu, S.: Mining top-k pairs of correlated subgraphs in a large network. In: VLDB, pp. 1511–1524 (2020)
25. Preti, G., De Francisci Morales, G., Riondato, M.: Maniacs: approximate mining of frequent subgraph patterns through sampling. In: SIGKDD, pp. 1348–1358 (2021)
26. Ryu, T., et al.: Scalable and efficient approach for high temporal fuzzy utility pattern mining. IEEE Trans. Cybern. **53**(12), 7672–7685 (2022)
27. Silva, A., Meira Jr, W., Zaki, M.J.: Mining attribute-structure correlated patterns in large attributed graphs. In: VLDB, p. 466–477 (2012)
28. Tian, S., et al.: GraphRPM: risk pattern mining on industrial large attributed graphs. In: Bifet, A., Krilavičius, T., Miliou, I., Nowaczyk, S. (eds.) ECML PKDD 2024. LNCS, vol. 14950, pp. 133–149. Springer, Cham (2024). https://doi.org/10.1007/978-3-031-70381-2_9
29. Valstar, L.D., Fletcher, G.H., Yoshida, Y.: Landmark indexing for evaluation of label-constrained reachability queries. In: SIGMOD, pp. 345–358 (2017)
30. Yano, Y., Akiba, T., Iwata, Y., Yoshida, Y.: Fast and scalable reachability queries on graphs by pruned labeling with landmarks and paths. In: CIKM, pp. 1601–1606 (2013)

Rumor Prevention: Approach of Minimizing the Competitive Influence of Unknown Rumors in Multi-layer Social Networks

Fei Gao[1], Qiang He[1], Xingwei Wang[1(✉)], Lin Qiu[1], and Min Huang[2]

[1] The College of Computer Science and Engineering, Northeastern University, Shenyang 110819, China
wangxw@mail.neu.edu.cn
[2] The College of Information Science and Engineering, Northeastern University, Shenyang 110819, China

Abstract. In the current development of social networks, technological advancements have provided users various platforms for acquiring and disseminating information. During the process of information propagation in social networks, users can participate in multiple social networks simultaneously. Existing rumor control methods typically consider scenarios where the rumors are already known, and releasing official refutation information results in a spread of competitive information. Addressing these phenomena, we first define an Unknown Rumor Competitive Influence Minimization (UR-CIM) problem, which aims to minimize the influence of rumor propagation in a multi-layer social network environment. Subsequently, we propose an Unknown Rumor Competitive Propagation Model (URCPM), which incorporates evolutionary game theory into the linear threshold model and designs a payoff matrix to calculate the user's benefits. Furthermore, to solve the UR-CIM problem, we design a <u>G</u>raph <u>C</u>onvolutional <u>N</u>etwork method based on a <u>M</u>ulti-layer network and <u>M</u>ulti-<u>I</u>nformation fusion (denoted as MMI-GCN) to select protective nodes, thereby achieving rumor control effectively. Finally, we conduct experiments on three real datasets, and the results demonstrate that MMI-GCN outperforms baseline algorithms in terms of minimizing the final number of individuals influenced by rumors.

Keywords: Unknown Rumor · Social networks · Evolutionary Game · Influence minimization · Graph Convolutional Network

1 Introduction

Social networks have become essential for people to obtain information, express themselves, promote marketing, etc. With the continuous innovation of internet technology, the way information is disseminated has gradually changed, and many dangerous elements are fabricating online rumors. Internet rumors rely on

H. Jung et al. (Eds.): DASFAA 2026, LNCS 16536, pp. 300–316, 2026.
https://doi.org/10.1007/978-981-92-0366-6_19

social platforms to generate many new characteristics, which are different from traditional rumors, such as anonymity, group communication, high communication efficiency, and short communication life cycle. Mainly after an emergency occurs, some people will fabricate that certain disasters will continue to occur or exaggerate harmful information that has already happened, causing public panic and disrupting social order. For example, some netizens spread false information about an earthquake in Jishishan County, Gansu Province, and a strong aftershock magnitude of 8.2. This attracted a large amount of attention and forwarding from other netizens caused panic among citizens, and seriously disrupting social order. Therefore, the issue of rumor control has attracted widespread attention in recent years.

The issue of social network rumor control aims to prevent, limit, or reduce the spread of rumors, thereby minimizing the number of individuals affected by rumors and promoting societal harmony and stability. However, current research predominantly focuses on known rumor propagation phenomena, with limited investigation into scenarios involving unknown rumors in social networks. Furthermore, with the continuous advancement of technology, users have access to various platforms for acquiring and disseminating information [10]. Users may simultaneously participate in multiple social networks, and rumor spreaders can propagate different rumors across various platforms. In such circumstances, failure to promptly curb the dissemination of network rumors could lead to heightened panic following emergency events. Therefore, there is an urgent need for research to detect [4, 11] and control [2, 5, 17] malicious network rumors in social networks rapidly.

In the current context of rumor control, the problem can be divided into propagation models and control algorithms. Commonly utilized propagation models include the linear threshold model and independent cascade model, alongside leveraging the similarity between rumor propagation and infectious disease transmission, using epidemic models to simulate rumor propagation. Research has recently explored the prediction of rumor propagation processes using game theory. Three main strategies exist for rumor control algorithms. The first involves blocking influential nodes in the social network to prevent the spread of rumors. The second strategy entails removing critical edges in the network, preventing the dissemination of information. The third approach focuses on selecting highly influential nodes to propagate positive information, thereby achieving the goal of rumor control.

Although many solutions to the rumor control problem exist, there are still two challenges: (1) Different users can join multiple social platforms simultaneously to form a multi-layer social network. At the same time, there may be many kinds of rumors on real social networks. After these rumors are generated, official information will be released to refute them. In this case, it is necessary to consider how to establish a process model of rumor information dissemination in a multi-layer social network where competing information exists. (2) Most existing research is based on known rumor control, but the generation of online rumors is usually unknown in the real world. However, relevant departments such as the government must be prepared in real-time to avoid unknown rumor information effectively. Therefore, our study considers the case of unknown rumor propagation in multi-layer social networks.

To address these two challenges, we propose the unknown rumor competition influence minimization problem in our research, which considers how to control rumors in the presence of competition among unknown rumors. Additionally, given the dispersed nature of information across multiple platforms in multi-layer social networks and the complex user interactions, we propose a method based on multi-information fusion using graph convolutional networks. This method enhances the comprehensive view of multi-layer social networks, effectively represents information propagation among users, and quickly selects seed nodes for rumor control in a competitive multi-layer social network environment. To sum up, the contributions of this paper are as follows:

1) We propose the unknown rumor competitive propagation model in multi-layer social networks, combined with evolutionary game theory. We design the game matrix and the benefits of both competing parties to simulate the spread of rumors in social networks.
2) We investigate the problem of minimizing the influence of unknown rumor competition and propose a multi-layer social network multi-information fusion graph convolutional network algorithm to solve this problem. The algorithm selects highly influential nodes as protection nodes for unknown rumors while spreading official information.
3) We verify the effectiveness of our method through three real social network datasets. The experimental results show that our proposed algorithm can effectively solve the impact of rumor spread when ultimately controlling it.

2 Preliminaries

2.1 Multi-layer Social Network Formulation

Assume a social network G_n consisting of n social platforms, each facilitating internal information propagation and enabling interaction between platforms. Moreover, nodes can join multiple social platforms simultaneously, contributing to intricate social relationships. Users' opinions can influence their responses to unknown rumor information. Consider a multi-layer social network $G_n = G_1, G_2, G_3, ..., G_n$, each layer $G_1 = (V, E)$, where V denotes the set of user nodes and E represents the associations between users. For example, we have a three-layer social network representing the WeChat, Sina Weibo, and Facebook networks, respectively. Users can influence one another within a multi-layer social network, and information can flow between layers.

2.2 The Factors Influencing Rumor Propagation

In analyzing the process of rumor propagation, both the users and the rumors themselves exert significant influence on the propagation of rumors. Therefore, definitions are provided based on the users and the rumor information.

Definition 1 (User Knowledge). User Knowledge (UK) is one of the critical factors affecting user attitudes. Users' awareness and understanding of rumors directly affect their attitude toward the information and the process of spreading rumors on social networks. The higher the user knowledge, the lower the influence of rumors. Its definition is as follows:

$$UK = \alpha * Edu + \beta * Exp \tag{1}$$

where α and β represent the influence of user education level and personal experience on user attitude, respectively. Edu represents education level, and Exp represents personal experience. α and β are parameters between 0 and 1.

Definition 2 (Real-time influence of information). The real-time influence of information (RII) is an essential factor affecting users' attitudes towards rumors. The closer the publisher's information is to current events, the higher the user's interest in the current information. The user may be influenced by the rumor to believe and continue to spread the rumor.

$$RII = \eta * Num(per) + \mu * Num(fans) \tag{2}$$

where $Num(per)$ represents the popularity of the information, $Num(fans)$ represents the influence of the information disseminator. η and $\mu \in [0,1]$ are parameters.

Definition 3 (User preference). User preference (UP) refers to the tendency of users to choose information in social networks. Users usually prefer a specific type of information, or they are more inclined to choose a specific type of information when it comes to information that affects each other. Therefore, user preference will have a more significant impact on the spread of rumors.

3 Model Formulation

In the interaction between rumor spreaders and official publishers, the information transmitted mutually influences each other. Upon receiving the information, rumor spreaders aim to maximize the dissemination of their message. In contrast, official publishers aim to maximize or minimize the dissemination of their message by rumor spreaders. There is a competitive-cooperative relationship between the two sides in the information propagation process, aiming to achieve their goals. Therefore, we will adopt the concept of evolutionary game theory to explain this process.

The propagation process is as follows: At time $t = 0$, unknown rumors begin to spread. When $t > 0$, the detection of the rumor's appearance occurs, but the unknown nature of the rumor's appearance remains. Therefore, at this time, the official needs to release information contrary to the rumor spreaders so that other nodes are no longer affected by the rumor spreaders. It is assumed that node u will activate it with probability $p(u, v)$ neighbor nodes; nodes activated by officials are no longer affected by rumors. Finally, the information dissemination stops when no new nodes are activated.

In disseminating unknown rumor information, it is necessary to select a fixed number of official publishers to mitigate the impact of rumor propagation. Thus, we assume two types of information propagators exist in the multi-layer social network, while the rest are normal social network users. Subsequently, an evolutionary game process between the information propagators and users over time is established based on the game process. The strategies of the users are divided into "believing in rumors" and "believing in truth", with different activation states for each strategy. Specifically, the "believing in rumors" strategy is activated as the rumor node set, while the "believing in truth" strategy is activated as the official truth node set. The model framework is shown in Fig. 1.

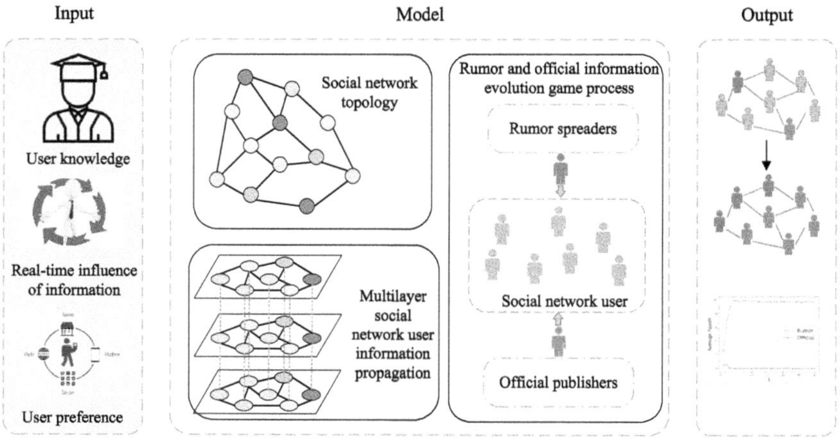

Fig. 1. A framework for minimizing the competitive influence of unknown rumors in multi-layer social networks.

Evolutionary Game Payoff Matrix. In rumor propagation, we consider three influencing factors: user knowledge, real-time influence of information, and user preference. The user payoff function incorporates coefficients corresponding to these factors, representing their respective impacts on rumor propagation. The formula is expressed as follows:

$$P = -a * UK + b * RII + c * UP \tag{3}$$

where P is user payoff function, $a + b + c = 1, 0 < a, b, c < 1$. In the process of rumor propagation, the user benefits generated by different communicators will change according to the chosen strategy. In practice, the following two game situations will occur: 1) Rumor Spreader (RS): Users believe the rumor spreader spreads the rumor, and the rumor spreader will receive corresponding rewards.

The payoff is $P + c_1 - p_1$ and c_1 is the reward for the spreader if the user believes the rumor, but the spreader will also be punished accordingly with p_1. 2) Official Publisher (OP): Users believe the truth information from official sources. To control the spread of rumors, users will receive a corresponding reward for spreading the true information. Therefore, the payoff for official information is $P + c_2$, where c_2 represents the additional reward received for disseminating the truth.

For users, receiving information from official sources yields a payoff of r_1, while receiving information from rumor spreaders yields a payoff of $r_2 - p_2$, where r_2 represents the reward for spreading rumor information, and p_2 represents the penalty for spreading rumors. It is impossible for users to choose to believe both the rumor and official information simultaneously. Thus, such cases do not exist. The payoff matrix is shown in Table 1 below.

Table 1. The game payoff matrix between information spreaders and users.

		Regular users	
		Trust Information (T)	Do not believe information (F)
Information disseminator	Rumor Spreader (R)	$P + c_1 - p_1, r_2 - p_2$	$P - p_1, 0$
	Official Publisher (O)	$P - p_1, 0$	$P, 0$

Game Analysis Between Rumor Spreaders and Official Publishers. Only the rumor information exists in the social network at the initial stage of the unknown rumor propagation. Therefore, the probability that users believe the information disseminated by rumor spreaders is x, and the probability that they do not believe it is $1 - x$. When a rumor propagation is detected, although it is unknown information, the information disseminated by the official publisher also carries a particular influence. Therefore, the probability that users choose to believe the information disseminated by official sources is y, and the probability that they do not believe it is $1 - y$.

(1) The payoff of rumor spreaders in the social network, whether users believe or do not believe the rumor, can be represented as:

$$W_{RS} = x * (P + c_1 + p_1) + (1 - x) * (P - p_1) \qquad (4)$$

(2) The payoff of official publishers in the social network, whether users believe or do not believe the official information, can be represented as:

$$W_{OP} = y * (P + c_2) + (1 - y) * P \qquad (5)$$

The Problem of Unknown Rumor Competitive Influence Minimization (UR-CIM Problem). Given a set of multi-layer social networks $G = (V, E, L)$, select the initial number of rumor spreaders as k. Under the influence of unknown rumor spread, and under the budget k, select effective R official publishers OP, which maximizes the payoff of the official publisher so that the activated nodes choose to believe the official news, thereby minimizing the impact of rumors. Therefore, minimizing the competitive impact of unknown rumors can be formulated as:

$$F(s) = \max \sum_{i=1}^{V} W_{OP}(s) \quad \text{or} \quad \min \sum_{i=1}^{V} W_{RS}(s), \quad s < k \tag{6}$$

4 Our Graph Convolutional Network-Based Algorithm

In this part, we design a multi-layer social network multi-information fusion graph convolutional network algorithm (MMI-GCN). As unknown rumor information spreads, we first select protectors, and these nodes will not be affected by rumors. Therefore, we design an algorithm that acts as a protector before rumors spread. When rumors are detected, the node acting as protector will act as the official publisher to spread official information. In this part, we use this algorithm to select more effective nodes as protectors to achieve more official publishers in the process of evolutionary game information transmission to achieve the purpose of rumor control.

Related Definitions and Descriptions. Because users join multi-layer social networks and can spread more information, we propose a node selection algorithm for MMI-GCN. This algorithm considers the importance of the user from the aspects of the user's adjacent neighbors, the relationship between different layers in the multi-layer network, and the node influence. This information is then fused through the attention mechanism to efficiently select protectors in the social network to achieve rumor control. Figure 2 is the node selection process of the graph convolution network for multi-layer social network multi-information fusion.

From Fig. 2, we can see that in addition to the traditional convolution process on the adjacency matrix A, the MMI-GCN model also convolves two other pieces of information. Among them, ID uses the centrality of the number of intermediaries as the metric of node influence to build the node influence matrix space and uses the relationship information between different social networks to construct a multi-layer relationship matrix space MD. After obtaining the MD and ID, two-layer graph convolutions are performed in three matrix spaces. $MGCN$, GCN, and $IGCN$ in Fig. 2 all represent two-layer graph convolutions with built-in regularization, which is the result of node selection based on the relationship information between different layers of the social network (different levels), user's neighbors (topological structure) and node influence (node characteristics).

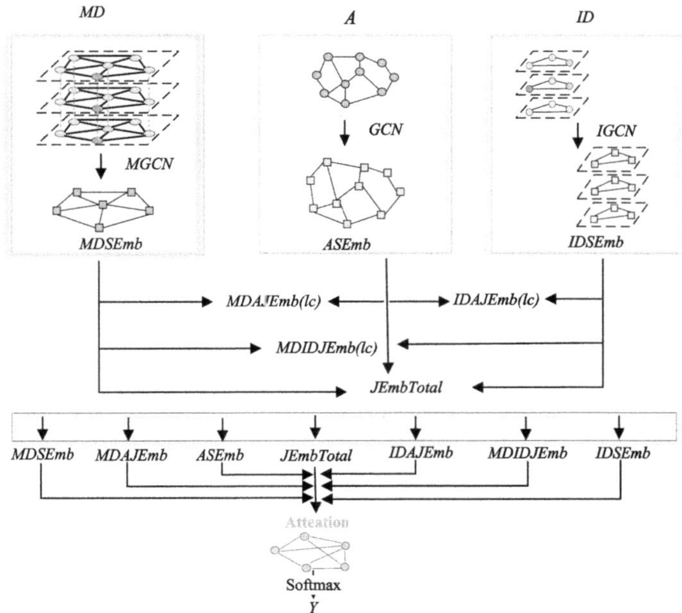

Fig. 2. A node selection framework for graph convolutional networks with multi-layer social network multi-information fusions.

Construct Matrix Space. To improve the node selection efficiency of unknown rumors control in social networks, we construct a corresponding relationship matrix according to the user's neighbors, the relationship information between different layers of social networks, and the node influence. The convolution operation in the constructed relational matrix space can better integrate the influence of multiple information on node selection.

MD is the relation matrix space of different layers. Since users can add multi-layer social networks, which significantly influence the propagation of unknown rumors, a multi-layer relationship information matrix is constructed according to users' social relations at different levels to represent the relations between different levels. If there is a connection relationship between nodes, the corresponding position in the matrix is set to 1; otherwise, it is 0. Finally, the relationship information between different levels is superimposed, and the integrated unified matrix MD space is obtained.

ID is the node influence matrix space. Firstly, the betweenness centrality is used to measure the influence of nodes. For multi-layer social networks, the influence of nodes needs to be calculated separately for each level. Then, the influence of nodes is constructed in the form of a matrix. The rows and columns of the matrix represent nodes in the network, and the elements in the matrix represent the influence relationship between nodes. At the same time, considering multi-layer social networks, we need to integrate the influence of different levels into a unified influence matrix, and the final spatial ID of the influence matrix of nodes is obtained.

Get Spatial Embedding. The usual two-layer graph convolution is performed for feature extraction for each spatial matrix. Before convolution, the space matrix is first regularized, such as the adjacency matrix $\hat{A} = D^{-1}\tilde{A}$, where $\hat{A} = A + I_n, \tilde{D}_{ij} = \sum_i \tilde{A}_{ij}$. In the same way, the different layer relationship information matrix $\hat{MD}$ and the node influence matrix $\hat{ID}$ after regularization preprocessing can be obtained. Then, we perform two convolution layers in each of the three matrix spaces after preprocessing with regularization.

The graph convolution is carried out in the relation information matrix space of different layers, the relation information between nodes of various layers is extracted and fused, and the embedded $K_{MD}(MDSEmb)$ after convolution of the relation matrix space is obtained, where W_{MD} is the weight matrix for learning in the relation matrix space. The calculation formula of K_{MD} is shown as follows:

$$K_{MD} = \hat{MD}\text{RELU}(\hat{MD}TW_{MD}^{(0)})W_{MD}^{(1)} \tag{7}$$

The graph convolution is carried out in the adjacency matrix space, node information of the neighbors is extracted and fused, and the embedded $K_A(ASEmb)$ after convolution of the adjacency matrix space is obtained, where W_A is the weight matrix for learning in the adjacency matrix space. The calculation formula of K_A is as follows:

$$K_A = \hat{A}\text{RELU}(\hat{A}TW_A^{(0)})W_A^{(1)} \tag{8}$$

The graph convolution is carried out in the node influence matrix space, the node influence information of each node is extracted and fused, and the embedded $K_{ID}(IDSEmb)$ after the convolution of the node influence matrix space is obtained, where W_{ID} is the weight matrix for learning in the node influence matrix space. The calculation formula of K_{ID} is shown as follows:

$$K_{ID} = \hat{ID}\text{RELU}(\hat{ID}TW_{ID}^{(0)})W_{ID}^{(1)} \tag{9}$$

Fusion Spatial Embedding. In a multi-layer social network, three kinds of information have different degrees of influence on node selection. Sometimes, two kinds of information may impact node selection, and sometimes, three kinds of information will also impact node selection. Therefore, given the embedding of three kinds of spatial convolution in relation matrix space, adjacency matrix space, and node influence matrix space, the different convolution information of three kinds of information fusion is obtained by adding, and the connected embedding is obtained.

We utilize an attention mechanism to integrate information from three different forms of embeddings to obtain the final embedding representation and determine the importance of each node. During this process, we concatenate all embedding information along the second dimension to obtain a multidimensional representation reflecting different perspectives. In our algorithm, we employ a two-layer fully connected neural network to learn the attention mechanism, using the *softmax* function to learn the weight distribution of each node across different embedding perspectives. The three combinations of combined three-space embeddings are shown in Eq. 10, and the three single-space embeddings are shown in Eq. 11:

$$K_{MDA} = \frac{K_{MD} + K_A}{2}$$
$$K_{IDA} = \frac{K_{ID} + K_A}{2} \tag{10}$$
$$K_{MDID} = \frac{K_{MDID} + K_A}{2}$$

$$K_{total} = \frac{K_{MD} + K_A + K_{ID}}{3} \tag{11}$$

To integrate spatial embedding, the formula is as follows:

$$\begin{aligned}(\alpha_{MD}, \alpha_A, \alpha_{ID}, \alpha_{MDA}, \alpha_{IDA}, \alpha_{MDID}, \alpha_{total}) \\ = att(K_{MD}, K_A, K_{ID}, K_{MDA}, K_{IDA}, K_{MDID}, K_{total})\end{aligned} \tag{12}$$

The embeddings are collectively incorporated into an attention mechanism, wherein the importance of each embedding for every node is determined. This is achieved through a vector representing the significance of each embedding. These importance-representing vectors integrate various embeddings, yielding the ultimate embedding. The formulation is shown as follows:

$$K = \sum \alpha_i \cdot K_j \tag{13}$$

Subsequently, for node selection, we must fully connect the final embed K through a layer activated by the *softmax* function. Herein, W denotes the weight matrix in the fully connected layer to be learned, b represents the bias value to be realized within the connection, and $\hat{Y}$ signifies the predicted category of the final output. The formulation is as follows:

$$\hat{Y} = softmax(W \cdot K + b) \tag{14}$$

Unknown Rumor Competition Minimization Algorithm. In this part, we aim to control rumors by maximizing the payoff of official publishers. Our algorithm is as follows: we begin by selecting protector nodes in a multi-layer social network. When a rumor is detected, these nodes spread the officially released truth. At this point, both rumor and official information compete within the network. To address the problem of minimizing the influence of unknown competing rumors, we select protector nodes using the MMI-GCN algorithm. These nodes then disseminate official information, thereby maximizing official publishers' utility and minimizing rumors' influence.

5 Experiment

In this part, we verify the unknown rumor competition game model and MMI-GCN method on three real data sets. All our experiments are carried out on the 12th Gen Intel(R) Core(TM) i7-12700 2.10 GHz processor on PyCharm Community Edition 2023.2.1 environment with 16GB RAM (Table 2).

Table 2. Statistics of three social network datasets.

Network	Nodes	Edges	Average degree	Maximum degree
fb-pages-food	620	2K	6	132
sb-Wiki-Vote	889	2.9K	6	102
email-univ	1133	5.5K	9	71

5.1 Experiment Setups

Dataset. We use three real social network datasets, which can be downloaded in http://networkrepository.com/soc.php. In these three datasets, "fb-pages-food" comprises page data collected from Facebook, representing relationships between different users regarding pages, with 620 nodes and 2.1K edges. "sb-Wiki-Vote" consists of data on Wikipedia who-votes-on-whom network, with a total of 889 nodes and 2.9K edges, where the edge from node i to node j signifies that user i voted for user j. "email-univ" denotes a social data network with 1133 nodes and 5.5K edges.

Experimental Setting. In the experiment, we set the probability of believing rumors and believing official information to be 0.3. To facilitate competitive game strategies, we choose the proportion of initial rumor spreaders not to exceed 30%. Additionally, we set the reward for the rumor spreader to be 15 if the user believes the rumor, the penalty for believing the rumor is 10, and the additional reward for spreading the official information is 20. At the same time, other parameters $r_1 = 50, r_2 = 40, p_2 = 15, c_2 = 15$.

Evaluation Metrics. We consider the minimization of rumor influence. Therefore, we examine the quantities of rumor and official information nodes, ultimately comparing the quantities of these two types of nodes. Additionally, we compare the proportions of rumor-influenced nodes and official information nodes when introducing a protector strategy under the influence of unknown rumors.

Comparison Algorithm. We compare the MMI-GCN algorithm with four algorithms with typical lines:

- The classical greedy algorithm (CGA) [6] is utilized to select a predetermined number of nodes for blocking. This algorithm serves as a baseline for illustrating the effects of the proposed strategy.

- The blocking algorithm (DRIMUX) [12] is based on a strategy that employs survival theory to identify nodes most susceptible to infection. In this approach, nodes can remain blocked indefinitely.
- The graph convolutional network algorithm (GCN) [7] obtains nodes with significance in the social network by training processes that involve acquiring node features and relationships between edges.
- The directed graph convolutional network algorithm (DGCN) [20] is based on the confidence-based opinion adoption model, selecting k most influential positive cascade nodes to suppress rumor propagation.

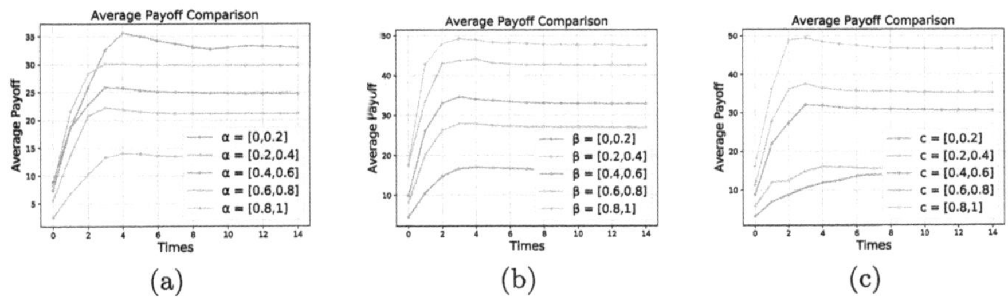

(a)   (b)   (c)

Fig. 3. Analysis of influencing factors.

5.2 Experimental Result

Propagation Model. Our experiments first analyze the rumor competition game model across three datasets. We select initial rumor spreaders not exceeding 30% of the total nodes and examine the influence process of rumor and refutation nodes during the competition game. We calculated the user benefit function P in the experiment according to Eq. 3, setting $a = 0.5, b = 0.3, c = 0.6$.

Next, we analyze the influence of UK, RII, and UP in competitive games. We calculated the user influence function according to Eq. 3 and set $UK = 100$, $RII = 200$, and $UP = 200$ in the fb-pages-food data set. As shown in Fig. 3, the control variable method was used to adjust the parameters of the three influential factors and analyze the income of rumor spreaders. Figure 3 show the benefits of the final rumor spread in the three datasets. As shown in Fig. 3(a), the larger the UK is, the fewer results will be affected by rumors, and rumor spreaders will earn less benefit. Similarly, by setting parameters that affect RII in the experiment in Fig. 3(b), we can see that the higher the RII, the higher the profit of the rumor spreaders, and the greater the final impact result. Finally, we analyze the influence of UP on the process of competitive games in Fig. 3(c). The larger the UP is, the more obvious the effect is. When users have a higher UP, they will judge the received information according to their history or their preferences, thus increasing the benefits of rumor spreaders. Therefore, the more users rely on this preference, the more susceptible they will be.

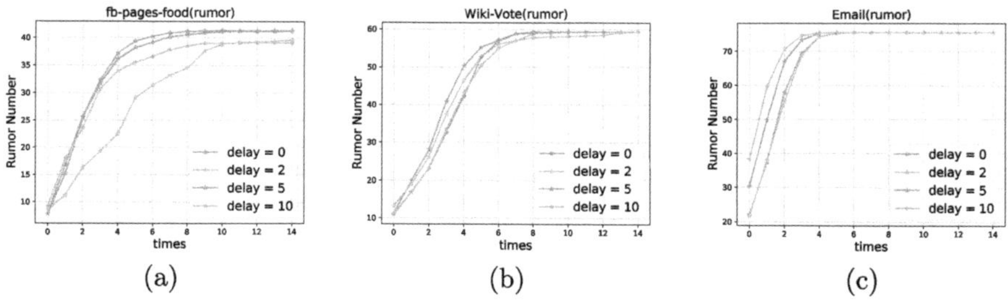

Fig. 4. Nodes activated by the rumors as the game progresses.

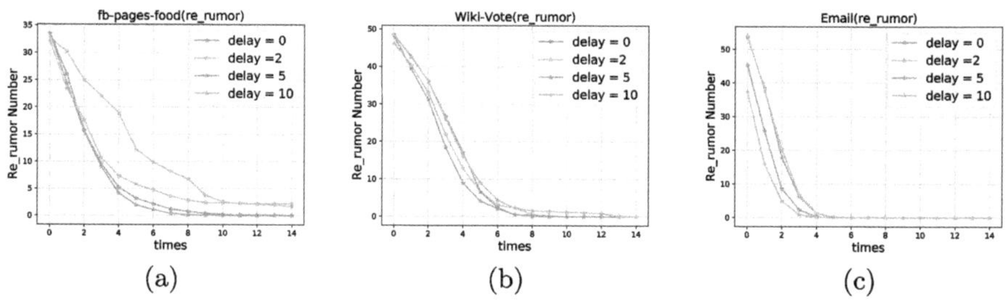

Fig. 5. Nodes activated by the official information as the game progresses.

Performance Evaluation of the Algorithm. We first considered the issue of protector delay. In real-world rumor dissemination scenarios, official publishers often detect the spread of rumors only after some time has passed since the rumors began spreading. Specifically, we examined the scenarios where official counter-rumor information was released during the competitive game's 2nd, 5th, and 10th iterations. The x-axis represents the number of rumor-spreading iterations, while the y-axis indicates the number of rumor-infected individuals at the end of the game. Figure 4 and 5 illustrate the number of nodes activated by rumors and official information as the number of iterations in the game increases. Similarly, releasing official information too late also yields limited effectiveness. The optimal strategy is to introduce official information as soon as rumors spread, thereby quickly reducing the impact of rumor propagation. Earlier dissemination of refutation information is more effective in controlling rumors. Given the influence of unknown rumors, our selected protectors, i.e., official publishers, continuously monitor information dissemination within the network. Consequently, they can react to rumors at the earliest possible moment.

Next, we analyze the performance of our proposed algorithm across three datasets. Initially, within these datasets, it is evident that introducing a protector mechanism, even without disseminating official information, significantly mitigates the ultimate impact of the rumor. The competitive game between rumor and official information begins when protectors act as official publishers. As the official information is propagated, the number of individuals believing the

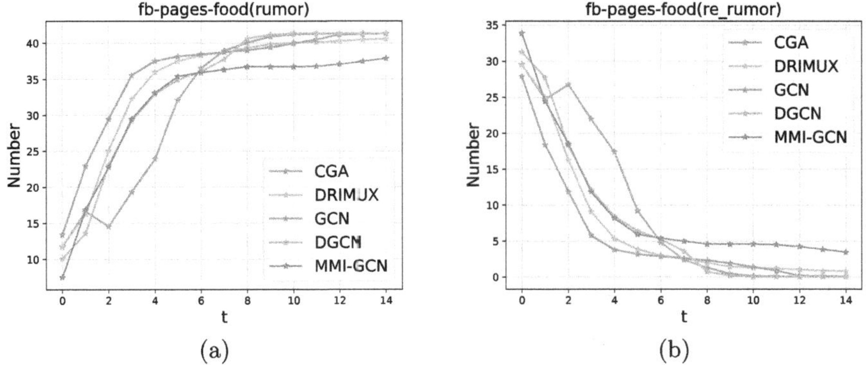

Fig. 6. Comparative results on rumor and official information.

rumor decreases, while the number of individuals believing the official information increases. In our experiments, we selected four comparison algorithms. From Fig. 6, it can be observed that the number of individuals believing the official information is higher in our MMI-GCN algorithm. In comparison, the number of individuals believing the rumors is lower, resulting in a reduced impact of rumors during the information dissemination process.

6   Related Work

Rumor Propagation Model. Rumor dissemination significantly impacts people's daily lives and can potentially lead to social unrest. Researchers commonly employ infectious disease models, independent cascade models, and linear threshold models to simulate rumor propagation. For example, Yu et al. [21] modified infectious disease models, Ding et al. [3] proposed a dynamic rumor propagation model with delays for non-network and network scenarios. Xiao et al. [15] introduced a rumor propagation prediction model based on image restoration techniques. In the current context of rumor propagation, which often involves multilingual environments and heterogeneous networks, Li et al. [8] proposed a rumor propagation model with an educational mechanism. Yu et al. [20] introduced the 2I2SR rumor propagation model based on multilingual environments. Thus, Xiao et al. [16] proposed a group behavior model for rumors and counter-rumors, considering the diversity and complexity of rumor propagation spaces. Wei et al. [14] considered the complex interactions present in information dissemination. Li et al. [9] designed a rumor propagation model based on a cognitive game involving rumors, counter-rumors, and stimulating rumors. Nonetheless, these methods ignore the competitive nature of rumors and refutations. Thus, this paper addresses this issue by proposing an unknown rumor competition rumor propagation model and corresponding solutions.

Rumor Control Strategy. Recently, there has been a growing body of research focusing on rumor control strategies under the competition between rumors and counter-rumors. In rumor control, it is essential to identify the most influential nodes for propagation. Wang et al. [13] proposed a node importance measurement algorithm based on multi-feature fusion to address issues in existing node importance analysis regarding network information and information dissemination accuracy. Ahad et al. [22] established a rumor propagation model based on network structure and designed decentralized community structure strategies to prevent rumor propagation. Xu et al. [17] considered the relationship between the impact block effect and the number of user impressions to minimize the impact count in rumor propagation. Similarly, Yao et al. [19] proposed five strategies to control rumors. Moreover, rapid responses to clarify rumors may be perceived as inaccurate in curbing rumors, Agarwal et al. [1] proposed two sequential game theory models to determine the optimal strategies for official institutions and social media, determining which rumors to address and when to clarify them. Yao et al. [18] addressed the simultaneous propagation of rumors and truths, proposing a multi-hop neighbor reinforcement algorithm.

However, these studies are unsuitable for addressing the problem of rumor control in the presence of unknown rumors. Based on these, this paper provides a solution to the rumor suppression problem in a competitive environment among unknown rumors in multi-layer social networks.

7 Conclusions

This paper addresses the phenomenon by investigating the problem of minimizing the influence of unknown rumor competition in multi-layer social networks. Initially, we propose a competitive rumor propagation model in multi-layer social networks and analyze the competition influence process. Simultaneously, we formulate the problem of minimizing the impact of unknown rumor competition and propose the graph convolutional network approach that integrates multiple pieces of information to select protector strategies, thereby achieving rumor control objectives. Experimental results also demonstrate the superiority of our proposed method over baseline algorithms in rumor control.

Acknowledgments. This work is supported by the National Natural Science Foundation of China under Grant No. 62432003, No. U25A20431; the Fundamental Research Funds for the Central Universities under Grant N2416002.

References

1. Agarwal, P., Al Aziz, R., Zhuang, J.: Interplay of rumor propagation and clarification on social media during crisis events-a game-theoretic approach. Eur. J. Oper. Res. **298**(2), 714–733 (2022)

2. Ding, L., Hu, P., Guan, Z.H., Li, T.: An efficient hybrid control strategy for restraining rumor spreading. IEEE Trans. Syst. Man Cybern. Syst. **51**(11), 6779–6791 (2020)
3. Ding, Y., Zhu, L.: Turing instability analysis of a rumor propagation model with time delay on non-network and complex networks. Inf. Sci. **667**, 120402 (2024)
4. Fang, L., Feng, K., Zhao, K., Hu, A., Li, T.: Unsupervised rumor detection based on propagation tree vae. IEEE Trans. Knowl. Data Eng. (2023)
5. He, Q., Zhang, D., Wang, X., Ma, L., Zhao, Y., Gao, F., Huang, M.: Graph convolutional network-based rumor blocking on social networks. IEEE Trans. Comput. Soc. Syst. (2022)
6. Hosni, A.I.E., Li, K.: Minimizing the influence of rumors during breaking news events in online social networks. Knowl.-Based Syst. **193**, 105452 (2020)
7. Huang, Q., Zhou, C., Wu, J., Wang, M., Wang, B.: Deep structure learning for rumor detection on twitter. In: 2019 International Joint Conference on Neural Networks (IJCNN), pp. 1–8. IEEE (2019)
8. Li, J., Jiang, H., Mei, X., Hu, C., Zhang, G.: Dynamical analysis of rumor spreading model in multi-lingual environment and heterogeneous complex networks. Inf. Sci. **536**, 391–408 (2020)
9. Li, Q., Xiang, T., Dai, T., Xiao, Y.: An information dissemination model based on the rumor and anti-rumor and stimulate-rumor and tripartite cognitive game. IEEE Trans. Cogn. Dev. Syst. **15**(2), 925–937 (2023)
10. Sun, M., Zhang, X., Ma, J., Xie, S., Liu, Y., Philip, S.Y.: Inconsistent matters: a knowledge-guided dual-consistency network for multi-modal rumor detection. IEEE Trans. Knowl. Data Eng. (2023)
11. Sun, X., et al.: Structure learning via meta-hyperedge for dynamic rumor detection. IEEE Trans. Knowl. Data Eng. (2022)
12. Wang, B., Chen, G., Fu, L., Song, L., Wang, X.: Drimux: dynamic rumor influence minimization with user experience in social networks. IEEE Trans. Knowl. Data Eng. **29**(10), 2168–2181 (2017)
13. Wang, Y.C., et al.: A multiple features fusion-based social network node importance measure for rumor control. Soft. Comput. **28**(3), 2501–2516 (2024)
14. Wei, L., Hu, D., Zhou, W., Wang, X., Hu, S.: Modeling the uncertainty of information propagation for rumor detection: a neuro-fuzzy approach. IEEE Trans. Neural Netw. Learn. Syst. (2022)
15. Xiao, Y., Huang, Z., Li, Q., Lu, X., Li, T.: Diffusion pixelation: a game diffusion model of rumor & anti-rumor inspired by image restoration. IEEE Trans. Knowl. Data Eng. **35**(5), 4682–4694 (2022)
16. Xiao, Y., Yang, Q., Sang, C., Liu, Y.: Rumor diffusion model based on representation learning and anti-rumor. IEEE Trans. Netw. Serv. Manage. **17**(3), 1910–1923 (2020)
17. Xu, P., Peng, Z., Wang, L.: Proactive rumor control: When impression counts. In: Pacific-Asia Conference on Knowledge Discovery and Data Mining, pp. 30–42. Springer (2023)
18. Yao, X., Gao, N., Gu, C., Huang, H.: Enhance rumor controlling algorithms based on boosting and blocking users in social networks. IEEE Trans. Comput. Soc. Syst. (2022)
19. Yao, X., Gu, Y., Gu, C., Huang, H.: Fast controlling of rumors with limited cost in social networks. Comput. Commun. **182**, 41–51 (2022)
20. Yu, S., Yu, Z., Jiang, H., Yang, S.: The dynamics and control of 2i2sr rumor spreading models in multilingual online social networks. Inf. Sci. **581**, 18–41 (2021)

21. Yu, Z., Lu, S., Wang, D., Li, Z.: Modeling and analysis of rumor propagation in social networks. Inf. Sci. **580**, 857–873 (2021)
22. Zehmakan, A.N., Out, C., Khelejan, S.H.: Why rumors spread fast in social networks, and how to stop it. In: proceedings of the 32nd International Joint Conference on Artificial Intelligence (IJCAI'23). International Joint Conferences on Artificial Intelligence Organization (2023)

Explainable Team Formation by Integrating Skill Evolution and High-Order Collaboration

Jiaming Pu[1], Yue Kou[1(✉)], Dong Li[2], Derong Shen[1], Tiezheng Nie[1], and Ge Yu[1]

[1] Northeastern University, Shenyang 110004, Liaoning, China
`2372038@stu.neu.edu.cn`,
`{kouyue,shenderong,nietiezheng,yuge}@cse.neu.edu.cn`
[2] Liaoning University, Shenyang 110036, Liaoning, China
`dongli@lnu.edu.cn`

Abstract. Explainable team formation has gained significant attention for its ability to enhance decision-makers' trust and team effectiveness. The dynamic evolution of experts' skills and their high-order collaborative relationships critically impact team performance and recommendation credibility. However, existing methods fail to capture the temporal decay of skill proficiency or model complex collaborative structures beyond direct relationships, while lacking sufficient justification for their recommendations. To address these limitations, we propose a novel Explainable Team Formation Model (ETFM) that incorporates skill evolution and higher-order collaboration. Specifically, we first construct a skill evolution-aware collaboration network enhanced with temporal decay factors to dynamically track proficiency levels. We then develop a hypergraph neural network-based representation learning method to capture latent synergy patterns from high-order relationships. Finally, we define two quantitative metrics ("skill familiarity" and "collaboration tightness") to generate transparent justifications. Extensive experiments on four real-world datasets demonstrate that our approach significantly outperforms state-of-the-art baselines across multiple evaluation metrics.

Keywords: Explainable Team Formation · Skill Evolution · Higher-Order Collaboration · Hypergraph Neural Network

1 Introduction

In mission-critical fields like open-source development, scientific research, and cross-organizational engineering, complex projects rely on experts from diverse backgrounds collaborating through social platforms. Explainable team formation

This work was supported by the National Natural Science Foundation of China (62472204).

acts as a critical bridge between system recommendations and human decision-makers, boosting trust in proposed teams and improving effectiveness. By offering intuitive explanations for team compositions, it helps organizations understand why certain configurations work best, supporting smarter resource allocation.

Team performance is fundamentally shaped by the evolving capabilities of members, where continuous learning and skill development directly impact collective outcomes. Experts follow distinct growth trajectories: some may experience skill decay due to prolonged inactivity, while others reinforce expertise through recent practice. These divergent paths create unique synergy patterns that significantly influence team cohesion and performance. By accounting for these dynamic capability pathways, we aim to enhance both team effectiveness and long-term sustainability.

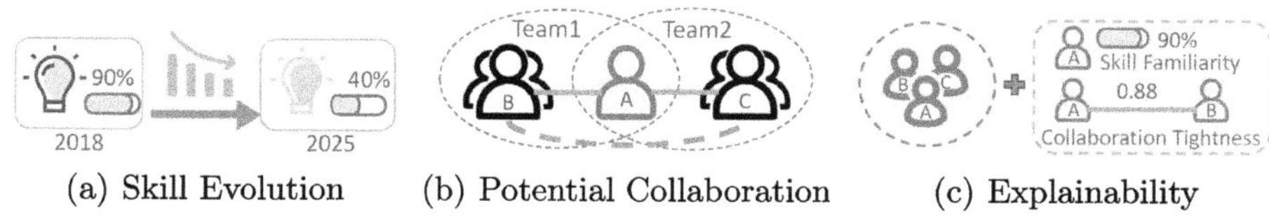

(a) Skill Evolution   (b) Potential Collaboration   (c) Explainability

Fig. 1. Motivation Examples.

We study the problem of skill evolution-aware explainable team formation. Three key issues need to be addressed. First, we need to design a temporal decay mechanism that dynamically captures experts' evolving skill proficiency. Skills naturally decay over time without practice. As shown in Fig. 1(a), in the film industry where VFX technology evolves rapidly, an expert who mastered GANs in 2018 but hasn't applied this skill recently would experience significant skill attenuation. A robust model should capture such temporal dynamics by considering both skill acquisition time and practice duration. Second, we need to design a representation learning approach capable of effectively modeling high-order collaboration patterns beyond direct pairwise relationships. Relying solely on direct historical collaborations while ignoring crucial indirect connections will lead to severe underestimation of team synergy potential. As shown in Fig. 1(b), in open-source software development, if expert A has collaborated with both B and C separately, B and C may demonstrate significant collaborative potential through their common connection A, even without direct collaboration history. A good model should accurately capture such high-order patterns. Finally, we need to develop an adaptive solution capable of generating comprehensive explanations for team recommendations. As shown in Fig. 1(c), when forming an interdisciplinary quantum computing research team, if the system not only recommends a team comprising Algorithm Expert A, Hardware Engineer B, and Physicist C, but also provides detailed justifications: "Expert A maintains 90% skill familiarity in quantum error correction algorithms, demonstrates 0.88 collaboration tightness with Hardware Engineer B across three previous projects, and the team members have established stable triangular collaboration patterns through shared project experiences", decision-makers can clearly understand the team's dual advantages in technical complementarity and collaborative experience, thereby gaining confidence in the team's capability.

Previous research in team formation primarily falls into three categories: text-based [1,2,18,19,22],topic-based [8,11,12], and graph-based [12–16] methods. However, these approaches ignore the temporal decay of experts' skill proficiency, fail to model higher-order collaborative structures beyond direct relationships, and provide limited justification for their recommendations. To overcome these limitations, we propose a novel Explainable Team Formation Model (ETFM) that incorporates skill evolution and higher-order collaboration. Specifically, a skill evolution-aware collaboration network is first constructed, enhanced with temporal decay factors where skill weights are dynamically adjusted according to recency and frequency of practice to accurately capture evolving skill proficiency. Next, a global representation learning method based on hypergraph neural networks is developed to model high-order collaborative relationships, enabling latent synergy patterns beyond direct pairwise connections to be effectively captured. Finally, an explainability analysis mechanism is proposed, in which both skill familiarity and collaboration tightness are quantified to generate intuitive justifications. We summarise our contributions as follows.

- We design a skill evolution-aware collaboration network with temporal decay factors that dynamically adjust skill weights based on skill acquisition time, usage frequency, and decay patterns. This enables accurate tracking of skill proficiency and real-time capability assessment, ensuring timely and reliable team recommendations that align with task requirements.
- We propose a HGNN-based representation learning method that effectively captures high-order collaborative relationships and latent synergy patterns. By integrating multidimensional expert information into unified representations, our approach enables the identification of teams with strong synergistic potential, significantly enhancing team formation quality.
- We define two quantitative metrics ("skill familiarity" and "collaboration tightness") to systematically assess experts' capability states and team collaboration potential. By leveraging these two metrics, we provide quantitative justification for team recommendation decisions, thereby enhancing the explainability and trustworthiness of the selection process.
- We evaluate our model on four real-world datasets. Experimental results demonstrate that our approach outperforms state-of-the-art team formation methods across multiple evaluation metrics.

2   Related Works

Text-Based Team Formation. Text-based methods identify experts by analyzing the semantic similarity between task descriptions and experts' textual data, such as publications or online profiles. Early approaches [1,3] employ document similarity metrics, while advanced techniques leverage document embeddings [22] and graph-based representations [2] to enhance matching accuracy. Although effective in capturing explicit skill requirements, these methods typically rely on shallow semantic analysis and overlook collaborative dynamics. Their performance is also constrained by data quality and availability, limiting practical utility in dynamic environments.

Topic-Based Team Formation. Topic-based approaches abstract specific task requirements into broader research themes, using statistical language models to match experts to predefined topics. Methods employ topic-conditioned probability models [11,12] and enhanced semantic representations [8] to rank candidates. While reducing dependency on raw text processing, these approaches are limited by the expressiveness of their topic models and fundamentally assume static skill profiles. This neglect of expertise evolution over time poses significant challenges in dynamic research landscapes.

Graph-Based Team Formation. Graph-based approaches explicitly model collaborative relationships to form teams with high synergy, typically framed as an optimal connected subgraph search in collaboration networks. Research has focused on minimizing communication costs via Group Steiner Tree problems [16], with extensions incorporating weight constraints [13], weighted node-labeled graphs [15], and constraint pattern graphs. Recent work reformulates the task as a QUBO problem solvable with quantum-inspired solvers [21].

To address data sparsity, semantic embedding methods learn low-dimensional expert representations using variational Bayesian networks [12] or heterogeneous collaboration networks (HCN) [14]. Neural approaches like node2vec [10] generate network embeddings but typically lack update mechanisms for dynamic environments. Recent work enhances interpretability: ExES [9] generates post-hoc explanations via feature perturbation, while [4] employs time-series prediction for dynamic environments. Complementary approaches [5] use retrieval-augmented generation to balance skill coverage and collaboration dynamics. Similarly, Dara et al. [5] introduce a retrieval-augmented neural team formation model that leverages a custom RAG architecture to retrieve historical team data and generate expert sequences.

Despite their strengths in modeling collaboration structures, graph-based methods exhibit limitations in dynamic adaptability, computational efficiency, and high-order synergy modeling. They often assume complete graph structures, scale poorly with $O(N^2)$ complexity, and perform suboptimally with incomplete information. These limitations motivate our approach, which incorporates temporal decay factors for skill evolution and hypergraph neural networks for complex collaboration patterns, while providing explainable justifications through quantitative metrics.

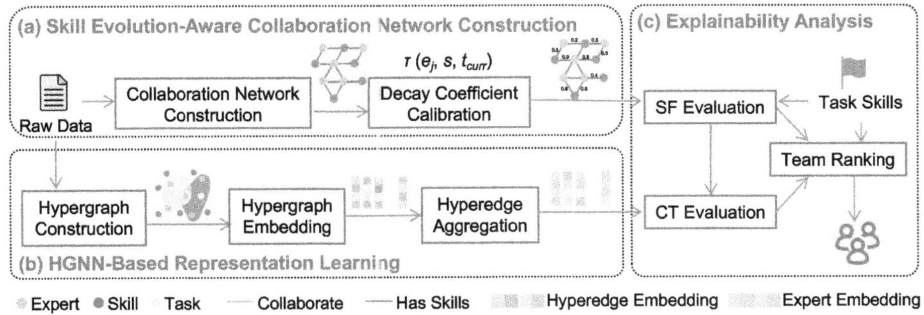

Fig. 2. Framework of ETFM.

3   Framework of Our Solution

Let $T = \{t_i\}$ and $E = \{e_j\}$ be the set of tasks and experts, respectively. Each task $t_i \in T$ has a specific skill requirement $S_{t_i} \subseteq S$, while each expert $e_j \in E$ possesses a set of skills. The team formation function $f : T \to 2^E$ maps a given task to an optimal subset of experts $E_{\text{opt}} \subseteq E$ that collectively satisfy the task's skill requirements while maintaining strong collaborative relationships. The notations used in this paper are listed in Table 1.

Table 1. Notation.

Symbols	Definitions and Descriptions
$T = \{t_i\}$	Task set
$E = \{e_j\}$	Expert set
$S = \{s_k\}$	Skill set
$\mathcal{G} = (\mathcal{V}, \mathcal{E}, \mathcal{T}, \mathcal{W})$	Heterogeneous collaboration network
$\mathcal{G}_{\text{hyper}} = (\mathcal{V}, \mathcal{E}_{\text{hyper}}, \mathcal{W})$	Temporal hypergraph
λ	Decay coefficient
$\tau(e_j, s, t_{\text{curr}})$	Current skill validity of e_j on s
$\text{HR}(e_a, e_b)$	High-order relationship strength
$f : T \to 2^E$	Team formation function
SF	Skill familiarity
CT	Collaboration tightness

In this work, we propose an Explainable Team Formation Model (ETFM), as illustrated in Fig. 2. The framework of ETFM consists of three key components: **(a) Skill Evolution-Aware Collaboration Network Construction.** We construct a heterogeneous collaboration network incorporating temporal decay factors, where skill weights are dynamically adjusted based on recency and frequency of practice. The decay coefficient is calibrated by domain characteristics, and skill validity is computed using an exponential decay model (see Sect. 4.1). **(b) HGNN-Based Representation Learning.** We build a hypergraph structure to model high-order collaborations and employ HGNN for embedding learning. This includes node-to-hyperedge and hyperedge-to-node aggregation (see Sect. 4.2). **(c) Explainability Analysis.** We define two metrics, *Skill Familiarity* (SF) and *Collaboration Tightness* (CT), to evaluate individual expertise and team synergy. They are used for candidate filtering and optimal team ranking. They are also combined to generate transparent justifications, with detailed rationality analysis ensuring their reliability (see Sect. 4.3).

4   Our Proposed Model: ETFM

4.1   Skill Evolution-Aware Collaboration Network Construction

To dynamically track expert skills and collaborations, we construct a heterogeneous collaboration network $\mathcal{G} = (\mathcal{V}, \mathcal{E}, \mathcal{T}, \mathcal{W})$, where $\mathcal{V} = E \cup T$ represents expert and task nodes, $\mathcal{E}$ denotes edges (e.g., expert-task relationships and expert-expert collaborations), $\mathcal{T}$ specifies node and edge types, and $\mathcal{W}$ assigns weights based on skill proficiency and temporal factors. This network captures both explicit collaborations and implicit skill relationships, enabling real-time assessment of expert capabilities.

Decay Coefficient Calibration. The decay coefficient λ is calibrated to reflect skill attenuation over time, varying by domain to account for different knowledge evolution rates. The coefficient is derived from the half-life period of skill relevance t_{half} using the relationship $\lambda = -\ln(0.5)/t_{\text{half}}$. Based on empirical data, we set $\lambda = 0.25$ for fast-evolving domains (e.g., Android, half-life ≈ 2.77 years) and $\lambda = 0.05$ for stable domains (e.g., History, half-life ≈ 13.86 years). This calibration ensures that skill weights prioritize recent usage, aligning with practical expertise decay.

Skill Validity Calculation Algorithm. The validity of skill s for expert e_j at current time t_{curr} is computed using an exponential decay model:

$$\tau(e_j, s, t_{\text{curr}}) = \sum_{k=1}^{N} \alpha_k \cdot \exp\left(-\lambda \cdot (t_{\text{curr}} - t_k)\right) \tag{1}$$

where α_k is the proficiency level demonstrated in the k-th application of skill s (e.g., based on answer scores), t_k is the timestamp of skill application, and N is the total number of skill applications. This formula quantifies the current effectiveness of an expert's skill, giving higher weight to recent activities. The output τ is used to weight edges in the collaboration network, ensuring that network connections reflect up-to-date expertise.

4.2 HGNN-Based Representation Learning

To capture high-order collaboration patterns beyond pairwise relationships, we employ a hypergraph neural network (HGNN) that learns latent representations of experts, skills, tasks, and teams. This approach models teams as hyperedges, allowing for the encapsulation of complex group dynamics.

Hypergraph Construction. We define a hypergraph $\mathcal{G}_{\text{hyper}} = (\mathcal{V}, \mathcal{E}_{\text{hyper}}, \mathcal{W})$, where $\mathcal{V}$ is the set of nodes (experts and tasks), with skills serving as attributes of expert nodes rather than separate entities, and $\mathcal{E}_{\text{hyper}}$ is the set of hyperedges representing historical teams. Each hyperedge h_i corresponds to a team instance, connecting a task node t_i and its answer experts $E_i = \{e_1, e_2, \ldots, e_m\}$. The weight w_h of a hyperedge is based on team performance metrics (e.g., aggregate vote scores of answers), which reflects historical success but does not include skill time decay (as that is handled separately). The incidence matrix H is defined such that $H(v, e) = 1$ if node v belongs to hyperedge e, and 0 otherwise.

Hypergraph Embedding & Hyperedge Aggregation. HGNN learns d-dimensional embeddings for nodes and hyperedges through layer-wise propagation. As shown in Fig. 3, this process employs two key aggregations that capture complex team formation relationships: node-to-hyperedge (Eq. 2) and hyperedge-to-node (Eq. 3) aggregation.

For each hyperedge h representing a historical team, we aggregate the embeddings of all expert nodes $\{e_1, e_2, \ldots, e_m\}$ and the task node t_i connected by this hyperedge.

$$\mathbf{h}^{(l+1)} = \sigma \left(\sum_{v \in h} \mathbf{W}_{\text{node}}^{(l)} \mathbf{v}^{(l)} \right) \tag{2}$$

where $\mathbf{W}_{\text{node}}^{(l)}$ is a learnable weight matrix and σ is the ReLU function. This aggregates individual expert features into a team representation that captures the collective expertise pattern.

For each expert node v, we aggregate information from all hyperedges (teams) that contain this expert:

$$\mathbf{v}^{(l+1)} = \sigma \left(\sum_{h \ni v} \mathbf{W}_{\text{hyper}}^{(l)} \mathbf{h}^{(l)} \right) \tag{3}$$

where $\mathbf{W}_{\text{hyper}}^{(l)}$ is a learnable weight matrix for hyperedge embeddings. For example, as shown in Fig. 3, in node-to-hyperedge aggregation, expert nodes $\{e_1, e_2, e_3\}$ and task node t_1 are aggregated to form a hyperedge h_1, representing "Team 1". Subsequently, during hyperedge-to-node aggregation, an expert node like e_2 receives integrated information from all the teams (h_1, h_2, h_3) it belongs to, enriching its representation with a holistic view of its collaborations. This enables each expert to learn from their diverse team experiences, enriching their individual representation with collaborative knowledge. This operation synthesizes individual expert characteristics into a unified team representation, capturing the collective expertise pattern. The model is trained using negative sampling and the Skip-gram loss function. For each positive sample (expert, team), we generate 5 negative samples by randomly replacing the expert with another expert not in the team. The loss function is defined as:

$$\mathcal{L} = - \sum_{(v,h)\in\mathcal{P}} \log \sigma(\mathbf{v} \cdot \mathbf{h}) - \sum_{(v,h')\in\mathcal{N}} \log \sigma(-\mathbf{v} \cdot \mathbf{h}') \tag{4}$$

where $\mathcal{P}$ is the set of positive samples (real teams) and $\mathcal{N}$ is the set of negative samples (randomly generated teams). Training uses the Adam optimizer with a learning rate of 0.001 and batch size of 256, typically for 50–100 epochs. The output is a set of embeddings that encode collaboration patterns and team structures.

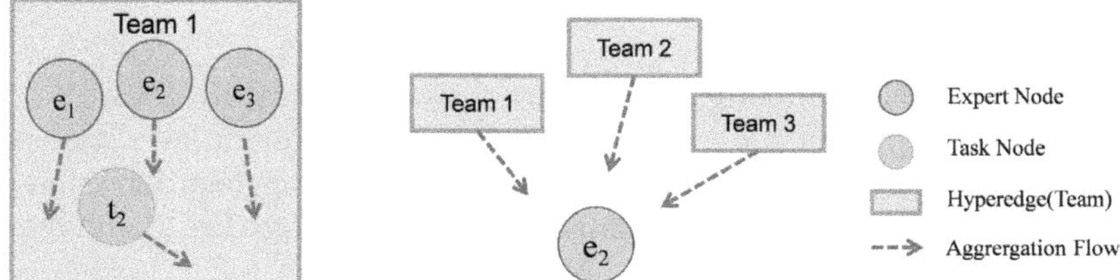

(a) Node-to-hyperedge Aggregation   (b) Hyperedge-to-node Aggregation

Fig. 3. Illustration of hypergraph aggregation process.

4.3   Explainability Analysis

ETFM generates transparent team recommendations by providing explanations based on the following two quantitative metrics:

Skill Familiarity. Skill familiarity quantifies an expert's current proficiency in a specific skill. It is formally defined by Eq. 5:

$$SF(e_j, s) = \frac{\tau(e_j, s, t_{\mathrm{curr}})}{\max_{e_k \in E} \tau(e_k, s, t_{\mathrm{curr}})} \tag{5}$$

This equation yields a normalized [0,1] score, indicating an expert's familiarity with a skill relative to the best performer in the pool. The score is computed independently using a skill validity mechanism, thereby providing an up-to-date justification for individual expert selection based on current expertise.

Collaboration Tightness. Collaboration tightness quantifies the historical collaboration strength between two experts, a metric directly derived from the hypergraph structure. It is formally defined as:

$$CT(e_a, e_b) = \frac{\sum_{h \in \mathcal{E}_{\text{hyper}}} I e_a, e_b \in h \cdot w_h}{\sqrt{\deg(e_a) \cdot \deg(e_b)}} \tag{6}$$

where I is the indicator function, w_h is the hyperedge weight (team performance), and $\deg(e)$ is the expert's hypergraph degree. The normalized score [0,1] indicates collaboration intensity, with higher values explaining better team cohesion and lower communication costs.

For a recommended team, the overall score is computed by combining the above metrics. To enhance transparency, a structured explanation is generated, which includes: 1) SF values for each expert's key skills (e.g., "Expert A (SF=0.9 in skill s_1) has demonstrated high proficiency through recent tasks."). 2) CT values between member pairs (e.g., "Experts B & C (CT=0.8) have collaborated effectively on 5 prior projects."). Presenting these quantitative justifications alongside the team recommendation builds user trust in the system's output.

Team Ranking. Given a new task t_{new} with required skills S_{new}, we first filter experts based on their SF values. An expert e_j is admitted to the candidate pool C only if $SF(e_j, s) \geq \theta_s$ for every required skill $s \in S_{\text{new}}$, ensuring all candidates meet the minimum proficiency. Teams are then ranked by a multi-objective function that balances skill coverage against collaboration potential:

$$\text{Score}(T) = w_1 \cdot \left(\frac{1}{|S_{\text{new}}|} \sum_{s \in S_{\text{new}}} \max_{e_j \in T} SF(e_j, s) \right) + w_2 \cdot \min_{e_a, e_b \in T} CT(e_a, e_b) \tag{7}$$

Here, w_1 and w_2 are tunable weights, SF is the average skill familiarity, and CT is the minimum collaboration tightness within the team. The top-k teams according to this score are recommended.

Rationality Analysis. The rationality of the SF and CT metrics is justified based on three key properties: 1) Monotonicity: Both metrics exhibit a positive correlation with team performance. Specifically, a higher SF consistently leads to better task outcomes, while a higher CT effectively reduces communication overhead. 2) Explainability: These metrics are derived from observable data, including skill usage history and collaboration records, which ensures their traceability and provides intuitive insights. 3) Adaptability: Key parameters (e.g., the coefficient λ and other weighting factors) can be calibrated for different application domains. This tunability helps prevent overfitting to specific scenarios and enhances the generalizability of the metrics. This analysis confirms that the SF and CT metrics offer reliable and actionable guidance for team formation.

5　Experiments

5.1　Experiments Setup

Datasets. We evaluate our model on four real-world datasets from Stack Exchange: Android, History, DBA, and Physics. The statistical details of these datasets are presented in Table 2. In our experimental setup, we define the "gold standard team" for a question as the group of all users who answered that question. The objective is to predict the team composition for each new question in the test set. To assess performance, we partition the questions in each dataset, using 90% for training and the remaining 10% for testing.

Table 2. The statistics of datasets.

Dataset	# Questions	# Answers	# Experts	# Teams
Android	882	2136	1018	857
History	1697	4702	1164	1575
DBA	2096	6976	1794	2554
Physics	5010	12559	2938	4613

Implementation Details. For all methods, the embedding dimension and batch size are set to 128 and 256, respectively. Node embedding vectors are randomly initialized. All other hyperparameters in the baseline models retain the default values as proposed in their original publications. Our framework is implemented using PyTorch, and all experiments are conducted on a server equipped with an Intel(R) Xeon(R) Silver 4310 CPU @ 2.10GHz, an NVIDIA A5000 GPU, and 256 GB of RAM. Model performance is evaluated using a suite of metrics commonly adopted in the literature [7,13,15]: Skill Coverage (SC), Communication Cost Level (CL), Precision at N (PN), Gold Standard Team Match (GM) and Expertise Level (EL). Specifically, for overall team quality assessment, SC calculates the percentage of required skills covered by the team. CL estimates the potential communication cost based on the number of prior collaborations among team members, whereas PN quantifies the proportion of correctly identified expert members, with N representing the number of test tasks. GM measures the exact match accuracy of the predicted team. And EL reflects the team's collective capability to solve the given task. The code and dataset are available at: https://github.com/Rose-JM/ETFM.

Baselines. We compare ETFM against a comprehensive set of baselines, including NeRank [17], Seq [20], metapath2vec [6], CCR [16], SA-CA-CC [23], NCCO [13], TOSA [8], TAPG [22], and E2T [14]. These methods represent diverse team formation strategies. Among them, the first three are employed in ablation studies to replace the HGNN-based representation learning module in our approach, while the latter six are used for overall performance comparison. Specifically, NeRank, Seq, and metapath2vec leverage textual information and network structures to identify experts and aggregate them into teams. For instance, NeRank utilizes both question titles and bodies, while Seq relies exclusively on tags. CCR, SA-CA-CC, NCCO, TOSA, and TAPG employ graph-based optimization techniques to form teams by solving problems related to skill coverage and collaboration efficiency. E2T learns low-dimensional representations to model expert compatibility and team synergy.

5.2 Overall Performance Comparison

We evaluate ETFM and the baseline methods using the following metrics: SC, CL and PN. The results, averaged over all test questions and team sizes (ranging from 1 to 6), are presented in Fig. 4 and Fig. 5. These results demonstrate that our ETFM model achieves strong performance across all metrics. The following observations are drawn from the experiments.

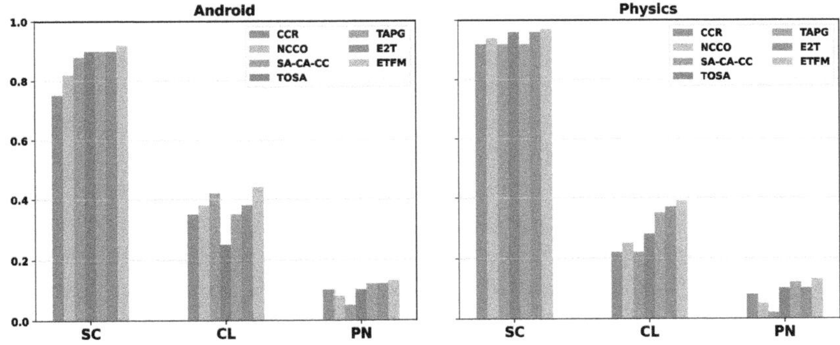

Fig. 4. Performance comparison on Android and Physics datasets.

SC Metric Analysis. ETFM consistently achieves the highest SC values across all datasets, with scores of 0.97 on Physics, DBA, and History, and 0.92 on Android. This superior performance can be attributed to its skill evolution-aware mechanism, which dynamically tracks proficiency levels through temporal decay factors. In contrast, methods such as CCR and NCCO exhibit lower SC values (ranging from 0.75 to 0.94), primarily due to their reliance on static skill representations that fail to adapt to evolving expertise requirements. While TOSA and E2T demonstrate competitive performance, they are still limited by their inability to effectively model skill decay dynamics.

CL Metric Analysis. ETFM achieves the highest CL scores across all datasets, reaching 0.50 on History and 0.39 on Physics and DBA. This improvement stems from its hypergraph neural network architecture, which effectively models high-order collaboration patterns beyond simple pairwise relationships. In contrast,

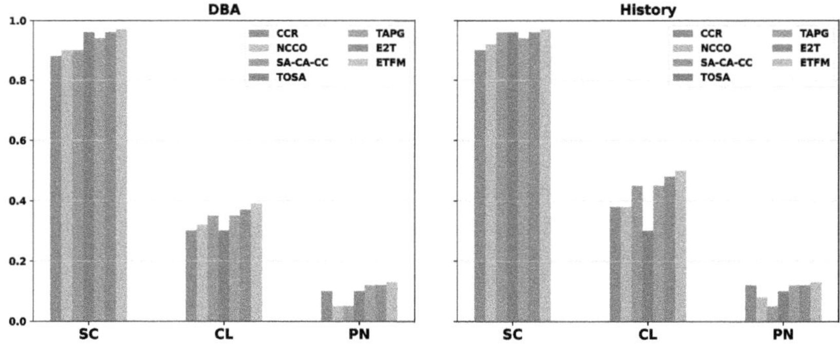

Fig. 5. Performance comparison on DBA and History datasets.

traditional graph-based methods like CCR and NCCO yield lower CL values (ranging from 0.22 to 0.45), as they primarily rely on direct historical collaborations. While E2T demonstrates better performance than these traditional methods, it remains constrained by its dependence on predefined metapaths for team formation.

PN Metric Analysis. ETFM achieves the best PN performance across all datasets. This consistent superiority can be attributed to its integration of explainability metrics, Skill Familiarity (SF) and Collaboration Tightness (CT), which guide the model toward more accurate and context-aware team recommendations. In contrast, methods like SA-CA-CC yield the lowest PN values, as their optimization primarily targets communication cost reduction at the expense of recommendation precision. While TAPG and E2T achieve moderate performance, their text-centric matching strategies often fail to capture the nuanced semantic relevance between task requirements and expert profiles, limiting their overall effectiveness.

The superior performance of ETFM across all evaluation metrics substantiates the effectiveness of its core architectural innovations: the dynamic modeling of skill evolution trajectories and the explicit encoding of high-order collaboration relationships. By capturing how individual competencies evolve over time and how complex team synergies emerge from multi-agent interactions, ETFM achieves a more nuanced and realistic representation of team dynamics. Furthermore, the model's inherently explainable nature significantly enhances its practical utility in real-world settings. This explicitness not only builds trust among decision-makers but also provides actionable insights for organizational resource allocation and team optimization.

5.3   Ablation Study

To gain deeper insights into ETFM's design, we conduct ablation studies by systematically replacing its core components. Specifically, we substitute the HGNN-based representation learning module with NeRank [17], Seq [20], metapath2vec [6], and E2T [14], respectively. Here, we adopt the set of metrics (SC,

CL, GM, and EL) to enable a precise diagnostic analysis of the model's core components. Specifically, SC and CL directly reflect the effectiveness of the skill evolution module and the high-order collaboration module, respectively. The strict exact-match criterion of GM is sensitive to changes in the overall output quality. The inclusion of EL serves as a targeted probe to evaluate the contribution of the skill familiarity mechanism to the team's collective capability, thereby providing complementary insights to the fundamental metrics. Due to space limitations, we present only the results on the Android dataset in Fig. 6 and Fig. 7. The results demonstrate that ETFM achieves the best performance across all metrics. The reasons for this are analyzed below. A detailed analysis of the Android dataset across all metrics reveals that ETFM maintains superior performance from team size 1 to 6.

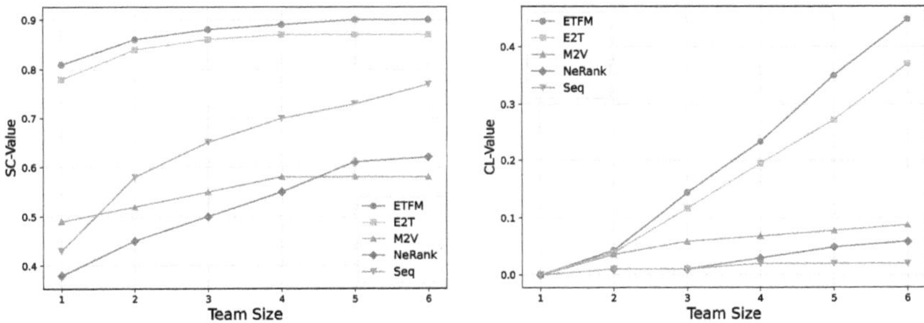

Fig. 6. Performance of variants on Android dataset in terms of SC and CL.

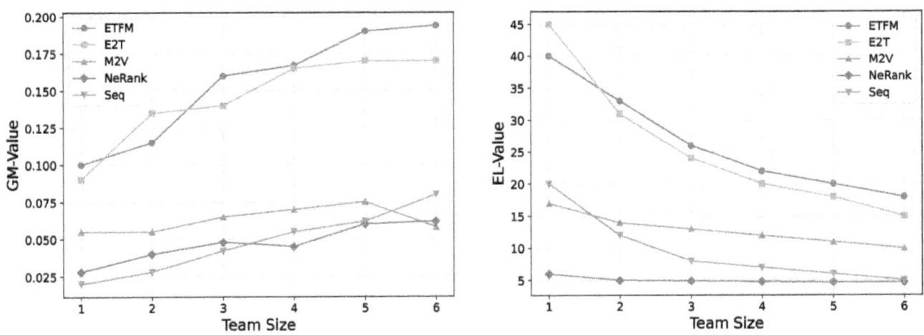

Fig. 7. Performance of variants on Android dataset in terms of GM and EL.

SC Metric Analysis. ETFM achieves the highest SC values (e.g., 0.9 at team size 6), demonstrating its superior capability in comprehensive skill coverage. This advantage stems from its proactive candidate filtering mechanism, which employs skill familiarity thresholds to guarantee that all required skills are assigned to sufficiently proficient experts. In contrast, variants such as NeRank and M2V exhibit lower SC due to their reliance on coarse-grained skill matching, an approach that risks overlooking critical skill requirements. Although Seq shows some improvement over time, it remains inferior because its tag-based matching lacks the necessary depth to assess true expertise.

CL Metric Analysis. ETFM consistently achieves the highest CL values across all epochs, significantly outperforming the variants. This sustained superiority is attributed to its hypergraph neural network, which effectively captures high-order collaborative relationships and models complex team structures beyond direct pairwise connections. In contrast, variants like NeRank and Seq rely on textual similarity or simple graph embeddings, failing to account for latent synergy patterns and thus yielding lower CL values. Although metapath2vec incorporates meta-paths, it still does not fully encapsulate team-level interactions, leading to suboptimal collaboration estimates.

GM Metric Analysis. ETFM consistently achieves high GM values (e.g., 0.194 at team size 6), underscoring its accuracy in team recommendation. This performance advantage stems from ETFM's skill evolution-aware mechanism, which dynamically adjusts skill proficiency via temporal decay to ensure expert selection reflects current capabilities. In contrast, variants such as E2T and NeRank rely on static skill representations that can become outdated, consequently reducing match accuracy. Although E2T shows competitive GM in specific cases (e.g., 0.135 at team size 2), its overall performance is hampered by the absence of skill decay modeling.

EL Metric Analysis. ETFM maintains high EL values (e.g., 18 at team size 6), reflecting its strong collective team capability. This is attributed to the integration of its skill familiarity metric, which prioritizes experts with recent practice, thereby elevating the overall team expertise. In contrast, variants such as M2V and Seq exhibit lower EL because they lack mechanisms to incorporate temporal dynamics. For instance, Seq's tag-based retrieval may include experts with obsolete skills, which diminishes the collective expertise level. While a general decreasing trend in EL is observed across all methods over epochs, potentially due to dataset characteristics, ETFM demonstrates greater robustness by exhibiting a notably slower decline.

5.4   Case Study

This case study demonstrates the efficacy of the SF and CT metrics in team formation. As illustrated in Fig. 8, for an Android memory optimization task requiring skills in Memory Management (s_1), Java Programming (s_2), and Android SDK (s_3), we evaluate four candidates: Expert A: Memory specialist (3 years inactive, SF=0.40). Expert B: Java expert (recently active, SF=0.79). Expert C: Android SDK expert (moderately active, SF=0.62). Expert D: Generalist (outdated skills, SF=0.26). Collaboration analysis shows: A–B (CT=0.85, strong), B–C (CT=0.72, good), A–C (CT=0.35, moderate). Team evaluation results confirm that Team A, B, C is recommended due to its optimal balance between skill coverage (leveraging B's Java proficiency at 79%, C's SDK at 62%, and A's Memory at 40%) and collaboration synergy (the strong A–B partnership minimizes integration overhead). This case quantitatively validates how SF and CT metrics enhance the transparency and trustworthiness of team recommendations.

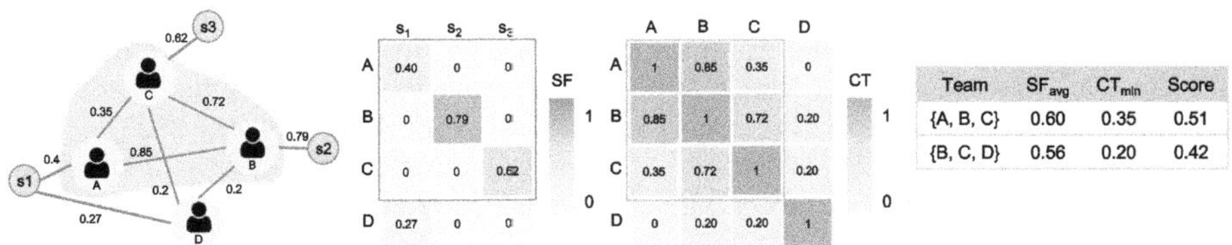

Fig. 8. Case study on explainable team formation.

6   Conclusion

In this work, we propose a novel Explainable Team Formation Model (ETFM) that dynamically integrates skill evolution and high-order collaboration for credible team recommendation. First, we construct a skill evolution-aware collaboration network enhanced with temporal decay factors to dynamically track proficiency levels. Then, we develop a HGNN-based representation learning method to capture latent synergy patterns from high-order relationships. We also define two quantitative metrics to generate transparent justifications for team recommendations. For future work, we will incorporate broader contextual factors, such as project dynamics and organizational constraints, to enhance the practical applicability of the model. Beyond that, we will explore the integration of large language models to enrich the semantic representation of skills and tasks, thereby improving alignment accuracy.

References

1. Berger, M., Zavrel, J., Groth, P.: Effective distributed representations for academic expert search. In: Proceedings of the 1st Workshop on Scholarly Document Processing (SDP&EMNLP), pp. 56–71 (2020)
2. Brochier, R., Gourru, A., Guille, A., Velcin, J.: New datasets and a benchmark of document network embedding methods for scientific expert finding. In: Proceedings of the 10th International Workshop on Bibliometric-enhanced Information Retrieval (BIR&ECIR), vol. 2591, pp. 16–29 (2020)
3. Brochier, R., Guille, A., Rothan, B., Velcin, J.: Impact of the query set on the evaluation of expert finding systems. In: Proceedings of the 41st International ACM SIGIR Conference on Research and Development in Information Retrieval, pp. 32–45 (2018)
4. Chen, Y., Zhang, L., Ding, Y., Guo, L., Bian, K.: Multi-task oriented team formation in online collaborative learning. Expert Syst. Appl. **259**, 125289 (2025)

5. Dara, M., Rad, R.H., Zarrinkalam, F., Bagheri, E.: Retrieval-augmented neural team formation. In: Proceedings of the 47th European Conference on Information Retrieval (ECIR), pp. 362–371 (2025)
6. Dong, Y., Chawla, N.V., Swami, A.: metapath2vec: Scalable representation learning for heterogeneous networks. In: Proceedings of the 23rd ACM SIGKDD International Conference on Knowledge Discovery and Data Mining, pp. 135–144 (2017)
7. Etemadi, R., Zihayat, M., Feng, K., Adelman, J., Bagheri, E.: Embedding-based team formation for community question answering. Inf. Sci. **623**, 671–692 (2023)
8. Gao, X., Wu, S., Xia, D., Xiong, H.: Topic-sensitive expert finding based solely on heterogeneous academic networks. Expert Syst. Appl. **213**, 119241 (2023)
9. Golzadeh, K., Golab, L., Szlichta, J.: Explaining expert search and team formation systems with exes. In: Proceedings of the 41st International Conference on Data Engineering (ICDE), pp. 224–237 (2025)
10. Grover, A., Leskovec, J.: node2vec: scalable feature learning for networks. In: Proceedings of the 22nd ACM SIGKDD International Conference on Knowledge Discovery and Data Mining, pp. 855–864 (2016)
11. Hamidi Rad, R., Bagheri, E., Kargar, M., Srivastava, D., Szlichta, J.: Retrieving skill-based teams from collaboration networks. In: Proceedings of the 44th International ACM SIGIR Conference on Research and Development in Information Retrieval, pp. 2015–2019 (2021)
12. Hamidi Rad, R., Fani, H., Bagheri, E., Kargar, M., Srivastava, D., Szlichta, J.: A variational neural architecture for skill-based team formation. ACM Trans. Inf. Syst. (TOIS) **42**(1), 1–28 (2023)
13. Kargar, M., Golab, L., Srivastava, D., Szlichta, J., Zihayat, M.: Effective keyword search over weighted graphs. IEEE Trans. Knowl. Data Eng. (TKDE) **34**(2), 601–616 (2020)
14. Kou, Y., et al.: Experts2team: Task relevance-induced team formation by combining global cohesion with local decoupling. In: Proceedings of the 30th International Conference on Database Systems for Advanced Applications (DASFAA) (2025)
15. Kou, Y., et al.: Efficient team formation in social networks based on constrained pattern graph. In: Proceedings of the 36th International Conference on Data Engineering (ICDE), pp. 889–900 (2020)
16. Lappas, T., Liu, K., Terzi, E.: Finding a team of experts in social networks. In: Proceedings of the 15th ACM SIGKDD International Conference on Knowledge Discovery and Data Mining, pp. 467–476 (2009)
17. Li, Z., Jiang, J.Y., Sun, Y., Wang, W.: Personalized question routing via heterogeneous network embedding. In: Proceedings of the AAAI Conference on Artificial Intelligence, vol. 33, pp. 192–199 (2019)
18. Reimers, N., Gurevych, I.: Sentence-bert: sentence embeddings using siamese bert-networks. In: Proceedings of the 2019 Conference on Empirical Methods in Natural Language Processing and the 9th International Joint Conference on Natural Language Processing (EMNLP&IJCNLP), pp. 3980–3990 (2019)
19. Rostami, P., Shakery, A.: A deep learning-based expert finding method to retrieve agile software teams from cqas. Inf. Process. Manag. **60**(2), 103144 (2023)
20. Sun, J., Zhao, J., Sun, H., Parthasarathy, S.: Endcold: An end-to-end framework for cold question routing in community question answering services. In: Proceedings of the 29th International Conference on International Joint Conferences on Artificial Intelligence (IJCAI), pp. 3244–3250 (2021)
21. Vombatkere, K., Lappas, T., Terzi, E.: A qubo framework for team formation. In: Proceedings of Machine Learning and Knowledge Discov. Databases (ECML&PKDD) (2025)

22. Xu, X., Liu, J., Wang, Y., Ke, X.: Academic expert finding via $(k, \mathcal{P})$-core based embedding over heterogeneous graphs. In: Proceedings of the 38th International Conference on Data Engineering (ICDE), pp. 338–351 (2022)
23. Zihayat, M., An, A., Golab, L., Kargar, M., Szlichta, J.: Authority-based team ciscovery in social networks. In: Proceedings of the 20th International Conference on Extending Database Technology (EDBT), pp. 498–501 (2017)

SP-GCRL: Influence Maximization on Incomplete Social Graphs

Haohua Niu[1], Yuxuan Yang[2], Lingfeng Zhang[3], Hao Li[4], Jiao Liang[1],
Zongfu Luo[1(✉)], and Luca Rossi[3(✉)]

[1] School of Systems Science and Engineering, Sun Yat-sen University, Guangzhou,
China
luozf@mail.sysu.edu.cn

[2] School of Finance, Jiangxi University of Finance and Economics, Jiangxi, China

[3] Department of Electrical and Electronic Engineering, The Hong Kong Polytechnic
University, Hong Kong, China
luca.rossi@polyu.edu.hk

[4] School of Computer Science and Engineering, Anhui University of Science and
Technology, Anhui, China

Abstract. Influence maximization (IM) in real platforms is challenged
by incomplete, noisy social graphs and non-stationary diffusion dynamics. We propose SP-GCRL, a social-propagation–aware graph contrastive
reinforcement learning framework that learns end-to-end seed selection
under partial observability. We first introduce a social-propagation-aware
nonlinear diffusion function to model reinforcement/diminishing effects
and probability drift under repeated exposure; we then construct dual
structural views and perform contrastive learning to obtain node representations robust to missing edges and weak ties, while replacing expensive strategy metrics with a GAT-based regression surrogate to improve
efficiency and scalability; finally, we use DDQN to learn an end-to-end
seed selection policy on top of these representations. Experiments on multiple real-world networks show that SP-GCRL achieves significant gains
over heuristic and learning-based baselines across budgets and topologies, while maintaining strong large-scale scalability.

Keywords: Social Networks · Influence Maximization

1 Introduction

Social networks have transformed interpersonal communication and the diffusion of information at scale, yielding rich observational traces for modeling user behavior and preferences. Within this landscape, Influence Maximization (IM) is a foundational problem with broad applications in viral marketing [6], recommender systems [32], and policy intervention [14]: given a budget k, select a seed set of k users that maximizes the expected cascade size under a predefined

---

H. Niu and Y. Yang—Equal contribution.

H. Jung et al. (Eds.): DASFAA 2026, LNCS 16536, pp. 334–350, 2026.
https://doi.org/10.1007/978-981-92-0366-6_21

diffusion mechanism. The identification and selection of users with global influence can substantially augment the efficacy and reach of social campaigns and targeted interventions [12].

The classical line of work initiated by [12] formalizes IM under the Independent Cascade (IC) and Linear Threshold (LT) models, proving NP-hardness while establishing a $(1-1/e)$-approximation via a greedy algorithm that exploits the submodularity of the expected spread. Building on this theory, a large literature has refined scalable heuristics and sampling-based estimators for IM [17,26,27]. Yet, real social systems exhibit heterogeneous and context-dependent communication patterns that deviate from these stylized models. Empirical studies document mechanisms such as selective exposure and homophily that induce systematic biases in who is exposed and who adopts [3], while communication unfolds through multi-stage, feedback-coupled dynamics shaped by network structure and context [1]. These observations motivate diffusion models that are both more expressive and more robust to data imperfections.

Recent learning-based approaches address part of this need by leveraging Graph Neural Networks (GNNs) and Reinforcement Learning (RL) to directly approximate influence dynamics from data [5,7,15,18]. However, in many practical settings the complete network and diffusion histories are not observable. Privacy constraints, API limitations, sampling bias, missing or delayed logs, and incomplete cross-platform linkages lead to partially observed graphs and cascades, with unobserved edges, unrecorded exposures, and censored adoptions. Such incompleteness fundamentally challenges both parametric diffusion modeling and representation learning, thus degrading IM performance.

To address these challenges, we propose SP-GCRL, a framework that combines self-supervised graph contrastive representation learning with RL for seed selection under partial observability. SP-GCRL learns node embeddings from incomplete graphs by contrasting structure-aware augmented views that emphasize information pathways most predictive of diffusion, and then trains a Double Deep Q-Network (DDQN) to produce seed sets from the learned representations. Our main contributions are:

(1) We introduce a unified probability formulation that captures both social reinforcement and social weakening, yielding a non-monotonic exposure–response curve aligned with observed diffusion behavior and accommodating deviations from IC/LT assumptions.

(2) We design two structural augmentations tailored to incomplete data, one that concentrates on high-information propagation corridors while being resilient to missing edges (Sect. 3.2(a)) and one that emphasizes nodes with high dynamical reachability under linearized propagation (Sect. 3.2(b)). Using Graph Contrastive Learning (GCL) we maximize agreement between these views producing embeddings robust to partial network and cascade observations.

(3) Leveraging the learned node embeddings, we train a DDQN to derive an end-to-end seed selection policy that operates effectively when only fragmentary graph or diffusion information is available, yielding the SP-GCRL framework.

(4) Across eight real-world datasets, under controlled masking and subsam-

pling regimes that emulate practical incompleteness, SP-GCRL consistently outperforms heuristic and contemporary learning-based baselines in achieved influence spread and generalization.

2   Related Work

Research on IM spans classical diffusion modeling and learning-based algorithms. The foundational IC and LT models formalize diffusion as, respectively, independent stochastic activations and thresholded aggregations of neighbor influence [6,12,30]. The triggering model unifies these paradigms by allowing node-specific triggering sets [26]. Heuristics such as PageRank are frequently employed for seed pre-selection [23]. These models confer analytical tractability and underpin scalable IM techniques, but they rely on strong independence and stationarity assumptions that can misalign with empirical propagation, motivating methods that accommodate heterogeneous, context-dependent dynamics [26].

To address these limitations, recent work leverages RL and GNNs. Early RL formulations cast IM as sequential decision making [2,19], followed by agents that optimize seed sets via rewards from simulated cascades [4,21,28] and frameworks that improve deployability through pretraining and model pools [18]. In parallel, GNN-based approaches encode structural signals to predict node influence or generate diversified diffusion patterns: regression-style predictors [15], end-to-end deep RL with GNN encoders [5], generative IM with knowledge distillation [20], and hybrids of graph convolution with neural bandits for exploration–exploitation under limited topology knowledge [9], alongside scalable architectures [35]. Beyond IM, GNNs have proved effective across chemistry, social analysis, and multi-agent systems [8,31,34], motivating self-supervision. GCL constructs augmented views (e.g., node dropping, edge perturbation, attribute masking) and maximizes agreement to learn label-efficient, robust embeddings [10,33,36], thereby improving generalization under noisy or incomplete observations.

3   The Proposed Framework

Figure 1 presents the overall SP-GCRL pipeline. Starting from an observed (possibly incomplete) social graph, we compute an exposure-aware diffusion signal, build two structural views (backbone and controllability), learn robust node embeddings via contrastive learning with a lightweight GAT approximation, and finally select seeds using a DDQN policy to maximize expected spread. Full details are provided in the following sections.

3.1   Social Information Propagation

In social networks, the probability of information propagation depends on multiple complex factors. Traditional models, such as the IC model, typically assume

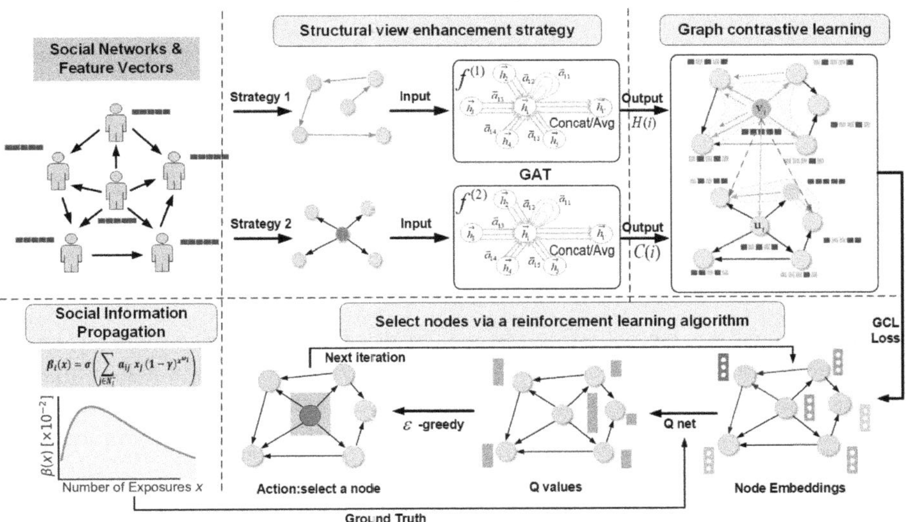

Fig. 1. The SP-GCRL framework.

that nodes propagate information with fixed probabilities, making it challenging to accurately capture the nonlinear effects of social relationships on user forwarding behaviors in real-world scenarios. To address this issue, inspired by the dynamic spreading equation proposed by [22], this study develops a nonlinear propagation model that incorporates social relationships and the "interest threshold" effect, which is expressed as

$$\beta_i(\mathbf{x}) = \sum_{j \in \mathcal{N}_i^-} a_{ij} x_j \left(1 - \gamma\right)^{f_i(x)} , \tag{1}$$

where $\beta_i(x)$ is the probability of the x-th attempt of the surrounding nodes to activate the user to forward the information, a_i represents the intrinsic spreading capability of the information for node i, defined as the edge weight $1/d_{in}(i)$ between nodes, where d_{in} represents the in-degree of the node. $\sigma(\cdot)$ denotes the sigmoid function, $f_i(x)$ is is an exposure-adjustment factor that modulates the effective influence of repeated views by accounting for users' uncertainty about the marginal benefit of retweeting, and γ is the average proportion of common neighbors shared between a user and its neighbors, which is calculated by the equation

$$\gamma = \frac{1}{N} \sum_{u=1}^{N} \frac{1}{|N_u|} \sum_{v \in N_u} \frac{|N_u \cap N_v|}{|N_u|} , \tag{2}$$

where N is the total number of nodes in the network, N_u denotes the neighbor set of node u, and $|N_u \cap N_v|$ represents the number of common neighbors shared between nodes u and v. The γ in Eq. (2) is computed from an ego-centric viewpoint: it divides the neighbour overlap by the size of the source node's neighbour set $|N_u|$ and then averages the result over all neighbours and all users.

Further, in order to determine the explicit form of the function $f_i(x)$, boundary conditions should be considered. Specifically, when a user first encounters the information ($x = 1$) and decides to forward it, a proportion γ of its neighbors are already aware of this information. Consequently, the proportion of effective new audience at this moment is $1 - \gamma$. According to this boundary condition, we have $\beta_i(1) = \alpha_i(1 - \gamma)$. Then, for $x > 1$, we plug $\alpha_i = \frac{\beta_i(1)}{1-\gamma}$ into Eq. (2) and obtain

$$f_i(x) = \frac{\ln\left[\frac{\beta_i(x)}{\beta_i(1)}\right] - \ln x}{\ln(1 - \gamma)} + 1 \,. \tag{3}$$

Analyzing data from large-scale real-world social networks shows that the most suitable form of function to approximate $f_i(x)$ is the power-law function [22], expressed specifically as $f_i(x) = x^{\omega_i}$ where ω_i is a parameter reflecting users' uncertainty or deviation in evaluating forwarding benefits of the information. Therefore, the final propagation probability formula becomes

$$\beta_i(\mathbf{x}) = \sigma\left(\sum_{j \in \mathcal{N}_i^-} a_{ij} x_j (1 - \gamma)^{x^{\omega_i}}\right), \tag{4}$$

where the sigmoid layer $\sigma(\cdot)$ bounds the activation probability strictly within $[0, 1]$ and produces a natural saturation effect, which better mirrors the empirical observation that users' forwarding willingness plateaus after a certain number of exposures. In order to show the form of the propagation equation more intuitively, Fig. 2 shows how the propagation probability beta varies as the number of exposures x increases, for different choices of the function parameters.

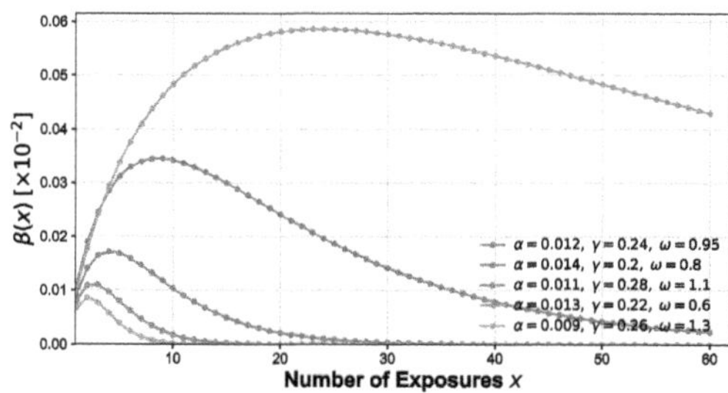

Fig. 2. For different choices of function parameters, the propagation probability varies as the number of exposures x increases.

3.2   Graph Contrastive Representation Learning

Large-scale recommendation scenarios present two major challenges, namely: 1) Label scarcity and incomplete graph structures—due to platform and policy restrictions, the visible social graph available to researchers is often only a sampled subset of the full network. In addition, privacy and access constraints frequently result in partially observed or structurally incomplete graphs, making it difficult to obtain reliable supervision signals; 2) Network noise and dynamic diffusion—weak social ties and machine accounts exist in social edges, while the propagation probability continuously drifts with both time and topic.

Traditional unsupervised methods based on reconstructed adjacency matrices or explicit probability estimation tend to rely excessively on local structure, struggle to generalize to multi-hop diffusion, or lack robustness with respect to noisy edges. To tackle these challenges, we design a GCL approach to learn node representations by maximizing the agreement of node embeddings across two subgraphs without labels. The following subsections therefore focus on the two core components of this pipeline: 1) how two types of subgraphs are constructed through topology- and feature-level perturbations, and 2) how the contrastive loss is formulated given these subgraphs.

(a) **Path-Entropy-Steiner Backbone Augmentation.** The propagation graph is estimated through n-round Monte Carlo sampling and denoted as $\mathcal{T}_r = (V_r, E_r)$. The path set of all nodes is defined as $\mathcal{P}_{u \to v} = \{p = (u, \dots, v) \subseteq \mathcal{T}_r\}$, and the path probability can be obtained as $\Pr(p) = \prod_{(i,j) \in p} \alpha_{ij}$. From an information theoretic perspective, we define the path inverse entropy for each propagation chain p as

$$\mathrm{H}(u, v) \; = \; \frac{1}{- \sum_{p \in \mathcal{P}_{u \to v}} \Pr(p) \log \Pr(p) + \epsilon} . \tag{5}$$

Recall that a Steiner tree in a weighted graph $G = (V, E)$ with a terminal set $T_{term} \subseteq V$ is a tree subgraph that spans all terminals and minimises the total edge cost, while being allowed to introduce additional non-terminal vertices called Steiner nodes [11]. The Steiner tree node set T_{term} is constructed by selecting nodes with the top k inverse entropy scores as $s_v = |V|^{-1} \sum_u \mathrm{H}(u, v)$. We build a minimum Steiner tree and choose the set of edges with minimum cost (most stable propagation) to make the set T_{term} of critical nodes connected, i.e.,

$$\mathcal{B} = \arg \min_{L \subseteq E} \left\{ \sum_{i,j \in T_{term}} \mathrm{H}(i, j) \mid T_{term} \subseteq \bigcup_{(i,j) \in L} \{i, j\} \right\} . \tag{6}$$

As the objective in Eq. (6) is NP-hard, in practice we employ the KMB heuristic to obtain a 2-approximate Steiner backbone. Specifically, we first build a metric closure of the Steiner tree node set T_{term} with edge weight $\tilde{c}_{uv} = 1/\mathrm{H}(u, v)$, compute all-pairs shortest paths, and extract a minimum spanning tree in the closure. The tree is then unfolded back to the original graph, and non-terminal leaves are pruned so that the resulting backbone $\mathcal{B}$ remains sparse. The overall complexity of this procedure is $\mathcal{O}(|T_{term}|(m + n \log n))$ in time and at most $2|T_{term}| - 2$ real edges in space, where n is the number of nodes and m is the number of edges.

We refer to the resulting Steiner backbone as a stable graph for information diffusion. To simulate perturbations in the network structure while preserving its core connectivity, we apply noise exclusively to the non-backbone component. Specifically, we randomly eliminate a subset of edges from the non-backbone graph by sampling a random masking matrix $\tilde{R} \in {0,1}^{k \times k}$, where each entry is independently drawn from a Bernoulli distribution as $\tilde{R}_{ij} \sim$ Bernoulli$(1 - p_r)$ if $A_{ij} = 1$ for the non-backbone graph and $\tilde{R}_{ij} = 0$ otherwise. Here p_r is the probability of each edge being removed. The resulting adjacency matrix can be computed as

$$\tilde{A} = A \circ \tilde{R}, \tag{7}$$

where $(x \circ y)_i = x_i y_i$ is the Hadamard product.

Apart from removing edges, we randomly mask a fraction of dimensions with zeros in node features. Formally, we first sample a random vector $\widetilde{m} \in \{0,1\}^F$ where each dimension of it independently is drawn from a Bernoulli distribution with probability $1 - p_m$, i.e., $\widetilde{m}_i \sim$ Bernoulli$(1 - p_m), \forall i$. Then, the generated node features $\widetilde{X}$ are computed as

$$\widetilde{X} = [x_1 \circ \widetilde{m}; x_2 \circ \widetilde{m}; \cdots ; x_N \circ \widetilde{m}]^{\top}, \tag{8}$$

where $[\cdot ; \cdot]$ denotes the concatenation operator. Finally, following the sequence of operations outline above, we obtain the subgraph $\tilde{G}_1$.

(b) Gramian Control Matrix Augmentation. In the analysis of information diffusion networks, it is essential not only to identify aggregation nodes where information converges but also to accurately locate influential "propagation source" nodes. These source nodes significantly enhance diffusion efficiency, enabling messages initiated from them to rapidly reach extensive areas of the network. To effectively characterize this source-driven diffusion behavior, we adopt the Gramian matrix concept from graph controllability theory. Based on this approach, we propose a subgraph generation strategy that identifies nodes exhibiting strong diffusion capabilities within the network.

We start by defining the propagation transition matrix P with elements $P_{uv} = \beta_{uv}$ if $(u,v) \in E$, 0 otherwise. We then transform P_{uv} into a stochastic transition matrix byperforming the row normalization

$$P_{uv} = D^{-1}A, \quad D_{uu} = \sum_v A_{uv}, \quad A_{uv} = w_{uv}. \tag{9}$$

Each element P_{uv} in $P \in \mathbb{R}^{|V| \times |V|}$ represents the probability of propagating from node u to v in one step. Given P, the propagation controllability Gramian matrix of graph structure is defined as

$$W_c = \sum_{t=0}^{J} P^t (P^t)^{\top}, \tag{10}$$

where P^t is the t-th power of the propagation matrix P, representing the path probability of the t steps propagation of between nodes and J is the truncation parameter of the number of propagation steps, indicating that we only consider the information diffusion ability up to J steps.

In order to accurately identify nodes with strong diffusion ability in the network, we define the propagation controllability score $C(i)$ of the nodes as the sum of the elements of row i in W_c

$$C(i) = \sum_{j \in V} [W_c]_{ij} \,, \tag{11}$$

where $[W_c]_{ij}$ indicates the total propagation energy (or influence ability) that node i can exert on node j over 0 to T propagation steps.

Accordingly, the top k nodes in terms of propagation controllability are selected to form the information diffusion source node set $\{v_1, v_2, ..., v_k\}$. After determining the set of diffusion source nodes, in order to construct a structural view that is complementary to the one obtained with the Path-Entropy-Steiner Backbone Augmentation strategy (information aggregation structure), we further construct an enhanced view that can highlight the information diffusion capability of the network by centering on these source nodes.

Finally, following the same idea as the Path-Entropy-Steiner Backbone Augmentation strategy, we perform feature masking on non-critical nodes and random edge dropping on the global graph to finally obtain the subgraph $\tilde{G}_2$.

(c) **Scaling to Large Graphs.** On a large-scale propagation graph, computing the metrics required by the above two strategies will incur a large computational overhead. To alleviate this bottleneck, we recast the computation as a supervised regression task and employ a Graph Attention Network (GAT) [29] to predict $H(u, v)$ and $C(i)$, as explained below.

We start by randomly initializing the node vectors S_v and T_v to describe the ability of node v to diffuse information outward when it is an "initiator" and its susceptibility to be activated by its neighbors when it is a "receiver", respectively.

For each layer k and direction $d \in \{\mathrm{out}, \mathrm{in}\}$, we define the directional neighbor set $N^d(v)$ and a shared scoring function

$$\psi_{u \to v}^{(k,d)} = \boldsymbol{a}_d^{(k)\top} \left(\boldsymbol{W}_{s,d}^{(k)} \boldsymbol{S}_u^{(k)} \,\|\, \boldsymbol{W}_{t,d}^{(k)} \boldsymbol{T}_v^{(k)} \right), \tag{12}$$

where $\|$ denotes concatenation and $\boldsymbol{a}_d^{(k)}$ is trainable. The normalized attention is the softmax over the corresponding directional neighborhood, i.e.,

$$\omega_{u \to v}^{(k,d)} = \frac{\exp\!\left(\mathrm{LeakyReLU}(\psi_{u \to v}^{(k,d)})\right)}{\sum\limits_{w \in N^c(v)} \exp\!\left(\mathrm{LeakyReLU}(\psi_{w \to v}^{(k,d)})\right)}. \tag{13}$$

We then form a direction-aware aggregation

$$\boldsymbol{m}_v^{(k,d)} \;=\; \sum_{u \in N^d(v)} \left(\eta_d^{(k)} \, \beta_{uv}^{(d)} \;+\; \rho_d^{(k)} \, \omega_{u \to v}^{(k,d)} \right) \boldsymbol{H}_u^{(k)}, \tag{14}$$

where $\beta_{uv}^{(d)}$ is the diffusion prior on edge (u,v) in direction d (e.g., p_{uv} for in), $\boldsymbol{H}_u^{(k)}$ denotes the neighbor-side representation used for message passing ($\boldsymbol{H}=\boldsymbol{S}$ for $d=$ out, $\boldsymbol{H}=\boldsymbol{T}$ for $d=$ in), and $\eta_d^{(k)}, \rho_d^{(k)}$ are learnable scalars.

With this setting to hand, the node representations are updated by combining self-state and directional messages, i.e.,

$$\boldsymbol{S}_v^{(k+1)} = \sigma\!\left(\lambda_s^{(k)} \boldsymbol{S}_v^{(k)} + \lambda_q^{(k)} \boldsymbol{m}_v^{(k,\text{out})} \right), \boldsymbol{T}_v^{(k+1)} = \sigma\!\left(\mu_t^{(k)} \boldsymbol{T}_v^{(k)} + \mu_x^{(k)} \boldsymbol{m}_v^{(k,\text{in})} \right), \tag{15}$$

where $\sigma(\cdot)$ is a nonlinearity (sigmoid by default) and all $\lambda_*^{(k)}, \mu_*^{(k)}$ are scalars. Finally, after K layers, two MLP heads make edge-level predictions using the paired representations,

$$\widehat{H}(u,v) \;=\; \mathrm{MLP}_{\theta_1}\!\left([\boldsymbol{S}_u^{(K)}, \boldsymbol{T}_v^{(K)}]\right), \qquad \widehat{C}(i) \;=\; \mathrm{MLP}_{\theta_2}\!\left([\boldsymbol{S}_u^{(K)}, \boldsymbol{T}_v^{(K)}]\right). \tag{16}$$

(d) Graph Contrastive Node Representations. Let $\widetilde{G}_1$ and $\widetilde{G}_2$ be the two subgraphs generated using the strategies outlined above. We denote their node embeddings as $U = f(\tilde{X}_1, \tilde{A}_1)$ and $V = f(\tilde{X}_2, \tilde{A}_2)$, where $\tilde{X}_*$ and $\tilde{A}_*$ are the feature matrices and adjacency matrices of the subgraphs.

We employ a contrastive objective (i.e., a discriminator) that distinguishes the embeddings of the same node in these two different views from other node embeddings. For any node v_i, its embedding generated in one view, $\boldsymbol{u}_i$, is treated as the anchor, its embedding generated in the other view, $\boldsymbol{v}_i$, forms the positive sample, and embeddings of nodes other than v_i in the two views are naturally regarded as negative samples. Formally, we define the critic $\theta(\boldsymbol{u}, \boldsymbol{v}) = s(g(\boldsymbol{u}), g(\boldsymbol{v}))$, where s is the cosine similarity and g is a nonlinear projection to enhance the expressive power of the critic. The projection g is implemented with a two-layer MLP. Let $Q = \sum_{k=1}^{N} \mathbb{1}_{[k \neq i]} e^{\theta(\boldsymbol{u}_i, \boldsymbol{v}_k)/\tau}, Z = \sum_{k=1}^{N} \mathbb{1}_{[k \neq i]} e^{\theta(\boldsymbol{u}_i, \boldsymbol{u}_k)/\tau}$, We define the pairwise objective for each positive pair $(\boldsymbol{u}_i, \boldsymbol{v}_i)$ as

$$\ell(\boldsymbol{u}_i, \boldsymbol{v}_i) = \log \frac{e^{\theta(\boldsymbol{u}_i, \boldsymbol{v}_i)/\tau}}{e^{\theta(\boldsymbol{u}_i, \boldsymbol{v}_i)/\tau} + Q + Z}, \tag{17}$$

where $\mathbb{1}_{[k \neq i]} \in \{0, 1\}$ is an indicator function that equals to 1 if $k \neq i$, and τ is a temperature parameter. The overall objective to be maximized is then defined as the average over all positive pairs, formally given by

$$\mathcal{J} = \frac{1}{2N} \sum_{i=1}^{N} [\ell(\boldsymbol{u}_i, \boldsymbol{v}_i) + \ell(\boldsymbol{v}_i, \boldsymbol{u}_i)] . \tag{18}$$

3.3 Reinforcement Learning

Based on the obtained node embeddings, the score function to measure the marginal gain of a node $u \in \bar{\mathcal{S}}_t = V \setminus \mathcal{S}_t$ with respect to the current seed set $\mathcal{S}_t$ is defined as

$$\hat{Q}(u, \mathcal{S}_t; \Theta) = \theta_1^\top \mathrm{ReLU}(\theta_2 z_u^{(K)}) \,, \tag{19}$$

where $\Theta = (\theta_1, \theta_2)$ are model parameters and z_u is the node embedding generated by GCL. We train all the parameters end-to-end using RL.

Specifically, we use a DDQN to perform end-to-end learning of the parameters in $\hat{Q}(u_t, \mathcal{S}_t; \Theta)$, as this allows us to avoid the over-optimistic issue of a simple DQN by adopting two networks, a behavior network and a target network, parameterized with Θ and Θ', respectively. The target network provides Q-values estimation of future states during training of the behavior network, and only updates parameters Θ' from the behavior network Θ every episodes. We use the term episode to represent a complete sequence of node additions starting from an empty set until termination, and a single action (node addition) within an episode is referred to as a step. To collect a more accurate estimate of future rewards, n-step Q-learning is utilized to update the parameters, which is to wait for n steps before updating parameters. Additionally, we apply the fit Q-iteration with experience replay for faster learning convergence. Formally, the update is performed by minimizing the square loss

$$(y - \hat{Q}(u_t, \mathcal{S}_t; \Theta))^2 \,, \tag{20}$$

where $y = \sum_{i=0}^{n-1} \rho^i r(\mathcal{S}_{t+i}, u_{t+i}) + \rho^n \max_v \hat{Q}(v, \mathcal{S}_{t+n}; \Theta')$, and $\rho \in [0, 1]$ is the discount rate, determining the importance of future rewards.

4 Experiments

4.1 Experiment Setup

The proposed SP-GCRL is compared with other approaches over eight real-world datasets, including Petster-hamster [16], Tv-show [24], Politician [24], Advogato [24], Public [24], Epinions [16], Twitter [9] and Weibo [9]. We also adopted a mini-dataset radoslaw-email [24] with 167 nodes to provide a qualitative evaluation of a case study. See Table 1 for a summary of dataset statistics . We evaluate our method against a range of representative baselines for influence maximization, encompassing random selection, traditional heuristics, and learning-based approaches. The traditional baseline used is PageRank [23], while learning-driven models comprise gIM [25], S2V-DQN [13], ToupleGDD [5], DeepIM [7], and BIGDN [35].

Edge weights on validation datasets and testing datasets have the same setting. To simulate incompletely observed networks, we process all datasets as follows: we randomly drop 50% of the edges and mask 50% of the node features to capture missing links and partially observed/noisy attributes. All models are evaluated under this setting. For each testing dataset, we vary the budget b such

Table 1. Statistics of the datasets used in this study.

Category	Dataset	Nodes	Edges
Social network	Petster-hamster	921	4,032
Social media pages	Tv-show	3,892	17,262
Social media pages	Politician	5,908	41,729
Trust network	Advogato	6,551	51,317
Social network	Public	11,565	67,114
Review network	Epinions	26,588	100,120
Social media	Twitter	11.6M	341.8M
Social media	Weibo	1.8M	23.8M
Organizational email network	radoslaw-email[†]	167	4,250

[†]For data visualization experiments only

that $b \in \{10, 20, 30, 40, 50\}$. All experiments are conducted on a machine with an Intel(R) Xeon(R) Platinum 8350C CPU (2.60 GHz, 48 cores), 384 GB of DDR4 RAM, an Nvidia RTX 4090 GPU with 24-GB memory, and running Ubuntu 20.04.

4.2 Propagation Equation Fitting

We fit the proposed social propagation equation on six datasets and find that it captures both global trends and local fluctuations of retweet probability with respect to exposure. As shown in Fig. 3 (a)–(f), the model closely matches empirical traces within the 95% confidence intervals, indicating robustness to sampling noise. Parameter analysis shows that α increases from 0.0003 to 0.0009, reflecting substantial variation in baseline diffusion strength, while ω ranges from 0.59 to 1.11, revealing heterogeneous exposure sensitivity. When $\omega \leq 1$, marginal gains from additional exposures diminish, whereas for $\omega > 1$ (Fig. 3 (e)), saturation occurs near $x \approx 6$, suggesting information overload. Overall, the equation provides accurate fitting across diverse networks while offering interpretable parameters for downstream analysis.

4.3 Influence Diffusion

Table 2 reports the expected influence spread achieved by all competing methods under budgets ranging from 10 to 50 on six real-world networks. Across all datasets and budgets, SP-GCRL achieves the highest influence spread and maintains a consistent lead over all baselines. On the small and medium graphs in Table I, the advantage is steady on Petster-hamster and Tv-show and expands with the seed budget; the gap becomes pronounced on Politician, indicating that contrastive representation learning benefits decision quality under sparse and heterogeneous topologies. Table II shows that the lead remains on Advogato,

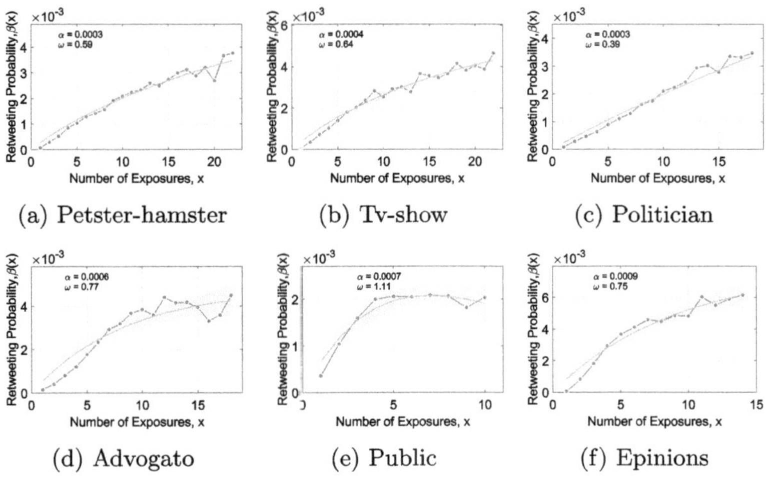

(a) Petster-hamster   (b) Tv-show   (c) Politician

(d) Advogato   (e) Public   (f) Epinions

Fig. 3. Fitting of the exposure–response curve on six datasets.

Public, and Epinions, where spreads are higher overall and fluctuate across budgets, yet SP-GCRL dominates most operating points, reflecting stable gains under differing graph densities and community structures. Table III further demonstrates scalability: on the large-scale Twitter and Weibo networks, SP-GCRL consistently ranks first for every budget, and the separation from the next best methods is maintained or amplified, suggesting that the learned local-to-global invariances improve marginal-gain estimation at scale. Taken together, the results support that coupling GCL with DDQN yields robust, budget-amplified improvements across networks of varying sizes and connectivity patterns.

4.4 Impact of GAT Approximation

We evaluate the impact of the GAT approximation to compute the subgraphs underpinning SBV (Steiner-Backbone View) and CGV (Controllability-Gramian View) on eight graphs of increasing size. Bars report end-to-end speedup when replacing the original strategy with a GAT approximation (higher is better). Lines with shaded bands show the approximation error relative to the non-approximate reference under the same view and its 95% confidence interval.

The speedup increases monotonically with graph size—from small on medium graphs (900–26K nodes) to substantial gains on large graphs (1.8M and 11.6M). Note also that the speedup obtained by the CGV approximation consistently exceeds the one obtained on SBV, with a widening gap at larger scales, suggesting more effective computational

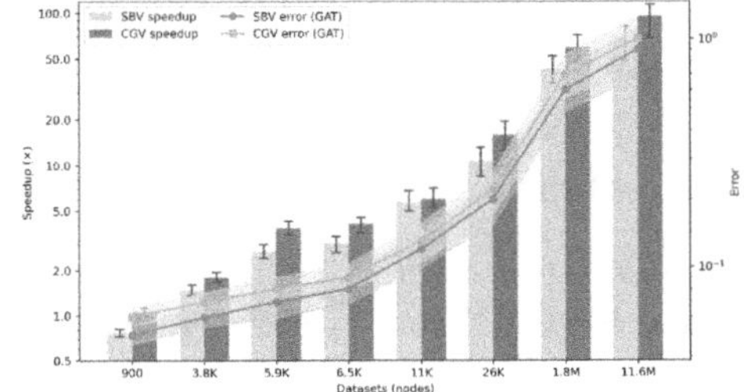

Fig. 4. Speedup and error under the GAT approximation.

Table 2. Performance comparison (I–III): Petster-hamster, Tv-show, Politician; Advogato, Public, Epinions; Twitter and Weibo. Best is highlighted in bold.

Methods (budget)	Petster-hamster					Tv-show					Politician				
	10	20	30	40	50	10	20	30	40	50	10	20	30	40	50
Random	35.9	38.1	92.2	130.8	94.2	23.8	64.5	91.5	79.5	172.2	52.6	41.2	74.7	175.2	124.4
PageRank	12.9	45.2	64.7	75.6	94.2	17.4	27.4	43.4	53.8	63.7	17.7	28.4	90.6	105.4	120.3
gIM	151.6	220.4	249.8	261.6	274.3	82.3	165.9	228.1	264.8	361.0	508.5	649.8	923.5	1128.6	1204.9
S2V-DQN	168.0	177.8	231.5	232.9	274.2	164.3	191.0	234.0	274.4	300.8	160.2	756.3	1,044.6	1,082.5	1,303.5
ToupleGDD	417.7	442.8	423.4	483.0	495.1	533.7	811.3	988.4	1,081.8	1,140.4	2,581.2	2,762.7	2,975.5	3,010.1	3,032.3
DeepIM	360.4	422.2	463.5	483.7	503.5	630.8	749.2	780.5	1,030.5	1,102.3	2,622.0	2,805.5	3,028.7	3189.2	3,312.1
BIGDN	381.4	416.7	450.7	450.2	468.8	751.6	1,114.6	1,208.6	1,362.4	1,424.3	2,627.1	2,870.5	3,063.7	3,176.2	3,251.3
SP-GCRL	**429.0**	**467.6**	**485.2**	**501.4**	**542.6**	**831.0**	**1,186.4**	**1,267.4**	**1,452.2**	**1,511.3**	**2,987.6**	**3,102.8**	**3,178.6**	**3,245.7**	**3,329.0**

Methods (budget)	Advogato					Public					Epinions				
	10	20	30	40	50	10	20	30	40	50	10	20	30	40	50
Random	28.8	116.0	216.0	102.6	181.9	17.3	48.7	111.9	118.1	251.6	20.8	58.4	134.3	141.7	301.9
PageRank	95.7	166.4	286.8	372.7	434.4	26.0	58.5	68.8	93.5	106.7	31.2	70.2	82.6	112.2	228.0
gIM	2,845.7	2,855.7	2,865.6	2,870.2	2,874.3	904.6	1,069.7	2,039.1	2,245.1	2,291.4	597.4	701.3	1,350.9	1,493.4	1,510.7
S2V-DQN	3,669.4	3,669.0	3,668.5	36,69.0	3,666.4	785.1	977.8	1,535.4	1,530.3	1,578.2	1,179.8	1,179.7	1,201.9	1,605.0	2,157.1
ToupleGDD	3,670.1	3,672.6	3,673.3	3,676.4	3,681.0	5,397.8	5,440.4	5,502.9	5,583.6	5,705.1	3,526.1	3,585.1	3,658.8	3,659.8	3,696.7
DeepIM	3,674.1	3,679.4	3,681.1	3,686.4	3,689.6	1,600.0	2,272.5	3,779.3	3,854.9	3,879.7	3,570.3	3,607.1	3,752.3	3,817.0	3,899.2
BIGDN	3,669.4	3,669.1	3,669.4	3,668.9	3,668.1	5,348.9	5,550.7	5,670.6	5,726.7	5,940.7	3,506.3	3,612.4	3,732.5	3,876.6	3,960.8
SP-GCRL	**3,702.1**	**3,698.0**	**3710.7**	**3,713.5**	**3,696.4**	**5,514.6**	**5,766.5**	**5,820.2**	**5,932.2**	**6,035.7**	**3,641.7**	**3,780.6**	**3,856.0**	**3,946.0**	**3,979.2**

Methods (budget)	Twitter					Weibo				
	10	20	30	40	50	10	20	30	40	50
Random	28,886	116,348	216,648	102,908	182,446	2,699	7,597	17,456	18,424	39,250
PageRank	95,987	166,899	287,660	373,818	435,703	4,056	9,126	10,733	14,586	16,645
gIM	2,850,943	2,880,517	2,910,268	2,940,733	2,970,184	140,736	170,284	300,912	340,587	360,941
S2V-DQN	3,380,642	3,401,213	3,423,567	3,446,118	3,478,905	120,438	150,217	235,906	242,511	249,873
BIGDN	3,452,831	3,472,409	3,503,276	3,534,882	3,566,341	824,728	865,903	896,447	914,205	936,582
ToupleGDD	3,521,764	3,542,395	3,563,981	3,594,257	3,621,803	782,315	813,604	842,997	861,732	883,419
DeepIM	3,602,417	3,623,058	3,645,221	3,666,774	3,688,932	241,368	332,741	523,518	582,906	624,177
SP-GCRL	**3,683,549**	**3,705,194**	**3,728,611**	**3,744,973**	**3,755,627**	**864,932**	**905,711**	**934,268**	**963,754**	**989,321**

reuse and parallelism. Errors grow mildly with size but remain acceptable; confidence bands are narrow on small graphs and broaden slightly at the million-node scale, with substantial overlap between views and no systematic bias. All results are averaged over multiple independent runs with fixed Monte-Carlo settings and identical random seeds; confidence intervals are obtained via bootstrap over run-level errors. Overall, the GAT approximation delivers scale-amplified runtime benefits for both views without noticeable accuracy loss, with CGV exhibiting the more stable efficiency advantage (Fig. 4).

4.5   Ablation Study

To assess the roles of GCL and view augmentation, we construct an ablation model SP-GNN that is identical to SP-GCRL in diffusion module, reward definition, DDQN decision head, and training configuration, but replaces the encoder with an L-layer GNN and removes both structural view augmentations and the local–global contrastive loss. As shown in Fig. 5, SP-GCRL achieves larger diffusion across all budgets and the advantage widens with increasing budget: on Tv-show and Petster-hamster the gain is about 3–5%; on Politician it reaches roughly twenty percent at budget 50; and on Advogato both methods fluctuate around 3.68–3.72×10^3 yet SP-GCRL maintains a 1% edge at most points with a notable jump at budget 30. The shaded bands indicate that these improvements are stable under randomness.

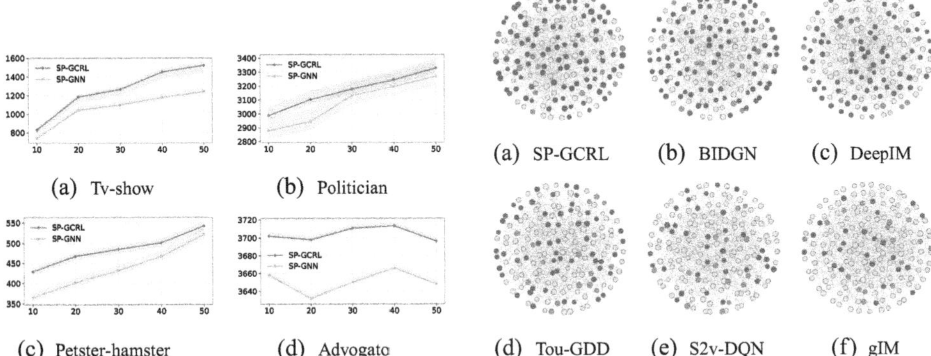

Fig. 5. Comparison of SP-GCRL and SP-GNN on four networks.

Fig. 6. Visualization on radoslaw-email (red: seeds; blue: influenced; grey: unaffected). (Color figure online)

4.6 Graph Diffusion Visualization

Finally, we conduct a case study to demonstrate the distribution of a selected set of seed nodes as well as the final propagation status of all nodes in Fig. 6, where red nodes indicate the initial seed node, blue nodes indicate the infected node during the influence spread, and grey nodes represent uninfected nodes. Visually, SP-GCRL disperses seeds more evenly and achieves the densest, most homogeneous cascade, whereas BIDGN, DeepIM and ToupleGDD cluster seeds in local communities, S2V-DQN and gIM leave sizeable peripheral regions uninfected, resulting in sparser overall spread.

5   Conclusion

In this work, we present SP-GCRL for influence maximization on social graphs with missing edges and noisy features. The framework combines a data-fitting propagation function, two structure-view augmentations for contrastive representation learning, a lightweight GAT surrogate that replaces costly subgraph construction, and a Double-DQN policy for seed selection. Evaluations on real networks from thousands to millions of nodes show that SP-GCRL consistently improves expected spread while reducing computation and memory. These results indicate that SP-GCRL is a practical and scalable solution for influence maximization on large, imperfect graphs.

Acknowledgments. This research is supported by the National Laboratory of Space Intelligent Control (No. HTKJ2023KL502003) and the Key Laboratory of Intelligent Space TTC&O (Space Engineering University), Ministry of Education (No. CYK2025-01-07).

Disclosure of Interests. The authors have no competing interests to declare that are relevant to the content of this article.

References

1. Aiyappa, R., Flammini, A., Ahn, Y.Y.: Emergence of simple and complex contagion dynamics from weighted belief networks. Sci. Adv. **10**(15), eadh4439 (2024)
2. Ali, K., Wang, C.Y., Chen, Y.S.: Boosting reinforcement learning in competitive influence maximization with transfer learning. In: 2018 IEEE/WIC/ACM International Conference on Web Intelligence (WI), pp. 395–400. IEEE (2018)
3. Bail, C.A., et al.: Exposure to opposing views on social media can increase political polarization. Proc. Natl. Acad. Sci. **115**(37), 9216–9221 (2018)
4. Chen, H., Qiu, W., Ou, H.C., An, B., Tambe, M.: Contingency-aware influence maximization: a reinforcement learning approach. In: Uncertainty in Artificial Intelligence, pp. 1535–1545. PMLR (2021)
5. Chen, T., Yan, S., Guo, J., Wu, W.: Touplegdd: a fine-designed solution of influence maximization by deep reinforcement learning. IEEE Trans. Comput. Soc. Syst. **11**(2), 2210–2221 (2023)
6. Chen, W., Yuan, Y., Zhang, L.: Scalable influence maximization in social networks under the linear threshold model. In: 2010 IEEE International Conference on Data Mining, pp. 88–97. IEEE (2010)
7. Chen, X., Lei, P.I., Sheng, Y., Liu, Y., Gong, Z.: Social influence learning for recommendation systems. In: Proceedings of the 33rd ACM International Conference on Information and Knowledge Management, pp. 312–322 (2024)
8. Du, W., Zhang, S., Di Wu, J.X., Zhao, Z., Fang, J., Wang, Y.: Mmgnn: a molecular merged graph neural network for explainable solvation free energy prediction. In: Proceedings of the Thirty-Third International Joint Conference on Artificial Intelligence, pp. 5808–5816 (2024)
9. Feng, Y., Tan, V.Y., Cautis, B.: Influence maximization via graph neural bandits. In: Proceedings of the 30th ACM SIGKDD Conference on Knowledge Discovery and Data Mining, pp. 771–781 (2024)
10. Hassani, K., Khasahmadi, A.H.: Contrastive multi-view representation learning on graphs. In: International Conference on Machine learning, pp. 4116–4126. PMLR (2020)
11. Hwang, F.K., Richards, D.S.: Steiner tree problems. Networks **22**(1), 55–89 (1992)
12. Kempe, D., Kleinberg, J., Tardos, É.: Maximizing the spread of influence through a social network. In: Proceedings of the Ninth ACM SIGKDD international Conference on Knowledge Discovery and Data Mining, pp. 137–146 (2003)

13. Khalil, E., Dai, H., Zhang, Y., Dilkina, B., Song, L.: Learning combinatorial optimization algorithms over graphs. Adv. Neural Inf. process. Syst. **30** (2017)
14. Klages-Mundt, A., Minca, A.: Optimal intervention in economic networks using influence maximization methods. Eur. J. Oper. Res. **300**(3), 1136–1148 (2022)
15. Kumar, S., Mallik, A., Khetarpal, A., Panda, B.S.: Influence maximization in social networks using graph embedding and graph neural network. Inf. Sci. **607**, 1617–1636 (2022)
16. Kunegis, J.: Konect: the koblenz network collection. In: Proceedings of the 22nd International Conference on World Wide Web, pp. 1343–1350 (2013)
17. Leskovec, J., Krause, A., Guestrin, C., Faloutsos, C., VanBriesen, J., Glance, N.: Cost-effective outbreak detection in networks. In: Proceedings of the 13th ACM SIGKDD International Conference on Knowledge Discovery and Data Mining, pp. 420–429 (2007)
18. Li, H., Xu, M., Bhowmick, S.S., Rayhan, J.S., Sun, C., Cui, J.: Piano: influence maximization meets deep reinforcement learning. IEEE Trans. Comput. Soc. Syst. **10**(3), 1288–1300 (2022)
19. Lin, S.C., Lin, S.D., Chen, M.S.: A learning-based framework to handle multi-round multi-party influence maximization on social networks. In: Proceedings of the 21th ACM SIGKDD International Conference on Knowledge Discovery and Data Mining, pp. 695–704 (2015)
20. Ling, C., et al.: Deep graph representation learning and optimization for influence maximization. In: International Conference on Machine Learning, pp. 21350–21361. PMLR (2023)
21. Ma, L., et alK.: Influence maximization in complex networks by using evolutionary deep reinforcement learning. IEEE Trans. Emerg. Top. Comput. Intell. **7**(4), 995–1009 (2022)
22. Meng, F., Xie, J., Sun, J., Xu, C., Zeng, Y., Wang, X., Jia, T., Huang, S., Deng, Y., Hu, Y.: Spreading dynamics of information on online social networks. Proc. Natl. Acad. Sci. **122**(4), e2410227122 (2025)
23. Page, L., Brin, S., Motwani, R., Winograd, T.: The pagerank citation ranking: Bringing order to the web. Tech. rep., Stanford infolab (1999)
24. Rossi, R.A., Ahmed, N.K.: The network data repository with interactive graph analytics and visualization. In: AAAI (2015). https://networkrepository.com
25. Shahrouz, S., Salehkaleybar, S., Hashemi, M.: gim: Gpu accelerated ris-based influence maximization algorithm. IEEE Trans. Parallel Distrib. Syst. **32**(10), 2386–2399 (2021)
26. Tang, Y., Shi, Y., Xiao, X.: Influence maximization in near-linear time: a martingale approach. In: Proceedings of the 2015 ACM SIGMOD International Conference on Management of Data, pp. 1539–1554 (2015)
27. Tang, Y., Xiao, X., Shi, Y.: Influence maximization: near-optimal time complexity meets practical efficiency. In: Proceedings of the 2014 ACM SIGMOD International Conference on Management of Data, pp. 75–86 (2014)
28. Tian, S., Mo, S., Wang, L., Peng, Z.: Deep reinforcement learning-based approach to tackle topic-aware influence maximization. Data Sci. Eng. **5**, 1–11 (2020)
29. Velickovic, P., Cucurull, G., Casanova, A., Romero, A., Lio, P., Bengio, Y., et al.: Graph attention networks. Stat **1050**(20), 10–48550 (2017)
30. Wang, C., Chen, W., Wang, Y.: Scalable influence maximization for independent cascade model in large-scale social networks. Data Min. Knowl. Disc. **25**, 545–576 (2012)
31. Wu, S., Sun, F., Zhang, W., Xie, X., Cui, B.: Graph neural networks in recommender systems: a survey. ACM Comput. Surv. **55**(5), 1–37 (2022)

32. Ye, M., Liu, X., Lee, W.C.: Exploring social influence for recommendation: a generative model approach. In: Proceedings of the 35th International ACM SIGIR Conference on Research and Development in Information Retrieval, pp. 671–680 (2012)
33. You, Y., Chen, T., Sui, Y., Chen, T., Wang, Z., Shen, Y.: Graph contrastive learning with augmentations. Adv. Neural. Inf. Process. Syst. **33**, 5812–5823 (2020)
34. Zhang, G., et al.: G-designer: Architecting multi-agent communication topologies via graph neural networks. arXiv preprint arXiv:2410.11782 (2024)
35. Zhu, W., Zhang, K., Zhong, J., Hou, C., Ji, J.: Bigdn: an end-to-end influence maximization framework based on deep reinforcement learning and graph neural networks. Expert Syst. Appl. 126384 (2025)
36. Zhu, Y., Xu, Y., Liu, Q., Wu, S.: An empirical study of graph contrastive learning. arXiv preprint arXiv:2109.01116 (2021)

ANS-CD: A Novel Label-Efficient Community Detection Approach via Active Node Selection

Junchang Xin[1,2], Sihan Liu[1], Mingcan Wang[1], Kaifu Long[1],
Chenxi Yao[1], and Zhiqiong Wang[3]($\boxtimes$)

[1] School of Computer Science and Engineering, Northeastern University, Shenyang,
China
xinjunchang@mail.neu.edu.cn, {liush7,longkf1}@mails.neu.edu.cn,
chenxiyao@stumail.neu.edu.cn
[2] Key Laboratory of Big Data Management and Analytics (Liaoning Province),
Northeastern University, Shenyang, China
[3] College of Medicine and Biological Information Engineering, Northeastern
University, Shenyang, China
wangzq@bmie.neu.edu.cn

Abstract. Community detection is a fundamental task that aims to partition a network into cohesive node groups. In real-world networks, partial prior information is available. However, existing community detection methods that utilize prior information ignore the sensitivity to different prior information and the dynamic evolution of node relevance to the expanding community, which lead to unstable results and an imbalance between annotation cost of prior information and performance. To address these issues, we propose ANS-CD, a community detection method via active node selection. Specifically, ANS-CD adopts a multi-perspective strategy to identify valuable nodes that exhibit high representativeness or uncertainty as core community seeds, ensuring efficient use of limited annotations. It then expands communities around them through an innovative conflict triggered labeling mechanism, which corrects the conflicts between communities and nodes during the expansion process. Therefore, compared with traditional methods that rely on prior information, ANS-CD can extract valuable prior information and dynamically capture high-valuable information with low annotation costs. Extensive experiments on real-world attributed networks demonstrate that ANS-CD outperforms comparison methods with fewer labeled nodes.

Keywords: Community detection · Active node selection ·
Annotation cost · Prior information

1 Introduction

Many real-world task relationships can be modeled as graphs of nodes and edges
[21]. In real networks, nodes are characterized topology and attribute infor-

H. Jung et al. (Eds.): DASFAA 2026, LNCS 16536, pp. 351–366, 2026.
https://doi.org/10.1007/978-981-92-0366-6_22

mation [20], highlighting the importance of community detection in attributed networks. While some studies focus on local community detection efficiently find communities for query nodes, they ignore global structures.

Diverse community detection methods have recently been developed. Methods for optimization-based methods, such as modularity optimization [1] find community structures by objective maximization. The methods based on representation learning, such as NMF [10] and graph embeddings like DeepWalk [18], effectively capture latent representations. Probabilistic inference models, typified by MRF [7] and SBM families [8], interpret community formation via generative processes and inference mechanism with strong theoretical interpretability. In real-world networks, partial prior information is available. However, traditional methods generally ignore the mining and utilization of this information, thus limiting the further improvement of model performance.

Recent semi-supervised methods have attempted to integrate prior information into community detection [9,14]. However, most of them depend on randomly partitioning the training set or randomly sampling priors. This means that they fail to balance prior information and performance. Some studies introduce active node selection strategies to actively identify valuable prior information. AC [5] iteratively query pairwise constraints for high-utility nodes, favoring high-degree nodes but overlook boundary or small communities. Our work tends to handle these problems and proposes a multi-perspective strategy to identify valuable nodes with efficient use of labels and high performance.

Therefore, an ANS-CD method is proposed, in which graph active selection is integrated with community detection to maximize model performance while minimize manual annotation costs. Specifically, ANS-CD actively selects nodes by jointly considering representativeness, uncertainty, and structuralattribute inconsistency during the preprocessing, as well as the conflicts between nodes and communities during the community expansion process.

The key contributions of this work are summarized as follows in four points:

1. Proposing a multi-perspective active node selection strategy that labeling nodes exhibiting high representativeness or uncertainty.
2. Designing a DivRank-based label coverage strategy to ensure full coverage of information in the network while reducing labeled nodes required.
3. Devising a progressive active selection strategy to manually annotate nodes conflicting with the current community during community expansion, thus correcting node-community conflicts in the expansion process.
4. Demonstrating the superiority of ANS-CD on four real-world attributed networks, where it outperforms both unsupervised and semi-supervised methods with fewer labeled nodes.

2   Related Work

2.1   Prior Information-Based Community Detection

Compared with unsupervised community detection, the incorporation of prior information can lead to significant performance gains. For instance, DSSC employs PMI to construct a pairwise constraint matrix, thereby introducing structured prior information for clustering [3]. DEGCN applies k-core filtering for graph reduction and leverages dual representations of topology and attribute associations, while proportionally sampling labeled communities [19]. In semi-supervised settings for community detection, random or proportional sampling strategies are usually adopted, which makes it impossible to fully explore the potential information of the training set. In addition, semi-supervised methods based on community-concept transfer depend on high-quality known communities [17], but are restricted by local structures, making it difficult to capture potential patterns beyond known community patterns. In general, these approaches have a fundamental weakness: the lack of active selection strategy for prior information. To address this weakness, ANS-CD employs a multi-perspective active node selection strategy to effectively identify valuable prior information.

2.2   Active Node Selection in Community Detection

In graph domains, manual annotation is costly, a practical way to reduce labeling effort is to actively select informative nodes that can most enhance model performance. Early studies such as AGE [4] integrate multiple criteria including uncertainty, representativeness, and node centrality to select highly informative nodes for manual annotation, thereby improving semi-supervised embedding and node classification performance. In semi-supervised clustering and community detection with pairwise constraints, active link or node-pair querying strategies have been proposed [12], which actively selects the most valuable pairwise constraints or links to manual annotation. However, these methods are limited to the preprocessing stage and fail to identify informative nodes during dynamic community expansion. To overcome this drawback, ANS-CD integrates a conflict triggered Labeling(CTL) mechanism that resolves nodecommunity conflicts and actively selects informative nodes for manual annotation during community expansion.

3   Proposed Framework

In this section, the problem definitions and notation descriptions are first provided. Then, the proposed ANS-CD framework is presented, which achieves high-performance community detection by actively selecting informative nodes during both preprocessing stage and community expansion.

3.1   Notations and Problem Definitions

A graph $G = (\mathcal{V}, \mathcal{E}, \mathbf{X})$ is defined by a set of nodes $\mathcal{V} = (v_1, ..., v_n)$, a set of edges $\mathcal{E}$, and an attribute matrix $\mathbf{X} \in \mathbb{R}^{n \times d}$. Let N_v denotes the neighborhood of a node $v \in \mathcal{V}$. Each edge $e_{ij} \in \mathcal{E}$ represents a link between v_i and v_j, while the row $\mathbf{x}_i$ in $\mathbf{X}$ represents the feature vector of node v_i.

The goal of community detection is to partition all nodes in a network into groups such that nodes within the same group are densely connected, while the connections between nodes in different groups are sparse. This study focuses on non-overlapping and global community detection. Formally, the community structure is defined as $\mathcal{CS} = \{C_1, C_2, \ldots, C_K\}, K \geq 1$, where K denotes the number of communities in the graph. To ensure the non-overlapping property, any two distinct communities must be disjoint, $\forall q \neq l,\ C_q \cap C_l = \emptyset$. At the same time, to ensure complete coverage of the network, the union of all communities must include every node: $\bigcup_{q=1}^{K} C_q = V$.

3.2   General Introduction of ANS-CD

To achieve global non-overlapping community detection in attribute graphs, a innovatively ANS-CD method is proposed. The overall workflow of ANS-CD is shown in Fig. 1. Nodes that are not assigned a community are outlined in gray; nodes that are assigned a community are outlined in black. Four types of nodes are distinguished in the framework: ordinary nodes, representative nodes, uncertain nodes, and CTL-satisfying nodes. The ANS-CD method first extracts adjacency matrix and node attribute information from the input attributed network. Based on the topological structure, the DivRank algorithm [15] can be used to identify representative nodes. Combining the topological structure and node attributes, the new ONE method can be used to mine uncertain nodes. The union of these two node sets constitutes the preprocessed selected nodes, which are subsequently queried and labeled by an oracle. These nodes are used as the initialization for each community. Next, the community expansion process iterates r rounds, each of which consists of three steps: (1) screening and expanding the community among candidate nodes, (2) finding nodes that meet the CTL conditions and manually querying them, and (3) manually labeling the nodes that meet the CTL. Finally, nodes with duplicate community labels and missed nodes are post-processed to ensure the completeness and consistency of global community detection.

3.3   Preprocessing for Community Node Selection

Uncertain Nodes Based on NEW-ONE. The Outlier-Aware Network Embedding (ONE) is an unsupervised learning algorithm tailored for attributed networks [2] that jointly integrates structural and attribute information to learn compact low-dimensional embeddings while reducing the impact of outliers. Specifically, ONE algorithm finds to minimize a cost function L, which projects all nodes into a k-dimensional embedding space, where k denotes the embedding dimension. The function L is specified as follows: minimize the following cost function.

$$L = L_{str} + \alpha L_{att} + \beta L_{union} \tag{1}$$

where L_{str} ,L_{att} , L_{union} denote the loss functions learned from the structure, attribute, and disagreement in structure and attributes, respectively. The

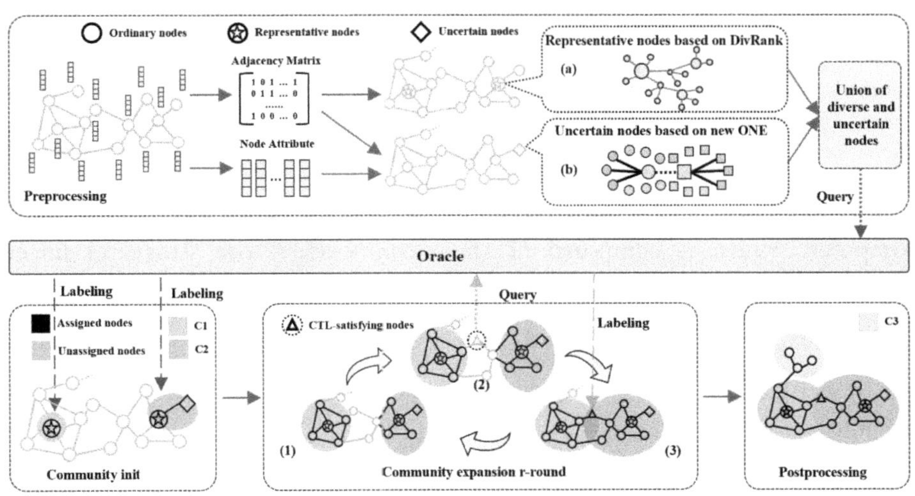

Fig. 1. The overall process of ANS-CD.

detailed formulation is not repeated. In this study, the principle of ONE method is utilized as the foundation for identifying uncertain nodes in attributed graphs.

As discussed above, identifying nodes with high uncertainty is crucial in active node selection. Such nodes often deviate from the overall pattern within a community and exhibit characteristics similar to the outliers defined in the ONE method. To improve computational efficiency, outlier scores are not incorporated into the loss function. Based on the ONE method, the New-ONE method is further developed. After obtaining low-dimensional embeddings from ONE, New-ONE evaluates node's uncertainty by jointly considering structural reconstruction error, attribute reconstruction error, and the inconsistency between structural and attribute embeddings, thereby enabling more effective identification of uncertain nodes. Formula 1 shows the comprehensive cost function. In New-ONE, the $\mathcal{L}_{str}$, $\mathcal{L}_{att}$ and $\mathcal{L}_{dis}$ function is specified as follows:

$$\mathcal{L}_{str} = \sum_{i=1}^{N}\sum_{j=1}^{N}\left(A_{ij} - G_{i\cdot}\cdot H_{\cdot j}\right)^2 \tag{2}$$

$$\mathcal{L}_{att} = \sum_{i=1}^{N}\sum_{d=1}^{N}\left(X_{id} - U_{i\cdot}\cdot V_{\cdot d}\right)^2 \tag{3}$$

$$\mathcal{L}_{dis} = \sum_{i=1}^{N}\sum_{k=1}^{K}\left(G_{ik} - U_i\cdot(W^T)_{\cdot k}\right)^2 \tag{4}$$

Let $A \in \mathbb{R}^{N \times N}$ denotes the adjacency matrix, which is factorized into $G \in \mathbb{R}^{N \times K}$ and $H \in \mathbb{R}^{K \times N}$. The matrix G, derived from this decomposition, represents the embedding that captures only structural information. Similarly, the attribute matrix X is factorized into $U \in \mathbb{R}^{N \times K}$ and $V \in \mathbb{R}^{K \times D}$, where U serves as the embedding reflecting attribute information. To address the potential misalignment between G and U, a linear transformation matrix $W \in \mathbb{R}^{K \times K}$ (Formula 4) is employed to align the two spaces, thereby minimizing the disagreement loss L union. The update rules for G, H, U, V and W follow those in ONE, except that the contribution of the outlier scores in the loss function is disregarded, which is equivalent to setting them to 1.

Simultaneously, a comprehensive uncertain score O_i (Formula 5) is computed, which is composed of three components: the structural uncertain score $O1_i = \exp\left(-\frac{\|A_{i,:}-G_iH\|_2^2}{2\sigma_1^2}\right)$, the attribute uncertain score $O2_i = \exp\left(-\frac{\|C_{i,:}-U_iV\|_2^2}{2\sigma_2^2}\right)$, and the structuralattribute disagreement score $O3_i = \exp\left(-\frac{\|G_i-U_iW\|_2^2}{2\sigma_3^2}\right)$ that captures inconsistencies between structure and attributes.

$$O_i = \frac{w_1\,O1_i + w_2\,O2_i + w_3\,O3_i}{w_1 + w_2 + w_3} \tag{5}$$

Here w_1, w_2, and w_3 are weighting factors controlling the contributions of structure, attributes, and their disagreement, respectively. Consequently, nodes with larger outlier scores O are considered greater uncertainty in the network.

However, focusing solely on nodes with high uncertainty in the network is insufficient, as this approach may overlook potential communities that already exhibit clear structural patterns. Therefore, it is necessary to further identify representative nodes that exhibit significant structural cohesion and have the potential to become community cores.

Representative Nodes Based on DivRank. Ma et al. proposed DivRank [15], a novel ranking algorithm that automatically balances centrality and diversity in the top-ranked items, ensuring that the selected top-k nodes are both influential and non-redundant. The central idea of DivRank is the "rich-gets-richer" principle: transition probabilities between vertices are dynamically adjusted based on the cumulative number of visits (or expected visits) to each vertex over time. This reinforcement process fosters competition among adjacent vertices—vertices with higher accumulated weights "absorb" weights from their neighbors, reducing redundancy and enhancing diversity by reducing overlap among the top-ranked nodes.

DivRank [15] is applied to select the top-k representative nodes while maintaining diversity. The parameter k is initialized to the number of labels and is increased until the representatives achieve complete label coverage. Let S denote the set of selected representative nodes and L the set of all labels in the graph. Each node $v \in S$ is associated with a label $\ell(v) \in L$. The label coverage $c(S)$ is then defined as:

$$c(S) = \frac{|\{\ell(v) \mid v \in S\}|}{|L|} \tag{6}$$

Here, $|\{\ell(v) \mid v \in S\}|$ is the number of distinct labels covered by S, and $|L|$ is the total number of labels in the graph. As a result, this strategy minimizes $cost_{re}$

in the manual annotation cost while inherently balancing representativeness and diversity among the chosen nodes.

Algorithm 1. Preprocessing for Initial Community Node Selection

---

Require: Graph $G = (\mathcal{V}, \mathcal{E}, \mathbf{X})$, label set L, budget of uncertainty B_{un}
Ensure: Selected nodes set s_nodes, number of selected representative nodes k
1: Initialize $k \leftarrow |L|$
2: $S \leftarrow \emptyset$
3: $s\_nodes \leftarrow \emptyset$
4: **while** $c(S) < 100\%$ **do**
5:     $S \leftarrow$ Select B_{un} nodes with the highest scores
6:     $c(S) \leftarrow$ Compute label coverage $c(S)$ based on L by Formula 6.
7:     $k \leftarrow k + 1$
8: **end while**
9: Compute uncertain scores using new ONE
10: Let $C_{un} \leftarrow$ top-k nodes with highest uncertain scores
11: $s\_nodes \leftarrow S \cup C_{un}$
12: **return** s_nodes, k

Algorithm 1 provides a preprocessing process for community node selection. This method merges high uncertain nodes with top-k DivRank nodes that can guarantee 100% label coverage to obtain the initial community node set.

3.4 Community Expansion

After selecting valuable nodes, including both representative and uncertain ones, as community cores, the next step is to perform community expansion around them.

Objective Function. The objective is to enhance intra-community connectivity and attribute homogeneity while simultaneously preserving inter-community sparsity and attribute diversity. An objective function that jointly exploits attribute and structural information is formulated for the expansion step:

$$C_o = m\, C_a^c + (1 - m)\, C_s^c \tag{7}$$

where m denotes a trade-off parameter that balances the contributions of attribute cohesiveness C_a^c and structural cohesiveness C_s^c.

Intuitively, C_a^c is the average attribute similarity among all pairs of nodes within community c, which is defined as follows:

$$cosine(vec_i, vec_j) = \frac{vec_i \cdot vec_j}{\|vec_i\|\|vec_j\|} \tag{8}$$

where $\|vec_i\|$ means the norm of vec_i

$$C_a^c = \frac{2\left(\sum_{i \in c}\sum_{j \in c, i < j} \cos(\boldsymbol{vec}_i, \boldsymbol{vec}_j) - |c|\right)}{|c| \times (|c| - 1)} \tag{9}$$

where $|c|$ denotes the number of nodes in community c, S_{ij} the cosine similarity of the attribute vectors of nodes i and j, $i, j \in V$. Then, the structural cohesiveness of a community is considered, which is evaluated from multiple perspectives, including the average internal degree of its nodes $d_c = 2e_{in}/|c|$ as well as the distribution of internal and external edges, represented by $M_c = e_{in}/e_{out}$ and $\rho_c = 2e_{in}/(|c| - 1)$. Then, d_c and M_c are normalized to the interval $[0, 1]$, giving $\bar{d}_c^{\text{norm}} = \bar{d}_c/(\bar{d}_c + 1)$ and $\bar{M}_c^{\text{norm}} = \bar{M}_c/(\bar{M}_c + 1)$.

$$C_s^c = \begin{cases} 1 & |c| = 1 \\ \frac{\rho_c + \bar{d}_c^{\text{norm}} + \bar{M}_c^{\text{norm}}}{3} & |c| > 1 \end{cases} \tag{10}$$

where, e_{in} and e_{out} denote the numbers of internal and external edges of a community c, respectively. Specifically, $e_{in} = |\{(u, v) \in E \mid u \in c \text{ and } v \in c\}|$, $e_{out} = |\{(u, v) \in E \mid u \in c \text{ and } v \notin c\}|$.

Another widely adopted objective function in community detection is modularity, which is proposed by Mark Newman and Michelle Girvan to measure the quality of community structures in the complex networks [16]. The modularity Q is the sum of quality scores over all communities in the network, which means that the quantity of a single community Q_c can be defined by:

$$Q_c = \left(\frac{E_c}{e} - \left(\frac{D_c}{2e}\right)^2\right) \tag{11}$$

where e denotes the total number of edges in the entire graph, E_c/e denotes the fraction of edges inside community c, and $D_c/2e$ represents the expected fraction of such edges in a random graph. A higher Q_c indicates a stronger and more cohesive community structure.

Conflict-Triggered Labeling. During community expansion, a conflict triggered labeling (CTL) mechanism is proposed to adapt to the dynamically changing characteristics of unassigned nodes. A node is considered potentially anomalous or part of a heterogeneous structure if, upon joining a new community, it reduces the community's structural or attribute cohesiveness, or if it is structurally connected but attribute-wise divergent, or attribute-similar but structurally disconnected. Those with the highest conflict scores relative to the current community are selected for manual labeling. CTL is formula by:

$$\text{CTL}(v) = \left\{v \,\middle|\, \left|C_a^c(S \cup \{v\}) - C_s^c(S \cup \{v\})\right| > \delta \ \lor \ \Delta O_c(v) < 0\right\} \tag{12}$$

where δ is the threshold used to determine whether adding node v to community S would create a significant discrepancy between the attribute cohesion C_a^c and structural cohesion C_s^c.

Algorithm 2. Community Expansion from Selected Nodes

Require: Initial community node set s_nodes, graph $G = (\mathcal{V}, \mathcal{E}, \mathbf{X})$, B_{CTL}, number of expansions r

Ensure: Expanded community c, $cost_{CTL}$

1: Initialize $c \leftarrow s\_nodes$, $cost_{CTL} = 0$
2: Compute initial C_a^c and C_s^c by Formula 9 and Formula 10
3: **for** $iter = 1$ to L **do**
4:     $cand \leftarrow \{u \in N_c \mid \Delta Co(u) = (C_a^{c \cup u} - C_a^c) + (C_s^{c \cup u} - C_s^c) > 0\}$
5:     **if** $cand = \emptyset$ **then**
6:         **break**
7:     **end if**
8:     Sort $cand$ in descending order by Co
9:     $C_{opt} \leftarrow$ sequentially add $u \in cand$ maximizing Co
10:    **if** u satisfies CTL condition Formula 12 **and** u is unlabeled **then**
11:       Query by oracle, $cost_{CTL}{+}{+}$
12:    **end if**
13:    **if** $C_h = \emptyset$ **then**
14:       $c \leftarrow C_{opt}$, add c to C_h
15:       **continue**
16:    **end if**
17:    $C_{max} \leftarrow$ historical community in C_h with highest Co
18:    $c \leftarrow \arg\max \{Co(C_{opt}), Co(C_{max})\}$, $C_h \leftarrow C_h \cup c$, update N_c
19: **end for**
20: **return** c, $cost_{CTL}$

Algorithm 2 presents the process of selecting candidate nodes based on the objective function C_o using a greedy strategy, followed by community expansion through CTL.

Post-processing of Community Assignments. To guarantee that the detected communities are both non-overlapping and completely covering the network, Algorithm 3 conducts the post-processing of community assiginments. It consists of two steps, one for handling nodes with duplicated community assignments (line 2–6), and the other for addressing nodes with missing community labels(line 7–20). As a result, the final community detection output is obtained.

3.5 Minimizing the Cost of Manual Annotation

To minimize manual annotation costs under a given budget constraint while ensuring that all nodes in the graph are assigned to non-overlapping communities and achieving optimal performance, a costbudget scheme comprising three components is devised. The total budget can be formulated as follows:

$$B = B_{re} + B_{un} + B_{CTL} \tag{13}$$

where B_{re}, B_{un} and B_{CTL} represent the portions of the budget allocated to representativeness, uncertainty, and the inconsistency condition in dynamic expan-

---

Algorithm 3. Post-processing of Community Assignments

Require: Communities $\mathcal{CS}$, graph $G = (\mathcal{V}, \mathcal{E}, \mathbf{X})$
Ensure: Refined communities $\mathcal{CS}$
1: $dup\_nodes \leftarrow \{\, v \mid |L(v)| \geq 2 \,\}$, $missing\_nodes \leftarrow \{\, v \mid |L(v)| = 0 \,\}$
2: **Duplicate handling:**
3: **for** each $v \in dup\_nodes$ **do**
4:     Keep v in the unique community $c^* = \arg\max_{c \in \mathcal{C}(v)} \Delta C_o$ by Formula 7
5:     Remove v from all other communities
6: **end for**
7: **Missing node assignment:**
8: Pre-compute $Q_{before} \leftarrow$ modularity of $\mathcal{C}$
9: **for** each $u \in missing\_nodes$ **do**
10:     **if** u is isolated **then**
11:         Assign u to community with highest content similarity
12:     **else**
13:         $c^* \leftarrow \arg\max_c (Q_{\text{after}} - Q_{\text{before}})$ ▷ c: communities of neighbors of u
14:         **if** $c^* \neq \emptyset$ **then**
15:             Assign u to c^*, update Q_{before}
16:         **else**
17:             Assign u to community with highest content similarity
18:         **end if**
19:     **end if**
20: **end for**
21: **return** $\mathcal{CS}$

---

sion, respectively. To guarantee 100% label coverage for representative nodes, the total budget B must satisfy the constraint $B \geq B_{re}$. Based on this requirement, the residual budget, expressed as $B = B - B_{re}$ is subsequently distributed between B_{un} and B_{CTL}. $B_{un} = \gamma * B$, $B_{CTL} = (1 - \gamma)B$, where $\gamma \in [0,1]$. In practice, during the dynamic community expansion process, B_{CTL} may not be fully consumed; hence, the actual manual annotation cost can be formulated $cost = cost_{re} + cost_{un} + cost_{CTL}$, where $cost_{re}$, $cost_{un}$, and $cost_{CTL}$ denote the actual manual annotation costs associated with representativeness, uncertainty, and inconsistency in dynamic community expansion, respectively. Clearly, each of these costs does not exceed the corresponding allocated budget. Representative nodes are identified by employing the DivRank method. The cost for representativeness, $cost_{re}$, is defined as the minimum value required to ensure full coverage of the network by the selected representative nodes, namely $B_{re} = cost_{re}$. To achieve a balanced and dynamic allocation of B_{CTL} during community expansion, the per-community budget is assigned as follows:

$$B_c = \left\lfloor \frac{B_{\text{CTL}} - B_{\text{used}}}{k - C_e} \right\rfloor \tag{14}$$

where B_{CTL} denotes the total budget for CTL, B_{used} the portion already consumed for CTL, k the total number of communities, and C_e is the number of communities that have been expanded.

Algorithm 4. ANS-CD

Require: Graph $G = (\mathcal{V}, \mathcal{E}, \mathbf{X})$, label set L, B_{un}
Ensure: Communities CS
 1: $k \leftarrow |L|, CS \leftarrow \emptyset$
 2: $s\_nodes, k \leftarrow$ Preprocessing for initial community node selection by Algorithm 1
 3: $CS \leftarrow$ Group s_nodes by oracle labeling
 4: $B_{CTL} \leftarrow$ Compute budget of CTL by Formula 13
 5: **for** $c \in CS$ **do**
 6:     $B_c \leftarrow$ CTL budget allocated to community c by Formula 14
 7:     $c \leftarrow$ Community expansion to c by Algorithm 2
 8: **end for**
 9: Post-processing of Community Assignments by Algorithm 3
10: **return** CS

3.6 Algorithm Implementation

Algorithm 4 outlines the overall workflow of ANS-CD. Given an attributed graph $G = (\mathcal{V}, \mathcal{E}, \mathbf{X})$, label set L and B_{un}, the algorithm first employs Algorithm 1 to obtain the node union set s_nodes, which includes both representative and uncertain nodes, along with the number of communities k that ensures full network coverage. Next, the oracle labels s_nodes to initialize the community set CS, where each community core is established. Based on the k and B_{un}, the balanced CTL budget B_{CTL} is computed. For each community in CS, an adaptive sub-budget B_c is allocated using Formula 14, and community expansion is performed via Algorithm 2. Finally, Algorithm 3 conducts post-processing to resolve duplicate assignments and unassigned nodes, producing the final global non-overlapping community detection result CS.

4 Experiments

4.1 Datasets and Evaluation Metrics

Our experiments are conducted on four public datasets for attributed community detection: Cora, Citeseer, Amazon-Photo (AP), and BlogCatalog (BC) [22]. Each dataset contains both node features and graph topologies, ranging from sparse to dense, which allows for a comprehensive evaluation. The number of communities in each dataset is known in advance. Detailed statistics of these datasets are summarized in Table 1. To comprehensively evaluate the performance of the ANS-CD method, four widely used metrics for attributed network community detection are employed: Accuracy (ACC), Normalized Mutual Information (NMI), Adjusted Rand Index (ARI), and F1-score (F1). All metrics range from 0 to 1, with higher score indicating better performance. The code of ANS-CD is available at https://github.com/lsh66/ANS-CD.

Table 1. Benchmark graph datasets. $|V|$: number of nodes; $|E|$: number of edges; $|X|$: dimensions of node features; $|C|$: number of communities

Dataset	$|V|$	$|E|$	$|X|$	$|C|$
Cora	2,708	5,429	1,433	7
Citeseer	3,327	4,732	3,703	6
AP	7,650	238,162	745	8
BC	5,196	343,486	8,189	6

4.2   Baseline Methods

The effectiveness of **ANS-CD** is evaluated by comparing it with several state-of-the-art community detection methods, including unsupervised (CDNMF [13], DANMF [23], HACD [24]) and semi-supervised (GM [11], SMACD [6], ND [9], GANSE [14]) methods on attributed networks.

- CDNMF combines NMF with contrastive learning, treating network topology and node attributes as dual perspectives, and introduces debiased negative sampling.
- DANMF fuses non-negative NMF with deep autoencoders to extract hierarchical latent features through a multi-layer encoder-decoder architecture, thus outperforming traditional shallow NMF methods.
- HACD utilizes the heterogeneous graph attention mechanism to introduce attributes into the model as independent nodes, and learns the semantic associations between nodes through attribute-level attention to achieve joint representation of structure and attributes.
- GM combines the feature extraction capabilities of GCN with the global dependency modeling capabilities of MRF.
- SMACD detects overlapping and non-overlapping communities by multi-view graph information and semi-supervised learning, requiring few node labels and enabling automated parameter adjustment for stable performance.
- ND is based on the nonlinear graph diffusion mechanism, introducing nonlinear dynamic processes, while taking into account the simplicity and efficiency of traditional diffusion models.
- GANSE combines generative adversarial networks with semi-supervised learning to generate community detection through vertex similarity rewiring, clustering coefficient rewards, and isolated node redistribution.

4.3   Experimental Setup

During community expansion, the maximum iteration counts r are set to 50, 60, 100 and 70 for Cora, CiteSeer, AP, and BC, respectively, to match their varying scales and sparsities. To highlight the low-annotation advantage of ANS-CD, we restrict the annotation budget to 5% of nodes when comparing with unsupervised methods and adopt parameter settings from the corresponding references for fair evaluation. When comparing ANS-CD with semi-supervised methods, to ensure fairness and avoid missing communities in global detection, the annotation budget of ANS-CD is set equal to the training set proportion used in semi-supervised methods. The training sets for semi-supervised methods are sampled proportionally across all node's classes. In our experiments, the manual annotation budget (or training set proportion) is configured to 5%, 10%, 20%, and 30% of the total number of nodes.

Additionally, an ablation study of ANS-CD is conducted on the Cora and CiteSeer datasets under a 5% total annotation budget. The annotation cost includes $cost_{re}$, $cost_{un}$, and $cost_{CTL}$, for representativeness, uncertainty, and CTL in dynamic community expansion, respectively. Each parts is set to zero in turn to examine the performance contribution of the remaining two.

4.4 Experimental Results

Figure 2 presents the comparison between ANS-CD and unsupervised baselines in terms of ACC, NMI, ARI, and F1 on the Cora, CiteSeer, AP, and BC datasets, showing that ANS-CD achieves superior performance. Table 2 shows the results of ANS-CD against semi-supervised baselines. Here, prior label information (PLI) refers to the proportion of manual annotation cost, that is, labeling nodes: 5%, 10%, 20%, and 30% of the total number of nodes, respectively. In most cases, our method ANS-CD achieves better performance than semi-supervised

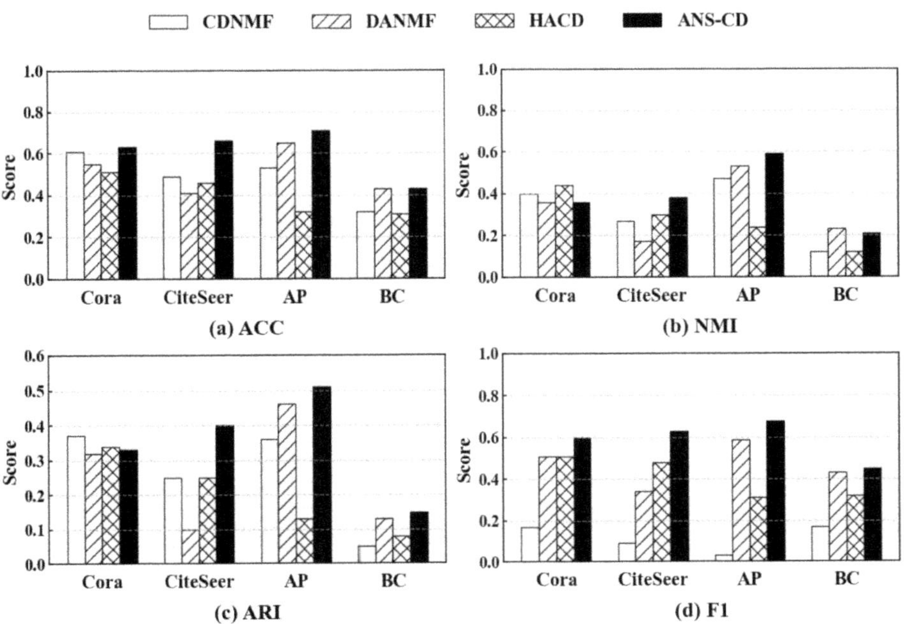

Fig. 2. Comparison of ANS-CD with unsupervised baselines.

methods even at lower PLI, clearly indicating that active node selection can reduce annotation cost. Although the results on BC are slightly weaker, this is mainly due to the dataset's highly overlapping community structure, substantial attribute noise, and low homophily.

Table 3 presents the ablation results: (1) Each of the three annotation costs contributes to the performance improvement of ANS-CD, as removing any one of them will cause the model to fail to achieve optimal results. (2) Excluding $cost_{re}$ results in the lowest scores on all evaluation metrics, indicating that representative annotation is the most critical factor. (3) Excluding $cost_{CTL}$ results in performance closest to that of ANS-CD, indicating that manual labeling of uncertain and representative nodes significantly improves the performance of community detection.

Table 2. Comparison under 5%, 10%, 20% and 30% annotation budgets, where PLI denotes the ratio of annotated nodes to the total number of nodes within the dataset, expressed as a percentage (%). Best results are in bold, and second best results are underlined.

Dataset	Metric	5% Budgets					10% Budgets				
		GM	SD	ND	GE	**ANS-CD**	GM	SD	ND	GE	**ANS-CD**
Cora	*PLI*	5.00				4.43	10.00				9.08
	ACC	0.458	0.390	<u>0.561</u>	0.265	**0.627**	<u>0.490</u>	0.339	0.440	0.283	**0.619**
	NMI	0.172	0.232	0.180	<u>0.236</u>	**0.362**	0.243	0.256	0.199	<u>0.285</u>	**0.344**
	ARI	0.164	0.167	0.165	<u>0.243</u>	**0.330**	0.255	0.178	0.130	<u>0.285</u>	**0.321**
	F1	0.344	0.254	0.458	<u>0.552</u>	**0.599**	0.402	0.276	0.434	<u>0.587</u>	**0.593**
CS	*PLI*	5.00				4.65	10.00				7.97
	ACC	0.432	0.389	0.435	<u>0.463</u>	**0.663**	0.470	0.323	0.300	<u>0.494</u>	**0.665**
	NMI	0.124	0.237	0.151	<u>0.341</u>	**0.383**	0.158	0.188	0.084	<u>0.315</u>	**0.384**
	ARI	0.117	0.090	0.127	<u>0.302</u>	**0.397**	0.134	0.043	0.018	<u>0.324</u>	**0.399**
	F1	0.373	0.297	0.416	<u>0.445</u>	**0.627**	0.399	0.227	0.253	<u>0.435</u>	**0.631**
AP	*PLI*	5.00				4.99	10.00				10.00
	ACC	0.398	<u>0.578</u>	0.365	0.501	**0.715**	0.385	<u>0.622</u>	0.236	0.522	**0.745**
	NMI	0.247	<u>0.436</u>	0.164	0.560	**0.588**	0.151	0.511	0.236	<u>0.593</u>	**0.619**
	ARI	0.161	0.385	0.050	<u>0.423</u>	**0.512**	0.057	0.398	0.089	<u>0.472</u>	**0.549**
	F1	0.191	0.301	0.235	<u>0.651</u>	**0.676**	0.144	0.316	0.213	<u>0.672</u>	**0.708**
BC	*PLI*	5.00				4.98	10.00				9.99
	ACC	**0.528**	0.392	0.389	0.345	<u>0.427</u>	**0.458**	0.413	0.358	0.398	<u>0.450</u>
	NMI	**0.317**	0.202	0.266	0.185	<u>0.311</u>	<u>0.276</u>	0.216	**0.280**	0.196	0.219
	ARI	<u>0.283</u>	0.163	0.189	0.126	**0.353**	<u>0.234</u>	0.164	0.174	0.123	**0.262**
	F1	<u>0.425</u>	0.356	0.131	0.410	**0.433**	0.330	0.392	0.133	<u>0.425</u>	**0.449**

Dataset	Metric	20% Budgets					30% Budgets				
		GM	SD	ND	GE	**ANS-CD**	GM	SD	ND	GE	**ANS-CD**
Cora	*PLI*	20.00				19.42	30.00				29.06
	ACC	<u>0.540</u>	0.376	0.398	0.307	**0.621**	<u>0.496</u>	0.443	0.430	0.304	**0.619**
	NMI	<u>0.384</u>	0.381	0.189	0.303	**0.451**	<u>0.342</u>	0.255	0.223	0.286	**0.344**
	ARI	**0.475**	0.256	0.093	0.284	<u>0.323</u>	<u>0.387</u>	0.347	0.123	0.277	**0.391**
	F1	<u>0.735</u>	0.293	0.395	0.677	**0.794**	0.259	0.324	0.427	<u>0.566</u>	**0.593**
CS	*PLI*	20.00				19.54	30.00				28.80
	ACC	<u>0.623</u>	0.413	0.358	0.502	**0.664**	**0.686**	0.462	0.257	0.541	<u>0.667</u>
	NMI	0.295	0.289	0.121	<u>0.372</u>	**0.383**	0.377	0.376	0.076	<u>0.382</u>	**0.387**
	ARI	0.304	0.156	0.080	<u>0.341</u>	**0.398**	<u>0.395</u>	0.235	0.041	0.375	**0.402**
	F1	<u>0.590</u>	0.330	0.314	0.512	**0.628**	<u>0.664</u>	0.427	0.185	0.561	**0.670**
AP	*PLI*	20.00				18.97	30.00				29.06
	ACC	0.268	<u>0.630</u>	0.360	0.542	**0.766**	0.366	<u>0.652</u>	0.386	0.559	**0.775**
	NMI	0.208	0.532	0.246	<u>0.620</u>	**0.651**	0.096	0.501	0.242	<u>0.671</u>	**0.692**
	ARI	0.104	0.423	0.152	<u>0.441</u>	**0.578**	0.053	0.436	0.090	<u>0.456</u>	**0.598**
	F1	0.075	0.348	0.184	<u>0.692</u>	**0.730**	0.129	0.352	0.168	<u>0.715</u>	**0.795**
BC	*PLI*	20.00				20.00	30.00				29.98
	ACC	<u>0.556</u>	0.433	0.391	0.375	**0.570**	0.406	0.471	<u>0.486</u>	0.435	**0.564**
	NMI	**0.350**	0.221	<u>0.327</u>	0.200	0.252	0.233	0.201	**0.408**	0.238	<u>0.389</u>
	ARI	**0.326**	0.176	0.213	0.166	<u>0.296</u>	0.188	0.192	**0.265**	0.220	<u>0.245</u>
	F1	<u>0.457</u>	0.421	0.167	0.403	**0.511**	0.322	0.441	0.232	<u>0.542</u>	**0.567**

Table 3. Ablation study results on Cora and CiteSeer datasets.

Method	Cora				CiteSeer			
	ACC	NMI	ARI	F1	ACC	NMI	ARI	F1
$cost_{un} + cost_{CTL}$	0.248	0.186	0.124	0.231	0.530	0.261	0.240	0.423
$cost_{re} + cost_{CTL}$	0.581	0.317	0.272	0.565	0.628	0.368	0.385	0.591
$cost_{un} + cost_{re}$	0.605	0.341	0.297	0.593	0.634	0.371	0.390	0.593
ANS-CD	**0.627**	**0.362**	**0.330**	**0.599**	**0.663**	**0.383**	**0.397**	**0.627**

5   Conclusions

To address the limitations of traditional community detection methods in exploring informative priors and balancing labeling costs, we propose ANS-CD, a novel label-efficient method. Specifically, it integrates a label coverage strategy, a novel ONE method for identifying high-uncertainty nodes, and a node-community conflict mechanism, enabling global community detection and dynamic optimization of communities during the expansion process. Experiments on datasets validate ANS-CD's superiority, and ablation studies confirm all three labeling costs enhance performance.

Acknowledgment. The work is supported by National Natural Science Foundation of China (62432003) and the Fundamental Research Funds for the Central Universities (N25BJD013).

Disclosure of Interests. The authors declare that they have no conflicts of interest.

References

1. Aref, S., Mostajabdaveh, M., Chheda, H.: Heuristic modularity maximization algorithms for community detection rarely return an optimal partition or anything similar. In: International Conference on Computational Science, pp. 612–626. Springer, Cham (2023). https://doi.org/10.1007/978-3-031-36027-5_48
2. Bandyopadhyay, S., Nagalapatti, L., Murty, M.N.: Outlier aware network embedding for attributed networks. In: AAAI, pp. 12–19. AAAI Press (2019)
3. Berahmand, K., Li, Y., Xu, Y.: A deep semi-supervised community detection based on point-wise mutual information. IEEE Trans. Comput. Soc. Syst. **11**(3), 3444–3456 (2023)
4. Cai, H., Zheng, V.W., Chang, K.C.C.: Active learning for graph embedding. arXiv preprint arXiv:1705.05085 (2017)
5. Cheng, J., Leng, M., Li, L., Zhou, H., Chen, X.: Active semi-supervised community detection based on must-link and cannot-link constraints. PLoS ONE **9**(10), e110088 (2014)
6. Gujral, E., Papalexakis, E.E.: SMACD: semi-supervised multi-aspect community detection. In: SDM, pp. 702–710. SIAM (2018)

7. He, D., You, X., Feng, Z., Jin, D., Yang, X., Zhang, W.: A network-specific markov random field approach to community detection. In: AAAI, pp. 306–313. AAAI Press (2018)

8. Holland, P.W., Laskey, K.B., Leinhardt, S.: Stochastic blockmodels: first steps. Soc. Netw. **5**(2), 109–137 (1983)

9. Ibrahim, R., Gleich, D.F.: Nonlinear diffusion for community detection and semi-supervised learning. In: WWW, pp. 739–750. ACM (2019)

10. Jin, D., He, J., Chai, B., He, D.: Semi-supervised community detection on attributed networks using non-negative matrix tri-factorization with node popularity. Front. Comp. Sci. **15**(4), 154324 (2021)

11. Jin, D., Liu, Z., Li, W., He, D., Zhang, W.: Graph convolutional networks meet markov random fields: Semi-supervised community detection in attribute networks. In: AAAI, pp. 152–159. AAAI Press (2019)

12. Li, Y., Jia, C., Li, J., Wang, X., Yu, J.: Enhanced semi-supervised community detection with active node and link selection. XXPhys. A **510**, 219–232 (2018)

13. Li, Y., Chen, J., Chen, C., Yang, L., Zheng, Z.: Contrastive deep nonnegative matrix factorization for community detection. In: ICASSP, pp. 6725–6729. IEEE (2024)

14. Liu, X., et al.: Semi-supervised community detection method based on generative adversarial networks. J. King Saud Univ. Comput. Inf. Sci. **36**(3), 102008 (2024)

15. Mei, Q., Guo, J., Radev, D.R.: Divrank: the interplay of prestige and diversity in information networks. In: KDD, pp. 1009–1018. ACM (2010)

16. Newman, M.E., Girvan, M.: Finding and evaluating community structure in networks. Phys. Rev. E **69**(2), 026113 (2004)

17. Ni, L., Ge, J., Zhang, Y., Luo, W., Sheng, V.S.: Semi-supervised local community detection. IEEE Trans. Knowl. Data Eng. **36**(2), 823–839 (2023)

18. Perozzi, B., Al-Rfou, R., Skiena, S.: Deepwalk: Online learning of social representations. In: Proceedings of the 20th ACM SIGKDD International Conference on Knowledge Discovery and Data Mining, pp. 701–710 (2014)

19. Rashnodi, O., Rastegarpour, M., Moradi, P., Zamanifar, A.: Community detection in attributed social networks using deep learning. J. Supercomput. **80**(18), 25933–25973 (2024)

20. Wang, M., Wang, Z., Qu, L., Long, K., Xin, J.: Bfmddt: A decision-tree-based gene regulatory network inference from multi-type datasets. IEEE Trans. Comput. Biol. Bioinform. (2025)

21. Xin, J., Wang, M., Qu, L., Chen, Q., Wang, W., Wang, Z.: Bic-lp: A hybrid higher-order dynamic bayesian network score function for gene regulatory network reconstruction. IEEE/ACM Trans. Comput. Biol. Bioinf. **21**(1), 188–199 (2023)

22. Yang, R., Shi, J., Xiao, X., Yang, Y., Liu, J., Bhowmick, S.S.: Scaling attributed network embedding to massive graphs. Proc. VLDB Endow. **14**(1), 37–49 (2020)

23. Ye, F., Chen, C., Zheng, Z.: Deep autoencoder-like nonnegative matrix factorization for community detection. In: CIKM, pp. 1393–1402. ACM (2018)

24. Zhang, A., Wang, X., Zhao, Y.: HACD: harnessing attribute semantics and mesoscopic structure for community detection. In: WSDM, pp. 616–624. ACM (2025)

Skyline Community Search over Edge-Attributed Bipartite Graphs

Fangda Guo[1]([📧]) [iD], Xuanpu Luo[1] [iD], Shiyuan Xu[2]([📧]) [iD], Haowen Gao[1,3]([📧]) [iD], Yanghao Liu[1,3] [iD], Huawei Shen[1] [iD], and Xueqi Cheng[1] [iD]

[1] State Key Lab of AI Safety, Institute of Computing Technology, CAS, Beijing, China
`{guofangda,seuviplxp,gaohaowen23s,liuyanghao19s,shenhuawei,cxq}@ict.ac.cn`
[2] School of Computing and Data Science, The University of Hong Kong, Hong Kong, China
`syxu2@cs.hku.hk`
[3] University of Chinese Academy of Sciences, Beijing, China

Abstract. Bipartite graphs, modeling relationships between two types of entities, are widely used in practical applications. Community search, a fundamental problem in bipartite graphs, has gained significant attention. However, existing studies focus on measuring structural cohesiveness between vertex sets while either ignoring edge attributes or considering only one-dimensional importance. In this paper, we introduce a novel community model, named edge-attributed skyline community (ESC), which preserves structural cohesiveness and captures the inherent dominance of multi-dimensional edge attributes in bipartite graphs. To search for ESCs, we developed an efficient peeling algorithm that iteratively deletes edges with the minimum attribute in each dimension. Additionally, we devised an expanding algorithm to reduce the search space and speed up the filtering of unpromising vertices using a proven upper bound. Extensive experiments on large-scale real-world datasets demonstrate the efficiency, effectiveness, and scalability of our approach. A case study compared with prior arts demonstrates that our design improves the precision and diversity of results.

Keywords: Bipartite graph · Cohesive subgraph · Community search

1 Introduction

Numerous real-world networks consist of community structures, making their discovery crucial for network analysis [1,34,39,40]. Community search aims to find the densest connected subgraphs containing specific input queries. It has garnered significant attention from database professionals in recent years [25,32,36]. Traditional search focuses primarily on the network topology [7,35,37,38]. However, in practical applications, it needs to consider not only the connections but also their attribute characteristics, as these can enhance the cohesion of identified communities [15,16]. This makes community search over attributed graphs

H. Jung et al. (Eds.): DASFAA 2026, LNCS 16536, pp. 367–383, 2026.
https://doi.org/10.1007/978-981-92-0366-6_23

particularly significant. Most current research addresses community constraints in terms of structure and attributes but typically only for scenarios involving a single type of entity [14,24]. However, in scenarios involving two different types of entities, such as collaboration networks [18], customer-product networks [27,33], and user-location networks [5], the focus tends to be on structural cohesiveness, often neglecting attributes. Approaches like the (α, β)-core [9], bitruss [26,28], and biclique [29,42] exemplify this trend by primarily concentrating on structure.

Some researchers considered one-dimensional importance for attribute information [31], but interactions often involve multiple dimensions. They missed the multidimensional aspects and fail to capture all cohesive communities. We aim to develop a model using both structural and multi-dimensional attribute information to identify all cohesive communities containing the queries. By designating U_2 as the query vertex and setting structural constraints of $\alpha = \beta = 2$, we extract two mutually exclusive communities[1] with the largest connected subgraphs based on the multidimensional attributes.

In this study, our goal is to identify vertices in the bipartite graph that are structurally closest to the query vertex and have similar attribute values, along with the corresponding edges, to find high-importance communities that meet these criteria. To achieve this, we propose a skyline community search problem for bipartite graphs with multi-dimensional edge attributes, named Edge-attributed Skyline Community Search Problem **(ESCS-Problem)**, along with its model termed Edge-attributed Skyline Community **(ESC)**.

Our initial effort targets bipartite graphs with multidimensional edge attributes, presenting a new challenge. A recent study [21] focuses on communities with multidimensional attributes in simple graphs, while another [30] deals with single-dimensional attributes, making it a subset of our scope. Additionally, the design [43] using Pareto optimality for one-dimensional weights differs in context and approach. Other works [23,45] on attribute community discovery in heterogeneous graphs handle only one-dimensional attributes and vertex-associated values, unsuitable for our scenario. We face challenges in balancing structural compactness with multidimensional attribute advantages, marking a creative endeavor. Due to space limitations, please refer to the full version[2]. Our contributions are summarized as follows.

- **Novel community model.** We introduce a new model for community search in bipartite graphs, considering both structural cohesion and multi-dimensional edge attributes.
- **New algorithms.** We propose an efficient peeling algorithm that removes edges with the minimum attribute and an expanding algorithm to reduce search space and speed up filtering. We also give lemmas to improve search performance, and verify computational complexity and completeness.
- **Extensive experiments.** We conduct extensive experiments over six real-world datasets to demonstrate the high efficiency and scalability of our proposed algorithms. We also give an extensive case study, demonstrating our

---

[1] Strictly speaking, communities may overlap.
[2] https://arxiv.org/abs/2401.12895.

ESC model's capability to match dense subgraphs of bipartite graphs with greater accuracy and personalization.

2 Problem Statement

Definition 1 (Bipartite Graph). *A bipartite graph $G = (U, L, E, X)$ is a graph whose vertices can be divided into two disjoint sets such that every edge connects a vertex from one set to a vertex from the other set. Note that U and L denote the sets of upper and lower vertices, respectively, and E represents the set of edges in G. The edges in G have d numerical attributes, which are denoted by X_i, where $X_i \in X$ and X_i is a d-dimensional vector. Let $|*|$ denote the modulo operation. We have $|U| + |L| = n$ and $|E| = m$.*

Definition 2 ((α, β)-core). *Given a bipartite graph $G = (U, L, E, X)$, an (α, β)-core is a connected subgraph where the degree of vertices in the upper layer is at least α, and the degree of vertices in the lower layer is at least β.*

Definition 3 (Significance). *Given a bipartite graph $G = (U, L, E, X)$ with edge attributes, we define the significance of G on the ith dimension (for $i = 1, 2, \cdots, d$) as: $f_i(G) = \min_{e \in |E|} X_i^e$.*

Definition 4 (Domination). *Given two bipartite graphs $G = (U, L, E, X)$ and $G' = (U', L', E', X')$, if $f_i(G) \leq f_i(G')$ (for $i = 1, 2, \cdots, d$) and there exists $f_i(G) < f_i(G')$ for certain i, we call G' dominates G, denoted by $G \preceq G'$.*

Definition 5 (ESC). *Given a bipartite graph $G = (U, L, E, X)$, α, β and query vertex q, where X is the multi-dimensional significance on edges [21,30,43]. An ESC is a connected subgraph $H = (U_H, L_H, E_H, X_H)$ of G, where:*

1) Cohesive property: H is an (α , β)-core contains q. It ensures that the connected subgraph contains the query and follows the (α, β)-core structure.
2) Skyline property: it does not exist a subgraph H' of G such that H' is an (α, β)-core and $H \prec H'$. It guarantees that the subgraph is not dominated by others, making it an optimal solution.
3) Maximal property: it does not exist a subgraph H' of G such that (1) H' is an (α, β)-core, (2) H' contains H, and (3) $f_i(H) = f_i(H')$ for all $i \in [1, d]$. It ensures the largest possible subgraph under constraints.

ESCS-Problem: Given a bipartite graph G, a query vertex q, and parameters α and β for structural constraints, the problem is aimed to search all ESCs containing q, where each ESC H cannot be dominated by any other ESC H'.

3   Peeling-Based Algorithm

3.1   Peeling for $d = 1$

In simple bipartite graphs with one-dimensional attributes, the algorithm follows a straightforward approach. It first sorts edges in ascending order of significance, then iteratively removes the least significant edge until the structure no longer maintains an (α, β)-core or the queried vertex is excluded. The resulting (α, β)-core is called an ESC. In the operational phase, peeling one edge at a time is inefficient. To address this, we've introduced DFSCom, as depicted in Algorithm 1, which recursively traverses adjacent vertices to ensure all structural constraints are met. Once these criteria are satisfied, the algorithm ends, removing all non-compliant vertices and their edges.

Algorithm 1: PDim1

Input: $G, \alpha, \beta, q, I, d, F$
Output: f_d

1 $G \leftarrow G \setminus \{e | e \text{ violates } I\}$; $G(q) \leftarrow$ maximal (α, β)-core containing q;
   $f_d \leftarrow f_d(G(q))$;
2 **while** $G(q) \neq \emptyset$ **do**
3 $\quad$ $w_{min} \leftarrow \min_{d\text{-dim}} G(q)$; **foreach** $(u, v) \in G(q)$ **do**
4 $\quad$ $\quad$ $visited(u, v) \leftarrow 0$;

5 $\quad$ **foreach** $(u, v) \in G(q)$ *with* $w(u, v) = w_{min}$ **do**
6 $\quad$ $\quad$ **if** $(u, v) \in F$ **then break**;
7 $\quad$ $\quad$ $S \leftarrow \emptyset, visited(u, v) \leftarrow 1$;
8 $\quad$ $\quad$ **if** $DFSCom(u, q, \alpha, \beta, S, F) = 0$ **then break**;
9 $\quad$ $\quad$ **if** $DFSCom(v, q, \beta, \alpha, S, F) = 0$ **then break**;
10 $\quad$ $\quad$ $G(q) \leftarrow G(q) \setminus \{(u, v)\}$; $f_d \leftarrow \max(f_d, f_d(G(q)))$;

11 **return** f_d;
12 **Function** DFSCom$(u, q, \alpha, \beta, S, F)$:
13 $\quad$ **if** $u = q \wedge deg(u, G(q)) = \alpha$ **then return** *0*;
14 $\quad$ Update $deg(u, G(q))$;
15 $\quad$ **if** $deg(u, G(q)) < \alpha$ **then**
16 $\quad$ $\quad$ **foreach** $v' \in N(u)$ *with* $visited(u, v') = 0$ **do**
17 $\quad$ $\quad$ $\quad$ $visited(u, v') \leftarrow 1$; $S \leftarrow S \cup \{(u, v')\}$; **if**
       $(u, v') \in F \vee DFSCom(v', q, \beta, \alpha, S, F) = 0$ **then** $G(q) \leftarrow G(q) \cup S$;
       return *0* ;
18 $\quad$ $\quad$ $S \leftarrow \emptyset$;
19 $\quad$ **return** *1*;

Completeness: The integrity of the algorithm is elucidated as follows. Assuming there exists another ESC H' that contains edges not present in the current

ESC H. It must satisfy at least one of the following two conditions: (1) it contains edges with significance smaller than $f(H)$, which contradicts the definition of $f(H')$ and is not possible; or (2) it contains edges with significance greater than $f(H)$, which were removed because the two vertices they connect do not satisfy the structural constraints. When H' adds these edges to the community, it must also add edges with significance smaller than $f(H)$ in order to satisfy the (α, β)-core constraint, which contradicts the definition of $f(H')$ and is not possible. Therefore, it is not possible to have another ESC.

3.2 Peeling for $d = 2$

As edge significance dimensionality grows, a backtracking-based dimensionality reduction strategy decomposes two-dimensional significance into one-dimensional sets, enabling existing methods to tackle new problems. Based on Lemma 1, we propose the Algorithm 2.

Lemma 1. *Without loss of generality, assuming that there exist n ESCs H_i, $(i = 1, 2, ..., n)$ in the bipartite graph G with two dimensions. Then, the significance of each dimension corresponding to different ESCs can be related as:* $f_1(H_1) < f_1(H_2) < ... < f_1(H_i) < ... < f_1(H_n), f_2(H_1) > f_2(H_2) > ... > f_2(H_i) > ... > f_2(H_n).$

Algorithm 2: PDim2

   Input: $G, \alpha, \beta, q, I, F$
   Output: R

1 $f_2 \leftarrow \mathrm{PDim1}(G, \alpha, \beta, q, I, 2, F)$;
2 $R \leftarrow \varnothing$;
3 **while** $f_2 > 0$ **do**
4     $I' \leftarrow I.update(\{X_2^i \geq f_2\})$;
5     $f_1 \leftarrow \mathrm{PDim1}(G, \alpha, \beta, q, I', 1, F)$;
6     $R \leftarrow R \cup \{(f_1, f_2)\}$;
7     $I' \leftarrow I.update(\{X_1^i > f_1\})$;
8     $f_2 \leftarrow \mathrm{PDim1}(G, \alpha, \beta, q, I', 2, F)$;
9 **return** R;

Completeness: Suppose an undiscovered ESC H. According to Lemma 3, we have $f_1(H_i) < f_1(H) < f_1(H_{i+1})$ and $f_2(H_j) > f_2(H) > f_2(H_{j+1})$. Assuming $i < j$, we have $f_1(H_i) < f_1(H) < f_1(H_j)$. Also, since $f_2(H_j) > f_2(H)$, H_j dominates H, thus H cannot exist. The same can be proven when $i > j$. Therefore, Algorithm expanding$(d = 2)$ is complete.

3.3   Peeling for $d = 3$

We implement an alternate Algorithm expending ($d = 3$). For obtaining $f_3(H_i)$, the process of being transferred function CandSig is as follows: in the process of gradually removing the edge with the minimum X_d, recording each X_d that makes the subgraph contain (α,β)-core. These recorded values constitute all possible significance for ESCs on d_{th} dimension. Then, we iterate through the set V of all possible values $v_i \in V$ during each iteration, edges with third-dimensional attribute values $X_3^e \le v_i$ are removed.

Algorithm 3: CandSig

Input: G, α, β, q, d
Output: T_d

1   $T_d \leftarrow \emptyset$; $G(q) \leftarrow$ maximal (α, β)-core containing q;
2   **while** $G(q) \ne \emptyset$ **do**
3      $f_d \leftarrow \min_{d\text{-dim}} G(q)$; $T_d \leftarrow T_d \cup \{f_d\}$;
4   Remove $e = (u, v) \in G(q)$ with $X_d^e = f_d$;
5   DFS($G(q), u, \alpha, \beta$); DFS($G(q), v, \beta, \alpha$);
6      $G(q) \leftarrow$ maximal (α, β)-core containing q;

7   **return** T_d;
8   **Function** DFS(G, u, α, β)**:**
9   **if** $deg(G, u) > \alpha$ **then**
10   **foreach** $v \in N(u)$ **do**
11          $G \leftarrow G \setminus \{(u, v)\}$; DFS($G, v, \beta, \alpha$);

Completeness: If there is any other $f_3(H_i)$ present, it is impossible to form an (α, β)-core. When fixing the $f_3(H_i)$, assume the existence of an ESC H' in the

---

Algorithm 4: PDim3

Input: $G, \alpha, \beta, q, I, F$
Output: R

1   $F_3 \leftarrow$ CandSig($G, \alpha, \beta, q, 3$); $R, S \leftarrow \emptyset$; Sort F_3 descending;
2   **foreach** $f_3 \in F_3$ **do**
3      $e \leftarrow \{e \in G \mid X_3^e = f_3\}$; $F \leftarrow F \cup e$;
4   **if** $(X_1^e, X_2^e) \preceq S$ **then continue**;
5      $I' \leftarrow I \cup \{X_3 \ge f_3\}$; $T \leftarrow$ PDim2($G, \alpha, \beta, q, I', F$);
6      $S \leftarrow S \cup T$; **foreach** $(f_1, f_2) \in T$ **do**
7        $R \leftarrow R \cup \{(f_1, f_2, f_3)\}$;

8   **return** R;

subgraph on the other two dimensions that Algorithm expending(d=2) did not find. As H' must be dominated by the already identified ESCs, contradicting the definition of ESC. Therefore, the algorithm does not overlook ESC solutions. For $(d > 3)$ cases, we also design a universal peeling approach[3] based on dimensionality reduction and backtracking, resembling mathematical induction.

4   Expanding-Based Algorithm

4.1   Expanding for $d = 1$

As shown in Algorithm 5, we tailor for the scenario where dimensionality $d = 1$. A straightforward method is to sort the edges of the initial graph by their significance. The iteration involves progressively incorporating the edge with the maximal value into an empty graph until it evolves into an (α,β)-core encompassing all queries. However, its efficiency is curtailed due to the addition of merely one edge per iteration. To mitigate this, we introduce Lemma 2 to establish a pseudo-upper bound for f_1 within the ESC, allowing batch processing.

Algorithm 5: EDim1

Input: G, α, β, q, d
Output: R

1 $G^*, C^*, R \leftarrow \varnothing$, $count \leftarrow \alpha$ or β; $f \leftarrow count^{th}$-max X_1^i of q's edges;
2 **while** $|E(G)| \geq |E(G^*)|$ **do**
3      $G^* \leftarrow \{e|X_1^i > f\}$; $C^* \leftarrow$ connected component in G^*;
4   **if** C^* *changed* $\wedge$ C^* *satisfies Lemmas 3 & 4* **then**
5   **if** $q \in C^*$ **then**
6            $H \leftarrow (\alpha, \beta)$-core of C^*; **if** $q \in H$ **then**
7               $R \leftarrow H$; **break**;
8      $count \leftarrow count + 1$; $f \leftarrow count^{th}$-max X_1^i in G;
9 **return** R;

Completeness: The existence of an ESC H' with $f_1(H') \geq f_1(H)$, H' must belong to one of the following two cases: (1) it contains edges that are not in H, or (2) the edge set of H' is a subset of the edge set of H. Since the algorithm continuously adds edges to an empty graph, the resulting community is the first (α, β)-core that includes the query. If H' contains edges that are not in H, these edges must have significance less than $f_1(H)$, and if H' contains such edges, it contradicts the definition of $f_1(H')$. So, the first case does not exist. As for (2), if the edge set of H' is a subset of the edge set of H, then H' is not the largest (α,β)-core, contradicting the definition of ESC, so there are no other ESCs.

---

[3] https://anonymous.4open.science/r/ESC-DC3F.

Lemma 2. *Given the query vertex q, the degree on the layer of q is α, the set of significance for edges connected to q store in N_E. The pseudo upper bound δ of the significance of the ESC is the αth maximal value in the sorted set N_E.*

By applying Lemma 2 to determine the pseudo upper bound δ, it used to incorporate only those edges where $X_1^e \geq \delta$, thereby markedly enhancing time efficiency. If the edges recently integrated fail to constitute an (α,β)-core encompassing the queries, the process necessitates continued iterations, sequentially incorporating remaining edges with the maximum significance into the current graph. Furthermore, we integrate insights from Lemma 3 and Lemma 4 [30], to further curtail the number of iterations and amplify efficiency. This culminates in the introduction of Algorithm peeling (d=1), a novel approach specifically devised to tackle the ESC search problem.

Lemma 3. *If a connected subgraph C^* contains an ESC H, then there must be: $\alpha\beta - \alpha - \beta \leq |E(C^*)| - |U(C^*)| - |L(C^*)|$.*

Lemma 4. *If a connected subgraph C^* contains an ESC H, then C^* has at least β vertices of degree α and α vertices of degree β.*

4.2   Expanding for $d = 2$

When the edge significance dimension is set to 2, multiple ESCs may exist. Identifying all ESCs requires both edge addition and deletion. We employ a dimensionality reduction approach, processing the two significance dimensions separately. The initial ESC community H is derived using Algorithm expending($d = 1$) based on the second-dimensional attribute. By Lemma 1, $f_1(H)$ is the lower bound of the first-dimensional significance for all ESCs. A new ESC H' with $f_1(H') > f_1(H)$ is then found by removing edges with $X_1^e = f_1(H)$ and adding those with $X_1^e > f_1(H)$. This iterative process continues until all ESCs are identified, forming the basis of Algorithm expending ($d = 2$). As shown in Algorithm 6, it initializes key variables: G^* (accumulated edges), f_1 (current ESC's first-dimensional significance), R (set of ESCs), *count* (tracking edge order by second-dimensional significance), and D (operating dimension). ESC identification consists of two steps: finding the first ESC and generating subsequent ones.

4.3   Expanding for $d = 3$

Expanding strategies can still be extended to three-dimensional situations. Utilizing a dimensionality reduction strategy, the values of the three-dimensional attributes are efficiently decomposed into two distinct sets: one set of one-dimensional values and another set of two-dimensional values. A preliminary and simplistic strategy entails enumerating all feasible values of f_3, setting each of these values consecutively, and subsequently conducting a subgraph analysis to locate ESCs within the context of the remaining two-dimensional attributes.

However, this method fails to guarantee finding ESCs for every f_3 value. Fixing f_3 constrains the search domain to $C_0 = \{X_1^e > 0, X_2^e > 0\}$.

Algorithm 6: EDim2

Input: G, α, β, q, F
Output: R

1 $G^*, R \leftarrow \varnothing,\ f_1 \leftarrow -1,\ count \leftarrow \alpha$ or $\beta,\ D \leftarrow 1$;
   $f \leftarrow count^{th}$-max X_2^i of q's edges;
2 **while** $|E(G^*)| < |E(G)|$ **do**
3   **if** $f_1 = -1$ **then**
4        $G^* \leftarrow \{e | X_2^i > f\};\ D \leftarrow 1,\ count \leftarrow count + 1$;
       $f \leftarrow count^{th}$-max X_2^i in G;
5   **else**
6        $G^* \leftarrow (G^* \setminus \{e | X_1^i = f_1\}) \cup \{e | X_1^i > f_1\};\ D \leftarrow 2$;
7     $S \leftarrow \varnothing$; **while** $H \leftarrow (\alpha, \beta)$-core of $G^* \neq \emptyset$ and $q \in H$ and $F \subseteq H$ **do**
8        $S \leftarrow H$; Remove edge with min X_D^i from G^*;
9   **if** $S = \varnothing$ **then**
10   **if** $f_1 = -1$ **then** continue;
11   **break**
12   **else if** $\exists ESC \in R \preceq S$ **then**
13   Remove dominated ESC from R
14    $(f_1, f_2) \leftarrow S;\ R \leftarrow R \cup \{(f_1, f_2)\}$;
15 **return** R;

To optimize the search, we use recursive decomposition, pruning unnecessary computations. An ESC must lie outside the shaded area, but its irregular shape prevents direct application of Algorithm peeling($d{=}2$). Thus, we partition the solution space into $C_1 = \{X_1^e > 0, X_2^e > f_2(H)\}$ and $C_2 = \{X_1^e > f_1(H), X_2^e > 0\}$, enabling separate applications of Algorithm peeling($d{=}2$). Each region's lower bounds are represented as $(0, f_2(H))$ for C_1 and $(f_1(H), 0)$ for C_2.

Lemma 5. *Suppose it exists n ESCs: $H_i = (f_1(H_i), f_2(H_i))$, $i \in [1, n]$, arranging all the horizontal and vertical coordinates as: $0 < f_1(H_1) < f_1(H_2) < ... < f_1(H_n), f_2(H_1) > f_2(H_2) > ... > f_2(H_n) > 0$. By combining the coordinates of the upper and lower inequalities, the solution space is: $C_1 = (0, f_2(H_1)), C_2 = (f_1(H_1), f_2(H_2)),...,C_{n+1} = (f_1(H_n), 0)$.*

As shown in Algorithm 7, we combine the aforementioned recursive decomposition manner for solving the ESC search problem in three-dimensional attribute bipartite graphs, which initially creates an empty graph G^* for each solution space. Subsequently, it iteratively adds a batch of edges to G^* that meet the significance requirements of the solution space. The algorithm then calls Algorithm peeling($d{=}2$) on G^* to compute ESCs. Following this, leveraging Lemma 5, it updates the solution space using the generated ESCs and calculates f_3 for each solution space. This process iterates until all ESCs are identified. In addition, a universal expanding approach is proposed for high-dimensional cases ($d > 3$), extending the solution space update in recursive decomposition based on Lemma 5.

5 Experiment

We validate the efficiency by varying the dimension d and structural parameters α and β. To assess scalability, we change the dataset size. Additionally, we demonstrate the algorithm's effectiveness through a case study. All experiments are conducted on an Ubuntu server with a 2.40GHz Intel(R) Xeon(R) Gold 6240R CPU and 512GB memory in Python. Table 1 provides the statistics of datasets, of which d_{1m}, d_{2m}, d_{1a} and d_{2a} denote the maximal degree of the upper vertices, the maximal degree of the lower vertices, the average degree of the upper vertices and the average degree of the lower vertices, respectively.

Algorithm 7: EDim3

Input: G, α, β, q, F
Output: R

1   $G^*, R, Q, P \leftarrow \varnothing$; $Q \leftarrow Q \cup (0, 0, \text{EDim1}(G, \alpha, \beta, q, 3))$;
2   **while** $Q \neq \emptyset$ **do**
3     $f_3 \leftarrow \max_{\text{3-dim}} Q$; $S \leftarrow \varnothing$;
4   **foreach** $(p_1, p_2, f_3) \in Q$ **do**
5       $G^* \leftarrow \{e | x_1^e > p_1, x_2^e > p_2, x_3^e \geq f_3\}$; Remove (p_1, p_2, f_3) from Q; **if** $x_3^e = f_3$ **then** $F \leftarrow F \cup e$;
6       $S \leftarrow S \cup \text{EDim2}(G^*, \alpha, \beta, q, F)$;
7   **foreach** $(f_1, f_2) \in S$ **do**
8   **if** $(f_1, f_2) \succeq P$ **then**
9   Remove dominated from P;
10   **else if** $(f_1, f_2) \preceq P$ **then**
11   continue;
12        $Q \leftarrow Q \cup (f_1, f_2, f_3)$; $P \leftarrow P \cup (f_1, f_2)$;
13   **foreach** $(point_1, point_2) \in Lemma5$ **do**
14        $G^* \leftarrow \{e | x_1^e > point_1, x_2^e > point_2\}$;
         $Q \leftarrow Q \cup (point_1, point_2, \text{EDim1}(G^*, \alpha, \beta, q, 3))$;

15 **return** R;

Datasets. We collected seven datasets, i.e., DBpedia, BookCrossing, IMDB, TV Tropes, arXiv, Crime, and MovieLens from the KONECT website. These datasets are all undirected bipartite graphs, and neither edges nor vertices have associated significance. As the currently available datasets lack edge weight attributes, we generated non-negative numerical multidimensional attributes for all attribute-less bipartite graph datasets, ensuring these attributes were independent. We used a uniform distribution for this purpose because it generates values with equal probability and avoids the correlation biases inherent in the Gaussian distribution. In real-world scenarios, edges with high significance in one dimension rarely exhibit high significance in others, making *anti-correlation* more common and justifying the use of uniform distribution.

Parameters. We vary three parameters: dimension (d), structural parameter (α), and dataset completeness (σ). Since α and β similarly constrain the community structure, we vary only α. Table 1 lists the parameter ranges and default values. For each experiment, we randomly selected 20 query vertex sets from each graph that ensured the existence of the maximal (α, β)-core, and averaged the time required for these sets.

Table 1. Datasets and Parameters used in our experiments

Dataset	Vertices	Edges	d_{1m}	d_{2m}	d_{1a}	d_{2a}
Crime	1K	1K	25	18	1.8	2.7
arXiv	39K	59K	116	18	3.5	2.6
DBpedia	225K	294K	28	12K	1.7	5.5
BookCrossing	446K	1.1M	14K	2.5K	10.9	3.4
IMDB	872K	2.7M	654	1.3K	3.9	14.6
TV Tropes	152K	3.2M	6.5K	12K	50.2	36.9
MovieLens	222K	25.0M	32.2K	81.4K	154.3	423.7

Para.	Tested values
d	1, 2, 3, 4
$\alpha(\beta)$	1, 2, 3, 4
σ	20, 40, 60, 80, 100%

5.1 Efficiency Evaluation

We investigate the time required to search for ESCs across various datasets by varying the attribute dimensionality d while keeping other variables constant, and please refer to footnote see footnote(2) for detailed explanation. As shown in Fig. 1, we fix the structural constraint parameters α and β at 2 and use the average time taken for 20 query vertices as the search time for each dimensionality. Note that, we use DB to denote DBpedia, BC to denote BookCross, TVT to denote TVTropes, ML to denote MovieLen. Figure 1(a)–(g) indicates that as attribute dimensionality increases, the time required for the peeling and expanding algorithms also increases.

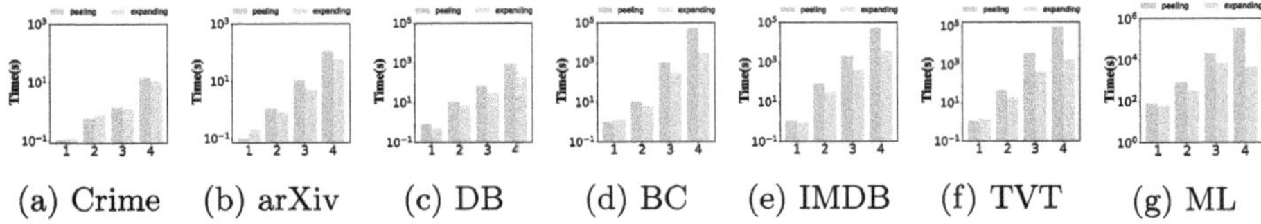

(a) Crime   (b) arXiv   (c) DB   (d) BC   (e) IMDB   (f) TVT   (g) ML

Fig. 1. The efficiency of the algorithms w.r.t dimensions change.

5.2 Effectiveness Evaluation

Scalability Evaluation. Without loss of generality, we take the expanding algorithm as the experimental object. To assess the scalability of Algorithm peeling ($d = 3$), we randomly select 20%, 40%, 60%, 80%, and 100% of the edges from the input dataset (vary σ) to form subgraphs for our experiments. The experimental results confirm that the algorithm consistently produces correct outcomes. As shown in Fig. 3, the algorithm's runtime generally rises with an increasing number of edges. However, deviations occur due to query vertex selection. For example, if a 20% edge subgraph fully covers a query vertex's ESC, adding more edges doesn't raise the processing time for that vertex.

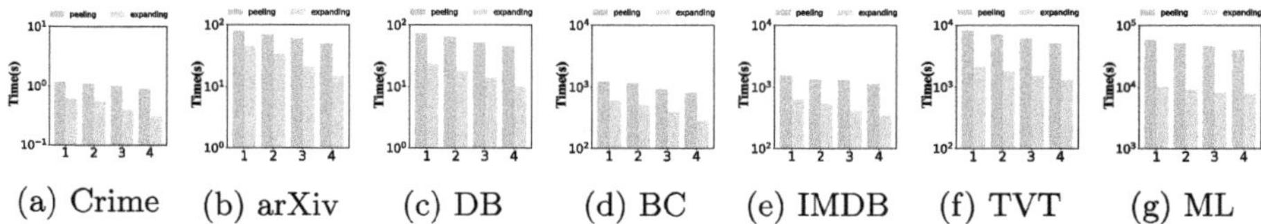

(a) Crime   (b) arXiv   (c) DB   (d) BC   (e) IMDB   (f) TVT   (g) ML

Fig. 2. The effectiveness of the algorithms w.r.t α changes.

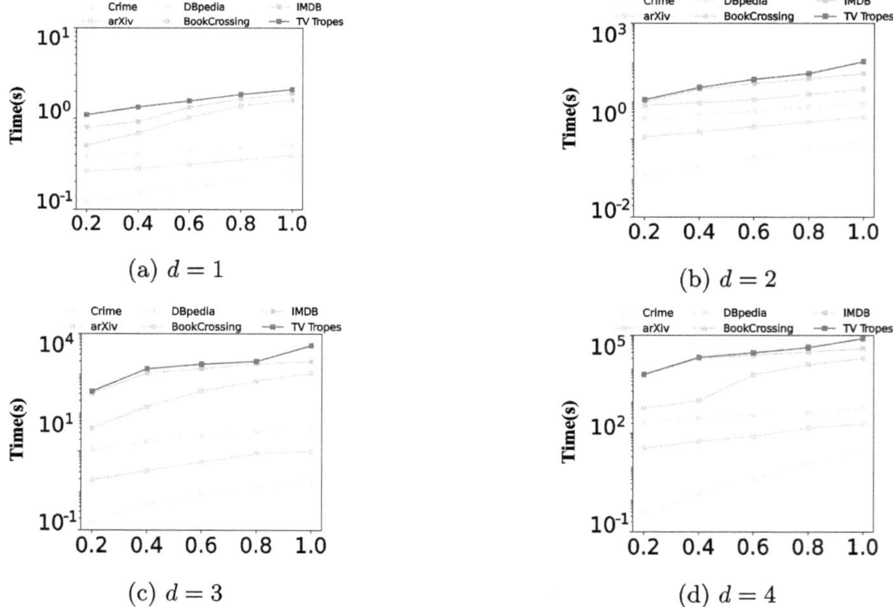

(a) $d = 1$

(b) $d = 2$

(c) $d = 3$

(d) $d = 4$

Fig. 3. Scalability test w.r.t varing dimensions.

Evaluating the Effect of α**.** We maintain other variables constant and vary the value of α across different datasets. As depicted in Fig. 2(a)–(g), both the peeling and expanding algorithms exhibit a decreasing trend in search time as α increases. Note that, we use DB to denote DBpedia, BC to denote BookCross, TVT to denote TVTropes, ML to denote MovieLen. This trend is due to the fact that as α increases, the size of the initial maximal (α,β)-core decreases, which in turn reduces the time required by the algorithms. Additionally, we observe that the efficiency of the expanding algorithm gradually surpasses that of the peeling algorithm. This is because a higher α leads to a larger discrepancy between the size of the (α,β)-core and the original graph, resulting in the expanding algorithm requiring less time.

Edge		e_{11}	e_{12}	e_{14}	e_{15}	e_{21}	e_{22}	e_{24}	e_{26}	e_{31}	e_{33}	e_{35}	e_{36}	e_{43}	e_{44}	e_{45}	e_{46}	e_{51}	e_{52}	e_{55}	e_{56}	e_{61}	e_{63}	e_{64}	e_{66}
Attribute	$d=1$	10	8	3	1	9	8	5	10	10	8	3	4	6	4	5	3	3	8	8	11	5	7	6	10
	$d=2$	13	10	4	3	11	10	4	11	14	15	4	3	17	12	2	6	3	10	16	11	4	6	9	18
	$d=3$	12	9	4	5	10	9	4	13	8	9	1	3	10	9	6	1	4	16	17	9	5	2	1	8

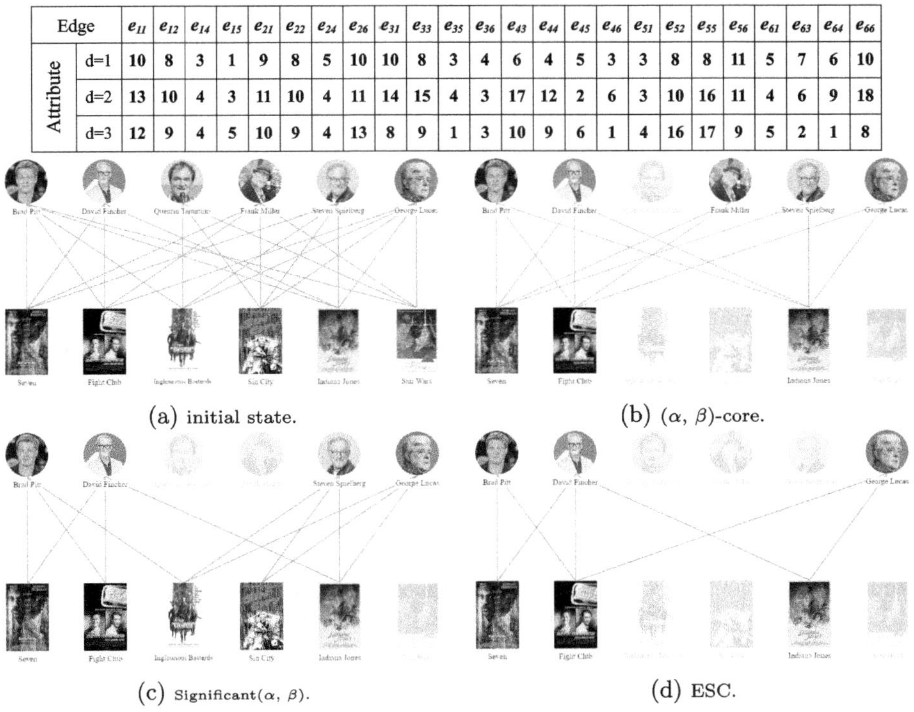

(a) initial state.　(b) (α,β)-core.

(c) Significant(α,β).　(d) ESC.

Fig. 4. Case study: comparison with different models.

5.3 Case Study

Comparison with Different Models. As shown in Fig. 4, we carried out a case study on a subset of the IMDB dataset. Given that each edge in the dataset lacks attribute information, "e_{ij}" represents the edge between the i-th movie and the j-th director, we assigned three-dimensional attribute values to each edge. The first dimension signifies the duration of a director's performance, the second dimension denotes the director's remuneration, and the third dimension represents the director's rating. All three dimensions encompass synthetic attribute information. Figure 4(a) showcases the initial state of the community. When we focus on the performance of a single dimension (where $d = 1$) and query vertex for *Brad Pitt* with parameters $\alpha = 2$ and $\beta = 4$, the communities identified by the algorithm proposed by [22] are illustrated in Fig. 4(b). In contrast, the communities identified by the algorithm proposed by [30] for the first dimension alone are shown in Fig. 4(c). The ESCs found by our algorithm are displayed in Fig. 4(d). The case study's outcomes clearly indicate that our algorithm's ESCs have a denser structure compared to those identified by [22]. Our algorithm effectively filters out edges with minimal attribute values in the second dimension, such as e_{35} and e_{36}, which highlights its precision in identifying more relevant solutions. Moreover, while [30] identifies an ESC with $f(H_1)$, our algorithm not only recognizes H_1 but also uncovers an additional ESC with $f(H_2) = (8, 10, 9)$. They are non-dominated, each excels in one dimension, underscoring that our design can discover more precise solutions in bipartite graphs with multi-dimensional attribute values than prior art [30].

6   Related Work

Community Search (CS). The objective of CS is to identify cohesive subgraphs related to queries [8], which can be divided into precise algorithms and heuristic algorithms. Precise often involves complex mathematical modeling and calculations, such as [3], [41]. Heuristics use approximate methods to quickly find better communities. For instance, in scheme [8], the minimum degree of the k-core is utilized as a constraint for community discovery [2,4].

Community Search on Attributed Graphs. Attributed graphs, categorized into k-core [6,21], k-truss [17,44], and k-clique [19] structures. [12] addressed CS on keyword attribute graphs by defining attribute communities based on structural cohesion and keyword aggregation, introducing the CL-tree index and algorithms. [11] developed a maintenance algorithm for the CL-tree index. For location graphs, [10] introduced spatial-aware communities. [13] presented three fast algorithms for continuous spatial-aware CS on dynamic spatial graphs. [2,4,20] introduced algorithms for large graphs and small j values. [14] proposed multi-attributed joint community search in road social networks, while [23] introduced the HSC model based on meta-path cohesiveness in heterogeneous graphs.

Community Search on Bipartite Graphs. [30] addressed the CS problem in edge-weighted bipartite graphs by introducing the significant (α, β)-community concept. [43] then explored the CS problem in influence bipartite graphs, introducing the Pareto-optimal (α, β)-community model, which considers both subgraph cohesion and vertex importance. It yields Pareto-optimal (α, β)-communities with vertex degree constraints and integrates Pareto optimality with $O(p \cdot m)$.

7   Conclusion

In this paper, we address the challenge of finding edge-attributed skyline communities (ESCs) in large bipartite graphs. We introduce a new model and propose an efficient peeling approach by iteratively removing edges with the least significant attributes across dimensions. A more effective expanding approach is designed to reduce search space, featuring a novel upper bound. Experimental results on real-world large bipartite graphs confirm the effectiveness and efficiency.

Acknowledgments. This work was supported by the NSFC (No. 62302485), and the Key Research Project of Chinese Academy of Sciences (No. RCJJ-145-24-21).

References

1. Bai, Z., et al.: Secmdp: towards privacy-preserving multimodal deep learning in end-edge-cloud. In: ICDE, pp. 1659–1670. IEEE (2024)
2. Bi, F., Chang, L., Lin, X., Zhang, W.: An optimal and progressive approach to online search of top-k influential communities. arXiv:1711.05857 (2017)
3. Bonchi, F., Khan, A., Severini, L.: Distance-generalized core decomposition. In: SIGMOD, pp. 1006–1023 (2019)
4. Chen, S., Wei, R., Popova, D., Thomo, A.: Efficient computation of importance based communities in web-scale networks using a single machine. In: CIKM, pp. 1553–1562 (2016)
5. Chen, X., Wang, K., Lin, X., Zhang, W., Qin, L., Zhang, Y.: Efficiently answering reachability and path queries on temporal bipartite graphs. PVLDB **14**(10), 1845–1858 (2021)
6. Chen, Y., Fang, Y., Cheng, R., Li, Y., Chen, X., Zhang, J.: Exploring communities in large profiled graphs. IEEE (2018)
7. Chen, Z., Zhao, Y., Yuan, L., Lin, X., Wang, K.: Index-based biclique percolation communities search on bipartite graphs. In: ICDE, pp. 2699–2712. IEEE (2023)
8. Cui, W., Xiao, Y., Wang, H., Wang, W.: Local search of communities in large graphs. In: SIGMOD, pp. 991–1002 (2014)
9. Ding, D., Li, H., Huang, Z., Mamoulis, N.: Efficient fault-tolerant group recommendation using alpha-beta-core. In: CIKM, pp. 2047–2050 (2017)
10. Fang, Y., Cheng, C., Luo, S., Hu, J., Li, X.: Effective community search over large spatial graphs. PVLDB (2017)
11. Fang, Y., Cheng, R., Li, X., Luo, S., Hu, J.: Effective community search over large spatial graphs. PVLDB **10**(6), 709–720 (2017)
12. Fang, Y., Cheng, R., Luo, S., Hu, J.: Effective community search for large attributed graphs. PVLDB **9**(12), 1233–1244 (2016)
13. Fang, Y., et al.: On spatial-aware community search. IEEE TKDE **31**(4), 783–798 (2018)
14. Guo, F., Yuan, Y., Wang, G., Zhao, X., Sun, H.: Multi-attributed community search in road-social networks. In: ICDE, pp. 109–120. IEEE (2021)
15. Guo, Y., Xi, Y., Wang, H., Wang, M., Wang, C., Jia, X.: Fededb: building a federated and encrypted data store via consortium blockchains. IEEE TKDE **36**(11), 6210–6224 (2023)

16. Guo, Y., Zhao, Y., Hou, S., Wang, C., Jia, X.: Verifying in the dark: verifiable machine unlearning by using invisible backdoor triggers. IEEE TIFS **19**, 708–721 (2023)
17. Huang, X., Lakshmanan, L.V.: Attribute-driven community search. PVLDB **10**(9), 949–960 (2017)
18. Ley, M.: The DBLP computer science bibliography: evolution, research issues, perspectives. In: International Symposium on String Processing and Information Retrieval, pp. 1–10 (2002)
19. Li, J., Wang, X., Deng, K., Yang, X., Yu, J.X.: Most influential community search over large social networks. In: ICDE (2017)
20. Li, R.-H., Qin, L., Yu, J.X., Mao, R.: Finding influential communities in massive networks. VLDB J. **26**(6), 751–776 (2017). https://doi.org/10.1007/s00778-017-0467-4
21. Li, R., et al.: Skyline community search in multi-valued networks. In: SIGMOD, pp. 457–472 (2018)
22. Liu, Q., Liao, X., Huang, X., Xu, J., Gao, Y.: Distributed (α, β)-core decomposition over bipartite graphs. In: IEEE ICDE, pp. 909–921. IEEE (2023)
23. Liu, Y., Guo, F., Xu, B., Bao, P., Shen, H., Cheng, X.: SACH: significant-attributed community search in heterogeneous information networks. In: ICDE, pp. 3283–3296. IEEE (2024)
24. Luo, J., Cao, X., Xie, X., Qu, Q., Xu, Z., Jensen, C.S.: Efficient attribute-constrained co-located community search. In: ICDE, pp. 1201–1212 (2020)
25. Meng, L., et al.: A survey of distributed graph algorithms on massive graphs. ACM Comput. Surv. **57**(2), 1–39 (2024)
26. Sarıyüce, A.E., Pinar, A.: Peeling bipartite networks for dense subgraph discovery. In: WSDM, pp. 504–512 (2018)
27. Wang, J., De Vries, A.P., Reinders, M.J.: Unifying user-based and item-based collaborative filtering approaches by similarity fusion. In: SIGIR, pp. 501–508 (2006)
28. Wang, K., Lin, X., Qin, L., Zhang, W., Zhang, Y.: Efficient bitruss decomposition for large-scale bipartite graphs. In: ICDE, pp. 661–672 (2020)
29. Wang, K., Zhang, W., Lin, X., Qin, L., Zhou, A.: Efficient personalized maximum biclique search. In: ICDE, pp. 498–511 (2022)
30. Wang, K., Zhang, W., Lin, X., Zhang, Y., Qin, L., Zhang, Y.: Efficient and effective community search on large-scale bipartite graphs. In: ICDE, pp. 85–96. IEEE (2021)
31. Wang, K., Zhang, W., Zhang, Y., Qin, L., Zhang, Y.: Discovering significant communities on bipartite graphs: an index-based approach. IEEE TKDE **35**(3), 2471–2485 (2023)
32. Xu, G., et al.: A model value transfer incentive mechanism for federated learning with smart contracts in aiot. IEEE IoTJ (2024)
33. Xu, S., et al.: Post-quantum searchable encryption supporting user-authorization for outsourced data management. In: ACM CIKM, pp. 2702–2711 (2024)
34. Xu, S., et al.: Towards efficient multi-user access control encrypted search for web data management. IEEE TDSC 1–16 (2025)
35. Xu, S., et al.: Lattice-based forward secure multi-user authenticated searchable encryption for cloud storage systems. IEEE Trans. Comput. (2025)
36. Xu, S., et al.: Lattice-based forward secure certificateless encryption scheme for cloud data management. In: International Conference on Database Systems for Advanced Applications, pp. 441–456. Springer, Heidelberg (2025). https://doi.org/10.1007/978-981-95-4155-3_31

37. Xu, S., Chen, X., Zhao, Y., Gao, S., Wang, J., Yiu, S.M.: Mulshare: a multi-receiver encrypted search scheme to secure cloud-assisted EMRS sharing. In: 2025 IEEE Conference on Communications and Network Security (CNS), pp. 1–9 (2025)
38. Xu, T., Xu, S., Chen, X., Chen, F., Li, H.: Multi-core token mixer: a novel approach for underwater image enhancement. Mach. Vis. Appl. **36**(2), 1–16 (2025)
39. Xu, Z., et al.: Effective community search on large attributed bipartite graphs. Int. J. Pattern Recogn. Artif. Intell. **37**(02), 2359002 (2023)
40. Yuan, L., Xu, J., Chen, Z., Ma C., Xu, J., Qin, L.: Efficient maximum balanced k-biplex search over bipartite graphs. In: ICDE, pp. 3098–3112. IEEE (2025)
41. Zhang, Y., Yu, J.X.: Unboundedness and efficiency of truss maintenance in evolving graphs. In: SIGMOD, pp. 1024–1041 (2019)
42. Zhang, Y., Phillips, C.A., Rogers, G.L., Baker, E.J., Chesler, E.J., Langston, M.A.: On finding bicliques in bipartite graphs: a novel algorithm and its application to the integration of diverse biological data types. BMC Bioinf. **15**, 1–18 (2014)
43. Zhang, Y., Wang, K., Zhang, W., Lin, X., Zhang, Y.: Pareto-optimal community search on large bipartite graphs. In: CIKM, pp. 2647–2656 (2021)
44. Zheng, Z., Ye, F., Li, R.H., Ling, G., Jin, T.: Finding weighted k-truss communities in large networks. Inf. Sci. **417** (2017)
45. Zhou, Y., Fang, Y., Luo, W., Ye, Y.: Influential community search over large heterogeneous information networks. PVLDB **16**(8), 2047–2060 (2023)

SPGRF: A Structure-Preserving Graph Reduction Framework

Xin Wang[1], Haiyang Pan[1], Zhenhao Tong[1], Ji Zhang[2], Bin Hu[1],
and Wenbo Xie[1(✉)]

[1] Southwest Petroleum University, Chengdu, China
{xinwang,wenboxie}@swpu.edu.cn,
{202422000626,202522000799,202422000621}@stu.swpu.edu.cn
[2] University of Southern Queensland, Toowoomba, Australia
ji.zhang@unisq.edu.au

Abstract. Large-scale graphs are prevalent. Their vast scale poses formidable challenges for analytical tasks. Graph reduction techniques, which can significantly shrink graph size while preserving key information, have thus become one of the key technologies for large graph analysis. However, existing methods mostly focus on low-order graph properties or task-specific objectives, omitting the high-order structures of a graph. The reduced graphs thus do not meet the expectation on downstream tasks. To address this, we propose the **Structure-Preserving Graph Reduction Framework (SPGRF)** to retain *crucial* structure of a large graph. Firstly, a local-to-global edge importance assessment method is developed to guide the construction of an initial graph skeleton. Secondly, a neighborhood-based enhancement mechanism is designed to compensate for the structural loss, producing a reduced graph G_s, that is better suited for downstream tasks. Extensive experiments on real graphs demonstrate that SPGRF excels across multiple metrics. The reduced graphs effectively preserve the original backbone structure and provide strong support for various downstream tasks, *e.g.*, when the reduction ratio is only 10%, the accuracy on the top-k (k=800) frequent patterns can even reach 96%. The source code is available at https://github.com/oceanphy/SPGRF.

Keywords: Graph Reduction · Graph Neural Network · Reduced Skeleton · Structured-Preserving Reduction

1 Introduction

Graphs have become a fundamental tool for modeling complex systems such as social networks, knowledge graphs, graph neural networks (GNNs) and so on. However, the overwhelm size hinders various analytical tasks, *e.g.*, learning of GNNs and its variant Graph Convolutional Networks (GCNs), frequent pattern mining (FPM), and others, on large graphs. In recent years, graph reduction has emerged as an effective approach to large-scale graph analysis. By retaining

H. Jung et al. (Eds.): DASFAA 2026, LNCS 16536, pp. 384–402, 2026.
https://doi.org/10.1007/978-981-92-0366-6_24

essential information, graph reduction effectively decreases the data scale of the original graph, and substantially improves performances of various downstream tasks [7]. However, existing techniques mostly engineered for specific downstream tasks, limiting their applicability. These include methods tailored for preserving specific graph properties, such as spectral characteristics [15], creating concise graph summaries via node aggregation [12], or distilling knowledge for GNNs training through graph synthesis [9]. Broadly, these task-specific methods operate by deleting, blurring (*i.e.*, aggregation), or artificially synthesizing parts of a graph. Such strategies inherently neglect the comprehensive, multi-scale structural information essential for robust performance across diverse analytical tasks.

In contrast, we propose a task-agnostic graph reduction framework SPGRF, applicable for various downstream tasks. Unlike prior methods, SPGRF uniquely integrates a local-to-global importance assessment with a neighborhood-guided enhancement mechanism to holistically preserve multi-scale structural information, ensuring its robustness and broad applicability without relying on task-specific heuristics.

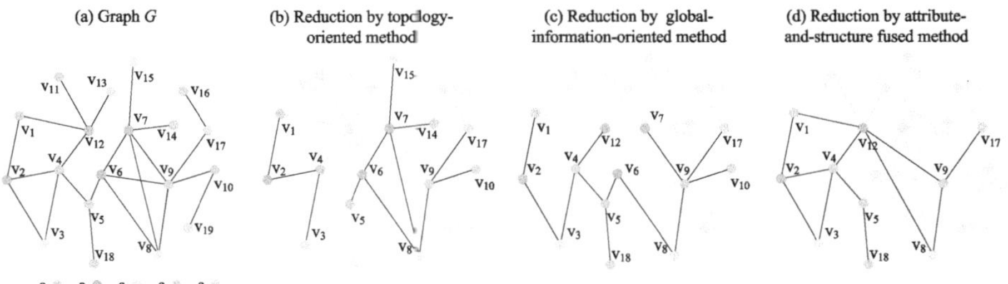

Fig. 1. Comparison of graph reduction methods.

Example 1. Given a graph G in Fig. 1 (a), node colors indicate their attributes, *i.e.*, a_1-a_5. Two representative graph reduction techniques, *i.e.*, topology-based method [17] and global-information-based method [11] generate reduced graphs G_s, as shown in Fig. 1 (b) and (c), respectively. They either ignore the feature information on nodes or fail to preserve crucial structure from a global perspective, resulting in degraded performance on downstream tasks like FPM, community detection and so on. Contrarily, when node features and higher-order structural features are considered simultaneously, the reduced graph, shown in Fig. 1 (d), exhibits its unique capability in maintaining both typical sub-structures and community boundaries of the original graph, *e.g.*, it well preserves substructures, *e.g.*, $\{v_2, v_3, v_4\}$, that corresponds to a frequent pattern with edge set $\{(a_1, a_2), (a_1, a_3), (a_2, a_3)\}$ in G and retains key bridge nodes such as v_{12} for the connection of communities.

However, constructing such a high-quality reduced graph G_s introduces two technical challenges: (1) how to quantify importance of nodes and edges and generate a concise skeleton, capturing crucial information of the original graph? (2) how to enhance this skeleton to compensate for structural loss and well support downstream tasks?

Contributions. The main contributions of this paper are summarized as follows:

(1) We propose a multi-stage structure-preserving graph reduction framework (SPGRF). SPGRF evaluates edge importance through a local-to-global mechanism. Leveraging the local structure for initial weighting, it subsequently employs a GCN-based model to infer global importance for edges. This process culminates in the construction of a graph skeleton comprising the most critical nodes and edges.

(2) We introduce a neighborhood-aware mechanism to enhance the graph skeleton through restoring essential missing structures. By identifying and reinstating critical indirect connections, the enhancement mechanism significantly improves the structural diversity and completeness of the final reduced graph.

(3) Extensive experiments on real graphs demonstrate that SPGRF consistently outperforms existing methods in terms of downstream performance, structure preservation, and efficiency. Notably, SPGRF achieves up to 96% accuracy on the top-k ($k = 800$) FPM task with reduction ratio as low as 10%, while also benefiting tasks like community detection and others.

2   Related Work

Existing techniques can be broadly categorized into three directions: sparsification, coarsening, and condensation.

Graph Sparsification. This line of work reduces graph size by eliminating redundant edges or nodes while maintaining key properties. Early methods focused on distance preservation, such as Spanner [2], while later works emphasize spectral similarity, as in GRASS [5]. With the rise of Graph Neural Networks (GNNs), sparsification methods have shifted toward model performance-driven objectives. For instance, WIS [18] scores edge importance based on multi-hop propagation, while GraphFS [26] incorporates fairness into edge removal. Additionally, sampling-based approaches like Forest Fire [25] and CNARW [13] optimize exploration efficiency in large graphs, and Wang et al. [22] leverage motif structures for higher-order preservation during node sampling, offering semantic-aware reductions for downstream tasks.

Graph Coarsening. Classical methods, *e.g.*, RSS [16], focus on minimizing reconstruction loss. Recent works integrate node features for better semantic consistency, *e.g.*, F-GCN [11]. Similarly, SsAG [1] merges nodes via weighted sampling to balance structural fidelity and degree of attribute retention. To avoid producing dense graphs, SSumM [12] integrates node merging with a sparsification mechanism guided by the minimum description length principle, yielding more compact summary graphs. For learning efficiency, SCAL [8] incorporates coarsening into GNN training pipelines with knowledge transfer. Bacciu et al. [3] introduces neural-inspired graph downsampling to support hierarchical design in irregular graph structures.

Graph Condensation. Unlike sparsification or coarsening, condensation synthesizes entirely new small graphs that mimic the training dynamics of the original. GCond and DosCond [9] use gradient matching and model decoupling for scalable graph distillation. SFGC [27] introduces long-term training trajectory guidance. For efficiency, GC-SNTK [20] uses neural tangent kernels to yield closed-form approximations. SimGC [24] and CGC [6] abandon gradient dependencies, using structural priors and distribution alignment for fast generation with high utility in dynamic or incremental settings.

3   Preliminaries

Definition 1 Graph [1]. *Given an undirected attributed graph $G = (V, E, L)$, where V and E are sets of nodes and edges, respectively. $L(v_i)$ is the attribute value for $v_i \in V$. For a node v, $N_G(v)$ denotes the neighborhood of v in graph G.*

Definition 2 Pattern [21]. *A pattern $Q = (V_p, E_p, f_v)$ is a graph where $f_v(u)$ for each node $u \in V_p$ is a conjunction of atomic formulas "$A = a$" over node attributes.*

Definition 3 Pattern Matching [22]. *Given a graph G and pattern Q, a node $v \in G$ matches $u \in Q$ (denoted $v \sim u$) if all atomic conditions in $f_v(u)$ are satisfied by $L(v)$. A match of Q in G is an isomorphic mapping f such that: (1) $f_v(u) \sim L(f(u))$ for all $u \in V_p$; (2) $(u, u') \in E_p$ iff $(f(u), f(u')) \in E$. Denote all matches by $\mathcal{M}(Q, G)$ and matching nodes of u by $img(u)$.*

Definition 4 Support [22]. *The support of a pattern Q in graph G, denoted by $Sup(Q, G)$, indicates how frequently Q occurs in G. We adopt the Minimum Image-based support (MNI), that is defined as:*

$$Sup(Q, G) = \min\{|img(u)| \mid u \in V_p\}, \tag{1}$$

where $img(u)$ denotes the set of nodes in G matching u in Q.

Definition 5 Graph Reduction [7]. *Given a graph $G = (V, E, L)$, graph reduction refers to the process of constructing a smaller-scale graph $G_s = (V_s, E_s, L_s)$ that preserves the critical structural and semantic information of G. We adopt the node reduction ratio $r = \frac{|V_s|}{|V|}$ to quantify the effectiveness of graph reduction, where $|V|$ and $|V_s|$ denote the total number of nodes in G and G_s, respectively.*

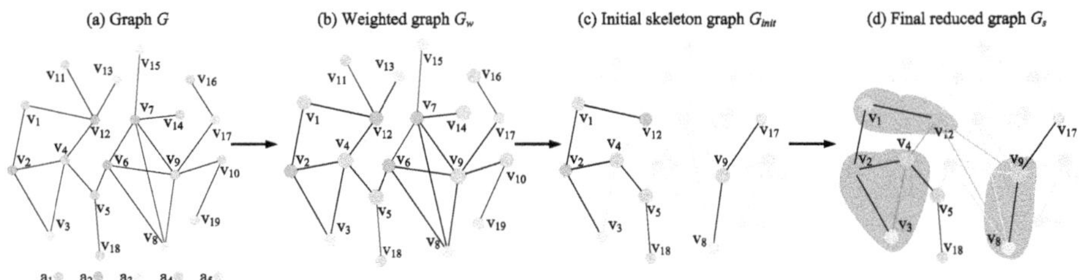

Fig. 2. Overview workflow of SPGRF.

4 Structure-Preserving Graph Reduction Framework

In this section, we introduce the SPGRF algorithm for graph reduction.

Algorithm 1. SPGRF

Input: Graph G, reduction ratio r, parameters d_r and τ.
Output: Reduced Graph G_s.
 1: initialize G_s, $\mathcal{M}$;
 2: $G_w := \mathsf{InitW}(G)$;
 3: $scores, Z := \mathsf{EdgeImp}(G_w)$;
 4: $G_{init} := \mathsf{GetInitGraph}(G_w, r, scores, \mathcal{M})$;
 5: $G_s := \mathsf{ConnectEdges}(G, G_{init}, scores, Z, d_r, \tau)$;
 6: **return** G_s;

Overall Framework. The framework SPGRF, detailed in Algorithm 1, takes a graph G, a reduction ratio r, and two hyperparameters: a edge retention ratio d_r, and a node similarity threshold τ as input and reduces a graph G into a smaller one G_s that meets the reduction ratio r. The algorithm begins with an initialization step (line 1) that generates a set of matches $\mathcal{M}(Q, G)$ for all single-edge patterns. It next works in four main phases. (a) SPGRF employs the InitW function to initialize weights for nodes and edges through a fused metric that incorporates node label frequency with pattern support. This step generates a weighted graph G_w, which encapsulates foundational yet localized importance. (b) To overcome the limitations caused by relying solely on local information, a GCN-based model EdgeImp is leveraged to obtain the context-aware edge importance scores and structure-aware node representations. (c) Utilizing the context-aware importance scores, SPGRF applies the procedure GetInitGraph to construct an initial graph skeleton G_{init}, which is a sparse yet structurally critical backbone of the original graph. (d) On skeleton G_{init}, the procedure ConnectEdges is developed to further enrich it, thereby obtaining a fine reduced graph. Figure 2 illustrates the overall workflow of SPGRF. We next proceed to elaborate on its key algorithmic components.

4.1   A Reduction Algorithm

Weight Initialization. The InitW function (not shown) utilizes both structural and attribute information of graph G to compute weights for its nodes and edges, thereby producing a weighted graph G_w. It is important to note that while G_w incorporates weights on nodes and edges, it is structurally identical to the original graph G. Therefore, the pattern matches computed on G are directly applicable to G_w for subsequent steps. We first quantify weights for node labels and single-edge patterns, respectively.

The weight $\omega_L(a)$ of a label a, is determined by its relative frequency in G:

$$\omega_L(a) = \frac{count(a)}{\sum_{l \in \Sigma_G} count(l)}, \tag{2}$$

where Σ_G represents the set of all labels in graph G, and the $count()$ function returns the number of nodes associated with a specific label.

The weight $\omega_Q(Q_e)$ of a single-edge pattern Q_e is defined as Eq. (3):

$$\omega_Q(Q_e) = \beta \cdot \frac{\sum_{v \in V_{Q_e}} \omega_L(L(v))}{2} + (1 - \beta) \cdot \frac{Sup(Q_e, G)}{\sum_{Q \in S_Q} Sup(Q, G)}, \tag{3}$$

where S_Q is a set consisting of all the single-edge patterns in graph G, and β is a parameter introduced to balance the weights of two components.

With these foundational weights established, the weight of a node v is defined as Eq. (4). The underlying design principle is that a node's importance is influenced by both its attributes and the characteristics of its local neighborhood.

$$\omega_N(v) = \omega_L(L(v)) \cdot \left(1 + \sum_{l \in S_N^l(v)} \omega_L(l) \right), \tag{4}$$

where $S_N^l(v)$ represents the set of labels from all the first-order neighbors of node v, excluding its own label $L(v)$.

Similarly, the weight $\omega_E(e)$ of an edge $e = (u, v)$ is defined as Eq. (5). Intuitively, $\omega_E(e)$ is calculated by aggregating the weights of its endpoints and the weight of its corresponding single-edge pattern Q_e.

$$\omega_E(e) = \omega_N(u) + \omega_N(v) + \omega_Q(Q_e). \tag{5}$$

Example 2. To illustrate the weighting process, we take the edge (v_6, v_8) from the graph in Fig. 1 (a) as a running example. First, the label weights are determined by their occurrence frequency. For instance, $\omega_L(a_2) = 0.21$ and $\omega_L(a_3) = 0.21$. Assuming the total support of all single-edge patterns is 14 and $\beta = 0.5$, we then compute the weight for the single-edge pattern $Q_1 : a_2 - a_3$ with a match (v_6, v_8). Via Eq. (3), its weight $\omega_Q(Q_e)$ equals to $0.5 \cdot \frac{0.21 + 0.21}{2} + 0.5 \cdot \frac{4}{14} = 0.248$. Subsequently, for node v_6, whose neighbors have labels $\{a_1, a_2, a_3\}$, its weight $\omega_N(v_6)$ is calculated via Eq. (4) as $0.21 \cdot (1 + 0.37 + 0.21) = 0.3318$. Similarly, the weight for node v_8 is calculated as $\omega_N(v_8) = 0.3318$. Finally, the weight of the edge (v_6, v_8) is computed using Eq. (5) by aggregating these components: $\omega_E(v_6, v_8) = \omega_N(v_6) + \omega_N(v_8) + \omega_Q(Q_1) = 0.3318 + 0.3318 + 0.248 = 0.9116$.

Algorithm 2. GetInitGraph

Input: Weighted graph G_w, reduction ratio r, and *scores*, $\mathcal{M}$.
Output: Graph skeleton G_{init}.

```
 1: initialize G_init, B_labels, S_Q, S_c, θ_t = 0;
 2: for each label l in B_labels do
 3:     k_l := max(1, ⌈r · |B_labels(l)|⌉); V'_s := Top-k_l(B_labels(l), ≻);
 4:     V_s := V_s ∪ V'_s;
 5: end for
 6: for each pattern Q in S_Q do
 7:     θ_t := max(1, ⌈r · Sup(Q, G_w)⌉);
 8:     S_c := ∅;
 9:     for each edge e = (u, v) ∈ M(Q, G_w) with {u, v} ⊂ V_s do
10:         S_c := S_c ∪ {e};
11:     end for
12:     sort S_c by the scores of each edge in descending order;
13:     for each edge e in S_c do
14:         E_s := E_s ∪ {e};
15:         if Sup(Q, G_init) ≥ θ_t then
16:             break;
17:         end if
18:     end for
19: end for
20: return G_init;
```

Skeleton Generation. The GetInitGraph procedure, outlined in Algorithm 2, constructs an initial skeleton G_{init} of graph G. The function takes a weighted graph G_w, a reduction ratio r, edge importance scores *scores* and a dictionary $\mathcal{M}$ that maps a single-edge pattern to its matches as input. It begins by initializing several data structures, *i.e.*, the target graph skeleton $G_{init} = (V_s, E_s)$, a dictionary B_{labels} which maps a label to all the nodes taking that label in G_w, set S_Q consisting of all unique single-edge patterns in G_w, set S_c for maintaining candidate edges. Additionally, a variable, θ_t, is initialized as 0. After initialization, the process unfolds in two sequential stages.

First, in the node selection stage (lines 2–5), GetInitGraph populates the node set V_s. A core design principle in this stage is to preserve the label distribution of the original graph. To achieve this, for each label l in the pre-computed B_{labels}, the procedure calculates the target number of nodes k_l based on the reduction ratio r. It then selects the top-k_l nodes with the highest weights from the group corresponding to label l and adds them to V_s.

Next, in the edge selection stage (lines 6–19), the function constructs the edge set E_s with the core objective of aligning supports of single-edge patterns with that of G_w. For each single-edge pattern Q in the pre-computed S_Q, the function calculates θ_t based on its support in G_w (*i.e.*, $Sup(Q, G_w)$) and the reduction ratio r. It then identifies a set S_c of candidate edges by filtering matches of the pattern $\mathcal{M}(Q, G_w)$, to include only those edges whose endpoints are both within the node set V_s. These candidate edges are sorted in descending order based on their importance scores. Finally, the function iteratively adds the top ranked edges to E_s until the support of pattern Q in the evolving skeleton G_{init} reaches the predefined θ_t. Notably, the computation of $Sup(Q, G_{init})$ does not require rescanning G_{init} in each iteration. Instead, we maintain a set consisting

---

Algorithm 3. ConnectEdges

Input: Graph G, skeleton $G_{init} = (V_s, E_s)$, *scores*, Z, parameters d_r and τ.
Output: Reduced graph G_s.

1: initialize *groupNodes*, S_m, *candNodes*;
2: **for each** node v in V_s **do**
3:     $S_m := \emptyset$;
4:     **for each** node u_1 in $N_G(v) \cap V_s$ **do**
5:         **if** $(v, u_1) \notin G_{init}$ **then**
6:             $S_m := S_m \cup \{(v, u_1)\}$;
7:         **end if**
8:     **end for**
9:     $k_e := \mathtt{max}(1, \lceil d_r \cdot |S_m| \rceil)$;
10:    select top-k_e edge set E_s' from S_m with highest *scores*;
11:    $E_s := E_s \cup E_s'$;
12:    **for each** node u_2 in $N_G(v) \setminus V_s$ **do**
13:        $candNodes := groupNodes(L(u_2))$;
14:        pick a node u_2' in *candNodes* with highest $cos\_sim(Z(u_2), Z(u_2'))$;
15:        **if** $cos\_sim(Z(u_2), Z(u_2')) \geq \tau$ **then**
16:           $E_s := E_s \cup \{(v, u_2')\}$;
17:        **end if**
18:     **end for**
19: **end for**
20: **return** updated skeleton G_{init} as reduced graph G_s;

---

of $img(u)$ for each pattern node u, and the support equals to the size of the smaller set. The resulting graph skeleton $G_{init} = (V_s, E_s)$ is then returned.

Example 3. With a reduction ratio of $r = 0.5$, the function first performs node selection. For instance, the a_2 label has four nodes in the original graph, so the target count is $4 \times 0.5 = 2$. Based on node weights, the top two nodes, v_2 and v_{12}, are selected. This process is repeated for all labels to form the node set V_s, as shown by the colored nodes in Fig. 2 (b). Next, edge selection aims to meet a target support for each single-edge pattern. For the $a_1 - a_2$ pattern with an original support of 4, the target becomes $4 \times 0.5 = 2$. The function iterates through candidate edges, sorted by their importance scores. An edge like (v_9, v_6) is skipped because $v_6 \notin V_s$. Then, (v_1, v_{12}) and (v_1, v_2) are added to E_s as their endpoints are in V_s. Once the support for the $a_1 - a_2$ pattern in G_{init} reaches the target of 2, the algorithm stops adding edges for this pattern and moves to the next. This procedure is applied to all single-edge patterns to form the initial skeleton G_{init}.

Skeleton Enhancement. The function ConnectEdges takes as input the graph G, the initial graph skeleton G_{init} with node and edge sets V_s, E_s, the edge importance scores *scores*, the node representations Z, a rate d_r for controlling edge connection, and a similarity threshold τ, and returns the final reduced graph G_s. This function comprises two main components: direct neighbor connections (lines 3–11) and indirect neighbor connections (lines 12–18). First, the function initializes a dictionary-based auxiliary structure *groupNodes* to store the set of nodes corresponding to each label in G, two sets S_m and *candNodes* to keep track of candidate edges and nodes, respectively (line 1). In the direct neighbor connection phase, for each node v in the node set V_s of G_{init}, if its neighboring node u_1 in G also belongs to V_s and the edge (v, u_1) does not exist in G_{init} (line 4), then (v, u_1) is added to the missing edge set S_m (line 5). To prevent over-connection, which could result in an excessively dense reduced graph and disrupt pattern support rankings, potentially impairing downstream tasks, the function introduces a direct connection rate d_r to control the number of direct connections k_e (line 9). The top k_e edges with the highest scores are then selected from the missing edge set S_m and added to the edge set E_s of the reduced graph (lines 10). Next, in the indirect neighbor connection phase, for neighboring nodes u_2 that are not present in V_s, nodes with the label $L(u_2)$ are first selected from *groupNodes* to form the *candNodes* (line 13). Since the node representations Z aggregate neighborhood information via the GCN, the similarity between node representations reflects their structural similarity. Based on this, the function identifies a substitute node u_2' from the nodes in V_s that share the same label as u_2, selecting the one with the highest cosine similarity to u_2 (line 14). If this similarity exceeds the threshold τ, the new edge (v, u_2') is added to E_s (line 15). After completing these edge connection operations, the function returns the final reduced graph G_s.

Example 4. The ConnectEdges function enhances the initial skeleton G_{init} from Fig. 2 (b) through two mechanisms. First, it performs direct connection recovery. For example, nodes v_3 and v_4 are neighbors in the original graph and both exist in V_s, but the edge between them is missing in G_{init}. The rule restores this edge, shown as the red line in Fig. 2 (c). Second, it handles indirect connections for nodes whose neighbors were removed. For node v_8, its neighbor v_7 (label a_2) is not in V_s. The function searches for a substitute within V_s that shares the same label, identifying v_{12} as the node with the highest representation similarity to v_7. Since this similarity exceeds the threshold τ, a new edge (v_8, v_{12}) is added, shown as the green line. Conversely, when processing v_5, the most similar substitute for its neighbor v_6 (v_2) has a similarity below τ, so no edge is added. After processing all nodes in V_s, the final reduced graph G_s is produced.

4.2   Complexity Analysis

Time Complexity. SPGRF consists of three sequential stages: *(i) Weight Initialization, (ii) Skeleton Generation,* and *(iii) Skeleton Enhancement.* In *(i)*, computing weights involves linear scans of nodes and edges, costing $O(|V| + |E|)$. In *(ii)*, grouping nodes and sorting edges requires $O(|V| \log |V| + |E| \log |E|)$. In *(iii)*, the reconnection process takes $O(|V| \cdot |E|)$ in the worst case due to neighborhood checks. Dominating the lower-order terms, the overall time complexity simplifies to $O(|V| \cdot |E|)$.

Space Complexity. Graph storage requires $O(|V| + |E|)$ space. Node representations (dimension d, used by GEEM and will be introduced in Sect. 5) incur $O(d \cdot |V|)$ cost. As d is small, the space is bounded by $O(|V| + |E|)$.

5   A Prediction Model

In this section, we introduce a GCN-based Model (GEEM) for edge importance evaluation. The model serves as the core of the function EdgeImp in Algorithm 1. As shown in Fig. 3, GEEM consists of three cascaded modules for node initial feature construction, node representation learning and the edge importance evaluation.

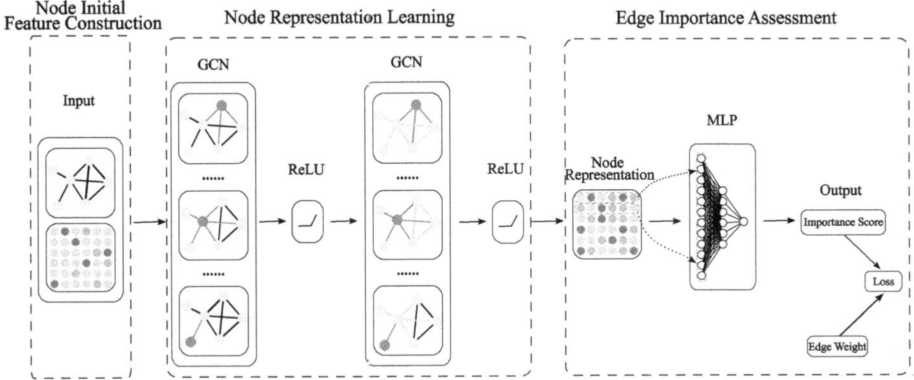

Fig. 3. Architecture of the GCN-based Edge Evaluation Model (GEEM). The dark blue node represents the target node that requires information aggregation from its surrounding neighbors. The red and yellow lines indicate the information flow from first-order neighbors and second-order neighbors, respectively. (Color figure online)

Node Initial Feature Construction. For each node v in a graph G, we construct an initial feature vector $h_v^{(0)} \in \mathbf{R}^d$ that encodes both semantic and structural importance. Specifically, discrete node labels are first mapped to dense semantic vectors e_v via word embedding techniques such as Word2Vec. These semantic vectors are then concatenated with the node importance weights $\omega_N(v)$ obtained from the InitW function. The initialization process is defined as Eq. (6).

$$h_v^{(0)} = [e_v \oplus \omega_N(v)].\tag{6}$$

Node Representation Learning. GEEM employs a two-layer GCN to update the initial node features $H^{(0)} = \{h_1^{(0)}, h_2^{(0)}, ..., h_v^{(0)}\}$. The GCN propagates and aggregates information from local neighborhoods through successive layers. At the l-th layer, the features are updated by

$$H^{(l+1)} = \sigma \left(\tilde{D}^{-\frac{1}{2}} \tilde{A} \tilde{D}^{-\frac{1}{2}} H^{(l)} W^{(l)} \right),\tag{7}$$

where $\tilde{A}$ denotes the adjacency matrix augmented with self-loops, $\tilde{D}$ is the corresponding degree matrix, $W^{(l)}$ are the learnable weight parameters, σ is a nonlinear activation function, and $H^{(l)}$ represents the node features at the l-th layer.

After two GCN layers, the matrix $Z = H^{(2)}$ is obtained. Each row vector z_v in Z (equivalently, $h_v^{(2)}$) is a learned representation of node v that aggregates structural and semantic information from its two-hop neighborhood. This multi-hop aggregation is essential for understanding the broader context in which each node resides.

Edge Importance Assessment. With the context-aware node representations Z, the edge importance can be calculated as follows. For any edge $e = (u, v)$ in the graph, its feature vector h_{uv} can be obtained by concatenating the representation of its two endpoints:

$$h_{uv} = [z_u \oplus z_v]. \tag{8}$$

The edge representation is then fed into a two-layer multilayer perceptron (MLP) to predict the importance score s_{uv}. The MLP performs the following computation,

$$s_{uv} = W_2 \sigma \left(W_1 h_{uv} + b_1 \right) + b_2, \tag{9}$$

where W_1, b_1 and W_2, b_2 denote the weights and biases of the first and second layers, respectively. The first layer projects the concatenated edge features into a 32-dimensional hidden space with ReLU activation, and the second layer maps this hidden representation to a scalar score. The collection of all predicted scores forms the set *scores* used for graph reduction.

Model Learning. To train GEEM, the edge weights $\omega_E(e)$ returned from the InitW function are used as the ground-truth supervision. Accordingly, the training objective is defined as minimizing the Mean Squared Error (MSE) between predicted scores s_{uv} and their ground-truth values $\omega_E(u, v)$ (Eq. (10)).

$$\text{Loss} = \frac{1}{|E|} \sum_{(u,v) \in E} \left(s_{uv} - \omega_E(u, v) \right)^2 . \tag{10}$$

The gradients of this objective are computed with respect to all learnable parameters, including the GCN weights $W^{(l)}$ and MLP parameters, and optimized using the Adam algorithm until convergence.

Inference by EdgeImp(). The procedure EdgeImp (line 3 in Algorithm 1) performs inference on a weighted graph G_w by using a well-trained GEEM model. It first constructs an initial feature vector for each node by combining its semantic and weight-based attributes. These features, along with the adjacency information of the graph, are then processed by the GEEM in a single forward pass. During this pass, the GCN layers of the model generate the structure-aware node representations Z, which are subsequently used by its MLP head to compute the final importance scores *scores*, for all edges. The function finally returns both Z and *scores*, for further applications.

6 Experiments

6.1 Experimental Setup

Datasets. We evaluate reduction performance on eight real-world datasets[1], covering diverse domains including social/e-commerce (YouTube, Amazon, Flickr), scholarly (DBLP, Cora), biological (PDB), internet topology (Skitter), and web graphs (Wiki). These datasets vary significantly in scale (2K–1.7M nodes) and average degree (6.62–31.83), presenting distinct structural patterns and sparsity levels. The first six datasets are used for main experiments, while Cora and Flickr serve for one downstream task *i.e.,* GNN node classification. Dataset statistics are summarized in Table 1.

Table 1. Data Statistics

Dataset	#Nodes	#Edges	Ave. Degree	Dataset	#Nodes	#Edges	Ave. Degree
YouTube	154,817	1,055,572	13.63	DBLP	317,080	1,049,866	6.62
PDB	20,226	83,356	8.24	Amazon	410,236	3,356,824	16.36
Skitter	1,696,415	11,095,298	13.08	Wiki	1,791,489	28,511,807	31.83
Cora	2,708	10,556	7.79	Flickr	89,250	899,756	20.16

Baselines. We compare our method against six state-of-the-art graph reduction approaches: (1) **SSumM** [12], which iteratively merges nodes and sparsifies hyperedges using the Minimum Description Length principle; (2) **WIS** [18], which preserves GNN node interactions by ranking edges based on walk contributions; (3) **CNARW** [13], a random walk approach prioritizing paths through nodes with many common neighbors; (4) **FF** [25], which simulates stochastic fire propagation from seed nodes; (5) **SS** [19], an enhanced breadth-first search that expands layer-by-layer from seed nodes; and (6) **SsAG** [1], which merges nodes based on joint structural and attribute similarity. For the hypergraph-based methods (SSumM, SsAG), we materialize their summaries into standard graphs for fair comparison. All experiments were conducted on a Windows 11 machine with a 3.70 GHz CPU and an NVIDIA GeForce RTX 4060Ti GPU. The code was implemented in Python 3.9.

Evaluation Metrics. To evaluate reduction quality, we employ a comprehensive suite of metrics assessing both structural and attribute fidelity. Following the methodology in [4], we measure structural properties including approximate diameter (Diameter), and both mean (MCC) and global (GCC) clustering coefficients. Additionally, we assess attribute deviation (AD) [14], average node reachability via closeness centrality (AvgCC) [28], information-theoretic entropy loss (EL) [10], and the preservation of eigenvalue structure via spectral distance (SD) [23]. The interpretation of each metric is detailed within its corresponding experimental results.

---

[1] https://github.com/oceanphy/SPGRF.

Implementation. The proposed framework involves three parameters: the reduction ratio r, the direct connection rate d_r (equivalent to r by default, unless otherwise specified), and the similarity threshold τ (fixed at 1). The complete parameter configurations are summarized in Table 2, where the third column specifies k values for the top-k FPM task. To ensure fair comparison, all edge-based reduction methods are calibrated to produce equivalent edge counts at each reduction level. For hypergraph-based methods (SSumM, SsAG), we convert their outputs into standard graphs: SSumM selects the edge with the highest support from each hyperedge, while SsAG determines representative nodes based on dominant attributes with degree-based tie-breaking, preserving all original edges among these representatives.

Table 2. Parameter settings across experiments

ID	r	k	Exp-#
P_1	[0.1, 0.3], Δ0.05	–	1, 3, 5
P_2	0.1	–	2, 3
P_3	0.2 (d_r: [0.1, 0.5], Δ0.1)	[300,2000,80,300,2000,1200]	4
P_4	[0.1, 0.3], Δ0.05	[300,300,300,300,200,800]	5

Note: order of k values: YouTube, DBLP, PDB, Amazon, Skitter, Wiki

6.2 Experimental Results

Exp-1: Overall Quality Metrics. Figure 4 (a) shows that SPGRF achieves superior AD performance on four datasets (YouTube, DBLP, PDB, and Skitter) while remaining competitive on Amazon and Wiki. The slightly lower performance on these two datasets stems from their high attribute dimensionality and skewed attribute distributions. For structural preservation, Fig. 4 (b) demonstrates that as r increases, the AvgCC of SPGRF-reduced graphs progressively converges toward the original graphs (black line). The moderate deviation on YouTube and Wiki occurs because SPGRF prioritizes high-frequency nodes, which affects global shortest-path distributions after skeleton enhancement. Figure 4 (c) reveals that SPGRF maintains consistently low EL across most datasets, with EL generally decreasing as r increases. The marginally higher EL on Amazon is attributable to its imbalanced label distribution, since SPGRF preserves high-frequency structural nodes, rare labels may be discarded during reduction, affecting this entropy-based metric.

Exp-2: Computational Efficiency. We assess runtime and memory usage under configuration P_2, as shown in Tables 3. Our approach demonstrates superior time efficiency, consistently securing the top two ranks across nearly all

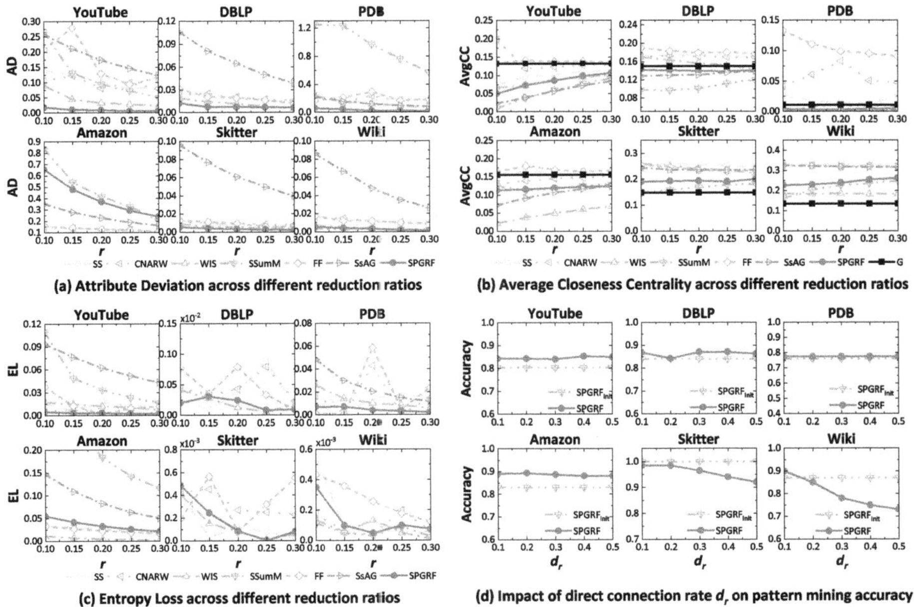

(a) Attribute Deviation across different reduction ratios

(b) Average Closeness Centrality across different reduction ratios

(c) Entropy Loss across different reduction ratios

(d) Impact of direct connection rate d_r on pattern mining accuracy

Fig. 4. Graph reduction quality and ablation study.

datasets. Specifically, $\mathrm{SPGRF}_{\mathrm{init}}$ achieves the fastest runtime, while SPGRF ranks a close second. While memory usage is higher than some baselines, mainly due to storing graph weights and node representations. This trade-off enables SPGRF to retain richer structural information. The additional cost is justified by its superior performance on downstream tasks, striking a favorable balance between efficiency and reduction quality.

Table 3. Reduction time consumption and memory usage. For each row, the best result is in **bold** and the second-best is underlined.

	Time Consumption (s)						Memory Usage (MB)					
	YouTube	DBLP	PDB	Amazon	Skitter	Wiki	YouTube	DBLP	PDB	Amazon	Skitter	Wiki
SPGRF	<u>2.87</u>	<u>5.21</u>	0.22	20.02	<u>63.11</u>	<u>160.82</u>	582.66	629.45	<u>23.21</u>	3498.52	13750.95	23102.4
$\mathrm{SPGRF}_{\mathrm{init}}$	**2.31**	**3.91**	**0.12**	16.95	**40.78**	**132.31**	<u>539.94</u>	<u>627.21</u>	**20.31**	3487.95	13550.79	23030.65
WIS	575.14	716.30	62.34	2096.95	17302.01	81563.39	1718.64	773.23	934.10	3065.45	10835.38	23732.77
SSumM	3.52	5.65	0.40	**10.81**	124.41	560.43	**314.52**	**72.47**	268.86	**553.43**	**1205.47**	**2096.21**
FF	4.08	6.52	0.23	19.94	68.98	164.80	1159.20	1738.09	1365.84	10898.48	10663.68	<u>21239.41</u>
SS	4.46	7.85	<u>0.17</u>	<u>15.86</u>	73.08	200.60	1290.25	1753.37	1192.21	<u>2858.20</u>	<u>6670.33</u>	21539.41
CANRW	5.20	9.77	0.20	18.03	413.17	2730.83	1727.09	1773.83	1396.30	2879.02	10898.48	21273.80
SsAG	5.12	6.11	0.41	22.13	93.41	252.02	1180.34	2376.9	2549.27	6379.53	16016.23	27392.74

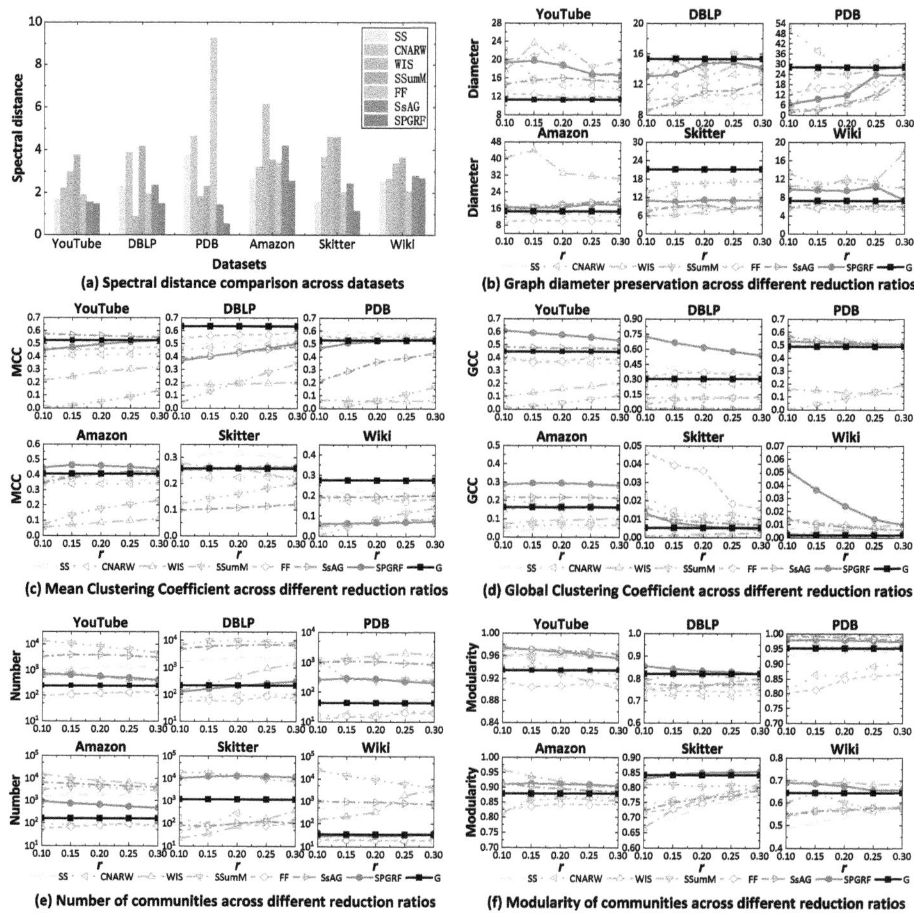

Fig. 5. Structural Property Preservation.

Exp-3: Structural Property Preservation. We conduct a comprehensive evaluation of how well different reduction algorithms preserve core structural properties of the original graphs. This includes spectral characteristics, global distances, local clustering patterns, and community structures. We next present four representative perspectives.

Spectral Distance. To assess overall structural similarity, we compute the Spectral Distance (SD) between original and reduced graphs under parameter configuration P_2. Lower SD indicates better preservation. As shown in Fig. 5 (a), SPGRF consistently achieves the lowest SD on all six datasets, with up to 2.6× improvement over the best baseline on PDB. This demonstrates that our method effectively retains global structural signals by jointly leveraging topological and attribute features, enhanced through GCN-based context modeling.

Diameter Preservation. We evaluate the preservation of graph diameter using an approximate diameter algorithm under configuration P_1, with results shown in Fig. 5 (b). SPGRF closely approximates the original graph diameter on datasets like DBLP, Amazon, and Wiki, especially as more nodes are retained (*i.e.*, larger r). Even on more challenging datasets, SPGRF exhibits a clear convergence trend toward the original diameter, showing its robustness in maintaining global reachability.

Clustering Coefficients. We assess clustering preservation using Mean Clustering Coefficient (MCC) and Global Clustering Coefficient (GCC) under P_1, with Fig. 5 (c) and (d) reporting the results. SPGRF achieves near-optimal MCC on YouTube, PDB, and Amazon, and excels in GCC on PDB and Skitter. The metric discrepancy reflects their different focuses—MCC emphasizes local density, while GCC captures global triangle distributions. Overall, SPGRF demonstrates superior ability to preserve both local and global clustering patterns.

Community Structure. We measure community structure preservation using the Louvain method, evaluating both the number of communities and modularity under P_1. Figure 5 (e) and (f) show that SPGRF consistently aligns closer to the original graph than all baselines. This improvement becomes more evident as the reduction ratio increases. These results validate our method's ability to retain key local structures, which in turn supports the emergence of accurate community formations.

Exp-4: Ablation Studies. To isolate the impact of our skeleton enhancement module, we compare the full SPGRF algorithm against $SPGRF_{init}$, a variant that omits this phase entirely using parameter configuration P_3. Figure 4 (d) shows that SPGRF consistently outperforms $SPGRF_{init}$ on YouTube, DBLP, PDB, and Amazon, validating the effectiveness of skeleton enhancement for preserving frequent patterns. However, accuracy tends to decline as d_r increases, occasionally falling below the performance $SPGRF_{init}$ at higher d_r values. This degradation occurs because excessive edge connections introduce structural noise that distorts the original pattern distribution. A notable exception appears in the Skitter dataset, where $SPGRF_{init}$ consistently outperforms SPGRF. This counterintuitive result stems from the inherently sparse structure and predominance of simple local patterns in Skitter, where additional edges from skeleton enhancement can create spurious patterns that interfere with mining genuine simple patterns.

Exp-5: Task-Specific Performance Evaluation. We evaluate performances on two representative downstream tasks, *i.e.*, frequent pattern mining and graph neural network (GNN)-based node classification, by using reduced graphs, to show their utilities. Experiments are conducted under parameter configurations P_4 and P_1, respectively.

Frequent Pattern Mining. We adopt a typical top-k FPM algorithm BMiner [22], which extracts frequent patterns without specifying a predefined support threshold. As shown in Fig. 6 (a), SPGRF consistently achieves the highest accuracy

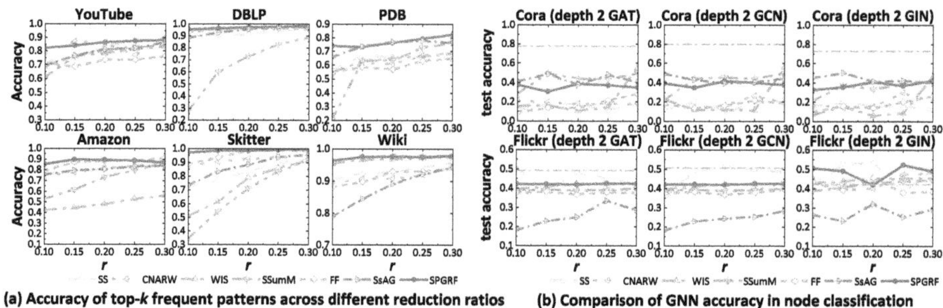

Fig. 6. Task-Specific Performance Evaluation.

across all six datasets. Its accuracy generally improves with increased reduction ratio r, indicating that higher values preserve richer structural pattern information. A slight dip on Amazon at $r > 0.2$ may stem from excessive edge connections introduced during skeleton enhancement, which could distort certain original patterns. These highlight the strong power of SPGRF in retaining subgraph-level structural fidelity even under substantial reduction.

GNN-Based Node Classification. We evaluate the utility of reduced graphs on node classification using Graph Attention Network (GAT), GCN, and Graph Isomorphism Network (GIN) under configuration P_1. As shown in Fig. 6 (b), SPGRF achieves stable and competitive performance, generally ranking among the top two to three methods across different models and datasets. Although it does not always outperform all baselines, SPGRF remains close to the performance of the original graph (green line), indicating its effectiveness in retaining useful structural and semantic information. Compared to other reduction methods that exhibit larger fluctuations, SPGRF shows higher robustness and consistency across varying GNN architectures and reduction ratios.

7   Conclusion

We present SPGRF to facilitate general-purpose graph analysis. By integrating GCN-based edge importance assessment and topology-aware enhancement, SPGRF retains crucial structures while significantly reducing graph size. Extensive experiments on diverse tasks, including FPM, node classification and others, demonstrate the excellent performance and broad applicability of SPGRF. In future, we plan to explore adaptive parameter tuning for evolving graphs through incremental learning, and skeleton enhancement strategies to allow SPGRF achieving even better generalization ability.

References

1. Ali, S., Ahmad, M., Beg, M.A., Khan, I.U., Faizullah, S., Khan, M.A.: SsAG: summarization and sparsification of attributed graphs. ACM Trans. Knowl. Discov. Data **18**(6), 141:1–141:22 (2024)
2. Althöfer, I., Das, G., Dobkin, D., Joseph, D., Soares, J.: On sparse spanners of weighted graphs. Discrete Comput. Geometry **9**(1), 81–100 (1993). https://doi.org/10.1007/BF02189308
3. Bacciu, D., Conte, A., Landolfi, F.: Generalizing downsampling from regular data to graphs. In: Proceedings of the AAAI Conference on Artificial Intelligence. vol. 37, pp. 6718–6727 (2023)
4. Chen, Y., et al.: Demystifying graph sparsification algorithms in graph properties preservation. Proc. VLDB Endowment **17**(3), 427–440 (2023)
5. Feng, Z.: GRASS: graph spectral sparsification leveraging scalable spectral perturbation analysis. IEEE Trans. Comput. Aided Des. Integr. Circuits Syst. **39**(12), 4944–4957 (2020)
6. Gao, X., Ye, G., Chen, T., Zhang, W., Yu, J., Yin, H.: Rethinking and accelerating graph condensation: a training-free approach with class partition. In: Proceedings of the ACM on Web Conference 2025, pp. 4359–4373 (2025)
7. Hashemi, M., Gong, S., Ni, J., Fan, W., Prakash, B.A., Jin, W.: A comprehensive survey on graph reduction: sparsification, coarsening, and condensation. In: Proceedings of the Thirty-Third International Joint Conference on Artificial Intelligence, pp. 8058–8066 (2024)
8. Huang, Z., Zhang, S., Xi, C., Liu, T., Zhou, M.: Scaling up graph neural networks via graph coarsening. In: Proceedings of the 27th ACM SIGKDD Conference on Knowledge Discovery & Data Mining, pp. 675–684 (2021)
9. Jin, W., Zhao, L., Zhang, S., Liu, Y., Tang, J., Shah, N.: Graph condensation for graph neural networks. In: International Conference on Learning Representations (2022)
10. Kiouche, A.E., Baste, J., Haddad, M., Seba, H., Bonifati, A.: Neighborhood-preserving graph sparsification. Proc. VLDB Endowment **17**(13), 4853–4866 (2024)
11. Kumar, M., Sharma, A., Saxena, S., Kumar, S.: Featured graph coarsening with similarity guarantees. In: International Conference on Machine Learning, pp. 17953–17975 (2023)
12. Lee, K., Jo, H., Ko, J., Lim, S., Shin, K.: SSumM: sparse summarization of massive graphs. In: Proceedings of the 26th ACM SIGKDD International Conference on Knowledge Discovery & Data Mining, pp. 144–154 (2020)
13. Li, Y., Wu, Z., Lin, S., Xie, H., Lv, M., Xu, Y., Lui, J.C.: Walking with perception: efficient random walk sampling via common neighbor awareness. In: 2019 IEEE 35th International Conference on Data Engineering, pp. 962–973 (2019)
14. Lin, M., Li, W., Lu, S.: Balanced influence maximization in attributed social network based on sampling. In: Proceedings of the 13th International Conference on Web Search and Data Mining, pp. 375–383 (2020)
15. Liu, Y., Qiu, R., Huang, Z.: CaT: balanced continual graph learning with graph condensation. In: 2023 IEEE International Conference on Data Mining, pp. 1157–1162 (2023)
16. Loukas, A., Vandergheynst, P.: Spectrally approximating large graphs with smaller graphs. In: International Conference on Machine Learning, pp. 3237–3246 (2018)
17. Meng, Y., Li, R.H., Lin, L., Li, X., Wang, G.: Topology-preserving graph coarsening: an elementary collapse-based approach. Proc. VLDB Endowment **17**(13), 4760–4772 (2024)

18. Razin, N., Verbin, T., Cohen, N.: On the ability of graph neural networks to model interactions between vertices. Adv. Neural. Inf. Process. Syst. **36**, 26501–26545 (2023)
19. Ricaud, B., Aspert, N., Miz, V.: Spikyball sampling: exploring large networks via an inhomogeneous filtered diffusion. Algorithms **13**(11), 275 (2020)
20. Wang, L., Fan, W., Li, J., Ma, Y., Li, Q.: Fast graph condensation with structure-based neural tangent kernel. In: Proceedings of the ACM Web Conference 2024, pp. 4439–4448 (2024)
21. Wang, X., Lan, Z., He, Y.A., Wang, Y., Liu, Z.G., Xie, W.B.: A cost-effective approach for mining near-optimal top-k patterns. Expert Syst. Appl. **202**, 117262 (2022)
22. Wang, X., et al.: Supports estimation via graph sampling. Expert Syst. Appl. **240**, 122554 (2024)
23. Xia, S., Wang, G., Xu, G., Zhao, S., Wang, G.: GBGC: efficient and adaptive graph coarsening via granular-ball computing. In: Proceedings of the Thirty-Fourth International Joint Conference on Artificial Intelligence, pp. 3489–3497 (2025)
24. Xiao, Z., Wang, Y., Liu, S., Wang, H., Song, M., Zheng, T.: Simple graph condensation. In: Joint European Conference on Machine Learning and Knowledge Discovery in Databases, pp. 53–71 (2024)
25. Zhang, X., et al.: A survey of large graph sampling techniques. J. Comput.-Aided Design Comput. Graph. **34**(12), 1805–1814 (2022)
26. Zhao, J., et al.: FS-GNN: improving fairness in graph neural networks via joint sparsification. Neurocomputing 130641 (2025)
27. Zheng, X., Zhang, M., Chen, C., Nguyen, Q.V.H., Zhu, X., Pan, S.: Structure-free graph condensation: from large-scale graphs to condensed graph-free data. Adv. Neural. Inf. Process. Syst. **36**, 6026–6047 (2023)
28. Zhou, Z., et al.: Context-aware sampling of large networks via graph representation learning. IEEE Trans. Visual Comput. Graphics **27**(2), 1709–1719 (2020)

Hop-Constrained s-t Simple Path Enumeration: Towards Reducing Repeated Vertex Checks

Tong Pei[1], Bin Wang[1(✉)], Hengzhao Ma[1], Xiaochun Yang[1], Rui Ding[1],
Jiayi Qu[1], and Baoyan Song[2]

[1] Northeastern University, Shenyang, China
{peit,qujy}@mails.neu.edu.cn, {binwang,yangxc}@mail.neu.edu.cn,
mahz@neu.edu.cn, ruiding.neu@outlook.com
[2] Liaoning University, Shenyang, China
bysong@lnu.edu.cn

Abstract. Hop-constrained s-t simple path enumeration (**HcPE**) is a fundamental problem in graph analysis and has broad applications in the real life. Given a graph G, two vertices s and t, and a hop constraint k, HcPE is to enumerate all simple paths from s to t with length not larger than k. Although there are some existing methods for solving HcPE, they still suffer from costly repeated vertex checks and large search space, leading to poor practical performance. To address the challenges above, we propose SimpleJoin, an efficient join-based method for HcPE which involves the following novel ideas. First, we define and utilize the one-way edge to reduce the number of repeated vertex checks. Second, we propose a new greedy strategy to guide the search for simple paths and reduce the search space. Third, an efficient four-step algorithm is proposed to solve HcPE, which consists of graph reduction to remove redundant vertices and edges, finding one-way edges for reducing repeated vertex checks, bidirectional partial path search guided by greedy strategy, and partial path join to produce the final results. Extensive experiments are conducted on nine real-world datasets to show the efficiency of our proposed method.

Keywords: Simple Path Enumeration · Hop-Constrained

1 Introduction

Given a graph G, a pair of vertices s, t and a hop constraint k, the hop-constrained s-t simple path enumeration (**HcPE**) is to enumerate all simple paths (i.e., paths without repeated vertices) from s to t with length not larger than k. HcPE has a wide range of application scenarios, as described in the following.

In financial transaction networks, HcPE can help detect anomalous fund flows which are transferred through multiple intermediate accounts or banks without

H. Jung et al. (Eds.): DASFAA 2026, LNCS 16536, pp. 403–420, 2026.
https://doi.org/10.1007/978-981-92-0366-6_25

any reasonable reason [1–3]. Since long paths increase the risk of fraudsters being detected, it is necessary to introduce the hop constraint [5]. In communication networks, HcPE assists in identifying potential causes of network failures by enumerating communication paths [6,7]. Constraining the hop helps capture realistic network behaviors since data transmission usually occurs over short and stable routes [8]. HcPE can also be used to obtain hop-constrained paths between entities in knowledge graphs as features to train the model [9–11]. Since long paths often represent weak associations, applying the hop constraint ensures that the extracted features are more meaningful [12].

Some methods have been proposed to address HcPE. DFS-based methods provide theoretical guarantees of polynomial delay [13–16], and join-based methods can share the computation by divide and conquer [4,14,15,20]. However, they still suffer from costly repeated vertex checks that are ignored in prior studies and the inherently large search space, leading to limited practical efficiency.

Repeated vertex check is a basic operation for obtaining simple paths in HcPE. Given two simple paths $\{s, \cdots, v_{join}\}$ and $\{v_{join}, \cdots, t\}$, where v_{join} is called the join vertex, repeated vertex check is defined as the operation to check whether the two simple paths share at least one common vertex except the join vertex v_{join}.

The repeated vertex check is inevitable in HcPE. Since HcPE aims to enumerate simple paths, the path expansion is necessary to solve HcPE. Assuming that an intermediate simple path p from s is obtained, when we expand p towards t, for each visited v we need to traverse all the vertices of p to check whether v already exists in p to ensure a new simple path is obtained by adding v into p. Moreover, for join-based methods which collect partial simple paths from s and t and join them to obtain simple paths from s to t, the partial path join operation also involves repeated vertex checks to prevent non-simple paths. The repeated vertex check of partial path join is similar with path expansion. When joining the path from s with the path from t, we still need to traverse all vertices of the two paths. We denote the number of vertices visited in the process of checking repeated vertices as the number of repeated vertex checks. Although repeated vertex check is an inevitable and costly operation in HcPE, existing methods do not propose effective techniques to reduce the number of repeated vertex checks.

The large search space to enumerate paths is an inherent problem for HcPE. Since the number of paths grows exponentially with k, the number of vertices traversed during the search for simple paths is enormous even for small values of k, especially in dense graphs. Existing methods still suffer from the redundant search space. For example, in the join-based method [19], a vertex may still be expanded even when all simple paths from it to the source or target have already been collected.

The central contributions of this paper are new techniques for addressing the above two challenges. To reduce the number of repeated vertex checks, we define the one-way edge which can be utilized to narrow down the scope of checking repeated vertices. To reduce the large search space of path enumeration, we propose a new greedy strategy to guide the search during path expansion. We also propose a new pruning technique to further reduce the search space. Summarizing these ideas, we propose SimpleJoin, a join-based method for HcPE, which reduces the number of repeated vertex checks and the search space as much as possible. SimpleJoin consists of four modules: graph reduction to remove redundant vertices and edges, finding one-way edges to reduce the number of repeated vertex checks, bidirectional partial path search guided by greedy strategy, and partial path join to obtain all simple paths from s to t with length not larger k.

Contributions. In this paper, our contributions are summarized as follows.

- We propose an efficient join-based method, **SimpleJoin**, which consists of four modules and achieves low query time.
- We define the one-way edge, which can be utilized to narrow down the scope of checking repeated vertices, which significantly reduces the number of repeated vertex checks.
- We propose a novel greedy strategy and pruning technique based on the longest path length to reduce the search space.
- Extensive experiments are conducted on nine real-world datasets, and the results demonstrate the efficiency of our proposed method.

Organization. The rest of this paper is organized as follows. Section 2 introduces the preliminaries and presents the problem of HcPE. Section 3 first provides an overview of SimpleJoin and then details its four modules. Section 4 evaluates our method. Section 5 reviews the related work. Finally, Sect. 6 concludes this paper.

2   Preliminaries

Table 1. A summary of notations frequently used

Notations	Descriptions
s, t, k	source, target and hop constraint
$G = (V, E), G^* = (V^*, E^*)$	original graph and reduced graph
n, m, n^*, m^*	the number of vertices and edges of G and G^*
$q(s, t, k)$	a query from s to t in G
$E_{out}(v), E_{in}(v), E^*_{out}(v), E^*_{in}(v)$	the edge sets of v in G and G^*
$deg_{out}(v), deg_{in}(v), deg^*_{out}(v), deg^*_{in}(v)$	the degrees of v in G and G^*
$dist_s(v), dist_t(v), dist^*_s(v), dist^*_t(v)$	the distance from s to v and from v to t in G and G^*
$w(s, t), p(s, t)$	a walk and a simple path from s to t
$w[x], p[x]$	the vertex at position x in walk w and simple path p
$w[x : y], p[x : y]$	the sub-walk (resp., sub-path) of walk w (resp., simple path p) from position x to y

Let $G = (V, E)$ denote a directed graph, where $V = \{v_1, v_2, \ldots, v_n\}$ is the set of vertices and $E = \{e_1, e_2, \ldots, e_m\}$ is the set of edges. Each $v_i \in V$ represents a vertex in G, and each $e_i = \langle u, v \rangle \in E$ represents a directed edge from u to

v in G. We use n and m to denote the number of vertices and edges in G, respectively. $E_{out}(v) = \{\langle v, u \rangle \in E\}$ and $E_{in}(v) = \{\langle u, v \rangle \in E\}$ denote the set of out-edges and in-edges of v, respectively. Similarly, the out-neighbor and in-neighbor set of v are denoted as $N_{out}(v) = \{u \mid \langle v, u \rangle \in E\}$ and $N_{in}(v) = \{u \mid \langle u, v \rangle \in E\}$, respectively. The out-degree and in-degree of v are denoted by $deg_{out}(v) = |N_{out}(v)|$ and $deg_{in}(v) = |N_{in}(v)|$. Moreover, we denote the distance (minimal hop number) from s to v and from v to t as $dist_s(v)$ and $dist_t(v)$, respectively.

A walk $w(s, t) = \{v_0 = s, v_1, \ldots, v_{l-1}, v_l = t\}$ is defined as a sequence of vertices from s to t with potentially repeated vertices, s.t. $\forall i \in [1, l]$, $\langle v_{i-1}, v_i \rangle \in E$. We denote by $|w|$ the number of vertices in a walk w, so the hop number (length) of w is $|w| - 1$. $w[x]$ denotes the vertex at position x, counting from 0. $w[x : y]$ denotes the sub-walk from position x to y. Additionally, $w[: y]$ is an abbreviation of $w[0 : y]$, $w[x :]$ is an abbreviation of $w[x : |w| - 1]$. Furthermore, we define a simple path from s to t as $p(s, t)$, which is a walk without any repeated vertex. The notation definition of p is similar to w. We use $\circ$ to denote the join operator of two simple paths. The notations frequently used in this paper are summarized in Table 1. In the following, we use the path to denote the simple path if no ambiguity is involved.

Problem Statement. Given a directed graph $G = (V, E)$, a pair of vertices s, t, and a hop constraint k, the *hop-constrained s-t simple path enumeration problem* (HcPE) aims to find all simple paths from s to t in G s.t. the length of each path is at most k. We use $q(s, t, k)$ to denote the query.

3   Our Join-Based Method

In this section, we first present an overview of our method, followed by a detailed description of its individual modules.

3.1   Overview

In this paper, we use a running example shown in Fig. 1 to introduce our method. The overall framework of our method consists of four modules, which are outlined

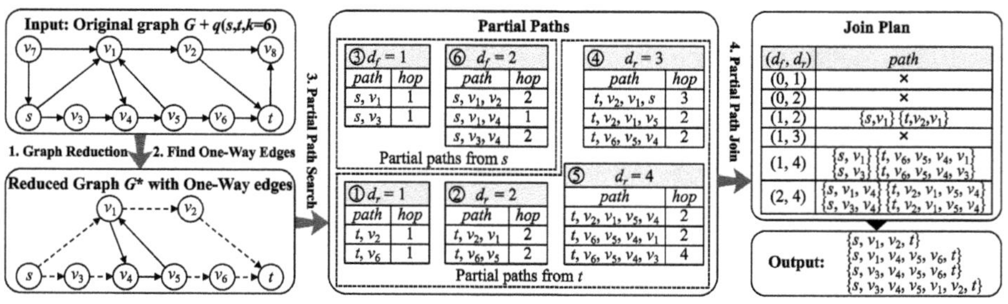

Fig. 1. An overview of SimpleJoin.

as follows. We first perform graph reduction to remove vertices and edges that do not satisfy the hop constraint. Next, we find the one-way edges on the reduced graph, which can be utilized to reduce the number of repeated vertex checks. We then perform bidirectional breadth-first search (BFS) from s and t to collect partial paths, recording the hop number of the farthest one-way edge from s or t in each path. Finally, we join the collected partial paths to output the results. When checking repeated vertices, it is sufficient to determine whether the sub-walk delineated by the recorded hop numbers contains any repeated vertices.

3.2 Graph Reduction

Inspired by [11], we reduce the graph by removing the edges that are definitely not in results. The vertex is removed when all the edges of the vertex is removed. We first perform a k-layer BFS from s to compute distances from s to vertices. Then a k-layer reverse BFS is performed to compute distances from vertices to t. During the reverse BFS process, we remove each edge $\langle u, v \rangle$ s.t. $dist_s(u) + dist_t(v) > k - 1$. After BFS, we get the reduced graph and reverse reduced graph. We denote the reduced graph as $G^* = (V^*, E^*)$, and its reverse graph as $\overline{G^*} = (V^*, \overline{E^*})$, where each edge $\langle u, v \rangle \in E^*$ corresponds an edge $\langle v, u \rangle \in \overline{E^*}$. For G^* (resp., $\overline{G^*}$), we sort the neighbors of each vertex by their distance to t (resp., s) in ascending order for an effective pruning strategy during partial path search. Given a query $q(s, t, k)$, the graph reduction preserves all k hop-constrained s-t simple paths, i.e., it is still correct to study HcPE in the (reverse) reduced graph.

Example. For the original graph in Fig. 1, given a query $q(s, t, 6)$, the edge $\langle v_7, s \rangle$ is removed since $dist_s(v_7) = \infty$, $dist_t(s) = 3$, and $dist_s(v_7) + dist_t(s) > 5$. Similarly, the edges $\langle v_7, v_1 \rangle$, $\langle v_2, v_8 \rangle$ and $\langle t, v_8 \rangle$ are also removed.

Complexity. Graph reduction performs BFS twice in G, so the time complexity of graph reduction is $O(n + m)$.

We use notations with the superscript * to denote their counterparts in the reduced graph, e.g., n^* is denoted as the number of vertices, m^* is denoted as the number of edges, and the same with $E_{out}^*(v)$, $E_{in}^*(v)$, $N_{out}^*(v)$, $N_{in}^*(v)$, $deg_{out}^*(v)$, $deg_{in}^*(v)$, $dist_s^*(v)$, and $dist_t^*(v)$. The following three modules work on G^*.

3.3 Find One-Way Edges

To reduce the number of repeated vertex checks, we define the one-way edge and propose the following theorems.

Definition 1 (One-Way Edge). *Given a directed graph $G = (V, E)$ and an edge $\langle u, v \rangle \in E$, if there is no path from v to u, then $\langle u, v \rangle$ is a one-way edge.*

Lemma 1. *Given a directed graph $G = (V, E)$ and an edge $\langle u, v \rangle \in E$, $\langle u, v \rangle$ is a one-way edge iff u and v belong to different strongly connected components (SCCs).*

Proof. ($\Rightarrow$) Assuming that u and v belong to the same SCC, then u and v are mutually reachable, which contradicts that $\langle u, v \rangle$ is a one-way edge.

($\Leftarrow$) Assuming that $\langle u, v \rangle$ is not a one-way edge, then v can reach u, which contradicts that u and v belong to different SCCs.

Lemma 2. *Given a directed graph G and a walk $w(s, t) = \{v_0 = s, v_1, \ldots, v_i, v_{i+1}, \ldots, v_l = t\}$ in G, where $\langle v_i, v_{i+1} \rangle$ is a one-way edge s.t. $i \in [0, l-1]$, assume that there exist repeated vertices in $w(s, t)$. $w(s, t)$ is divided into $w(s, v_i)$ and $w(v_{i+1}, t)$. For any repeated vertex v_{rep} in $w(s, t)$, v_{rep} does not simultaneously exists in both $w(s, v_i)$ and $w(v_{i+1}, t)$.*

Proof. Assume that there exists the repeated vertex v_{rep} in both $w(s, v_i)$ and $w(v_{i+1}, t)$. Since v_{i+1} can reach v_{rep}, and v_{rep} can reach v_i, i.e., v_{i+1} can reach v_i, which contradicts that $\langle v_i, v_{i+1} \rangle$ is a one-way edge.

Theorem 1. *Given a directed graph G and two simple paths $p(s, u_l) = \{u_0 = s, u_1, \ldots, u_i, u_{i+1}, \ldots, u_l\}$, $p(u_l, t) = \{v_0 = u_l, v_1, \ldots, v_j, v_{j+1}, \ldots, v_{l'} = t\}$ in G, where $\langle u_i, u_{i+1} \rangle$ and $\langle v_j, v_{j+1} \rangle$ are one-way edges s.t. $i \in [0, l-1] \wedge j \in [0, l'-1]$, join $p(s, u_l)$ with $p(u_l, t)$ to obtain the walk $w(s, t) = \{s, \ldots, u_i, u_{i+1}, \ldots, v_j, v_{j+1}, \ldots, t\}$. For any repeated vertex v_{rep} in $w(s, t)$, v_{rep} can only exists in the sub-walk $w(u_{i+1}, v_j)$ of $w(s, t)$.*

Algorithm 1: Find One-Way Edges

Input: Reduced graph $G^* = (V^*, E^*)$ and source s
Output: A set of all one-way edges E_{owe}

```
 1  Order, E_owe = ∅, scc ← 0, S ← an empty stack;
 2  foreach v ∈ V* do  SCC(v) ← n* + 1 ;
 3  perform DFS post-order traversal from s in G*;
 4  Order ← the ranks of all vertices in DFS post-order traversal;
 5  foreach v ∈ Order in reverse order do
 6      if SCC(v) > n* then
 7          scc ← scc + 1;
 8          S.push(v);
 9          while S ≠ ∅ do
10              u ← S.pop();
11              SCC(u) ← scc;
12              foreach ⟨u', u⟩ ∈ E*_in(u) do
13                  if SCC(u') > n* then  S.push(u') ;
14                  else
15                      if SCC(u') ≠ scc then  E_owe ← E_owe ∪ {⟨u', u⟩} ;
16  return E_owe
```

Proof. Assume that repeated vertices do not only exist in $w(u_{i+1}, v_j)$. Since $p(s, u_l)$ and $p(u_l, t)$ are simple paths, there are three kinds of existence of repeated vertices (the walks mentioned below are all sub-walks of $w(s, t)$): (1) $w(s, u_i)$ and $w(v_1, v_j)$; (2) $w(u_{i+1}, u_{l-1})$ and $w(v_{j+1}, t)$; (3) $w(s, u_i)$ and $w(v_{j+1}, t)$. According to Lemma 2, since $\langle u_i, u_{i+1} \rangle$ is a one-way edge, repeated vertices cannot exist simultaneously in both $w(s, u_i)$ and $w(u_{i+1}, t)$. Since $w(v_1, v_j)$ is a sub-walk of $w(u_{i+1}, t)$, repeated vertices cannot exist simultaneously in both $w(s, u_i)$ and $w(v_1, v_j)$. The proofs for cases (2) and (3) are similar.

Corollary 1. *Let $w(s, t)$ be the walk obtained by joining $p(s, u_l)$ with $p(u_l, t)$ in Theorem 1. The number of vertices in sub-walk $w(u_{i+1}, v_j)$ satisfies*

$$|w(u_{i+1}, v_j)| = l - i + j. \tag{1}$$

If $\langle u_i, u_{i+1} \rangle$ is the one-way edge with the maximum hop number in $p(s, u_l)$ and $\langle v_j, v_{j+1} \rangle$ is the one-way edge with the minimum hop number in $p(u_l, t)$, then $|w(u_{i+1}, v_j)|$ is minimized.

According to Corollary 1, assuming that two partial simple paths $p(s, u)$ and $p(u, t)$ are to be joined to get a potential simple path from s to t, the one-way edges can be utilized to reduce the number of repeated vertex checks. Assuming that $\langle u_1, v_1 \rangle$ is the one-way edge with maximum hop number in $p(s, u)$, and $\langle u_2, v_2 \rangle$ is the one-way edge with minimum hop number in $p(u, t)$, we can only check repeated vertices in sub-walk $w(v_1, u_2)$.

Inspired by Kosaraju algorithm [21], we propose Algorithm 1 to find one-way edges. We can perform depth-first search (DFS) twice to label the edges connecting different SCCs as one-way edges. In lines 14–15, we determine whether u and its in-neighbor u' are in the same SCC, if not, we label $\langle u', u \rangle$ as the one-way edge. As shown in Fig. 1, the one-way edges of G^* are dashed arrows.

Complexity. Since G^* is traversed twice, the time complexity of Algorithm 1 is $O(n^* + m^*)$.

3.4   Partial Path Search

We use bidirectional BFS to find simple paths from s to t, whose potential power to improve the performance of HcPE has been proved [20]. During bidirectional BFS, we obtain intermediate results such as simple paths $p(s, u)$ and $p(v, t)$, which we call partial paths. A final simple path from s to t $\{s, \cdots, v_{join}, \cdots, t\}$ is obtained by joining a partial path from s $\{s, \cdots, v_{join}\}$ with a partial path from t $\{t, \cdots, v_{join}\}$, where v_{join} is the join vertex. Note that the partial paths from vertices to t are stored in reverse order.

When performing bidirectional search to obtain the partial paths, a straightforward idea is to set the depth of the search from s to $\lfloor k/2 \rfloor$, and set the depth of the search from t to $k - \lfloor k/2 \rfloor$. However, since different graphs have different distributions, such setting may lead to poor performance. In order to reduce

---

Algorithm 2: Partial Path Search

Input: Reduced graph G^*, reverse reduced graph $\overline{G^*}$, source s, target t, hop constraint k and a set of all one-way edges E_{owe}

Output: Partial paths I_f and I_r, the search depth d_f and d_r, a sequence of join plan S_{join}, a set of join vertices V_{join}

1 $d_f \leftarrow 0, d_r \leftarrow 0, deg_f \leftarrow deg^*_{out}(s), deg_r \leftarrow deg^*_{in}(t), p_s \leftarrow \{s\}, p_t \leftarrow \{t\};$

2 $p_s.hop \leftarrow 0, p_t.hop \leftarrow 0, P_f.push(p_s), P_r.push(p_t);$

3 **while** $P_f \neq \emptyset \wedge P_r \neq \emptyset$ **do**

4   **if** $d_f + d_r \geq k$ **then break**;

5   **if** $deg_f < deg_r$ **then** Expand$(G^*, t, k, I_f, E_{owe}, P_f, deg_f, d_f, dist^*_t, V_{join})$;

6   **else** Expand$(\overline{G^*}, s, k, I_r, E_{owe}, P_r, deg_r, d_r, dist^*_s, V_{join})$;

7 $S_{join} \leftarrow S_{join} \cup (d_f, d_r);$

8 **Procedure** Expand$(G, t, k, I, E_{owe}, P, deg, d, dist, V_{join})$

9 $d \leftarrow d + 1, deg \leftarrow 0, next \leftarrow \emptyset;$

10   **foreach** $p \in P$ **do**

11   **foreach** $\langle u, v \rangle \in E_{out}(p[|p| - 1])$ **do**

12   **if** $dist(v) + d > k$ **then break** ;

13   **if** $v \notin p[p.hop :]$ **then**

14 $p' \leftarrow p \circ v;$

15 $V_{join}(d) \leftarrow V_{join}(d) \cup \{v\};$

16   **if** $\langle u, v \rangle \in E_{owe}$ **then** $p'.hop \leftarrow d;$

17   **else** $p'.hop \leftarrow p.hop$;

18 $I(d)(v).push(p');$

19   **if** $v \neq t$ **then** $next.push(p'), deg \leftarrow deg + deg_{out}(v)$;

20 $P.clear(), swap(P, next);$

---

the search space, we propose a new greedy strategy to choose from the forward direction and reverse direction to expand.

To utilize the intermediate results of bidirectional BFS, we store partial paths from both s and t, and employ the greedy strategy for partial path search. Algorithm 2 presents the pseudocode of partial path search.

Algorithm 2 performs BFS from s in G^* and from t in $\overline{G^*}$, and uses the procedure Expand to search layer by layer to collect partial paths. The queues P_f and P_r store the paths of current layer in the search process. The last vertex of each path in P_f and P_r is called the forward frontier and reverse frontier, and deg_f and deg_r denote the total out-degrees of forward frontiers and the total in-degrees of reverse frontiers, respectively. Our greedy strategy is to search for the direction with smaller sum of degrees each time. If deg_f is smaller, we search in the forward direction in line 5; otherwise, we search in the reverse direction in line 6. In line 4, the search ends when the sum of the forward depth d_f and the reverse depth d_r reaches k; otherwise, the search continues.

Next, we describe two important points of the procedure Expand that expands the frontiers and collects the partial paths.

First, a new pruning technique is used in line 12. If a neighbor v of u satisfies $dist(v) + d > k$ where d is the depth of BFS, v is impossible to form a result path with p. Since we sort the neighbors of the vertices in G^* (resp., $\overline{G^*}$) in ascending order by their distance to t and (resp., from s) (see Sect. 3.2), all the rest neighbors v' of u can be skipped since $dist(v') \geq dist(v)$. The pruning technique applies similarly to the reverse expand procedure.

Second, an important technique for reducing the number of repeated vertex checks based on the one-way edge is used in line 13. Based on Definition 1 and Lemma 2, we only need to check whether v appears in the sub-path $p[p.hop :]$, where $p.hop$ denotes the hop number of the one-way edge that is closest to u in p. If $v \notin p[p.hop :]$, we join the current path p with the vertex v to get a new path p' in line 14. In lines 16–17, we record the hop number of the one-way edge that is closest to v in p'. If the expanded edge is a one-way edge, the current expansion depth (i.e., the hop number of the one-way edge $\langle u, v \rangle$ in p') is recorded; otherwise, the hop number remains unchanged.

We store the join vertex v in V_{join} according to the expansion depth d in line 15 for use during partial path join. Note that V_{join} is collected only in the forward direction. In line 18, we collect the expanded path. At the end of the whole algorithm, the join plan is stored into S_{join} in line 7, where the pair (d_f, d_r) denotes joining partial paths $I_f(d_f)$ with $I_r(d_r)$ during partial path join.

Example. We search for partial paths on G^* with $q(s, t, 6)$, as illustrated in Fig. 1. During the search phase, we collect partial paths, denoted by *path*, and record for each *path* the maximum hop number of all one-way edges it contains, denoted by *hop*. The expansion order of Algorithm 2 is given in the upper-left corner of each table in the partial path search phase. Assuming that we have performed the third expansion (i.e., the forward search with $d_f = 1$), the corresponding degrees are $deg_f = deg^*_{out}(v_1) + deg^*_{out}(v_3) = 3$ and $deg_r = deg^*_{in}(v_1) + deg^*_{in}(v_5) = 3$. Since $deg_f \geq deg_r$, a reverse search with $d_r = 3$ is performed. We first expand $p_1 = \{t, v_2, v_1\}$ whose *hop* is 2. Since $s \notin p_1[2 :] = \{v_1\}$, we get $\{t, v_2, v_1, s\}$. Then we record *hop* 3 of one-way edge $\langle s, v_1 \rangle$ in this path. Additionally, since $v_5 \notin p_1[2 :]$, we get $\{t, v_2, v_1, v_5\}$. However, $\langle v_5, v_1 \rangle$ is not a one-way edge, so *hop* 2 remains unchanged. Similarly, $\{t, v_6, v_5\}$ is expanded to obtain $\{t, v_6, v_5, v_4\}$, and *hop* 2 remains unchanged.

Complexity. The time complexity of Algorithm 2 is $O(d_f \times |W_f| + d_r \times |W_r|) = O(k \times |W|)$, s.t. $d_f + d_r = k$. $|W_f|$ (resp., $|W_r|$) represents the number of walks from s (resp., t) with at most d_f (resp., d_r) hops in G^*, and $|W|$ denotes the total number of walks from s to t in G^*. The space complexity is $O(d_f \times |SP_f| + d_r \times |SP_r|) = O(k \times |W|)$. $|SP_f|$ (resp., $|SP_r|$) denotes the number of collected simple paths from s (resp., t) with at most d_f (resp., d_r) hops.

Optimization via Pruning. A pruning strategy is proposed to optimize partial path search. During partial path search, assume that the forward search from a vertex u reaches a vertex v. If we know that all simple paths from v to t have been collected, there is no need to continue expanding v in the forward direction, since all k hop-constrained simple paths through v from s to t can be obtained by joining the collected paths. Formally, let $l_s^*(v)$ and $l_t^*(v)$ denote the hop number of the longest simple path from s to v and from v to t in G^*, respectively. Assuming that the reverse (resp., forward) search has expanded d_r (resp., d_f) layers, the pruning strategy is to skip v in the forward (resp., reverse) search if $l_t^*(v) \leq d_r$ (resp., $l_s^*(v) \leq d_f$).

In the following, we only explain the pruning process for forward search; reverse search is similar. When the reverse search has already expanded d_r layers, all simple paths from target t with length no more than d_r have been collected. If $l_t^*(v) \leq d_r$, then the longest simple path must have been collected, i.e., all simple paths from v to t have been collected. However, determining whether all simple paths from v to t have been collected is equivalent to determining whether there is a simple path with length $d_r + 1$ or more from v to t in the graph, which has been proven to be NP-complete [22].

Nevertheless, we provide a special case where $l_s^*(v)$ and $l_t^*(v)$ can be easily computed: when the reduced graph is DAG. Now we explain how to compute $l_t^*(v)$. We visit each vertex u in reverse topological order (continuously remove vertices with out-degree 0 from the reduced graph), traverse $\langle v, u \rangle \in E_{in}^*(u)$, and compute $l_t^*(v) = \max(l_t^*(v), l_t^*(u) + 1)$. The expression for $l_t^*(v)$ is as follows:

$$l_t^*(v) = \begin{cases} 0, & v = t, \\ \max_{\langle v,u \rangle \in E_{out}^*(v)} \left(l_t^*(u) + 1\right), & v \neq t. \end{cases} \tag{2}$$

The above pruning operation is added in line 19 of Algorithm 2, updating frontiers during forward (resp., reverse) expansion only if G^* is a DAG and $l_t^*(v) > d_r$ (resp., $l_s^*(v) > d_f$). Determining whether the graph is DAG is straightforward by using Algorithm 1. If all edges in the reduced graph G^* are one-way edges, then G^* is DAG.

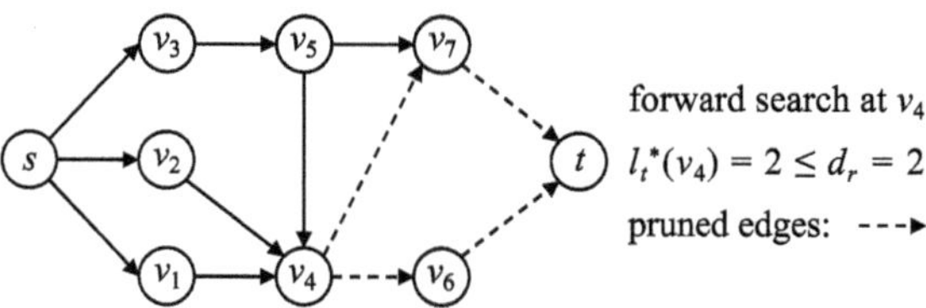

Fig. 2. An illustration of pruning.

Example. Assume that we search for partial paths with $q(s, t, 6)$ on the DAG shown in Fig. 2 and have performed the reverse search for two layers (i.e., $d_r = 2$). When the forward search reaches v_4, we find that all simple paths from v_4 to t

have been collected (i.e., $l_t^*(v_4) \leq d_r$). Therefore, there is no need to continue to expand v_4 in the forward direction. That is, after expanding from v_1, v_2 and v_5 to v_4, respectively, we stop the search, reducing the search space. The pruned edges are shown as dashed arrows.

Complexity. The computations $l_s^*(v)$ and $l_t^*(v)$ are performed after graph reduction and before partial path search. This process needs to traverse G^* twice, so the time complexity is $O(n^* + m^*)$.

3.5 Partial Path Join

Based on the partial paths collected by Algorithm 2, we can join the simple paths to obtain the complete results according to the sequence of join plan S_{plan}. $S_{plan} = \{(a_1, b_1), (a_2, b_2), \ldots, (a_k, b_k)\}$, where $a_i + b_i = i, \forall i \in [1, k]$. Our method is correct since S_{plan} stores all plans of length from 1 to k, i.e., all simple paths from s to t with length not larger than k are obtained by partial path join.

Algorithm 3: Partial Path Join

> **Input:** Partial paths I_f and I_r, the final search depth d_f and d_r, a sequence of join plan S_{join} and a set of join vertices V_{join}
> **Output:** All k hop-constrained simple paths from s to t

1 **foreach** $(i, j) \in S_{join}$ **do**
2   **if** $i = 0$ **then** Output the simple paths $I_r(j)(s)$;
3   **else**
4   **if** $j = 0$ **then** Output the simple paths $I_f(i)(t)$;
5   **else**
6   **foreach** $v \in V_{join}(i)$ **do**
7   **foreach** $p_f \in I_f(i)(v)$ **do**
8   **foreach** $p_r \in I_r(j)(v)$ **do**
9   **if** $p_f[p_f.hop : |p_f| - 2] \cap p_r[p_r.hop : |p_r| - 2] = \emptyset$ **then**
10   Output the simple path $p_f \circ p_r$;

Algorithm 3 presents the pseudocode of partial path join. We enumerate the join plan from length 1 to length k in line 1. In lines 2–4, if $i = 0 \vee j = 0$, we directly output the simple paths from s to t. If $i \neq 0 \wedge j \neq 0$, we join the partial simple paths according to S_{join} in lines 6–10. In line 9, we only check whether the sub-walk in Corollary 1 has repeated vertices, and if not, output the joined simple path in line 10. Note that the join vertex v (the last vertex in the partial path) needs not be checked, since p_f and p_r are both simple paths, each existing only one v. Additionally, if the graph G^* is DAG, we directly join the paths without the check in line 9.

Example. The join plans are illustrated in Fig. 1. Taking the join plan $(2,4)$ as an example, we join the paths $p_1 = \{s, v_1, v_4\}$ and $p_2 = \{s, v_3, v_4\}$ with $p_3 = \{t, v_2, v_1, v_5, v_4\}$ according to V_{join}, respectively. Since $p_1.hop = 1$, $p_3.hop = 2$, and $(p_1[1:1] = \{v_1\}) \cap (p_3[2:3] = \{v_1, v_5\}) = \{v_1\}$, $p_1 \circ p_3$ has a repeated vertex v_1 and therefore is not a simple path. In contrast, there is no need to check repeated vertices for $p_2 \circ p_3$ since $p_2.hop = 2$ (i.e., the last edge of p_2 is the one-way edge), so $p_2 \circ p_3 = \{s, v_3, v_4, v_5, v_1, v_2, t\}$ is a simple path.

Complexity. Since at most $|W|$ walks are checked for repeated vertices, the time complexity of Algorithm 3 is $O(k|W|)$. Here, $|W|$ denotes the total number of walks from s to t in G^*.

Table 2. Network statistics

Dataset	Name	Type	$|V|$	$|E|$	d_{avg}
SFHH-conf-sensor	SC	Interaction	403	70,261	348.69
Reactome	RE	Metabolic	6,327	147,547	46.64
Com-Amazon	CA	Miscellaneous	334,863	925,872	5.53
Enron	EN	Email	87,273	1,148,072	26.31
Wikipedia Links (jv)	WL	Hyperlink	72,453	1,831,665	50.56
Web-Google	WG	Web	875,713	5,105,039	11.66
TREC WT10g	TW	Hyperlink	1,601,787	8,063,026	10.07
Wiki-User-Edits-Page	WU	Interaction	2,094,546	8,998,641	8.59
Human-Jung2015	HJ	Brain	763,149	40,258,003	105.50

4   Experimental Evaluation

Settings. We conduct experiments on a Linux server with an Intel Xeon 2.20GHz CPU and 256GB RAM. All methods are implemented in C++ and compiled using a g++ compiler at -O3 optimization level. For each dataset, we randomly generate 1000 queries where the distance from s to t is at most k, ensuring that each query has at least one result. If a method cannot answer all queries within 100 h, the corresponding results are not reported in experiments.

Datasets. Detailed statistics of the datasets are summarized in Table 2. d_{avg} denotes the average degree of vertices in the graph. 9 real-world datasets are used to evaluate our proposed method. These datasets are from a variety domains such as web graphs, social networks and biology graphs. All graphs are obtained from KONECT[1] and Network Repository[2].

---

[1] http://konect.cc/.

[2] https://networkrepository.com/index.php.

Comparisons. We compare the following three state-of-the-art HcPE methods with SimpleJoin. We obtain the source code of BC-JOIN [14] and CPE [19] from the original authors, and PathEnum [20] is open-source.

- BC-JOIN: A bidirectional search method based on middle vertices.
- PathEnum: An adaptive method based on cardinality estimation.
- CPE: A partial path-based indexing method based on bidirectional BFS.

Table 3. Query time and #checks comparison with $k = 6$

Dataset	BC-JOIN		PathEnum		CPE		SimpleJoin	
	time (ms)	#checks	time (ms)	#checks	time (ms)	#checks	time (ms)	#checks
SC	1.67e+4	5.14e+8	6.64e+0	1.24e+6	5.94e+0	1.27e+6	**3.54e+0**	**2.68e+4**
RE	1.72e+4	9.55e+9	5.17e+0	1.56e+6	6.74e+0	2.02e+6	**1.59e+0**	**1.88e+4**
CA	4.81e−1	1.03e+3	9.90e−2	3.61e+1	9.85e−2	9.00e+1	**4.37e−2**	**1.45e+1**
EN	1.09e+5	1.34e+11	1.10e+1	1.31e+6	1.05e+1	1.28e+6	**7.46e+0**	**3.24e+4**
WL	-	-	8.35e+0	9.95e+5	9.15e+0	1.57e+6	**6.37e+0**	**2.19e+4**
WG	4.65e+2	2.99e+6	1.17e+0	4.96e+3	7.40e−1	5.69e+3	**3.78e−1**	**2.49e+2**
TW	-	-	2.37e+0	2.75e+5	1.85e+0	2.81e+5	**1.04e+0**	**2.24e+3**
WU	6.96e+3	2.48e+8	3.04e+1	1.61e+4	2.22e+1	2.21e+4	**1.89e+1**	**1.78e+3**
HJ	-	-	5.07e+2	3.25e+8	1.18e+3	3.87e+8	**7.36e+1**	**5.88e+5**

4.1 Efficiency of HcPE

Efficiency on Different Datasets. We report the average query time and the average number of repeated vertex checks (#checks) with $k = 6$ in Table 3. We can observe that SimpleJoin consistently achieves the best performance on all datasets. For example, SimpleJoin outperforms BC-JOIN, PathEnum, and CPE by 10832.85X, 3.25X, and 4.23X on RE, respectively. This performance advantage primarily stems from SimpleJoin's significant reduction in repeated vertex checks. In contrast, BC-JOIN incurs substantial overhead from barrier updates despite providing the polynomial delay per output. The query time of PathEnum is influenced by its search space cardinality estimation. The greedy strategy of CPE, which considers the number of partial paths, explores a larger search space compared to our degree-based strategy. Even if the paths from the vertex to s or t have been collected, CPE still continues the search. However, the pruning strategy of our partial path search prevents this situation in DAG. In addition, SimpleJoin achieves the minimum number of repeated vertex checks. Regarding the number of repeated vertex checks on RE, it is reduced by 99.9%, 98.8%, and 99.1% in SimpleJoin compared to BC-JOIN, PathEnum, and CPE, respectively. Obviously, the one-way edge and the corresponding theorems reduce the number of repeated vertex checks, which significantly improves the performance.

Effect of k. We evaluate four methods on the RE, CA, WG and HJ datasets by varying k from 2 to 8, comparing their average query time and number of repeated vertex checks, and reporting the average number of results for each k.

Query Time Comparison. In Fig. 3(a)-(d), we evaluate the average query time of methods over 1000 queries on four datasets by varying k. SimpleJoin performs best on all datasets and all k. The query time for all methods increases with k, as the number of results grows exponentially. As shown in Fig. 3(d), other methods cannot answer all queries with $k = 8$ within 100 h on HJ. The reason is as mentioned above: other methods suffer from excessive repeated vertex checks.

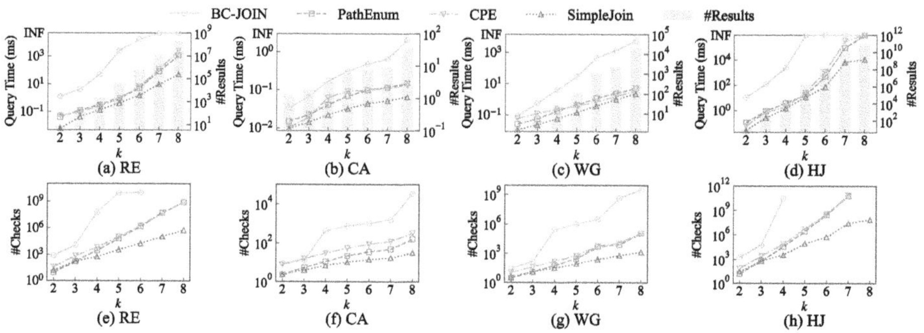

Fig. 3. Query time and #checks for varying k.

Number of Repeated Vertex Checks Comparison. We count the average number of repeated vertex checks over 1000 queries on four datasets by varying k, as shown in Fig. 3(e)-(h). The number of repeated vertex checks grows exponentially with k on all datasets. When $k \leq 3$, the length of each path is relatively short, and the number of repeated vertex checks performed by each method is similar on all datasets. When $k > 3$, the gap in the number of repeated vertex checks between SimpleJoin and other methods widens as k increases on all datasets. Therefore, the proposed one-way edge significantly reduces the number of repeated vertex checks, thereby ensuring lower query time.

Number of Results Comparison. We report the average number of results over 1000 queries on four datasets by varying k, as shown in Fig. 3(a)-(d). The number of results increases with k on each dataset. When k is the same, the number of results is smallest on CA, followed by WG, then RE, and largest on HJ. For the same k and the same method, the query time follows the same trend. This suggests that the query time is positively correlated with the number of results.

4.2 Effect of Distances from s to t

When $k = 6$ we generate 500 queries for each distance $2, 3, \cdots, 6$ on each of the four datasets. Figure 4 presents the average query time with different distances. For RE, WG and HJ, as the distance between s and t approaches k, the number of simple paths decreases, resulting in shorter query time. In contrast, the query time on CA becomes longer as the distance increases. This situation arises since the number of simple paths is small and graph reduction becomes the primary factor. The performance gap between SimpleJoin and other methods becomes more pronounced at smaller distances. This is because queries between s and t in closer proximity always have a larger number of paths. For example, on HJ, when the distance is set to 2, SimpleJoin is 5.02X faster than PathEnum and 15.24X faster than CPE. BC-JOIN fails to answer all queries within 100 h.

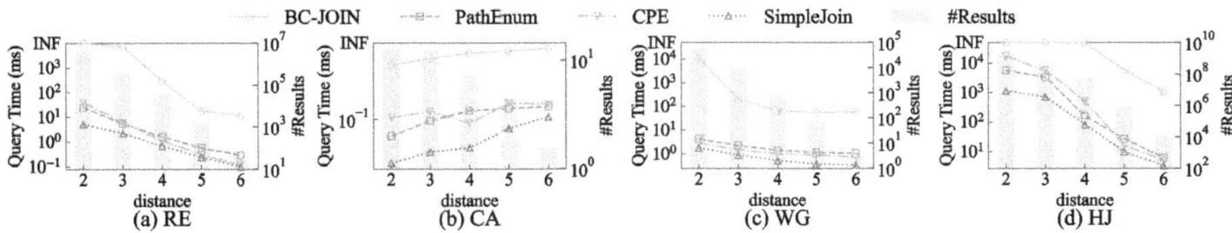

Fig. 4. Query time for varying distances between query vertices s and t.

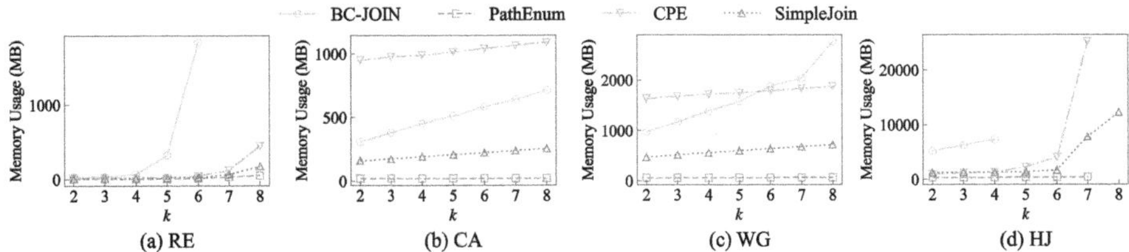

Fig. 5. Memory usage for varying k.

4.3 Memory Usage

We report average memory usage over 1000 queries on four datasets by varying k as shown in Fig. 5. Compared to BC-JOIN and CPE, SimpleJoin always consumes less memory on RE, CA and WG. On HJ, the memory usage of SimpleJoin is comparable to CPE when $k \leq 5$, but becomes significantly lower when $k > 5$. PathEnum consumes the least memory on four datasets, since its index stores neighbors that satisfy the hop constraint instead of paths. In future work, we will aim to design a more light-weight method without compromising query efficiency.

5   Related Work

DFS-based Methods. Several backtracking-based methods [13–16] perform DFS from s to enumerate simple paths, outputting a path whenever t is reached. These methods achieve polynomial delay per output, while their pruning strategies are different. T-DFS and T-DFS2 [13,16] reduce the search space utilizing the distance from vertices to t, which guarantees that there must exist a result for each search branch. Peng et al. [14,15] proposed BC-DFS based on barrier, which initially sets the barrier of each vertex as the distance from t. In the search process, if the result is not found in the subtree rooted at a vertex, the barrier will be increased to prevent the search from falling into the same subtree. BC-DFS is much faster than TDFS and T-DFS2 [14,15], while it still has large search space.

Join-Based Methods. In order to reduce the search space, BC-DFS is further optimized to propose BC-JOIN [14,15]. BC-JOIN first finds the middle vertices of the paths, then applies BC-DFS twice to obtain the paths from s to the middle vertices and from the middle vertices to t, and finally joins these paths to produce the results. HP-Index is a real-time constrained cycle detection method that can be applied in HcPE [17]. It enumerates simple paths between high-degree vertices in advance and stores them. During online enumeration, once the search reaches a high-degree vertex, it skips further search and directly joins the partial paths searched online with precomputed paths. However, HP-Index has high memory overhead due to the simple paths between high-degree vertices. PathEnum [20] first constructs the online index and performs a preliminary cardinality estimation of the search space. If the estimation does not exceed the threshold, it directly conducts DFS to enumerate paths; otherwise, a full-fledged cardinality estimation is performed to determine whether to adopt the join-based strategy, and finally partial path join is performed. However, the cardinality estimation process introduces additional computational overhead, which affects the query time. Zhang et al. [19] proposed a join-based method called CPE. To reduce the overhead of partial path join, CPE adopts bidirectional search and always expands in the direction with fewer paths, aiming to reduce the number of collected paths. Although bidirectional search helps reduce the search space compared with unidirectional search, CPE focuses only on the number of collected paths during the search and overlooks the overall search efficiency.

Most importantly, the above methods do not consider the repeated vertex check. In order to obtain the simple paths, they traverse all the vertices of paths to check repeated vertices, which significantly limits their practical efficiency.

6   Conclusion

In this paper, we study the hop-constrained s-t simple path enumeration (HcPE) problem. To the best of our knowledge, all existing HcPE methods suffer from excessive repeated vertex checks and large search space. To address the above challenges, we propose SimpleJoin, a join-based method consisting of four modules: graph reduction, finding one-way edges, partial path search, and partial path join. Based on the one-way edge, SimpleJoin significantly reduces the number of repeated vertex checks. We employ a greedy strategy to guide the search for simple paths and introduce a pruning operation to prevent unnecessary expansion of vertices and edges, thereby effectively reducing the search space. Extensive experiments on nine real-world datasets demonstrate the superior efficiency compared to state-of-the-art methods.

Acknowledgments. The work is partially supported by the National Key Research and Development Program of China (No. 2024YFF0617702), the National Natural Science Foundation of China (No. U23A20309, U22A2025, 62232007), and the 111 Project (No. B16009).

References

1. Yue D., Wu X., Wang Y., et al.: A review of data mining-based financial fraud detection research. In: International Conference on Wireless Communications, Networking and Mobile Computing, pp. 5519–5522 (2007)
2. Jedrzejek C., Bak J., Falkowski M.: Graph mining for detection of a large class of financial crimes. In: International Conference on Conceptual Structures, vol. 46 (2009)
3. F. A. T. Force: Fatf report: Money laundering and terrorist financing vulnerabilities of legal professionals. In: FATF (2013)
4. Zhang J., Yang S., Ouyang D., et al.: Hop-constrained s-t simple path enumeration on large dynamic graphs. In: ICDE, pp. 762-775 (2023)
5. Li X., Liu S., Li Z., et al.: FlowScope: spotting money laundering based on graphs. In: AAAI, pp. 4731–4738 (2020)
6. Misra, R.B., Misra, K.B.: Enumeration of all simple paths in a communication network. Microelectron. Reliab. **20**(4), 419–426 (1980)
7. Vass B., Tapolcai J., Hay D., et al.: How to model and enumerate geographically correlated failure events in communication networks. Guide to Disaster-Resilient Communication Networks, pp. 87–115 (2020)
8. Fontugne R., Pelsser C., Aben E., et al.: Pinpointing delay and forwarding anomalies using large-scale traceroute measurements. In: IMC, pp. 15–28 (2017)
9. Zhu, Z., Yuan, X., Galkin, M., et al.: A*Net: a scalable path-based reasoning method for knowledge graphs. Adv. Neural. Inf. Process. Syst. **36**, 59323–59336 (2023)
10. Tan X., Wang X., Liu Q., et al.: Paths-over-graph: knowledge graph empowered large language model reasoning. In: WWW, pp. 3505–3522 (2025)
11. Shi, B., Weninger, T.: Discriminative predicate path mining for fact checking in knowledge graphs. Knowl.-Based Syst. **104**, 123–133 (2016)
12. Zeng Y., Fang Y., Ma C., et al.: Efficient distributed hop-constrained path enumeration on large-scale graphs. In: SIGMOD, pp. 1–25 (2024)
13. Grossi R., Marino A., Versari L.: Efficient algorithms for listing k disjoint st-Paths in graphs. In: LATIN, pp. 544–557 (2018)
14. Peng, Y., Zhang, Y., Lin, X., et al.: Hop-constrained s-t simple path enumeration: towards bridging theory and practice. Proc. VLDB Endow. **13**(4), 463–476 (2019)

15. Peng, Y., Lin, X., Zhang, Y., et al.: Efficient hop-constrained s-t simple path enumeration. VLDB J. **30**(5), 799–823 (2021)
16. Rizzi R., Sacomoto G., Sagot M F.: Efficiently listing bounded length st-paths. In: International Workshop on Combinatorial Algorithms, pp. 318–329 (2014)
17. Qiu, X., Cen, W., Qian, Z., et al.: Real-time constrained cycle detection in large dynamic graphs. Proc. VLDB Endow. **11**(12), 1876–1888 (2018)
18. Hao, K., Yuan, L., Zhang, W.: Distributed hop-constrained s-t simple path enumeration at billion scale. Proc. VLDB Endow. **15**(2), 169–182 (2021)
19. Zhang J., Yang S., Ouyang D., et al.: Hop-constrained s-t simple path enumeration on large dynamic graphs. In: ICDE, pp. 762–775 (2023)
20. Sun S., Chen Y., He B., et al.: PathEnum: towards real-time hop-constrained s-t path enumeration. In: SIGMOD, pp. 1758–1770 (2021)
21. Sharir, M.: A strong-connectivity algorithm and its applications in data flow analysis. Comput. Math. Appl. **7**(1), 67–72 (1981)
22. Johnson D. S., Garey M. R.: Computers and intractability: A guide to the theory of NP-completeness. WH Freeman (1979)

Constrained Reachability Queries on Hypergraphs

Faming Li[(✉)], Han Guo, Shengli Qiu, Xiaochun Yang, and Bin Wang

School of Computer Science and Engineering, Northeastern University, Shenyang 110819, China

{lifaming,yangxc,binwang}@mail.neu.edu.cn, guohan@stumail.neu.edu.cn, 2401891@stu.neu.edu.cn

Abstract. The hypergraph, which captures group-wise relationships among multiple entities, effectively addresses the limitation of conventional graphs that only represent pairwise relationships. It has been widely applied in querying and mining collaboration networks, social networks, protein interaction networks, and so on. In this paper, we study the reachability on hypergraphs. We formally define the reachability problem with a specified constraint on the edges of the hypergraphs. To overcome the time inefficiency of traversal-based methods, and the space overhead and poor update support of union-find and 2-hop-based approaches, we propose a novel tree-based index HRTree for efficient hypergraph reachability queries, achieving $O(\log(|E|))$ query time and $O(|E|)$ space complexity, where E denotes the hyperedge set of a hypergraph. Furthermore, we design an incremental maintenance method for HRTree to support hypergraph updates without index reconstruction. The maintenance time is only related to the influence region of updates. Extensive experiments on 8 real-world datasets demonstrate that our algorithm outperforms baseline methods in both time and space efficiency. Moreover, HRTree efficiently supports updates on hypergraphs.

Keywords: Hypergraph · Reachability · Index · Update

1 Introduction

The graph serves as a fundamental model for representing entities and their many-to-many relationships. Owing to its strong representational capacity, the graph has been widely adopted for modeling, managing, and mining various types of data, including social networks, communication networks, and biological information networks [19]. However, in conventional graph models, each edge can connect only two vertices, representing a binary relationship between two entities, which fundamentally limits their ability to model complex interactions among multiple entities. To address this limitation, *hypergraph* has been introduced, where a single hyperedge can incorporate multiple vertices to capture group-wise relationships among entities. Notable hypergraphs include collaboration networks, social networks, and protein-protein interaction networks [12].

H. Jung et al. (Eds.): DASFAA 2026, LNCS 16536, pp. 421–437, 2026.
https://doi.org/10.1007/978-981-92-0366-6_26

Querying and mining hypergraphs can enable us to gain deeper insights into the hidden knowledge behind graph data. The motifs on hypergraphs can be used to construct characteristic profiles of real-world hypergraphs and distinguish hypergraphs from different domains [13]. Hyperpaths between different hyperedges can be modeled as interactions between proteins in bioinformatics networks [17]. In integrated circuit design, representing circuit components and their relationships as a hypergraph enables an optimal circuit layout through hypergraph partitioning [9]. Similar applications of hypergraphs include text classification [7], product recommendation [23], and so on [12].

Although extensive studies on hypergraphs have been conducted, such as shortest paths, representation learning, and motifs, numerous representative queries on hypergraphs have not been investigated. Among these queries, *reachability query* on hypergraphs, as one of the most important types of queries on graph data, has not yet been systematically studied. Given a hypergraph G, two hyperedges e_i and e_j, and a threshold τ, the reachability query on G determines if there is a hyperpath from e_i to e_j, where a hyperpath is a sequence of hyperedges such that the cardinality of the intersection of any two successive hyperedges is not less than τ. Similar to the practical value of reachability queries on traditional graphs, reachability queries on hypergraphs facilitate the discovery of latent knowledge and the efficient management of hypergraph data. For example, in co-authorship networks, hyperedges represent multi-author papers where intersections correspond to authors contributing to multiple publications [15]. The reachability between two hyperedges quantifies the connectivity of research groups. In social networks, hyperedges model interest communities with intersections that indicate individuals with multiple interests [11,18]. Reachability between two hyperedges measures the information propagation potential between communities [2].

Compared with reachability on traditional graphs, reachability on hypergraphs becomes more complex, mainly due to the following two reasons. First, the adjacency relationships between the edges on hypergraphs are complex. This is because the adjacency between two edges is defined based on the cardinality of intersecting vertex sets. Queries often involve constraints on the intersection size to cater to different applications, which poses challenges for efficiently executing queries and constructing indexes on hypergraphs. Second, the dynamic nature of hypergraphs presents challenges for both the maintenance and query processing on hypergraphs. Any single vertex update on a hypergraph may propagate to multiple hyperedges, thereby affecting these hyperedges and the adjacency relationships among them. Efficiently executing queries on dynamically changing hypergraphs is a non-trivial task.

Given a hypergraph G, two target hyperedges e_i and e_j, and a threshold τ constraining the set intersection size, an intuitive approach to handle reachability problems on hypergraphs is to traverse G using Breadth-First Search (BFS) or Depth-First Search (DFS) to determine whether e_i and e_j are reachable through a path where the intersection size between any two adjacent edges on the path is greater than τ. This method requires traversing the entire graph data, which results in poor time efficiency. Another intuitive approach is to construct a union-find data structure for each possible τ to support reachability queries on hypergraphs, which requires storing multiple union-find structures based on τ values and updating multiple associated structures upon updates. Both methods obviously fail to meet the requirements of efficient queries on large-scale hypergraph data in terms of both time and space efficiency.

To address the above limitations of the two intuitive approaches, we propose a novel index HRTree for hypergraph reachability queries. HRTree adopts a tree structure that organizes connected components with different parameters into distinct subtrees, eliminating redundancy through hierarchical containment relationships between different connected components. Meanwhile, we design an efficient query algorithm on HRTree to support reachability queries, which can complete the query in $O(h)$ time via a bottom-up approach on HRTree, where $h = \log(|E|)$ denotes the height of HRTree and $|E|$ is the number of edges in the hypergraph. To handle hypergraph updates, we systematically categorize update types and develop incremental maintenance algorithms for HRTree that only modify the influence regions, avoiding full index reconstruction.

In this paper, we make the following key contributions.

- We formally define and study the constrained reachability problem on hypergraphs.
- We propose a novel index HRTree to support reachability queries on hypergraphs.
- We develop incremental maintenance algorithms to handle updates to hypergraphs on HRTree, avoiding rebuilding the index.
- We conduct extensive experiments on 8 real-world hypergraphs to validate the proposed index and algorithms.

The rest of the paper is organized as follows. Section 2 introduces some concepts related to hypergraphs and formally defines the reachability problem on hypergraphs. Section 3.1 introduces the proposed index and query processing on the index, Sect. 3.2 details the index construction algorithm, and Sect. 3.3 shows how to maintain the index under updates. Section 4 presents the experimental evaluation of our index and algorithms. Section 5 discusses related work, and Sect. 6 concludes the paper.

2   Preliminaries

In this section, we introduce some fundamental concepts related to hypergraphs and formally define the constrained reachability problem on hypergraphs.

$G = (V, E)$ is a hypergraph, where V is the set of vertices and $E = \{e_1, e_2, \ldots, e_{|E|}\}$ is the set of hyperedges. Each hyperedge $e_i \in E$ is a non-empty subset of V, that is, $e_i \subsetneq V$ for any $i \in \{1, 2, \ldots, |E|\}$. e_i and e_j are *adjacent* if $e_i \cap e_j \neq \emptyset$. e_i and e_j are called *τ-adjacent* if $|e_i \cap e_j| \geq \tau$. Let $[n]$ be the set of natural numbers from 1 to n. A *hyperpath* P is a sequence of hyperedges $< e_1, e_2, \ldots, e_p >$ such that 1) $e_i \cap e_{i+1} \neq \emptyset$ for any $i \in [p-1]$ and 2) $e_j \neq e_k$

for any $j \neq k \in [p]$. P is called a τ-*hyperpath* if $|e_i \cap e_{i+1}| \geq \tau$ for any $i \in [p-1]$ and a τ-hyperpath between e_i and e_j is represented as $P^\tau_{e_i,e_j}$. For any two edges e_i and e_j, e_i is *reachable* to e_j if there is a hyperpath from e_i to e_j. Similarly, e_i is τ-*reachable* to e_j if there is a τ-hyperpath from e_i to e_j. Notice that the reachability between vertices in V can be determined by querying the reachability among the hyperedges to which the vertices are incident. For the reachability between e_i and e_j, the following lemma holds.

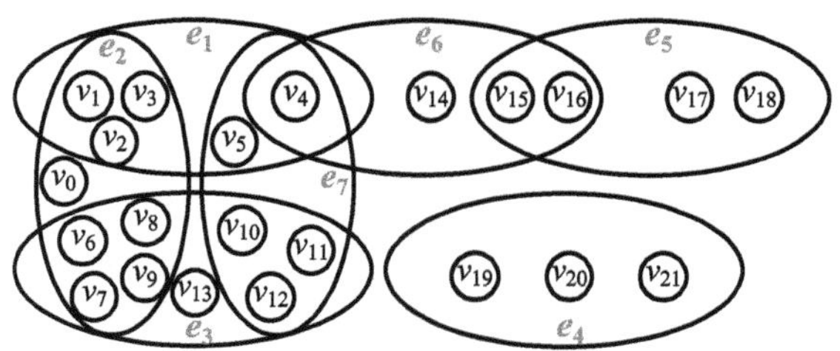

Fig. 1. A hypergraph with 22 vertices and 7 hyperedges.

Lemma 1. *If e_i is τ-reachable to e_j on G, e_i is $(\tau-1)$-reachable to e_j.*

A subgraph $S = (V' \subset V, E' \subset E)$ of G is a τ-connected component (τ-CC) if 1) for any two hyperedges $e_i, e_j \in E'$, e_i is τ-reachable to e_j in S, 2) there is no $E'' \supset E'$ that satisfies the condition 1) [1]. A τ-CC is a maximal connected subgraph that ensures that every two hyperedges are connected by at least a τ-hyperpath. If the subgraph S only satisfies condition 1), then S is called a τ-connected subgraph.

Lemma 2. *For any $\tau \in \mathbb{N}$ and $\tau > 1$, τ-CC $\subset (\tau-1)$-CC $\subset \cdots \subset 1$-CC.*

Example 1. Figure 1 illustrates a hypergraph comprising 22 vertices and 7 hyperedges. e_5 and e_6 are 2-adjacent as $e_5 \cap e_6 = \{v_{15}, v_{16}\}$. e_1 is 3-reachable to e_3 as there is a 3-hyperpath $< e_1, e_2, e_3 >$ from the hyperedge e_1 to the hyperedge e_3. The subgraph induced by the hyperedge set $\{e_1, e_2, e_3, e_7\}$ is a 3-CC.

Now, we formally define the constrained reachability problem on hypergraphs.

Constrained Reachability on Hypergraphs
Input: A hypergraph $G = (V, E)$, two hyperedges e_i and e_j, a threshold τ.
Output: true if $P^\tau_{e_i,e_j}$ exists, otherwise false.

3   Index for Hypergraph Reachability

In this section, we present the index HRTree, a novel index specifically designed to support efficient constrained reachability queries on hypergraphs. Given two hyperedges e_i, e_j, and a threshold τ, the reachability between e_i and e_j can be directly and efficiently answered on HRTree.

3.1 Overview of HRTree

HRTree is a tree that contains two kinds of nodes, leaf nodes and non-leaf nodes. Each leaf node n_L corresponds to a hyperedge in the hypergraph G. Each non-leaf node n_N, as a root of the tree T', connects multiple subtrees. A reachability factor α is stored in n_N and indicates that all hyperedges corresponding to leaf nodes in T' are mutually α-reachable. For each non-leaf node in HRTree, its reachability factor is greater than its parent's reachability factor. The HRTree constructed for the hypergraph in Fig. 1 is shown in Fig. 2.

Given a query $Q(e_i, e_j, \tau)$ with two hyperedges e_i, e_j, and a threshold τ, the τ-reachability between the two hyperedges can be immediately answered using HRTree. Let $n_L^{e_i}$ and $n_L^{e_j}$ be the leaf nodes corresponding to e_i and e_j in HRTree. Starting from $n_L^{e_i}$ and $n_L^{e_j}$, we search for their *lowest common ancestor* in HRTree, the root node n_R of the minimal subtree containing $n_L^{e_i}$ and $n_L^{e_j}$. The lowest common ancestor n_R and the reachability factor α in n_R reveal that 1) e_i and e_j are in a α-CC, and 2) e_i and e_j are not in a $(\alpha+1)$-CC. If n_R is found and the reachability factor α in the node is greater than or equal to τ, then the two hyperedges are returned as τ-reachable. Otherwise, e_i is not τ-reachable to e_j. To accelerate the query process, if α in any searched non-leaf node is less than τ, the search can be stopped immediately.

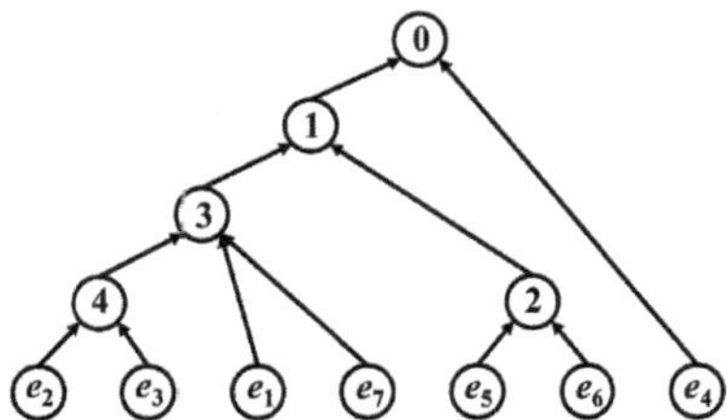

Fig. 2. HRTree constructed for the hypergraph in Fig. 1.

Example 2. Given a query $Q(e_1, e_3, 3)$, the lowest common ancestor of $n_L^{e_1}$ and $n_L^{e_3}$ is the non-leaf node with a reachability factor 3 in Fig. 2. Therefore, **true** is returned. This indicates that e_1 is 3-reachable to e_3 on the hypergraph in Fig. 1. For another query $Q(e_3, e_6, 3)$, the lowest common ancestor of $n_L^{e_3}$ and $n_L^{e_6}$ is the non-leaf node with a reachability factor 1 in Fig. 2. Therefore, e_3 is not 3-reachable to e_6. Indeed, when a non-tree node with a reachability factor of 2 is found when searching for its lowest common ancestor, the query can be terminated.

Time complexity. Let h be the height of HRTree. The time complexity of searching for the lowest common ancestor is $O(h)$. More details on the height and space complexity of HRTree will be introduced in Sect. 3.2.

3.2 Constructing **HRTree**

Now, we will introduce how to construct **HRTree** to support the constrained reachability queries on the hypergraph.

Algorithm 1 constructs **HRTree** in a bottom-up manner. Initially, the algorithm generates a corresponding leaf node (which also serves as a root node) for each hyperedge. All these leaf nodes form a forest $\mathcal{F}$ (Line 1). Each tree T' in $\mathcal{F}$ represents a α-connected subgraph formed by $E(T')$, where α is the reachability factor in the root of T' and $E(T')$ is the set of hyperedges corresponding to all leaf nodes in T'. For every pair of hyperedges with non-empty intersection, Algorithm 1 constructs a triple to store the hyperedge pair and their intersection cardinality, and aggregates these generated triples into a set L (Line 2). According to Lemma 1, to prioritize the processing of hyperedge pairs with larger intersections, the algorithm sorts L in descending order based on intersection cardinalities within the triples (Line 3). The algorithm subsequently processes the triples in L in turn to progressively construct **HRTree** based on the trees to which the hyperedges in each triple belong. For each triple $< k, e_i, e_j >$, it first retrieves the root nodes of the trees containing hyperedges e_i and e_j (Line 4–5). Depending on whether the root nodes are null, the reachability factors within the roots, and k, distinct processing strategies are applied (Line 6–21).

- *Case 1: Both roots are null (Line 6–7).* $n_L^{e_i}$ and $n_L^{e_j}$ do not belong to any tree in $\mathcal{F}$, and the reachability between e_i and e_j depends on the cardinality of the intersection of the two hyperedges. Therefore, a new non-leaf node is initialized as the parent of $n_L^{e_i}$ and $n_L^{e_j}$, and its reachability factor is set to the intersection cardinality k.
- *Case 2: One of roots is null (Line 8–14).* If one of the leaf nodes (assume that it is hyperedge $n_L^{e_j}$) has been in a tree T_{e_j}, the algorithm evaluates the reachability between e_i and $E(T_{e_j})$. Due to the descending order of L, the subgraph S_{e_j} induced by $E(T_{e_j})$ is guaranteed to be at least k-connected. If S_{e_j} is exactly k-connected, $n_L^{e_i}$ can be appended as a child of r_{e_j} to form a new tree (Line 9–10). Otherwise, the algorithm initializes a new non-leaf node to represent a new tree whose reachability factor is k and sets it as the parent of r_{e_j} and $n_L^{e_i}$ (Line 13–14). The new tree represents that the subgraph induced by $e_i \cup E(T_{e_j})$ forms a k-connected subgraph.
- *Case 3: None of the roots is null (Line 15–21).* Let $n_L^{e_i}$ and $n_L^{e_j}$ be the leaf nodes in T_{e_i} and T_{e_j}, respectively. The algorithm evaluates the reachability between subgraph induced by $E(T_{e_i})$ and $E(T_{e_j})$. If one of subgraph induced by $E(T_{e_i})$ (or $E(T_{e_j})$) is a k-connected subgraph, T_{e_j} (or T_{e_i}) is connected to the root of T_{e_i} (or T_{e_j}) as a subtree (Line 18–21). Otherwise, the algorithm initializes a new non-leaf node to represent a new tree whose reachability factor is k and sets it as the parent of T_{e_i} and T_{e_j} (Line 16–17). The subgraph induced by $E(T_{e_j}) \cup E(T_{e_i})$ is a new k-connected subgraph.

If the forest does not ultimately coalesce into a single tree, this indicates the presence of disconnected subgraphs in G. The algorithm initializes a non-leaf

---

Algorithm 1: Constructing HRTree

Input: A hypergraph $G = (V, E)$
Output: HRTree T

1 Initialize each hyperedge in E as a leaf node to form a forest $\mathcal{F}$
2 $L \leftarrow \{< |e_i \cap e_j|, e_i, e_j > \mid |e_i \cap e_j| > 0, e_i \in E, e_j \in E, i \neq j\}$
3 Sort L in descending order based on the size of intersections of hyperedges
4 **for** *each* $< k, e_i, e_j >$ *in* L **do**
5 $\quad$ $r_{e_i} \leftarrow$ GetRoot$(e_i, \mathcal{F})$, $r_{e_j} \leftarrow$ GetRoot$(e_j, \mathcal{F})$
6 $\quad$ **if** $r_{e_i} =$ null **and** $r_{e_j} =$ null **then**
7 $\quad\quad$ Initialize a node n_N, $n_N.\alpha \leftarrow k$, $n_L^{e_i}.parent \leftarrow n_N$, $n_L^{e_j}.parent \leftarrow n_N$
8 $\quad$ **else if** $r_{e_i} =$ null **then**
9 $\quad\quad$ **if** $r_{e_j}.\alpha = k$ **then**
10 $\quad\quad\quad$ $n_L^{e_i}.parent \leftarrow r_{e_j}$
11 $\quad\quad$ **else**
12 $\quad\quad\quad$ Initialize a node n_N $n_N.\alpha \leftarrow k$, $n_L^{e_i}.parent \leftarrow n_N$, $r_{e_j}.parent \leftarrow n_N$
13 $\quad$ **else if** $r_{e_j} =$ null **then**
14 $\quad\quad$ // Process $n_L^{e_j}$ and r_{e_i} in the same manner as the case $r_{e_i} =$ null
15 $\quad$ **else**
16 $\quad\quad$ **if** $r_{e_i}.\alpha > k$ **and** $r_{e_j}.\alpha > k$ **then**
17 $\quad\quad\quad$ Initialize a node n_N. $n_N.\alpha \leftarrow k$, $r_{e_i}.parent \leftarrow n_N$, $r_{e_j}.parent \leftarrow n_N$
18 $\quad\quad$ **else if** $r_{e_i}.\alpha = k$ **and** $r_{e_j}.\alpha \geq k$ **then**
19 $\quad\quad\quad$ $r_{e_j}.parent \leftarrow r_{e_i}$
20 $\quad\quad$ **else if** $r_{e_i}.\alpha > k$ **and** $r_{e_j}.\alpha = k$ **then**
21 $\quad\quad\quad$ $r_{e_i}.parent \leftarrow r_{e_j}$

22 **return** GenTree$(\mathcal{F})$

---

node with a reachability factor $\alpha = 0$, which serves as a root node to interconnect all subtrees in $\mathcal{F}$ (using Procedure GenTree). Finally, a HRTree is returned (Line 22).

Example 3. Given the hypergraph in Fig. 1, 7 leaf nodes are initialized and added to $\mathcal{F}$. 7 elements are stored in L and they are $\{< 4, e_2, e_3 >, < 3, e_1, e_2 >, < 3, e_3, e_7 >, < 2, e_1, e_7 >, < 2, e_5, e_6 >, < 1, e_1, e_6 >, < 1, e_6, e_7 >\}$ after sorting. $< 4, e_2, e_3 >$ is first considered by the algorithm and the root r_{e_2} and r_{e_3} are both empty. Thus, a new non-leaf node n_N^4 is initialized with $n_N^4.\alpha = 4$ and n_N^4 is set as the parent of $n_L^{e_2}$ and $n_L^{e_3}$. Next, $< 3, e_1, e_2 >$ is selected. The root of e_1 is empty and the root of e_2 is n_N^4. Because $n_N^4.\alpha = 4 > 3$, a new non-leaf node n_N^3 is initialized with $n_N^3.\alpha = 3$ and n_N^3 is set as the parent of n_N^4 and $n_L^{e_1}$. The procedure continues as all elements in L are processed. $n_L^{e_4}$ has not been processed because it is not adjacent to any hyperedge. Therefore, there are two trees in $\mathcal{F}$, *i.e.*, the tree with n_N^0 as the root and the one with $n_L^{e_4}$. The algorithm finally initializes a non-leaf node n_N^0 as the root and connects the trees in $\mathcal{F}$.

Time and space complexity. According to the definition of the tree structure, when the number of leaf nodes is $|E|$, the maximum number of non-leaf nodes is $|E| - 1$. Therefore, the space complexity of HRTree is $O(|E|)$. Therefore, the average height of the tree is $O(\log|E|)$, while the worst-case height is $O(|E|)$. Let d denote the average number of 1-adjacent hyperedges to each hyperedge. Thus, the size of L is $O(d|E|)$ and the time complexity of sorting L is $O(d|E|\cdot\log(d|E|))$. The time complexity of GetRoot is $O(\log|E|)$. So, Algorithm 1 takes $O(d|E| \cdot \log(d|E|))$ time to construct HRTree from L (Line 4–22). The time complexity to obtain a tree from a forest is $O(|E|)$. The total time complexity of Algorithm 1 is $O(d|E| \cdot \log(d|E|))$.

3.3 Processing Dynamic Hypergraphs

This part will introduce how to maintain HRTree when the hypergraph updates. Before introducing the specific method, we first present the definition of updates on hypergraphs.

Definition 1. *An update operation on a hypergraph $G = (V, E)$ is a triple $\Delta = (op, E_\Delta, v)$, where $op \in \{+, -\}, E_\Delta \subset E \cup \{\emptyset\}, v \in V$. The "+" represents adding the vertex v to the hyperedges in E_Δ, and the "−" represents removing the vertex v from the hyperedges in E_Δ.*

The update operations can be categorized according to the number of hyperedge sets in which the vertex v is located.

- *Case 1:* $|E_\Delta| = 1$. This case indicates that the update operation only adds a single vertex to one hyperedge or removes a single vertex from one hyperedge, and this vertex does not affect the adjacency relationship between any pair of hyperedges, thereby leaving the reachability of the hypergraph unchanged.
- *Case 2:* $|E_\Delta| = 2$. This case indicates that the update operation either adds the same vertex to two hyperedges or removes it from both. The adjacency relationship between the two hyperedges is modified, and the reachability among the hyperedges of G may be altered.
- *Case 3:* $|E_\Delta| > 2$. This case indicates that the update operation either adds the same vertex to multiple hyperedges or removes it from them simultaneously. The adjacency relationships between the $|E_\Delta|$ hyperedges are modified, and the reachability among the hyperedges of G may also be altered.

In Case 1, the update operation does not affect the reachability of the hyperedges on the hypergraph and only requires updating the corresponding hyperedge. The process of adding or removing a hyperedge with multiple vertices can be achieved by repeatedly executing the operations defined in Case 1. In Case 3, it is necessary to process each pair of hyperedges in E_Δ and consider the reachability of other edges affected by the update of each pair of hyperedges. The scenario in which an update occurs between an arbitrary pair of hyperedges corresponds to Case 2. It can be seen that Case 2 serves as the foundation for Case 3. Consequently, in the following section, we will focus on describing how to process updates on HRTree under Case 2.

It is observed that when $|E_\Delta| = 2$, the update operation exhibits a "continuous" impact on the connectivity of subgraphs. Specifically, such an operation

---

Algorithm 2: Adding a vertex to two hyperedges

Input: HRTree T, $\Delta = (+, \{e_i, e_j\}, v)$
Output: HRTree T' after update

1   $T_{inf}, T_{inf}^1, T_{inf}^2 \leftarrow$ GetInfRegion(T, e_i, e_j)
2   **if** $T_{inf}.root.\alpha = |e_i \cap e_j|$ **then**
3   **if** $T_{inf}.root$ *has two children in* HRTree **then**
4       $\llcorner$ $T_{inf}.root.\alpha \leftarrow T_{inf}.root.\alpha + 1$

5   **else**
6   Initialize a node n_N and set $n_N.\alpha \leftarrow T_{inf}.root.\alpha + 1$
7   set T_{inf}^1 and T_{inf}^2 be the subtrees of n_N
8   set n_N be the children of $T_{inf}.root$
9       $\llcorner$ set the tree rooted at n_N be T_{inf}

10   **if** $T_{inf}^1.root.\alpha = T_{inf}.root.\alpha$ **then**
11       $\llcorner$ set the children of $T_{inf}^1.root$ be the children of T_{inf} and delete $T_{inf}^1.root$

12   **if** $T_{inf}^2.root.\alpha = T_{inf}.root.\alpha$ **then**
13       $\llcorner$ set the children of $T_{inf}^2.root$ be the children of T_{inf} and delete $T_{inf}^2.root$

14 **return** T

---

can only transition a τ-CC to a $(\tau - 1)$-CC, $(\tau + 1)$-CC, or leave it unchanged, rather than changing it to a $(\tau - 2)$-CC, $(\tau + 2)$-CC, or any other connected component. This continuity property ensures minimal structural modifications to HRTree, which will be detailed below.

Before introducing how to perform update operations on HRTree, we first define the concept of *influence region*.

Definition 2. *Given two hyperedges e_i and e_j, their influence region is a tree T_{inf} in* HRTree *such that T_{inf} is the subtree of* HRTree *rooted at the lowest common ancestor of the leaf nodes $n_L^{e_i}$ and $n_L^{e_j}$.*

The influence region defines the minimum region in HRTree that needs to be considered for changes when an update occurs. For example, in Fig. 3(a), the subtree within the blue dashed triangle is the influence region T_{inf} in HRTree when an update occurs in e_1 and e_2. That is, the update of HRTree can be achieved by merely modifying T_{inf} when e_1 and e_2 undergo changes.

Adding a vertex to two hyperedges. We first consider the case where $op = +$, *i.e.*, adding a vertex to two hyperedges. The algorithm is described in Algorithm 2. It first identifies the influence region T_{inf} of e_i and e_j in T, along with the subtree T_{inf}^1 containing $n_L^{e_i}$ and the subtree T_{inf}^2 containing $n_L^{e_j}$, whose roots are children of $T_{inf}.root$ (Line 1). Subsequently, it processes the root nodes of T_{inf}, T_{inf}^1, and T_{inf}^1 based on the reachability factor in the nodes (Line 2–13). The algorithm modifies HRTree only when $T_{inf}.root.\alpha = |e_i \cap e_j|$. If $T_{inf}.root.\alpha > |e_i \cap e_j|$, $n_L^{e_i}$ and $n_L^{e_j}$ are still in the $(T_{inf}.root.\alpha)$-CC when a vertex is added to the two hyperedges. Therefore, the structure of HRTree remains unchanged. If $T_{inf}.root.\alpha = |e_i \cap e_j|$, $n_L^{e_i}$ and $n_L^{e_j}$ should be in a $(T_{inf}.root.\alpha + 1)$-CC when a vertex is added to the two hyperedges. When $T_{inf}.root$ has two children in HRTree, the subgraph induced by $E(T_{inf}^1) \cup E(T_{inf}^2)$

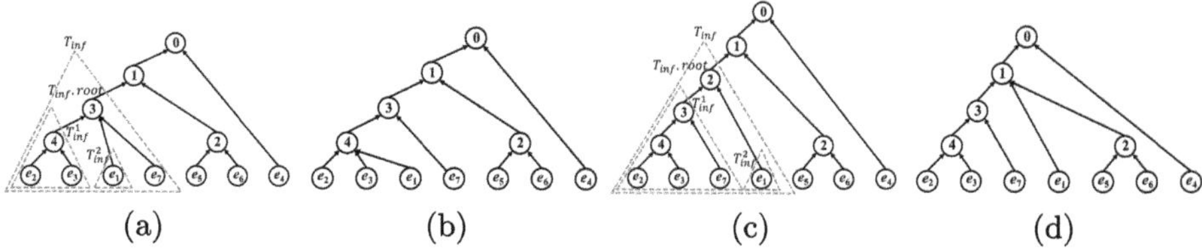

Fig. 3. Examples for updates on HRTree. (a) T_{inf} when $\Delta = (+, \{e_1, e_2\}, v_{22})$ and $\Delta = (-, \{e_1, e_2\}, v_1)$. (b) HRTree after applying update operation $\Delta = (+, \{e_1, e_2\}, v_{22})$ on HRTree in Fig. 3(a). (c) HRTree after applying update operation $\Delta = (-, \{e_1, e_2\}, v_1)$ on HRTree in Fig. 3(a). (d) HRTree after applying update operation $\Delta = (-, \{e_1, e_2\}, v_2)$ and $\Delta = (-, \{e_1, e_7\}, v_5)$ on HRTree in Fig. 3(c).

can be a $(T_{inf}.root.\alpha + 1)$-CC and the reachability factor can be immediately modified (Line 3–4). Otherwise, a new non-leaf node n_N is initialized and used to represent the reachability of the subgraph induced by $E(T_{inf}^1) \cup E(T_{inf}^2)$ (Line 5–9). The algorithm then removes the non-leaf nodes with the identical reachability factor from T_{inf} (Line 10–13). Finally, a new HRTree after update is returned (Line 14).

Example 4. Given an update $\Delta = (+, \{e_1, e_2\}, v_{22})$ applied on G and the HRTree in Fig. 2, the subtree within the blue dashed triangle is the influence region T_{inf} in HRTree in Fig. 3(a). The two subtrees T_{inf}^1 and T_{inf}^2 of T_{inf} are shown within the red dashed triangle. Since $|e_1 \cap e_2| = 3$ and $T_{inf}.root$ has three children, a new node n_N is initialized with its reachability factor set to $3+1 = 4$. Meanwhile, T_{inf}^1 and T_{inf}^2 are set as subtrees of n_N, n_N is set as a child of $T_{inf}.root$, and the subtree rooted at n_N is established as the influence region. As $n_N.\alpha = 4$ is the same as $T_{inf}^1.root.\alpha$, the redundant node $T_{inf}^1.root$ is removed and its children are connected to n_N. The new HRTree is shown in Fig. 3(b).

Time and space complexity. The time complexity of Procedure GetInfRegion is $O(h)$, where h is the height of HRTree. All operations in Line 2–9 in Algorithm 2 take $O(1)$ time. The operation in Line 11 or in Line 13 takes $O(D)$ time, where D is the maximum degree of the nodes in HRTree. Therefore, the time complexity of Algorithm 2 is $O(h + D)$.

Removing a Vertex from Two Hyperedges. We next process the case where $op = -$, *i.e.*, removing a vertex from two hyperedges. Algorithm 3 shows the details. It first identifies the influence region T_{inf} of e_i and e_j in T (Line 1). Subsequently, the parent of T_{inf} and the set E' of edges corresponding to the leaf nodes in T_{inf} are obtained and used to construct a new subtree T'' (Line 2–3). Algorithm 1 is called to generate the subtree T'' on E' (Line 4). For the newly returned tree T'', it is necessary to compare the reachability factor stored in its root node with that in the parent node of T_{inf}. If they are the same, all children of T'' are linked to $P_{T_{inf}}$ as the children of $P_{T_{inf}}$ to eliminate redundant nodes (Line 5–6). Otherwise, T'' is linked to $P_{T_{inf}}$ as a child (Line 7–8). Finally, T_{inf} is deleted and the updated T is returned (Line 9–10).

Algorithm 3: Removing a vertex from two hyperedges

Input: HRTree T, $\Delta = (-, \{e_i, e_j\}, v)$
Output: HRTree T' after update

1 $T_{inf}, T_{inf}^1, T_{inf}^2 \leftarrow$ GetInfRegion(T, e_i, e_j)
2 $P_{T_{inf}} \leftarrow$ Parent(T_{inf}, T)
3 $E' \leftarrow$ Edges(T_{inf}, E)
4 apply Algorithm 1 on $G' = (V, E')$ to get a tree T''
5 **if** $P_{T_{inf}}.\alpha = T''.root.\alpha$ **then**
6 $\quad\lfloor$ set the children of $T''.root$ be the children of $P_{T_{inf}}$ and delete $T''.root$

7 **else**
8 $\quad\lfloor$ set $T''.root$ be the child of $P_{T_{inf}}$

9 delete T_{inf} from T
10 **return** T

Example 5. Given an update $\Delta = (-, \{e_1, e_2\}, v_1)$ applied on G and the HRTree in Fig. 2, the subtree within the blue dashed triangle is the influence region T_{inf} in HRTree in Fig. 3(a). Due to the deletion of v_1, e_1 is 2-reachable to e_2. The subgraph containing e_2, e_3 and e_7 is still a 3-CC and 2-reachable to e_1. Therefore, the reachability factor in the root of T'' is 2. T'' is shown in Fig. 3(a), which is within the blue dashed triangle. Since the reachability factor in the parent $P_{T_{inf}}$ is 1, T'' is linked to $P_{T_{inf}}$ as a child. Given another update $\Delta = (-, \{e_1, e_2\}, v_2)$ applied on G, the subtree within the blue dashed triangle is the influence region T_{inf} in HRTree. As e_1 is 2-reachable to e_7, HRTree in Fig. 3(c) is unchanged when $\Delta = (-, \{e_1, e_2\}, v_2)$ is applied on G. When an update $\Delta = (-, \{e_1, e_2\}, v_2)$ is applied on G, the subtree within the blue dashed triangle is the influence region T_{inf} in Fig. 3(c). The subgraph containing e_2, e_3 and e_7 is 1-reachable to e_1. Because the reachability factor in the root of T'' is 1 and is the same as that of $P_{T_{inf}}$, the children of T'' are linked to $P_{T_{inf}}$ as children. The new HRTree is shown in Fig. 3(d).

Time and Space Complexity. Procedure GetInfRegion takes $O(h)$ time, where h is the height of HRTree. Procedure Parent takes $O(1)$ time and Procedure Edges takes $O(E')$ time. It takes $O(d|E'| \cdot \log(d|E'|))$ time to call Algorithm 1 to construct a tree on the set E' of edges. All operations in Line 5–8 in Algorithm 2 take $O(D)$ time, where D is the maximum degree of the nodes in HRTree. Therefore, the time complexity of Algorithm 2 is $O(d|E'| \cdot \log(d|E'|))$. In the worst case, the size of E' is comparable to that of E. However, in most scenarios, given two adjacent hyperedges, their influence regions are often minimal, *i.e.*, $|E'| \ll |E|$. Thus, Algorithm 3 can update HRTree efficiently.

4 Experiments

4.1 Experimental Setup

Settings. All experiments are conducted on a high-performance computing machine consisting of an 8-core Intel E5-1660v4 processor running at 3.2 GHz. The machine is equipped with 144 GB of RAM. All algorithms are implemented in C++17 and compiled using g++ 9.4.0 compiler on an Ubuntu 20.04 system. The source code is available at GitHub[1].

Table 1. Statistics of the real-world datasets.

Dataset	Abbr.	#Vertices	#Hyperedges	#Converted Edges
Geometry Questions	GMQ	585	1,193	212,237
Cooking Recipes	CRP	6,714	39,774	355,979,539
Drug Abuse Warning	DAW	2,559	87,104	460,821,691
NDC-substances	NDC	5,311	112,405	36,365,883
Threads-Ask-Ubuntu	TAU	125,602	192,947	27,363,251
Trivago Hotel Clicks	THC	172,738	233,202	5,955,227
Congress Bills	CGB	1,719	260,851	546,359,117
Threads-Maths-SX	TMS	201,864	719,792	546,359,117

Datasets. All datasets are downloaded from a public data warehouse[2] and the details are listed in Table 1. The last column shows the number of edges after the hypergraph is represented as a line graph. The set of vertices in a line graph corresponds to the edge sets of the hypergraph. If the size of the intersection of any two edge sets in the hypergraph is greater than 1, a weighted edge exists between the corresponding vertices in the line graph, and the weight of this edge is the size of this intersection.

Queries. The threshold k for the intersection size between two hyperedges is taken from $\{2, 4, 6, 8, 10\}$. For each hypergraph, we randomly select 1,000 distinct vertex pairs to evaluate the query performance of different algorithms. For updates, we also select 1,000 distinct vertex pairs and vary the ratio of vertex additions and vertex deletions. That is, the proportion r of vertex additions is taken from $\{0\%, 25\%, 50\%, 75\%, 100\%\}$, and the corresponding proportion of vertex deletions is $1 - r$. In the experiments, each query is executed 10 times, and the average time and memory usage are reported. For the evaluation of updates on hypergraphs, we set its time limit to 10^4 milliseconds and use "!" to denote that an algorithm fails to complete the update within this time limit.

---

[1] https://anonymous.4open.science/r/ReachabilityInHypergraph-85BD.
[2] https://www.cs.cornell.edu/~arb/data/.

Algorithms. We compare our algorithm, which is denoted as HRTree, with four classical algorithms. The details of the algorithms are as follows.

- BiBFS. A graph traversal algorithm that explores edges using breadth-first search.
- WeightedPLL [24]. An index-based algorithm utilizing the 2-hop index constructed on the transformed line graph of a hypergraph.
- HierarchicalUF [22]. An algorithm implemented using the union find set. For different thresholds k, HierarchicalUF constructs multiple union-find sets to support reachability queries on a hypergraph.
- DWR. It is adapted from DLCR [3] and supports reachability queries on the corresponding line graph when the hypergraph is updated.

Table 2. Memory usage of algorithms.

Dataset	BiBFS (MB)	HierarchicalUF (MB)	WeightedPLL (MB)	HRTree (MB)
GMQ	5.07	10.82	5.57	6.71
CRP	10.42	11941.09	8574.05	22.72
DAW	13.51	11785.38	10885.95	47.57
NDC	10.48	2594.10	1035.64	26.95
TAU	35.41	799.51	619.96	107.78
THC	42.68	367.42	174.02	137.66
CGB	42.11	14343.95	12529.15	142.15
TMS	101.92	17732.07	17112.27	404.06

4.2 Experimental Results

Memory Usage. Table 2 illustrates the memory usage of the algorithms. On all datasets, HRTree consistently consumes less memory. This is because HRTree does not need to convert the hypergraph into a more space-intensive line graph (adopted by WeightedPLL) nor does it require constructing multiple indexes for different values of k (adopted by HierarchicalUF). This ensures that HRTree achieves favorable efficiency during query and update operations. BiBFS has the minimum memory usage, as it only needs to store the hypergraph itself without constructing additional indexes. However, its query efficiency is poor, as shown in Fig. 4. When updates are taken into account, the memory usage of the algorithms does not undergo significant changes. This is because the changes to the hypergraph itself or the indices are not substantial when the hypergraph is updated.

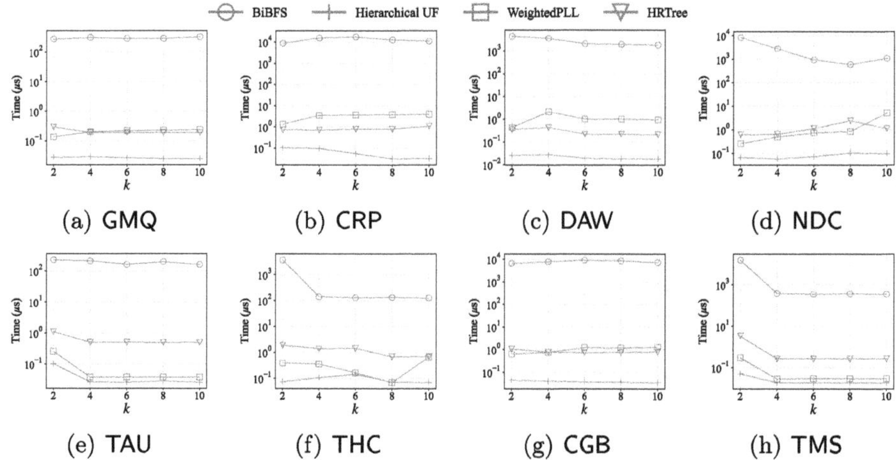

Fig. 4. Query time for algorithms on different hypergraphs.

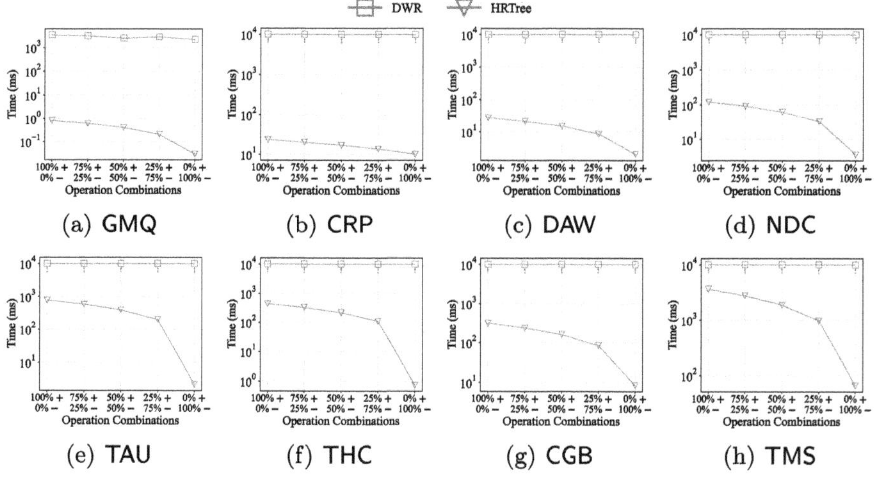

Fig. 5. Index update time for different hypergraphs (+ denotes vertex insertion; − denotes vertex deletion).

Query Performance. Figure 4 shows the execution time of algorithms on different hypergraphs. On all hypergraphs, HRTree can consistently return query results in approximately microseconds. As k increases, the query time decreases in most cases. This is because the number of k-reachable vertices among the generated 1,000 vertex pairs reduces, allowing the algorithm to terminate early. BiBFS takes the longest time, as it needs to traverse a large space on the hypergraph to determine the reachability between vertices. WeightedPLL performs slightly better than HRTree during querying due to its adoption of the 2-hop index. However, when these two sets are large, the query performance of Weight-edPLL deteriorates, as shown in Fig. 4(c) and (g).

Update Evaluation. We compare the update times of HRTree and DWR when the ratio of vertex additions to vertex deletions in updates varies. The results are shown in Fig. 5. Compared with DWR, HRTree is faster than DWR by more than three orders of magnitude in processing updates. This benefits from the concise index structure of HRTree and the fact that updates only occur within the affected regions in the index, while DWR requires extensive searching and index reorganization. For large hypergraphs, from CRP to TMS in Table 1, DWR fails to complete updates within 10^4 milliseconds, indicating that its scalability is poor. As the proportion of deletions in updates increases, the update processing time of HRTree gradually decreases. This is because HRTree only needs to delete a small number of nodes (or even none) within the affected regions of the index.

5   Related Work

Reachability Queries on Graphs. Recent reachability query methods can be broadly categorized into dimension labeling and set labeling approaches. Dimension labeling methods, such as Chain-Cover [4], Path-Tree [10], MGTag [25], assign labels or intervals to vertices and determine reachability through label comparison or interval inclusion, resorting to path search only when inconclusive. Grail constructs k spanning trees to generate multiple intervals using a min-post strategy, while Ferrari merges adjacent intervals to reduce query time. In contrast, set labeling methods, such as TF-label [5], BFL [21], etc. It leverages set containment and intersection to evaluate reachability. TF-label recursively folds the graph and derives L_{in} and L_{out} sets based on level filters, while BFL maps transitive closure indexes into hashed sets, pruning unreachable queries via containment checks before exploring successors.

Queries on Hypergraphs. Early studies on s-reachability are largely conceptual. Lu et al. [14] and Aksoy et al. [1] define s-walks and s-connected components to generalize classical graph notions such as distance and centrality. Cooley et al. [6] further examine the emergence of s-connected components in random hypergraphs from a probabilistic perspective. Later works explore shortest-path algorithms [8,20] that can indirectly answer reachability queries on hypergraphs, though they are either limited to 1-connectivity or produce redundant results. The recent HYPED framework [16] provides an approximate index for shortest s-connect paths by enumerating s-connected components and applying landmark-based indexing.

6   Conclusions

In this paper, we study the constrained reachability problem on hypergraphs. By analyzing the inefficiencies of intuitive approaches in terms of space and time complexity, we propose a novel index HRTree to support reachability queries on hypergraphs. HRTree supports reachability queries in logarithmic time while maintaining space complexity linear to the number of hyperedges. Furthermore, HRTree enables incremental updates where maintenance time depends solely on the influence regions of the updates. Experimental results demonstrate that our algorithm simultaneously handles updates with low time delay and enables fast reachability query processing. In the future, we will further investigate more efficient distributed reachability algorithms on hypergraphs, as well as reachability queries with richer semantics, such as label-constrained reachability queries.

Acknowledgments. The work is partially supported by the National Key Research and Development Program of China (2024YFF0617702), the National Natural Science Foundation of China (Nos. 62232007, U22A2025, U23A20309, 62202091), the Fundamental Research Funds for the Central Universities (Nos. N2316008, N25LPY012), the Liaoning Provincial United Natural Science Foundation (No. 2023-BSBA-129), the CCF-ApsaraDB Research Fund, and the CCF Alibaba Cloud Yaochi Research Fund (No. CCF-Aliyun2024006).

References

1. Aksoy, S.G., Joslyn, C., Ortiz Marrero, C., Praggastis, B., Purvine, E.: Hypernetwork science via high-order hypergraph walks. EPJ Data Sci. **9**(1), 1–34 (2020). https://doi.org/10.1140/epjds/s13688-020-00231-0
2. Antelmi, A., Cordasco, G., Spagnuolo, C., Szufel, P.: Social influence maximization in hypergraphs. Entropy **23**(7), 796 (2021)
3. Chen, X., Peng, Y., Wang, S., Yu, J.X.: Dlcr: efficient indexing for label-constrained reachability queries on large dynamic graphs. VLDB **15**(8), 1645–1657 (2022)
4. Chen, Y., Chen, Y.: An efficient algorithm for answering graph reachability queries. In: ICDE, pp. 893–902 (2008)
5. Cheng, J., Huang, S., Wu, H., Fu, A.W.C.: Tf-label: A topological-folding labeling scheme for reachability querying in a large graph. In: SIGMOD, pp. 193–204 (2013)
6. Cooley, O., Kang, M., Koch, C.: Evolution of high-order connected components in random hypergraphs. Electron. Notes Discrete Math. **49**, 569–575 (2015)
7. Ding, K., Wang, J., Li, J., Li, D., Liu, H.: Be more with less: Hypergraph attention networks for inductive text classification. In: EMNLP, pp. 4927–4936 (2020)
8. Gao, J., Zhao, Q., Ren, W., Swami, A., Ramanathan, R., Bar-Noy, A.: Dynamic shortest path algorithms for hypergraphs. TON
9. Grzesiak-Kopec, K., Oramus, P., Ogorzalek, M.: Hypergraphs and extremal optimization in 3D integrated circuit design automation. Adv. Eng. Inf. **33**, 491–501 (2017)
10. Jin, R., Xiang, Y., Ruan, N., Wang, H.: Efficiently answering reachability queries on very large directed graphs. In: SIGMOD, pp. 595–608 (2008)
11. Kovács, B., Benedek, B., Palla, G.: Community detection in hypergraphs through hyperedge percolation. Sci. Rep. **15**(1), 36032 (2025)
12. Lee, G., Bu, F., Eliassi-Rad, T., Shin, K.: A survey on hypergraph mining: patterns, tools, and generators. ACM Comput. Surv. **57**(8), 1–36 (2025)
13. Lee, G., Ko, J., Shin, K.: Hypergraph motifs: concepts, algorithms, and discoveries. VLDB **13**(11), 2256–2269 (2020)

14. Lu, L., Peng, X.: High-order random walks and generalized laplacians on hypergraphs. Internet Math. **9**(1), 3–32 (2013)
15. Luo, Q., Yu, D., Cai, Z., Lin, X., Wang, G., Cheng, X.: Toward maintenance of hypercores in large-scale dynamic hypergraphs. VLDB J. **32**(3), 647–664 (2023)
16. Preti, G., De Francisci Morales, G., Bonchi, F.: Hyper-distance oracles in hypergraphs. VLDB J. **33**(5), 1333–1356 (2024)
17. Ritz, A.M., Avent, B., Murali, T.M.: Pathway analysis with signaling hypergraphs. IEEE ACM Trans. Comput. Biol. Bioinform. **14**(5), 1042–1055 (2017)
18. Ruggeri, N., Contisciani, M., Battiston, F., De Bacco, C.: Community detection in large hypergraphs. Sci. Adv. **9**(28), eadg9159 (2023)
19. Sahu, S., Mhedhbi, A., Salihoglu, S., Lin, J., Özsu, M.T.: The ubiquity of large graphs and surprising challenges of graph processing: extended survey. VLDB J. **29**(2–3), 595–618 (2020)
20. Shun, J.: Practical parallel hypergraph algorithms. In: PPoPP, pp. 232–249 (2020)
21. Su, J., Zhu, Q., Wei, H., Yu, J.X.: Reachability querying: can it be even faster? TKDE **29**(3), 683–697 (2016)
22. Tarjan, R.E., van Leeuwen, J.: Worst-case analysis of set union algorithms. J. ACM **31**(2), 245–281 (1984)
23. Xia, X., Yin, H., Yu, J., Wang, Q., Cui, L., Zhang, X.: Self-supervised hypergraph convolutional networks for session-based recommendation. In: AAAI, pp. 4503–4511 (2021)
24. Yano, Y., Akiba, T., Iwata, Y., Yoshida, Y.: Fast and scalable reachability queries on graphs by pruned labeling with landmarks and paths. In: CIKM, pp. 1601–1606 (2013)
25. Yuan, P., You, Y., Zhou, S., Jin, H., Liu, L.: Providing fast reachability query services with mgtag: a multi-dimensional graph labeling method. TSC **15**(2), 1000–1011 (2022)

Toward Generalizable Multi-domain Graph Foundation Models via Dynamic Domain-Aware Feature Alignment

Dahyun Jeong and Heasoo Hwang[✉]

Department of Computer Science, University of Seoul, Seoul, Republic of Korea
{dahyun21,hwang}@uos.ac.kr

Abstract. Graphs are pervasive across diverse domains, requiring a Graph Foundation Model (GFM) pre-trained to capture generalizable knowledge across heterogeneous domains. However, multi-domain pre-training poses two challenges: (i) domain discrepancy—as graphs from different domains possess distinct feature dimensions and semantics—and (ii) adaptation difficulty to unseen target domains with limited supervision. Many prior studies relied on PCA-based preprocessing to unify multi-domain feature dimensions. However, such static feature alignment risks the loss of information essential for the model optimization and the acquisition of generalizable knowledge. Moreover, since it does not exploit structural information and inter-domain relationships, it often yields suboptimal alignment. We propose Dynamic Domain-aware Feature Alignment (DDFA), a novel framework that jointly considers intra-domain dimension alignment and inter-domain semantic alignment. We also introduce WB regularization to reduce domain discrepancy efficiently, thus guaranteeing upper-bound of generalization error. In the adaptation stage, we design a target-specific dual-space dimension encoder with a novel node sampling strategy and a prompting method to fully exploit limited target supervision and pre-trained knowledge. Extensive experiments on seven datasets demonstrate the superior performance on most datasets under few-shot node classification scenarios.

Keywords: Graph Foundation Models · Multi-Domain Pre-Training · Prompt Learning · Domain Adaptation

1 Introduction

Graphs model complex relationships among entities in domains such as social networks, citation networks, and e-commerce networks. We encode graphs using GNNs to exploit the structure and semantics of graphs. Inspired by the success of foundation models in CV and NLP, a Graph Foundation Model (GFM) capable of generalizing to diverse domains is highly demanded [23]. During graph pre-training, a GNN learns intrinsic graph properties via self-supervised objectives. However, the pre-trained GNN is effective only when a target graph and a pre-train graph share the same or similar domains [16,24]. Cross-domain transfer

H. Jung et al. (Eds.): DASFAA 2026, LNCS 16536, pp. 438–454, 2026.
https://doi.org/10.1007/978-981-92-0366-6_27

methods learn domain-invariant representations but still rely on knowledge from a single source domain [3,5,17,19]. To obtain a GFM, it is essential to perform multi-domain graph pre-training that integrates generalizable knowledge across diverse domains [23,26,27].

However, multi-domain pre-training poses two main challenges. The first is domain discrepancy. For instance, in the citation network *Cora* [9], a paper is represented as a 1,433-dimensional binary vector indicating the presence of specific words. In the e-commerce graph *Photo* [14], a product is represented as a 745-dimensional tf-idf vector. Aligning such heterogeneous features—which differ in both dimensionality and semantics—into a common space is non-trivial. A GFM needs to balance domain-invariant knowledge and domain-specific characteristics for generalization [18,21,22,27]. For dimension alignment, recent methods mainly use linear decomposition such as PCA or SVD during preprocessing [20,25–27], which maintain domain-specific characteristics but may remove information crucial for model optimization. Moreover, they cannot exploit graph structure and inter-domain relationships, leading to suboptimal alignment. The second is the difficulty of adapting a pre-trained model to unseen target domains with limited supervision. For this, prompts that can enable target domains to effectively leverage all the multi-domain knowledge are crucial. However, existing methods [20,26] mainly perform semantic alignment separately as in Fig. 1b, instead of jointly optimizing dimension alignment and semantic alignment.

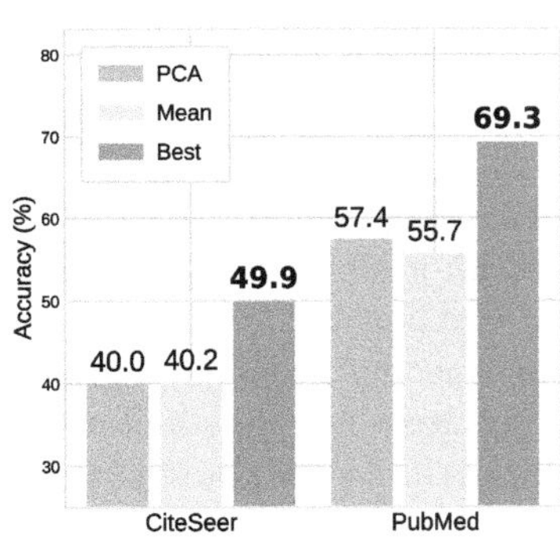

(a) PCA vs. reordered dimensions.

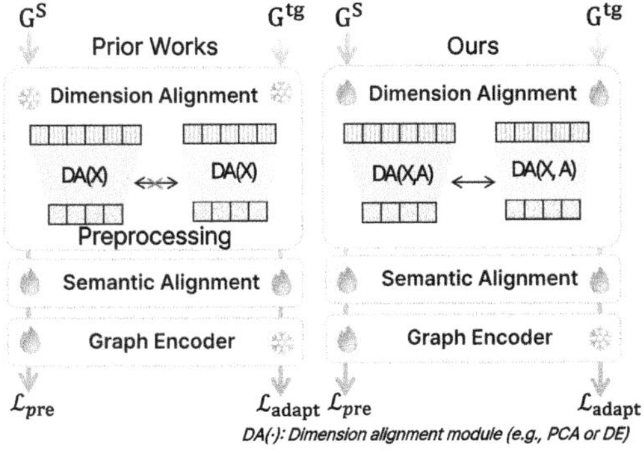

(b) Static vs. dynamic dimension alignment.

Fig. 1. (a) Limitation of PCA (b) Dimension alignment of prior works and ours.

To demonstrate the potential of dynamic dimension alignment, we examine the impact of a simple dimension alignment method, dimension reordering in Fig. 1a

Given a source domain, *Cora*, we perform PCA and pre-train a GCN. For a target domain, *Citeseer* or *Pubmed*, we freeze the pre-trained GNN and train a classifier after aligning target domain features using PCA and random dimension reordering. We repeat random dimension reordering 100 times to get the mean and the best accuracy. The accuracy obtained with the original PCA dimensions is almost identical to the mean accuracy. While PCA is a commonly used feature alignment method, the order of the original PCA dimensions is not particularly better than random orderings. Surprisingly, the best accuracy is almost 10–12%p higher than the accuracy with PCA dimensions. This indicates that even a dimension reordering can significantly improve the adaptation performance if the order of dimensions is consistent with a pre-trained model.

Motivated by this, we propose **D**ynamic **D**omain-aware **F**eature **A**lignment (**DDFA**), a novel multi-domain feature alignment method that dynamically aligns source domain features during pre-training as in Fig. 1b. DDFA jointly considers both intra-domain dimension alignment and inter-domain semantic alignment to preserve diversity. Specifically, for **intra-domain dimension alignment**, we optimize domain-specific dimension encoders to maximize the diversity of basis transformation vectors within each domain. For **inter-domain semantic alignment**, we generate a domain token per source domain to capture prominent domain-specific semantics. After dimension and semantic alignment, we perform **WB regularization** to obtain a graph encoder capable of effective adaptation to unseen target domains. We use the Wasserstein Barycenter of source domains, μ_{wb}, as a common domain, and minimize the W1 distance between each source domain and μ_{wb}. This guarantees a theoretical upper-bound of generalization error [18] and improves pre-training efficiency. In the adaptation stage, we use a dual-space dimension alignment and target-specific and source-mixture prompts to enable effective few-shot transfer to new domains. We use both labeled and randomly sampled nodes to fully exploit limited supervision. Our main contributions are summarized as follows:

- We propose DDFA, a novel GFM framework based on dynamic feature alignment that jointly considers dimension- and domain-level diversity as well as domain generalization during multi-domain pre-training.
- We suggest a target-specific dual-space dimension encoder and a novel prompt method that effectively uses weak supervision for cross-domain adaptation.
- We conduct extensive experiments on seven datasets and show the superiority of DDFA under few-shot scenarios.

2   Related Work

Graph Pre-training. Graph pre-training aims to learn generalizable representations via self-supervised learning [16,24]. GraphPrompt [8] unified pre-training and downstream tasks under a common subgraph similarity template, while GPF [4] introduced a universal prompt framework applicable regardless of the pre-training strategy. Although they enable parameter-efficient adaptation, they assume that pre-training and downstream graphs share the same or similar domains, limiting cross-domain generalization. To address domain gaps in cross-domain settings, graph transfer learning methods aim to learn domain-invariant representations [3,5,17,19]. While such methods effectively transfer knowledge from a source to a target graph, they rely on a single source domain, not leveraging multi-domain knowledge. Moreover, they are usually adapted to a specific domain or task, which hinders their ability to generalize to unseen domains.

Multi-domain Graph Pre-training. Recent studies have extended graph pre-training to multi-domain scenarios to build GFMs. GCOPE [27] introduced domain-specific virtual nodes to interconnect multi-domain graphs and align feature and topology, but it does not explicitly align semantic features, limiting its cross-domain generalization. MDGPT [26] proposed domain tokens to unify the semantic spaces of multiple source domains and introduced dual prompts to align unseen domains. MDGFM [20] leveraged adaptive graph structure learning to align feature and topology distributions, effectively extracting domain-invariant knowledge with robust generalization. SAMGPT [25] incorporated structure tokens in the aggregation stage and adopted a dual prompt for structure alignment, allowing target domains to adapt to domain-specific structural knowledge. However, they still rely on static dimension alignment methods such as PCA.

Feature Unification. PCA projects features into a lower-dimensional space by maximizing variance. While simple and effective, it cannot perform task-specific alignment since it is performed before training. Also, it tends to overemphasize node-wise feature differences, losing homophily information. FUG [28] suggested a dimension encoder that exploits the distribution of each source dimension to generate a basis transformation vector, which achieves superior transferability. However, FUG is limited to single-source pre-training, making its multi-domain extension non-trivial. We extend the dimension encoding mechanism to a multi-domain setting, enabling dynamic feature alignment.

3 Preliminary

3.1 Notations

We consider a set of m source domain graphs $\mathcal{G}^{\mathcal{S}} = \{G^{(i)}\}_{i=1}^{m}$ from varied domains $\mathcal{D}^{\mathcal{S}} = \{\mathcal{D}^{(i)}\}_{i=1}^{m}$ and a target domain graph G^{tg} from unseen domain $\mathcal{D}^{tg}$ where $\mathcal{D}^{tg} \notin \mathcal{D}^{\mathcal{S}}$. Each graph $G^{(i)}$ is $(V^{(i)}, \mathbf{A}^{(i)}, \mathbf{X}^{(i)})$ where $V^{(i)}$ is the node set, $\mathbf{A}^{(i)}$ is the adjacency matrix, and $\mathbf{X}^{(i)} \in \mathbb{R}^{|V^{(i)}| \times d^{(i)}}$ is feature matrix. Self-loops are added to obtain $\widetilde{\mathbf{A}}^{(i)} = \mathbf{A}^{(i)} + \mathbf{I}_{|V^{(i)}|}$ and normalized it as $\widehat{\mathbf{A}}^{(i)} = (\widetilde{\mathbf{D}}^{(i)})^{-1/2}\widetilde{\mathbf{A}}^{(i)}(\widetilde{\mathbf{D}}^{(i)})^{-1/2}$ where $\widetilde{\mathbf{D}}_{jj}^{(i)} = \sum_{k}\mathbf{A}_{kk}$. To enable representation learning across heterogeneous domains, each feature matrix $\mathbf{X}^{(i)}$ is transformed to have a unified feature dimension, k.

Once feature alignment is done, a GNN encoder with L layers is applied to update node representations by aggregating neighborhood information. The l-th layer of a standard GNN updates $\mathbf{H}_{(l-1)}^{(i)}$ as $\mathbf{H}_{(l)}^{(i)} = \sigma(\widehat{\mathbf{A}}^{(i)}\mathbf{H}_{(l-1)}^{(i)}\mathbf{W}_{(l)})$, where $\mathbf{H}^{(i)}_{(0)} = \mathbf{X}^{(i)}$ and $\sigma(\cdot)$ is a non-linear activation. $\mathbf{H}^{(i)}$ is the final representation after L-th layer, and the j-th row $\mathbf{h}^{(i)}_j$ corresponds to the embedding of $v^{(i)}_j$. The train set for K-shot node classification is denoted as $\Omega_{train} = \{(x_1, y_1), (x_2, y_2), \dots\}$ where x_i is a node and y_i is the corresponding label.

3.2 Dimension Encoder DE($\cdot$)

The dimension encoder (DE) is proposed by FUG [28] for cross-domain adaptation. It is a *dimension-specific* encoder that differs from conventional MLPs that are sample-specific. It takes the values of a specific feature dimension j as input and outputs a basis transformation vector $\mathbf{t}_j$. To reuse a pre-trained DE for a target graph, FUG samples ns nodes on both graphs and pre-trains a DE with a fixed input dimensionality, ns. Let $\widehat{\mathbf{X}}$ denote Sampling$(\mathbf{X}, ns)$.

$$\mathbf{t}_j = \mathbf{DE}(\mathbf{X}_{:,j}) = \mathrm{Norm}(\mathrm{MLP}(\widehat{\mathbf{X}}^{\top}_{:,j})) \tag{1}$$

Unlike PCA, a DE performs a non-linear projection to obtain a basis transformation matrix $\mathbf{T} \in \mathbb{R}^{d \times k}$ through MLPs. To reflect distinctions among feature dimensions, a DE is optimized by $\mathcal{L}_{de}$ in Eq. (2). It relaxes PCA's strict orthogonality constraints to preserve inter-dimension correlation.

$$\mathcal{L}_{de}(\mathbf{T}) = \| \frac{1}{d} \sum_{j=1}^{d} \mathbf{t}_j - \mathbf{0} \|_2^2 \tag{2}$$

The transformed feature matrix $\widetilde{\mathbf{X}} = \mathbf{XT} \in \mathbb{R}^{|V| \times k}$ is then used as input to the graph encoder to obtain node embedding $\mathbf{H}$. To reduce the cost of negative sample selection and the sampling bias, a global regularization loss, $\mathcal{L}_{uniform}$, in Eq. (3) is proposed to enhance the uniformity of overall representations:

$$\mathcal{L}_{uniform}(\mathbf{H}) = \| \frac{1}{|V|} \sum_{v=1}^{|V|} \mathrm{Norm}(\mathbf{h}_v) - \mathbf{0} \|_2^2 \tag{3}$$

Instead of training a DE on a source domain to reuse it for a target domain [28], we jointly train domain-specific DEs during pre-training, and a target-specific DE during domain adaptation. Although we do not reuse the pre-trained DE, we still employ node sampling to constrain parameter size and apply a unified DE architecture across source graphs. Sampled nodes are also used for efficient approximation of the WB and W1 distances during pre-training. Moreover, with a DE, we encode smoothed features ($\widehat{\mathbf{A}}\mathbf{X}$) for more stable and structure-aware dimension alignment, while [28] encodes raw features ($\mathbf{X}$) (Fig. 2).

4   Methodology

4.1   Preserving Domain Diversity

Before pre-training a graph encoder, we need to project heterogeneous features of source domain graphs onto a shared space of k dimensions. To improve the performance and the generalizability of a pre-trained graph encoder, we want to maximally preserve the diversity of dimensions of each domain and domain-specific semantic characteristics. For this, we perform **intra-domain dimension alignment** and **inter-domain semantic alignment** with jointly optimizing dimension-level diversity and domain-level diversity using $\mathcal{L}_{diversity}$.

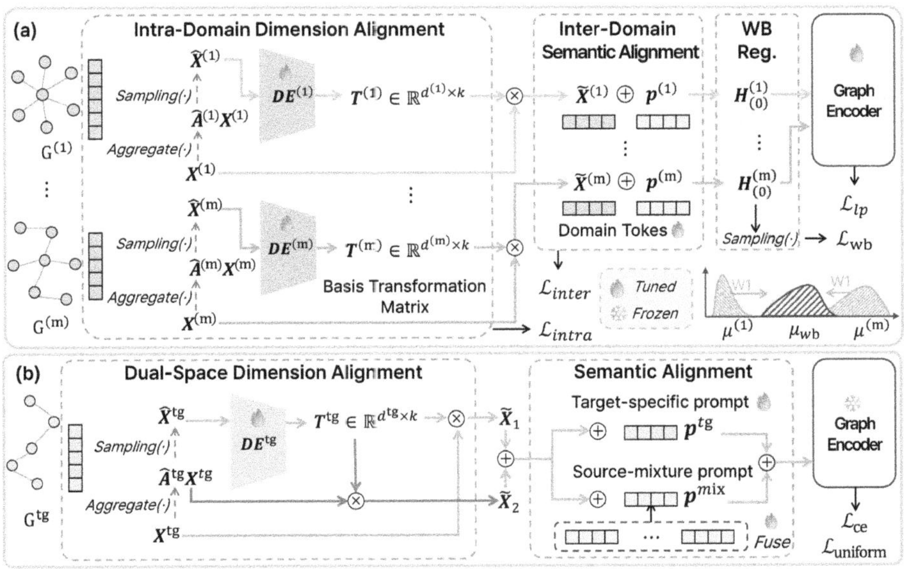

Fig. 2. Overview of our proposed DDFA. (a) Multi-domain pre-training stage. (b) Target-domain adaptation stage.

Intra-domain Dimension Alignment. To unify feature dimensions, we use structure-aware domain-specific DEs. The DE of the i-th source graph, $G^{(i)}$, captures dimension-specific distribution of features in $\widehat{\mathbf{A}}^{(i)}\mathbf{X}^{(i)}$. We encode smoothed features, $\widehat{\mathbf{A}}^{(i)}\mathbf{X}^{(i)}$, to mitigate domain-specific noises in raw features, $\mathbf{X}^{(i)}$, while enabling a more stable and structure-aware dimension alignment. To ensure efficiency and architectural consistency, we sample ns nodes per DE, maintaining a fixed input dimension after transposing $\widehat{\mathbf{X}}^{(i)}$.

$$\widehat{\mathbf{X}}^{(i)} = \mathrm{Sampling}(\widehat{\mathbf{A}}^{(i)}\mathbf{X}^{(i)}, ns) \in \mathbb{R}^{ns \times d_i} \tag{4}$$

Then, $\mathbf{DE}^{(i)}$ produces a basis transformation matrix $\mathbf{T}^{(i)} \in \mathbb{R}^{d_i \times k}$ from $\widehat{\mathbf{X}}^{(i)}$ via Eq. (1). The *intra-domain loss*, $\mathcal{L}_{intra}$, ensures basis transformation vectors in $\mathbf{T}^{(i)}$ are uniformly distributed on a hypersphere, preserving dimension diversity and the intrinsic characteristics of each domain.

$$\mathcal{L}_{intra} = \sum_{i=1}^{m} \mathcal{L}_{de}(\mathbf{T}^{(i)}) \tag{5}$$

Using $\mathbf{T}^{(i)}$, we perform dimension alignment of $\mathbf{X}^{(i)}$ and obtain $\widetilde{\mathbf{X}}^{(i)}$ of k dimensions as in Eq. (6). This transformation enables not only dimensional unification but structure-aware dimension alignment optimized jointly across domains.

$$\widetilde{\mathbf{X}}^{(i)} = \mathbf{X}^{(i)}\mathbf{T}^{(i)} \in \mathbb{R}^{|V^{(i)}| \times k} \tag{6}$$

Inter-domain Semantic Alignment. While intra-domain dimension alignment unifies heterogeneous feature spaces and preserves dimension-level diversity, it may risk losing domain-level semantic diversity. To preserve domain-specific semantics and thus enhance inter-domain diversity, we introduce domain tokens $\mathbf{p}^{(i)} \in \mathbb{R}^{1 \times k}$ for each domain $\mathcal{D}^{(i)}$.

$$\mathbf{H}_{(0)}^{(i)} = \widetilde{\mathbf{X}}^{(i)} + \mathbf{p}^{(i)} \tag{7}$$

Each domain token can be seen as a domain-specific bias that represents unique semantic characteristics of the corresponding domain within the shared latent space. We define the *inter-domain loss*, $\mathcal{L}_{inter}$, for domain-level diversity. Instead of imposing strict orthogonality constraints between domain tokens, we distribute domain tokens on a hypersphere to preserve inter-domain correlation.

$$\mathcal{L}_{inter} = \| \frac{1}{m} \sum_{i=1}^{m} \mathbf{p}^{(i)} - \mathbf{0} \|_2^2 \tag{8}$$

We define the *diversity loss*, $\mathcal{L}_{diversity}$, that jointly optimizes dimension-level diversity and domain-level diversity.

$$\mathcal{L}_{diversity} = \mathcal{L}_{intra} + \mathcal{L}_{inter} \tag{9}$$

4.2   Improving Domain Generalization

To ensure generalization to unseen domains, we want to reduce the discrepancy between source domain graphs and capture domain-invariant information. Inspired by the theoretical analysis in [18], we want to reduce the first-order Wasserstein (W1) distance between source domains to guarantee the upperbound of generalization error to unseen domains. For training efficiency, we consider the Wasserstein Barycenter (WB) of source domains as a common domain and define $\mathcal{L}_{wb}$, **a Wasserstein barycenter regularization**, that reduces the discrepancy between WB and each source domain within a unified latent space.

By Lemma 1 and Theorem 1 in [18], we conclude that if we minimize W1 distances between source domains in the shared space during pre-training, we can obtain a graph encoder that guarantees the upper-bound of generalization error on unseen target domains. Let $P(G)$ be the probability distribution of a source domain graph G and $W_1(\cdot, \cdot)$ be the W1 distance between two probability distributions. If $W_1(P(G^i), P(G^j)) \leq \xi$ holds for all G^i, G^j pairs in $\mathcal{G}^S$, the set of source domain graphs, the pre-trained graph encoder can perform effectively on an unseen target domain graph within guaranteed upper-bound of generalization error. As the inter-domain discrepancy reduces, ξ decreases and thus the generalization error bound becomes tighter.

However, instead of computing all pairwise W1 distances, we compute W1 distances from the WB of source domains to each source domain for training efficiency. Notice that W1 distance computation is costly, and computing all pairwise W1 distances becomes even more expensive ($\mathcal{O}(m^2)$). Instead, we use the WB of source domains, μ_{wb}, in Eq. (10), for efficient pre-training. WB is defined as the central distribution that minimizes the sum of W1 distances to all source domain distributions [7].

$$\mu_{wb} = \arg\min_{\mu} \sum_{i=1}^{m} W_1(P(G^{(i)}), \mu) \tag{10}$$

WB Regularization. Considering μ_{wb} as a common domain distribution in the aligned space, we reduce the W1 distances between μ_{wb} and each source domain distribution $P(G^{(i)})$. To approximate $P(G^{(i)})$, we use the node representations obtained after dimension and semantic alignment, $\mathbf{H}_{(0)}^{(i)}$, as empirical vectors. For efficiency, we use only ns sampled empirical vectors. Each source distribution is approximated by an empirical measure $\mu_i = \frac{1}{ns} \sum_{j=1}^{ns} \delta_{\mathbf{H}_{(0),j}^{(i)}}$, where δ denotes the Dirac measure, assigning a unit mass at each empirical vector [7]. The empirical vectors of μ_{wb} are derived from the empirical measures of source domains, $\{\mu_i\}_{i=1}^{m}$. We randomly initialize the μ_{wb} and compute the W1 distance $W_1(\mu_i, \mu_{wb})$ using the Sinkhorn-based entropic regularized optimal transport algorithm [1]. μ_{wb} is updated via the fast barycenter approximation method [2] every q epochs. Then, the WB regularization loss, $\mathcal{L}_{wb}$, is defined as follows.

$$\mathcal{L}_{wb} = \sum_{i=1}^{m} W_1(\mu_i, \mu_{wb}) \tag{11}$$

While we minimize the W1 distances between μ_{wb} and each source domain, we can still achieve a low upper-bound of generalization error. We do this by optimizing W1 distances between WB and source domains to be evenly distributed so that the maximum W1 distance between WB and source domains can be reduced. Let $\xi' = \max_{1 \leq i \leq m} W_1(P(G^{(i)}), \mu_{wb})$ denote the maximum W1 distance to μ_{wb}. Since $W_1(\cdot, \cdot)$ satisfies the triangle inequality [10], $W_1(P(G^{(i)}), P(G^{(j)})) \leq W_1(P(G^{(i)}), \mu_{wb}) + W_1(P(G^{(j)}), \mu_{wb})$ holds. Thus, since $W_1(P(G^{(i)}), \mu_{wb}) \leq \xi'$ for every source domain graph $G^{(i)}$,

$W_1(P(G^{(i)}), P(G^{(j)})) \leq 2\xi'$. Since the pairwise W1 distance between source domains is bounded by twice of the maximum distance between WB and each source domain, ξ', we can reduce the variance of W1 distances between WB and each source domain during pre-training by adding $\eta \sum_{i=1}^{m} \| W_1(\mu_i, \mu_{wb}) \|_2^2$ to $\mathcal{L}_{wb}$ in Eq. (11). We do not include this term in our experiments.

4.3 Pre-training Objectives

The graph encoder $f(\cdot)$ receives aligned representation $\mathbf{H}_{(0)}^{(i)}$ and graph structure $\mathbf{A}^{(i)}$ of the i-th source domain as input to generate $\mathbf{H}^{(i)}$, i.e., $\mathbf{H}^{(i)} = f_\Theta(\mathbf{H}_{(0)}^{(i)}, \mathbf{A}^{(i)})$. As in [8,26], we pre-train with link prediction loss ($\mathcal{L}_{lp}$) to leverage abundant edge information in large multi-domain graphs. We employ a universal task template of node similarity, which unifies link prediction and classification objectives. For each anchor node u, the model increases similarity to its positive neighbor v^+ and decreases it to negative samples v^-. $g(\cdot)$ denotes cosine similarity, and $\mathbf{H}^S = \{\mathbf{H}^{(i)}\}_{i=1}^{m}$.

$$\mathcal{L}_{lp}(\mathbf{H}^S) = - \sum_{(u,v+,v-)} \ln \frac{\exp(g(\mathbf{h}_u^{(i)}, \mathbf{h}_{v+}^{(i)})/\tau)}{\sum_{v \in \{v+,v-\}} \exp(g(\mathbf{h}_u^{(i)}, \mathbf{h}_v^{(i)})/\tau)} \tag{12}$$

$$\mathcal{L}_{pre}(\Theta) = \mathcal{L}_{lp} + \alpha\mathcal{L}_{wb} + \beta\mathcal{L}_{diversity} \tag{13}$$

In the pre-training loss, $\mathcal{L}_{pre}(\Theta)$, Θ denotes all learnable parameters. The coefficients α and β are hyperparameters that control the relative contributions of the WB regularization and the diversity loss, respectively.

4.4 Cross-Domain Adaptation

Dual-Space Dimension Alignment. For effective cross-domain adaptation, we ensure architectural consistency by using the same dimension alignment method for the target domain. To align the target with the common latent space of multiple source domains, we use a target-specific dimension encoder, $\mathbf{DE}^{tg}$.

Unlike fully random sampling during pre-training, we train $\mathbf{DE}^{tg}$ using samples reflecting weak supervision of few-shot scenarios. Among ns samples, we include all labeled nodes in Ω_{train}. Then, we randomly sample the remaining $ns - |\Omega_{train}|$ nodes. To consider the structural context of sampled nodes, we use representations in the aggregated feature space $\widehat{\mathbf{A}}^{tg}\mathbf{X}^{tg}$.

$$\widehat{\mathbf{X}}^{tg} = \{(\widehat{\mathbf{A}}^{tg}\mathbf{X}^{tg})_{x_i} | (x_i, y_i) \in \Omega_{train}\} \cup \text{Sampling}(\widehat{\mathbf{A}}^{tg}\mathbf{X}^{tg}, ns - |\Omega_{train}|) \tag{14}$$

The sampled features, $\widehat{\mathbf{X}}^{tg} \in \mathbb{R}^{ns \times d^{tg}}$, are fed into $\mathbf{DE}^{tg}$ to obtain the target basis transformation matrix $\mathbf{T}^{tg}$. To maximize the use of limited supervision in the target domain, we perform dual-space dimension alignment by projecting not only the raw features but also the aggregated structural features.

$$\widetilde{\mathbf{X}}_1 = \mathbf{X}^{tg}\mathbf{T}^{tg} \qquad \widetilde{\mathbf{X}}_2 = \widehat{\mathbf{A}}^{tg}\mathbf{X}^{tg}\mathbf{T}^{tg} \tag{15}$$

The representation after dual-space dimension alignment is $\widetilde{\mathbf{X}}^{tg} = \widetilde{\mathbf{X}}_1 + \widetilde{\mathbf{X}}_2$, which incorporates both local feature and structural information.

Algorithm 1. Pre-training Stage for DDFA

Input: (1) $\mathcal{G}^{\mathcal{S}}$: a set of m source domain graphs, $\{G^{(i)} = (V^{(i)}, \mathbf{A}^{(i)}, \mathbf{X}^{(i)})\}_{i=1}^{m}$; (2) $d^{(i)}$: the number of feature dimensions of $G^{(i)}$; (3) k: unified feature dimensionality; (4) ns: the number of samples; (5) q: barycenter update interval;

Output: Optimized model parameters Θ

1: Initialize parameters Θ, including graph encoder $f(\cdot)$, dimension encoders $\mathbf{DE}$ and domain tokens $\{\mathbf{p}^{(i)}\}_{i=1}^{m}$. Initialize the Wasserstein barycenter μ_{wb}.

2: Normalize adjacency matrix with self-loop $\{\widehat{\mathbf{A}}^{(i)}\}_{i=1}^{m}$, and generate $\{\widehat{\mathbf{A}}^{(i)}\mathbf{X}^{(i)}\}_{i=1}^{m}$

3: **for** until converge **do**

4:      Initialize all losses: $\mathcal{L}_{intra} = \mathcal{L}_{inter} = \mathcal{L}_{wb} = \mathcal{L}_{lp} = 0$

5:      **for all** $G^{(i)} \in \mathcal{G}^{\mathcal{S}}$ **do**

6:          // **Intra-Domain Dimension Alignment**:

7:          $\widehat{\mathbf{X}}^{(i)} \leftarrow \text{Sampling}(\widehat{\mathbf{A}}^{(i)}\mathbf{X}^{(i)}, ns) \in \mathbb{R}^{ns \times d_i}$

8:          **for** $j = 1 \cdots d^{(i)}$ **do**

9:              $\mathbf{t}_j^{(i)} \leftarrow \text{DE}^{(i)}(\mathbf{X}_{:,j}^{(i)}) = \text{Norm}[\text{MLP}^{(i)}(\widehat{\mathbf{X}}_{:,j}^{(i)\top})]$

10:          **end for**

11:          Basis transformation matrix $\mathbf{T}^{(i)} \leftarrow [\mathbf{t}_1^{(i)}, \cdots, \mathbf{t}_{d^{(i)}}^{(i)}] \in \mathbb{R}^{d^{(i)} \times k}$

12:          $\widetilde{\mathbf{X}}^{(i)} \leftarrow \mathbf{X}^{(i)}\mathbf{T}^{(i)} \in \mathbb{R}^{|V^{(i)}| \times k}$

13:          Update $\mathcal{L}_{intra}$ by Eq.(5)

14:          // **Inter-Domain Semantic Alignment**:

15:          $\mathbf{H}_{(0)}^{(i)} \leftarrow \widetilde{\mathbf{X}}^{(i)} + \mathbf{p}^{(i)}$

16:          Update $\mathcal{L}_{inter}$ by Eq.(8)

17:          // **Wasserstein Barycenter Regularization**:

18:          Update $\mathcal{L}_{wb}$ by Eq.(11)

19:          // **Graph Embedding**:

20:          $\mathbf{H}^{(i)} \leftarrow f(\mathbf{H}_{(0)}^{(i)}, \mathbf{A}^{(i)})$

21:          Update $\mathcal{L}_{lp}$ by Eq.(12)

22:      **end for**

23:      $\mathcal{L}_{diversity} \leftarrow \mathcal{L}_{intra} + \mathcal{L}_{inter}$

24:      $\mathcal{L}_{pre} \leftarrow \mathcal{L}_{lp} + \alpha L_{wb} + \beta \mathcal{L}_{diversity}$

25:      Update all learnable parameters Θ by minimizing $\mathcal{L}_{pre}$

26:      Update μ_{wb} every q rounds

27: **end for**

Semantic Alignment. To leverage multi-domain knowledge, we use two types of prompts: a target-specific prompt $\mathbf{p}^{tg} \in \mathbb{R}^{1 \times k}$ to exploit the unified knowledge in a pre-trained graph encoder and a source-mixture prompt $\mathbf{p}^{mix} \in \mathbb{R}^{1 \times k}$ incorporating domain-specific semantics. The target-specific prompt is adapted to domain-invariant knowledge shared across source domains. The source-mixture prompt aggregates domain-specific semantics from multiple source domains, as in Eq. (16). α_i is a learnable weight that controls the importance of source domain token $\mathbf{p}^{(i)}$ according to its relevance to the target domain. These two prompts allow the target domain to align within the shared latent space while adaptively referencing source domain priors.

$$\mathbf{p}^{mix} = \sum_{i=1}^{m} \alpha_i \mathbf{p}^{(i)} \tag{16}$$

The representation after semantic alignment is defined as a weighted combination of two prompt-guided representations. λ_1 and λ_2 are the learnable parameters that adaptively control the contribution of prompts.

$$\mathbf{H}_{(0)}^{tg} = \lambda_1(\widetilde{\mathbf{X}}^{tg} + \mathbf{p}^{tg}) + \lambda_2(\widetilde{\mathbf{X}}^{tg} + \mathbf{p}^{mix}) \tag{17}$$

Adaptation Loss. For node-level classification, we use a universal task template based on node similarity, which is consistent with the pre-training objective. Given the labeled set Ω_{train}, the supervised contrastive loss is defined as:

$$\mathcal{L}_{ce} = -\sum_{(x_i,y_i)\in\Omega_{train}} \ln \frac{\exp(g(\mathbf{h}_{x_i}, \mathbf{h}_{y_i})/\tau)}{\sum_{y\in Y} \exp(g(\mathbf{h}_{x_i}, \mathbf{h}_{y})/\tau)} \tag{18}$$

To mitigate overfitting under weak supervision and promote representation uniformity, we add the uniformity loss term, $\mathcal{L}_{uniform}$, in Eq. (3). The overall adaptation loss is then formulated as:

$$\mathcal{L}_{adapt}(\mathbf{H}; \phi, \mathcal{A}, \mathbf{p}^{tg}) = \mathcal{L}_{ce} + \gamma\mathcal{L}_{uniform} \tag{19}$$

where γ is a hyperparameter that balances loss terms, and ϕ denotes the parameters of the target dimension encoder $\mathbf{DE}^{tg}$.

5 Experiments

5.1 Experimental Settings

Datasets and Baselines. We evaluate our model on seven datasets in diverse domains. *Cora* [9], *Citeseer* [13], and *Pubmed* [13] are paper citation networks. *Photo* [14] and *Computers* [14] are e-commerce networks from Amazon co-purchasing data. *Facebook* [11] is a Facebook web page network. *LastFM* [12] is a social network. The statistics of the datasets are presented in Table 1.

We compare our method with three groups of baselines: **(1) Vanilla GNNs:** GCN [6] and GAT [15], which are trained from scratch under a supervised setting. **(2) Graph Pre-training Models:** DGI [16] and GraphCL [24], which follow the pre-train-then-fine-tune pipeline. GraphPrompt [8] and GPF [4], which adopt a pre-train-then-prompt-tune framework. **(3) Multi-Domain Pre-training Models:** We follow the original training protocol in each paper: GCOPE [27] employs *fine-tuning*, whereas MDGPT [26] and SAMGPT [25] adopt *prompt-tuning*. MDGFM [20] was excluded due to out-of-memory errors.

Table 1. Statistics of datasets (*Avg.ND denotes the average node degree).

Dataset	# Nodes	# Edges	# Features	# Labels	Avg.ND
Cora	2,708	10,566	1,433	7	3.89
Citeseer	3,327	9,104	3,703	6	2.73
Pubmed	19,717	88,648	500	3	4.49
Photo	7,650	238,162	745	8	31.13
Computers	13,752	491,722	767	10	35.75
Facebook	22,470	342,004	128	4	15.22
LastFM	7,624	55,612	128	18	7.29

Multi-domain Pre-training and Downstream Adaptation. Following previous studies [25–27], we treat a dataset as a distinct domain. All datasets except the target are used as source domains. For evaluation, we perform K-shot node classification, where K labeled nodes per class are randomly sampled as supervision, excluding query nodes. 100 few-shot episodes are constructed on a fixed set of 1,000 query nodes for consistency. The performance measure is the accuracy of 500 evaluations using five random seeds $\{39, 40, 41, 42, 43\}$.

Implementation Details. All experiments are conducted using PyTorch 2.5.1 and PyTorch Geometric 2.6.1 with CUDA 12.1 on an NVIDIA RTX 6000 Ada GPU (48GB VRAM). We use a 3-layer GCN backbone with 256 hidden dimensions and unified dimension $k = 50$. DE is a 2-layer MLP with 100 hidden dimensions. Our method is trained with a sample size $ns = 256$ for 10,000 epochs with early stopping. Default loss weights are $\alpha = 3$, $\beta = 100$, and $\gamma = 100$. For target adaptation, learning rate$\in \{0.01, 0.001, 0.0001, 0.00005\}$ and weight decay$= 1 \times 10^{-4}$ are used with temperature $\tau = 0.2$. For fair comparison, we follow the settings of the official implementations of all baselines.

5.2   Performance Evaluation

We evaluate our model under K-shot node classification, where K ranges from 1 to 10. Table 2 presents the accuracy under the most challenging one-shot setting. Firstly, our method consistently outperforms baselines across most datasets, demonstrating the effectiveness of our method. Secondly, compared to baselines with static feature alignment such as GCOPE, MDGPT, and SAMGPT, our method achieves an average improvement of 6.2%p, with the most significant gain of 10%p compared to the second-best on *Cora*. This suggests that static feature alignment leads to suboptimal alignment. The superiority of our dynamic feature alignment method originates from exploiting not only intra-domain diversity but inter-domain relationships and the pre-training objective, as well as structural information.

In Fig. 3, our method consistently outperforms other baselines across various K-shot settings. The gap narrows on *Cora* and *Pubmed* as K increases, but our

Table 2. Accuracy(%) of one-shot node classification. **Bold** and <u>underlined</u> values indicate the best and second-best results, respectively.

Target →	Cora	Citeseer	Pubmed	Photo	Computers	Facebook	LastFM
GCN	28.25±10.17	30.04±8.25	46.87±10.52	43.53±15.34	43.95±8.72	<u>42.18</u>±9.16	32.01±6.67
GAT	29.97±11.04	24.68±7.12	47.66±8.11	43.81±9.02	38.79±8.90	34.61±8.00	19.57±4.99
DGI	32.64±6.65	34.52±6.21	45.17±7.78	48.12±8.56	41.78±7.69	34.42±6.04	23.46±5.21
GraphCL	38.13±7.50	41.89±7.70	47.38±7.97	51.97±9.16	44.10±8.56	40.26±6.37	28.60±5.56
GraphPrompt	29.25±8.01	29.52±2.80	46.08±3.46	42.93±7.36	43.92±5.76	35.95±1.06	22.12±2.65
GPF	31.26±7.55	34.61±6.97	39.78±6.14	40.52±7.08	30.60±5.51	36.53±6.29	21.73±5.26
GCOPE	34.64±9.22	31.01±7.19	<u>52.04</u>±9.30	56.63±11.00	49.95±10.72	32.62±6.28	29.06±6.98
MDGPT	42.09±8.47	<u>42.05</u>±8.42	50.37±8.65	**63.96**±10.97	**52.64**±10.10	41.90±8.35	30.82±6.03
SAMGPT	<u>42.62</u>±7.99	37.78±8.39	47.65±8.93	57.64±9.94	46.63±7.75	40.03±7.27	<u>32.27</u>±5.31
DDFA	**53.04**±8.62	**47.03**±8.81	**52.54**±10.82	<u>61.25</u>±12.64	<u>51.08</u>±10.77	**47.68**±10.32	**35.59**±7.99

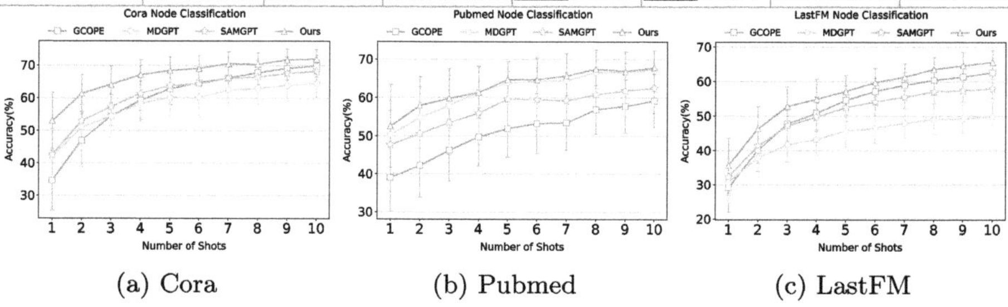

(a) Cora   (b) Pubmed   (c) LastFM

Fig. 3. Few-shot Node Classification.

model achieves notably higher accuracy in low-shot settings, demonstrating its strong ability to utilize limited supervision effectively.

5.3 Ablation Study

Data Ablation. To analyze the effect of domain diversity, we increase the number of source graphs and measure accuracy. As shown in Fig. 4, accuracy improves consistently as more source domains are used during pre-training. This suggests that our model effectively learns transferable knowledge and benefits from multi-domain pre-training, emphasizing the importance of domain diversity for enhanced generalization. For target *Cora*, source graphs are added in the order of {*LastFM, Citeseer, Facebook, Pubmed*}, and {*Citeseer, Pubmed, Photo, Computers*} for target *Facebook*.

Model Ablation. We perform a component-wise ablation in Table 3. Firstly, removing the uniformity loss causes the largest average drop ($-11.28\%p$), especially on *Cora*, *Citeseer*, and *Facebook*, showing the impact of $\mathcal{L}_{uniform}$ in preventing overfitting under weak supervision. Secondly, replacing the aggregated feature sampling, Sampling($\hat{\mathbf{A}}\mathbf{X}$), with raw feature sampling also degrades accuracy ($-3.74\%p$). This demonstrates that structure-aware sampling provides more stable feature distributions for dimension alignment. Lastly, all the key components of our method contribute to higher accuracy except for $\mathbf{p}^{mix}$. The effect of $\mathbf{p}^{mix}$ varies across datasets, suggesting that mixing source domain semantics directly during target feature alignment is not effective in some datasets.

Table 3. Model ablation study. "All" is the full model; others are results without the specified component. "Avg. (%, ↓)" denotes average accuracy drop.

Dataset	All	$\mathcal{L}_{intra}$	$\mathcal{L}_{inter}$	$\mathcal{L}_{wb}$	$\widetilde{\mathbf{X}}_2$	$\mathbf{p}^{tg}$	$\mathbf{p}^{mix}$	$\mathcal{L}_{uniform}$	Sampling($\widehat{\mathbf{A}}\mathbf{X}$)	Avg. (%, ↓)
Cora	53.04	52.74	52.28	52.26	52.72	52.48	53.02	39.96	51.16	−4.17
Citeseer	47.03	43.73	45.82	45.35	46.41	45.18	46.07	28.97	44.32	−8.08
Pubmed	52.54	53.10	52.68	53.10	52.68	52.42	52.97	52.53	52.66	0.43
Photo	61.25	60.55	60.51	60.76	61.80	62.33	61.64	60.62	60.76	−0.21
Computers	51.08	51.28	48.76	50.69	50.75	51.73	52.37	50.33	49.41	−0.81
Facebook	47.68	48.62	48.60	48.95	46.62	45.15	48.63	43.49	45.61	−1.51
LastFM	35.39	34.27	34.28	35.49	33.57	31.58	34.98	33.96	32.50	−4.95
Avg. (%, ↓)	−	−1.29	−1.61	−0.45	−1.33	−2.68	0.31	−11.28	−3.74	−

5.4 Parameter Sensitivity

In Fig. 5, we use target *Citeseer*, and examine the sensitivity of our model to loss weights, α, β, and γ, of L_{wb}, $L_{diversity}$, and $L_{uniform}$. For all weights, accuracy improves as non-zero weight values are used, indicating all the corresponding loss terms are effective. The best accuracy is achieved when $\alpha = 1$, $\beta = 150$. For γ, accuracy drastically improves as γ increases from 0 to 5, and then saturates. This indicates that the diversity loss is more important than other loss terms, and the uniformity loss is particularly effective on *Citeseer*.

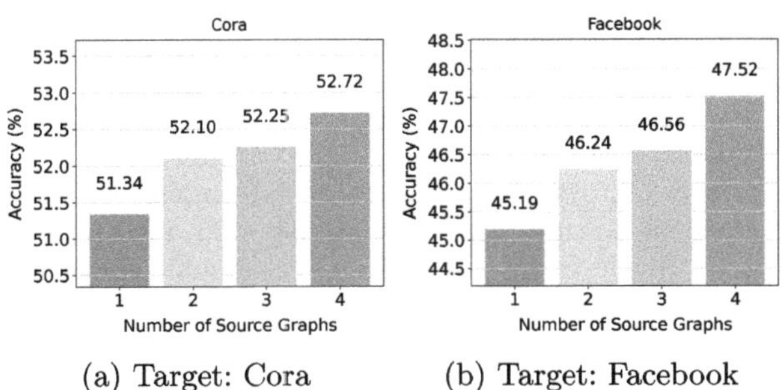

(a) Target: Cora   (b) Target: Facebook

Fig. 4. Data Ablation Study.

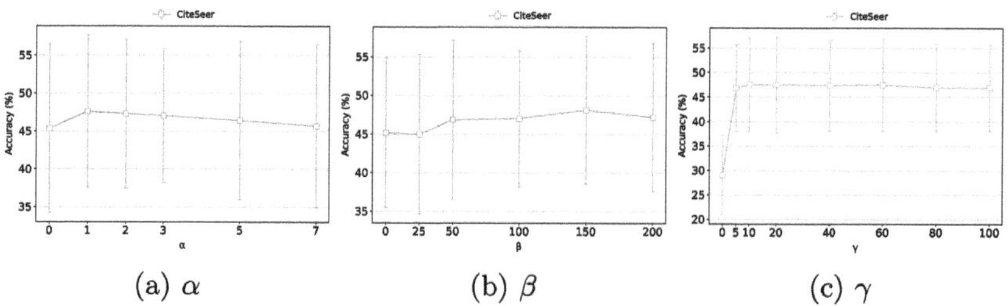

Fig. 5. Hyperparameter Sensitivity.

6   Conclusions

We propose DDFA, a GFM based on dynamic feature alignment that balances dimension/domain diversity and domain generalization during multi-domain pre-training and adaptation. During pre-training, DDFA learns diverse yet consistent representations through intra-domain dimension and inter-domain semantic alignment, with WB regularization enhancing generalization. In the adaptation stage, dual-space dimension alignment and prompts effectively leverage weak supervision signals. Extensive experiments demonstrate the superiority of DDFA in few-shot scenarios. As future work, we plan to improve WB regularization to minimize the variance of W1 distances, thereby directly tightening the generalization error bound. We also want to extend our method to dynamic structure alignment.

Acknowledgments. This work was supported by the National Research Foundation of Korea (NRF) grant funded by the Korea government (MSIT) (NRF-2022R1A2C1011637).

References

1. Cuturi, M.: Sinkhorn distances: Lightspeed computation of optimal transport. Adv. Neural Inf. Process. Syst. **26** (2013)
2. Cuturi, M., Doucet, A.: Fast computation of wasserstein barycenters. In: International Conference on Machine Learning, pp. 685–693. PMLR (2014)
3. Ding, K., Shu, K., Shan, X., Li, J., Liu, H.: Cross-domain graph anomaly detection. IEEE Trans. Neural Netw. Learn. Syst. **33**(6), 2406–2415 (2022). https://doi.org/10.1109/TNNLS.2021.3110982
4. Fang, T., Zhang, Y., Yang, Y., Wang, C., Chen, L.: Universal prompt tuning for graph neural networks. Adv. Neural. Inf. Process. Syst. **36**, 52464–52489 (2023)
5. Hassani, K.: Cross-domain few-shot graph classification. In: Proceedings of the AAAI Conference on Artificial Intelligence, vol. 36, pp. 6856–6864 (2022)
6. Kipf, T.: Semi-supervised classification with graph convolutional networks. arXiv preprint arXiv:1609.02907 (2016)
7. Lin, M., Li, W., Li, D., Chen, Y., Li, G., Lu, S.: Multi-domain generalized graph meta learning. In: Proceedings of the AAAI Conference on Artificial Intelligence, vol. 37, pp. 4479–4487 (2023)

8. Liu, Z., Yu, X., Fang, Y., Zhang, X.: Graphprompt: Unifying pre-training and downstream tasks for graph neural networks. In: Proceedings of the ACM Web Conference 2023, pp. 417–428. WWW '23, Association for Computing Machinery, New York, NY, USA (2023). https://doi.org/10.1145/3543507.3583386
9. McCallum, A.K., Nigam, K., Rennie, J., Seymore, K.: Automating the construction of internet portals with machine learning. Inf. Retrieval **3**(2), 127–163 (2000)
10. Panaretos, V.M., Zemel, Y.: Statistical aspects of wasserstein distances. Ann. Rev. Stat. Appl. **6**(1), 405–431 (2019)
11. Rozemberczki, B., Allen, C., Sarkar, R.: Multi-scale attributed node embedding. J. Complex Netw. **9**(2), cnab014 (2021)
12. Rozemberczki, B., Sarkar, R.: Characteristic functions on graphs: Birds of a feather, from statistical descriptors to parametric models. In: Proceedings of the 29th ACM International Conference on Information & Knowledge Management, pp. 1325–1334 (2020)
13. Sen, P., Namata, G., Bilgic, M., Getoor, L., Galligher, B., Eliassi-Rad, T.: Collective classification in network data. AI Mag. **29**(3), 93–93 (2008)
14. Shchur, O., Mumme, M., Bojchevski, A., Günnemann, S.: Pitfalls of graph neural network evaluation. arXiv preprint arXiv:1811.05868 (2018)
15. Velickovic, P., Cucurull, G., Casanova, A., Romero, A., Lio, P., Bengio, Y., et al.: Graph attention networks. Stat **1050**(20), 10–48550 (2017)
16. Veličković, P., Fedus, W., Hamilton, W.L., Liò, P., Bengio, Y., Hjelm, R.D.: Deep graph infomax. arXiv preprint arXiv:1809.10341 (2018)
17. Wang, C., Liang, Y., Liu, Z., Zhang, T., Yu, P.S.: Pre-training graph neural network for cross domain recommendation. In: 2021 IEEE Third International Conference on Cognitive Machine Intelligence (CogMI), pp. 140–145 (2021). https://doi.org/10.1109/CogMI52975.2021.00026
18. Wang, P., Bai, L., Yang, X., Wang, X., Liang, J.: Generalization error bounds for graph neural networks in unseen domains. IEEE Trans. Artif. Intell. **6**(10), 2712–2721 (2025). https://doi.org/10.1109/TAI.2025.3556978
19. Wang, Q., Pang, G., Salehi, M., Buntine, W., Leckie, C.: Cross-domain graph anomaly detection via anomaly-aware contrastive alignment. In: Proceedings of the AAAI Conference on Artificial Intelligence, vol. 37, pp. 4676–4684 (2023)
20. Wang, S., Wang, B., Shen, Z., Deng, B., Kang, Z.: Multi-domain graph foundation models: robust knowledge transfer via topology alignment. arXiv preprint arXiv:2502.02017 (2025)
21. Wang, S., Zhou, J., Chen, Q., Zhang, Q., Gui, T., Huang, X.: Domain generalization via causal adjustment for cross-domain sentiment analysis. arXiv preprint arXiv:2402.14536 (2024)
22. Wang, Y., Wang, L., Shi, S., Li, V.O., Tu, Z.: Go from the general to the particular: multi-domain translation with domain transformation networks. In: Proceedings of the AAAI Conference on Artificial Intelligence, vol. 34, pp. 9233–9241 (2020)
23. Wang, Z., Liu, Z., Ma, T., Li, J., Zhang, Z., Fu, X., et al.: Graph foundation models: a comprehensive survey. arXiv preprint arXiv:2505.15116 (2025)
24. You, Y., Chen, T., Sui, Y., Chen, T., Wang, Z., Shen, Y.: Graph contrastive learning with augmentations. Adv. Neural. Inf. Process. Syst. **33**, 5812–5823 (2020)
25. Yu, X., Gong, Z., Zhou, C., Fang, Y., Zhang, H.: Samgpt: text-free graph foundation model for multi-domain pre-training and cross-domain adaptation. In: Proceedings of the ACM on Web Conference 2025, pp. 1142–1153. WWW '25, Association for Computing Machinery, New York, NY, USA (2025)
26. Yu, X., Zhou, C., Fang, Y., Zhang, X.: Text-free multi-domain graph pre-training: toward graph foundation models. arXiv preprint arXiv:2405.13934 (2024)

27. Zhao, H., Chen, A., Sun, X., Cheng, H., Li, J.: All in one and one for all: a simple yet effective method towards cross-domain graph pretraining. In: Proceedings of the 30th ACM SIGKDD Conference on Knowledge Discovery and Data Mining, pp. 4443–4454 (2024)
28. Zhao, J., Jin, D., Ge, M., Shan, L., Wang, X., He, D., et al.: Fug: Feature-universal graph contrastive pre-training for graphs with diverse node features. Adv. Neural. Inf. Process. Syst. **37**, 4003–4034 (2024)

G^{2}rammar: Bilingual Grammar Modeling for Enhanced Text-Attributed Graph Learning

Heng Zheng[1], Haochen You[2], Zijun Liu[3], Zijian Zhang[4], Lubin Gan[5], Hao Zhang[6], Wenjun Huang[7], and Jin Huang[1]([✉])

[1] South China Normal University, Guangzhou, China
`{2024025450,huangjin}@m.scnu.edu.cn`
[2] Columbia University, New York, USA
`hy2854@columbia.edu`
[3] iAUTO, Taipei, Taiwan
`liuzijun@iauto.com`
[4] University of Pennsylvania, Philadelphia, USA
`zzjharry@umich.edu`
[5] University of Science and Technology of China, Hefei, China
`ganlubin@mail.ustc.edu.cn`
[6] University of Chinese Academy of Sciences, Beijing, China
`zhang_hao1999@yeah.net`
[7] Sun Yat-sen University, Guangzhou, China
`huangwj98@mail2.sysu.edu.cn`

Abstract. Text-attributed graphs require models to effectively integrate both structural topology and semantic content. Recent approaches apply large language models to graphs by linearizing structures into token sequences through random walks. These methods create concise graph vocabularies to replace verbose natural language descriptions. However, they overlook a critical component that makes language expressive: grammar. In natural language, grammar assigns syntactic roles to words and defines their functions within sentences. Similarly, nodes in graphs play distinct structural roles as hubs, bridges, or peripheral members. Current graph language methods provide tokens without grammatical annotations to indicate these structural or semantic roles. This absence limits language models' ability to reason about graph topology effectively. We propose **G^{2}rammar**, a bilingual grammar framework that explicitly encodes both structural and semantic grammar for text-attributed graphs. Structural grammar characterizes topological roles through centrality and neighborhood patterns. Semantic grammar captures content relationships through textual informativity. The framework implements two-stage learning with structural grammar pretraining followed by semantic grammar fine-tuning. Extensive experiments on real-world datasets demonstrate that G^{2}rammar consistently outperforms competitive baselines by providing language models with the grammatical context needed to understand graph structures.

Keywords: Graph Neural Networks · Large Language Models · Text-Attributed Graphs · Graph Representation Learning

1   Introduction

Graph neural networks have made significant progress in text-attributed graph learning tasks, including applications in areas such as social networks and citation networks [14,20,30]. Accurately classifying nodes in these graphs requires the effective integration of both structural details from graph topology and semantic information embedded in node attributes [17,19]. This dual dependency raises a primary challenge: how can models effectively encode both the connectivity patterns defining network structure and the content information residing in node features? Traditional GNNs address this through message passing mechanisms that aggregate neighbor features according to graph structure [11]. However, their limited expressiveness and scalability constraints motivate exploration of more powerful alternatives [24,32].

Recent advances in large language models have motivated researchers to leverage their exceptional text understanding capabilities for graph learning tasks [1,3]. Two main paradigms have emerged. The first describes graph structures using natural language by enumerating nodes and connections in textual form [10,34]. The second employs LLMs to generate embeddings for node attributes and aggregates these embeddings through GNN layers [6,40]. Recent work such as GDL4LLM [18] proposes treating graphs as a new language. This approach replaces verbose natural language descriptions with concise graph tokens and forms graph sentences through random walks. By creating graph vocabulary and pre-training LLMs on graph language corpus, these methods successfully address the redundancy and verbosity of natural language descriptions (Fig. 1).

The answer lies in grammar. Language models process graphs through linearized token sequences. Current graph language methods sample paths primarily via random walks to represent local neighborhoods. This approach mirrors how we read sentences word by word. Yet natural language achieves expressiveness not just through vocabulary but through grammar. Grammar assigns syntactic roles to words and defines their functions within sentences. Consider this parallel in graphs. A random walk might produce nodeA → nodeB → nodeC. Another yields nodeD → nodeE → nodeF. Both appear as three-token sequences, yet they may encode different structural meanings. The first path might traverse from a highly-cited hub through a bridge to a peripheral node. The second might connect three nodes within a tightly-coupled cluster. NodeA could be coordinating multiple communities while nodeD remains within a single cluster. Current graph language representations treat these distinctions as vocabulary differences rather than grammatical ones. The language model observes token transitions without understanding the syntactic roles these nodes play. This perspective suggests graphs need grammatical annotation just as sentences do.

The observation reveals a critical insight. Linearization preserves node connections but loses their grammatical roles in network topology. Token sequences

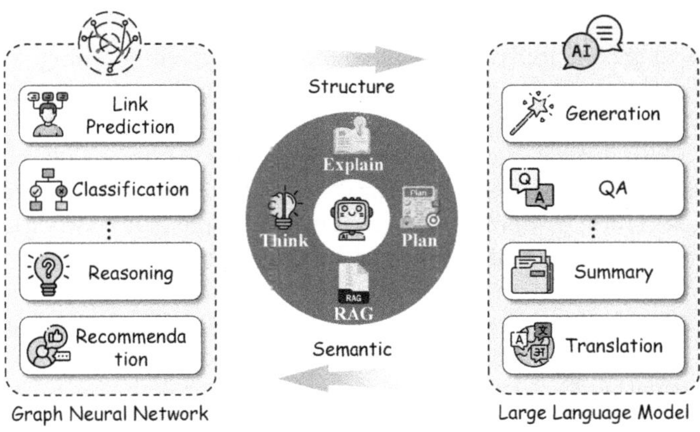

Fig. 1. Across diverse graph domains, the integration of graphs and LLMs has proven to be highly effective in tackling a variety of downstream tasks.

record transitions from nodeA to nodeB without indicating whether nodeA acts as a hub coordinating communities or a member within a single cluster. This parallels reading words without knowing if they are subjects or objects. Graph neural networks avoid this limitation by computing on adjacency matrices during message passing. They aggregate information with access to complete neighborhood structures. Language models instead process linearized paths through sequential windows, seeing one path at a time without broader grammatical context. Text-attributed graphs add another layer. Structural roles and textual content form a semantic grammar. Central hubs may discuss specialized topics while peripheral nodes contain broadly applicable information. This creates rich interactions between structural and semantic roles. Current graph language methods create token vocabularies but lack grammatical annotations. The model receives bare tokens without markers for their structural or semantic functions. This absence of graph grammar limits understanding just as removing grammar from natural language would impair comprehension.

Inspired by this linguistic perspective, we propose G^{2}rammar, a novel framework for text-attributed graphs that incorporates structural and semantic grammar into a language model's representation. This framework follows a two-stage learning approach inspired by cognitive processes. In the initial stage, structural grammar pre-training emphasizes learning graph topology through annotated paths obtained via random walks. Nodes and edges are explicitly labeled based on centrality and proximity measures, respectively. In the fine-tuning stage, semantic grammar complements structural annotations by incorporating informativity measures derived from textual attributes. This design allows the model to jointly reason over both structural and semantic dimensions, overcoming potential limitations of purely sequential representations. Our contributions can be summarized as follows:

- We identify the loss of structural role information during graph linearization as a critical gap in applying language models to graph learning. We establish the connection between graph structural roles and linguistic grammar as a principled solution.
- We introduce G^{2}rammar, a bilingual grammar system that characterizes topological roles through structural grammar and content relationships through semantic grammar. The framework implements two-stage learning with structural pre-training followed by semantic fine-tuning.
- Extensive
  experiments on three real-world datasets demonstrate that G^{2}rammar consistently outperforms competitive baselines including GDL4LLM. The results validate effective high-order neighbor modeling through concise grammatical annotations.

2 Related Works

2.1 Graph Representation Learning

Graph representation learning aims to encode graph-structured data into low-dimensional vector spaces for various downstream tasks [15]. Early approaches rely on random walk-based methods to capture neighborhood information. Deep-Walk [25] samples node sequences through random walks and applies Skip-Gram models to learn node embeddings. Node2vec [12] extends this idea by introducing biased random walks to balance local and global structural patterns. Graph Neural Networks emerge as a more powerful paradigm by aggregating neighborhood features through message passing mechanisms [26]. Graph Convolutional Networks [20] operate on spectral graph theory to perform localized feature aggregation. GraphSAGE [14] introduces sampling-based aggregation to improve scalability on large graphs. Graph Attention Networks [30] employ attention mechanisms to assign different weights to neighboring nodes during aggregation. Despite their success in capturing structural patterns, GNNs face challenges in expressiveness and scalability [32]. Recent efforts explore self-supervised learning [16,35] and contrastive learning [35,38] on graphs to enhance representation quality, significantly improving learned representations across various domains including social networks, molecular structures, and citation networks.

2.2 Language Models for Graphs

Large language models demonstrate remarkable capabilities in natural language understanding and generation [3,29]. Two main paradigms emerge in applying language models to graphs. The first describes graph structures using natural language by enumerating nodes and connections in textual form [10,13]. Instruct-GLM [34] designs templates to describe local ego-graph structures and conducts instruction tuning for node classification. The second paradigm employs language models to generate embeddings for node attributes and aggregates these embeddings through graph neural network layers [6]. SPECTER [7] fine-tunes SciBERT on citation networks to learn paper representations. GIANT [6] constructs neighborhood-based pretraining objectives to align textual features with graph structures. Recent work proposes treating graphs as a new language rather than describing them in natural language. GDL4LLM [18] creates graph vocabularies by treating nodes as tokens and forms graph sentences through random walks, replacing verbose natural language descriptions with concise graph tokens. Despite these advances, current approaches primarily focus on vocabulary and sequencing without explicitly encoding the structural roles of nodes and edges in the graph topology.

3   Preliminary

3.1   Graph and Language Notations

We define a text-attributed graph as $\mathcal{G} = (\mathcal{V}, \mathcal{E}, \mathcal{X})$ where $\mathcal{V} = \{v_1, \ldots, v_{|\mathcal{V}|}\}$ denotes nodes, $\mathcal{E} \subseteq \mathcal{V} \times \mathcal{V}$ represents edges, and $\mathcal{X} = \{x_1, \ldots, x_{|\mathcal{V}|}\}$ contains textual attributes with x_i corresponding to v_i. The adjacency matrix $\mathbf{A} \in \{0,1\}^{|\mathcal{V}| \times |\mathcal{V}|}$ encodes structure where $\mathbf{A}_{ij} = 1$ indicates an edge. We denote the k-hop neighborhood as $\mathcal{N}^k(v_i)$.

Treating graphs as language, nodes become tokens and paths become sentences. A path $p = \langle v_{i_1}, \ldots, v_{i_l} \rangle$ represents connected nodes where consecutive pairs form edges. $\mathcal{P}(v_i)$ denotes paths from v_i, and the corpus is $\mathcal{C} = \bigcup_{v_i \in \mathcal{V}} \mathcal{P}(v_i)$.

3.2   Graph Grammar as Structural Roles

We introduce a bilingual grammar system operating on two dimensions. The structural grammar $\mathcal{G}_s$ defines node roles via $\phi : \mathcal{V} \to \{\text{HUB}, \text{BRIDGE}, \text{REGULAR}\}$ where HUB nodes exhibit high degree centrality, BRIDGE nodes demonstrate high betweenness centrality connecting communities, and REGULAR nodes represent ordinary members. Edge relations are characterized by $\psi : \mathcal{E} \to \{\text{STRONG}, \text{WEAK}\}$ based on Jaccard similarity:

$$J(v_i, v_j) = \frac{|\mathcal{N}^1(v_i) \cap \mathcal{N}^1(v_j)|}{|\mathcal{N}^1(v_i) \cup \mathcal{N}^1(v_j)|}, \tag{1}$$

measuring structural proximity through shared neighbors. The semantic grammar $\mathcal{G}_m$ enriches structural annotations with mutual information between textual attributes to capture semantic relatedness (Fig. 2).

4   Methodology

4.1   Structural Grammar Construction

Node and Edge Role Assignment. For each node v_i, we compute degree centrality $d_i = \sum_{j=1}^{|\mathcal{V}|} \mathbf{A}_{ij}$ and betweenness centrality. Nodes are assigned roles via $\phi : \mathcal{V} \to \{\text{HUB}, \text{BRIDGE}, \text{REGULAR}\}$ where HUB requires $d_i > \tau_d$ (80th

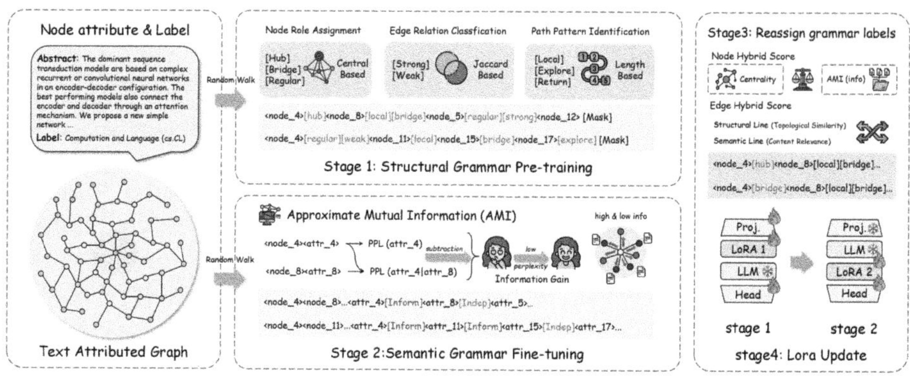

Fig. 2. The G^2rammar Framework for Bilingual Grammar Learning on Text-Attributed Graphs. Our framework provides language models with grammatical context by explicitly encoding structural and semantic grammar through two-stage learning: structural grammar pre-training followed by bilingual grammar fine-tuning.

percentile), BRIDGE requires betweenness $> \tau_b$ (75th percentile), and others are REGULAR. These thresholds identify structurally significant nodes while maintaining diversity.

Edges are classified via $\psi : \mathcal{E} \rightarrow \{\text{STRONG}, \text{WEAK}\}$ where STRONG is assigned when $J(v_i, v_j) > 0.5$. We efficiently compute Jaccard similarity using MinHash functions $h_r : \mathcal{V} \rightarrow [B]$ for $r = 1, \ldots, R$:

$$\Pr[h_r(\mathcal{N}^1(v_i)) = h_r(\mathcal{N}^1(v_j))] = J(v_i, v_j), \tag{2}$$

where $h_r(\mathcal{N}^1(v_i)) = \min_{u \in \mathcal{N}^1(v_i)} h_r(u)$. This provides linear-time approximation versus quadratic exhaustive comparison.

Path Pattern Identification. We define the path patterns via $\rho : \mathcal{C} \rightarrow \{\text{LOCAL}, \text{EXPLORE}, \text{RETURN}\}$. A path $p = \langle v_{i_1}, \ldots, v_{i_l} \rangle$ receives LOCAL when $l \leq 2$, EXPLORE when containing a BRIDGE node, and RETURN when start and end nodes are identical. These capture fundamental information diffusion modes: homophilous clustering, heterophilous bridging, and feedback mechanisms.

4.2 Pretraining with Structural Grammar

We extend the tokenizer vocabulary with eight grammar tokens: [HUB], [BRIDGE], [REGULAR], [STRONG], [WEAK], [LOCAL], [EXPLORE], [RETURN]. For each node v_i, we sample k random walks of length l following normalized adjacency:

$$P(v_j|v_i) = \frac{\mathbf{A}_{ij}}{\sum_{t \in \mathcal{N}^1(v_i)} \mathbf{A}_{it}} \tag{3}$$

Each path $p = \langle v_{i_1}, \ldots, v_{i_l} \rangle$ is transformed into a grammar-annotated sequence:

$$s = \langle v_{i_1}, \phi(v_{i_1}), \psi(e_{i_1 i_2}), v_{i_2}, \phi(v_{i_2}), \ldots, v_{i_l}, \phi(v_{i_l}) \rangle \oplus [\rho(p)], \tag{4}$$

where $e_{ij i_{j+1}}$ denotes edges between consecutive nodes.

We adopt LoRA with trainable matrices $\mathbf{B}_1 \in \mathbb{R}^{d \times r}$ and $\mathbf{A}_1 \in \mathbb{R}^{r \times d'}$ where $r \ll \min(d, d')$, forming $\Delta\mathbf{W}_1 = \mathbf{B}_1 \mathbf{A}_1$ added to frozen $\mathbf{W}_0$. The pretraining objective maximizes:

$$\mathcal{L}_{\text{pretrain}} = -\sum_{s \in \mathcal{C}} \sum_{t=1}^{|s|} \log P(s_t | s_{<t}; \mathbf{W}_0, \Delta\mathbf{W}_1) \tag{5}$$

A projection matrix $\mathbf{W}_p \in \mathbb{R}^{d \times d}$ maps node embeddings from textual attributes into the model's hidden space.

4.3   Semantic Grammar Enhancement

We quantify semantic relatedness through approximate mutual information from conditional perplexity. For nodes (v_i, v_j) with texts x_i, x_j:

$$\text{AMI}(v_i, v_j) = \log \text{PPL}(x_j) - \log \text{PPL}(x_j | x_i), \tag{6}$$

where $\text{PPL}(x_j) = \exp\left(-\frac{1}{|x_j|} \sum_{t=1}^{|x_j|} \log P(x_{j,t} | x_{j,<t})\right)$ is unconditional perplexity and $\text{PPL}(x_j | x_i)$ is conditional perplexity given x_i. High AMI indicates strong semantic informativeness.

4.4   Two-Stage Semantic Retrieval

To balance efficiency and precision, we employ two-stage retrieval. First, FAISS rapidly retrieves top K candidates (typically $K = 100$) based on cosine similarity of sentence embeddings. Second, we compute AMI only for these K candidates and select top K' nodes ($K' \ll K$) with highest mutual information. This reduces cost from $O(|\mathcal{V}|)$ to $O(K)$ language model evaluations per node.

We define semantic edge relations $\psi_{\text{sem}} : \mathcal{E} \cup \mathcal{E}_{\text{sem}} \to \{\text{INFORM}, \text{INDEP}\}$ where $\mathcal{E}_{\text{sem}}$ denotes semantic edges. INFORM is assigned when $\text{AMI}(v_i, v_j) > \tau_{\text{AMI}}$ (median AMI value), enabling distinction between cases where structural proximity coincides with or diverges from semantic similarity.

4.5   Fine-Tuning with Bilingual Grammar

Hybrid Grammar Score Computation. We integrate both grammars through hybrid scores. For node v_i:

$$\sigma_i = \alpha \cdot \frac{d_i}{\max_j d_j} + (1 - \alpha) \cdot \frac{\sum_{v_j \in \mathcal{V}} \text{AMI}(v_i, v_j)}{\max_k \sum_{v_j \in \mathcal{V}} \text{AMI}(v_k, v_j)}, \tag{7}$$

where $\alpha \in [0, 1]$ balances structural and semantic importance. Edge relations are fused as:

$$\omega_{ij} = \alpha \cdot J(v_i, v_j) + (1 - \alpha) \cdot \frac{\text{AMI}(v_i, v_j)}{\max_{(u,v) \in \mathcal{E}} \text{AMI}(u, v)} \tag{8}$$

Typically $\alpha = 0.5$ equally weights both dimensions.

Classification with Bilingual Annotations. For each target node, we sample k paths encoded as:

$$\tilde{s} = x_{i_1} \oplus \langle v_{i_1}, \phi_{\text{hybrid}}(v_{i_1}), \psi_{\text{hybrid}}(e_{i_1 i_2}), v_{i_2}, \phi_{\text{hybrid}}(v_{i_2}), \ldots \rangle \oplus [\rho(p)], \tag{9}$$

where ϕ_{hybrid} and ψ_{hybrid} use hybrid scores, merging semantic content with grammatical structure. We introduce a second LoRA module $\Delta\mathbf{W}_2 = \mathbf{B}_2\mathbf{A}_2$ yielding $\mathbf{W} = \mathbf{W}_0 + \Delta\mathbf{W}_1 + \Delta\mathbf{W}_2$ where $\Delta\mathbf{W}_1$ remains frozen.

The classification head uses $\hat{\mathbf{W}}_h \in \mathbb{R}^{d \times C}$ where C is the number of classes. For node v_i with hidden representation $\mathbf{t}_i \in \mathbb{R}^d$:

$$P(y_i = c | \tilde{s}) = \frac{\exp(\hat{\mathbf{W}}_{h,c}^\top \mathbf{t}_i + b_c)}{\sum_{c'=1}^{C} \exp(\hat{\mathbf{W}}_{h,c'}^\top \mathbf{t}_i + b_{c'})} \tag{10}$$

The fine-tuning objective minimizes cross-entropy:

$$\mathcal{L}_{\text{finetune}} = - \sum_{v_i \in \mathcal{V}_{\text{train}}} \sum_{c=1}^{C} y_i^c \log P(y_i = c | \tilde{s}_i), \tag{11}$$

where $y_i^c \in \{0, 1\}$ indicates true class membership. This bilingual integration leverages structural patterns from pretraining and semantic relationships from fine-tuning for comprehensive graph understanding.

5   Experiments

5.1   Datasets

We evaluate G^{2}rammar on three widely-used text-attributed graph datasets spanning different domains. ACM [28] is an academic citation network containing 3,025 papers categorized into three research areas, where nodes represent papers and edges denote citation relationships. Wiki [27] is a knowledge graph extracted from Wikipedia with 2,405 documents classified into 17 categories, where nodes correspond to documents and edges represent hyperlink connections. Amazon [23] is a product co-purchasing network comprising 13,752 items across 10 product categories, where nodes represent products and edges indicate frequent co-purchase patterns. Each dataset provides rich textual attributes including titles and abstracts for ACM, article content for Wiki, and product descriptions for Amazon. These datasets exhibit diverse structural properties with ACM showing strong citation clustering, Wiki demonstrating semantic connectivity, and Amazon reflecting collaborative filtering patterns.

Table 1. Node classification performance comparison among baselines w.r.t. micro classification accuracy across three datasets.

NLP Models	GNNs	ACM		Wiki		Amazon	
		Val.	Test	Val.	Test	Val.	Test
Fine-tuned LMs + GNNs							
Bert	-	74.4	73.2	69.5	68.8	86.2	87.0
	GCN	77.6	77.1	69.4	68.4	92.3	92.8
	GAT	77.9	78.0	70.5	69.8	92.5	92.4
	GraphSAGE	77.3	76.8	73.1	72.7	92.0	92.3
Roberta	-	78.1	76.6	67.8	68.1	84.9	85.9
	GCN	80.1	79.4	68.5	68.0	92.3	92.5
	GAT	79.7	78.9	70.1	71.0	92.5	92.4
	GraphSAGE	78.5	78.3	72.7	72.1	92.2	92.1
Fine-tuned Large Language Models +/- GNNs							
GraphAdapter	-	80.8	80.4	71.9	71.7	94.1	93.4
Llama3-8b	-	80.7	80.6	71.9	71.2	92.0	91.6
Llama3-8b	GraphSAGE	<u>82.0</u>	81.3	72.8	<u>73.4</u>	93.1	92.8
Qwen2-7B	-	81.2	80.9	72.3	71.8	92.6	92.1
Qwen2-7B	GraphSAGE	81.8	<u>81.5</u>	<u>73.1</u>	72.6	93.5	93.2
Specialized Frameworks for Text-Attributed Graphs							
GLEM		81.4	79.8	72.6	71.2	92.5	93.3
GraphFormers		75.3	75.1	66.8	67.5	85.6	86.4
LLAGA		77.2	77.5	71.7	72.0	90.1	90.8
InstructGLM		75.4	74.5	72.2	70.6	94.3	94.2
TEA-GLM		76.8	75.9	71.5	71.8	93.2	93.5
GraphAny		78.3	77.2	70.9	70.3	92.8	93.1
GraphCLIP		79.5	78.6	72.8	72.4	93.7	93.9
GraphInsight		77.9	76.8	71.3	70.9	92.4	92.7
UniGLM		78.7	77.9	72.1	71.6	93.5	93.8
GDL4LLM		81.9	81.4	<u>74.3</u>	73.2	<u>94.6</u>	<u>94.6</u>
G^{2}rammar		**83.6**	**83.7**	**78.5**	**76.1**	**95.5**	**96.1**

5.2   Baselines

We compare G^{2}rammar against three categories of competitive baselines. First, we evaluate fine-tuned language models combined with GNNs, including BERT [8] and RoBERTa [22] paired with GCN [20], GAT [30], and Graph-

Table 2. Comparison w.r.t. the number of used tokens and the order of graph structure modeled (Token/(order)).

LLMs-	ACM		Wiki		Amazon	
	Val.	Test	Val.	Test	Val.	Test
InstructGLM	146(1)	149(1)	1024(2)	319(1)	532(2)	538(2)
LLAGA-HO	155(4)	155(4)	130(4)	130(4)	140(4)	140(4)
GDL4LLM	54(4)	54(4)	72(4)	72(4)	72(4)	72(4)
G^{2}rammar	**48(4)**	**51(4)**	**65(4)**	**68(4)**	**75(4)**	**69(4)**
Reduction (%)	**67.12**	**65.77**	**50.00**	**47.69**	**46.43**	**50.71**

Table 3. Training and test time comparison across LLMs for three datasets. Times are presented in 'hh:mm:ss' format.

LLMs-	ACM		Wiki		Amazon	
	Train	Test	Train	Test	Train	Test
InstructGLM	5:59:04	0:06:41	4:30:16	0:05:01	6:07:02	0:07:08
LLAGA-HO	0:47:26	0:04:41	0:35:18	0:01:57	0:49:29	0:04:04
GDL4LLM	1:05:23	0:02:06	0:34:03	0:01:17	**0:32:15**	0:01:45
G^{2}rammar	**0:52:17**	**0:01:58**	**0:31:45**	**0:01:12**	0:35:42	**0:01:38**

SAGE [14]. Second, we test large language models with optional GNN integration, specifically Llama3-8B [29] and Qwen2-7B [2] both standalone and combined with GraphSAGE, as well as GraphAdapter [21] which adapts LLMs for graph structures. Third, we include specialized frameworks designed for text-attributed graphs: GLEM [36] which combine language models with graph learning, GraphFormers [33] using transformer architectures, LLAGA [5] and Instruct-GLM [34] employing instruction tuning, TEA-GLM [31] and GraphAny [37] focusing on text-enhanced aggregation, GraphCLIP [39] using contrastive learning, GraphInsight [4] with interpretable representations, UniGLM [9] providing unified graph-language modeling, and GDL4LLM [18] treating graphs as a new language. This comprehensive comparison validates G^{2}rammar's effectiveness against state-of-the-art methods.

5.3 Metrics

Following standard protocols, we adopt micro-averaged classification accuracy as the primary metric, which treats each node prediction equally for balanced assessment across class sizes. We report both validation and test accuracy to demonstrate generalization. Computational efficiency is measured through token count combined with neighborhood order (Token/(order)) and actual training/inference time. These metrics collectively evaluate predictive performance and resource utilization for comprehensive comparison.

5.4   Implementation Details

We implement G^{2}rammar using PyTorch with Qwen2-72B as backbone. For structural grammar, we set $k = 8$ walks per node with length $l = 4$. Node role thresholds use 80th percentile for HUB and 75th for BRIDGE based on centrality distributions. Edge classification employs Jaccard threshold 0.5 with MinHash using $R = 128$ hash functions. For semantic grammar, we retrieve top $K = 100$ candidates and select $K' = 10$ neighbors via AMI scores. Hybrid fusion weight $\alpha = 0.5$ balances structural and semantic contributions. We employ LoRA with rank $r = 8$, AdamW optimizer with learning rate 5×10^{-4}, and batch size 16. Pretraining runs 3 epochs followed by 5 fine-tuning epochs. Experiments use 8 NVIDIA A100 80G GPUs with mixed precision training.

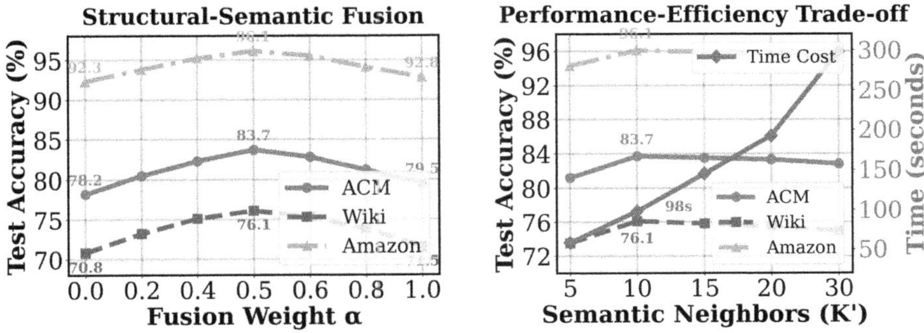

Fig. 3. Left: Impact of fusion weight α on node classification accuracy across three datasets. Right: Performance-efficiency trade-off with semantic neighbor count K'.

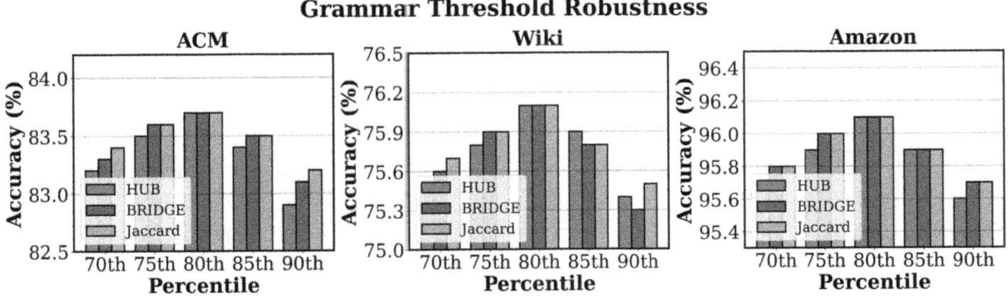

Fig. 4. Sensitivity analysis of grammar threshold configurations across three datasets. We evaluate HUB centrality threshold, BRIDGE betweenness threshold, and Jaccard similarity threshold. Consistent performance across different threshold settings (standard deviation $< 0.4\%$ within each group) demonstrates G^{2}rammar's robustness to hyperparameter choices.

5.5 Main Result

Table 1 demonstrates that G²rammar consistently achieves the best performance across all three datasets. On ACM, our model reaches 83.7% test accuracy, surpassing the previous best method GDL4LLM at 81.4% by 2.3%. On Wiki, we obtain 76.1% accuracy with 2.9% improvement over GDL4LLM at 73.2%. On Amazon, G²rammar achieves 96.1% accuracy, outperforming GDL4LLM at 94.6% by 1.5%. These results validate our core hypothesis that providing language models with explicit grammatical annotations for structural and semantic roles enables better graph understanding than treating graphs as vocabulary alone. The bilingual grammar framework successfully addresses the critical gap identified in existing graph language methods.

5.6 Model Analysis

Table 2 and Table 3 demonstrate G²rammar's efficiency advantage through concise grammatical annotations. While maintaining 4-hop neighborhood modeling, our approach reduces token usage by 65.77% on ACM with only 48 tokens compared to InstructGLM's 146 tokens, by 47.69% on Wiki, and by 50.71% on Amazon. This compression comes from replacing verbose path descriptions with compact grammar tokens such as [HUB], [BRIDGE], and [STRONG]. G²rammar requires only 52 min training on ACM versus GDL4LLM's 65 min, with even larger speedups over InstructGLM at 6 h. The grammar-based representation enables high-order structural modeling without the computational burden of exhaustive path enumeration, making our framework practical for large-scale graph applications.

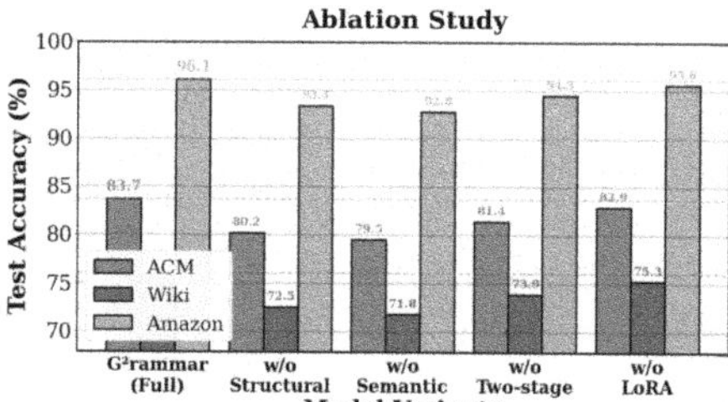
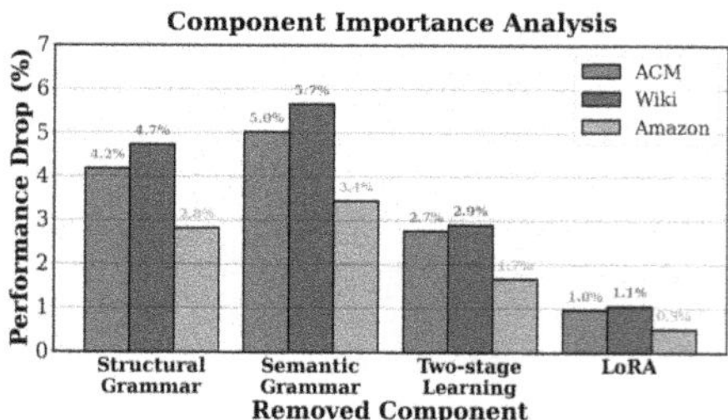

Fig. 5. Left: Ablation study comparing G²rammar variants across three datasets. **Right**: Component importance ranking. Semantic grammar is most critical (5% drop when removed).Semantic grammar is most critical (5% drop when removed).

5.7 Hyper-Parameter Analysis

Figure 3 and Fig. 4 examine G²rammar's sensitivity to key hyperparameters. The left panel of Fig. 3 shows that fusion weight α, which balances structural and semantic grammar, performs stably across 0.2–0.8, with optimal results at $\alpha = 0.5$ for all datasets. This indicates both grammar types contribute meaningfully and equally. The right panel reveals a performance-efficiency trade-off for semantic neighbors K' where accuracy peaks around $K' = 10$ while inference time grows linearly, suggesting this value balances quality and speed. Figure 4 demonstrates robustness to grammar threshold choices including HUB centrality, BRIDGE betweenness, and Jaccard similarity thresholds, with standard deviation under 0.4% across 70th-90th percentiles. This stability stems from our two-stage learning design, which separates grammar extraction from classification and allows the model to adapt to different threshold configurations during training.

5.8 Ablation Study

Figure 5 validates the necessity of each component in G^{2}rammar through systematic ablation. The left panel shows that removing the entire grammar module causes the largest performance drop of 11.8% on ACM, 2.3% on Wiki, and 6.0% on Amazon, confirming that grammatical annotations are the core innovation. The right panel quantifies individual component contributions through performance degradation when each is removed. Semantic grammar proves most critical with 5.0% average drop, followed by structural grammar at 4.3%, two-stage learning at 2.5%, and LoRA at 1.0%. Removing structural information alone decreases accuracy by 3.0% on ACM, demonstrating that graph topology complements textual semantics rather than competing with it. The two-stage learning strategy consistently brings 2–3% gains by enabling better grammar extraction before fine-tuning. These results confirm that G^{2}rammar's design choices work synergistically, with both structural and semantic grammar essential for comprehensive graph understanding.

6 Conclusion

We introduced G^{2}rammar, a bilingual grammar framework that bridges language models and text-attributed graphs by explicitly encoding structural and semantic grammar through node roles, edge relations, and path patterns. A two-stage approach combines structural pre-training with semantic fine-tuning for comprehensive graph understanding. Extensive experiments demonstrate consistent improvements over baselines with superior efficiency, opening new directions for graph-language integration by treating grammatical structures as first-class components.

Acknowledgments. This work was supported by the Natural Science Foundation of Guangdong Province, China. "Research on Key Theories and Technologies for Nano-learning"

References

1. Achiam, J., et al.: Gpt-4 technical report. arXiv preprint arXiv:2303.08774 (2023)
2. Bai, J., et al.: Qwen technical report. arXiv preprint arXiv:2309.16609 (2023)
3. Brown, T., et al.: Language models are few-shot learners. Adv. Neural. Inf. Process. Syst. **33**, 1877–1901 (2020)
4. Cao, Y., Han, S., Gao, Z., Ding, Z., Xie, X., Zhou, S.K.: Graphinsight: unlocking insights in large language models for graph structure understanding. arXiv preprint arXiv:2409.03258 (2024)
5. Chen, R., Huang, T., Zhao, S., Liu, S., Yu, H., Ma, Y.: Llaga: large language and graph assistant. arXiv preprint arXiv:2402.08170 (2024)
6. Chien, E., et al.: Node feature extraction by self-supervised multi-scale neighborhood prediction. arXiv preprint arXiv:2111.00064 (2021)
7. Cohan, A., Feldman, S., Beltagy, I., Downey, D., Weld, D.: Specter: document-level representation learning using citation-informed transformers. In: Proceedings of the 58th Annual Meeting of the Association for Computational Linguistics, pp. 2270–2282 (2020)
8. Devlin, J., Chang, M.W., Lee, K., Toutanova, K.: Bert: Pre-training of deep bidirectional transformers for language understanding. In: Proceedings of NAACL-HLT, pp. 4171–4186 (2019)
9. Fang, Y., Fan, D., Ding, S., Liu, N., Tan, Q.: Uniglm: training one unified language model for text-attributed graph embedding. arXiv preprint arXiv:2406.12052 (2024)
10. Fatemi, B., Halcrow, J., Perozzi, B.: Talk like a graph: encoding graphs for large language models. arXiv preprint arXiv:2310.04560 (2023)
11. Gilmer, J., Schoenholz, S.S., Riley, P.F., Vinyals, O., Dahl, G.E.: Neural message passing for quantum chemistry. In: International conference on machine learning. pp. 1263–1272 (2017)
12. Grover, A., Leskovec, J.: node2vec: Scalable feature learning for networks. In: Proceedings of the 22nd ACM SIGKDD International Conference on Knowledge Discovery and Data Mining, pp. 855–864 (2016)
13. Guo, J., Du, L., Liu, H., Zhou, M., He, X., Han, S.: Gpt4graph: Can large language models understand graph structured data? an empirical evaluation and benchmarking (2023)
14. Hamilton, W., Ying, Z., Leskovec, J.: Inductive representation learning on large graphs. Adv. Neural Inf. Process. Syst. 1024–1034 (2017)
15. Hamilton, W.L.: Graph representation learning. Synth. Lect. Artif. Intell. Mach. Learn. **14**(3), 1–159 (2020)
16. Hu, W., et al.: Strategies for pre-training graph neural networks. arXiv preprint arXiv:1905.12265 (2020)
17. Hu, Z., Dong, Y., Wang, K., Chang, K.W., Sun, Y.: Gpt-gnn: Generative pre-training of graph neural networks. In: Proceedings of the 26th ACM SIGKDD International Conference on Knowledge Discovery & Data Mining, pp. 1857–1867 (2020)
18. Huang, J., Zhang, X., Mei, Q., Ma, J.: Can LLMs effectively leverage graph structural information through prompts, and why? arXiv preprint arXiv:2309.16595 (2024)
19. Jin, B., Liu, G., Han, C., Jiang, M., Ji, H., Han, J.: Large language models on graphs: a comprehensive survey. arXiv preprint arXiv:2312.02783 (2023)

20. Kipf, T.N., Welling, M.: Semi-supervised classification with graph convolutional networks. In: International Conference on Learning Representations (2017)
21. Li, X., Lian, D., Lu, Z., Bai, J., Chen, Z., Wang, X.: Graphadapter: tuning vision-language models with dual knowledge graph. arXiv preprint arXiv:2309.13625 (2023)
22. Liu, Y., et al.: Roberta: A robustly optimized bert pretraining approach. arXiv preprint arXiv:1907.11692 (2019)
23. McAuley, J., Targett, C., Shi, Q., Van Den Hengel, A.: Image-based recommendations on styles and substitutes. In: Proceedings of the 38th International ACM SIGIR Conference on Research and Development in Information Retrieval, pp. 43–52 (2015)
24. Morris, C., Ritzert, M., Fey, M., Hamilton, W.L., Lenssen, J.E., Rattan, G., Grohe, M.: Weisfeiler and leman go neural: higher-order graph neural networks **33**, 4602–4609 (2019)
25. Perozzi, B., Al-Rfou, R., Skiena, S.: Deepwalk: Online learning of social representations. In: Proceedings of the 20th ACM SIGKDD International Conference on Knowledge Discovery and Data Mining, pp. 701–710 (2014)
26. Scarselli, F., Gori, M., Tsoi, A.C., Hagenbuchner, M., Monfardini, G.: The graph neural network model. IEEE Trans. Neural Netw. **20**(1), 61–80 (2008)
27. Sen, P., Namata, G., Bilgic, M., Getoor, L., Galligher, B., Eliassi-Rad, T.: Collective classification in network data. In: AI magazine, vol. 29, pp. 93–93 (2008)
28. Tang, J., Zhang, J., Yao, L., Li, J., Zhang, L., Su, Z.: Arnetminer: extraction and mining of academic social networks. In: Proceedings of the 14th ACM SIGKDD International Conference on Knowledge Discovery and Data Mining, pp. 990–998 (2008)
29. Touvron, H., et al.: Llama 2: open foundation and fine-tuned chat models. arXiv preprint arXiv:2307.09288 (2023)
30. Veličković, P., Cucurull, G., Casanova, A., Romero, A., Lio, P., Bengio, Y.: Graph attention networks. In: International Conference on Learning Representations (2018)
31. Wang, D., Zuo, Y., Li, F., Wu, J.: Llms as zero-shot graph learners: alignment of GNN representations with llm token embeddings. arXiv preprint arXiv:2408.14512 (2024)
32. Xu, K., Hu, W., Leskovec, J., Jegelka, S.: How powerful are graph neural networks? In: International Conference on Learning Representations (2019)
33. Yang, J., Liu, Z., Xiao, S., Li, C., Lian, D., Agrawal, S., Singh, A., Sun, G., Xie, X.: Graphformers: GNN-nested transformers for representation learning on textual graph. Adv. Neural. Inf. Process. Syst. **35**, 28798–28810 (2022)
34. Ye, R., Zhang, C., Wang, R., Xu, S., Zhang, Y.: Language is all a graph needs. arXiv preprint arXiv:2308.07134 (2024)
35. You, Y., Chen, T., Sui, Y., Chen, T., Wang, Z., Shen, Y.: Graph contrastive learning with augmentations. Adv. Neural Inf. Process. Syst. **33**, 5812–5823 (2020)
36. Zhao, J., et al.: Learning on large-scale text-attributed graphs via variational inference. In: International Conference on Learning Representations (2023)
37. Zhao, J., Zhu, Z., Galkin, M., Mostafa, H., Bronstein, M., Tang, J.: Graphany: a foundation model for node classification on any graph. arXiv preprint arXiv:2405.20445 (2024)
38. Zhu, Y., Xu, Y., Liu, Q., Wu, S.: An empirical study of graph contrastive learning. In: NeurIPS 2021 Workshop on Self-Supervised Learning-Theory and Practice (2021)

39. Zhu, Y., et al.: Graphclip: Enhancing transferability in graph foundation models for text-attributed graphs. arXiv preprint arXiv:2410.10329 (2024)
40. Zhu, Z., Zhang, Z., Xhonneux, L.P., Tang, J.: Neural bellman-ford networks: a general graph neural network framework for link prediction. Adv. Neural Inf. Process. Syst. **34**, 29476–29490 (2021)

Unleashing the Power of Pre-trained Graph Models in Federated Graph Learning

Huabin Sun[1], Bo Yan[1], Yaoqi Liu[1], Shaohua Fan[2], Yang Cao[3], and Chuan Shi[1]([⊠])

[1] Beijing University of Posts and Telecommunications, Beijing, China
`{sunhuabin,boyan,yaoqiliu,shichuan}@bupt.edu.cn`
[2] Inner Mongolia University, Hohhot, China
`shfan@imu.edu.cn`
[3] Institute of Science Tokyo, Tokyo, Japan
`cao@c.titech.ac.jp`

Abstract. Federated graph learning (FGL) enables collaborative model training without sharing local data, but suffers from severe client heterogeneity, such as shifts in label distributions and graph sizes. Pre-trained graph models (PGMs), learned from large-scale graph data, encode general structural and semantic priors, offering a promising solution to alleviate heterogeneity in FGL. To this end, we first explores the potential of PGMs for FGL. We empirically reveal that directly fine-tuning PGMs in federated settings often results in degraded performance due to misalignment between the global prior and heterogeneous local distributions. To address this misalignment, we introduce PeFGL, a PGM-enhanced FGL framework that effectively adapts pre-trained priors to heterogeneous local graphs. PeFGL is built on two key components. First, we employ conditional generative diffusion models to augment local graphs, enriching structural diversity and compensating for missing patterns that hinder PGM adaptation. We further design a pre-trained-knowledge-guided filtering mechanism to select generated samples that align with both the PGM prior and local distribution. Then, based on the augmented and filtered data, we extract invariant subgraphs that can be shared across clients to guide PGM fine-tuning, which suppresses noisy substructures while reducing cross-client heterogeneity. Extensive experiments on four real-world datasets demonstrate that our framework consistently achieves state-of-the-art performance.

Keywords: Federated graph learning · Pre-trained graph models · Heterogeneous graph data

1 Introduction

Graph Neural Networks (GNNs) [5,10,20] have become a powerful framework to learn representations from structured graph data. Through neighborhood aggre-

H. Sun and B. Yan—Equal contribution.

gation, they have demonstrated outstanding performance in diverse applications such as node and graph classification. Traditional GNNs hold a basic assumption that the graph data is centrally stored. In practice, however, the graph data is owned by different parties, and due to privacy concerns, it is hard to collect data centrally, which makes the centralized training of GNNs infeasible.

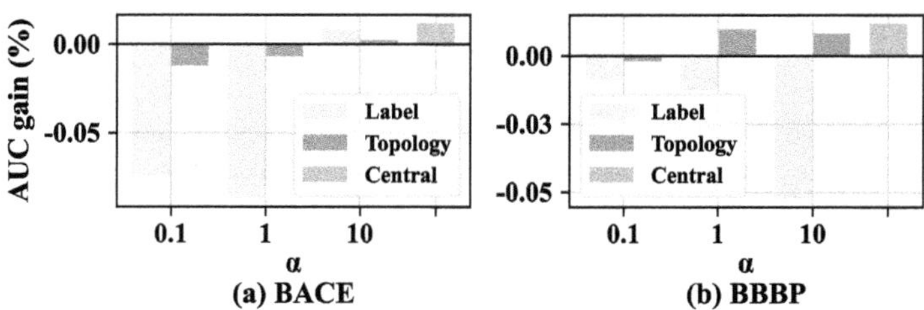

Fig. 1. The performance gains by fine-tuning PGMs under different types (label, topology) and degrees (0.1, 1, 10) of heterogeneity.

Federated learning (FL) [14,27] provides a paradigm for collaboratively training models by multiple parties without sharing raw data. To enable graph data to enjoy the benefits, researchers have developed federated graph learning (FGL) [22,28,32], where distributed graph owners jointly train GNNs under privacy constraints. A major challenge of FGL is data heterogeneity, i.e., the graph data of different clients is not independent and identically distributed (non-IID), which dramatically undermines the performance of FGL. Existing studies addressing heterogeneity can be divided into knowledge sharing and data augmentation. Knowledge sharing methods aim to share common knowledge (e.g., structural and spectral knowledge) across clients and leave personalized knowledge locally [18,19,23]. In contrast, data augmentation methods aim to augment local graphs to align the data distribution across clients [3,33]. However, these approaches largely operate within client-side data, where shared signals can be weak, and augmentation alignment is hard to control. This leaves open how to acquire stronger transferable knowledge to narrow client gaps.

Recently, public pre-trained knowledge, which captures transferable knowledge from large-scale datasets, has shown great potential in mitigating heterogeneity of FL in computer vision (CV) [15,31] and natural language processing (NLP) [1,13] domains. Typically, it has been shown that simply fine-tuning pre-trained models can yield significant improvements under heterogeneous settings [9,15]. On the other hand, pre-trained graph models (PGMs), such as GraphCL [29] and GCC [16], are trained on large-scale graph datasets to encode transferable structural and semantic priors, significantly boosting the performance of various downstream tasks. This naturally raises a question: *Can we directly fine-tune PGMs to mitigate the heterogeneity in FGL?* Answering this question helps to clarify whether PGMs' transferable priors can be easily adapted to FGL, or whether new designs are required to unleash their potentials.

Therefore, we conduct empirical studies to show the performance gains after fine-tuning PGMs under various degrees of heterogeneity, controlled by the Dirichlet parameter α. We adopt the typical PGM provided by [7] and consider two major types of graph heterogeneity (i.e., label and topology heterogeneity). The results on BACE and BBBP datasets are shown in Fig. 1. We observe that PGMs improve the model performance in the centralized setting but suffers from negative transfer in the federated setting, which deviates from common observations in CV and NLP domains. This raises another question: *How to unleash the power of PGMs to tackle the heterogeneity issue in FGL?* Answering this question is non-trivial, since the complex structural priors captured by PGMs often conflict with heterogeneous local topologies and label distributions, making effective adaptation particularly challenging in federated settings.

In this paper, we propose a novel framework named PeFGL to effectively adapt PGMs to tackle the heterogeneity issue of FGL and further enhance the performance of FGL. Specifically, to mitigate the mismatch between the pre-trained knowledge and the heterogeneous local graph distributions, we employ a conditional generative diffusion model to augment local graphs, enriching their structural diversity and compensating for missing local patterns that impede PGM adaptation. We further introduce a pre-trained-knowledge-guided filtering mechanism that selectively retains generated samples consistent with both the PGM prior and the client-specific distribution. This mechanism measures the representation similarity between the original and generated samples via the PGM and preserves the top-ranked ones whose labels are consistently predicted by the PGM. Based on these augmented local graphs, we extract invariant subgraphs through local gradient regularization, which captures shared graph semantics across clients. These invariant structures provide reliable guidance for fine-tuning PGMs, enabling robust cross-client knowledge alignment while mitigating the adverse impact of noisy or domain-specific substructures.

The contribution of this paper is summarized as follows:

- To the best of our knowledge, this is the first work to explore the potential of PGMs for FGL, enabling more consistent and transferable graph representation learning across heterogeneous clients.
- We propose PeFGL, a novel framework that adapts pre-trained graph priors to heterogeneous local graphs. PeFGL first augments minority local classes via conditional diffusion and then aligns client updates by extracting invariant subgraphs shared across clients.
- Experiments on four datasets demonstrate that PeFGL outperforms existing FGL methods, achieving up to 6.69% improvements compared to baselines.

2   Related Work

2.1   Federated Graph Learning

Briefly speaking, FGL can be divided into inter-graph and intra-graph FGL [30]. Intra-graph FGL assumes that each client owns a part of the entire graph while inter-graph FGL assumes that multiple entire graphs exist in each client [3,23,25,26]. This work considers the inter-graph FGL scenario. To tackle the heterogeneity issue in inter-graph FGL, GCFL [23] designs a sequence-based gradient clustering method to identify homogeneous clients and aggregate gradients from the same clusters. FedStar [18] introduces a decoupled GNN that only shares the parameters which capture graph structure patterns. FedSSP [19] shares generic spectral knowledge among clients. Other studies seek to generate common graphs [3,19]. FedVN [3] introduces virtual nodes locally, and FedGOG [33] further explores to generate out-of-distributions (OOD) samples to achieve better generalization. Despite the success, none of the above methods consider leveraging external knowledge (e.g., PGMs) to enhance the inter-graph FGL.

2.2   Graph Pre-training

The key insight of graph pre-training is to learn transferable knowledge from large unlabeled graphs, enabling efficient adaptation to downstream tasks via fine-tuning. [2]. Typical graph pre-training methods can be divided into contrastive-based and predictive-based. The contrastive-based methods aim to maximize the mutual information (MI) between two augmented graph views, thereby learning discriminative graph features [16,17,21,29]. For example, DGI [21] learns the node embeddings by maximizing the MI between global and local graph representations. GraphCL [29] designs different graph views by perturbing the graph information. Different from contrast-based methods, predictive-based methods pre-train the model by predicting the graph properties [6–8]. Among them, AttrMasking and ContextPred [7] predict masked node/edge attributes and subgraph-based contexts respectively. GPT-GNN [8] further generates the whole node attributes based on the graph structure. Nevertheless, when adapting these PGMs to downstream tasks, all these efforts assume that the downstream data is centrally stored, without considering the real-world privacy requirements.

3   Preliminary

In this section, we first present some basic definitions and concepts related to our work, including pre-trained graph models and federated graph learning, then we formalize our problem.

Let $G = (X, A)$ denote a graph G with node features $X \in \mathbb{R}^{n \times d}$ and adjacency matrix A. We define the pre-trained graph models as follows.

Definition 1. *Pre-trained Graph Model (PGM).* *A PGM refers to a graph neural network f_ϕ that is first trained on a large-scale source graph dataset $\mathcal{D}_{pre}$ using supervised or self-supervised objectives to learn generalizable graph representations. Formally, the pre-training process can be expressed as:*

$$\phi^* = \arg \min_\phi \mathcal{L}_{pre}(f_\phi, \mathcal{D}_{pre}), \tag{1}$$

where $\mathcal{L}_{pre}$ denotes the pre-training loss (e.g., contrastive, predictive, or generative). The obtained parameters ϕ^* encode transferable structural and semantic priors that can be fine-tuned on downstream graph tasks.

Our graph model is composed of a pre-trained graph encoder $f_\phi : (X, A) \rightarrow \mathbb{R}^{d_z}$ and a classifier head $c_\psi : \mathbb{R}^{d_z} \rightarrow \mathbb{R}^C$, denoting as:

$$F_\theta = c_\psi \circ f_\phi, \tag{2}$$

where $\theta = (\phi, \psi)$ is the total parameters. Given a graph G, the graph model first encodes G into a vector $z = f_\phi(G)$ and then obtains the class probabilities $p_\psi(y \mid G) = \mathrm{softmax}(c_\psi(z))$. In practice, ϕ or ψ can be fine-tuned with downstream graph data to adapt to specific tasks and enhance performance.

Definition 2. *Federated Graph Learning (FGL).* FGL *aims to collaboratively train a global graph model θ across multiple clients $\mathcal{C} = \{1, 2, \ldots, K\}$ without sharing raw data. The overall objective can be formalized as:*

$$\min_\theta \sum_{k=1}^{K} w_k \frac{1}{|\mathcal{D}_k|} \sum_{(G,y)\in\mathcal{D}_k} \ell(F_\theta(G), y), \quad s.t. \sum_{k=1}^{K} w_k = 1. \tag{3}$$

Specifically, at each communication round t, each client k locally optimizes its model parameters θ_k based on its private graph dataset $\mathcal{D}_k = \{(G_k^i, y_k^i)\}_{i=1}^{n_k}$. The global model is then updated by aggregating the local parameters:

$$\theta^{t+1} = \mathrm{Agg}(\{\theta_k^t\}_{k=1}^{K}). \tag{4}$$

A typical aggregation function $\mathrm{Agg}(\cdot)$ is Fedavg [14], which is defined as follows:

$$\theta^{t+1} = \sum_{k=1}^{K} \frac{n_k}{\sum_{j=1}^{K} n_j} \theta_k^t, \tag{5}$$

where n_k denotes the number of samples in client k.

Based on the above preliminaries, we define our problem as follows:

Definition 3. *PGM-enhanced FGL.* Given *a PGM with parameters θ^* stored in the server, and a set of clients $\mathcal{C} = \{1, 2, \ldots, K\}$ with private dataset $\mathcal{D}_k$ on client k, the objective of PGM-enhanced FGL is:*

$$\min_\theta \sum_{k=1}^{K} w_k \frac{1}{|\mathcal{D}_k|} \sum_{(G,y)\in\mathcal{D}_k} \ell(F_\theta(G), y) + \Omega(\theta, \theta^*), \quad s.t. \sum_{k=1}^{K} w_k = 1, \tag{6}$$

where ℓ is the loss function and $\Omega(\theta, \theta^*)$ is a regularization that aligns the global model θ with the pre-trained prior θ^*.

$\Omega(\theta, \theta^*)$ enables the model to retain global transferable knowledge while adapting to heterogeneous client distributions. A naive solution is directly setting $\theta^{(0)} = \theta^*$ and performing federated fine-tuning. However, it leads to optimization conflicts and even negative transfer, as shown before. Thus, the core challenge of PGM-enhanced FGL lies in mitigating the misalignment between $p(\theta^*)$ and the heterogeneous $\{p(\theta_k)\}_{k=1}^{K}$.

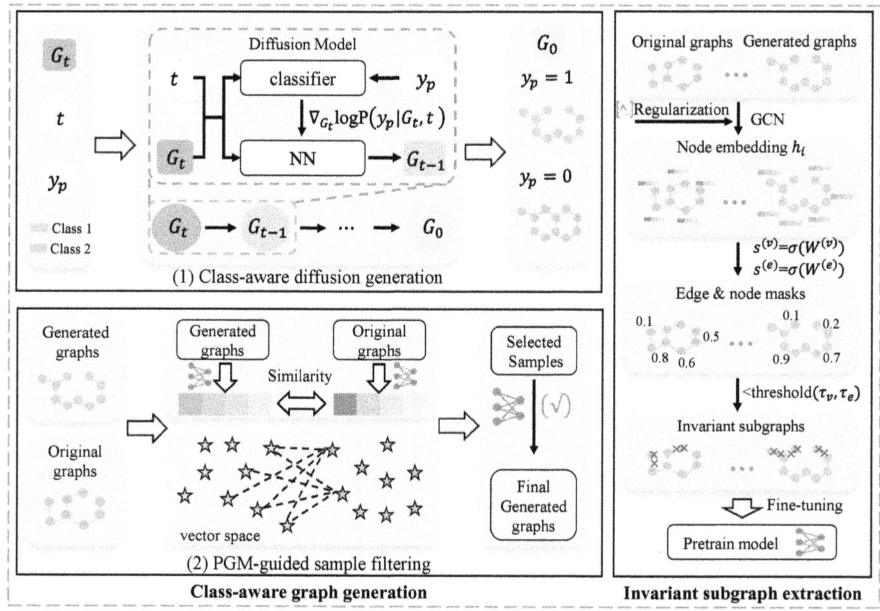

Fig. 2. The overall framework of PeFGL.

4   Methodology

In this section, we first present an overview of the proposed PeFGL, then detail the two key components of PeFGL, including class-aware graph generation and invariant subgraph extraction.

4.1   Overall Framework

The overall framework of PeFGL is shown in Fig. 2. PeFGL consists of two stages, graph generation and subgraph extraction. In the graph generation stage, all the clients first collaboratively train a class-aware diffusion model and then each client generates their minority-class graphs locally, therefore enhancing the data diversity and mitigating heterogeneity. To align the generated graph distribution with the local distribution and prior knowledge, a PGM-guided filter mechanism is further applied. First, the generated graphs are fed into the PGM encoder to obtain their representations, and the top-k samples with the closest distances to the client's original graphs are selected as candidates. Among these candidates, only those whose predicted labels by the current PGM classifier match the true label are retained. In the subgraph extraction stage, based on the original and generated data, a collaboratively trained graph masker automatically extracts subgraphs that are invariant across clients and consistent with the prior knowledge. To achieve this, a global gradient regularizer is applied to the masker, aligning its local updates with global gradients. Finally, the extracted subgraphs are used to fine-tune PGM, thus mitigating negative transfer.

4.2 Class-Aware Graph Generation

As discussed, naive fine-tuning of the PGM will harm performance due to the mismatch between PGM's prior and the heterogeneous local data distributions. This mismatch stems from their disparity: the PGM, trained on large-scale graph datasets, provides a general unbiased knowledge, whereas client data exhibits distributional bias due to class imbalance. To alleviate the resulting bias and mismatch, we adopt a generate-and-select strategy: we first use class-aware diffusion to generate minority-class samples; subsequently, recognizing that synthetic graphs may be noisy or inconsistent with the prior, we introduce a PGM-guided filter that retains only samples consistent with both the frozen prior and the local data statistics.

Class-Aware Diffusion Generation. Diffusion-based graph generation has two key phases. In the forward phase, a clean graph $G = (X, A)$ is gradually perturbed by a time-dependent noise kernel q_t. This produces noisy graphs $G_t = (X_t, A_t)$ at different times t, where larger t means more noise has been added to the graph. In the reverse phase, a score-based diffusion model starts from noise and iteratively denoises G_t. It learns a time-dependent score function $s_\theta(G_t) \approx \nabla_{G_t} \log p_t(G_t)$ that approximates the score of the perturbed graph distribution p_t at diffusion time t and provides update directions that move G_t toward higher-density regions. This reverse process defines an unconditional generative model that does not use label information.

We next extend this framework to class-conditional graph generation. Given a target class y_i, we construct a class-conditional score function that guides the reverse process according to both the data distribution and the label information. At diffusion time t, we define the class-guided score as

$$\tilde{s}_\theta(G_t; y_i) = s_\theta(G_t) + \Lambda \nabla_{G_t} \log p(y_i \mid G_t, t),$$
$$\Lambda = \mathrm{diag}(\lambda_X I_X, \lambda_A I_A). \tag{7}$$

where $\nabla_{G_t} \log p(y_i \mid G_t, t)$ is the gradient of the class-posterior, and $\lambda_X, \lambda_A \geq 0$ control the guidance strength on node and edge components, respectively. This construction is consistent with Bayes' rule, since the score of the class-conditional distribution can be decomposed into a marginal term and a class-posterior term.

To instantiate the class-posterior term in Eq. (7), we require estimates of $\nabla_{G_t} \log p(y_i \mid G_t, t)$ across different noise levels t. We obtain these estimates by federatively training a noise-aware classifier $f_{noi}(G_t, t)$. On client k, the classifier is trained on the local dataset $\mathcal{D}_k$. For each sample $(G, y) \in \mathcal{D}_k$, we first sample a time t, apply the same forward kernel $q_t(\cdot \mid G)$ as in the diffusion process to obtain a noisy graph G_t, and then compute loss on G_t. The classification objective on client k is defined as

$$\mathcal{L}_k^{\mathrm{clf}}(f) = \ell_{\mathrm{CE}}(f_{noi}(G_t, t), y), \tag{8}$$

where ℓ_{CE} denotes the cross-entropy loss between the softmax of $f_{\mathrm{noise}}(G_t, t)$ and the label y. During federated training, each client uploads only classifier parameters for weighted aggregation on the server, and the raw graphs remain on local devices.

After obtaining $\tilde{s}_\theta$, we use it in the reverse-time stochastic differential equation (SDE) that defines the denoising process. For the graph state G_t, the reverse SDE is written as

$$\mathrm{d}G_t = [f_t(G_t) - g_t^2\, \tilde{s}_\theta(G_t; y_i)]\mathrm{d}t + g_t\, \mathrm{d}\bar{W}, \tag{9}$$

where f_t is a drift coefficient, g_t is a diffusion coefficient, and $\bar{W}$ is a standard Wiener process (Brownian motion). Starting from pure noise and integrating Eq. (9) down to $t = 0$ yields, for each client k, a set of generated graphs $\widetilde{\mathcal{D}}_k$ corresponding to the target classes, which augments minority labels and enhances structural diversity.

PGM-Guided Sample Filtering. Class-aware diffusion, however, does not guarantee that all generated graphs are useful. In practice, the generated graphs vary in quality and may deviate from the local data distribution. Furthermore, some graphs may be incompatible with the graph priors encoded in the PGM, leading to negative transfer when used directly for fine-tuning. To address this issue, we perform PGM-guided sample filtering: we measure proximity in the PGM's representation space and retain only those generated graphs that are closest to real graphs of the same class and correctly predicted by the downstream classifier.

Based on the original graphs $\mathcal{D}_k = \{(X, A)\}$ and the generated graphs $\widetilde{\mathcal{D}}_k = \{(\widetilde{X}, \widetilde{A})\}$ on client k, we compute graph representations $z = f_\phi(X, A)$ and $\tilde{z} = f_\phi(\widetilde{X}, \widetilde{A})$ by the PGM encoder f_θ. For each generated graph, we define its minimum distance to all local graphs:

$$d_{\mathrm{min}}^{\mathrm{pre}}(\widetilde{X}, \widetilde{A}) = \min_{(X,A)\in\mathcal{D}_k} \|\, \tilde{z} - z\,\|_2. \tag{10}$$

Then, we rank all generated graphs in $\widetilde{\mathcal{D}}_k$ by $d_{\mathrm{min}}^{\mathrm{pre}}$ in ascending order, and, after sorting, we write the ordered set as $\widetilde{\mathcal{D}}_k = \{\widetilde{G}^{(1)}, \widetilde{G}^{(2)}, \ldots\}$. We select those with the k smallest distances. The resulting candidate set is $\mathcal{C}_k = \{\widetilde{G}^{(m)}\}_{m=1}^k$. After obtaining the candidate set $\mathcal{C}_k$, we further apply a label-consistency filter. Let $\hat{y}_\psi(\widetilde{X}, \widetilde{A})$ denote the predicted label produced by the classifier head c_ψ. We retain only those candidates whose predicted labels match their target labels:

$$\mathcal{S}_k = \{\hat{y}_\psi(\widetilde{X}, \widetilde{A}) = \widetilde{y}\}. \tag{11}$$

After filtering, we construct the final local training set $T_k = D_k \cup S_k$.

This distance-and-consistency-based selection acts as a generation correction step. By measuring how close each generated graph is to the real local data in the PGM representation space, it filters out samples that deviate from the client distribution and the prior, preventing noisy augmentations from misleading the model. The label-consistency criterion further enforces semantic alignment with the task, favoring generated graphs near the decision boundary but on the correct side, thus providing informative, margin-enlarging examples. Overall, this strategy improves the alignment between the prior and the local data distribution, leading to more stable fine-tuning and reduced cross-client discrepancy.

4.3   Invariant Subgraph Extraction

In FGL, graphs often exhibit topology heterogeneity, i.e., differences in graph size across clients, which may hinder the adaptation of PGMs. Moreover, many substructures may be irrelevant or noisy to the downstream tasks. To mitigate task-irrelevant variations while retaining label-relevant signals, we extract invariant predictive subgraphs that are stable across clients. Making predictions on these subgraphs rather than full graphs suppresses client-specific nuisances and promotes more consistent PGM adaptation under heterogeneity.

Subgraph Extraction. Given a graph $G = (X, A)$, we first adapt a vanilla GCN [10] with parameters $\omega^{(gcn)}$ to encode each node i into an embedding h_i. Based on these embeddings, we compute a node score for each node and an edge score for each pair of nodes:

$$
\begin{aligned}
{}_i &= \sigma(\mathrm{W}^{(v)} h_i) \in [0, 1], \\
[s^{(e)}(X, A)]_{ij} &= \sigma(\mathrm{W}^{(e)}[\, h_i \| h_j \,]) \in [0, 1],
\end{aligned}
\tag{12}
$$

where $\mathrm{W}^{(v)}$ and $\mathrm{W}^{(e)}$ are trainable parameters that project node and edge embeddings to scalar scores, and σ is the sigmoid function. With fixed thresholds $\tau_v, \tau_e \in (0, 1)$, we obtain mask matrices:

$$
\begin{aligned}
\mathrm{M}^{(v)}(X, A) &= \mathbb{I}[s^{(v)}(X, A) \geq \tau_v], \\
\mathrm{M}^{(e)}(X, A) &= \mathbb{I}[s^{(e)}(X, A) \geq \tau_e].
\end{aligned}
\tag{13}
$$

where $\mathbb{I}[\cdot]$ is the indicator function. Using these masks, we construct the masked graph inputs:

$$
\widehat{X} = X \odot \mathrm{M}^{(v)}(X, A), \qquad \widehat{A} = A \odot \mathrm{M}^{(e)}(X, A),
\tag{14}
$$

where $\odot$ denotes the Hadamard product. Thus, the subgraph extraction can be seen as a trainable masker $\omega = (\omega^{(gcn)}, W^{(v)}, W^{(e)})$ applied to the original graphs. The obtained subgraph $\widehat{G} = (\widehat{X}, \widehat{A})$ is used to fine-tune the PGM F_θ.

Invariant Regularizer. To further ensure the extracted subgraphs are invariant, i.e., they should preserve a consistent relationship with the target labels across clients, we draw inspiration from [4] and propose to add a global gradient penalty to the parameter of subgraph extraction.

Let the supervised fine-tuning loss on the extracted subgraphs be

$$\ell_{\text{task}}(X, A, y; \theta, \omega) = \text{CE}\left(\text{softmax}(F_\theta(\widehat{X}, \widehat{A})), y\right). \tag{15}$$

Define the local masker gradient of client k and the global masker gradient as

$$g_w^{(l)} = \mathbb{E}_{(X,A,y)\in T_k} \nabla_\omega \ell_{\text{task}}(X, A, y; \theta, \omega),$$
$$g_w^{(g)} = \text{Agg}(\{g_j\}_{j\in \mathcal{C}_r}), \tag{16}$$

where $\mathcal{C}_r$ denotes the participating clients in round r. Then we add a L_2 regularizer between these two gradients:

$$\mathcal{R}_k^{\text{inv}}(\omega) = \| g_w^{(l)} - g_w^{(g)} \|_2. \tag{17}$$

In this way, the local objective is

$$\mathcal{L}_k(\theta, \omega) = \mathbb{E}_{(X,A,y)\in T_k} \ell_{\text{task}}(X, A, y; \theta, \omega) + \gamma \mathcal{R}_k^{\text{inv}}(\omega), \tag{18}$$

where the penalty weight γ scales the local-global gradient alignment strength. Note that $\mathcal{R}_k^{\text{inv}}$ only affects the masker parameters ω. As a result, the masker can mask irrelevant or noisy substructures and preserve subgraphs that exhibit invariant relationships with the target labels across clients.

5   Experiment

5.1   Experimental Settings

Datasets. We evaluate our method on four representative graph datasets from [7], namely BACE, BBBP, HIV, and PPI. Cross-client heterogeneity is introduced via label-based and topology-based partitioning [12], where the degree of heterogeneity is controlled by a Dirichlet distribution with parameter α.

Baseline. We select seven representative baselines from two categories: (1) FL methods including FedAvg [14], FedProx [11], and FedIIR [4]; (2) FGL methods including GCFL+ [23], FedStar [18], FedSSP [19], and FedVN [3]. For all FGL baselines, we use GIN [24] as the backbone.

Implementation Details. The Dirichlet parameter α is selected from $\{0.1, 1, 10\}$, where a smaller α indicates stronger heterogeneity. For class-aware diffusion, we set the conditional control weights to $\lambda_X = 0.4$ and $\lambda_A = 0$ on BACE and BBBP, and $\lambda_X = 0.4$ and $\lambda_A = 0.2$ on HIV. We set $k = 15$ in the filtering step. The masking module is a two-layer GCN with 128 hidden dimensions. The node and edge thresholds are the same. We optimize the model with Adam using a learning rate of 10^{-3} and use ROC-AUC as the evaluation metric.

5.2 Overall Performance

As is shown in Table 1, under different degrees of label heterogeneity, our method consistently outperforms all baselines. Under severe heterogeneity (i.e., $\alpha = 0.1$), our method achieves substantial performance gains, indicating that the class-conditional data generation effectively replenishes minority-class samples, while invariant subgraph extraction suppresses structural conflicts during aggregation. Moreover, PeFGL generally exhibits smaller standard deviations, demonstrating robustness under severe heterogeneity. Table 2 presents results under various degrees of topology heterogeneity and PeFGL also achieves the SOTA performance. When topology heterogeneity is severe (e.g., in HIV), traditional FGL models tend to capture topology-specific spurious correlations. In contrast, our invariant-subgraph constraint concentrates on learning structures shared among different topologies and thus shows great generalization capabilities. These findings further verify that aligning clients to the pre-trained prior and then fine-tuning the PGM allows the model to fit local distributions while maintaining global consistency.

Table 1. Performance (%) of baselines under label heterogeneity.

Centralized	BACE			BBBP			HIV			PPI		
	86.39±0.86			95.83±0.37			77.87±1.95			65.15±1.88		
	0.1	1	10	0.1	1	10	0.1	1	10	0.1	1	10
Fedavg	82.74±4.47	81.58±3.84	81.87±1.81	88.32±7.77	95.68±0.92	96.30±0.72	77.01±5.67	81.67±3.90	77.14±3.31	79.17±6.08	66.25±0.88	69.53±2.40
FedProx	80.25±2.97	80.17±5.66	83.32±2.28	89.04±7.89	95.53±0.77	96.47±0.32	75.28±5.89	74.23±1.13	72.60±3.56	64.20±1.45	64.54±2.87	64.65±0.84
FedIIR	73.63±2.13	80.35±1.96	82.84±0.08	75.61±1.16	94.17±1.67	95.02±0.37	67.68±8.41	73.73±3.74	74.84±4.25	59.80±1.29	60.81±1.70	61.40±1.36
GCFL+	79.65±1.84	83.94±3.14	84.14±1.60	89.50±7.16	96.23±0.68	96.38±0.40	68.73±7.79	74.47±3.85	70.38±1.04	70.88±2.02	64.54±4.60	64.81±2.03
Fedstar	59.59±0.61	62.29±2.92	70.04±3.16	78.73±5.36	85.18±4.40	86.18±3.43	69.33±6.78	68.66±5.07	69.34±5.86	51.71±0.81	51.99±1.18	55.61±2.46
Fedssp	60.45±7.36	65.85±9.23	56.02±4.50	57.75±2.34	75.18±6.43	79.79±2.48	53.10±5.73	53.33±6.41	54.91±0.50	79.64±2.84	73.86±1.52	67.21±0.87
Fedvn	72.88±4.92	76.45±2.54	80.78±2.22	89.07±4.22	89.59±5.25	95.54±0.98	71.72±1.91	73.20±5.26	69.16±4.10	73.48±7.66	74.88±4.82	66.51±7.23
PeFGL	**84.25±2.07**	**84.33±3.54**	**87.83±0.92**	**93.39±4.10**	**98.11±0.67**	**97.52±0.45**	**79.34±0.56**	**87.10±1.69**	**82.30±3.54**	**79.29±4.88**	**75.89±3.84**	**76.19±2.32**

Table 2. Performance (%) of baselines under topology heterogeneity.

Centralized	BACE			BBBP			HIV			PPI		
	86.39±0.86			95.83±0.37			77.87±1.95			65.15±1.88		
	0.1	1	10	0.1	1	10	0.1	1	10	0.1	1	10
Fedavg	82.43±0.57	83.61±0.63	83.58±0.22	94.97±0.77	93.46±2.97	95.96±0.42	70.73±1.34	71.62±1.52	71.63±2.84	60.43±2.02	63.95±1.15	63.57±1.92
FedProx	84.59±2.81	84.33±2.52	84.18±1.17	96.03±0.98	92.35±5.52	95.43±0.79	69.10±0.86	71.89±2.67	69.58±1.87	59.26±4.56	62.40±2.12	62.45±1.45
FedIIR	80.99±1.09	79.66±1.28	80.74±1.60	95.05±0.52	93.65±0.38	94.73±0.67	74.58±3.54	73.53±2.70	76.75±4.82	58.61±2.35	61.98±0.46	62.03±0.67
GCFL+	82.53±3.37	84.37±1.73	84.29±2.57	95.95±0.76	92.23±5.71	96.10±0.13	72.40±3.01	72.93±2.99	74.16±4.40	59.82±3.30	64.25±1.93	64.44±1.91
Fedstar	70.79±1.52	66.69±1.18	69.95±2.30	84.37±2.08	84.10±2.61	85.91±1.23	71.21±3.67	71.58±3.41	70.59±2.34	54.98±2.04	56.37±4.48	56.27±3.31
Fedssp	63.14±4.78	62.95±3.20	62.63±4.50	83.18±2.69	78.03±1.96	80.74±1.57	61.92±1.30	58.08±1.84	61.00±1.42	56.49±0.64	56.39±1.15	59.32±1.19
Fedvn	75.93±0.64	79.24±2.36	75.20±3.40	93.64±1.15	92.62±1.70	94.12±1.29	74.80±3.43	69.72±2.38	67.41±2.53	57.43±0.61	56.58±1.51	53.74±1.76
PeFGL	**86.79±0.11**	**87.12±0.88**	**88.75±1.02**	**96.37±0.12**	**94.18±2.58**	**96.42±0.84**	**76.77±3.39**	**77.48±2.65**	**79.03±3.75**	**68.41±2.79**	**71.76±3.07**	**71.96±1.53**

5.3 Ablation Study

To show the effect of each component in PeFGL, we conduct ablation studies with five variants: **ours-g-m** (federated fine-tuning of the pre-trained model

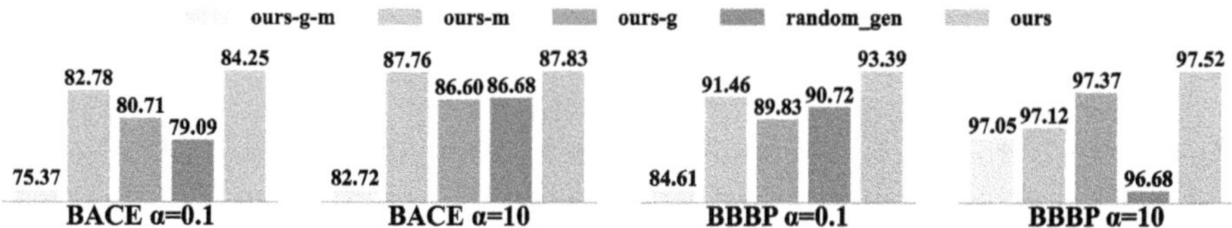

Fig. 3. Ablation study under label heterogeneity.

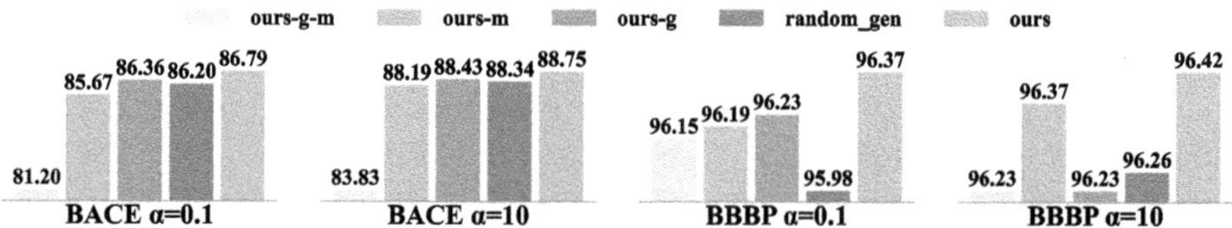

Fig. 4. Ablation study under topology heterogeneity.

only), `ours-g` (invariant subgraph extraction only), `ours-m` (minority-class generation only), `random_gen` (randomly selecting generated samples), and `ours` (PeFGL). Figure 3 and Fig. 4 report the results under different types of heterogeneity. Among different α, PeFGL achieves the best AUC. For `ours-g-m`, adding either single component brings consistent gains, indicating that minority-class generation and invariant structural regularization are both effective. `random_gen` performs worse than PeFGL and even inferior to the single-component variants, and in a few cases slightly worse than `ours-g-m`, showing that prior-guided selection is necessary to avoid including noisy synthetic samples. Under label heterogeneity, `ours-m` tends to yield larger improvements, whereas under topology heterogeneity, `ours-g` contributes more. It suggests that when label imbalance is severe, addressing the data-coverage gap is most useful, whereas when class coverage is sufficient, structural constraints become more critical.

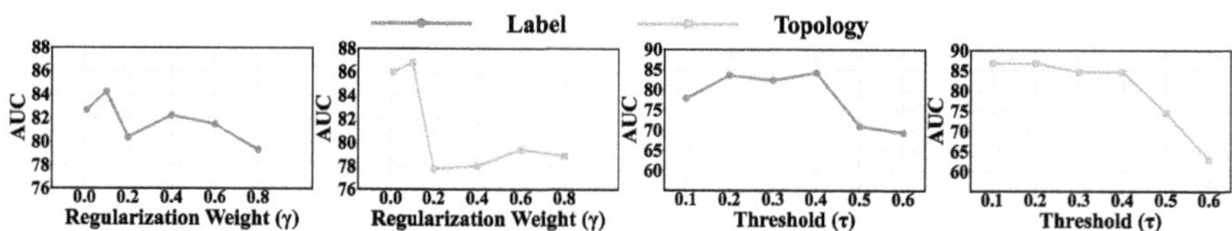

Fig. 5. Parameter sensitivity on BACE ($\alpha = 0.1$).

5.4   Parameter Analysis

We conduct parameter analysis regarding the mask threshold τ and the regularization weight γ, which is reported in Fig. 5.

Table 3. Results w/ and w/o PGM ($\alpha = 0.1$)

Method	BACE				BBBP			
	Label		Topology		Label		Topology	
	w/o	w/	w/o	w/	w/o	w/	w/o	w/
FedIIR	73.63±2.13	65.11±9.83	80.99±1.09	77.87±0.82	75.61±1.16	78.84±12.61	95.05±0.52	94.21±0.16
FedProx	80.25±2.97	70.77±9.87	84.59±2.81	81.59±0.64	89.04±7.89	81.71±16.52	96.03±0.98	96.20±0.61
GCFL+	79.65±1.84	76.36±4.22	82.53±3.37	81.07±0.60	89.50±7.16	85.63±12.47	95.95±0.76	96.23±0.54
Fedavg	82.74±4.47	75.37±4.95	82.43±0.57	81.20±1.58	88.32±7.78	84.61±10.35	94.97±0.77	96.15±0.08

The Threshold of Edge/Node Masks. Since we set the node and edge thresholds to be identical, we use a unified threshold τ ($\tau_v = \tau_e = \tau$) to control the sparsity of the invariant subgraph. As shown in Fig. 5, under label heterogeneity, the AUC first increases and then decreases as τ grows. When τ is too small, the mask retains many client-specific edges and introduces noise, whereas overly large τ removes too many informative edges. Under topology heterogeneity, small thresholds are more stable, and progressively increasing τ leads to a clear degradation in AUC, consistent with the intuition that cross-client shared structures are relatively sparse and can be easily damaged by aggressive pruning.

The Weight of the Regularization Term. For the regularization weight γ, performance under both label and topology heterogeneity is highest when γ remains small and declines as γ increases. Within this range, increasing γ tends to improve performance, whereas larger values lead to a clear performance drop, with a sharper decline under topology heterogeneity. These results suggest that a small amount of regularization suffices to encourage cross-client alignment of invariant subgraphs, whereas larger penalties over-constrain the model, blur class-discriminative structure, and diminish the benefit of the pre-trained priors.

5.5 Results of Existing Methods with PGM

Table 3 reports results for existing heterogeneity-mitigation methods combined with a PGM, evaluated on three datasets with $\alpha = 0.1$. Under label heterogeneity, integrating the pre-trained PGM generally yields worse performance than training the same backbone without the pre-trained prior. Under topology heterogeneity, we observe minor gains only on BBBP. These results indicate that addressing heterogeneity alone does not resolve the negative transfer when adapting pre-trained models in federated settings.

5.6 Results of Different Fine-Tuning Strategies

To assess the difference between classifier-only fine-tuning and full fine-tuning, we compare the two strategies in Fig. 6 (α=0.1). Full fine-tuning yields substantial improvements over classifier-only fine-tuning. This observation is consistent with our hypothesis: the pre-trained encoder carries source-domain priors that are

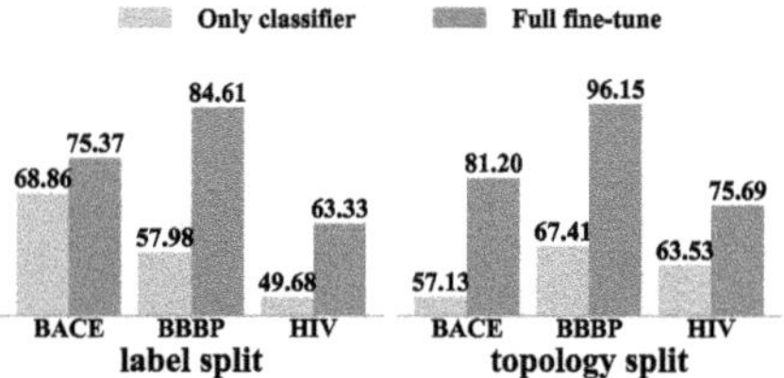
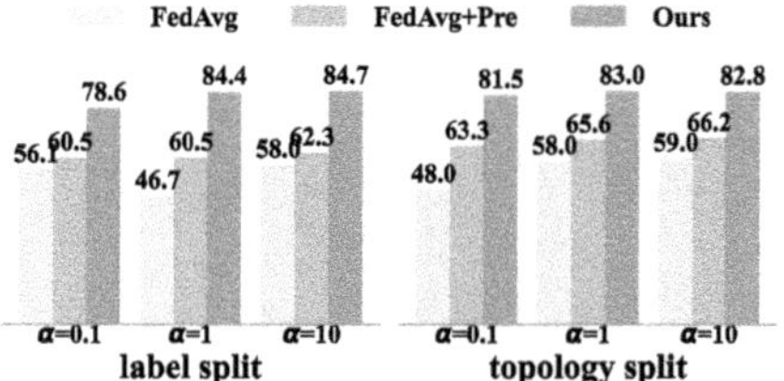

Fig. 6. Fine-tuning strategy comparison. **Fig. 7.** Results on unseen graphs.

misaligned with heterogeneous client distributions. Simply optimizing the classifier provides merely shallow adjustments and hardly mitigates the representation drift induced by label imbalance or topology shifts. These findings motivate our framework that builds on full fine-tuning and we additionally employ class-aware generation, prior-guided sample selection, and invariant subgraph constraints to enable more effective encoder adaptations.

5.7 Results on Unseen Graphs

As shown in Fig. 7, we evaluate PeFGL on unseen graphs to show whether PeFGL truly learns transferable structural regularities rather than overfitting to client-specific biases. The results reveal that vanilla FedAvg generalizes poorly to unseen graphs, and naively federated fine-tuning a pre-trained graph model can be brittle, exhibiting unstable gains and even negative transfer. In contrast, PeFGL consistently shows stronger robustness, indicating that it effectively mitigates the prior–local mismatch during federated optimization and encourages the model to rely on more transferable patterns.

Fig. 8. Visualization of generated graphs.

5.8 Visualization

Figure 8 visualizes several molecular graphs generated by our diffusion model on the BACE dataset. The generated samples are chemically valid and include functional motifs, such as hydroxyl, amine, ether, and fluorinated groups, that are closely associated with BACE inhibitory activity. These functional groups serve as key determinants of molecular bioactivity. The generated graphs enrich the diversity of ring structures while preserving chemically meaningful substructures. Such augmentation enhances the representation of activity-related patterns and facilitates cross-client alignment in federated settings.

6   Conclusion

In this work, we propose PeFGL, a framework that adapts pre-trained graph priors to heterogeneous clients by coupling class-aware graph generation with invariant subgraph extraction. We apply the diffusion model to generate local graphs, which are then filtered by both the PGM prior and local distributions. Then, we design the invariant subgraphs extraction for fine-tuning. Evaluations on several real-world datasets against strong baselines achieve consistent improvements, and further analysis also demonstrates the effectiveness of PeFGL.

Acknowledgments. This work is supported in part by the National Natural Science Foundation of China (No. 62550138, 62192784, 62572064, 62472329, 62402263), JST PRESTO JPMJPR23P5, JST CREST JPMJCR21M2, and JST NEXUS JPMJNX25C4.

Disclosure of Interests. The authors have no competing interests to declare that are relevant to the content of this article.

References

1. Agarwal, A., Rezagholizadeh, M., Parthasarathi, P.: Practical takes on federated learning with pretrained language models. In: Findings of the Association for Computational Linguistics: EACL 2023, pp. 454–471 (2023)
2. Cao, Y., et al.: When to pre-train graph neural networks? from data generation perspective! In: Proceedings of the 29th ACM SIGKDD Conference on Knowledge Discovery and Data Mining, pp. 142–153 (2023)
3. Fu, X., et al.: Virtual nodes can help: Tackling distribution shifts in federated graph learning. In: Proceedings of the AAAI Conference on Artificial Intelligence, vol. 39, pp. 16657–16665 (2025)
4. Guo, Y., Guo, K., Cao, X., Wu, T., Chang, Y.: Out-of-distribution generalization of federated learning via implicit invariant relationships. In: International Conference on Machine Learning, pp. 11905–11933. PMLR (2023)
5. Hamilton, W., Ying, Z., Leskovec, J.: Inductive representation learning on large graphs. Adv. Neural Inf. Process. Syst. **30** (2017)
6. Hou, Z., et al.J.: Graphmae: Self-supervised masked graph autoencoders. In: Proceedings of the 28th ACM SIGKDD Conference on Knowledge Discovery and Data Mining, pp, 594–604 (2022)
7. Hu, W., et al.: Strategies for pre-training graph neural networks. arXiv preprint arXiv:1905.12265 (2019)
8. Hu, Z., Dong, Y., Wang, K., Chang, K.W., Sun, Y.: Gpt-gnn: Generative pre-training of graph neural networks. In: Proceedings of the 26th ACM SIGKDD International Conference on Knowledge Discovery & Data Mining, pp. 1857–1867 (2020)
9. Jhunjhunwala, D., Sharma, P., Xu, Z., Joshi, G.: Initialization matters: Unraveling the impact of pre-training on federated learning. arXiv preprint arXiv:2502.08024 (2025)
10. Kipf, T.: Semi-supervised classification with graph convolutional networks. arXiv preprint arXiv:1609.02907 (2016)

11. Li, T., Sahu, A.K., Zaheer, M., Sanjabi, M., Talwalkar, A., Smith, V.: Federated optimization in heterogeneous networks. Proc. Mach. Learn. Syst. **2**, 429–450 (2020)
12. Li, X., et al.: Openfgl: A comprehensive benchmark for federated graph learning. arXiv preprint arXiv:2408.16288 (2024)
13. Lit, Z., Sit, S., Wang, J., Xiao, J.: Federated split bert for heterogeneous text classification. In: 2022 International Joint Conference on Neural Networks (IJCNN), pp. 1–8. IEEE (2022)
14. McMahan, B., Moore, E., Ramage, D., Hampson, S., Arcas, B.A.: Communication-efficient learning of deep networks from decentralized data. In: Artificial Intelligence and Statistics, pp. 1273–1282. PMLR (2017)
15. Nguyen, J., Wang, J., Malik, K., Sanjabi, M., Rabbat, M.: Where to begin? on the impact of pre-training and initialization in federated learning. arXiv preprint arXiv:2206.15387 (2022)
16. Qiu, J., et al.: Gcc: Graph contrastive coding for graph neural network pre-training. In: Proceedings of the 26th ACM SIGKDD International Conference on Knowledge Discovery & Data Mining, pp. 1150–1160 (2020)
17. Sun, F.Y., Hoffmann, J., Verma, V., Tang, J.: Infograph: unsupervised and semi-supervised graph-level representation learning via mutual information maximization. arXiv preprint arXiv:1908.01000 (2019)
18. Tan, Y., Liu, Y., Long, G., Jiang, J., Lu, Q., Zhang, C.: Federated learning on non-iid graphs via structural knowledge sharing. In: Proceedings of the AAAI Conference on Artificial Intelligence, vol. 37, pp. 9953–9961 (2023)
19. Tan, Z., Wan, G., Huang, W., Ye, M.: Fedssp: federated graph learning with spectral knowledge and personalized preference. Adv. Neural. Inf. Process. Syst. **37**, 34561–34581 (2024)
20. Veličković, P., Cucurull, G., Casanova, A., Romero, A., Lio, P., Bengio, Y.: Graph attention networks. arXiv preprint arXiv:1710.10903 (2017)
21. Veličković, P., Fedus, W., Hamilton, W.L., Liò, P., Bengio, Y., Hjelm, R.D.: Deep graph infomax. arXiv preprint arXiv:1809.10341 (2018)
22. Wan, C., Li, Y., Li, A., Kim, N.S., Lin, Y.: Bns-gcn: Efficient full-graph training of graph convolutional networks with partition-parallelism and random boundary node sampling. Proc. Mach. Learn. Syst. **4**, 673–693 (2022)
23. Xie, H., Ma, J., Xiong, L., Yang, C.: Federated graph classification over non-iid graphs. Adv. Neural. Inf. Process. Syst. **34**, 18839–18852 (2021)
24. Xu, K., Hu, W., Leskovec, J., Jegelka, S.: How powerful are graph neural networks? arXiv preprint arXiv:1810.00826 (2018)
25. Yan, B., Cao, Y., Wang, H., Yang, W., Du, J., Shi, C.: Federated heterogeneous graph neural network for privacy-preserving recommendation. In: Proceedings of the ACM Web Conference 2024, pp. 3919–3929 (2024)
26. Yan, B., He, S., Yang, C., Liu, S., Cao, Y., Shi, C.: Federated graph condensation with information bottleneck principles. In: Proceedings of the AAAI Conference on Artificial Intelligence, vol. 39, pp. 12990–12998 (2025)
27. Yang, Q., Liu, Y., Chen, T., Tong, Y.: Federated machine learning: concept and applications. ACM Trans. Intell. Syst. Technol. (TIST) **10**(2), 1–19 (2019)
28. Yao, Y., Jin, W., Ravi, S., Joe-Wong, C.: Fedgcn: convergence-communication tradeoffs in federated training of graph convolutional networks. Adv. Neural. Inf. Process. Syst. **36**, 79748–79760 (2023)
29. You, Y., Chen, T., Sui, Y., Chen, T., Wang, Z., Shen, Y.: Graph contrastive learning with augmentations. Adv. Neural. Inf. Process. Syst. **33**, 5812–5823 (2020)

30. Zhang, H., Shen, T., Wu, F., Yin, M., Yang, H., Wu, C.: Federated graph learning–a position paper. arXiv preprint arXiv:2105.11099 (2021)
31. Zhang, J., Liu, Y., Hua, Y., Cao, J.: An upload-efficient scheme for transferring knowledge from a server-side pre-trained generator to clients in heterogeneous federated learning. In: Proceedings of the IEEE/CVF Conference on Computer Vision and Pattern Recognition, pp. 12109–12119 (2024)
32. Zhang, K., Yang, C., Li, X., Sun, L., Yiu, S.M.: Subgraph federated learning with missing neighbor generation. Adv. Neural. Inf. Process. Syst. **34**, 6671–6682 (2021)
33. Zhou, P., et al.: Fedgog: Federated graph out-of-distribution generalization with diffusion data exploration and latent embedding decorrelation. In: Proceedings of the AAAI Conference on Artificial Intelligence, vol. 39, pp. 22965–22973 (2025)

Friend-Aware Contrastive Learning with Aligned Social Graph for Social Recommendation

Zaisheng Ruan, Yanmin Zhu(✉), Wenze Ma, and Xuhao Zhao

Shanghai Jiao Tong University, Shanghai, China
{zeasonjuan,yzhu,mawenze991226,zhaoxuhao}@sjtu.edu.cn

Abstract. Social recommendation leverages friends' preferences to improve user recommendation quality. However, recent studies reveal an inherent misalignment between the social and interaction graphs. Directly using these raw graphs leads to suboptimal performance. Recently, contrastive learning (CL) has been introduced to mitigate this issue. It treats the same user's representations from social and interaction views as a positive pair, while representations of different users across the two views as negative pairs, aligning views by maximizing the similarity of positive pairs and minimizing that of negative pairs. Nevertheless, existing approaches treat all different users uniformly when constructing sample pairs, regardless of their behavioral similarity. As a result, highly similar users are pushed away like dissimilar ones, weakening alignment effectiveness. In this paper, we propose Friend-aware Contrastive Learning with Aligned Social Graph for Social Recommendation (FACAR). Specially, to alleviate the graph-level misalignment and facilitate subsequent processing, FACAR first performs preference-aware social relation mining by integrating user preferences into social relations to refine the social graph. Then, to avoid uniformly sampling pairs in CL so as to better align the two views in representations-level, we construct a Trust Friend Set for each user and design a friend-aware contrastive learning framework based on relations characterized above. Experiments on three benchmarks demonstrate that FACAR outperforms state-of-the-art baselines. Our implementation is available at: https://github.com/ZeasonJuan/FACAR.

Keywords: Social Recommendation · Contrastive Learning

1 Introduction

Social recommendation (SR) has gradually emerged as a crucial scenario in the recommendation domain, driven by the increasing ubiquity of social networks. Its effectiveness primarily relies on two key theories: social influence [10], a user's preferences can be affected by those of their friends, and social homophily [12], users with similar preferences are more likely to become friends.

H. Jung et al. (Eds.): DASFAA 2026, LNCS 16536, pp. 488–503, 2026.
https://doi.org/10.1007/978-981-92-0366-6_30

Despite the benefits, recent studies [8] indicate an inherent misalignment between the social graph and the interaction graph, i.e., social edges do not reliably reflect preference similarity [7,9,24]. Therefore, directly leveraging these raw graphs in recommendation may yield sub-optimal performance. To alleviate this misalignment, recent works adopt contrastive learning (CL) to align representations between the social and interaction views [7,17,26]. In this paradigm, the representations of the same user from the social and interaction graphs forms a positive pair, while embeddings of different users taken across the two views constitute negative pairs. By maximizing agreement between positive pairs formed across views and minimizing agreement of negative pairs, CL captures the shared semantics of two views and reduces the gap between social and interaction data.

However, most contrastive learning approaches in social recommendation field [9,26,27] treat all different users uniformly when constructing sample pairs, ignoring the varying degrees of behavioral similarity among users. As a result, the information of behaviorally similar friends (or users with high social-level consistency) fails to align with the target user and is even pushed apart like dissimilar users. This may reduce the effectiveness of CL in aligning social and preference graphs. As illustrated in Fig. 1(a), user B is socially connected with users A, C, and D. Moreover, A and C exhibit highly similar item interaction patterns with B, users D shows entirely different preferences. When uniform sampling strategy is adopted, users like A and C, who share both strong social links and highly similar preferences with B, are not only unrecognized as similar, but are treated as negative samples just like D, whose behavior differs entirely from B. This indiscriminate treatment forces the model to push away behaviorally aligned representations, which runs counter to the goal of aligning social relations with users' true preference similarity.

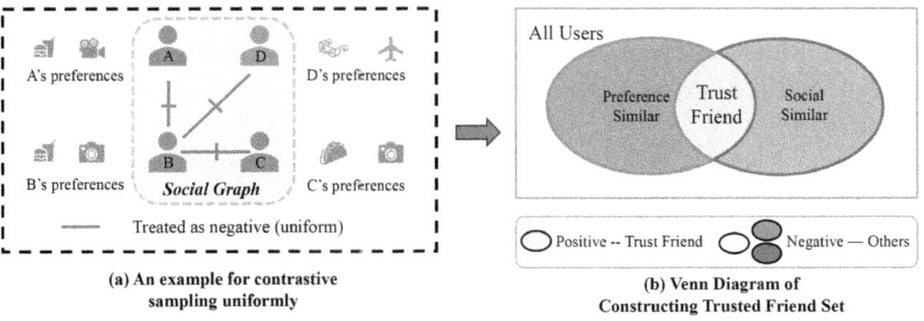

(a) An example for contrastive sampling uniformly

(b) Venn Diagram of Constructing Trusted Friend Set

Fig. 1. Comparison of sampling strategies for social recommendation: (a) uniform in-batch sampling that treats all non-identical users as negatives; (b) Venn-based friend-aware sampling (ours) where the intersection defines the Trusted Friends.

To address these issues, we propose Friend-aware Contrastive Learning with Aligned Social Graph for Social Recommendation (FACAR). FACAR first performs preference-aware social relation mining, integrating user preference information to refine social relations. Based on the refined social graph, FACAR then construct a Trust Friend Set as illustrated in Fig. 1(b), which identifies users similar at both the social and preference levels. Finally, we design a Friend-aware Contrastive Learning (FCL) framework that replaces uniform pair sampling strategy with a relation-aware sampling, based on the Trust Friend Set. This facilitates distinguishing users by similarity levels during contrastive learning and finally aligns the social and interaction spaces.

Our main contributions are summarized as follows:

- We perform preference-aware social relation mining to inject interaction information into social relation, thereby producing a refined social graph. The resulting graph serves as the foundation for constructing the Trust Friend Set, guiding positive pair sampling and optimizing the contrastive objective in the subsequent Friend-aware Contrastive Learning module.
- We construct a Trust Friend Set based on the refined social graph and design a Friend-aware Contrastive Learning objective upon it, aiming to distinguish behaviorally similar users from dissimilar ones, thereby mitigating the misalignment between social and preference data.
- We conduct comprehensive experiments on multiple benchmark datasets, validates the robustness of the proposed method in social recommendation.

2   Related Work

2.1   Graph-Based Social Recommendation and Contrastive Learning

Recently, Graph Neural Networks (GNNs) have gained traction in social recommendation [3,16,21], owing to their ability to capture complex structural patterns and higher-order dependencies within social networks. For example, Diffnet++ [23] employs a diffusion process to capture both direct and higher-order social influences, thereby improving representations. DESIGN [19] adopts knowledge distillation in graph learning, improving recommendation performance.

However, as social relationships often exhibit a semantic gap with users' true preferences [8,28,29], recent studies leverages contrastive learning (CL) to align social graph with behavioral signals [7,17,26]. For example, SEPT [26] constructs three graph views to conduct contrastive learning. It leverages tri-training to predict positive samples for each view. GDSSL [7] utilizes the refined social relation graph generated by diffusion model for contrastive learning.

Despite effectiveness, existing CL-based methods form positive pairs by pairing the same user across different views, while treating all other users as negatives, which uniformly weights all social relations. By contrast, we design a friend-aware contrastive learning module based on a Trust Friend Set, which adopts a relation-aware sampling strategy when constructing pairs.

2.2   Social Graph Refinement for Recommendation

Unlike contrastive learning which mitigate the gap between social and behavioral signals in representation level, social graph refinement methods addresses the graph itself. Two common routes are: graph sparsification and graph structure learning. The former removes relations showing low alignment to interaction signals [1,13], and the latter add informative links based on user preference [9,17]. On the sparsification side, GDMSR [13] scores relations using interaction signals and employs curriculum-style training with an adaptive filter to suppress weak ties. On the structure learning side, MADM [9] uses a mixture-of-experts to estimate user similarity and maximizes mutual information to guide link reconstruction, yielding a social topology consistent with user preferences.

In this paper, we not only refines the social graph, but also leverage the resulting relations to guide friend-aware contrastive learning, thereby further bridging the social–preference gap.

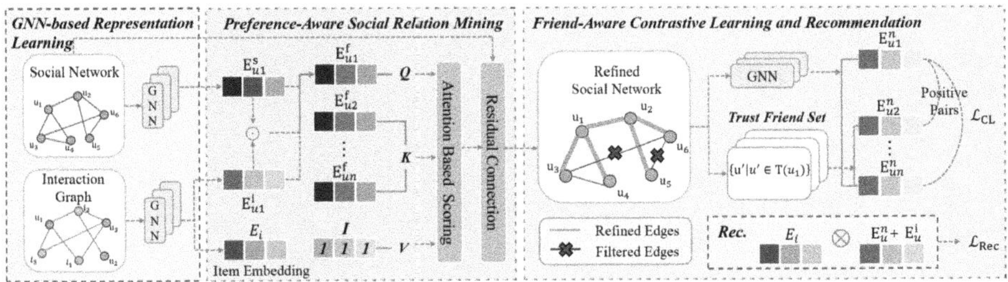

Fig. 2. Overall framework of Preference-Aware Social Graph Denoising for Social Recommendation.

3   Preliminary

In the social recommendation scenario, let $\mathcal{U} = \{u_1, u_2, ..., u_N\}$ and $\mathcal{I} = \{i_1, i_2, ..., i_M\}$ denote the sets of users and items, respectively, where N is the number of users and M is the number of items. The user–item interaction graph $\mathcal{G} = (\mathcal{U}, \mathcal{I}, \mathcal{E})$ is represented by a bipartite adjacency matrix, where an entry $r_{ui} = 1$ indicates an interaction between user u and item i. The social network $\mathcal{G}_S = (\mathcal{U}, \mathcal{E}^s)$ is represented by an adjacency matrix, where an entry $s_{uv} = 1$ denotes the existence of a relation between user u and user v. Each user and item embedding lies in a d-dimensional latent space.

Task Formulation. Given the interaction graph $\mathcal{G}$ and the social network $\mathcal{G}_S$, the task of social recommendation is to generate a ranked list of items for each user u. The ranking is based on the probability that user u will interact with the items, where items have higher interaction probabilities appearing earlier.

4   Methodology

In this section, we describe our FACAR model. The overall structure is illustrated in Fig. 2. Given $\mathcal{G}$ and $\mathcal{G}_S$, we learn user preference embeddings E_u^i, item embeddings E_i and user social embeddings E_u^s via two GNN encoders. Then, a preference-aware social relation mining module integrates user preferences into social relations to refine $\mathcal{G}_S$, producing a refined social graph $\hat{\mathcal{G}}_S$. Based on $\hat{\mathcal{G}}_S$, we further construct a Trust Friend Set T_u for each user and apply a friend-aware contrastive objective to enhance user representation learning. Finally, recommendations are generated by inner product scoring.

4.1   GNN-Based Representation Learning

To encode user preferences, item representations in $\mathcal{G}$, and user social relations in $\mathcal{G}_S$, a GNN is employed on both graphs. The resulting embeddings E_u^s, E_u^i, and E_i are computed as:

$$E_u^{s\,(0)} = E_u^{i\,(0)}, \tag{1}$$

$$E_u^{s\,(k)} = GNN(E_u^{s\,(k-1)}, \{E_{u'}^{s\,(k-1)} | u' \in S_u\}), \tag{2}$$

$$E_u^{i\,(k)} = GNN(E_u^{i\,(k-1)}, \{E_i^{(k-1)} | i \in R_u\}), \tag{3}$$

$$E_i^{(k)} = GNN(E_i^{(k-1)}, \{E_u^{i\,(k-1)} | u \in R_i\}), \tag{4}$$

where S_u denotes the set of social neighbors of user u; R_u denotes the set of items interacted by user u; and R_i denotes the set of users who interacted with item i. LightGCN [6] is adopted to be the GNN module, for its effectiveness and parameter efficiency for recommendation.

As shown in Eq. 1, the initial representations of users are shared between the social graph and the interaction graph. Then, as described in Eq. 2–4, user and item embeddings are iteratively updated through multiple layers of graph convolution. Specially, the representation at layer k is derived from a combination of the node's own embedding and the aggregated embeddings of its neighbors from layer $(k-1)$.

$$E = \frac{1}{K+1} \sum_{k=0}^{K} E^{(k)}, E \in \{E_u^s, E_u^i, E_i\}. \tag{5}$$

Finally, as shown in Eq. 5, the embeddings obtained from all K propagation layers are averaged to produce the final representations.

4.2   Preference-Aware Social Relation Mining

This module integrates both social similarity and preference similarity for relation characterization, and accordingly refines the original social graph.

Given preference embeddings E_u^i and social embeddings E_u^s of users, a comprehensive representation that reflects preference and social-level consistency of the user E_u^f is computed as:

$$E_u^f = [E_u^s, E_u^i, E_u^s \odot E_u^i], \tag{6}$$

$$E_u^f, E_u^s, E_u^i \in \mathbb{R}^d, \tag{7}$$

where $\odot$ denotes the element-wise (Hadamard) product, and $[\cdot, \cdot]$ denotes concatenation along the feature dimension. Compared to approaches that consider only social-level similarity [9] or only preference similarity [13], this design mitigates the bias toward characterizing real relations caused by relying solely on the distribution of users in a single view, thereby producing preference-aware relation for subsequent refinement.

After obtaining the fused representation E_u^f, attention-based relation scoring is applied to assess the task relevance of each edge. For clarity of presentation, the following formulations illustrate the derivation using a representative user u_1, while the procedure is identical for all users:

$$Q = \tanh(W_q E_{u_1}^f), \quad Q \in \mathbb{R}^d, \tag{8}$$

$$K = \tanh(W_k E_{u'}^f), \quad u' \in \mathcal{U}, \quad K \in \mathbb{R}^{N \times d}, \tag{9}$$

$$\hat{\mathcal{E}}_{u_1}^s = softmax(\frac{QK^\top}{\sqrt{d}}), \tag{10}$$

where W_q, W_k are the learnable weight matrices.

The fused representation of user u_1 are treated as the query, and all users serve as keys. Then, as shown in Eq. 10, a scaled dot-product operation is applied to characterize the relational strength between u_1 and all users. Unlike standard cross-attention [20], no value pathway V is employed; as the objective is to estimate relation strengths between u_1 and each user, rather than to generate a new sequence representation.

Residual Connection. From the above process, a attention-scored social graph is obtained, whose edges represent preference-aware relation strength. However, directly utilizing this refined graph in downstream tasks can lead to adverse effects, especially during the early stages of training, as the relation characterizing relies on E_u^s and E_u^i, which are not yet well-learned.

To address this challenge, following approaches [9,22], we introduce a residual connection that integrates the original social network S with the refined network $\hat{S}$. Specifically, the final refined social network is computed as:

$$\hat{\mathcal{E}}^s = \tau \hat{\mathcal{E}}^s + (1 - \tau)\mathcal{E}^s, \tag{11}$$

with the adaptive coefficient τ defined by

$$\tau = 0.5 \times (1 - cos(\frac{step}{r}\pi)), \tag{12}$$

where τ is the adaptive weight that balances the contributions of the original and refined networks. *step* denotes the current training iteration. r is the total number of training epochs, its default value is 100.

At the beginning of training, τ remains close to zero, meaning the model primarily relies on the original social network $\mathcal{G}_S$, which is more stable when the representations are immature. As training progresses, τ gradually increases, thereby allowing the model to place greater emphasis on the refined network $\hat{\mathcal{G}}_S$. This strategy mitigates the negative impact of inaccurate initial representations.

4.3 Friend-Aware Contrastive Learning and Recommendation

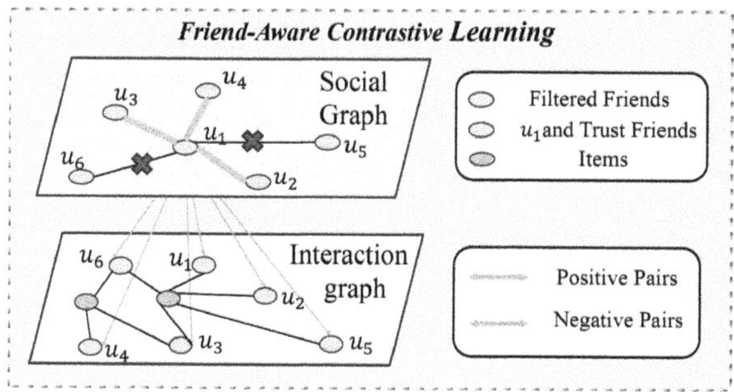

Fig. 3. Friend-aware Contrastive Learning.

Based on the refined social graph $\hat{\mathcal{G}}_S$, the resulting user representation E_u^n is derived in a manner similar to Eq. 2–Eq. 1. This representation is shared across two subsequent components, including friend-aware contrastive learning and recommendation.

Friend-Aware Contrastive Learning. To address the issues of data misalignment and uniform sampling, we introduce a friend-aware contrastive learning module that exploits refined relations to construct positive pairs and optimize the objective for CL; its architecture is shown in Fig. 3.

First, a Trust Friend Set of each user is constructed by relation-aware sampling strategy. Specifically, for user u_1, we select friends whose relation strengths $\hat{\mathcal{E}}_{u_1}^s$ exceed a dynamic threshold η_u, defined as:

$$\eta_u = percentile(\{\hat{\mathcal{E}}_{u_1}^s | \hat{\mathcal{E}}_{u_1}^s > 0\}, p), \tag{13}$$

$$p = p_0 - \tau \cdot (p_0 - p_r), \tag{14}$$

where the initial selection percentage p_0 (It is set to 90%) decays over training epochs. p_r is the final selection percentage which is set as 50%, and we have introduced τ in Eq. 12.

The size of each user's Trust Friend Set T_{u_1} is further limited by a predefined upper bound Q [2]:

$$
T_{u_1} = \begin{cases} \{u' | \hat{\mathcal{E}}^s_{u_1,u'} > \eta_u\} & \text{, if } |\{u' | \hat{\mathcal{E}}^s_{u_1,u'} > \eta_u\}| \leq Q \\ TopQ & \text{, otherwise,} \end{cases} \tag{15}
$$

where $TopQ$ refers to Q friends of u with the strongest relation weights $\hat{\mathcal{E}}^s_{u_1,u'}$.

Then, positive pairs in CL objective are constructed by matching u_1 with friends $u' \in T_{u_1}$. Each pair contributes to the objective with a confidence weight derived from the refined edge weight $\hat{\mathcal{E}}^s_{u_1}$. The objective is defined as:

$$
\mathcal{L}_{CLu_1} = \sum_{v \in T_{u_1}} -log \frac{G(E^f_{u_1}, E^i_{u_1}) + \hat{\mathcal{E}}^s_{u_1,v}(G(E^f_{u_1}, E^i_v) + G(E^i_{u_1}, E^f_v))}{\sum_{w \in \mathcal{U}}(G(E^f_{u_1}, E^i_w) + G(E^i_{u_1}, E^f_w))}, \tag{16}
$$

$$
G(a,b) = exp(cos(g(a), g(b))/t), \tag{17}
$$

where a, b denote representations, and $g(\cdot)$ is the projection head implemented as a two-layers MLP. The parameter t controls the temperature in contrastive scaling. Positive instances correspond to the Trust friends T_{u_1}, while negatives include all users. Each positive pair is assigned a confidence weight $\hat{\mathcal{E}}^s_{u1,u'}$.

As shown in Eq. 16, the self-view term $G(E^f_{u_1}, E^i_{u_1})$ and the friend-view terms $G(E^f_{u_1}, E^i_{u'}) + G(E^i_{u_1}, E^f_{u'})$ are optimized to maximize the similarity within each pair. The former term reinforces the consistency between the fused and preference representations of user u_1, while the latter term aligns the representations of u_1 and its Trust friends across different views, transferring reliable relational signals into representation learning and aligning u_1's representations with that of its friends across social and preference views.

The formulations above, illustrated with user u_1, are applied identically to all users. The overall contrastive loss of the model is obtained by summing the individual user-level losses, as follows:

$$
\mathcal{L}_{CL} = \sum_{u \in \mathcal{U}} \mathcal{L}_{CLu}, \tag{18}
$$

where $\mathcal{L}_{CLu}$ denotes the contrastive objective for user u.

Recommendation. Item representations E_i are derived from Eq. 4 and Eq. 5. As introduced above, the user representation E^n_u is obtained accordingly. With the final user and item representations, the preference score of user u over item i is computed as follows:

$$
\hat{y}_{ui} = \langle \mathbf{E}^n_u + \mathbf{E}^i_u, \mathbf{E}_i \rangle. \tag{19}
$$

We adopt the Bayesian Personalized Ranking (BPR) loss [14] to optimize the recommendation task. The BPR loss is designed to promote scores for observed (positive) user–item interactions while penalizing unobserved (negative) interactions, thereby producing a ranking that the user is likely to prefer. Specifically, the BPR loss is formulated as:

$$\mathcal{L}_{BPR} = -\sum_{u \in \mathcal{U}} \sum_{i \in \mathcal{E}_u} \sum_{j \notin \mathcal{E}_u} log(\sigma(\hat{y}_{ui} - \hat{y}_{uj})), \tag{20}$$

where $\hat{y}_{ui}$ and $\hat{y}_{uj}$ are the predicted scores for the positive and negative interactions, respectively; $\sigma(\cdot)$ is the sigmoid function.

After computing the BPR loss, overall loss function is formulated as follows:

$$\mathcal{L} = \mathcal{L}_{BPR} + \alpha \mathcal{L}_{CL} + \beta \|\Theta\|^2 . \tag{21}$$

Contrastive learning loss is weighted by the hyperparameter α, and the rest is the regularization term.

Table 1. Statistical information of datasets

	Ciao	Flickr	Yelp
Users	7,355	7,402	17,053
Items	17,867	39,906	21,904
Interactions	140,628	267,773	185,177
Relations	111,679	259,014	142,808
Interaction Density	0.11%	0.09%	0.05%
Relation Density	0.21%	0.47%	0.06%

4.4 Complexity Analysis

Space Complexity. Our model involves learning user and item embeddings. Given N users, M items, and latent dimension d, the embedding tables require $\mathcal{O}((N+M) \times d)$ parameters. In the preference-aware social relation mining module, the attention-based scoring employs two learnable weight matrices W_q, W_k, each of size $d \times d$, leading to an additional $\mathcal{O}(2d^2)$ parameters. In the friend-aware contrastive learning module, the projection head is a two-layer MLP with input and hidden size d and output size d. It thus introduces approximately $\mathcal{O}(2d^2 + 2d)$ parameters.

The total space complexity of FACAR is $\mathcal{O}((N+M)d+4d^2)$, which is dominated by the embedding tables and grows linearly with the number of users and items. Therefore, our model lighter than most existing social recommendation methods (e.g., [9,17,23]).

Time Complexity. The time complexity of FACAR mainly arises from three modules: (1) GNN-based propagation, (2) preference-aware social relation characterization, and (3) friend-aware contrastive learning. For (1), the propagation cost across K layers is $\mathcal{O}(K(|\mathcal{E}| + |\mathcal{E}^s|)d)$. For (2), relation modeling involves a dot-product and two linear projections, yielding $\mathcal{O}(|\mathcal{E}^s|d^2)$ as the dominant term. For (3), contrastive learning computes similarities among positive pairs (u, u'), $u' \in T_u$, and in-batch negatives, resulting in $\mathcal{O}(NQd + NBd)$.

Overall, the total training complexity is $\mathcal{O}\big(K(|\mathcal{E}| + |\mathcal{E}^s|)d + |\mathcal{E}^s|d^2 + NQd + NBd\big)$, which is primarily determined by the number of social and interaction edges since K, B, Q, and d are relatively small. In practice, FACAR achieves faster training than most social recommendation models [9,17].

5   Experiments

5.1   Experimental Setup

Datasets. To validate the effectiveness of our proposed FACAR model, we conduct experiments on three widely adopted datasets for social recommendation [5,11,18]: Ciao[1] Flickr[2], and Yelp[3]. Each dataset provides both user-item interactions and user-user social relations, making them suitable for our experimental evaluation. Following common practice, we randomly split each dataset into training, validation, and testing sets with a ratio of 80%, 10%, and 10%, respectively. The detailed statistics of these datasets are summarized in Table 1.

Baseline Methods. To assess FACAR, we compare it with GNN-based social recommenders (GraphRec [3], DESIGN [19]), CL methods with uniform pair construction (SEPT [26]), and social-graph refinement approaches (GDMSR [13], MADM [9], GBSR [25], SGIL [24]).

Evaluation. All models rank candidate items to produce Top-K recommendation lists for evaluation. To measure the effectiveness of these approaches, we employ two widely-used metrics: Recall@K and NDCG@K [4,15], with K set to 15 and 20, which reflect the performance of recommendation models in accurately retrieving relevant items and ranking them effectively, respectively.

Parameter Settings. All models are implemented using PyTorch. To ensure a fair comparison, all models are trained under the same experimental settings. Specifically, we set the embedding dimension to 8, batch size to 1024 and optimize models using the Adam optimizer with learning rate to be 1e-3. We initialize all embeddings with a Gaussian distribution with a mean value of 0 and a standard variance of 0.1. For all baselines, we follow the original settings and perform the

---

[1] https://www.ciao.co.uk.
[2] https://www.flickr.com.
[3] https://www.yelp.com.

grid search to find the best settings. The hyperparameter α is tuned within the set $0, 1e\text{-}5, 1e\text{-}4, \ldots, 1e\text{-}1, 0.2, 0.3, 0.5, 0.8, 1.0$, and the temperature parameter t in the contrastive learning objective is searched over $0.1, 0.3, 0.5, 0.8, 1, 10, 100$. The details of hyper-parameter study for key parameters are in Sect. 5.5.

5.2   Overall Performance

Table 2. Performance comparison of FACAR with different baselines in terms of Recall@15 (R@15), Recall@20 (R@20), and NDCG@20 (N@20). **Bold** indicates the best performance, <u>underline</u> denotes the second-best and * represents that the improvements compared with base models are statistically significant for $p < 0.05$.

Models	Ciao			Flickr			Yelp		
	R@15	R@20	N@20	R@15	R@20	N@20	R@15	R@20	N@20
GraphRec	0.3769	0.4358	0.3103	0.2719	0.3361	0.1999	0.4428	0.4994	0.3748
DESIGN	0.4391	0.5078	0.3408	0.3231	0.3943	<u>0.2322</u>	<u>0.6350</u>	<u>0.6711</u>	<u>0.5530</u>
SEPT	0.4459	0.5135	<u>0.3483</u>	0.3063	0.3819	0.2236	-	-	-
GDMSR	0.4433	0.5070	0.3412	0.3046	0.3809	0.2007	0.5867	0.6434	0.4290
MADM	0.4246	0.4923	0.3229	0.2811	0.3448	0.2018	-	-	-
GBSR	0.4594	0.5185	0.3263	0.3225	0.3847	0.2284	0.5854	0.6385	0.5299
SGIL	**0.4808**	<u>0.5284</u>	0.3369	<u>0.341</u>	<u>0.4177</u>	0.2319	0.5397	0.6016	0.4274
FACAR	<u>0.4759</u>	**0.5400***	**0.3711***	**0.3694***	**0.4381***	**0.2635***	**0.6644***	**0.6960***	**0.5822***

As shown in Table 2, FACAR consistently outperforms all baselines on three datasets. Compared with GNN-based social recommenders, FACAR yields better results by refining social relations and aligning social-preference graphs. Against contrastive-learning-based models, FACAR avoids uniform negative sampling through a friend-aware, weight-based sampling strategy, enabling finer alignment across social and interaction views and achieve better results. Compared with graph refinement methods, FACAR jointly performs graph-level refinement and representation-level alignment, achieving more robust recommendations.

5.3   Ablation Study

To further investigate the contributions of the individual components in FACAR model, we conduct an ablation study comparing the following variants:

- **FACAR-w/o PAM**: This variant removes the preference-aware social relation characterization. For each user u, the Trusted Friend Set T_u is formed by *randomly sampling* friends from the observed neighbor set S_u. All selected pairs are assigned equal weights. FCL is then applied on these randomly chosen friends in T_u.

Table 3. Ablation results of different components in terms of Recall@15 (R@15), Recall@20 (R@20), and NDCG@20 (N@20). **Bold** indicates the best performance, <u>underline</u> denotes the second-best and * represents that the improvements compared with base models are statistically significant for $p < 0.05$.

Models	Ciao			Flickr			Yelp		
	R@15	R@20	N@20	R@15	R@20	N@20	R@15	R@20	N@20
FACAR-w/o PAM	<u>0.4716</u>	<u>0.5384</u>	<u>0.3708</u>	0.3447	0.4187	0.2497	0.6259	0.6675	0.5446
FACAR-w/o FCL	0.4571	0.5232	0.3532	0.3489	<u>0.4197</u>	<u>0.2513</u>	<u>0.6540</u>	<u>0.6941</u>	<u>0.5593</u>
FACAR-w/o TF	0.4578	0.5247	0.3541	<u>0.3492</u>	0.4170	0.2500	0.6521	0.6902	0.5589
FACAR	**0.4759***	**0.5400***	**0.3711***	**0.3694***	**0.4381***	**0.2635***	**0.6644***	**0.6960***	**0.5822***

- **FACAR-w/o FCL**: This variant disables the friend-aware contrastive learning module while retaining the preference-aware relation mining. The refined social graph is still generated based on preference signals, but the friend-aware contrastive objective is omitted.
- **FACAR-w/o TF**: Built upon FACAR-w/o FCL, this variant replaces the proposed friend-aware contrastive learning with a conventional contrastive learning objective without considering trust-based relational confidence.

As shown in Table 3, removing preference-aware relation mining (FACAR-w/o PAM) consistently degrades performance, verifying that scoring relations before FCL is crucial for the subsequent contrastive learning stage. Eliminating friend-aware contrastive learning (FACAR-w/o FCL) further reduces accuracy, showing that omitting FCL-based alignment between the preference and social views truly harms performance. Deleting trust friends set and replacing FCL with vanilla CL (FACAR-w/o TF) also harms results, demonstrating the effectiveness of sharing information between trust friends and the necessity of trust friends set and the relation-aware sampling strategy.

5.4 Misalignment Stress Test

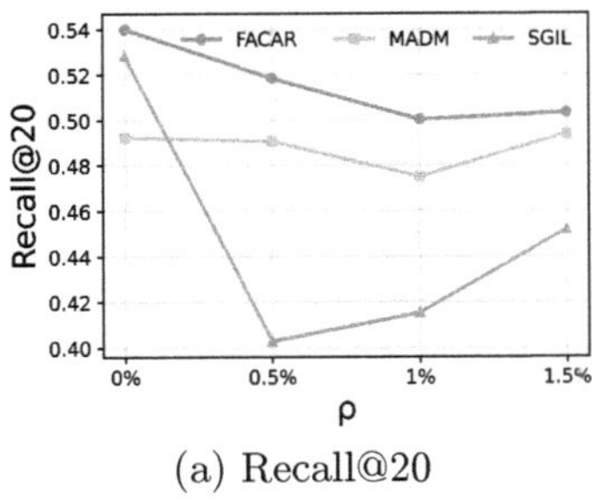

(a) Recall@20

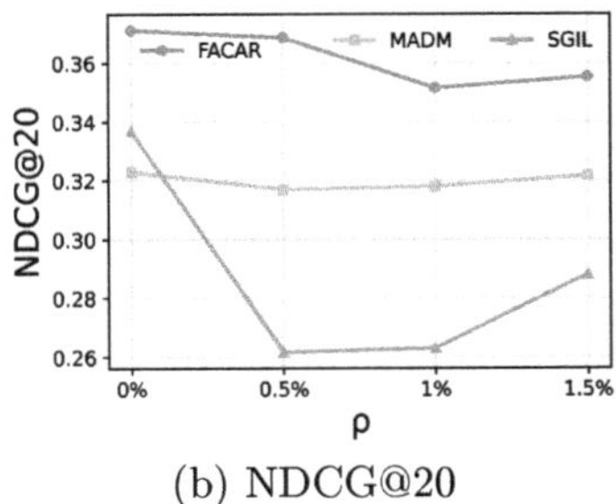

(b) NDCG@20

Fig. 4. Ciao: Misalignment stress test under noise injection.

To assess whether FACAR effectively aligns social signals with user preferences, we deliberately enlarge the social–preference gap by injecting random social links. Concretely, for a noise rate $\rho \in \{0\%, 0.5\%, 1\%, 1.5\%\}$, we uniformly sample ρ of all currently missing directed user–user pairs $(u \neq v)$ and add them as social edges. The interaction graph and evaluation splits are kept unchanged.

As Fig. 4 shows, FACAR exhibits a smooth, gradual degradation as noise increases, maintaining competitive performance throughout. In contrast, SGIL deteriorates rapidly even at modest noise levels, while MADM shows mild yet irregular variations. FACAR consistently retains superior ranking quality.

These trends suggest that FACAR's friend-aware pair construction with relation-aware sampling is better at resisting preference-agnostic links.

5.5 Hyperparameter Analysis

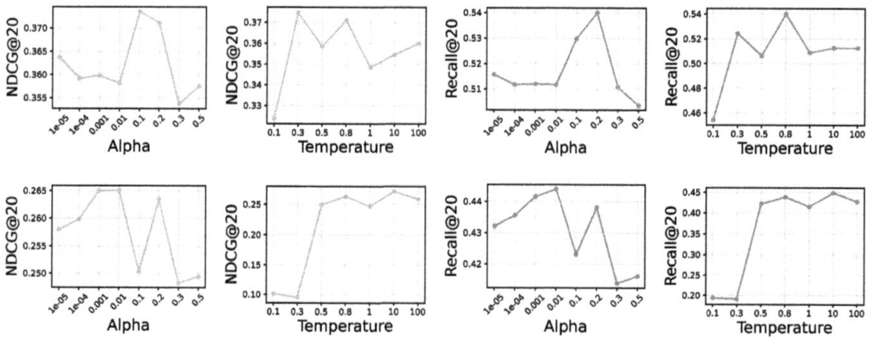

Fig. 5. Hyperparameter sensitivity analysis on different parameters.

To examine the robustness of FACAR with respect to key hyperparameters, we conduct a sensitivity analysis on two representative parameters: (1) the contrastive weight α, which balances the contribution of the friend-aware contrastive objective, and (2) the temperature t, which controls the sharpness of the softmax distribution in the contrastive loss. Experiments are performed on the Ciao and Flickr datasets using Recall@20 and NDCG@20 as evaluation metrics. The results are summarized in Fig. 5.

Effect of α. The α is varied within $10^{-5}, 10^{-4}, 10^{-3}, 10^{-2}, 10^{-1}, 0.2, 0.3, 0.5$. On Ciao, the reaches its peak around $\alpha = 0.1$–0.2, indicating that a moderate weighting of the FCL helps the model exploit social relations effectively. When α becomes too small, the contrastive objective has insufficient influence; conversely, excessively large values (e.g., $\alpha > 0.3$) lead to performance degradation, suggesting that overemphasizing contrastive alignment harms recommendation accuracy. On Flickr, the best results occur around $\alpha = 10^{-2}$–0.2, following a similar pattern but with a smaller optimal range, which may stem from the denser social relations and higher inherent noise level in Flickr.

Effect of T. The hyperparameter t is tuned over $0.1, 0.3, 0.5, 0.8, 1, 10, 100$. On Ciao, both Recall and NDCG improve as t increases from 0.1 to 0.8, reaching the best performance within the range of 0.3–0.8. Smaller temperatures make the contrastive loss overly sensitive to minor representation differences, while extremely large values (e.g., $t > 10$) smooth the softmax excessively, weakening discrimination between positive and negative pairs. On Flickr, however, the optimal temperature is higher (around $t = 10$), implying that a smoother contrastive objective better handles the noisier user–user relations in Flickr.

6   Conclusion

This work revisits the misalignment between social relations and user preferences in social recommendation and shows that uniform pair construction in contrastive learning may undermine user representations. We propose FACAR, which incorporate user preferences into social graph, and performs friend-aware contrastive learning with relation-aware sampling strategy drawn from a Trusted Friend Set, achieving social-preference alignment.

Extensive experiments on three benchmarks demonstrate gains over baselines. Ablation results verify that removing either relation mining or friend-aware contrast substantially degrades performance, and replacing the latter with vanilla contrastive learning reduces accuracy. Under misalignment stress test, FACAR maintains advantages, indicating robustness to spurious links.

Acknowledgments. This research is supported in part by National Natural Science Foundation of China (No. 62472277, No. 62072304, No. 62572309), and the Shanghai East Talents Program (2023-177).

Disclosure of Interests. The authors have no competing interests to declare that are relevant to the content of this article.

References

1. Dai, E., Jin, W., Liu, H., Wang, S.: Towards robust graph neural networks for noisy graphs with sparse labels In: Proceedings of the Fifteenth ACM International Conference on Web Search and Data Mining, pp. 181–191. WSDM '22, Association for Computing Machinery, New York, NY, USA (2022). https://doi.org/10.1145/3488560.3498408
2. Dunbar, R.: How many friends does one person need? Harvard University Press, Dunbar's number and other evolutionary quirks (2010)
3. Fan, W., et al.: Graph neural networks for social recommendation. In: Liu, L., White, R.W., Mantrach, A., Silvestri, F., McAuley, J.J., Baeza-Yates, R., Zia, L. (eds.) The World Wide Web Conference, WWW 2019, San Francisco, CA, USA, May 13–17, 2019, pp. 417–426. ACM (2019). https://doi.org/10.1145/3308558.3313488
4. Gunawardana, A., Shani, G.: A survey of accuracy evaluation metrics of recommendation tasks. J. Mach. Learn. Res. 10, 2935–2962 (2009). https://doi.org/10.5555/1577069.1755883

5. He, R., McAuley, J.: Ups and downs: Modeling the visual evolution of fashion trends with one-class collaborative filtering. In: Proceedings of the 25th International Conference on World Wide Web (WWW) (2016)
6. He, X., Deng, K., Wang, X., Li, Y., Zhang, Y., Wang, M.: Lightgcn: Simplifying and powering graph convolution network for recommendation. In: Huang, J.X., et al. (eds.) Proceedings of the 43rd International ACM SIGIR conference on research and development in Information Retrieval, SIGIR 2020, Virtual Event, China, July 25-30, 2020, pp. 639–648. ACM (2020). https://doi.org/10.1145/3397271.3401063
7. Li, J., Wang, H.: Graph diffusive self-supervised learning for social recommendation. In: Yang, G.H., Wang, H., Han, S., Hauff, C., Zuccon, G., Zhang, Y. (eds.) Proceedings of the 47th International ACM SIGIR Conference on Research and Development in Information Retrieval, SIGIR 2024, Washington DC, USA, July 14–18, 2024, pp. 2442–2446. ACM (2024). https://doi.org/10.1145/3626772.3657962
8. Ma, W., et al.: Towards effective and consistent information extraction for social recommendation: A minimum and sufficiency perspective. In: Proceedings of the 2025 International Conference on Multimedia Retrieval, pp. 991–999. ICMR '25, Association for Computing Machinery, New York, NY, USA (2025). https://doi.org/10.1145/3731715.3733452
9. Ma, W., et al.: Madm: A model-agnostic denoising module for graph-based social recommendation. In: Proceedings of the 17th ACM International Conference on Web Search and Data Mining, pp. 501–509. WSDM '24, Association for Computing Machinery, New York, NY, USA (2024). https://doi.org/10.1145/3616855.3635784
10. Marsden, P.V., Friedkin, N.E.: Network studies of social influence. Sociol. Methods Res. **22**, 127 – 151 (1993). https://api.semanticscholar.org/CorpusID:143267928
11. Massa, P., Avesani, P.: Trust-aware recommender systems. In: Proceedings of the 2007 ACM Conference on Recommender Systems (2007)
12. Mcpherson, M., Smith-Lovin, L., Cook, J.: Birds of a feather: Homophily in social networks. Ann. Rev. Sociol. **27**, 415 (2001). https://doi.org/10.3410/f.725356294.793504070
13. Quan, Y., Ding, J., Gao, C., Yi, L., Jin, D., Li, Y.: Robust preference-guided denoising for graph based social recommendation. In: Ding, Y., Tang, J., Sequeda, J.F., Aroyo, L., Castillo, C., Houben, G. (eds.) Proceedings of the ACM Web Conference 2023, WWW 2023, Austin, TX, USA, 30 April 2023–4 May 2023, pp. 1097–1108. ACM (2023). https://doi.org/10.1145/3543507.3583374
14. Rendle, S., Freudenthaler, C., Gantner, Z., Schmidt-Thieme, L.: BPR: bayesian personalized ranking from implicit feedback. In: Bilmes, J.A., Ng, A.Y. (eds.) UAI 2009, Proceedings of the Twenty-Fifth Conference on Uncertainty in Artificial Intelligence, Montreal, QC, Canada, June 18–21, 2009. pp, 452–461. AUAI Press (2009). https://www.auai.org/uai2009/papers/UAI2009_0139_48141db02b9f0b02bc7158819ebfa2c7.pdf
15. Steck, H.: Evaluation of recommendations: rating-prediction and ranking. In: Yang, Q., King, I., Li, Q., Pu, P., Karypis, G. (eds.) Seventh ACM Conference on Recommender Systems, RecSys '13, Hong Kong, China, October 12–16, 2013, pp. 213–220. ACM (2013). https://doi.org/10.1145/2507157.2507160
16. Sun, X., Zhang, J., Wu, X., Cheng, H., Xiong, Y., Li, J.: Graph prompt learning: a comprehensive survey and beyond (2023). https://arxiv.org/abs/2311.16534

17. Sun, Y., Sun, Z., Du, Y., Zhang, J., Ong, Y.S.: Self-supervised denoising through independent cascade graph augmentation for robust social recommendation. In: Proceedings of the 30th ACM SIGKDD Conference on Knowledge Discovery and Data Mining, pp. 2806–2817. KDD '24, Association for Computing Machinery, New York, NY, USA (2024). https://doi.org/10.1145/3637528.3671958

18. Tang, J., Hu, X., Liu, H.: Social recommendation: a review. Soc. Netw. Anal. Min. (2013)

19. Tao, Y., Li, Y., Zhang, S., Hou, Z., Wu, Z.: Revisiting graph based social recommendation: A distillation enhanced social graph network. In: Laforest, F., et al. (eds.) WWW '22: The ACM Web Conference 2022, Virtual Event, Lyon, France, April 25–29, 2022, pp. 2830–2838. ACM (2022). https://doi.org/10.1145/3485447.3512003

20. Vaswani, A., et al.: Attention is all you need. In: Proceedings of the 31st International Conference on Neural Information Processing Systems, pp. 6000–6010. NIPS'17, Curran Associates Inc., Red Hook, NY, USA (2017)

21. Wang, Q., Sun, X., Cheng, H.: Does graph prompt work? a data operation perspective with theoretical analysis (2025). https://arxiv.org/abs/2410.01635

22. Wei, C., Bai, B., Bai, K., Wang, F.: Gsl4rec: Session-based recommendations with collective graph structure learning and next interaction prediction. In: Laforest, F., et al. (eds.) WWW '22: The ACM Web Conference 2022, Virtual Event, Lyon, France, April 25–29, 2022, pp. 2120–2130. ACM (2022). https://doi.org/10.1145/3485447.3512085

23. Wu, L., Li, J., Sun, P., Hong, R., Ge, Y., Wang, M.: Diffnet++: A neural influence and interest diffusion network for social recommendation. CoRR abs/2002.00844 (2020). https://arxiv.org/abs/2002.00844

24. Yang, Y., et al.: Invariance matters: Empowering social recommendation via graph invariant learning. In: Proceedings of the 48th International ACM SIGIR Conference on Research and Development in Information Retrieval, pp. 2038–2047. SIGIR '25, Association for Computing Machinery, New York, NY, USA (2025). https://doi.org/10.1145/3726302.3730013

25. Yang, Y., Wu, L., Wang, Z., He, Z., Hong, R., Wang, M.: Graph bottlenecked social recommendation. In: Baeza-Yates, R., Bonchi, F. (eds.) Proceedings of the 30th ACM SIGKDD Conference on Knowledge Discovery and Data Mining, KDD 2024, Barcelona, Spain, August 25–29, 2024, pp. 3853–3862. ACM (2024). https://doi.org/10.1145/3637528.3671807

26. Yu, J., Yin, H., Gao, M., Xia, X., Zhang, X., Hung, N.Q.V.: Socially-aware self-supervised tri-training for recommendation. CoRR abs/2106.03569 (2021). https://arxiv.org/abs/2106.03569

27. Yu, J., Yin, H., Li, J., Wang, Q., Hung, N.Q.V., Zhang, X.: Self-supervised multi-channel hypergraph convolutional network for social recommendation. CoRR abs/2101.06448 (2021). https://arxiv.org/abs/2101.06448

28. Zhao, X., et alF.: Social relation-level privacy risks and preservation in social recommender systems. In: Ferro, N., Maistro, M., Pasi, G., Alonso, O., Trotman, A., Verberne, S. (eds.) Proceedings of the 48th International ACM SIGIR Conference on Research and Development in Information Retrieval, SIGIR 2025, Padua, Italy, July 13-18, 2025, pp. 1728–1737. ACM (2025). https://doi.org/10.1145/3726302.3730086

29. Zhao, X., et al.: Dual-adaptive update strategies-enhanced meta-optimization for user cold-start recommendation. ACM Trans. Inf. Syst. **43**(6) (2025). https://doi.org/10.1145/3746634

Machine Learning for Database

ActiveDiag: Dynamic Fusion of Discrepancy and Uncertainty in Active Learning for Database Anomaly Diagnosis

Peize Yuan[1], Zixuan Li[1], Xiyue Gao[1(✉)], Mingzhe Wang[1], Hui Li[1], Yanguo Peng[1], Yaofeng Tu[2], and Jiangtao Cui[1]

[1] School of Computer Science and Technology, Xidian University, Xi'an, Shaanxi, China
{pzyuan,lzx}@stu.xidian.edu.cn,
{xygao,wangmingzhe,hli,ygpeng,jtcui}@xidian.edu.cn
[2] ZTE Corporation, Nanjing, China
tu.yaofeng@zte.com.cn

Abstract. Performance anomalies in database systems degrade service quality, making efficient diagnosis essential. However, existing methods rely on large amounts of labeled data, which is costly to obtain in practice. Active learning tackles this by selectively labeling samples. Despite this, most active learning query strategies prioritize samples with high uncertainty while neglecting those with high discrepancy, hindering the rapid and accurate establishment of decision boundaries. To address these challenges, we propose ActiveDiag, an active learning-based framework for anomaly diagnosis in database systems. It incorporates a Discrepancy and Uncertainty Dynamic Fusion (DUDF) query strategy and a Transformer-based Siamese Network (TSN) classifier to enhance diagnostic performance. Specifically, DUDF considers both discrepancy and uncertainty, dynamically fusing them at different query stages. Additionally, TSN combines the powerful feature representation capability of transformers and the precise similarity measurement capability of siamese networks. Furthermore, we introduce a threshold-constrained cold start strategy to mitigate inaccurate uncertainty estimation of samples during the early query stage, particularly when certain predefined categories are absent from the labeled dataset. Extensive experiments show ActiveDiag achieves superior diagnostic performance, achieving a Macro-F1 of 0.8820 and a Micro-F1 of 0.9070 with 50 expert-labeled samples, outperforming the best previous framework by 9.31% and 9.69%.

Keywords: Active Learning · Anomaly Diagnosis · Database Systems

1 Introduction

With the rapid growth of data and complexity of database systems, performance anomalies such as query delays and throughput reduction have become more frequent. The immediate diagnosis of these anomalies is crucial for ensuring database reliability and enhancing user experience.

Traditional diagnosis methods often rely on manual monitoring and expert judgment, which are time-consuming and error-prone. In contrast, modern methods [20,21,24] leverage machine learning and deep learning techniques to automatically identify complex anomalies by analyzing the Key Performance Indicators (KPIs) from system monitoring, significantly improving the accuracy of diagnosis. Despite these advancements, existing methods for database anomaly diagnosis still face significant challenges, particularly the scarcity of high-quality labeled datasets [15]. The inherent complexity of database systems makes it challenging to construct sufficiently comprehensive and representative labeled datasets, necessitating considerable time, resources, and specialized domain expertise. Active learning has garnered attention as an effective solution to alleviate this problem. Query strategies are used to select the most valuable samples for labeling, improving model performance with minimal labeling cost. Existing query strategies are typically based on informativeness and representativeness, which have proven effective in many domains [2,12]. However, these strategies exhibit limitations in anomaly diagnosis tasks. Uncertainty-based query strategies, a classic approach within informativeness, prioritize samples with low model diagnostic confidence but overlook the significant discrepancies among different categories. Samples with high discrepancy are typically located at the periphery of the data distribution and often represent potential anomalies, playing an essential role in driving the rapid improvement of the model's performance. Similarly, representative-based strategies fail to address the class imbalance in anomaly detection, where normal samples are dense and numerous, while anomalous samples are sparse and few. As a result, these strategies focus on normal samples, hindering the model's ability to learn from anomalous categories. Additionally, a high-performance classifier is paramount for the success of anomaly diagnosis. Traditional classifiers for anomaly diagnosis perform well with sufficient labeled data, but perform poorly with limited labeled data, as they struggle to extract high-dimensional features and discern subtle inter-category distinctions under such conditions [6,36].

To address these challenges, we propose ActiveDiag, an active learning framework that combines a Discrepancy and Uncertainty Dynamic Fusion (DUDF) query strategy with a Transformer-based Siamese Network (TSN) classifier for diagnosing performance anomalies in database systems. DUDF integrates the discrepancy, which reflects the degree to which samples deviate from the distribution of the majority, with uncertainty, which reflects the classifier's confidence in its diagnosis for the sample. During the early stage of querying, DUDF assigns higher weight to discrepancy, selecting samples that are potential anomalies, and thereby facilitating the rapid establishment of decision boundaries. As querying progresses, DUDF shifts its focus to samples with high uncertainty, further refining these decision boundaries. This dynamic fusion ensures the continuous selection of high-value samples throughout the learning process, ultimately optimizing model's performance at all stages. Meanwhile, TSN combines the transformer with a siamese network structure, mapping similar samples closer together in the feature space while pushing dissimilar ones farther apart [3]. This design enhances the model's representation capacity, enabling accurate discrimination of subtle category differences and maintains high diagnostic performance under limited labeled data. We further propose a threshold-constrained cold start strategy to estimate the uncertainty of samples in the early query stage. By calculating the maximum similarity between the sample and known category samples, the strategy generates smaller uncertainty scores when the similarity exceeds the threshold and larger scores when it falls below, enabling more accurate and faster uncertainty estimation even with incomplete category coverage. Our specific contributions are as follows:

- We design DUDF, a novel strategy that dynamically fuses discrepancy and uncertainty to select the most valuable samples for expert labeling.
- We employ TSN for anomaly diagnosis, leveraging transformer and siamese network to learn category differences, ensuring high accuracy with limited labeled data.
- We propose a threshold-constrained cold start strategy to alleviate inaccurate uncertainty estimation in the early query stage, when predefined categories are incomplete.
- The experimental results show that ActiveDiag achieves superior diagnostic performance. Notably, it improves Macro-F1 and Micro-F1 by 9.31% and 9.69%, respectively, over the previous best method, using only 50 labeled samples.

The rest of the paper starts with an overview of the related work. Section 3 introduces our proposed framework ActiveDiag, Sect. 4 details experiments, and we conclude in Sect. 5.

2   Related Work

Database Anomaly Diagnosis. Existing methods for database anomaly diagnosis typically concentrate on attributing anomalies to specific KPIs, SQL queries (SQLs), or predefined categories. Both FluxInfer [24] and CauseRank [26] aim to attribute anomalies to specific KPIs, such as *queries_per_second*, *transactions_per_second*, *cpu_usage*, *memory_cache*, and *mem_buffer*. FluxInfer constructs a Weighted Undirected Dependency Graph to capture relationships between anomalous KPIs and applies a weighted PageRank algorithm to pinpoint KPIs linked to the root cause of performance issues. Similarly, CauseRank employs causal inference with a group-based causal discovery algorithm to build causal graphs from KPIs, to identify KPIs related to performance issues. It then ranks these KPIs by importance using the Causal Oriented Personalized PageRank algorithm, enabling accurate root cause identification. Pin-SQL [25] categorizes anomalous SQLs into high-impact SQLs correlated with anomalies and root cause SQLs that are the root causes of the performance issue. By analyzing KPIs and query logs and tracing the propagation chain of relevant SQLs, it precisely pinpoints the root cause SQLs, effectively addressing performance issues in cloud databases. Methods such as Tejo [34] and DistDiagnosis [37] treat anomaly diagnosis as a classification task. They use classifiers like XGBoost and Random Forest to attribute anomalies to specific categories, such as I/O bottlenecks, missing indexes, and heavy workloads. More recent work, such as OpDiag [14], achieves a more comprehensive anomaly diagnosis by simultaneously attributing anomalies to specific SQLs, KPIs, categories, and even query operators.

Active Learning Query Strategy. Active learning is a machine learning paradigm that selectively labels the most valuable samples, thereby enhancing model performance while minimizing labeling costs [2]. This approach is particularly effective in scenarios where labeled data is scarce or labeling is resource-intensive. Query strategies in active learning are typically categorized as informativeness-based and representativeness-based strategies [18]. Informativeness-based strategies encompass techniques such as uncertainty sampling and query-by-committee, which prioritize the selection of instances that provide the most information for model improvement. In contrast, representativeness-based strategies, including density-based and diversity-based approaches, focus on selecting instances that best represent the underlying data distribution, ensuring a more comprehensive model understanding. Active learning can be categorized based on application scenarios into three types [18]: membership query synthesis, stream-based selective sampling, and pool-based active learning. Monitoring systems generate vast unlabeled data that well reflect database behavior, making performance anomaly diagnosis suitable for pool-based active learning.

3   ActiveDiag Framework

As shown in Fig. 1, ActiveDiag starts with the system monitoring module to collect the original KPIs and transformed them into data in specific format. The data is then processed by the discrepancy and uncertainty calculation modules, which calculate samples' discrepancy and uncertainty scores, respectively. These scores are fused in the dynamic fusion module to prioritize the most valuable samples for expert labeling. Labeled samples are used to generate pairs for training the TSN. For online diagnosis, the TSN module computes the probability of each sample belonging to each category, thereby generating the final diagnosis results for database administrators to handle. The detailed description of each module is provided in the following content.

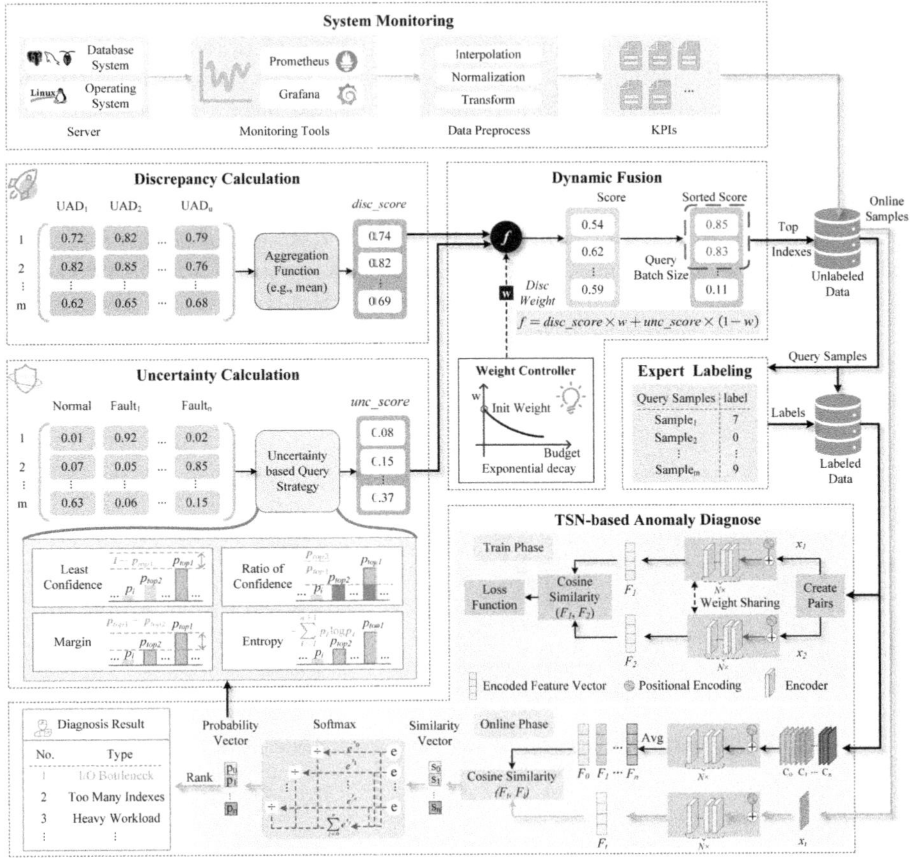

Fig. 1. The ActiveDiag framework, with orange arrows representing the online diagnosis process and black arrows representing the active learning and model training process.(Color figure online)

3.1 System Monitoring

Database systems typically operate in complex environments, coexisting with various applications and system services that share resources and mutually influence one another. To create a unified view of the system's performance, both database and operating system KPIs must be monitored [15]. The KPIs related to the operating system include I/O operations, CPU utilization, memory usage, network activity, as well as additional metrics such as interrupt counts, lock statuses, process counts, and socket statistics. The KPIs related to the database system include connection management, SQL query execution, transaction processing, and lock states. Tools such as Prometheus[1] and Grafana[2] are deployed

---

[1] https://prometheus.io/.
[2] https://grafana.com/.

to provide an efficient solution for continuous system monitoring and visualization. Prometheus supports monitoring various types of database systems and operating systems through Exporters[3]. As a result, ActiveDiag can be adapted for anomaly diagnosis across a wide range of mainstream database systems, such as MySQL, MariaDB, PostgreSQL, and CockroachDB. Meanwhile, Grafana is employed to visualize the KPIs collected by Prometheus. It provides a user-friendly interface for displaying the KPIs in real-time, allowing users to intuitively observe system performance and facilitating data-driven decision-making. Subsequently, we handle missing values by mean imputation and normalize the KPIs using Z-score standardization. The KPIs are then transformed into samples in specific format, where each sample consists of a sequence of KPIs recorded at multiple timestamps with a fixed interval between consecutive timestamps. Finally, these samples are stored, making them ready for anomaly diagnosis.

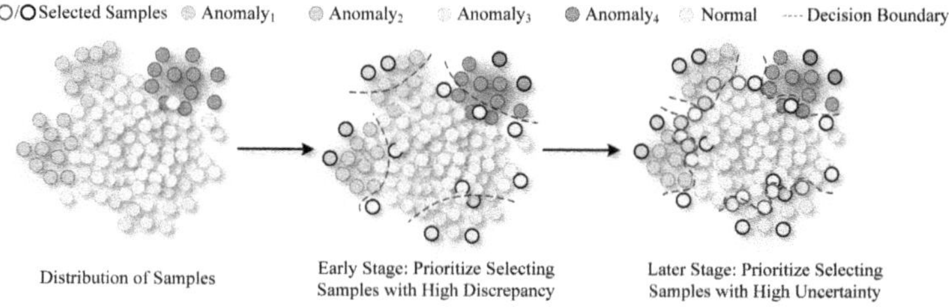

Fig. 2. The dynamic process of sample selection via DUDF Strategy

3.2 DUDF Query Strategy

DUDF simultaneously considers both discrepancy and uncertainty, with their focus adjusted dynamically at different query stages, rather than adopting a rigid strategy that treats them in isolation at each phase. As illustrated in Fig. 2, during the early stage of querying, DUDF prioritizes samples with high discrepancy, which are typically located at the periphery of the data distribution and are often indicative of potential anomalies. Labeling these samples facilitates the rapid establishment of an initial decision boundary between normal and anomalous samples. Moreover, the specific labels provided through expert labeling assist in defining decision boundaries between different categories. It is worth noting that samples with extremely high uncertainty may still be selected in this stage, as the samples are chosen based on the final weighted scores. As the querying process progresses, DUDF increasingly focuses on samples exhibiting higher uncertainty, which are often difficult to diagnose. Labeling these samples enables in precisely refining decision boundaries, further enhancing classification performance. Next, we provide a detailed description of the three core steps of DUDF: Discrepancy Calculation, Uncertainty Calculation, and Dynamic Fusion.

---

[3] https://prometheus.io/docs/instrumenting/exporters/.

Discrepancy Calculation. This module calculates discrepancy scores to identify samples with the highest discrepancy in the data distribution. Samples with high discrepancy scores are often located at the periphery of the data distribution and may correspond to potential anomalies. Unsupervised anomaly detection algorithms, which identify anomalous samples based on their deviation from the majority of the data, inherently align with discrepancy calculation methods, making anomaly scores a valid proxy for discrepancy scores. To obtain more robust and comprehensive discrepancy scores, we calculate anomaly scores from multiple unsupervised anomaly detection algorithms using aggregation functions such as *Max*, *Min*, *Mean*, or *Median*. This ensemble approach capitalizes on the complementary strengths of different methods. Additionally, databases and operating systems generate numerous KPIs, often in the hundreds or thousands, leading to high-dimensional data. In such spaces, many unsupervised anomaly detection algorithms perform poorly, making it essential to employ algorithms specifically designed for high-dimensional settings.

In this paper, we integrate three such algorithms: iForest [23], HBOS [10], and COPOD [22], each of which has been proven effective in high-dimensional data [42]. These algorithms are particularly well-suited to address the complexity and sparsity inherent in high-dimensional KPIs. First, iForest isolates samples by constructing random binary trees. For each sample x, we calculate the average path length $h(x)$ required to isolate it in the forest. The discrepancy score is defined as

$$d_{\text{iForest}}(x) = 2^{-\frac{h(x)}{c(m)}}, \tag{1}$$

where m denotes the total number of samples, $c(m)$ is a normalization constant approximated by $2H(m-1) - \frac{2(m-1)}{m}$, and $H(i)$ represents the i-th harmonic number, which can be further approximated as $\ln(i)+\gamma$, with γ being the Euler–Mascheroni constant. This formula converts the path length into a discrepancy score within the range $[0, 1]$; shorter paths (indicating greater discrepancy) yield scores closer to 1. Second, HBOS evaluates the sparsity of each sample based on the density of its features' histograms. For each sample x, the discrepancy score is defined as

$$d'_{\text{HBOS}}(x) = \sum_{i=1}^{d} \log \left(\frac{1}{\text{hist}_i(x)} \right), \tag{2}$$

where d denotes the number of dimensions, and $\text{hist}_i(x)$ represents the histogram-based density estimate of sample x for the i-th feature. To eliminate scale differences, we apply min-max normalization, yielding $d_{\text{HBOS}}(x)$ in the range $[0, 1]$, where higher values indicate greater discrepancy. Third, COPOD models the multivariate dependency structure of the samples using copula theory [22]. For each sample x, we compute its marginal Cumulative Distribution Function (CDF) values $V_i(x_i)$ in each dimension and then estimate the tail probability of the joint CDF. For each sample x, the discrepancy score is defined as

$$d_{\text{COPOD}}(x) = 1 - C(V_1(x_1), V_2(x_2), \ldots, V_d(x_d)), \tag{3}$$

where C is the copula function, and d is the feature dimension. Since $d_{\mathrm{COPOD}}(x)$ is inherently a probability value, it naturally falls within the range $[0, 1]$. Higher values indicate that the sample lies in the tail region of the distribution, reflecting higher discrepancy. For each sample x, the final discrepancy score is defined as

$$disc\_score(x) = AGG(d_{\mathrm{iForest}}(x), d_{\mathrm{HBOS}}(x), d_{\mathrm{COPOD}}(x)), \tag{4}$$

where AGG is the aggregation function, typically Max, Min, $Mean$, or $Median$ in practice.

Uncertainty Calculation. This module is designed to calculate uncertainty scores to identify samples with the highest uncertainty in the model's predictions, which are most likely to provide valuable insights for refining the model's decision boundaries. For each sample x, the module applies one of the classic uncertainty-based query strategies, such as Least Confidence [5], Entropy [32], Ratio of Confidence, or Margin [30], to compute an uncertainty score denoted as $unc\_score(x)$. The details of these strategies are as follows:

- Least Confidence. It calculates the uncertainty score based on the model's confidence in its top prediction, defined as $LC(x) = 1 - p_{\mathrm{top1}}(y|x)$, where p_{top1} is the probability of the most likely predicted category. A higher LC indicates greater uncertainty in the model's top prediction.
- Entropy. It calculates the uncertainty score using Shannon entropy of the model's predicted probability distribution, defined as $E(x) = -\sum_i p_i(y|x) \log p_i(y|x)$, where $p_i(y|x)$ represents the predicted probability that the model assigns to the i-th possible category for input x. A higher entropy indicates greater uncertainty.
- Ratio of Confidence. It calculates the uncertainty score based on the ratio of the model's top two predicted probabilities, defined as $RC(x) = \frac{p_{\mathrm{top2}}(y|x)}{p_{\mathrm{top1}}(y|x)}$, where p_{top1} and p_{top2} are the highest and second-highest predicted probabilities, respectively. A higher RC indicates greater uncertainty.
- Margin. It calculates the uncertainty score based on the difference between the model's top two predicted probabilities, defined as $M(x) = p_{\mathrm{top1}}(y|x) - p_{\mathrm{top2}}(y|x)$. A smaller margin indicates higher uncertainty, as the model has difficulty distinguishing between the top two predictions.

Dynamic Fusion. This module balances discrepancy and uncertainty in the sample selection process, allowing the model to dynamically adjust its focus between the two factors. During the early stage of querying, high-discrepancy samples, which are potential anomalies, are prioritized, facilitating the rapid establishment of decision boundaries and early performance improvements. However, as learning progresses, an overemphasis on high-discrepancy samples may hinder the refinement of intricate decision boundaries, ultimately leading to a decline in diagnostic performance. To address this, we introduce a dynamic decay mechanism for the discrepancy score weight w, controlled by an exponential decay function defined as follows:

$$w = \alpha \cdot e^{-\frac{r}{T}}, \tag{5}$$

where $\alpha \in (0, 1.0)$ is the initial weight, r denotes the current round of active learning, and T controls the decay rate. A higher α emphasizes high-discrepancy samples early, while a lower α accelerates the shift to high-uncertainty samples. Larger T slows the decay, maintaining the emphasis on high-discrepancy samples for a longer duration. We conducted extensive experiments and recommend $\alpha \in [0.6, 0.8]$ and $T \in [50, 250]$. Finally, for each sample x, the fusion score is defined as

$$fusion\_score(x) = w \cdot disc\_score(x) + (1 - w) \cdot unc\_score(x). \tag{6}$$

Following this, we select the sample with the highest fusion score for expert labeling, repeating the process until the budget (the upper limit on the number of samples allowed for labeling) is reached. This dynamic fusion ensures continuous selection of high-value samples throughout the learning process. The pseudocode for DUDF is shown in Algorithm 1.

Algorithm 1: The DUDF Query Strategy

Input: U (unlabeled dataset), $\mathcal{B}$ (budget), α (initial weight), g (uncertainty-based query strategy), T (decay rate constant), AGG (aggregation function), r (current round of active learning)

Output: Q (selected samples)

1   $Q \leftarrow \emptyset$;
2   **for** $r = 1$ **to** $\mathcal{B}$ **do**
3   Compute w via Eq. (5);
4   **for** $x \in U \setminus Q$ **do**
5         $unc\_score(x) \leftarrow g(x)$;
6   Compute $disc\_score(x)$ via Eq. (4);
7   Compute $fusion\_score(x)$ via Eq. (6);
8      $x \leftarrow \arg\max_{x \in U \setminus Q} fusion\_score(x)$;
9      $Q \leftarrow Q \cup \{x\}$;
10   **return** Q;

3.3   TSN-Based Anomaly Diagnosis

To enable accurate diagnosis with limited labeled data, ActiveDiag leverages a TSN-based classifier that combines a transformer for feature extraction and a siamese network for similarity learning.

During the training phase, samples from the same category are matched to form positive pairs, while samples from different categories are matched to form negative pairs. Two transformer encoders [41] with shared parameters are then used to represent these input pairs, generating encoded feature vectors. The following loss function is adopted to optimize the model:

$$\mathcal{L} = \frac{1}{2B} \sum_{i=1}^{B} \left(y_i \cdot (1 - cos(F_i^1, F_i^2))^2 + (1 - y_i) \cdot [\max(0, cos(F_i^1, F_i^2) - m)]^2 \right),$$

(7)

where B is the batch size of sample pairs, y_i is the binary label indicating whether the pair is positive (1) or negative (0), $cos(F_i^1, F_i^2)$ represents the cosine similarity between the encoded feature vectors F_i^1 and F_i^2 of sample pairs after encoding by the transformer, and m is a margin threshold, which ensures that the similarity of negative sample pairs stays below this value to encourage the model to better distinguish between different categories of samples during training. We conducted extensive experiments and recommend $m = 0.3$. The loss function minimizes distances between features of same-category samples and maximizes distances between those of different categories.

During the diagnosis phase, similar to Tejo and DistDiagnosis, we model the diagnosis problem as a multi-class classification task. The final classification is conducted by calculating the cosine similarity between the target sample's encoded feature vector F_t and the prototype vector F_i of each known category. F_i is computed by encoding all samples belonging to category i with the transformer and averaging the resulting feature vectors. These cosine similarity scores are then input into a softmax classifier, which normalizes them into probabilities. The category with the highest probability is selected as the diagnostic result.

This classifier leverages intra-class similarities and emphasizes inter-class differences during training, thereby improving sensitivity to category distinctions and addressing the challenges posed by data scarcity. The pseudocode for this diagnostic algorithm is shown in Algorithm 2.

3.4   Threshold-Constrained Cold Start

During the initialization of active learning, we select only a small number of random samples to be labeled by experts as the initial labeled dataset. These samples often cover a subset of categories, leaving the model without prior knowledge of the remaining category. Consequently, online diagnostic samples are evaluated with limited category information, leaving other categories unconsidered. The absence of category information severely impacts the accuracy of the sample uncertainty estimation.

To mitigate the cold-start problem, we propose a threshold-constrained strategy. Given an online diagnostic sample, we compute the cosine similarity between its encoded feature vector F_t and the prototype vector F_i of each known category. The highest similarity score across all categories is denoted as M, representing the model's most confident match. Based on this, we define the uncertainty score

---

Algorithm 2: TSN-based Anomaly Diagnosis

Input: U (unlabeled dataset), E_1, E_2 (transformer encoders), $\mathcal{L}$ (contrastive loss), $\mathcal{D}$ (dictionary of categories with corresponding samples), B (batch size), ξ (epochs), x_t (the online sample)

Output: $\hat{y}$ (predicted category label for the online sample)

   // **Training phase**

1 Initialize E_1 and E_2 with shared weights;

2 Initialize empty set P to store constructed sample pairs;

3 **for** $c_i \in \mathcal{D}.keys()$ **do**

4   **for** $x \in \mathcal{D}[c_i]$ **do**

5   **for** $x^+ \in \mathcal{D}[c_i]$ **do**

6   **if** $x^+ \neq x$ and $(x, x^-) \notin P$ and $(x^+, x) \notin P$ **then**

7   Add (x, x^+) to P;

8   **for** $c_j \in \mathcal{D}.keys()$ and $c_j \neq c_i$ **do**

9   **for** $x^- \in \mathcal{D}[c_j]$ **do**

10   **if** $x^- \neq x$ and $(x, x^-) \notin P$ and $(x^-, x) \notin P$ **then**

11   Add (x, x^-) to P;

12 **for** $j = 1$ **to** ξ **do**

13   **for** $i = 1$ **to** $\lceil |P|/B \rceil$ **do**

14   Select a batch P' with $|P'| = B$ from P (iteratively);

15   **for** $(x_i^1, x_i^2) \in P'$ **do**

16           $F_i^1 \leftarrow E_1(x_i^1)$, $F_i^2 \leftarrow E_2(x_i^2)$;

17   Compute the loss function $\mathcal{L}$ using Eq. (7), then backpropagate and update E_1 and E_2.

   // **Diagnosis phase**

18 $F_t \leftarrow E(x_t)$; // **Compute the feature vector of the sample to be diagnosed**

19 **for** $c_i \in \mathcal{D}.keys()$ **do**

20     $F_i \leftarrow \frac{1}{|\mathcal{D}[c_i]|} \sum_{x_j \in \mathcal{D}[c_i]} E(x_j)$; // **Compute the prototype vector of the category** c_i

21     $S_i \leftarrow$ Cosine Similarity(F_t, F_i);

22 $P \leftarrow$ Softmax(S) ; // **Compute probability vector**

23 $\hat{y} \leftarrow \arg\max_i P_i$

24 **return** $\hat{y}$;

---

unc_score as follows:

$$unc\_score = \begin{cases} \text{random}(0, u), & \text{if } M > \tau \\ \text{random}(v, 1.0), & \text{if } M \leq \tau \end{cases}, \tag{8}$$

where u determines the upper bound of the low uncertainty score, v determines the lower bound of the high uncertainty score; and τ represents a similarity threshold used to determine whether the online sample has a sufficiently strong association with a known category. If $M > \tau$, this indicates that the sample, despite not having encountered samples from other categories, still exhibits a high similarity with a known category. This strong similarity implies that the sample is likely to belong to that category, resulting in a low uncertainty score. Conversely, lower similarity suggests a weaker association with known categories and a higher likelihood of belonging elsewhere. This corresponds to a higher uncertainty score, increasing its probability of being selected for expert labeling. The strategy facilitates rapid and accurate uncertainty estimation even when predefined categories are incomplete in the early stage, mitigating the cold-start problem. The values of u and v can be set based on expert knowledge, with typical values being $u = 0.3$ and $v = 0.7$, while τ should be determined experimentally, with its optimal value being evaluated and justified in Subsect. 4.5.

4   Experiments

4.1   Experimental Setup

Experimental Environment. We simulate an Online Transaction Processing (OLTP) scenario in PostgreSQL running on a CentOS platform, using OLTP-Bench[4] to model database workloads and Prometheus to monitor KPIs from both PostgreSQL and CentOS. Each sample consists of a sequence of KPIs spanning 12 timestamps, with an interval of 5 s between consecutive timestamps. Meanwhile, we simulate various categories of anomalies, resulting in a dataset for anomaly diagnosis. Due to the application of the threshold-constrained cold start strategy, only a few samples (typically 10–20) need to be randomly labeled during initialization in active learning. Notably, DUDF adopts the "ratio of confidence" as the default method for uncertainty calculation.

Anomaly Category. Guided by prior studies, including DBPA [15] and DB-MAGS [33], we systematically define nine representative categories of database performance anomalies. These include CPU bottlenecks, I/O bottlenecks, network faults, highly concurrent inserts, highly concurrent commits, heavy workloads, missing indexes, excessive indexes and lock waits. Specifically, we use Chaosblade[5] to inject CPU bottlenecks, I/O bottlenecks, and network faults. The remaining anomalies are injected by specific SQL scripts, following the standards outlined in DBPA and DB-MAGS.

Evaluation Metrics. Given that anomaly diagnosis is considered a multi-class classification task, we use Macro-F1 (Ma-F) and Micro-F1 (Mi-F) as evaluation metrics. Ma-F calculates the average F1-score across all categories, reflecting the model's overall effectiveness across all categories. On the other hand, Mi-F is derived from global statistics, capturing the model's comprehensive performance across the entire dataset. To ensure the reliability of experiments and reduce random errors, each experiment is repeated three times and the mean value of these results is reported as the final result.

---

[4] https://github.com/oltpbenchmark/oltpbench.
[5] https://github.com/chaosblade-io/chaosblade.

4.2   Comparison Study

Comparison with Prior Frameworks. To evaluate the effectiveness of ActiveDiag, we compare it against prior classification-based anomaly diagnosis frameworks including Tejo [34], AnoDD [29], AutoMonitor [17], ISQUAD [27], FaultDR [9], FIRED [38], DistDiagnosis [37], ADSDN [39] and OpDiag [14]. As shown in Table 1, ActiveDiag achieves best performance across all different budgets. Particularly, it achieves a Ma-F of 0.8820 and a Mi-F of 0.9070 with only 50 budget, outperforming the best previous framework, FIRED, by 9.31% and 9.69%, respectively. As budget increases, the Ma-F and Mi-F of ActiveDiag steadily improve, surpassing 0.9 at 100 budget and reaching 0.9139 and 0.9187 at 250 budget, while the majority of other frameworks fail to approach 0.9. These results highlight the superior performance of ActiveDiag with limited labeled data.

Table 1. Comparison with Prior Database Anomaly Diagnosis Frameworks

Framework	Budget									
	50		100		150		200		250	
	Ma-F	Mi-F	Ma-F	Mi-F	Ma-F	Mi-F	Ma-F	Mi-F	Ma-F	Mi-F
Tejo [34]	0.7802	0.7863	0.8139	0.8155	0.8663	0.8841	0.8798	0.8917	0.8854	0.8974
AnoDD [29]	0.7033	0.7002	0.7892	0.7714	0.8276	0.8326	0.8531	0.8581	0.8601	0.8733
AutoMonitor [17]	0.7852	0.7920	0.8318	0.8168	0.8860	0.8787	0.9085	0.9073	0.9064	0.9085
ISQUAD [27]	0.6761	0.6668	0.7603	0.7628	0.8231	0.8439	0.8447	0.8686	0.8419	0.8680
FaultDR [9]	0.8026	0.8177	0.8772	0.8688	0.8839	0.8790	0.8872	0.8793	0.8992	0.8933
FIRED [38]	0.8069	0.8269	0.8403	0.8346	0.8473	0.8393	0.8612	0.8501	0.8811	0.8641
DistDiagnosis [37]	0.6864	0.6831	0.7490	0.7863	0.7858	0.8257	0.8190	0.8346	0.8275	0.8361
ADSDN [39]	0.8062	0.8219	0.8432	0.8527	0.8726	0.8735	0.8864	0.8881	0.8910	0.8922
ActiveDiag	**0.8820**	**0.9070**	**0.9002**	**0.9120**	**0.9042**	**0.9124**	**0.9115**	**0.9171**	**0.9139**	**0.9187**

Comparison with Prior Query Strategies. To demonstrate the superiority of DUDF, we compare it with several state-of-the-art active learning query strategies, including QUIRE [13], CoreSet [31], BADGE [1], NNAL [11], TypiClust [12], ProbCover [40], SBAL [35], AL-FEW [7], and Random, all using the same classifier (e.g., TSN). As shown in Table 2, DUDF achieves best performance across all different budgets. Particularly, it achieves a Mi-F of 0.9070 with only 50 budget, outperforming TypiClust by 4.16%. Additionally, DUDF achieves both Ma-F and Mi-F greater than 0.9 with 100 budget, while no other strategy reaches this level until 250 budget. These results highlight DUDF's superiority in enhancing early-stage performance and sustaining strong performance in later stages.

Comparison with Prior Classifiers. To demonstrate the superiority of TSN, we compare it against classical machine learning approaches such as XGBoost, LightGBM, GNB [6], and GBDT [8], as well as state-of-the-art deep learning

Table 2. Comparison with Prior Query Strategies

Query Strategy	Budget									
	50		100		150		200		250	
	Ma-F	Mi-F	Ma-F	Mi-F	Ma-F	Mi-F	Ma-F	Mi-F	Ma-F	Mi-F
Random	0.8422	0.8323	0.8695	0.8644	0.8721	0.8720	0.8884	0.8869	0.8883	0.8866
QUIRE [13]	0.7994	0.7958	0.8118	0.8133	0.8226	0.8311	0.8328	0.8434	0.9084	0.9012
CoreSet [31]	0.8634	0.8688	0.8840	0.8908	0.8977	0.8977	0.8985	0.8990	0.8985	0.9050
BADGE [1]	0.6408	0.8298	0.8760	0.8800	0.8870	0.8904	0.8909	0.8914	0.8909	0.8968
NNAL [11]	0.8278	0.8333	0.8578	0.8654	0.8641	0.8758	0.8839	0.8815	0.8945	0.8939
TypiClust [12]	0.8690	0.8708	0.8824	0.8871	0.8911	0.8882	0.8912	0.8895	0.8953	0.8930
ProbCover [40]	0.8577	0.8657	0.8697	0.8676	0.8801	0.8876	0.8852	0.8936	0.8901	0.8965
SBAL [35]	0.8471	0.8371	0.8745	0.8694	0.8771	0.8719	0.8935	0.8919	0.8934	0.8917
AL-FEW [7]	0.8453	0.8354	0.8727	0.8676	0.8768	0.8791	0.8917	0.8902	0.8916	0.8899
DUDF	**0.8820**	**0.9070**	**0.9002**	**0.9120**	**0.9042**	**0.9124**	**0.9115**	**0.9171**	**0.9139**	**0.9187**

Table 3. Comparison with Prior Classifiers

Classifier	Shots (Number of Labeled Samples per Category)									
	5		10		15		20		25	
	Ma-F	Mi-F	Ma-F	Mi-F	Ma-F	Mi-F	Ma-F	Mi-F	Ma-F	Mi-F
XGBoost	0.7241	0.7831	0.8338	0.8266	0.8151	0.8444	0.8556	0.8587	0.8982	0.8965
LightGBM	0.8656	0.8492	0.8764	0.8660	0.8903	0.8866	0.8941	0.8892	0.8940	0.8898
GNB [6]	0.8641	0.8466	0.8695	0.8536	0.8718	0.8625	0.8742	0.8644	0.8526	0.8444
GBDT [8]	0.7853	0.7898	0.8644	0.8523	0.8876	0.8844	0.8959	0.8869	0.9008	0.8990
Hydra [36]	0.8455	0.8625	0.8557	0.8758	0.8675	0.8889	0.8661	0.8876	0.8839	0.9011
LITETime [16]	0.7865	0.8066	0.8564	0.8577	0.8768	0.8800	0.8759	0.8806	0.8807	0.8800
HIVE-COTE v2 [28]	0.8566	0.8406	0.8768	0.8815	0.8872	0.8768	0.8893	0.8968	0.9023	0.9013
ShapeFormer [19]	0.7813	0.7817	0.8536	0.8562	0.8613	0.8644	0.8759	0.8791	0.8898	0.8915
ConvTimeNet [4]	0.7778	0.8079	0.8496	0.8422	0.8505	0.8603	0.8716	0.8739	0.8996	0.9001
TSN	**0.8851**	**0.8828**	**0.8991**	**0.9032**	**0.9126**	**0.9083**	**0.9202**	**0.9172**	**0.9344**	**0.9312**

approaches, including HIVE-COTE v2 [28], Hydra [36], LITETime [16], Shape-Former [19] and ConvTimeNet [4]. As shown in Table 3, TSN achieves best performance across all shots among the classifiers. Particularly, TSN achieves Ma-F and Mi-F of 0.9344 and 0.9312 with 25 shots, outperforming the best classifiers, HIVE-COTE v2, by 3.56% and 3.32%, respectively, while other classifiers can hardly approach 0.9.

4.3   Ablation Study

To evaluate the effectiveness of DUDF, we conduct an ablation study comparing three variants for uncertainty-based query strategies: (1) Q_1, which represents the original uncertainty-based query strategy; (2) Q_2, which combines uncertainty and discrepancy without dynamically adjusting their weights; and (3)

Q_3, which combines both uncertainty and discrepancy, dynamically adjusting their weights as budget increases, representing the proposed DUDF. As shown in Table 4, Q_3 consistently outperforms both Q_1 and Q_2 in different budgets. Specifically, for Least Confidence, Q_3 improves Mi-F by 6.54% over Q_1 and by 3.32% over Q_2 with only 50 budget. Similar improvements are observed in other strategies, where Q_3 remains superior the alternatives.

Table 4. Ablation Study on DUDF

Uncertainty Strategy	Variants	Budget									
		50		100		150		200		250	
		Ma-F	Mi-F	Ma-F	Mi-F	Ma-F	Mi-F	Ma-F	Mi-F	Ma-F	Mi-F
Least Confidence	Q_1	0.8485	0.8438	0.8624	0.8600	0.8714	0.8755	0.8757	0.8771	0.8786	0.8790
	Q_2	0.8442	0.8701	0.8782	0.8990	0.8808	0.9006	0.8961	0.9041	0.8987	0.8965
	Q_3	**0.8627**	**0.8990**	**0.8864**	**0.9085**	**0.8972**	**0.9117**	**0.9046**	**0.9130**	**0.9050**	**0.9165**
Entropy	Q_1	0.8742	0.8704	0.8733	0.8688	0.8870	0.8838	0.8980	0.8920	0.8977	0.8908
	Q_2	0.8732	0.8733	0.8776	0.8781	0.8881	0.8850	0.8948	0.8955	0.9001	0.8904
	Q_3	**0.8745**	**0.8815**	**0.8874**	**0.9016**	**0.8937**	**0.9063**	**0.8987**	**0.8955**	**0.9002**	**0.9025**
Ratio of Confidence	Q_1	0.8226	0.8431	0.8356	0.8920	0.8945	0.8962	0.8999	0.9009	0.9000	0.9025
	Q_2	0.8429	0.8625	0.8769	0.9025	0.8774	0.9060	0.8979	0.9025	0.8965	0.8933
	Q_3	**0.8820**	**0.9070**	**0.9002**	**0.9120**	**0.9042**	**0.9124**	**0.9115**	**0.9171**	**0.9139**	**0.9187**
Margin	Q_1	0.8256	0.8552	0.8772	0.8971	0.8820	0.9028	0.8987	0.9054	0.9108	0.9139
	Q_2	0.8624	0.9003	0.8759	0.9012	0.8768	0.9076	0.8857	0.9050	0.8971	0.9050
	Q_3	**0.8691**	**0.9019**	**0.8876**	**0.9155**	**0.9021**	**0.9047**	**0.9096**	**0.9146**	**0.9125**	**0.9162**

4.4 Generalization Study

Performance under Different Workloads. To assess the generalizability of Active-Diag across different workloads, we conduct experiments on four benchmarks from OLTPBench: TPC-C (W_1), TATP (W_2), Voter (W_3) and Smallbank (W_4). As illustrated in Fig. 3(a), both Ma-F and Mi-F exhibit consistent improvement, ultimately reaching around 0.92 at 250 budget across all the workloads. These results underscore the robust adaptability of ActiveDiag across a variety of workloads.

Performance under Different Databases. To evaluate the generalizability of ActiveDiag across different database systems, we conduct experiments on PostgreSQL (DB_1), MySQL (DB_2), CockroachDB (DB_3), and MariaDB (DB_4). The experimental procedures for MySQL, CockroachDB, and MariaDB follow the same steps as those for PostgreSQL, as detailed in Sect. 4.1. As shown in Fig. 3(b), both Ma-F and Mi-F demonstrate consistent improvement, surpassing

0.9 with a 100 budget across all the database systems, highlighting the strong portability of ActiveDiag.

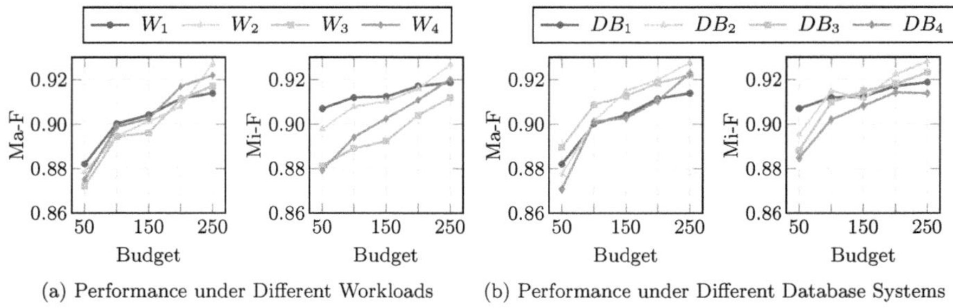

(a) Performance under Different Workloads   (b) Performance under Different Database Systems

Fig. 3. Generalization Study

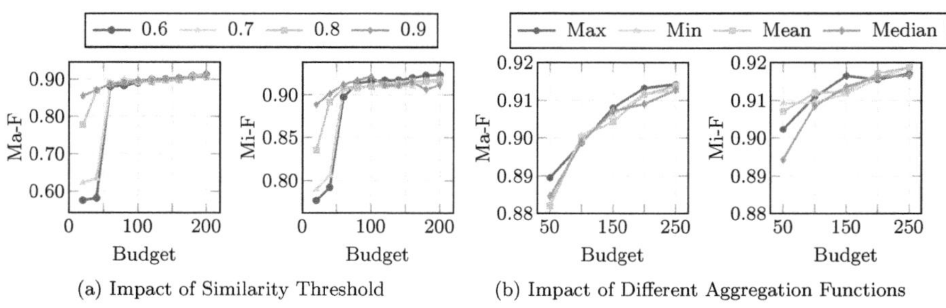

(a) Impact of Similarity Threshold   (b) Impact of Different Aggregation Functions

Fig. 4. Parameter Sensitivity Analysis

4.5   Parameter Sensitivity Analysis

Similarity Threshold in Cold-Start Strategy. To guide the selection of similarity thresholds, we compare various threshold settings. As shown in Fig. 4(a), setting a low threshold (e.g., 0.6) significantly degrades performance in the early stages, resulting in a noticeable lag compared to other thresholds. On the other hand, setting a high threshold (e.g., 0.9) leads to superior performance, with smooth improvement. As budget increases, the predefined categories are progressively covered, at which point the cold-start strategy is no longer necessary and all curves nearly overlap.

Aggregation Functions in Discrepancy Calculation. To assess the impact of aggregation functions on diagnostic performance, we compare four common aggregation functions: *Max, Min, Mean,* and *Median.* As shown in Fig. 4(b), all four functions demonstrate strong performance. As budget increases, both Ma-F and Mi-F show steady improvements, with all Ma-F exceeding 0.91 and Mi-F approaching 0.92 at a budget of 250. The choice of aggregation function has minimal impact on the final diagnostic performance, with the differences between the functions being negligible.

5 Conclusion

In this work, we introduce ActiveDiag, an active learning-based framework for database anomaly diagnosis. Extensive experiments demonstrate its superiority, effectiveness, and generalizability. Future work will focus on extending Active-Diag to distributed databases, exploring concurrency and causal relationships among anomalies, and refining the diagnosis method.

Acknowledgments. This study was funded by National Natural Science Foundation of China under Grant No. 62302370, 62272369, 62372352, 62306223, and ZTE Industry-University-Institute Cooperation Funds (No. IA20240710014).

Disclosure of Interests. The authors have no competing interests to declare that are relevant to the content of this article.

References

1. Ash, J.T., Zhang, C., Krishnamurthy, A., Langford, J., Agarwal, A.: Deep batch active learning by diverse, uncertain gradient lower bounds. In: International Conference on Learning Representations (2019)
2. Bae, W., Noh, J., Sutherland, D.J.: Generalized coverage for more robust low-budget active learning. In: Leonardis, A., Ricci, E., Roth, S., Russakovsky, O., Sattler, T., Varol, G. (eds.) Computer Vision. ECCV 2024. LNCS, vol. 15141. Springer, Cham (2025). https://doi.org/10.1007/978-3-031-73010-8_19
3. Bandara, W.G.C., Patel, V.M.: A transformer-based siamese network for change detection. In: IEEE International Geoscience and Remote Sensing Symposium, pp. 207–210. IEEE (2022)
4. Cheng, M., Yang, J., Pan, T., Liu, Q., Li, Z., Wang, S.: ConvTimeNet: a deep hierarchical fully convolutional model for multivariate time series analysis. In: Companion Proceedings of the ACM on Web Conference 2025, pp. 171–180 (2025)
5. Culotta, A., McCallum, A.: Reducing labeling effort for structured prediction tasks. In: The Association for the Advancement of Artificial Intelligence, vol. 5, pp. 746–751 (2005)
6. Duda, R.O., Hart, P.E.: Pattern Classification and Scene Analysis. A Wiley-Interscience Publication (1973)
7. Elokri, C., Ouaderhman, T., Chamlal, H.: AL-FEW: an enhanced approach for optimized query examples through feature weighting in active learning. Exp. Syst. Appl. **266**, 126045 (2025)
8. Friedman, J.H.: Greedy function approximation: a gradient boosting machine. Ann. Stat., 1189–1232 (2001)

9. Gadde, H.: AI-powered fault detection and recovery in high-availability databases. Int. J. Mach. Learn. Res. Cybersecur. Artif. Int. **15**(1), 500–529 (2024)
10. Goldstein, M., Dengel, A.: Histogram-based outlier score (HBOS): a fast unsupervised anomaly detection algorithm. In: KI-2012: Poster and Demo Track, vol. 1, pp. 59–63 (2012)
11. Gweon, H., Yu, H.: A nearest neighbor-based active learning method and its application to time series classification. Pattern Recogn. Lett. **146**, 230–236 (2021)
12. Hacohen, G., Dekel, A., Weinshall, D.: Active learning on a budget: opposite strategies suit high and low budgets. arXiv preprint arXiv:2202.02794 (2022)
13. Huang, S.J., Jin, R., Zhou, Z.H.: Active learning by querying informative and representative examples. In: Advances in Neural Information Processing Systems, vol. 23 (2010)
14. Huang, S., Wang, Z., Wu, Y., Tu, Y., et al.: OpDiag: unveiling database performance anomalies through query operator attribution. IEEE Trans. Knowl. Data Eng. (2025)
15. Huang, S., Wang, Z., Zhang, X., Tu, Y., Li, Z., Cui, B.: DBPA: a benchmark for transactional database performance anomalies. ACM Manage. Data **1**(1), 1–26 (2023)
16. Ismail-Fawaz, A., Devanne, M., Berretti, S., Weber, J., Forestier, G.: LITE: light inception with boosting techniques for time series classification. In: International Conference on Data Science and Advanced Analytics, pp. 1–10. IEEE (2023)
17. Jin, L., Li, G.: AI-based database performance diagnosis (in Chinese). J. Softw. **32**(3), 845–858 (2021)
18. Kumar, P., Gupta, A.: Active learning query strategies for classification, regression, and clustering: a survey. J. Comput. Sci. Technol. **35**, 913–945 (2020)
19. Le, X.M., Luo, L., Aickelin, U., Tran, M.T.: ShapeFormer: Shapelet transformer for multivariate time series classification. In: Proceedings of the 30th ACM SIGKDD Conference on Knowledge Discovery and Data Mining, pp. 1484–1494 (2024)
20. Li, G., Zhou, X., Sun, J., Yu, X., et al.: openGauss: an autonomous database system. Proc. VLDB Endow. **14**(12), 3028–3042 (2021)
21. Li, Y., Han, J., Sun, Y., Shi, B., Gong, Z.: A root cause analysis framework for microservice systems with multimodal data. ZTE Commun. **23**(4) (2025)
22. Li, Z., Zhao, Y., Botta, N., Ionescu, C., Hu, X.: COPOD: copula-based outlier detection. In: IEEE International Conference on Data Mining, pp. 1118–1123. IEEE (2020)
23. Liu, F.T., Ting, K.M., Zhou, Z.H.: Isolation-based anomaly detection. ACM Trans. Knowl. Discov. Data **6**(1), 1–39 (2012)
24. Liu, P., Zhang, S., Sun, Y., Meng, Y., Yang, J., Pei, D.: FluxInfer: automatic diagnosis of performance anomaly for online database system. In: IEEE International Performance Computing and Communications Conference, pp. 1–8. IEEE (2020)
25. Liu, X., Yin, Z., Zhao, C., Ge, C., et al.: PinSQL: pinpoint root cause SQLs to resolve performance issues in cloud databases. In: International Conference on Data Engineering, pp. 2549–2561. IEEE (2022)
26. Lu, X., Xie, Z., Li, Z., Li, M.O.: Generic and robust performance diagnosis via causal inference for OLTP database systems. In: IEEE International Symposium on Cluster, Cloud and Internet Computing, pp. 655–664. IEEE (2022)
27. Ma, M., Yin, Z., Zhang, S., Wang, S., et al.: Diagnosing root causes of intermittent slow queries in cloud databases. Proc. VLDB Endow. **13**(8), 1176–1189 (2020)
28. Middlehurst, M., Large, J., Flynn, M., Lines, J., Bostrom, A., Bagnall, A.: HIVE-COTE 2.0: a new meta ensemble for time series classification. Mach. Learn. **110**(11), 3211–3243 (2021)

29. Sauvanaud, C., Kaâniche, M., Kanoun, K., et al.: Anomaly detection and diagnosis for cloud services: practical experiments and lessons learned. J. Syst. Softw. **139**, 84–106 (2018)
30. Scheffer, T., Decomain, C., Wrobel, S.: Active hidden Markov models for information extraction. In: Hoffmann, F., Hand, D.J., Adams, N., Fisher, D., Guimaraes, G. (eds.) Advances in Intelligent Data Analysis, IDA 2001. LNCS, vol. 2189. Springer, Heidelberg (2001). https://doi.org/10.1007/3-540-44816-0_31
31. Sener, O., Savarese, S.: Active learning for convolutional neural networks: a core-set approach. In: International Conference on Learning Representations (2018)
32. Shannon, C.E.: A mathematical theory of communication. Bell Syst. Tech. J. **27**(3), 379–423 (1948)
33. Shen, Y., Li, S., Shen, M., Cai, P., et al.: DB-MAGS: multi-anomaly data generation system for transactional databases. Proc. VLDB Endow. **17**(12), 4497–4500 (2024)
34. Silvestre, G., Sauvanaud, C., Kaâniche, M., Kanoun, K.: Tejo: a supervised anomaly detection scheme for newSQL databases. In: Fantechi, A., Pelliccione, P. (eds.) Software Engineering for Resilient Systems, SERENE 2015. LNCS, vol. 9274, pp. 114–127. Springer, Cham (2015). https://doi.org/10.1007/978-3-319-23129-7_9
35. Sui, Q., Ghosh, S.K.: Similarity-based active learning methods. Exp. Syst. Appl. **251**, 123849 (2024)
36. Tan, C.W., Dempster, A., et al.: MultiRocket: multiple pooling operators and transformations for fast and effective time series classification. Data Min. Knowl. Disc. **36**(5), 1623–1646 (2022)
37. Xiang, Q., Shao, Y., Xu, Q., Yang, C.: Distributed diagnosis for compound anomalies (in Chinese). J. Softw. **15**(1), 115–137 (2024)
38. Xin, R., Chen, P., Grosso, P., Zhao, Z.: A fine-grained robust performance diagnosis framework for run-time cloud applications. Fut. Gener. Comput. Syst. **155**, 300–311 (2024)
39. Yan, L., Zheng, B., Qing, J., You, W., et al.: Anomaly diagnosis with Siamese Discrepancy Networks in distributed cloud databases. In: IEEE International Conference on Data Engineering, pp. 4053–4065. IEEE Computer Society (2025)
40. Yehuda, O., Dekel, A., Hacohen, G., Weinshall, D.: Active learning through a covering lens. In: Advances in Neural Information Processing Systems, vol. 35, pp. 22354–22367 (2022)
41. Yu, W., Liu, Y., Zhang, J., Ye, J., Ge, X.: Root cause analysis of poor FTTR quality based on transformer mechanisms. ZTE Commun. **23**(4) (2025)
42. Yuan, L. et al.: ATS: a fully automatic troubleshooting system with efficient anomaly detection and localization. In: Mikyška, J., de Mulatier, C., Paszynski, M., Krzhizhanovskaya, V.V., Dongarra, J.J., Sloot, P.M. (eds.) Computational Science, ICCS 2023. LNCS, vol. 14077, pp. 476–491. Springer, Cham (2023). https://doi.org/10.1007/978-3-031-36030-5_38

Q-Doctor: Retrieval-Augmented Diagnosis and Multi-agent Correction for Query Performance Anomalies

Yiwen Han, Zhicheng Pan[✉], Lixiang Chen, Chengcheng Yang, Rong Zhang, and Xuan Zhou

School of Data Science and Engineering, East China Normal University, Shanghai, China
{ywhan03,zcpan,lxchen}@stu.ecnu.edu.cn,
{ccyang,rzhang,xzhou}@dase.ecnu.edu.cn

Abstract. The stability and efficiency of database systems are critical to numerous data-intensive applications. However, it remains challenging to keep stable and high query performance, particularly under complex analytical workloads where subtle performance anomalies would cause significant latency and resource inefficiencies. Existing approaches often separate diagnosis from correction—focusing either on detecting execution anomalies or on black-box tuning techniques with limited interpretability and generality. In this paper, we propose **Q-Doctor**, a Query-level retrieval-augmented framework for Diagnosing and correcting database performance. First, Q-Doctor jointly encodes both query semantics and execution behaviors via a hybrid representation combining graph and tree neural encoders. This representation enables efficient retrieval of similar historical cases, which then guides an informed and fine-grained diagnosis. Moreover, a multi-agent correction module is introduced to collaboratively refine SQL hints and system configurations via reinforcement-guided iterations. Extensive experiments on well-established benchmarks demonstrate that Q-Doctor could accurately identify hidden performance anomalies and significantly improve query performance.

Keywords: Query Diagnosis · Knob Tuning · Query Hint · RAG · LLM

1 Introduction

With the proliferation of data-driven applications, effective diagnosis has become a critical capability for relational database systems (RDBMS). It focuses on identifying, analyzing and resolving anomalies, thereby ensuring high data availability and sustained workload performance [23]. Nevertheless, it remains a non-trivial task, as many factors, e.g., mis-configured system parameters or suboptimal execution plans, would lead to substantial performance degradations.

Recently, a growing body of research has been devoted to the anomaly *detection* or *correction* of query performance in RDBMS. In terms of anomaly detection, previous studies analyze system behaviors and query execution patterns to identify performance issues and diagnose their underlying causes [15,21]. In terms of anomaly correction, great progress has been made regarding knobs and plans [8,22]. Specifically, knob correction focuses on automatic system-level configuration adjustment [6,18]. Plan correction targets the optimizer and executor behaviors to generate more efficient execution plans [11,24]. However, these methods operate as black-box tuners, which optimize performance without explicitly identifying the root causes of query inefficiency.

Fortunately, recent studies have proposed to leverage large language models (LLMs) to automate the database performance improvement. Through in-context learning, LLMs can retrieve task-relevant knowledge and generate suggestions to address performance issues [2,3,16]. However, there remain three critical challenges to achieve accurate anomaly detection and efficient correction.

Performance Anomalies Are Not Explicit (C1). It is often difficult to detect performance anomalies, as there is no "gold standard" that defines the expected performance for a given query. Effective diagnosis requires integrating multiple heterogeneous sources of evidence [10] like: ❶ *key performance indicators* (KPIs) that reflect run-time performance metrics such as resource utilization and throughput, ❷ *system logs* that record execution signals, including warnings and errors, ❸ *historical query cases* that serve as analogical evidence for similar execution patterns, etc. However, these sources of information are both *complex* and *diverse*. On one hand, KPIs are time-series data, logs are unstructured natural language messages, and historical cases are structured records. It requires an integrated interpretation to effectively utilize these heterogeneous formats. On the other hand, each type of information provides a unique and essential perspective on performance issues—none can be ignored. Therefore, it also poses a significant challenge to process such diverse contents within a unified framework.

Useful References Are Critical But Hard to Identify (C2). To effectively diagnose query performance, it is crucial to leverage similar well-performed historical queries executed in the same system environment. Their execution states offer valuable insights into the current query, since we can identify the key factors that contribute to their superior performance, thereby enabling the detection of anomalies in current query behavior and facilitating informed corrective actions. A prerequisite for it is to measure the similarity between SQL queries, and such similarity should reflect the useful information of historical cases for the current query. Thus, the second challenge is to design informative query representations that could comprehensively characterize a given query.

Unification of Multi-dimensional Correction (C3). Effective correction involves two critical dimensions: *system knob* and *query plan*–each addressing distinct layers of the database system. Specifically, knob correction operates at the system level, managing runtime resources and execution environments. In contrast, plan correction focuses on the optimizer-side, determining the execution strategies for individual queries. Existing approaches often overlook the precise utilization of established optimization and tuning knowledge, ignoring the unification of these two dimensions. Although we can leverage the LLMs to bridge these two layers, it is insufficient to solely utilize the in-context learning capabilities of LLMs. A primary concern is *hallucination*, which refers to the tendency of models to generate incorrect or unsupported outputs. Moreover, they often lack domain knowledge of the database, which might lead to suboptimal or even misleading recommendations.

Solutions. To tackle the above challenges, we propose Q-Doctor, a Query-level retrieval-augmented framework for Diagnosing and correcting database performance. Our goal is to enable Q-Doctor to rapidly detect performance anomalies and efficiently restore normal execution behavior. To address C1, we first incorporate historical query cases, KPIs, and log information to support anomaly detection. Moreover, we design a component that could *semantically understand* these heterogeneous information and generate a comprehensive query profile. To address C2, we design a representation model, which captures statement-level and plan-level features of the given query. The statement-level feature encapsulates user intents, while the plan-level feature reflects the runtime behavior. With the representation, we leverage the vector similarity search to retrieve the most similar historical cases. To address C3, we design a multi-agent collaboration mechanism to handle the knob correction and plan correction. To enhance the interpretability, we propose the *diagnosis-based correction*, where each correction action is grounded in a detailed diagnostic analysis. Moreover, due to the intricacy of the task, we include a multi-round iterative mechanism to progressively improve the query performance.

In summary, we make the following contributions:

- We propose *Q-Doctor*, a RAG-based framework that focuses on query-level anomaly detection and correction.
- We design a sophisticated query representation method that incorporates both statement-level semantics and plan-level characteristics, enabling the integration of historically well-performing queries as end-to-end references.
- We propose a multi-agent collaborative framework and multi-round iterative mechanism, ensuring accurate detection and diagnosis-based correction.
- We conduct experiments on benchmarks including JOB, TPC-H and STATS. The experimental results show that Q-Doctor can accurately detect anomalies and correct them, resulting in substantial performance improvements.

2 Overview

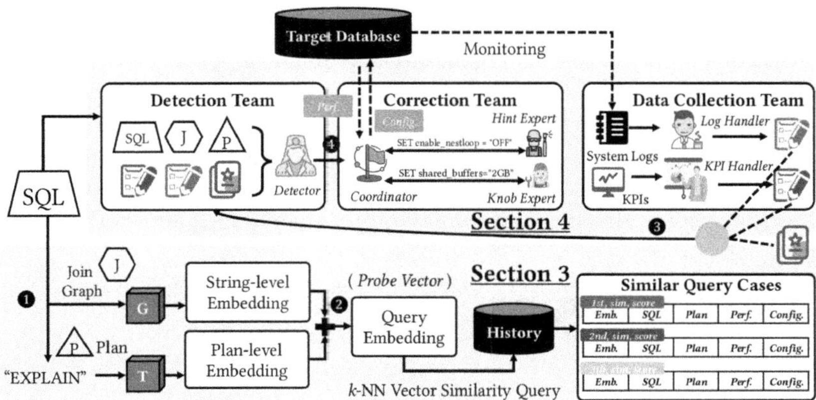

Fig. 1. The overview of Q-Doctor.

System Architecture. Q-Doctor aims to achieve accurate detection and effective correction of query performance. Figure 1 depicts the architecture of Q-Doctor, which contains the following steps: ❶ As the given SQL query q has been executed, we can easily obtain its join graph $J(q)$ and query plan $P(q)$; ❷ Then, $J(q)$ and $P(q)$ are input into the corresponding representation model, thus finally generating the query embedding $\mathcal{E}(q)$, which is then used to retrieve similar query cases for the downstream tasks; ❸ Q-Doctor utilizes the query statement $S(q)$ (i.e. static SQL text), plan $P(q)$, join graph $J(q)$, current configuration $\boldsymbol{\theta}$, query latency $perf(q)$, runtime metrics (from logs and KPIs), similar historical cases and knowledge to determine whether performance anomaly exists and form a diagnosis report; ❹ If the performance anomaly is detected, Q-Doctor initiates the correction process and iteratively refines the configuration $\boldsymbol{\theta}'$.

Query Representation. Historical query cases has been extensively validated, indicating that their execution states (e.g., plans, configurations) can offer valuable insights into current queries. To find such cases, Q-Doctor leverages a sophisticated query representation model to capture their features and embed them into a unified space. Subsequently, a k-NN similarity search is conducted over these embeddings to retrieve similar cases. The representation incorporates two types of information: *statement-level* feature and *plan-level* feature. Specifically, Q-Doctor uses *graph neural network* (GNN) [19] and *tree convolutional neural network* (TCNN) [12] to obtain the corresponding embeddings.

Detection and Correction. ① As the performance metrics provide valuable insights into the operational status of queries [10], we consider logs and KPIs to construct a query performance profile. ② Then, together with current query information and retrieved historical query cases, they are used for the anomaly detection. We also design a semantic-aware interpreter that jointly understands heterogeneous information and generates a diagnosis report to guide subsequent correction decisions. ③ Finally, the query performance is corrected via two aspects, i.e., knob correction and plan correction [8]. Accordingly, we design a delicate coordination mechanism with three different experts to integrate their configuration recommendations. Moreover, we iteratively apply new configurations to the system, driving continuous improvements in query performance.

3   Query Case Retrieval

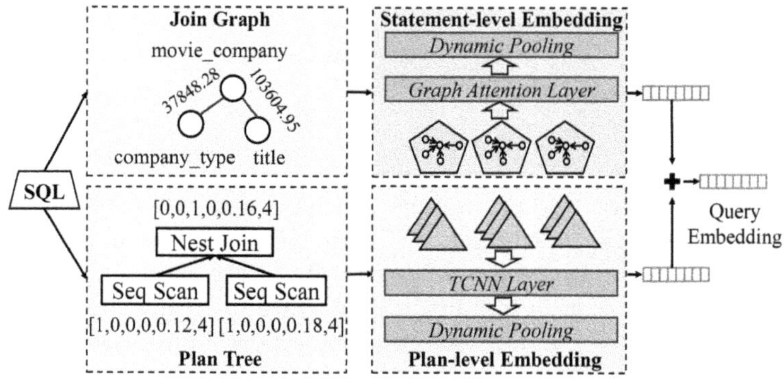

Fig. 2. Query representation model.

In this section, we design a representation model that maps queries into a unified embedding space for effective retrieval of insightful historical cases. As shown in Fig. 2, we consider two features: *statement-level feature* (see Sect. 3.1) and *plan-level feature* (see Sect. 3.2). To effectively capture these two features, we introduce different techniques to transform them into fixed-size vectors. Finally, we present a fusion strategy to combine these two features and derive a joint query representation (see Sect. 3.3).

3.1   Statement-Level Feature

The statement-level feature encapsulates the users' purposes, without the specific procedural aspects of *how* the query should be executed.

Statement-Level Encoding. We use the *join graph* that captures the join dependencies as the statement-level encoding. This is because for a query, join operations tend to be common sources of performance bottlenecks and indicators of users' purposes [3]. To explicitly encode the information, we parse the SQL statement and obtain its *join graph* [1]. In the join graph, each node corresponds to a table referenced by the query, and each edge indicates a join predicate; the edge weight encodes the estimated join cost derived from the database cardinality estimator [24]. For a given query q, we denote the encoding as $E_s(q)$.

Statement-level Embedding. After obtaining the join graph, we further obtain the statement-level embedding $\mathcal{E}_s(q)$ utilizing a multi-layer *graph neural network* (GNN) [19]. To incorporate the importance of nodes, an attention mechanism is employed based on *graph attention networks* (GAT) [19].

Specifically, let W denote a shared learnable weight matrix applied to all node features, and a represents the attention parameter vector. For a given node i, let $h_i^{(l)}$ denote its feature vector at the l-th GAT layer. The attention coefficient between node i and each of its neighboring nodes $j \in \mathcal{N}(i)$ is $e_{ij} = \text{LeakyReLU}(a^T[Wh_i^{(l)}|Wh_j^{(l)}])$. The coefficients e_{ij} are then normalized with the softmax function: $\alpha_{ij} = \frac{\exp(e_{ij})}{\sum_{k\in\mathcal{N}_i}\exp(e_{ik})}$. It reflects the importance of each neighboring node j in updating the embedding of node i. The updated node representation at the $(l+1)$-th GAT layer is computed as a weighted aggregation of its neighbors' features: $h_i^{(l+1)} = \sigma(\sum_{j\in\mathcal{N}_i}\alpha_{ij}Wh_j^{(l)})$. Moreover, to capture diverse aspects of neighborhood information and improve training stability, we adopt a multi-head attention mechanism, where K independent attention heads are employed in parallel. The outputs of these heads are concatenated to form the final node representation, i.e., $h_i^{(l+1)} = \|_{k=1}^{K}\sigma\left(\sum_{j\in\mathcal{N}(i)}\alpha_{ij}^{(k)}W^{(k)}h_j^{(l)}\right)$. Finally, a *dynamic pooling* mechanism is applied over the graph to aggregate node embeddings into a fixed-length embedding $\mathcal{E}_s(q)$.

3.2   Plan-Level Feature

Unlike statement-level features, plan-level features reflect the physical execution strategy chosen by the query optimizer, reflecting runtime behavior of a query.

Plan-Level Encoding. We use the *plan tree* as the plan-level encoding, since a query plan is a tree structure that describes how a query would be executed. Each node in this tree represents a *physical operator*, and the edges capture the execution dependencies between operators. Specifically, the node encoding includes: ❶ *Operator Type:* The physical operator type of the node specifies the execution strategy (e.g. `Seq Scan`, `Hash Join`, `Aggregate`, etc.). They are encoded by one-hot encoding. ❷ *Cardinality:* It refers to the estimated number of distinct values or tuples processed at the operator node. ❸ *Row Width:* It indicates the average size of each row in bytes, reflecting the physical cost of data accesses. Finally, by combining the above three types of key information, each operator node is encoded as a fixed-size vector, denoted as $E_p(q)$.

Plan-Level Embedding. After obtaining the plan tree, we obtain the plan-level embedding $\mathcal{E}_p(q)$ with the *tree convolutional neural network* (TCNN) [12]. The TCNN operates on the structure of the plan tree, extracting features by sliding a set of convolutional kernels over the tree nodes and their local substructures. It enables the model to capture the hierarchical dependencies among different operators. By stacking multiple TCNN layers, the model could effectively capture structural information from the execution plan. Finally, the output node features are also aggregated via a *dynamic pooling* mechanism, which compresses the entire plan tree into a fixed-size embedding $\mathcal{E}_p(q)$.

3.3 Put It Together

For a given query q, we incorporate both the statement-level and plan-level features, and finally obtain its unified representation: $\mathcal{E}(q) = \mathcal{E}_s(q) \oplus \mathcal{E}_p(q)$.

Training. After obtaining $\mathcal{E}(q)$, we input it into a feed-forward network (FFN) with its query latency as the prediction target. The model is trained by minimizing the mean squared error (MSE) loss: $\mathcal{L} = \frac{1}{n}\sum_{i=1}^{n}(y_i - \hat{y}_i)^2$, where n denotes the number of training samples, y_i represents the ground-truth query latency of the i-th sample, and $\hat{y}_i$ is the corresponding predicted query latency.

Retrieval. Given queries q_1 and q_2, their similarity is computed using *cosine* similarity between $\mathcal{E}(q_1)$ and $\mathcal{E}(q_2)$: $\frac{\mathcal{E}(q_1)\cdot\mathcal{E}(q_2)}{\|\mathcal{E}(q_1)\|\cdot\|\mathcal{E}(q_2)\|}$. With the query embedding and cosine similarity, we can conduct k-nearest neighbor similarity search over historical queries for the top-k similar cases.

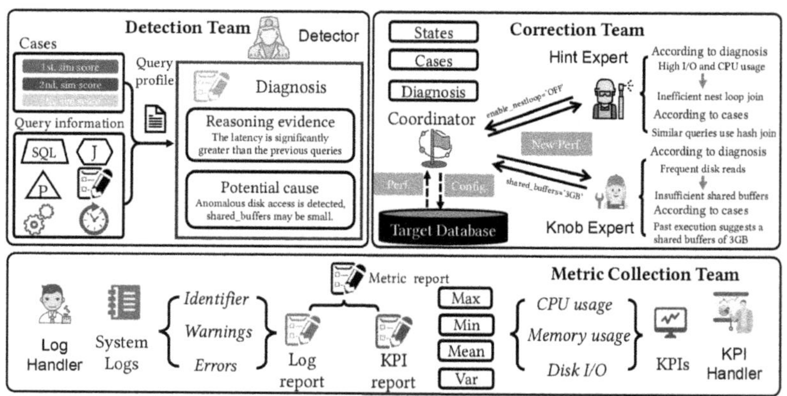

Fig. 3. Online detection and correction.

4   Detection and Correction

In this section, we design a collaborative multi-agent framework for anomaly detection and correction. As illustrated in Fig. 3, the framework comprises six domain experts that are organized into three functionally specialized teams.

❶ **Metric Collection Team** (Sect. 4.1). Metric collection team consists of Log_Handler and KPI_Handler. They collect two representative metrics: logs and KPIs to support the anomaly detection and correction.

❷ **Detection Team** (Sect. 4.2). Detection team is responsible for anomaly detection and consists of Detector. It handles heterogeneous information and generates a comprehensive *diagnosis report*.

❸ **Correction Team** (Sect. 4.3). Correction team performs the anomaly correction, which involves three experts: (*i*) Knob_Expert responsible for knob correction; (*ii*) Hint_Expert responsible for plan correction with query hint; and (*iii*) Coordinator that coordinates the whole correction process. These experts collaboratively resolve performance issues through an iterative correction loop.

4.1   Metric Collection

Analysis of query performance requires a thorough understanding of related metrics [10]. Here, we consider two representative metrics: *system logs* and *key performance indicators (KPIs)*, as they provide direct insights into the internal states of the system. These metrics are collected by the following two experts, respectively.

Log Handler. The Log_Handler is responsible for parsing native database system logs to extract structured information, including query identifiers, warnings, and error messages. These logs serve as *explicit* system responses from the database engine during query execution.

KPI Handler. The KPI_Handler captures *implicit* key performance indicators that could provide quantitative resource usages. Specifically, it continuously monitors various indicators, such as CPU usage, memory usage, etc. They are sampled at fixed time intervals and aggregated to compute representative statistics, including maximum, minimum, mean, and variance. Finally, the aggregated statistics are organized into a structured JSON format for downstream experts.

4.2   Anomaly Detection

To enhance the accuracy of anomaly detection, it is crucial to integrate heterogeneous information to form a holistic understanding of query behavior. To this end, we propose Detector, a diagnostic expert that takes advantage of runtime metrics and system states during query execution to construct the performance profile of the current query. Then, with in-context learning capability, it identifies performance anomalies and associates them with possible root causes.

More specifically, the Detector first receives two kinds of information, including: ❶ **Query state**: query statement $S(Q)$, execution plan $P(Q)$, join graph $J(Q)$, configuration θ, execution latency $Perf(Q)$, metrics including KPIs (from KPI_Handler) and system logs (from Log_Handler); ❷ **Historical cases**: execution information (e.g. latency) of similar historical queries (see Sect. 3.3).

Next, the Detector assesses whether a performance anomaly exists. Its reasoning process consists of two steps: (1) It performs semantic interpretation of these heterogeneous inputs to uncover anomalies, such as unusually high latency compared with historical executions. Based on this analysis, it decides whether corrective actions are required. (2) To provide a solid foundation for downstream performance correction, the Detector provides interpretable evidence along with potential root causes for the anomaly, i.e. diagnosis report.

Diagnosis Report. Detector synthesizes its analysis into a structured detection report: ❶ **State:** Current query Q, $P(Q)$, $J(Q)$, θ, $perf(Q)$, logs and KPIs; ❷ **Cases:** Historical query cases referenced during the anomaly detection phase; ❸ **Diagnosis:** The *evidence* for Detector's anomaly assessment, along with its initial diagnosis of *potential causes* of performance degradation.

4.3   Anomaly Correction

The anomaly correction involves knob correction and plan correction. Since they focus on different aspects of the database system, we set Knob_Expert and Hint_Expert for them, respectively. Moreover, we introduce a Coordinator to *coordinate* the entire correction process.

Coordinator. The Coordinator serves as the "leader", which orchestrates the information flow among agents. It integrates suggestions from different correction experts and interacts with the database to evaluate refined configurations. First, the Coordinator receives the diagnosis report from the Detector, including the *state*, *cases*, and *diagnosis*. Then, it determines which expert is the most appropriate to perform the correction process [13]. The report is forwarded to the selected expert (i.e., Knob_Expert or Hint_Expert). Through the process, experts refine their suggestions based on feedback from the database system.

Knob Expert. The Knob_Expert focuses on knob correction, which involves two aspects: ❶ *Diagnosis* Upon receiving diagnosis report from Detector, it identifies performance bottlenecks caused by misconfigured knobs and pinpoints the knobs most likely responsible for the performance issue; ❷ *Cases* It compares current configurations with historical cases to identify the reasons why similar queries previously exhibited better performance, whereas the current query suffers from performance degradation. Finally, Knob_Expert generates refined configurations to correct performance based on both empirical evidence from diagnosis and historical executions (e.g. `set shared_buffers to "3GB"`).

Hint Expert. Hint_Expert focuses on plan correction and also makes inferences from two aspects: ❶ *Diagnosis* Upon receiving diagnosis report from the Detector, it identifies performance bottlenecks caused by suboptimal plan; ❷ *Cases* It analyzes the discrepancies between the current query plan and the historical ones, including structural differences and physical operator differences, aiming to identify why previous plans achieved better performance while the current plan exhibits performance degradation. Based on the above analysis, Hint_Expert proposes explainable hints to control the generation of the execution plan (e.g. `set enable_nestloop to "off"`).

Correction Loop. Upon receiving suggestions from Knob_Expert and Hint_Expert, the Coordinator generates a new configuration θ' and applies it to the database system to obtain new execution latency $perf'(Q)$ and runtime metrics. The information is used to evaluate the effectiveness of the new configuration. It then selects a new expert and the updated query profile is forwarded to the selected expert for further analysis.

The correction loop follows the thinking–action–observation paradigm [20]: ❶ *thinking*: evaluating performance and analyzing strategies for further improvement; ❷ *action*: proposing a new configuration and ❸ *observation*: observing the effectiveness of the configuration. Through dynamic interaction with the database, experts continuously update their understanding of the environments, enabling them to progressively refine their recommendations.

5   Experimental Evaluation

5.1   Experimental Setup

Settings. Q-Doctor is implemented in Python 3.10, with PostgreSQL (PG) 14.0. All experiments are conducted on a machine running Ubuntu 22.04, equipped with a 16-core CPU, a 32 GB DRAM, and a 1 TB hard disk. Unless specified, GPT-4o-mini is used as the default LLM. Historical queries are automatically collected, and the top 15 queries with the best performance are stored in the vector database for efficient retrieval. During detection, a query is labeled as anomalous if it exhibits a significant latency regression or if resource utilization exhibits abrupt deviations. During correction, we consider both the query hints [11] and configurable knobs [6].

Benchmarks. We conduct experiments on the following three benchmarks:
① **Join Order Benchmark (JOB)** [7]: It consists of 21 tables, primarily designed to evaluate the query optimizer's effectiveness in join order.
② **TPC-H** [17]: It is mainly designed to evaluate database performance under complex queries across 8 tables. Here, we adopt scale factors (SF) of 1 and 10.
③ **STATS** [5]: It is designed to evaluate the end-to-end performance of the query optimizer. It emphasizes realistic workloads and diverse operator combinations.

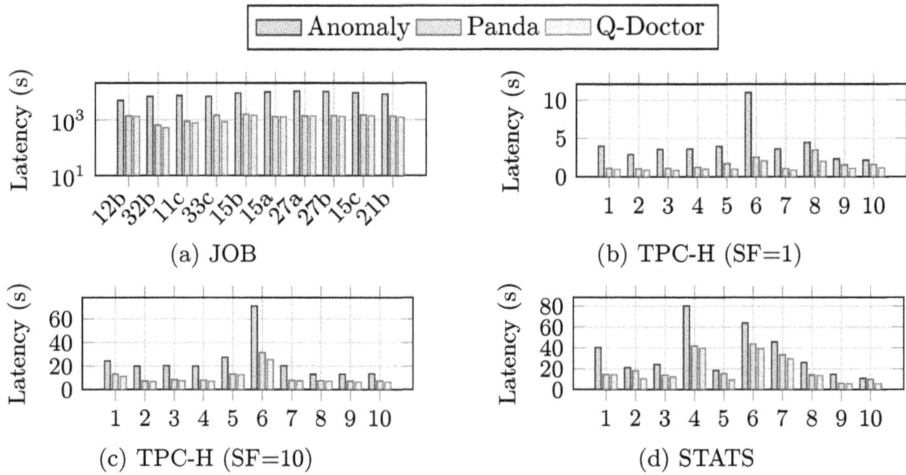

Fig. 4. Effect of Q-Doctor on different workloads.

5.2 End-to-End Result Analysis

Metric. For a given query q, in addition to its execution latency, we define its *improvement rate* Δ as:

$$\Delta = \frac{\text{Latency}_{\text{anom}}(q) - \text{Latency}_{\text{corr}}(q)}{\text{Latency}_{\text{anom}}(q)}, \tag{1}$$

where $\text{Latency}_{\text{anom}}(q)$ denotes the execution latency of the anomalous query q, and $\text{Latency}_{\text{corr}}(q)$ denotes the execution latency after correction.

Baseline. Q-Doctor is a diagnosis-driven correction framework and shares similarities with Panda [15], a performance debugger based on LLMs. However, Panda employs a single-agent correction process without iterative refinement. In the following experiments, we focus on the comparison between Q-Doctor and Panda, as it represents the state-of-the-art approach. Systems such as Bao [11] and GPTuner [6] *directly* search for configurations without explicit diagnostic reasoning, thus we do not select them as baselines.

Figure 4 illustrates the execution latency of a representative subset of queries that are most sensitive to anomalies in four different workloads. The horizontal axis represents the query, while the vertical axis denotes the execution latency. Among the four workloads, the JOB workload shows the highest average performance improvement, achieving an improvement rate of 93%. It exhibits high sensitivity to join operations, where suboptimal choices can lead to substantial performance degradation, and Q-Doctor effectively mitigates this issue. For the TPC-H workload, the Q-Doctor achieves the improvement rates of 71% and 62% under scale factors of 1 and 10, respectively. This demonstrates the generalization of our approach across different data scales. Moreover, we also observe that

Q-Doctor consistently achieves superior results in performance correction compared to Panda, which highlights the advantage of decoupling different correction strategies and the multi-round iteration mechanism.

5.3 Overhead Analysis

Table 1 summarizes the Q-Doctor's reasoning runtime and token usage statistics. The token usage is divided into two components: *prompt tokens* (#PT) and *completion tokens* (#CT). Q-Doctor demonstrates stable overhead in both reasoning time and monetary cost across various benchmarks. Specifically, the reasoning process is completed in approximately 80 s, which indicates that Q-Doctor scales well in diverse analytical workloads. Compared with manual efforts, Q-Doctor can significantly reduce the diagnosis latency. Moreover, given that a substantial fraction of queries in real-world workloads are repetitive, it effectively amortizes the reasoning cost in practice [4,14]. Q-Doctor also maintains consistent token usage, with prompt and completion tokens averaging around 500,000 and 2,000 respectively. The stability in token usage facilitates reliable scaling to other workloads without incurring unpredictable overhead.

Table 1. Reasoning time and monetary cost in the correction process of Q-Doctor.

Benchmark	Time (s)	# PT	# CT	$ Cost	AVG. Time (s)	AVG. $ Cost
JOB	82.50	513,226	2080	0.084USD	27.500	0.0168USD
TPC-H (1×)	80.72	503,262	1,992	0.081USD	16.144	0.0162USD
TPC-H (10×)	83.68	503,372	2,003	0.077USD	16.736	0.0154USD
STATS	81.42	502,987	2,059	0.076USD	16.284	0.0156USD

5.4 Hyper-parameters Sensitivity Analysis

The Number of Historical Query Cases. During the detection and correction phase, Q-Doctor leverages historical query cases as references. Figure 5(a) illustrates the impact of the number of retrieved cases k on performance improvement. As k increases, the overall improvement grows and gradually stabilizes. For example, under the JOB workload, the improvement rate reaches 88%, 90%, and 93% for $k = 1$, 3, and 5, respectively. However, increasing k beyond a certain point does not lead to additional performance gains, with the improvement rate even dropping to 92% and 90% when $k = 7$ and $k = 9$. It shows that introducing more historical queries increases the likelihood of incorporating less relevant cases. Besides, a larger k results in longer prompt token sequences, thus increasing the token usage. Based on the observations, we set k to 5.

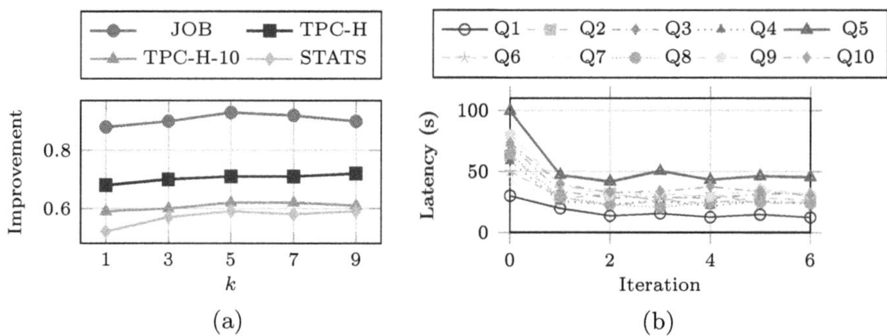

Fig. 5. (a) Relationship between improvement rate and the number of historical query examples k; (b) Latency trends during the correction process.

The Maximum Correction Iterations. The correction team of Q-Doctor iteratively searches for the optimal configuration parameters $\boldsymbol{\theta}_{knob}$ and $\boldsymbol{\theta}_{hint}$ to correct anomalous queries. Figure 5(b) illustrates the latency trend of parts of queries during the correction process. Take Q1 as an example, its initial anomalous latency is 30 s. As correction proceeds, Q-Doctor improves the performance in the early iterations, with latency decreasing from 19 to 13 s. Then, the latency tends to stabilize and fluctuate within a narrow range. The result shows that Q-Doctor typically achieves efficient performance correction within a few iterations. Here we set the maximum iterations to 5 in Q-Doctor, providing sufficient space for Q-Doctor to explore the configurations for the optimal performance and avoid unnecessary resource consumption.

5.5   Effectiveness of Anomaly Detection

In this section, we evaluate Q-Doctor's capability for anomaly detection. Here, we use four standard classification outcomes: *true positives* (TP), *false positives* (FP), *true negatives* (TN), and *false negatives* (FN). They serve as the basis for computing **Precision**, **Recall** and **F1-score**. As shown in Table 2, Q-Doctor achieves high precision, recall, and F1-score. Taking JOB as an example, it achieves a precision of 92%, a recall of 88%, and an F1-score of 89%, demonstrating its effectiveness in identifying performance anomalies under various conditions.

Table 2. Precision, Recall and F1-score of the detection process.

	JOB	TPC-H (1×)	TPC-H (10×)	STATS
Precision	0.92	0.89	0.86	0.92
Recall	0.88	0.83	0.80	0.89
F1-score	0.89	0.85	0.82	0.90

5.6   Generalization Analysis

Varying Hardware Configurations. We evaluate Q-Doctor on another machine with a 4-core CPU and 16 GB of RAM. As presented in Fig. 6, Q-Doctor can also achieve substantial performance improvements when applied to another machine with different settings.

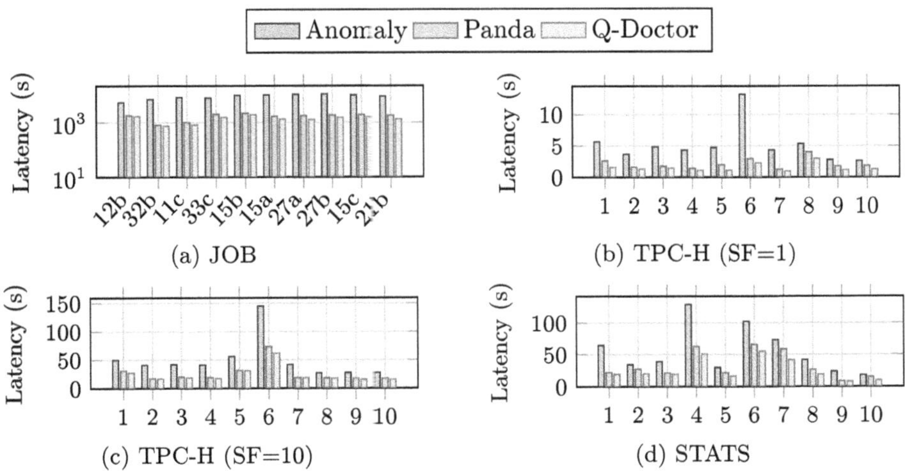

Fig. 6. Effect of Q-Doctor under another hardware environment.

Table 3. F1-score, improvement rate and reasoning time under other LLMs.

Model	JOB			TPC-H (1×)			TPC-H (10×)			STATS		
	F1-score	Δ	Time (s)	F1-score	Δ	Time (s)	F1-score	Δ	Time(s)	F1-score	Δ	Time (s)
GPT-4o-mini	0.92	0.93	82.50	0.88	0.71	80.72	0.86	0.62	83.68	0.92	0.59	81.42
GPT-4-1106	0.91	0.94	84.20	0.91	0.74	82.51	0.89	0.68	86.14	0.90	0.66	80.18
Deepseek-V3	0.92	0.93	83.50	0.88	0.72	79.52	0.88	0.65	82.19	0.91	0.62	78.27
Deepseek-R1	0.93	0.93	562.42	0.92	0.78	542.41	0.91	0.71	559.16	0.91	0.71	581.23

Varying LLMs. We evaluate different LLMs with the results reported in Table 3. Specifically, we evaluate F1-score, improvement rate Δ and reasoning time of different LLMs, including GPT-4-1106, Deepseek-V3 and Deepseek-R1.

We observe that our framework can integrate various LLMs and consistently achieve high F1-scores and improvement rates across different workloads. Notably, the Deepseek-R1 model, operating in *deep reasoning* mode, achieves the best overall performance. Taking TPC-H as an example, Deepseek-R1 achieves an F1-score of 0.92 and the improvement rate of 78%, outperforming GPT-4 (F1-score 0.91, improvement rate 74%) and Deepseek-V3 (F1-score 0.88, improvement rate 72%). It suggests that deeper analysis contributes to more effective detection and correction. However, it also incurs higher reasoning overhead, highlighting the trade-off between effectiveness and efficiency.

5.7 Ablation Study

Table 4. Ablation study on the effect of the reasoning contexts and different modules of Q-Doctor in performance improvement rate.

Benchmark	Q-Doctor	Representation Model		Reasoning Context			Expert	
		w/o $\mathcal{E}_s(q)$	w/o $\mathcal{E}_p(q)$	w/o Cases	w/o Metrics	w/o Report	w/o Knob_Expert	w/o Hint_Expert
JOB	0.93	0.87	0.89	0.86	0.85	0.88	0.92	0.14
TPC-H (1×)	0.71	0.65	0.61	0.64	0.61	0.66	0.59	0.65
TPC-H (10×)	0.62	0.55	0.50	0.55	0.53	0.56	0.48	0.56
STATS	0.59	0.52	0.49	0.51	0.49	0.55	0.51	0.54

Query Representation Model. As discussed in Sect. 3, Q-Doctor utilizes a fused representation model involving the statement-level feature $\mathcal{E}_s(q)$ and the plan-level feature $\mathcal{E}_p(q)$. Therefore, we evaluate the effect of these two types of information in the correction process. We observe declines in the performance improvement rate when both features are eliminated. Taking STATS workload as an example, the performance improvement rate decreases by 7% and 10%, respectively, as reported in Table 4. These results highlight the importance of both user intent and execution behavior to extract features of SQL queries.

Reasoning Contexts and Experts in the Correction Process. Q-Doctor leverages (1) historical query cases, (2) metrics including system log information and KPIs, and (3) diagnosis report from the Detector to achieve performance correction. Here, we discuss the effects of these three types of information. Moreover, we will discuss separately experts' contributions to the improvement rate.

In Table 4, we observe that the absence of different types of information leads to a decrease in the improvement rate. This can be explained as follows. ❶ The historical query cases serve as concrete examples in the current hardware environment. ❷ The metrics further enable experts to construct the query profile to identify performance bottlenecks. ❸ The diagnosis report serves as the basis for subsequent correction, enhancing the interpretability of the entire process. Moreover, we observe that both Knob_Expert and Hint_Expert contribute to the overall performance correction. In the JOB benchmark, Hint_Expert plays a more dominant role, as JOB queries are particularly sensitive to join order and join types in the execution plan. For the other three benchmarks, both experts lead to substantial improvements, with Knob_Expert contributing slightly more to the overall improvement rate. The results demonstrate that Q-Doctor could effectively determine the credit assignment between the two expert agents when dealing with multi-agent collaborations.

6   Related Work

LLM-based Database Diagnosis. Recent studies have explored LLMs for database performance diagnosis and correction [10,15,21,23]. Panda [15] employs LLMs as autonomous agents to analyze database performance issues and generate natural language recommendations. D-Bot [23] combines offline knowledge extraction with a collaborative diagnostic mechanism, allowing multiple agents to cooperatively analyze complex performance problems. However, it targets workload-level diagnosis and could not be applied to query-level diagnosis.

Knob Tuning and Query Optimization. Numerous studies have explored improving query performance via knob tuning and query optimization, using learned models or LLMs [3,6,9,11,18,24]. GPTuner [6] integrates domain knowledge from manuals and forums, and utilizes a Bayesian Optimization framework for knob tuning. Bao [11] reformulates query optimization as a contextual multi-armed bandit (CMAB) problem and employs the TCNN together with the Thompson sampling technique to improve optimization efficiency. However, these studies rely on black-box models, which hinder interpretability and transparency. Due to space limits, we refer the readers to [8,22] for detailed comparisons.

While prior studies address diagnosis or correction in isolation, none provides a unified framework that jointly leverages heterogeneous evidence, historical analogies, and multi-dimensional tuning—precisely the gap Q-Doctor fills.

7   Conclusion

This paper presents Q-Doctor, a RAG-based framework for query-level anomaly detection and correction. Q-Doctor leverages historically similar query cases together with a multi-agent collaborative mechanism to accurately identify performance anomalies and generate effective corrections. Experimental results demonstrate that Q-Doctor achieves high detection accuracy and delivers substantial performance improvements through diagnosis-driven correction.

Acknowledgement. This work is supported by the National Key R&D Program of China (2024YFC3308102), NSFC (62461146205, 92270202).

References

1. Chen, T., Gao, J., Tu, Y., Xu, M.: GLO: towards generalized learned query optimization. In: ICDE, pp. 4843–4855 (2024)
2. Fan, C., Pan, Z., Sun, W., Yang, C., Chen, W.N.: LATuner: an LLM-enhanced database tuning system based on adaptive surrogate model. In: Bifet, A., Davis, J., Krilavičius, T., Kull, M., Ntoutsi, E., Žliobaitė, I. (eds.) Machine Learning and Knowledge Discovery in Databases. Research Track. ECML PKDD 2024. LNCS, vol. 14945, pp. 372–388. Springer, Cham (2024). https://doi.org/10.1007/978-3-031-70362-1_22

3. Giannakouris, V., Trummer, I.: λ-tune: Harnessing large language models for automated database system tuning. Proc. ACM Manage. Data **3**(1), 1–26 (2025)

4. Han, Y., et al.: ByteCard: enhancing ByteDance's data warehouse with learned cardinality estimation. In: SIGMOD, pp. 41–54 (2024)

5. Han, Y., et al.: Cardinality estimation in DBMs: a comprehensive benchmark evaluation. PVLDB **15**(4) (2022)

6. Lao, J., et al.: GPTuner: a manual-reading database tuning system via GPT-guided Bayesian optimization. PVLDB **17**(8), 1939–1952 (2024)

7. Leis, V., Gubichev, A., Mirchev, A., Boncz, P., Kemper, A., Neumann, T.: How good are query optimizers, really? PVLDB **9**(3), 204–215 (2015)

8. Li, G., Zhou, X., Cao, L.: AI meets database: AI4DB and DB4AI. In: SIGMOD, pp. 2859–2866 (2021)

9. Li, G., Zhou, X., Li, S., Gao, B.: QTune: a query-aware database tuning system with deep reinforcement learning. PVLDB **12**(12), 2118–2130 (2019)

10. Ma, M., et al.: Diagnosing root causes of intermittent slow queries in cloud databases. PVLDB **13**(8), 1176–1189 (2020)

11. Marcus, R., Negi, P., Mao, H., Tatbul, N., Alizadeh, M., Kraska, T.: Bao: making learned query optimization practical. In: SIGMOD, pp. 1275–1288 (2021)

12. Mou, L., Li, G., Zhang, L., Wang, T., Jin, Z.: Convolutional neural networks over tree structures for programming language processing. In: AAAI, vol. 30 (2016)

13. Pan, Z., et al.: Hyper: hybrid physical design advisor with multi-agent reinforcement learning. In: ICDE, pp. 1565–1578 (2025)

14. Shankhdhar, P., Liu, F., Narale, J., Sun, J., Schlussel, R., Antova, L.: Presto's history-based query optimizer. PVLDB **17**(12), 4077–4089 (2024)

15. Singh, V.Y., et al.: Panda: performance debugging for databases using LLM agents. In: CIDR (2024)

16. Sun, W., et al.: Rabbit: retrieval-augmented generation enables better automatic database knob tuning. In: ICDE, pp. 3807–3820 (2025)

17. TPC-H (2021). http://www.tpc.org/tpc

18. Trummer, I.: DB-BERT: a database tuning tool that "reads the manual". In: SIGMOD, pp. 190–203 (2022)

19. Wu, L., Cui, P., Pei, J., Zhao, L., Guo, X.: Graph neural networks: foundation, frontiers and applications. In: SIGKDD, pp. 4840–4841 (2022)

20. Yao, S., et al.: ReACT: synergizing reasoning and acting in language models. In: ICLR (2023)

21. Yoon, D.Y., Niu, N., Mozafari, B.: DBSherlock: a performance diagnostic tool for transactional databases. In: SIGMOD, pp. 1599–1614 (2016)

22. Zhou, X., Chai, C., Li, G., Sun, J.: Database meets artificial intelligence: a survey. TKDE **34**(3), 1096–1116 (2020)

23. Zhou, X., et al.: D-Bot: database diagnosis system using large language models. PVLDB **17**(10), 2514–2527 (2024)

24. Zhu, R., et al.: Lero: a learning-to-rank query optimizer. PVLDB **16**(6), 1466–1479 (2023)

DBRooter: An Efficient Causal Root Cause Analysis Framework for Distributed Databases

Qingfeng Xiang[1], Yingxia Shao[1,2]([✉]), Chenglin Tian[1], Quanqing Xu[3], and Qiyao Luo[3]

[1] Beijing University of Posts and Telecommunications, Beijing, China
`{xiangqingfeng,shaoyx,tianchenglin}@bupt.edu.cn`
[2] Inspur Computer Technology Co., Ltd., Shandong, China
[3] Independent Researchers, Beijing, China

Abstract. Distributed databases are fundamental to modern infrastructure, but face complex performance anomalies due to their multi-node architecture. Existing root cause analysis methods, particularly those based on causal inference, struggle with the scale and complexity of distributed environments. They suffer from high computational costs and exhibit low accuracy in diagnosing high-dimensional runtime monitoring metrics. To address these challenges, we propose DBRooter, a diagnostic framework for distributed databases. DBRooter introduces dependency-guided causal modeling for distributed metrics that efficiently constructs local causal graphs in parallel and integrates them with a node dependency graph derived from query plans. In addition, it employs a workload-aware anomaly metric ranking method using structural causal models to quantitatively assess causal effects and identify root causes. The experimental results show that DBRooter effectively diagnoses anomalies in individual queries, transactions, and entire workloads and outperforms the state-of-the-art baselines. Compared to the second-best method, AC@1, AC@3, and AC@5 are improved up to 10%, 15%, and 12%.

Keywords: Distributed Database · Root Cause Analysis · Anomaly Diagnose · Structural Causal Model

1 Introduction

Distributed databases are fundamental to modern infrastructure, providing high availability, scalability, and massive data processing capabilities for critical domains like fintech and cloud computing [21]. However, their multiple node architecture introduces complex performance anomalies, including query latency spikes, replica synchronization failures, and resource hotspot contention [2,24,27]. These anomalies often exhibit inter-node and inter-metric correlations, making root cause analysis challenging for database administrators (DBAs) [31].

H. Jung et al. (Eds.): DASFAA 2026, LNCS 16536, pp. 543–560, 2026.
https://doi.org/10.1007/978-981-92-0366-6_33

In root cause analysis, traditional methods [19,29] rely on statistical features to identify critical metrics. Recently, causal inference-based techniques [4,12,14, 16,17] have introduced a new perspective for metric-based root cause analysis. Unlike purely correlation-driven, statistical, or AI-based approaches [30], causal graphs explicitly model causal dependencies between metrics, enabling the identification of key nodes in anomaly propagation paths [26]. For example, when an anomaly occurs, causal inference methods build metric causal graphs, followed by unsupervised ranking to identify root cause metrics [11,13]. Such methodologies have achieved significant progress in network fault diagnosis, microservice analysis, and database root cause analysis [6,7].

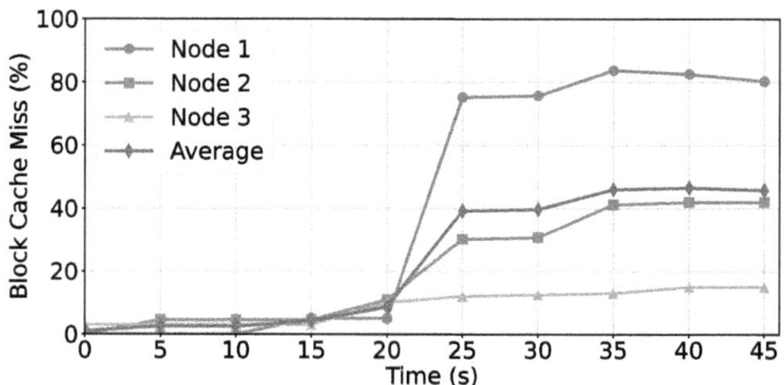

Fig. 1. The rate of block cache miss in a 3-node OceanBase cluster experiencing a small buffer anomaly. The X-axis represents the elapsed time (seconds) after the anomaly was triggered, while the Y-axis indicates the cache miss (percentage).

In distributed databases, the large number of nodes generates massive runtime metrics [23]. When performance anomalies occur, the patterns of metric variation across this vast set of nodes exhibit significant diversity. For example, Fig. 1 shows a 3-node OceanBase distributed database cluster experiencing a small buffer anomaly: while the average cache miss rate of the cluster surges, individual nodes exhibit divergent behaviors, nodes 1 and 2 experience severe degradation, while node 3 shows minimal impact. This divergence arises because cluster-level metrics are aggregated from the metrics of all underlying nodes, and the performance and anomaly patterns of individual nodes do not have to align with those of the cluster. An effective diagnostic method must therefore analyze metrics across all nodes jointly to identify root causes.

However, current causality-based diagnostic approaches face two critical challenges in distributed environments: **1) The large number of anomaly-irrelevant metrics impairs the performance of diagnosis.** Distributed databases provide exhaustive metric systems, where each node hosts a complete set of monitoring metrics, the total number of metrics increases linearly with the number of nodes. Existing methods [3,4,14,16] typically model the relationships among metrics of the entire system directly. However, distributed execution exhibits strong data locality: queries or transactions typically access only a subset of nodes. Modeling the metrics of the entire system is both inefficient and inaccurate when only a few of nodes and their metrics are relevant to the ongoing anomaly. This not only increases computational overhead but also conceals real inter-metric and inter-node causal effects, degrading the diagnostic precision.

2) Underutilization of Causality. Traditional root cause analysis methods [14,15,17,22] typically employ unsupervised graph node ranking (e.g., Page-Rank, RandomWalk), which primarily leverages two types of information from causal graphs: directional relationships between metrics (edges), and strength of the correlation effect between metrics (edge weights). However, these methods fail to quantify the intensity and mechanism of causal effects. When an anomaly such as the increase of query timeout rates occurs, a database might exhibit both thread pool exhaustion and an abnormal variation in CPU utilization. But only the former is the true root cause of the anomaly, while the latter is merely a side effect. The limitation of existing methods easily leads to ranking the critical root causes after other secondary metrics.

To address these challenges, we propose the DBRooter diagnostic framework. DBRooter introduces Dependency-Guided Causal Modeling for distributed metrics to effectively capture relationships between metrics. It begins with redundant node filtering: using the query plan to prune nodes not accessed by the target query and then applying an anomaly detector to exclude healthy nodes, thereby reducing the scope of the analysis. After redundant node filtering, DBRooter analyzes the query plan to discover runtime data flow and inter-node dependencies. Guided by this dependency structure, DBRooter constructs local causal graphs within each abnormal node and adds inter-node causal links only where data dependencies exist, avoiding costly exhaustive discovery and enabling parallel graph construction. These steps yield a more accurate and efficient causal model over distributed metrics. To pinpoint the root cause of anomalies and address Challenge 2, DBRooter proposes a Workload-Aware Anomaly Metric Ranking method. It employs Structural Causal Models(SCMs) [9] to model causal relationships. Beyond constructing causal paths, SCMs explicitly define functional relationships between variables to describe causal mechanisms, providing a quantitative basis for the root cause ranking. We further propose a workload-aware SQL selection mechanism, which can handle scenarios where a DBA needs to analyze individual queries, single transactions, and the entire workload.

We evaluated our approach on distributed database clusters comprising 3, 5, 7, and 9 nodes. The results show that DBRooter improves the average AC@1, AC@3, and AC@5 for root cause diagnosis by up to 10%, 15%, and 12%, respectively, and overall outperforms all state-of-the-art baselines. Specifically, the contributions of this work are summarized as follows:

- We propose an efficient causal root cause analysis framework, DBRooter, which accurately pinpoints root cause metrics in distributed databases.
- We propose dependency-guided causal modeling for distributed metrics, leading to a more accurate and efficient causal model.
- We propose workload-aware anomaly metric ranking based on structural causal models, supporting diagnosis for queries, transactions, and workloads.

- We evaluated our approach on distributed database clusters comprising up to 9 nodes. Experimental results demonstrate that DBRooter outperforms other SOTA baselines.

2   Problem Statements

Consider a distributed database $\mathcal{DB} = \{\text{Node}_i\}_{i=1}^{N}$ comprising N interconnected nodes. $\mathcal{DB}$ processes a workload $\mathcal{W} = \{\text{query}_j\}_{j=1}^{K}$ consisting of K distinct query statements. Each node generates comprehensive monitoring metrics, formally represented as a multivariate time series $\mathbf{M} = \{\mathbf{M}_i\}_{i=1}^{N}$, where $\mathbf{M}_i \in \mathbf{R}^{d \times T}$ denotes the d-dimensional metrics of Node_i sampled over T time steps.

When performance anomalies occur in $\mathcal{DB}$ (e.g., CPU saturation or Network Saturation), the DBRooter framework initiates automated root cause analysis (RCA). Given $\mathbf{M}$, DBRooter performs causal analysis to uncover latent fault propagation. The objective of DBRooter is to compute a diagnostic score for every metric on every node, thereby identifying the most likely root causes. Formally, for each metric k $(1 \leq k \leq d)$ at each node i $(1 \leq i \leq N)$, DBRooter computes a diagnostic score $\delta(\mathbf{M}_i^{(k)})$ using:

$$\delta(\mathbf{M}_i^{(k)}) = f_{\text{RCA}}(\mathbf{M}_i^{(k)}, \mathbf{M}), \tag{1}$$

where f_{RCA} is used to calculate the score $\delta(\mathbf{M}_i^{(k)})$ for a single metric, which quantifies the likelihood of the metric $\mathbf{M}_i^{(k)}$ being a root cause of the anomaly.

By applying root cause f_{RCA} to all $N \times d$ metrics, DBRooter finally generates a comprehensive set of diagnostic tuples

$$\mathcal{S} = \{\langle \text{Node}_i, \text{Metric}_k, \delta(\mathbf{M}_i^{(k)}) \rangle\}, \tag{2}$$

this set is then ranked by descending diagnostic scores to produce the final output: an ordered root cause sequence $\mathcal{D} = \{\langle \text{Node}_r, \text{Metric}_r \rangle\}_{r=1}^{m}$ of the top-m most probable root cause indicators. This diagnostic sequence provides database administrators (DBAs) with a causal road map for anomaly mitigation. Rather than inspecting all $N \times d$ metrics, DBAs prioritize remediation using $\mathcal{D}$. For instance, if $\langle \text{Node}_7, \texttt{disk\_queue\_depth} \rangle$ dominates $\mathcal{D}$, the DBA immediately investigates I/O saturation on Node 7. DBRooter thus transforms high-dimensional metrics into actionable suggestions for database operations.

3   Overview of DBRooter

Figure 2 shows the overview of DBRooter. When a performance alert occurs in the database, such as increased query latency or reduced throughput, DBRooter applies dependency-guided causal modeling to the distributed metrics. Starting from query execution plans and node-level performance metrics, DBRooter

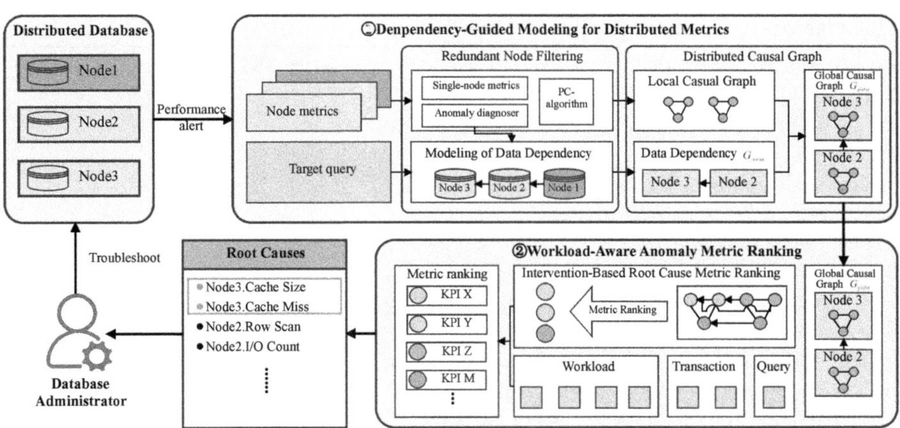

Fig. 2. The overview of DBRooter

introduces a redundant node filtering mechanism to prune nodes that are unrelated to the target query or exhibit normal operational behavior, narrowing the scope of diagnosis. Within the remaining anomalous nodes, it constructs a global causal graph of metrics by integrating intra-node metric dependencies with inter-node data flow relationships derived from distributed execution semantics. This unified causal structure captures how performance deviations propagate across both metrics and nodes.

Based on this causal graph, DBRooter then employs a workload-aware anomaly metric ranking technique to identify root causes. It builds structural causal models to model functional relationships between metrics and performs interventions on metrics to quantify the extent of their deviation as abnormal scores. In addition, to accommodate different diagnostic scenarios, DBRooter introduces a workload-aware SQL selection mechanism that performs multi-granularity analysis at the query, transaction, and workload levels, and computes corresponding anomaly scores for the metrics. Finally, by ranking metrics according to their abnormal scores, DBRooter identifies the most likely root causes and presents them to the DBA in an interpretable manner, enabling fast, automated, and insight-driven diagnosis even in complex environments.

4 Dependency-Guided Causal Modeling for Distributed Metrics

The dependency-guided causal modeling consists of two components: (1) *redundant node filtering*, which removes irrelevant nodes to reduce the complexity of diagnosis; and (2) *distributed causal graph*, which integrates the knowledge of query execution into the construction of a global causal graph. Together, these components allow DBRooter to efficiently construct a meaningful causal model tailored to the metrics of a distributed database.

Algorithm 1. Modeling of Data Dependency

Input: Query execution plan P, monitoring metrics $\mathbf{M} = \{\mathbf{M}_i\}_{i=1}^{N}$
Output: Data dependency graph $\mathcal{G}_d$
 1: // **Extract nodes accessed in query execution**
 2: $V \leftarrow \emptyset$
 3: **for** each operator op in execution plan P **do**
 4:     $node\_id \leftarrow \text{GetNodeId}(op)$
 5:     $V \leftarrow V \cup \{node\_id\}$
 6: **end for**
 7: // **Extract inter-node data dependency**
 8: $E \leftarrow \emptyset$
 9: **for** each operator op in execution plan P **do**
10:     $target\_node \leftarrow \text{GetNodeId}(op)$
11:     **for** each remote data access in op **do**
12:         $source\_node \leftarrow \text{GetRemoteNodeId}(remote\_access)$
13:         $E \leftarrow E \cup \{(source\_node, target\_node)\}$
14:     **end for**
15: **end for**
16: $G_d \leftarrow (V, E)$
17: // **Remove cycles while preserving sink constraint**
18: $cycles \leftarrow \text{FindCycles}(G_d)$
19: **while** $cycles \neq \emptyset$ **do**
20:     $C \leftarrow \text{SelectCycle}(cycles)$
21:     $edge \leftarrow \text{SelectEdge}(C, G_d)$ {Remove edge that preserves sink constraint}
22:     $E \leftarrow E \setminus \{edge\}$
23:     $G_d \leftarrow (V, E)$
24:     $cycles \leftarrow \text{FindCycles}(G_d)$
25: **end while**
26: **return** G_d

4.1   Redundant Node Filtering

In large-scale distributed databases, hundreds of metrics can be collected across numerous nodes. However, many of these metrics are redundant (e.g., from normal nodes or irrelevant nodes). Including such metrics in causal analysis increases both computational overhead and the complexity of diagnosis. DBRooter addresses this problem through a redundant node filtering mechanism, which consists of the modeling of data dependency and normal node filtering.

Modeling of Data Dependency. Algorithm 1 illustrates the modeling of data dependency. First, we filter out nodes that do not participate in the anomalous query's execution, as they are unlikely to contribute directly to the performance degradation. Given a target query, by parsing the query plan P, DBRooter identifies the subset of nodes actively accessed during execution, denoted as $V = \{\text{Node}_i\}$, (Algorithm 1, Lines 1–6).

Next, we construct a directed data dependency graph $G_d = (V, E)$ to model the data dependencies among the active nodes in V. In G_d, each edge $(u, v) \in E$ represents a data flow from node u to node v, indicating that the computation on node v depends on data residing on or produced by node u. Specifically, for any operator in the query plan (e.g., join or aggregation) executed on node v that requires remote data from node u, we add a directed edge $u \rightarrow v$ to E, (Algorithm 1, Lines 7–16). G_d denotes all nodes accessed during a query, with data flowing along the direction of the edges. If a node exhibits a performance anomaly that slows data production, it will consequently block downstream nodes. G_d therefore captures potential anomaly-propagation paths.

Since the final query result is returned to the client from a single node (typically the node that receives the query), the dependency graph contains exactly one *sink* node—defined as a node with zero out-degree. Formally, let out-deg(v) denotes the out-degree of node v, then

$$\exists!\, s \in V \text{ such that } \text{out-deg}(s) = 0 \quad \text{and} \quad \forall v \in V \setminus \{s\}, \text{ out-deg}(v) > 0. \quad (3)$$

To ensure G_d is a directed acyclic graph (DAG), a prerequisite for causal inference, we detect and break any cycles that may arise due to complex multi-hop data exchanges or optimizer-generated query plans. However, edges cannot be removed arbitrarily, as this could disrupt data dependencies and distort execution semantics. To preserve the logical flow of data, we impose the constraint defined in Eq. 3: an edge (u, v) may only be removed if doing so does not introduce a new sink node (i.e., out-deg$(v) > 0$ after removal, unless v is the original sink). If a cycle $C \subseteq V$ is detected, we iteratively remove one edge from C that satisfies this constraint until the graph becomes acyclic, (Algorithm 1, Lines 17–25). In real query execution, data flow outwards from distributed storage nodes to processing nodes responsible for aggregation and response generation. This selective edge removal ensures that the global directionality of data propagation is preserved throughout the transformation.

Normal Node Filtering. After obtaining $G_d = (V, E)$, we observe that although all nodes in V participate in query execution, not all of them exhibit anomalous behavior. To narrow down the root cause candidates, DBRooter employs a classifier to filter out normal nodes. We use an XGBoost [5] classifier trained on historical metrics. For each node $v_i \in V$, we extract its metrics $\mathbf{M}_i$ and predict its anomaly status as: $a_i = \text{XGBoostClassifier}(\mathbf{M}_i)$, where $a_i \in \{0, 1\}$, with $a_i = 1$ indicating that node v_i is anomalous.

We then prune G_d by removing all non-anomalous nodes. If a non-anomalous node has no incoming or outgoing edges, it lies at an endpoint of the data-flow chain, and we simply remove it. For other non-anomalous nodes, to preserve the data flow semantics, we reconnect the graph by bypassing the nodes: for a removed node v, we add direct edges from each of its predecessors to each of its successors. Formally, let pred$(v) = \{u \mid (u, v) \in E\}$ and succ$(v) = \{w \mid (v, w) \in E\}$. Upon removing v, we update the edge set as:

$$E \leftarrow E \cup \{(u, w) \mid u \in \text{pred}(v),\ w \in \text{succ}(v)\} \setminus \{(u, v), (v, w) \mid u, w \in V\}. \quad (4)$$

The final pruned graph, denoted $G_{\mathrm{anom}} = (V_{\mathrm{anom}}, E_{\mathrm{anom}})$, contains only anomalous nodes ($V_{\mathrm{anom}} = \{v_i \in V \mid a_i = 1\}$), and its edges represent data dependencies among them.

4.2   Distributed Causal Graph

To enable efficient and interpretable root cause localization in large-scale distributed databases, we construct a global distributed causal graph that captures both intra-node metric interactions and inter-node data-flow dependencies. This construction consists of two steps: (1) local metric modeling on each anomalous node, and (2) global integration guided by the data dependency structure. Algorithm 2 shows the procedure of distributed causal graph construction.

Algorithm 2. Distributed Causal Graph Construction

Input: Data dependency graph G_{anom}, monitoring metrics $\mathbf{M} = \{\mathbf{M}_i\}_{i=1}^{N}$
Output: Global causal graph $\mathcal{G}_{\mathrm{global}}$
1: // **Metrics Modeling for Each Node**
2: **for** each anomalous node $v_i \in V_{\mathrm{anom}}$ in parallel **do**
3:     $\mathcal{G}_i \leftarrow \mathrm{PCAlgorithm}(\mathbf{M}_i)$
4: **end for**
5: // **Distributed Causal Graph Construction**
6: $\mathcal{V}_{\mathrm{global}} \leftarrow \bigcup_{v_i \in V_{\mathrm{anom}}} \{(v_i, k) \mid k = 1, \ldots, d\}$
7: $\mathcal{E}_{\mathrm{global}} \leftarrow \emptyset$
8: **for** each local graph $\mathcal{G}_i$ **do**
9:     **for** each edge $M_i^{(k)} \rightarrow M_i^{(\ell)} \in \mathcal{E}_i$ **do**
10:        $\mathcal{E}_{\mathrm{global}} \leftarrow \mathcal{E}_{\mathrm{global}} \cup \{((v_i, k), (v_i, \ell))\}$
11:    **end for**
12: **end for**
13: **for** each edge $(v_u, v_v) \in E_{\mathrm{anom}}$ **do**
14:    **for** each metric dimension $k \in \{1, \ldots, d\}$ **do**
15:        $\mathcal{E}_{\mathrm{global}} \leftarrow \mathcal{E}_{\mathrm{global}} \cup \{((v_u, k), (v_v, k))\}$
16:    **end for**
17: **end for**
18: $\mathcal{G}_{\mathrm{global}} \leftarrow (\mathcal{V}_{\mathrm{global}}, \mathcal{E}_{\mathrm{global}})$
19: **return** $\mathcal{G}_{\mathrm{global}}$

Metric Modeling for Each Node. For each anomalous node $v_i \in V_{\mathrm{anom}}$, we model the causal relationships among its performance metrics. We apply the PC algorithm [10,20] to learn a local causal graph $\mathcal{G}_i = (\mathbf{M}_i, \mathcal{E}_i)$, where each node corresponds to a metric $M_i^{(k)}$, and a directed edge $M_i^{(k)} \rightarrow M_i^{(\ell)} \in \mathcal{E}_i$ indicates that $M_i^{(k)}$ is a direct cause of $M_i^{(\ell)}$, (Algorithm 2, Lines 1–4). Critically, this step is parallelizable: each node's causal graph $\mathcal{G}_i$ is learned independently using only its local metrics. This design ensures linear scalability with respect to the number of anomalous nodes and enables efficient deployment in large clusters.

Distributed Causal Graph Construction. Having obtained the local causal graphs $\{\mathcal{G}_i\}_{v_i \in V_{\text{anom}}}$, we now integrate them into a global distributed causal graph $\mathcal{G}_{\text{global}}$ by incorporating inter-node data dependencies from G_{anom}.

The integration proceeds as follows: we treat each $\mathcal{G}_i$ as a subgraph in $\mathcal{G}_{\text{global}}$, and for every directed edge $(v_u, v_v) \in E_{\text{anom}}$ (indicating data flows from v_u to v_v), we add inter-node causal edges between *corresponding metric types*, (Algorithm 2, Lines 5–18). Specifically, for each metric dimension $k \in \{1, \ldots, d\}$, we introduce a directed edge $M_u^{(k)} \longrightarrow M_v^{(k)}$. This reflects the intuition that performance degradation in a metric on a source node can directly impact the same downstream metrics on the destination node due to data propagation.

Formally, the global causal graph is defined as: $\mathcal{G}_{\text{global}} = (\mathcal{V}_{\text{global}}, \mathcal{E}_{\text{global}})$, where the node set is the disjoint union of all local metrics:

$$\mathcal{V}_{\text{global}} = \bigcup_{v_i \in V_{\text{anom}}} \{(v_i, k) \mid k = 1, \ldots, d\}, \tag{5}$$

and the edge set combines intra-node and inter-node causal relations:

$$\mathcal{E}_{\text{global}} = \underbrace{\bigcup_{v_i \in V_{\text{anom}}} \left\{ ((v_i, k), (v_i, \ell)) \mid M_i^{(k)} \to M_i^{(\ell)} \in \mathcal{E}_i \right\}}_{\text{intra-node causal edges}}$$
$$\cup \underbrace{\bigcup_{(v_u, v_v) \in E_{\text{anom}}} \left\{ ((v_u, k), (v_v, k)) \mid k = 1, \ldots, d \right\}.}_{\text{inter-node data dependency edges}} \tag{6}$$

Since both the local graphs $\mathcal{G}_i$ (learned via PC) and the inter-node dependency graph G_{anom} are directed acyclic graphs (DAGs), and inter-node edges strictly follow the direction of E_{anom}, the resulting $\mathcal{G}_{\text{global}}$ is guaranteed to be acyclic. By jointly encoding intra-node causal relationships and inter-node data dependency semantics, $\mathcal{G}_{\text{global}}$ provides a holistic view of how anomalies propagate across the distributed system. Moreover, this construction avoids explicit causality testing among metrics of different nodes by directly using inter-node dependency relationships. This substantially reduces computational overhead while preserving the capacity to capture inter-node causal relationships.

4.3 The Time Complexity Analysis of Dependency-Guided Causal Modeling for Distributed Metrics

First, DBRooter builds a data dependency graph G_d from the target query. Next, it leverages the normal node filtering to prune G_d to G_{anom}. DBRooter then models the causal relationship of metrics in each node and constructs $\mathcal{G}_{global}$. Its total time complexity is $O(d^3)$, as shown in Theorem 1:

Theorem 1. *The time complexity of dependency-guided causal modeling for distributed metrics is $O(d^3)$.*

Proof. The execution process of dependency-guided causal modeling for distributed metrics is as follows:

Phase 1: Modeling of Data Dependency. The modeling of data dependency requires parsing query execution plans, which takes $O(|P|)$ time where $|P|$ is the number of operators in the execution plan. Let $|V|$ be the number of active nodes. Building the data dependency graph G_d involves cycle detection and removal. Cycle detection in a directed graph takes $O(|V| + |E|)$ time, where $|E|$ is the number of edges. For each detected cycle, edge removal takes $O(|V|)$ time in the worst case. Since the number of cycles is bounded by $O(|V|)$ in practice, the total complexity for building G_d is $O(|V|^2)$ in the worst case.

Phase 2: Normal Node Filtering. DBRooter uses XGBoost for anomalous node detection, and the time complexity is approximately $O(d \cdot T)$ for classifying each node (omitting the tree depth and the number of trees). Since anomaly detection can be performed independently for each node, the total complexity for this phase is $O(|V| \cdot d \cdot T)$.

Phase 3: Metrics Modeling for Each Node. The PC algorithm for each anomalous node has complexity $O(d^3)$, where d is the number of metrics per node. Since local causal graphs are constructed in parallel across $|V_{\mathrm{anom}}|$ anomalous nodes, the effective complexity remains $O(d^3)$.

Phase 4: Distributed Causal Graph Construction. Integrating local graphs with inter-node dependencies takes $O(|V_{\mathrm{anom}}| \cdot d^2 + |E_{\mathrm{anom}}| \cdot d)$ time, where $|E_{\mathrm{anom}}|$ is the number of edges in the pruned dependency graph.

Overall Complexity. The total time complexity is $O(|P| + |V|^2 + |V| \cdot d \cdot T + d^3 + |V_{\mathrm{anom}}| \cdot d^2 + |E_{\mathrm{anom}}| \cdot d)$. In practice, since $|V_{\mathrm{anom}}| < |V|$, $|E_{\mathrm{anom}}| < |V|^2$ and $|V| < d$, the effective complexity is dominated by $O(d^3)$, which is significantly lower than traditional PC-based approaches that require $O((N \cdot d)^3)$ for global causal discovery across all N nodes in a distributed database.

5　Workload-Aware Anomaly Metric Ranking

While accurate causal metric modeling establishes the foundation for root cause analysis, effective diagnosis depends on the ability to prioritize and identify metrics most responsible for performance degradation. DBRooter introduces a workload-aware anomaly metric ranking, which consists of two key components: (1) *intervention-based root cause metric ranking*, which leverages structural causal models to extract functional causal relations between metrics; and (2) *workload-aware SQL selection*, which identifies the subset of queries in the current workload and reduces computational overhead.

5.1 Intervention-Based Root Cause Metric Ranking

First, DBRooter leverages SCMs to model the functional causal relationship between metrics. Given the global causal graph $\mathcal{G}_{\text{global}}$, we construct a structural causal model $\mathcal{M} = \langle \mathbf{U}, \mathbf{V}, \mathbf{F} \rangle$, where $\mathbf{V}$ denotes the set of observed metric variables (i.e., nodes in $\mathcal{G}_{\text{global}}$), $\mathbf{U}$ represents unobserved exogenous factors, and $\mathbf{F} = \{f_v \mid v \in \mathbf{V}\}$ defines the set of structural functions such that:

$$V_j = f_j\big(pa(V_j), U_j\big), \tag{7}$$

where V_j is $(v_i, k) \in \mathcal{V}_{\text{global}}$, $pa(V_j)$ denotes the parent variables of V_j in $\mathcal{G}_{\text{global}}$. These functions are estimated from historical normal-state data using regression, ensuring faithful representation of dependencies among metrics.

DBRooter's ranking mechanism uses intervention within the SCM framework [9,12]. Unlike correlation-based or topology-driven ranking schemes, this approach quantifies the actual impact of each anomalous metric on the target performance degradation through simulated interventions. To identify root cause metrics, we treat the observed performance anomaly as an unmeasured intervention that perturbs one or more system variables. For each candidate metric variable $V_j \in \mathbf{V}$, we perform a counterfactual intervention $\text{do}(V_j = v_j^{\text{obs}} + \delta_j)$, where v_j^{obs} is its observed value during the anomaly window and δ_i models the deviation from its expected behavior under normal conditions. We then propagate this intervention through $\mathcal{M}$ to compute the resulting change in downstream variables. Let $\overline{V_j}$ be the regression value for V_j. Assuming the residuals follow an $i.i.d.$ normal distribution $N(\mu_j, \sigma_j)$, we define the abnormal score of V_j as

$$\delta(\mathbf{M}_i^{(k)}) = \delta(V_j) = \frac{(V_j - \overline{V_j}) - \mu_j}{\sigma_j}. \tag{8}$$

By computing $\delta(V_j)$ for all $V_j (i.e., Metric\ M_i^{(k)})$ in $\mathcal{G}_{\text{global}}$, DBRooter generates a ranked list of metrics ordered by their counterfactual influence. This intervention-based approach goes beyond mere correlation or centrality, directly modeling how local anomalies propagate through the causal structure, enabling precise and interpretable root cause localization.

5.2 Workload-Aware SQL Selection

While the intervention-based ranking framework operates on individual queries, real-world database workloads often consist of multiple concurrent SQL statements. Given a workload $W = \{q_1, q_2, \ldots, q_K\}$, we extend the root cause metric score to the workload level by aggregating anomaly impacts across queries. DBRooter supports three diagnostic scenarios with different granularities:

Individual Query Analysis. For individual query diagnosis, DBRooter directly extracts the execution plan for the target query and computes root cause scores based on the specific nodes and metrics involved in that query's execution.

Single Transaction Analysis. For transaction-level diagnosis, DBRooter aggregates anomaly scores across all queries within the transaction. Given a transaction $T = \{q_1, q_2, \ldots, q_k\}$ consisting of k queries, the final score for each metric V_j is computed as the average impact across all queries in the transaction:

$$\delta_T(V_j) = \frac{1}{|T|} \sum_{q_i \in T} \delta_{q_i}(V_j). \tag{9}$$

This averaging approach ensures that the root cause ranking reflects the overall performance impact of the entire transaction.

Entire Workload Analysis. However, for large-scale workloads, not all queries are equally relevant to the observed performance issue. Including low-impact or irrelevant queries can increase the computation overhead. To address this, DBRooter leverages built-in database performance monitors (e.g., OceanBase ASH report) to extract the top-$m\%$ resource-consuming SQLs (e.g., those with the highest CPU time), and forms a critical subset $W_{\text{top}} \subset W$. We then compute

$$\delta(V_j) = \frac{1}{|W_{\text{top}}|} \sum_{q_k \in W_{\text{top}}} \delta_{q_k}(V_j). \tag{10}$$

It restricts causal analysis to the most influential queries, preserving diagnostic precision while substantially reducing computational overhead.

6   Experiments

Baselines. To evaluate the effectiveness of DBRooter, we compare DBRooter with five state-of-the-art root cause analysis baselines. In experiments, we simultaneously diagnose the metrics of all nodes in the distributed database.

- **MicroCause** [17] is based on causal graphs and unsupervised ranking of graph nodes, it introduces time delays into computations on the edges of the causal graph to model real-world anomaly impact patterns.
- **FluxInfer** [14] models relations among metrics using an undirected metric relation graph and uses PageRank to find the most influential metrics.
- **CauseInfer** [4] is also based on causal graphs and unsupervised ranking of graph nodes, but it ranks metrics using Deep First Search.
- **AutoMAP** [16] uses graph subtraction to capture differences between the system's normal and abnormal behaviors. It was originally designed for microservice diagnosis, and we adapt it for distributed database diagnosis.
- **Balance** [3] uses regression and attribution analysis, first fitting the effects of key metrics on performance anomalies and then evaluating metric importance based on the regression coefficients.

Environment. We conduct experiments on up to 10 Ubuntu 20.04 servers. Each server is equipped with 32 AMD EPYC 9K84 processor cores, 64 GB RAM, and a 500 GB SSD. We evaluate two database systems—OceanBase 4.3.0 [25] and MySQL 8.0—configured as clusters of 3, 5, 7, and 9 nodes. An additional server is used to generate benchmark requests. m is set to 20%.

Anomaly Scenarios. We constructed 13 concrete anomaly scenarios, including 8 individual anomalies, 4 compound anomalies and additionally a real-world workload imbalance case. Details are provided in Table 1. We use TPC-C and TPC-H workloads. The diagnosis results are evaluated by four database experts, including one senior engineer in a database development department.

Table 1. Anomaly Scenarios.

Anomaly Type	Anomaly Name			
Individual Anomaly	CPU Saturation	I/O Saturation	Network Saturation	Small Buffer
	Heavy Workload	Too many indexes	Missing Inexes	Dump
Compound Anomaly	CPU Saturation+I/O Saturation		CPU Saturation + Network Saturation	
	I/O Saturation+Network Saturation		Small Buffer+I/O Saturation	
Real World Anomaly	Workload Imbalance			

6.1 Root Cause Analysis Accuracy Comparison

Table 2. AC@k on OceanBase cluster.

# of Nodes	3			5			7			9		
AC@k	AC@1	AC@3	AC@5	AC@1	AC@3	AC@5	AC@1	AC@3	AC@5	AC@1	AC@3	AC@5
FluxInfer	0.27	0.32	0.51	0.26	0.33	0.47	0.20	0.27	0.33	0.17	0.23	0.31
MicroCause	0.22	0.34	0.41	0.18	0.31	0.35	0.15	0.24	0.29	0.13	0.23	0.30
Balance	<u>0.45</u>	<u>0.55</u>	<u>0.62</u>	**0.48**	<u>0.53</u>	<u>0.69</u>	<u>0.35</u>	<u>0.42</u>	<u>0.58</u>	<u>0.30</u>	<u>0.43</u>	<u>0.54</u>
CauseInfer	0.18	0.32	0.49	0.19	0.28	0.42	0.10	0.19	0.23	0.07	0.14	0.21
AutoMAP	0.23	0.33	0.47	0.22	0.29	0.41	0.18	0.31	0.41	0.21	0.27	0.38
DBRooter	**0.51**	**0.62**	**0.71**	<u>0.47</u>	**0.61**	**0.75**	**0.43**	**0.57**	**0.64**	**0.37**	**0.53**	**0.59**

Table 2 and Table 3 compare DBRooter's diagnostic performance against baselines on OceanBase and MySQL clusters. The bold and underlining indicate the best and second-best performance. DBRooter consistently outperforms all baselines across both database systems: compared with the second-best method, AC@1, AC@3, and AC@5 improve by 10%, 15%, and 12%, corresponding to relative gains of 30.3%, 35.7%, and 23.5%. This substantial improvement demonstrates DBRooter's superior ability to identify root causes accurately and rank them appropriately.

Table 3. AC@k on MySQL cluster.

# of Nodes	3			5			7			9		
AC@k	AC@1	AC@3	AC@5	AC@1	AC@3	AC@5	AC@1	AC@3	AC@5	AC@1	AC@3	AC@5
FluxInfer	0.11	0.14	0.21	0.10	0.15	0.18	0.11	0.14	0.15	0.13	0.15	0.16
MicroCause	0.13	0.20	0.24	0.14	0.21	0.23	0.13	0.19	0.22	0.14	0.18	0.28
Balance	0.34	0.42	0.51	0.33	0.45	0.47	0.29	0.37	**0.43**	0.25	**0.33**	**0.38**
CauseInfer	0.11	0.17	0.28	0.08	0.15	0.17	0.09	0.11	0.14	0.04	0.07	0.23
AutoMAP	0.14	0.22	0.30	0.15	0.16	0.22	0.08	0.12	0.15	0.05	0.17	0.09
DBRooter	**0.41**	**0.55**	**0.63**	**0.43**	**0.51**	**0.53**	**0.33**	**0.38**	0.41	**0.27**	0.31	0.35

On OceanBase clusters, DBRooter achieves up to 51% AC@1 accuracy, while on MySQL clusters, it reaches 41% AC@1 accuracy. This consistency across different database architectures validates DBRooter's design principles and its applicability to diverse distributed database environments.

6.2   Root Cause Analysis Comparison in Different Granularities

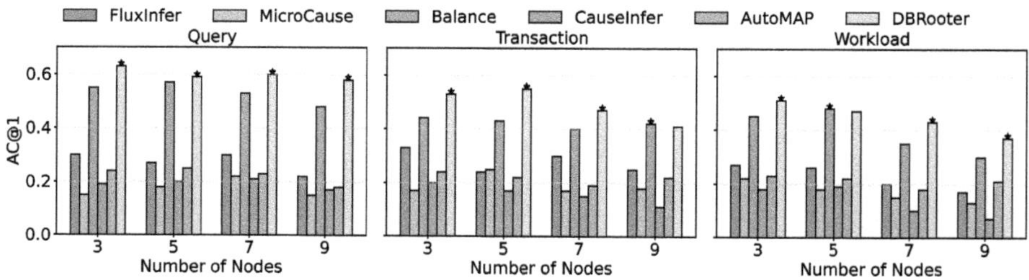

Fig. 3. The comparison of root cause analysis in different granularities. The star markers denote the best accuracy.

To further demonstrate DBRooter's effectiveness across diagnostic granularities, Fig. 3 compares its performance when diagnosing at the query, transaction, and workload levels on OceanBase. For query or transaction diagnosis, the database executes the corresponding single query or transaction. The results show that DBRooter achieves the best performance. The consistent high performance across all granularities highlights DBRooter's adaptive analysis capability. DBRooter's workload-aware SQL selection provides flexible and context-sensitive diagnosis that adapts to different operational requirements. This support is crucial for real-world database operations, where DBAs may need to diagnose issues at different levels depending on the scenarios.

Table 4. Ablation Study on OceanBase.

# of Nodes	3			5			7			9		
AC@k	AC@1	AC@3	AC@5	AC@1	AC@3	AC@5	AC@1	AC@3	AC@5	AC@1	AC@3	AC@5
DBRooter-w/o-DG	0.42	0.53	0.62	0.45	0.57	0.68	0.33	0.48	0.52	0.35	0.49	0.57
DBRooter-w/o-WR	0.21	0.44	0.52	0.23	0.38	0.45	0.19	0.25	0.38	0.17	0.23	0.33
DBRooter	**0.51**	**0.62**	**0.71**	**0.47**	**0.61**	**0.75**	**0.43**	**0.57**	**0.64**	**0.37**	**0.53**	**0.59**

6.3 Ablation Study

Table 4 presents the ablation experiments, where DBRooter-w/o-DG removes dependency-guided causal modeling and instead applies the PC algorithm, and DBRooter-w/o-WR removes the workload-aware anomaly metric ranking and replaces it with a simple PageRank. From the results, we have the following findings: First, dependency-guided causal modeling improves causal modeling accuracy through inter-node data dependencies; compared with DBRooter-w/o-DG, DBRooter achieves up to 10%, 9%, and 12% improvements on AC@1, AC@3, and AC@5, respectively. Second, replacing simple unsupervised ranking with workload-aware anomaly metric ranking substantially enhances DBRooter's diagnostic performance. Compared to DBRooter-w/o-WR, DBRooter achieves up to 0.34, 0.30, and 0.30 improvements on AC@1, AC@3, and AC@5, respectively. This significant improvement indicates that SCM can capture the influence relationships among database metrics and perform quantitative analysis.

6.4 Efficiency Comparison

Table 5. Root Cause Analysis Overhead Comparison (Seconds).

# of Nodes	FluxInfer	MicroCause	Balance	CauseInfer	AutoMAP	DBRooter
3	18.21	32.63	7.62	23.12	20.61	**5.68**
5	43.11	83.12	11.92	47.80	44.82	**7.12**
7	74.29	162.32	15.63	81.53	77.31	**13.26**
9	170.81	376.25	**21.71**	219.51	189.71	27.82

Table 5 presents the average time overhead of each method during diagnosis. DBRooter achieves the lowest overhead across 3–7 nodes. Since Balance's regression and attribution processes have lower time complexity, it is faster than DBRooter at 9 nodes. However, in this case, DBRooter effectively leverages causality while maintaining a comparable time overhead to Balance. Among the other baselines, FluxInfer is faster than MicroCause, CauseIner, and AutoMap because it does not require a complex causal discovery process. Compared with these methods, DBRooter's distributed causal graph construction yields lower and more gracefully scaling overhead as the cluster size increases.

7   Related Works

There already exist numerous root cause analysis methods, most of which can be adapted to databases. They can be roughly divided into three categories:

Data-driven statistical methods [1,19,29] identifies root causes by analyzing the characteristics of monitoring metrics. For example, ϵ-diagnosis [19] captures outliers via time-series statistics. However, these methods often employ simple statistical techniques, failing to capture complex causal patterns.

Causal graph–based methods [4,12,14,16–18,28] adopts a causality-driven analysis approach. For instance, FluxInfer [14] proposes an undirected metric relationship graph to model mutual influences among metrics and uses weighted PageRank to find the most influential metrics as root causes. Micro-Cause [17] models metrics with a directed causal graph and introduces lag into causal computation. However, current causal methods often rely on unsupervised node ranking, underutilizing causal information, and struggle to scale to large distributed database clusters due to the complexity of building causal graphs.

Attribution-based methods [3,8] designs a regression and attribution analysis paradigm, selecting as root-cause metrics those that contribute most to the modeled anomaly. For example, Balance [3] first builds a forward module to predict performance anomalies and then selects metrics with larger regression coefficients as root causes. However, such methods often handle all metrics simultaneously and struggle to perceive the complex interactions among metrics.

8   Conclusion

In this paper, we present DBRooter, an efficient causal root-cause analysis framework for distributed databases. DBRooter leverages dependency-guided causal modeling for distributed metrics to efficiently construct local causal graphs in parallel and integrate them with data dependencies. It further employs workload-aware anomaly metric ranking with structural causal models to quantitatively assess causal effects and accurately identify root causes. Experimental results show that DBRooter effectively diagnoses anomalies in individual queries, transactions, and entire workloads, outperforming state-of-the-art baselines.

Acknowledgments. This work is supported by the National Natural Science Foundation of China (Nos. 62272054, 62192784), Beijing Nova Program (No. 20230484319, 20250484968), State Key Laboratory of Multimedia Information Processing Open Fund (No. SKLMIP-KF-2025-07) and Shandong Key Laboratory of Advanced Computing. Yingxia Shao is the corresponding author.

References

1. Bhagwan, R., et al: Adtributor: revenue debugging in advertising systems. In: NSDI, pp. 43–55 (2014)
2. Cao, H., et al.: Aion: live migration for in-memory databases with zero downtime and reduced redundant data transfer. Data Sci. Eng. **10**, 212–229 (2025)
3. Chen, C., et al.: BALANCE: Bayesian linear attribution for root cause localization. Proc. ACM Manage. Data **1**(1), 1–26 (2023)
4. Chen, P., et al.: CauseInfer: automatic and distributed performance diagnosis with hierarchical causality graph in large distributed systems. In: INFOCOM, pp. 1887–1895 (2014)
5. Chen, T., et al.: XGBoost: a scalable tree boosting system. In: SIGKDD (2016)
6. Dang, Y., et al.: AiOps: real-world challenges and research innovations. In: ICSE-Companion, pp. 4–5 (2019)
7. Fang, A., et al.: A goal-driven survey on root cause analysis. arXiv preprint arXiv:2510.19593 (2025)
8. Huang, S., et al.: OpDiag: unveiling database performance anomalies through query operator attribution. IEEE TKDE **37**(6), 3613–3626 (2025)
9. Judea, P.: Causality: models, reasoning, and inference. Econ. Theor. (2003)
10. Kalisch, M., et al.: Causal inference using graphical models with the R package pcalg. J. Stat. Softw. **47**, 1–26 (2012)
11. Kim, M., et al.: Root cause detection in a service-oriented architecture. ACM Sigmetrics Perform. Eval. Rev. **41**(1), 93–104 (2013)
12. Li, M., et al.: Causal inference-based root cause analysis for online service systems with intervention recognition. In: SIGKDD, pp. 3230–3240 (2022)
13. Liu, D., et al.: MicroHECL: high-efficient root cause localization in large-scale microservice systems. In: ICSE-SEIP, pp. 338–347 (2021)
14. Liu, P., et al.: FluxInfer: automatic diagnosis of performance anomaly for online database system. In: IPCCC, pp. 1–8 (2020)
15. Ma, M., et al.: MS-Rank: multi-metric and self-adaptive root cause diagnosis for microservice applications. In: ICWS, pp. 60–67 (2019)
16. Ma, M., et al.: AutoMap: diagnose your microservice-based web applications automatically. WWW, pp. 246–258 (2020)
17. Meng, Y., et al.: Localizing failure root causes in a microservice through causality inference. In: IWQoS, pp. 1–10 (2020)
18. Qiu, J., et al.: A causality mining and knowledge graph based method of root cause diagnosis for performance anomaly in cloud applications. Appl. Sci. (2020)
19. Shan, H., et al.: ϵ-diagnosis: unsupervised and real-time diagnosis of small-window long-tail latency in large-scale microservice platforms. In: WWW, pp. 3215–3222 (2019)
20. Spirtes, P., et al.: An algorithm for fast recovery of sparse causal graphs. Soc. Sci. Comput. Rev. **9**(1), 62–72 (1991)
21. Wang, J., et al.: Big data service architecture: a survey. J. Internet Tech. **21**(2), 393–405 (2020)
22. Wang, P., et al.: CloudRanger: root cause identification for cloud native systems. In: CCGRID, pp. 492–502 (2018)
23. Wu, Q., et al.: DBPecker:a graph-based compound anomaly diagnosis system for distributed RDBMSs. Proc. VLDB Endow. **18**(12), 5383–5386 (2025)
24. Xiang, Q., et al.: Distributed database diagnosis method for compound anomalies. J. Softw. **36**(3), 1022–1039 (2025)

25. Yang, Z., et al.: OceanBase: a 707 million tpmC distributed relational database system. Proc. VLDB Endow. **15**(12), 3385–3397 (2022)
26. Yao, L., et al.: A survey on causal inference. ACM TKDD **15**(5), 1–46 (2021)
27. Zhang, G., et al.: DBCatcher: a cloud database online anomaly detection system based on indicator correlation. In: ICDE, pp. 1126–1139 (2023)
28. Zhang, Y., et al.: CloudRCA: a root cause analysis framework for cloud computing platforms. In: CIKM, pp. 4373–4382 (2021)
29. Zhao, Y., et al.: Multi-stage location for root-cause metrics in online service systems. In: NOMS, pp. 1–9 (2023)
30. Zhou, X., et al.: DB-GPT: large language model meets database. Data Sci. Eng. **9**, 102–111 (2024)
31. Zhu, X., et al.: CoLA: model collaboration for log-based anomaly detection. Proc. VLDB Endow. **18**(11), 3979–3987 (2025)

Accuracy-Aware Log Replay with Fine-Grained Prioritization for Real-Time Prediction Queries

Shanshan Huang[1], Jing Jiang[2], Peng Cai[1(✉)], Qiwen Dong[1], and Huiqi Hu[1]

[1] East China Normal University, Shanghai, China
`52195100004@stu.ecnu.edu.cn`, `{pcai,qwdong,hqhu}@dase.ecnu.edu.cn`
[2] Beijing Natural Original Number Technology Co., Ltd., Beijing, China
`jiangjing@obase.com.cn`

Abstract. Machine learning applications require timely access to fresh data from primary–backup databases to ensure accurate inference. Existing log replay strategies treat all updates equally and adhere to log-order dependencies, or only prioritize frequently accessed tables, resulting in high latency for prediction queries that usually access a small subset of attributes. Allowing immediate query execution can reduce latency but risks substantial accuracy degradation, as prediction models exhibit varying sensitivity even to minor data staleness. In this paper, we propose **AALR**, an accuracy-aware log replay strategy that accelerates data visibility for prediction queries while preserving inference accuracy. AALR prioritizes replay at the attribute level, enabling fine-grained replay aligned with query access patterns to avoid unnecessary delay. It leverages a learning-based model to quantify the relationship between data freshness and prediction accuracy, supporting adaptive replay decisions under diverse workloads. Furthermore, AALR introduces an epoch-based two-step replay mechanism, combining column-level parallel classification with row-level latest transaction retention to improve parallelism and resource utilization. Extensive experiments on multiple real-world datasets demonstrate that AALR significantly reduces data visibility latency for prediction queries while maintaining high prediction accuracy, outperforming state-of-the-art log replay strategies.

Keywords: Parallel Log Replay · Real-time Prediction · Data Freshness · Machine Learning

1 Introduction

Machine learning-based applications are widely used across various domains [15, 22], such as finance and e-commerce, supporting tasks like fraud detection and personalized recommendations, which rely on primary-backup databases for efficient data storage and access. The primary database processes a large volume of business transactions (i.e., read-write transactions) and propagates their modification logs to the backup database for synchronization [21]. The backup database

H. Jung et al. (Eds.): DASFAA 2026, LNCS 16536, pp. 561–577, 2026.
https://doi.org/10.1007/978-981-92-0363-6_34

primarily serves analytical queries (i.e., read-only transactions), retrieving feature data for prediction models to support efficient inference and decision-making. On one hand, existing log replay strategies [11,21,28] treat all log records equally and enforce a strict log-order dependency, requiring the replay of irrelevant updates before prediction-relevant data become visible, which delays decision-making. On the other hand, prediction models exhibit varying sensitivity to data freshness, where even slight staleness can lead to significant deviations in prediction results [23], making immediate query execution without timely data updates unreliable.

To better illustrate the limitations of existing log replay strategies in prediction scenarios, we consider a movie rating prediction task [6], as shown in Fig. 1. The upper-left part of the figure depicts the schema of *MoviesDB*, which consists of three tables: `user_info`, `rating_info`, and `search_info`, storing user profiles, movie ratings, and search histories, respectively. The primary database continuously processes frequent OLTP(Online Transaction Processing) updates on these tables. The upper-right part of Fig. 1 shows the *MoviesModel*, which predicts movie ratings based on user preferences and historical ratings. Prediction queries are executed on the backup database, where feature values are computed using columns such as `like_actor` and `rating`.

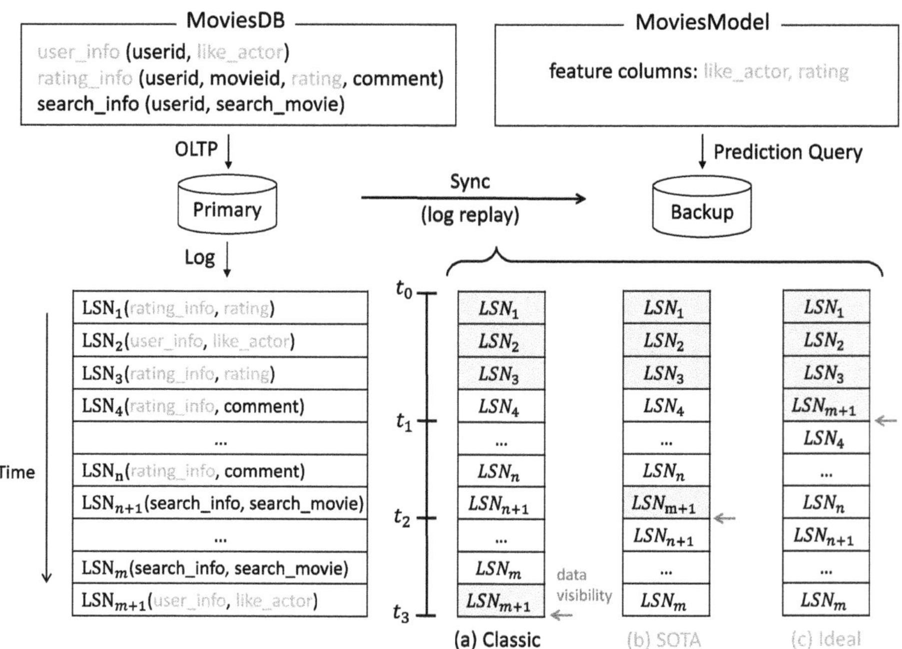

Fig. 1. Motivating example. The classic log replay method (a) and the SOTA method (b) both incur unnecessary delays by replaying modifications to data columns that are irrelevant to the *MoviesModel* prediction results, leading to data visibility later than the ideal case (c) at time t_1.

The lower-left part of Fig. 1 shows the log records generated by the primary database from time t_0 to t_3. Each record is denoted as LSN_i(`TableName`, `ColumnName`), indicating a modification to column `ColumnName` in table `TableName`. These logs are replayed on the backup database, as illustrated in the lower-right part of the figure, which compares three representative log replay strategies: (a) Classic, (b) state-of-the-art(SOTA), and (c) Ideal.

In the classic parallel log replay method (a) [11,19,27,28] shown in Fig. 1, all log records are treated as equally important and replayed strictly in log order dependency. As a result, prediction-relevant data such as the `like_actor` column only becomes available at time t_3, after replaying numerous irrelevant updates (e.g., `comment` and `search_movie`), introducing unnecessary delays for prediction queries. The SOTA method (b) [29] prioritizes frequently accessed tables (i.e., `user_info` and `rating_info`), reducing data visibility latency so that prediction-relevant data is available by time t_2. However, updates to irrelevant attributes, such as `comment`, are still replayed, limiting further reduction in latency. In contrast, the ideal method (c) ensures that prediction-relevant data becomes visible as early as time t_1 by eliminating delays from irrelevant updates.

Existing log replay strategies are inadequate for ensuring timely availability of fresh data, as they replay updates irrelevant to prediction queries. Method (a) neglects query access patterns, while method (b) lacks sufficient granularity in replay prioritization. Serving prediction queries without waiting for log replay reduces visibility latency, but can substantially degrade accuracy even with minor data staleness [23,26]. To achieve the ideal performance illustrated by method (c) and effectively support prediction workloads, three key challenges must be addressed: (1) designing a fine-grained replay prioritization mechanism that minimizes delays from irrelevant log entries; (2) quantifying the sensitivity of prediction models to data freshness to optimize data visibility while maintaining accuracy; (3) developing a log replay method that outperforms state-of-the-art approaches in both parallelism and overhead.

In this paper, we propose **AALR**, an _a_ccuracy-_a_ware _l_og _r_eplay strategy that enables fine-grained, attribute-level replay prioritization, accelerating data visibility while maintaining prediction accuracy. To address the first challenge, AALR restricts log replay to query-relevant attributes, eliminating unnecessary delays caused by irrelevant updates. For the second challenge, AALR introduces a data freshness metric [20] and leverages machine learning to model the relationship between data freshness and prediction accuracy, ensuring that only impactful log records are processed prior to query execution. Finally, to address the third challenge, AALR extends the epoch-based replay mechanism [21,29] with workload-aware optimizations, achieving higher replay parallelism and improved resource efficiency.

The main contributions of this paper are summarized as follows:

- We analyze the limitations of existing log replay strategies, highlighting their inefficiency and their neglect of the varying impact of data changes on model accuracy, which delays the availability of fresh data.

- We propose AALR, which integrates attribute-level replay prioritization with a model of data freshness impact on prediction accuracy, optimizing both data visibility and prediction reliability.
- We develop a two-step parallel replay strategy that adapts to query workloads, improving replay concurrency and resource utilization through column-level classification and row-level transaction retention.
- We experimentally demonstrate that AALR reduces data visibility latency and improves prediction accuracy compared with state-of-the-art log replay methods across multiple real-world datasets.

2   Related Work

Log replay is a core mechanism in database systems for maintaining data consistency and high availability. Existing techniques can be broadly categorized into serial replay and parallel replay strategies, with parallel methods further classified by log format into logical, physical, and hybrid approaches. Each category offers different considerations for efficiency, latency, and system overhead.

Serial Replay. Serial log replay is the earliest and simplest strategy, executing log records sequentially in commit order. It avoids transaction dependency issues and provides strong serializable consistency [4]. However, it quickly becomes a performance bottleneck in high-concurrency or large-scale databases, as even non-conflicting transactions cannot be processed in parallel, resulting in high replay latency [11]. Consequently, serial replay is now mainly used as a consistency baseline or for experimental comparison.

Logic-Based Parallel Replay. Logic-based parallel replay records only key operation information (e.g., SQL commands), reducing storage and transmission overhead while alleviating logging load on the primary node [19,21,25,27]. The main challenge is handling transaction dependencies. For example, command logging [19] partitions data to enable parallel replay and batches transactions targeting the same partition. PACMAN [27] combines static table-level dependency analysis with dynamic parameter-level checks to execute non-conflicting transactions concurrently. Deterministic replay methods [21] log only write sets and transaction identifiers, allowing replicas to re-execute transactions deterministically. Logic-based replay improves throughput and reduces storage cost while preserving logical consistency, but dependency analysis incurs computational overhead under high concurrency or complex schemas. Optimizations such as batching, priority queues, and dependency graph pruning mitigate this, yet trade-offs between scheduling latency and resource use remain.

Physical-Based Parallel Replay. Physical log replay records actual data modifications, allowing high parallelism without complex dependency analysis [11,12,16,29]. Unlike logic-based logs, physical logs capture changes at the page or row level, enabling concurrent execution of non-conflicting updates and reducing replay latency. For instance, C5 [11] uses row-level scheduling with dedicated FIFO queues, KuaFu [12] performs topological sorting based on write–write dependencies, and Lee et al. [16] employ row version identifiers (RVID) to accelerate validation. AETS [29] prioritizes hot tables under OLAP workloads to improve query performance. Physical-base replay is effective for latency-sensitive systems with low transaction contention, providing near real-time synchronization while preserving consistency. Its drawbacks include large log volumes, high storage and bandwidth requirements, and potential conflicts under complex transaction dependencies.

Hybrid Logic–Physical Replay. Hybrid replay strategies combine logic and physical logs to reduce storage overhead while supporting parallel execution and maintaining consistency. These approaches leverage logical logs to minimize storage and transmission costs, and use physical logs to accelerate replay of critical or conflicting transactions via adaptive scheduling. Systems switch between logic- and physical-based replay dynamically according to dependency patterns and runtime workload. For instance, Adaptive Logging [28] implements a dynamic logging and recovery framework that constructs a lightweight dependency analyzer and adjusts the ratio of logical to physical logs based on workload characteristics, effectively improving replay efficiency while preserving consistency.

3   Methodology

3.1   Overall Framework

The core idea of AALR is to learn the sensitivity of prediction accuracy to data freshness, guiding an attribute-level prioritized log replay strategy. Figure 2 presents the overall architecture, which comprises four key components:

Log Processor①: responsible for parsing and segmenting transaction logs obtained from the primary database and grouping them into query-driven batches, thereby establishing the foundation for subsequent parallel processing.

Accuracy-Aware Model②: this model quantifies the effect of changes in each attribute on prediction accuracy and ranks the attributes according to their importance, providing the basis for selecting high-priority attributes for replay.

Attribute Prioritizer③: Based on learned attribute importance, this component selects a subset of columns that satisfy prediction accuracy requirements and replays them first to achieve faster visibility of prediction-relevant data.

Parallel Replayer④: this component employs a two-step parallel replay strategy, further enhancing replay parallelism and improving resource utilization.

The overall workflow of AALR proceeds as follows: the Log Processor① continuously retrieves log streams from the primary database, parses and batches the logs, and passes the structured entries to the Parallel Replayer④. Based on the sensitivity rankings provided by the Accuracy-Aware Model②, the Attribute Prioritizer③ determines the set of prioritized attributes. Finally, the Parallel

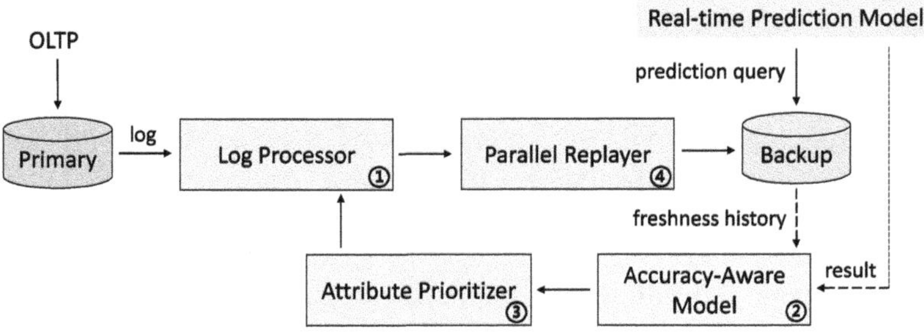

Fig. 2. Overall architecture of AALR.

Replayer④ executes the two-stage replay strategy, achieving both high replay parallelism and prioritized synchronization of critical data, thereby providing fresher and more accurate data for prediction queries.

3.2 Log Structure and Processing

This section first introduces the log structure, followed by a description of the Log Processor component, which batches log records for efficient parallel replay.

Log Record Structure. The log format adopted in our work is similar to that used in SiloR [25], where each log record contains both the before-image and after-image to facilitate accelerated replay. The detailed structure of the log fields is illustrated in Fig. 3:

LSN	Operation Type	Transaction ID	Table ID	Row ID	Modified Column List	Before Image	After Image

Fig. 3. Log record structure (Modification operations).

- LSN(Logical Sequence Number): A monotonically increasing identifier representing the order of log records.
- Operation Type: Specifies the modification type applied to the data item, including INSERT, UPDATE, and DELETE.
- Transaction ID: Identifies the transaction to which the log record belongs, ensuring transaction-level consistency.
- Table ID: Indicates the table to which the modified data item belongs.
- Row ID: Identifies the specific row being modified.
- Modified Column List: Records the set of columns affected by the operation.
- Before Image: The values of data items before modification, used for rollback or validation.
- After Image: The values of data items after modification, used for data recovery and replay.

Query-Driven Log Batching. To maximize replay parallelism while maintaining consistency, we adopt an epoch-based batching approach [17,25,29] that partitions transactions into multiple epochs. Each epoch is formed based on predefined criteria such as transaction count, time interval, or memory utilization. Transactions within the same epoch can be replayed in parallel, while the epoch boundaries serve as synchronization points that ensure consistent query execution and preserve the global ordering across epochs. However, traditional epoch-based batching methods neglect the query access patterns (e.g., query range, execution frequency, and concurrency level). As a result, they may replay many data versions that are never accessed by prediction queries, leading to unnecessary overhead. However, to overcome this limitation, we propose a query-driven batching strategy that dynamically adjusts epoch boundaries based on query timestamps. By aligning batching decisions with actual query arrival patterns, this approach effectively eliminates redundant log replays (further detailed in Step 2 of the two-phase replay strategy in Sect. 4), thereby improving replay efficiency and reducing resource consumption.

3.3 Accuracy-Aware Attribute Prioritization

In prediction scenarios, only a subset of attributes contributes to feature computation. To accelerate data visibility without compromising prediction accuracy, we introduce an attribute-level replay prioritization strategy composed of two key components: the Accuracy-Aware Model and the Attribute Prioritizer, which collaboratively determine the subset of attributes to prioritize during replay.

Training Accuracy-Aware Model. To quantify the relationship between data freshness and downstream prediction accuracy, we develop an Accuracy-Aware Model that estimates attribute importance scores, enabling the identification of attributes whose changes have the greatest impact on prediction accuracy, as formulated in:

$$PredResult\_gap = f(Freshness\_gaps) \tag{1}$$

Specifically, **Freshness** refers to the data freshness, which captures the frequency or timeliness of data modifications [8]. We track the number of modifications to each column used in the downstream model to better capture the complexity of data updates. The **Freshness_gap** represents the difference in data freshness between the primary and backup databases and is defined as:

$$Freshness\_gap_i = Freshness\_primary_i - Freshness\_backup_i \tag{2}$$

where i is the column index. The **Freshness_gaps** vector is derived from the raw columns directly or indirectly contributing to the model input features:

$$Freshness\_gaps = [Freshness\_gap_1, Freshness\_gap_2, ..., Freshness\_gap_n] \tag{3}$$

PredResult denotes the prediction result, whereas **PredResult_gap** measures the discrepancy between predictions based on fresh data (from the primary database) and those based on potentially stale data (from the backup database):

$$PredResult\_gap = |PredResult\_primary - PredResult\_backup| \quad (4)$$

When the $Freshness\_gap_i$ exhibits a significant impact on $PredResult\_gap$, it indicates that changes in the corresponding column strongly affect the accuracy of the real-time prediction model. To quantify this effect, we adopt a feature importance analysis, represented by the function $f(\cdot)$ in Eq. (1). By default, we use Random Forests [13] to evaluate attribute importance, as they provide robust, interpretable feature rankings with minimal hyperparameter tuning. Users can also adopt custom scoring methods via the provided API to accommodate domain-specific requirements. Alternative approaches, such as permutation importance [7], SHAP values [18], or Lasso coefficients [24], are also supported.

Attribute Prioritizer. Not all feature columns contribute equally to prediction accuracy. Based on the importance ranking provided by the Accuracy-Aware Model, a prioritized subset of columns is selected. To formally control the effect of replaying only this subset, we introduce a configurable threshold α that bounds the maximum allowable prediction discrepancy. For a prediction task t, let C denote the complete set of feature columns required by the model, and $C' \subseteq C$ represent the selected subset of prioritized columns. Let $Pred_t(C)$ and $Pred_t(C')$ denote the prediction results based on the complete column set and the prioritized subset, respectively. The following constraint must hold:

$$L(Pred_t(C), Pred_t(C')) \leq \alpha \quad (5)$$

where $L(\cdot)$ denotes a task-specific loss function used to quantify prediction discrepancy (e.g., Mean Absolute Error). In our experiments, we set $\alpha = 0.05$, which satisfies the accuracy requirements of downstream prediction tasks while enabling substantial improvements in log replay efficiency.

Model Validity Maintenance. A key concern is whether the accuracy-aware model remains effective as the system workload evolves. In practice, when the downstream prediction model is fixed and the workload changes gradually, the learned column importance tends to remain stable. For moderate data distribution shifts, the random forest model demonstrates sufficient robustness, avoiding frequent retraining. To handle long-term or substantial workload changes, AALR employs a lightweight periodic retraining mechanism based on recent logs and prediction feedback. This process incurs negligible overhead and operates independently of transactional and query processing, ensuring that the attribute prioritization remains accurate over time with minimal maintenance cost.

4 Parallel Replay Strategy

Two-Step Replay Strategy. In AALR, this strategy is designed to support column-grained parallel log replay and consists of two sequential steps:

- *Step 1: Column-level Parallel Transaction Classification.* To minimize data visibility delay for prediction queries, it is essential to identify and prioritize the replay of log entries corresponding to attribute columns that have a significant impact on prediction accuracy. Specifically, based on the prioritized column set provided by the Attribute Prioritizer, log entries are dispatched in parallel and grouped according to the column IDs modified by transactions, forming prioritized and other column groups.
- *Step 2: Row-level Latest Transaction Retention.* For each column group, log entries are processed to retain only the latest update for rows modified by multiple transactions, determined by comparing their LSNs. This eliminates redundant operations and ensures efficient log replay. Subsequently, logs that update prioritized columns are replayed and committed first, improving the timeliness of critical data visibility for prediction queries.

Consistency Guarantee. The proposed replay strategy ensures data consistency while accelerating log application. In Step 2, skipping outdated modifications does not compromise consistency, as the Log Processor adopts a query-driven batching mechanism (Sect. 3.2). Under this mechanism, queries are issued only at epoch boundaries, ensuring that omitted updates correspond to data versions never visible to queries. Consequently, replaying only the latest modifications within each epoch eliminates redundant log operations without affecting query correctness.

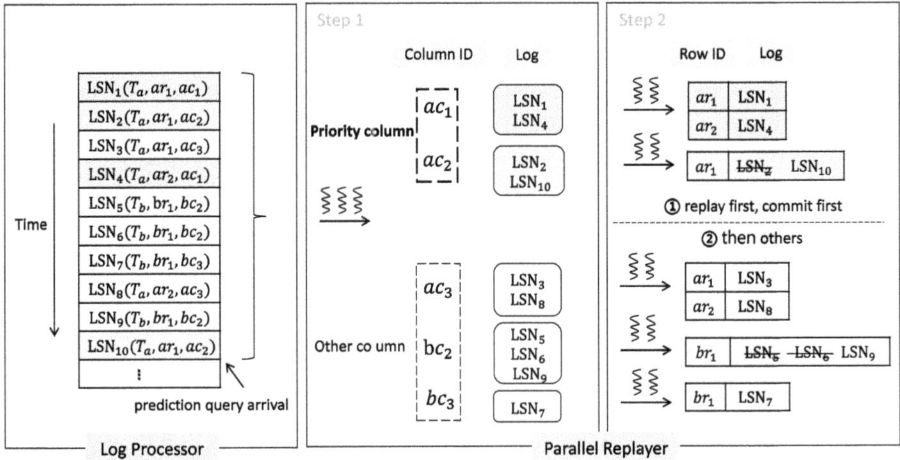

Fig. 4. An example of parallel replay workflow.

Potential Applications. Although AALR is designed for prediction-query workloads, its attribute-level prioritization extends to other data-intensive applications [9] where queries access only a subset of attributes. By prioritizing relevant updates and deferring irrelevant ones without compromising consistency, AALR reduces log processing and accelerates data visibility. Such scenarios include real-time reporting, interactive analytics, and IoT sensor monitoring, where timely access to fresh, query-relevant data is critical.

Example. In Fig. 4, we illustrate the parallel replay workflow of AALR. Within an epoch, the backup database replays operations from records 1 to 10, which modify tables T_a and T_b. For example, record $LSN_1(T_a, ar_1, ac_1)$ represents an update to column ac_1 of row ar_1 in table T_a. The Log Processor dispatches the log entries in parallel to multiple column groups based on the modified columns (e.g., ac_1, ac_2, ac_c, bc_2, bc_3), corresponding to Step 1 of the parallel replay. In Step 2, for each column group, only the most recent transaction for each modified row is retained. Specifically, the priority column groups ar_1 and ar_2, which have a greater impact on the prediction results, are processed first (see ①). In the ac_1 group, LSN_1 and LSN_4 modify rows ar_1 and ar_2, respectively; therefore, both transactions are executed and committed. In the ac_2 group, LSN_2 and LSN_{10} modify row ar_1; therefore, only the most recent transaction, LSN_{10}, is executed and committed. At this point, the prediction query data becomes visible. Finally, the same process is applied to other column groups that are unrelated to the prediction query (see ②).

5 Experiment

5.1 Experimental Setup

Environment. Experiments were conducted on a three-server cluster comprising a primary database, a backup database, and a load generator. Each server had 64 CPU cores, 128 GB memory, and a 10 Gbps network interface. The load generator issued modification transactions and prediction queries from real-world datasets. The primary server ran MySQL 8.0 [5] and handled updates, while the backup server, implemented in C++ as an in-memory database prototype based on a B+ tree, was synchronized via AALR.

Benchmarks. Three real-world open-source datasets, representing common real-time prediction scenarios, were selected. These datasets vary in data volume, number of features, and the prediction models applied:

- *Movies* [3]: the Movies Dataset records historical ratings and movie descriptions, which are used for movie rating prediction. The dataset consists of five tables, and an embedding-based DNN [10] is employed for prediction. The required features are derived from four columns across four tables.

- *Logistics* [2]: the Logistics Dataset captures changes in truck loading status, warehouse information, and cargo details, which are used to estimate the waiting time for trucks to enter the factory. The dataset consists of three tables, and a random forest regression model [13] is applied. The required feature data comes from seven columns across the three tables.
- *Credit* [1]: the Credit Card Dataset records customer information and credit card transaction history, which are used for credit risk assessment. The dataset consists of three tables, and a LightGBM model [14] is employed for prediction, utilizing feature data from ten columns across two tables.

Metrics. We primarily collect the following metrics for performance evaluation:

- *Replay Throughput*: the number of log entries applied per second during log replay on the backup server;
- *Replay Time*: the total time required to replay all logs on the backup;
- *Data Visibility Delay*: the total waiting time from the query timestamp until all data required by the query are committed and visible on the backup;
- *Model Accuracy*: quantified by the Mean Absolute Error (MAE). For a query workload of size n, MAE is defined as:

$$\text{MAE} = \frac{1}{n} \sum_{i=1}^{n} |\hat{y}_i - y_i| \tag{6}$$

where $\hat{y}_i$ is the predicted value and y_i is the true value for the i-th query.

Baselines. The following log replay strategies were implemented for comparison:

- **C5** [11]: A log replay strategy without prioritization. It is based on row-level scheduling, where each row maintains its own queue, and a global scheduling queue ensures transaction commit order.
- **AETS** [29]: A log replay strategy with table-level prioritization. Transactions within a table are executed in parallel while preserving their original commit order. In our implementation, we prioritize replaying logs for tables accessed by prediction queries, with an epoch size set to 2048 transactions.
- **TSLR**: A two-step replay strategy in AALR, where the second step does not prioritize the execution and commit of important attribute columns. This serves to evaluate the efficiency of our replay strategy in terms of parallelism.

5.2   Replay Efficiency

To compare the replay efficiency, we present the results of four replay strategies under three scenarios, with replay throughput and replay time shown in Fig. 5 and 6, respectively. We observed that compared with C5, AALR achieves nearly twice the replay throughput while reducing replay time to approximately 0.3× (a reduction of about 70%). In contrast, AETS improves performance to some extent through table-level parallel execution; however, its throughput remains constrained due to single-threaded log dispatching. Building upon this, AALR further increases throughput by approximately 1.4× and reduces replay time to about 0.4× (roughly a 60% reduction). Moreover, the results from TSLR confirm that the performance advantage of AALR primarily stems from its higher degree of parallelism. **In summary**, AALR outperforms existing methods in both log replay throughput and total replay time. This improvement stems from its parallelization at both log entry batching and transaction commit, rather than relying on single-threaded commit (as in C5) or sequential grouping with table-level parallel commit (as in AETS).

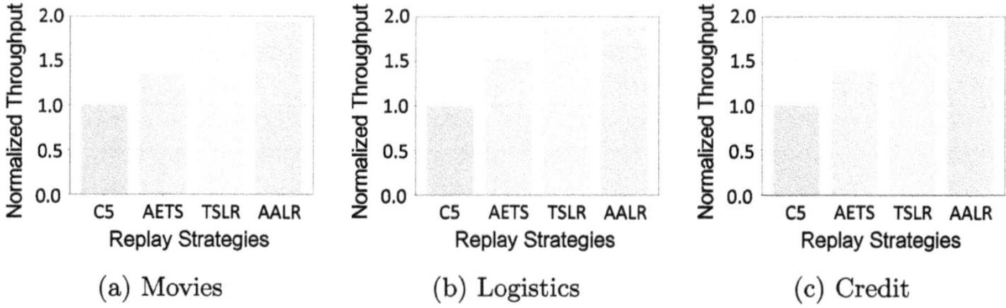

(a) Movies   (b) Logistics   (c) Credit

Fig. 5. Replay throughput under different workloads (Normalized by C5).

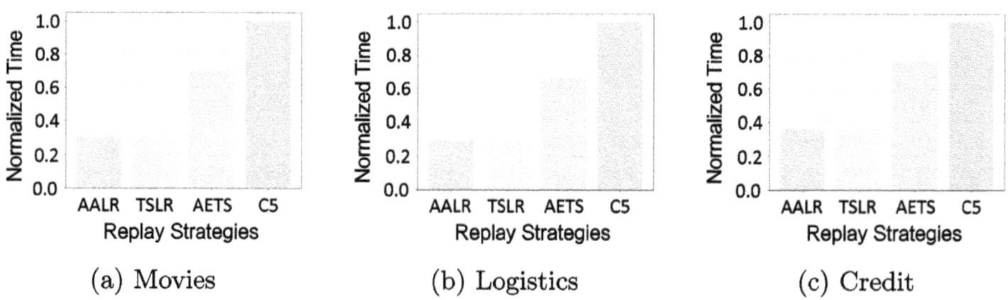

(a) Movies   (b) Logistics   (c) Credit

Fig. 6. Replay time under different workloads (Normalized by C5).

5.3   Data Visibility Delay

Log replay strategies significantly influence how quickly backup servers can deliver fresh data for prediction queries. To assess this effect, we measured data visibility latency under different strategies. The results are shown in Fig. 7. We observed that compared with C5, AALR reduces visibility latency to approximately 0.25× (a reduction of about 75%) Compared with AETS, AALR reduces latency to approximately 0.5× (about a 50% reduction), whereas AETS can only prioritize at the table level, still introducing delays from updates to irrelevant columns. For TSLR, although it improves parallelism, it does not effectively prioritize prediction-relevant columns, leading to weaker performance than AETS. **In summary**, AALR substantially reduces data visibility latency in prediction scenarios. This advantage arises not only from its parallel execution capability (as in TSLR) but also from its accuracy-aware prioritization. Unlike approaches that wait for all updates (C5) or all table updates (AETS), AALR prioritizes the replay of modifications to only those attributes accessed by prediction queries.

5.4   Scalability

To evaluate the scalability of AALR, we conducted two experiments: (1) multi-core scalability; (2) scalability under high update workloads. For the first setting, the Movies workload was executed on the primary server to generate logs, including a 1-minute warm-up and a 5-minute high-load phase. Replay throughput was measured while incrementally increasing the number of replay threads. In the second setting, replay threads were fixed at four while the primary update rate gradually increased to 10,000 tps to simulate dynamic high-load conditions. As shown in Fig. 8(a), AALR achieves near-linear throughput growth with increasing replay threads and consistently outperforms C5 and AETS Fig. 8(b) shows that under rising update loads, AALR maintains a stable data visibility latency, while AETS and C5 experience a sharp increase due to table-level replay granularity and lack of update prioritization. **In summary**, AALR demonstrates strong and stable scalability under both multi-core and high-load environments. Its prioritized replay of selected attributes and highly parallel replay operations enable efficient utilization of system resources, maintaining low latency and high throughput even under intensive workloads.

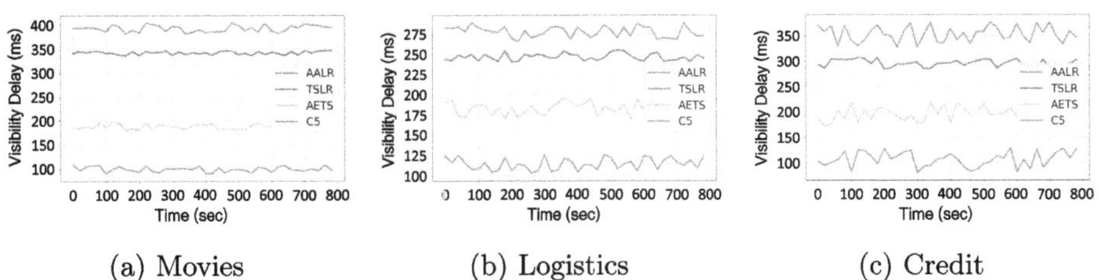

| (a) Movies | (b) Logistics | (c) Credit |

Fig. 7. Visibility delay under different workloads.

5.5   AALR Overhead

To evaluate the runtime overhead introduced by AALR, we measured both performance metrics and internal management overhead under three representative workloads, as shown in Table. 1. For performance overhead, the normalized CPU

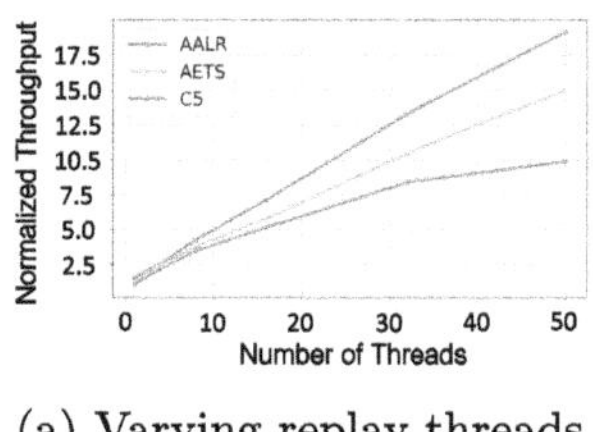
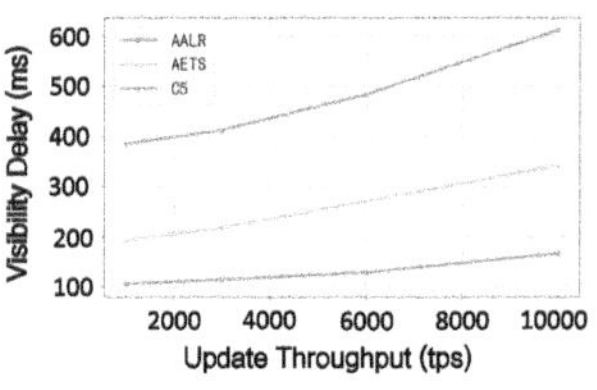

(a) Varying replay threads   (b) Varying update load

Fig. 8. Scalability evaluation of AALR.

and memory usage are reported relative to the baseline method C5. AALR consistently incurs lower costs, with CPU overhead reduced by approximately 12% and memory overhead reduced by around 23% compared with C5. For management overhead, we measured the time spent in each phase of the replay process: log entry dispatching, transaction retention within column groups, replay execution, and commit operations. AALR introduces only minimal management overhead (about 1%), with the vast majority of time spent on actual replay execution, demonstrating the efficiency of its log dispatching and transaction retention mechanisms. **In summary**, AALR achieves efficient and low-overhead log replay due to the high parallelism in log grouping and execution/commit of the latest transactions, while skipping updates for data that are not accessed. As a result, both CPU and memory consumption are reduced without compromising replay effectiveness, highlighting the scalability and practicality of AALR in high-frequency update scenarios.

Table 1. System overhead of AALR.

Workload	Performance Overhead		Management Overhead			
	CPU (norm.)	Memory (norm.)	Log Dispatching	Transaction Retention	Replay Execution	Commit Operation
Movies	0.87	0.76	0.69%	0.13%	99.06%	0.12%
Logistics	0.85	0.74	0.54%	0.09%	99.29%	0.08%
Credit	0.92	0.81	0.92%	0.17%	98.70%	0.21%

5.6   Impact of Attribute Replay Prioritization

To examine how attribute prioritization influences both data visibility and prediction accuracy, we conducted a one-hour experiment using the Movies workload. A total of 268,754 samples were collected to train the accuracy-aware model, which produced an importance ranking for four columns involved in rating prediction. Initially, all four columns were prioritized, achieving the freshest data and a minimum MAE of 0.047. We then progressively removed the least important columns and measured data visibility delay and prediction accuracy. When one minimally impactful column was excluded (prioritized columns = 3

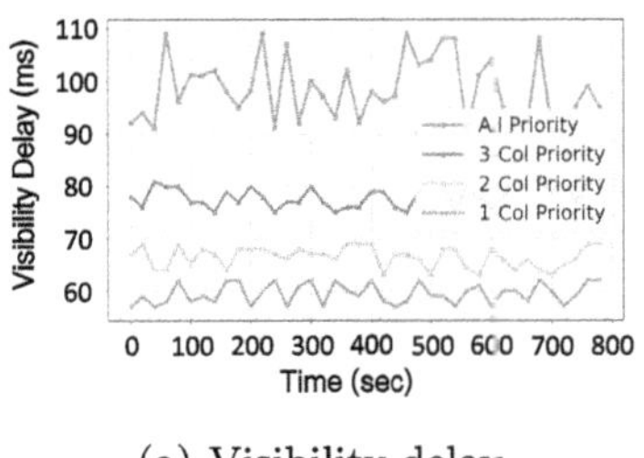

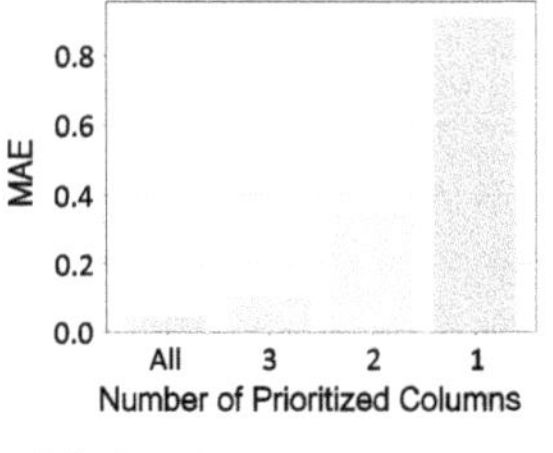

<table>
<tr><td>(a) Visibility delay</td><td>(b) Prediction accuracy</td></tr>
</table>

Fig. 9. Impact of column prioritization on prediction accuracy and data visibility.

in Fig. 9), visibility delay decreased by 25% while MAE increased by only 0.05. **In summary**, AALR reduces the prioritized replay subset to further optimize data visibility delay, while preserving prediction accuracy within the user-defined threshold. This is based on the observation that changes to some feature columns have minimal impact on prediction results, allowing their replay to be deferred.

6   Conclusion

This paper investigates the limitations of existing log replay strategies in ensuring timely and fresh data for prediction scenarios, which may lead to inaccurate or delayed decisions. We propose AALR, an accuracy-aware parallel log replay strategy that enables fine-grained, attribute-level prioritization. By learning the relationship between data freshness and prediction accuracy, AALR replays only prediction-relevant updates while deferring irrelevant ones. A two-step replay mechanism further enhances parallelism and resource efficiency. Experimental results show that AALR reduces data visibility latency by up to 65% while maintaining almost the same prediction accuracy (with less than 0.05 degradation in MAE) compared with state-of-the-art methods. These findings demonstrate the effectiveness of AALR in supporting real-time prediction workloads, and future work will extend it to more complex multi-model and cross-database synchronization scenarios.

Acknowledgments. We thank the anonymous reviewers for their insightful comments and feedback. This work is supported by grant from the National Natural Science Foundation of China No. U22B2020.

References

1. Credit card dataset. https://www.kaggle.com/datasets/rikdifos/credit-card-approval-prediction/data
2. Logistics dataset. https://github.com/Hss-477/Logistics-Dataset
3. Movies dataset. https://www.kaggle.com/datasets/rounakbanik/the-movies-dataset/data

4. MySQL 5.7 reference manual, chapter 16: Replication. https://dev.mysql.com/doc/refman/5.7/en/replication.html
5. MySQL 8.0 Reference Manual. https://dev.mysql.com/doc/refman/8.0/en/
6. Abarja, R.A., Wibowo, A.: Movie rating prediction using convolutional neural network based on historical values. Int. J. **8**, 2156–2164 (2020)
7. Altmann, A., Toloşi, L., Sander, O., Lengauer, T.: Permutation importance: a corrected feature importance measure. Bioinformatics **26**(10), 1340–1347 (2010)
8. Bouzeghoub, M.: A framework for analysis of data freshness. In: Proceedings of the 2004 International Workshop on Information Quality in Information Systems, pp. 59–67 (2004)
9. Chen, C.P., Zhang, C.Y.: Data-intensive applications, challenges, techniques and technologies: a survey on big data. Inf. Sci. **275**, 314–347 (2014)
10. He, X., Liao, L., Zhang, H., Nie, L., Hu, X., Chua, T.S.: Neural collaborative filtering. In: Proceedings of the 26th International Conference on World Wide Web, pp. 173–182 (2017)
11. Helt, J., Sharma, A., Abadi, D.J., Lloyd, W., Faleiro, J.M.: C5: cloned concurrency control that always keeps up. arXiv preprint arXiv:2207.02746 (2022)
12. Hong, C., Zhou, D., Yang, M., Kuo, C., Zhang, L., Zhou, L.: KuaFu: closing the parallelism gap in database replication. In: 2013 IEEE 29th International Conference on Data Engineering (ICDE), pp. 1186–1195. IEEE (2013)
13. Jaiswal, J.K., Samikannu, R.: Application of random forest algorithm on feature subset selection and classification and regression. In: 2017 World Congress on Computing and Communication Technologies (WCCCT), pp. 65–68. IEEE (2017)
14. Ke, G., et al.: LightGBM: a highly efficient gradient boosting decision tree. Adv. Neural. Inf. Process. Syst. **30** (2017)
15. Kumar, A., Boehm, M., Yang, J.: Data management in machine learning: challenges, techniques, and systems. In: Proceedings of the 2017 ACM International Conference on Management of Data, pp. 1717–1722 (2017)
16. Lee, J., et al.: Parallel replication across formats for scaling out mixed OLTP/OLAP workloads in main-memory databases. VLDB J. **27**, 421–444 (2018)
17. Lu, Y., Yu, X., Cao, L., Madden, S.: Epoch-based commit and replication in distributed OLTP databases (2021)
18. Lundberg, S.M., Lee, S.I.: A unified approach to interpreting model predictions. Adv. Neural. Inf. Process. Syst. **30** (2017)
19. Malviya, N., Weisberg, A., Madden, S., Stonebraker, M.: Rethinking main memory OLTP recovery. In: 2014 IEEE 30th International Conference on Data Engineering, pp. 604–615. IEEE (2014)
20. Peralta, V.: Data freshness and data accuracy: a state of the art. Instituto de Computacion, Facultad de Ingenieria, Universidad de la Republica2006 (2006)
21. Qin, D., Brown, A.D., Goel, A.: Scalable replay-based replication for fast databases. Proc. VLDB Endow. **10**(13), 2025–2036 (2017)
22. Sarker, I.H.: Machine learning: algorithms, real-world applications and research directions. SN Comput. Sci. **2**(3), 160 (2021)
23. Shisher, M.K.C., Sun, Y.: How does data freshness affect real-time supervised learning? In: Proceedings of the Twenty-Third International Symposium on Theory, Algorithmic Foundations, and Protocol Design for Mobile Networks and Mobile Computing, pp. 31–40 (2022)
24. Tibshirani, R.: Regression shrinkage and selection via the lasso. J. R. Stat. Soc. Ser. B Stat Methodol. **58**(1), 267–288 (1996)

25. Tu, S., Zheng, W., Kohler, E., Liskov, B., Madden, S.: Speedy transactions in multicore in-memory databases. In: Proceedings of the Twenty-Fourth ACM Symposium on Operating Systems Principles, pp. 18–32 (2013)
26. Wooders, S., et al.: RALF: accuracy-aware scheduling for feature store maintenance. Proc. VLDB Endow. **17**(3), 563–576 (2023)
27. Wu, Y., Guo, W., Chan, C.Y., Tan, K.L.: Fast failure recovery for main-memory DBMSs on multicores. In: Proceedings of the 2017 ACM International Conference on Management of Data, pp. 267–281 (2017)
28. Yao, C., Agrawal, D., Chen, G., Ooi, B.C., Wu, S.: Adaptive logging: optimizing logging and recovery costs in distributed in-memory databases. In: Proceedings of the 2016 International Conference on Management of Data, pp. 1119–1134 (2016)
29. Zhu, J.P., et al.: Log replaying for real-time HTAP: an adaptive epoch-based two-stage framework. In: 2024 IEEE 40th International Conference on Data Engineering (ICDE), pp. 2096–2108. IEEE (2024)

LLM-Driven Online Aggregation for Unstructured Text Analytics

Chao Hui[1], Weizheng Lu[2(✉)], Yanjie Gao[2], Lingfeng Xiong[2], Yunhai Wang[2], and Yueguo Chen[2]

[1] Shandong University, Qingdao, China
`chaohui@mail.sdu.edu.cn`
[2] Renmin University of China, Beijing, China
`{luweizheng,gaoyanjie,lenfeng2022,wang.yh,chenyueguo}@ruc.edu.cn`

Abstract. Large Language Models (LLMs) exhibit strong capabilities in text processing, and recent research has augmented SQL and DataFrame with LLM-powered semantic operators for data analysis. However, LLM-based data processing is hindered by slower token generation speeds compared to relational queries. To enhance real-time responsiveness, we propose OLLA, an LLM-driven online aggregation framework that accelerates semantic processing within relational queries. In contrast to batch-processing systems that yield results only after the entire dataset is processed, our approach incrementally transforms text into a structured data stream and applies online aggregation to provide progressive output. To enhance our online aggregation process, we introduce a semantic stratified sampling approach that improves data selection and expedites convergence to the ground truth. Evaluations show that OLLA reaches 1% accuracy error bound compared with labeled ground truth using less than 4% of the full-data time. It achieves speedups ranging from $1.6\times$ to $38\times$ across diverse domains, measured by comparing the time to reach a 5% error bound with that of full-data time. We release our code at https://github.com/olla-project/llm-online-agg. git.

Keywords: Large Language Model · Online Aggregation · Text Processing

1 Introduction

Extracting insights from the ever-growing volume of unstructured text has been a long-standing research problem [6,26]. Due to the strong semantic understanding capabilities, Large Language Models (LLMs) have recently emerged as a powerful paradigm for unstructured text analysis, and researchers are integrating LLM-based text analysis into relational queries, with recent examples including LOTUS [22] and UQE [8]. These systems employ LLMs to parse unstructured

---

C. Hui, W. Lu, and Y. Gao—Equal contribution.

© The Author(s), under exclusive license to Springer Nature Singapore Pte Ltd. 2026
H. Jung et al. (Eds.): DASFAA 2026, LNCS 16536, pp. 578–594, 2026.
https://doi.org/10.1007/978-981-92-0366-6_35

text into structured fields and then perform relational analysis on the resulting data. This approach enables users to perform complex statistical analysis on text using familiar interfaces such as SQL or DataFrame. However, converting text into structured fields requires row-by-row processing, and LLM token generation is significantly slower than the execution speed of relational queries [17]. For instance, processing a set of texts with an LLM might take several minutes, whereas a relational query may complete in seconds. This performance disparity hinders the large-scale adoption of LLM-based text analysis in production environments. Thus, simply integrating LLMs into relational engines is insufficient to meet the demand for low-latency text analysis.

Inspired by online aggregation in relational analysis, which provides progressively refined approximate results, we propose a novel approach that combines online aggregation with LLM-based text analysis. We present **OLLA** (**O**nline **L**arge **L**anguage model **A**ggregator), a novel LLM-driven online aggregation framework to support fast, interactive analytics over large-scale unstructured text. OLLA applies LLM-driven information transformation to convert unstructured data into a structured data stream and then performs online aggregation on the data stream. To ensure our online aggregation process converges rapidly to the ground truth, OLLA introduces a semantic stratified sampling strategy. It first converts unstructured text into a vector space using an embedding model, then clusters the embedding vectors, and finally samples and adjusts the resulting clusters.

Our evaluation demonstrates that OLLA reaches the 1% absolute error bound on accuracy against the labeled ground truth using less than 4% of the full data processing time. And it delivers speedups from 1.6× to 38×, where speedups are measured by comparing the time required to reach a 5% error bound of confidence interval with that of full data processing.

In summary, our contributions are as follows:

- OLLA builds upon the principles of online aggregation, enabling early, progressively refined query results. This allows users to obtain approximate insights in real time without waiting for full data processing to complete.
- We introduce a semantic indexing and stratified sampling mechanism where unstructured texts are converted into embedding vectors, clustered into strata, and then sampled uniformly from each stratum. This dynamic process improves both precision and efficiency over time.
- We implement a prototype and evaluate it on a range of representative queries over diverse real-world unstructured datasets. Experimental results show that our approach significantly outperforms baseline methods in terms of both response latency and convergence speed.

2 Background and Related Work

2.1 LLM for Data Processing

Large Language Models (LLMs) have recently shown strong capabilities in extracting and understanding semantics from unstructured text. Some recent studies extend SQL [8] or DataFrame [22] on unstructured data analysis by embedding semantic operators driven by LLMs. Likewise, production systems such as Google BigQuery[1] and Databricks[2] have integrated LLM capabilities directly into their SQL APIs. This type of query invokes the LLM on each row to process the text column. Users have to wait for long periods to get results when working with large datasets, and the latency limits practicality in large-scale settings.

2.2 Online Aggregation

Online aggregation provides users with approximate results early, refining results progressively as more data is processed [11,16]. Unlike batch queries that delay output until full completion, online aggregation supports incremental computation and interactive exploration, and enables users to observe query convergence over time and terminate early once results achieve the expected accuracy, which significantly reduces the time-to-insight. Systems like Ripple Joins [9], BlinkDB [1], and VerdictDB [21] have demonstrated the effectiveness in streaming and approximate query processing contexts.

3 LLM-Driven Online Aggregation

3.1 Problem Definition

Consider a table $T = \{x_i\}_{i=1}^{N}$ with mixed attributes: some columns contain unstructured text $\mathcal{D}_i$ (e.g., logs, reviews, conversations) while the rest are numeric, categorical, or otherwise structured data. A user query Q combines ordinary SQL with an LLM-driven function $LLM(\mathcal{D}_i)$ and then applies an aggregate g (e.g., COUNT, AVG). Because invoking the LLM for every row is expensive, we incrementally aggregate, and at each step, provide a running estimate along with an α-level confidence interval.

3.2 Workflow of OLLA

To realize this vision of fast and interactive text analytics, OLLA integrates several key components, as illustrated in Fig. 1.

Semantic Stratified Sampling. In OLLA, each text entry is embedded into a high-dimensional vector space using an embedding model. The embedding vectors are then clustered using K-means algorithm [2]. We then sample uniformly from the generated clusters. Since the initial embedding-based clusters may not align perfectly with the eventual LLM transformations, we introduce an adaptive adjustment algorithm. This algorithm iteratively adjusts and samples the strata, progressively purifying them to align with the LLM's output.

---

[1] https://cloud.google.com/blog/products/ai-machine-learning/llm-with-vertex-ai-only-using-sql-queries-in-bigquery.

[2] https://docs.databricks.com/aws/en/large-language-models/ai-functions.

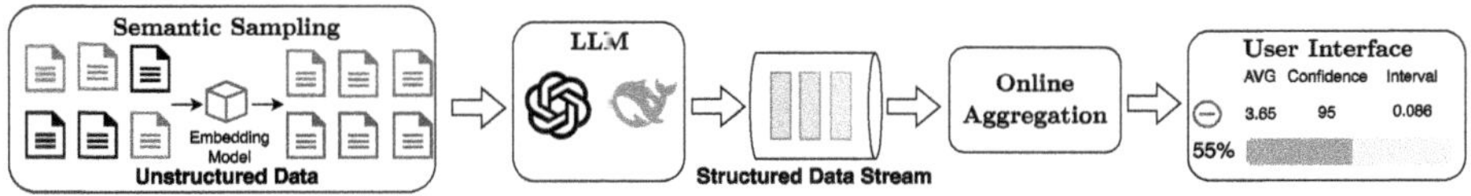

Fig. 1. OLLA system architecture: unstructured data is processed by the LLM module to produce streaming structured data, which is then incrementally aggregated by the online aggregation engine.

Unstructured Data to Structured Data Streams. Raw unstructured data is transformed into a structured data stream using LLMs.

Online Aggregation. OLLA adopts online aggregation to process incrementally arrived data streams, and the system delivers approximate answers with confidence bounds, enabling timely insights without waiting for full computation.

3.3   Implementation

We built a demonstration system to verify the viability of our approach. In the semantic sampling module, we adopt SentenceTransformers v4.1.0 [23] and Faiss v1.8.0 [12] to implement our sampling algorithm. We use the *all-MiniLM-L6-v2* model[3] as our embedding model to encode text into dense vectors. In the unstructured-to-structured transformation module, we deploy an LLM inference service based on vLLM v0.8.4 [15]. To ensure broad applicability, this module supports both on-premise (self-hosted) and cloud-hosted deployment configurations. The output of the LLM service is continuously streamed to a Kafka v4.0.0 [14] message broker, which acts as a buffer and transport layer between the text transformation module and the downstream online aggregation engine. For the online aggregation and query processing, we use Apache Spark Streaming v3.5.1 [3]. Spark Streaming allows us to apply complex transformations and aggregations over real-time structured data, such as group-by or filtering operations. It also computes the confidence intervals.

3.4   Query Examples

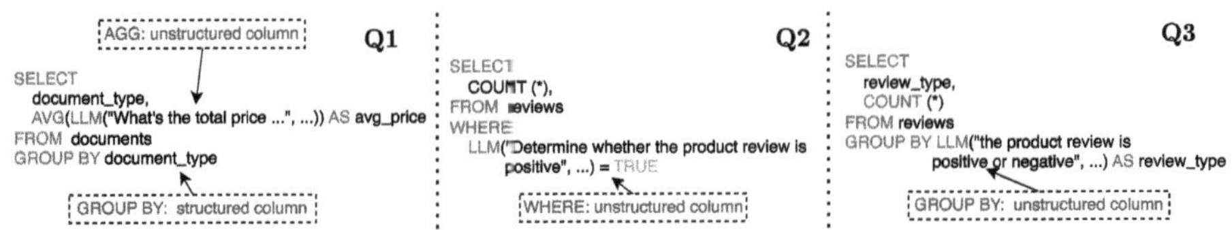

Fig. 2. Three types of user queries.

As shown in Fig. 2, we categorize the user queries that can be accelerated by our method into three types:

---

[3] https://huggingface.co/sentence-transformers/all-MiniLM-L6-v2.

Q1 - LLM in the `SELECT` clause: `GROUP BY` on structured columns with aggregation on an unstructured column using LLM. Traditional sampling is applied to the structured columns. We refer to this query type as `SELECT`.

Q2 - LLM in the `WHERE` clause: filters on an unstructured column and aggregation on structured columns. We call this query type `WHERE`.

Q3 - LLM in `GROUP BY` clause: `GROUP BY` an unstructured column and aggregates over structured columns. We denote this query type as `GROUP BY`.

We classify the three categories because online aggregation relies on data sampling, and the sample method directly affects the result's accuracy and the confidence interval's convergence speed [16]. Sampling on structured columns can adopt conventional techniques, such as uniform sampling [19] or stratified sampling [13]. Unstructured columns in `GROUP BY` or `WHERE` clauses cannot leverage traditional sampling techniques, and we will discuss them in the next Sect. 4

4   Semantic Sampling and Statistical Estimation

4.1   Problem Formulation

Given the text collection $\mathcal{D}$, we first construct a semantic embedding space through a mapping function $E : \mathcal{D} \rightarrow \mathbb{R}^d$, where each text is transformed into a d-dimensional dense vector. The stratification process partitions the embedding space into strata using K-means clustering. Each stratum S_h is characterized by $S_h = (\mathcal{D}_h, \mathcal{M}_h)$ where $\mathcal{D}_h \subset \mathcal{D}$ represents the contained texts and $\mathcal{M}_h$ maintains stratum-specific statistics. For online estimation, we employ the following equation to get the confidence interval:

$$\epsilon_n = (z_p^2 V_n / n)^{1/2} \tag{1}$$

where z_p is the quantile value determined by the confidence level, n is the sample size, and V_n represents the variance term. The specific computation of p and V_n varies between filtering and group aggregation scenarios.

The stratification achieves two primary objectives. For *filtering* cases, it helps identify and prioritize strata likely to contain valid samples (`WHERE TRUE`), thereby accelerating confidence interval convergence. For *group aggregation* cases, it uniformly samples from semantically coherent groups. The estimator is unbiased in *filtering* because it is derived from standard stratified random sampling with fixed strata. In *group aggregation*, it is asymptotically unbiased because any potential bias from the dynamic stratum adjustment diminishes to zero as the sample size increases.

4.2   Online Filtering with Semantic Stratified Sampling

In the *filtering* scenario, our goal is to estimate aggregations over records that satisfy the `WHERE` clause. Two key observations motivate our sampling strategy. First, only valid samples, i.e., records for $\text{LLM}(x) = \text{True}$, contribute to the final result, as data not satisfying the condition are discarded. Second, the confidence interval of the online aggregation narrows as more valid samples are collected, as described in Eq. 1. Therefore, by prioritizing early sampling from strata more likely to yield valid records, we can accelerate the convergence of the confidence interval.

Each stratum S_h maintains a set of statistics $\mathcal{M}_h = \{\hat{p}_h, \pi_h\}$ where $\hat{p}_h$ represents the estimated valid rate and π_h indicates the sampling priority based on this rate. To compute these statistics, we first draw an initial total sample of size m from the entire dataset $\mathcal{D}$, assuming it is representative of the overall data distribution. This sample is then allocated to each stratum S_h proportionally to its size. The set of records sampled from stratum S_h is denoted as $\mathcal{R}_h$, with a size of $m_h = |\mathcal{R}_h| = m \cdot |\mathcal{D}_h|/|\mathcal{D}|$. The valid rate is then estimated using this initial sample set:

$$\hat{p}_h = \frac{1}{m_h} \sum_{x \in \mathcal{R}_h} \mathbb{I}[\text{LLM}(x) = \text{True}], \tag{2}$$

where $\mathbb{I}[\cdot]$ denotes the indicator function. Strata are ranked in descending order of $\hat{p}_h$ to assign sampling priorities π_h, guiding the subsequent sampling process toward those more likely to yield valid records. Let $v(x)$ denote the value of the aggregation column for a record with text x. For example, if we aim to compute the average age of people whose textual descriptions meet certain criteria, then $v(x)$ returns the age value for the record containing text x. The variance term V_n in the confidence interval (Eq. 1) is computed as

$$V_n = 1/(n-1) \sum_{i=1}^{n} (v(x_i) - \bar{v})^2, \tag{3}$$

where $v(x_i)$ is the aggregation value of the i-th valid sample, and $\bar{v} = \frac{1}{n} \sum_{i=1}^{n} v(x_i)$ is their mean. The confidence level $p = \alpha$ is typically pre-specified (e.g., 95%).

4.3   Online Aggregation with Semantic Stratified Sampling

For GROUP BY on text, the main challenge is that group membership is unknown prior to LLM inference. Our approach leverages the semantic similarity principle: texts with similar embeddings are likely to be classified into the same group. Based on this insight, we develop an adaptive stratification strategy that iteratively refines strata through sampling, recording, and adjustment phases.

Let $\mathcal{D}$ be a dataset with N records, where each record belongs to one of K possible categories after LLM inference. At the initial stage (i.e., $t = 0$), the stratification of the dataset is generally performed based on prior knowledge or empirically. A common approach is to determine the number of strata H using the total sample size N and the number of categories K. For instance, the empirical rule $H_0 = K \log N$, $H_{\max} = 2K \log N$ can be applied, where H_0 is the initial estimated number of strata and $H_{\max}$ denotes a predefined maximum number of strata. At time t, each stratum $S_h^{(t)}$ maintains a set of statistics:

$$\mathcal{M}_h^{(t)} = \{m_h^{(t)}, X_{hk}^{(t)}, \hat{p}_{hk}^{(t)}, \tau_h^{(t)}, \hat{V}_h^{(t)}\}, \tag{4}$$

where $m_h^{(t)}$ is the number of samples collected by time t, $X_{hk}^{(t)}$ tracks category k's frequency, $\hat{p}_{hk}^{(t)} = X_{hk}^{(t)}/m_h^{(t)}$ estimates category k's proportion, $\tau_h^{(t)} = \arg\max_k X_{hk}^{(t)}$ identifies the dominant category, and $\hat{V}_h^{(t)}$ measures stratum heterogeneity, defined as

$$\hat{V}_h^{(t)} = \sum_{k=1}^{K} \hat{p}_{hk}^{(t)}(1 - \hat{p}_{hk}^{(t)}). \tag{5}$$

Our iterative process consists of three phases: Sampling, Recording, and Adjustment. Figure 3 depicts these phases, which we discuss in detail below.

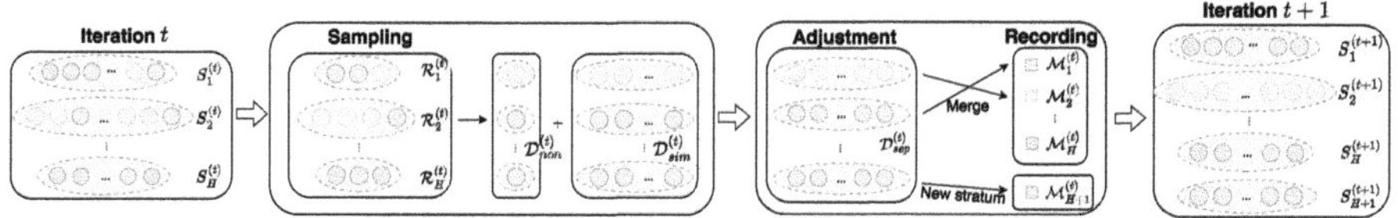

Fig. 3. The adjustment process that enforces homogeneity within each stratum and heterogeneity across strata.

Sampling. In each iteration t, we draw a total of $n^{(t)}$ samples across all strata, denoted as $\{\mathcal{R}_h^{(t)}\}_{h=1}^{H}$. For each stratum $S_h^{(t)}$, the sample size $n_h^{(t)} = |\mathcal{R}_h^{(t)}|$ is determined by Neyman allocation:

$$n_h^{(t)} = \begin{cases} n^{(t)} \cdot \dfrac{N_h^{(t)}}{N}, & \text{if } t = 0, \\[2ex] n^{(t)} \cdot \dfrac{N_h^{(t)}\sqrt{\hat{V}_h^{(t)}}}{\sum_{i=1}^{H} N_i^{(t)}\sqrt{\hat{V}_i^{(t)}}}, & \text{if } t > 0. \end{cases} \tag{6}$$

The Neyman allocation method minimizes the variance of stratified sampling theoretically by considering both stratum size and its variance, ensuring that strata with higher variability are allocated more samples to improve precision and efficiency of the estimates. [18] At initial iteration ($t = 0$), the sample size is allocated proportionally to the stratum size. In subsequent iterations ($t > 0$), the allocation considers both stratum size and its estimated variance $\hat{V}_h^{(t)}$.

Recording. After obtaining samples $\{\mathcal{R}_h^{(t)}\}_{h=1}^{H}$, we stream them to LLM for inference. As responses arrive, we update the stratum statistics $\mathcal{M}_h^{(t)}$ by recording frequency counts $X_{hk}^{(t)}$, computing proportions $\hat{p}_{hk}^{(t)}$, identifying dominant category $\tau_h^{(t)}$, and measuring heterogeneity $\hat{V}_h^{(t)}$.

Adjustment. Each stratum $S_h^{(t)}$'s normalized variance is

$$\tilde{V}_h^{(t)} = \hat{V}_h^{(t)}/(1 - 1/K). \tag{7}$$

Here, the denominator $1 - 1/K$ corresponds to the theoretical maximum variance under a uniform category distribution, serving as a normalization factor. If the normalized variance exceeds threshold θ, we trigger a refinement process to improve stratum homogeneity by first identifying heterogeneous groups:

$$\mathcal{D}_{non}^{(t)} = \{x \in \mathcal{R}_h^{(t)} \mid \text{label}(x) \neq \tau_h^{(t)}\}, \tag{8}$$

and then locating semantically similar, unsampled records:

$$\mathcal{D}_{sim}^{(t)} = \{x \in \mathcal{D}_h^{(t)} \backslash \mathcal{R}_h^{(t)} \mid \exists y \in \mathcal{D}_{non}^{(t)} : \text{sim}(E(x), E(y)) > \gamma\}, \tag{9}$$

where $\text{sim}(u, v)$ denotes the cosine similarity between embedding vectors u and v. These two sets are combined as $\mathcal{D}_{sep}^{(t)} = \mathcal{D}_{non}^{(t)} \cup \mathcal{D}_{sim}^{(t)}$, which represents potentially misclustered or outlier samples requiring reallocation. Based on the number of existing strata, we either split or merge $\mathcal{D}_{sep}^{(t)}$. When $H < H_{\max}$, we create a new stratum from $\mathcal{D}_{sep}^{(t)}$. Otherwise, we merge $\mathcal{D}_{sep}^{(t)}$ with an existing stratum sharing the same dominant category. The target stratum h^* is selected based on maximum embedding similarity:

$$h^* = \arg \max_{h'}\{\text{sim}(E(\mathcal{D}_{sep}^{(t)}), E(\mathcal{D}_{h'})) \mid \tau_{h'}^{(t)} = \tau_{\mathcal{D}_{sep}^{(t)}}\}, \tag{10}$$

where $E(\mathcal{D})$ represents the mean embedding vector of all texts in set $\mathcal{D}$. After each iteration, we compute the variance term V_n and confidence level p for Eq. 1:

$$V_n = \sum_{h=1}^{H} \left(\frac{N_h^{(t)}}{N}\right)^2 f_h^{(t)} \hat{V}_h^{(t)}, \tag{11}$$

where $f_h^{(t)} = \frac{N_h^{(t)} - m_h^{(t)}}{m_h^{(t)}(N_h^{(t)} - 1)}$ is the finite population correction term arising from the hypergeometric distribution due to sampling without replacement within each stratum, and $p = \alpha/K$ accounts for multiple comparisons across K categories.

This iterative process continues until convergence (all strata achieve sufficient purity ($\tilde{V}_h^{(t)} \leq \theta$)), with each cycle improving stratum purity ($V^{(t+1)} < V^{(t)}$) while maintaining bounded variance ($0 \leq V_h \leq 1 - \frac{1}{K}$). The complete procedure is summarized in Algorithm 1.

Algorithm 1. Online Aggregation with Semantic Stratified Sampling

---

Require: Dataset $\mathcal{D}$, LLM function LLM, Embedding function E, initial strata number H_0, maximum strata number $H_{\max}$, confidence level α, variance threshold θ, similarity threshold γ

1: **for** each data point $x \in \mathcal{D}$ **do**
2:     Compute embedding vector $e_x \leftarrow E(x)$
3: **end for**
4: Partition embeddings $\{e_x\}$ into H strata: $\{S_1^{(0)}, \ldots, S_{H_0}^{(0)}\}$
5: $t \leftarrow 0$
6: **while** $\exists h : \tilde{V}_h^{(t)} > \theta$ **do**
7:     **if** $t = 0$ **then**
8:         $n_h^{(t)} \leftarrow n^{(t)} \cdot N_h^{(t)}/N$ for each stratum $S_h^{(t)}$
9:     **else**
10:         $n_h^{(t)} \leftarrow n^{(t)} \cdot N_h^{(t)}\sqrt{\hat{V}_h^{(t)}} / \sum_{i=1}^{H} N_i^{(t)}\sqrt{\hat{V}_i^{(t)}}$
11:     **end if**
12:     Draw samples $\mathcal{R}_h^{(t)}$ of size $n_h^{(t)}$ from each stratum
13:     **for** each sample $x \in \mathcal{R}_h^{(t)}$ **do**
14:         Get category label $k \leftarrow \text{LLM}(x)$
15:         Update $\mathcal{M}_h^{(t)} \leftarrow \{m_h^{(t)}, X_{hk}^{(t)}, \hat{p}_{hk}^{(t)}, \tau_h^{(t)}, \hat{V}_h^{(t)}\}$
16:     **end for**
17:     **for** each stratum h with $\tilde{V}_h^{(t)} > \theta$ **do**
18:         Obtain $\mathcal{D}_{sep}^{(t)} \leftarrow \mathcal{D}_{non}^{(t)} \cup \mathcal{D}_{sim}^{(t)}$
19:         **if** $H < H_{\max}$ **then**
20:             Create new stratum from $\mathcal{D}_{sep}^{(t)}$
21:         **else**
22:             Merge $\mathcal{D}_{sep}^{(t)}$ with most similar existing stratum S_{h*}
23:         **end if**
24:     **end for**
25:     Update confidence intervals using V_n and $p = \alpha/K$
26:     $t \leftarrow t + 1$
27: **end while**
28: **return** Proportion estimates $\{\hat{p}_{hk}^{(t)}\}$ and confidence intervals

---

5 Evaluation

5.1 Experimental Setup

Datasets. We evaluate our system using a diverse set of datasets spanning product reviews, document understanding, and text classification. Due to the absence of standard benchmarks for LLM queries, we curated real-world datasets from diverse sources and formulated representative queries over them. Table 1 shows the dataset description, size, and the queries on them.

Metrics. We adopt the statistical metrics: *Absolute Error, Confidence Interval,* and *Cumulative Valids.* All reported times in our experiments refer to the average time per query. The *Absolute Error* between the progressive streaming output and the labeled ground truth can be referred to as accuracy. The *Confidence Interval* is computed following Eq. 1, with the specific computation of

Table 1. Datasets and queries used in the experiments. The *Size* column (n_{rows}) indicates the number of rows used. The *Query Type* column corresponds to the three categories illustrated in Fig. 2 and Sect. 3.4, as the LLM operator in the SELECT, FILTER, or GROUP BY clause.

Dataset	Description	Size (n_{rows})	Query Type
Company Documents [7]	A document dataset comprising business documents like invoices, purchase orders	10,000	Q1
BBC News [5]	A multi-class news classification dataset with synthetically added numerical fields (e.g., view counts) for aggregation	15,000	Q2,Q3
arXiv [25]	Contains metadata for research papers from arXiv, including titles and abstracts	10,000	Q2
Amazon Product [10]	Product reviews and metadata, with a sentiment distribution balanced for analysis. For our experiments, we created a balanced dataset with an equal number of positive and negative reviews	15,000	Q2
Movie [20]	A collection of movie reviews from critics, paired with metadata such as scores	15,000	Q2,Q3
Chinese Resume [24]	A collection of Chinese resumes where the self-introduction section includes personal information such as name and age	20,000	Q1

the confidence level p and variance term V_n varying between filtering and group aggregation scenarios as described in Sect. 4. The *Cumulative Valids* refers to the total number of outputs that satisfy predefined validity constraints (e.g., syntactic correctness).

Environment. Our experiments are conducted on a machine equipped with 2 Intel Xeon 6438M CPUs (2.2 GHz, 60 MB cache), 512 GB of main memory, and 8 NVIDIA A800 GPUs (80 GB HBM memory), running Rocky Linux 8.9.

5.2   Accuracy of OLLA

We evaluate the accuracy of OLLA along two dimensions: (1) the correctness of the LLM's unstructured-to-structured data transformation, and (2) the precision of the streaming aggregation result compared to the ground truth from the fully labeled dataset.

Table 2. Accuracy on *BBC News* classification and *Movie* sentiment analysis.

Model	BBC News	Movie
Qwen2.5 7B	0.90	0.84
DeepSeek-V3.1 671B	0.95	0.88
Ground Truth	1.00	1.00

Table 2 compares model accuracies on the *BBC News* and *Movie* datasets against their ground truth labels, which provide complete ground truth labels for quantitative evaluation. The results indicate that LLMs performance is model-dependent and context-dependent, and LLMs' outputs do not 100% align with the pre-defined labels. The DeepSeek-V3.1 671B (no-thinking) model achieves the best performance, but it incurs a high inference cost. While the much smaller compact Qwen2.5-7B model is a compelling and cost-effective alternative, achieving accuracy rates of over 80% and even 90%.

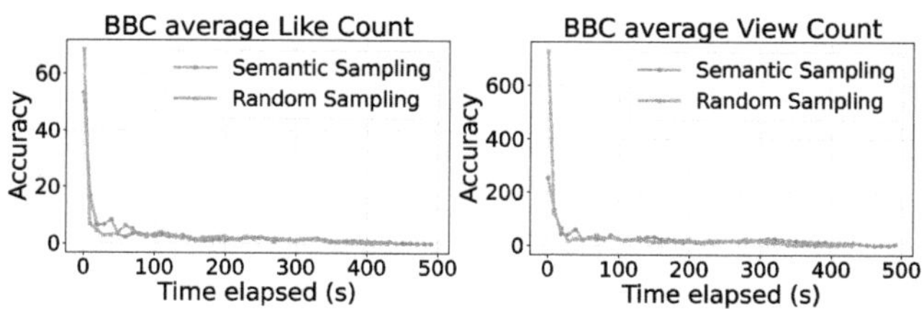

Fig. 4. Convergence of accuracy over time for online aggregation. Accuracy: the absolute error between the streaming aggregate result and the ground truth. Left: average *like_count* of the *BBC News*. Right: average *view_count* of the *BBC News*.

We compare the online aggregation results against the ground truth from the full dataset to directly measure OLLA's empirical performance, rather than relying solely on its confidence interval estimates. Figure 4 plots the absolute error between the streaming aggregated result and the ground truth computed over the entire dataset, showing that the online aggregation process converges remarkably quickly. OLLA reaches the 1% error bound comparing the final labeled ground truth using only 2.43% and 3.49% of the entire dataset processing time in our two settings. This figure illustrates the fundamental trade-off in online aggregation: as execution time (x-axis) increases, the error decreases. What's more, our semantic sampling method reduces error more quickly than random sampling, particularly during the initial stages of the aggregation process.

5.3   Efficiency of OLLA

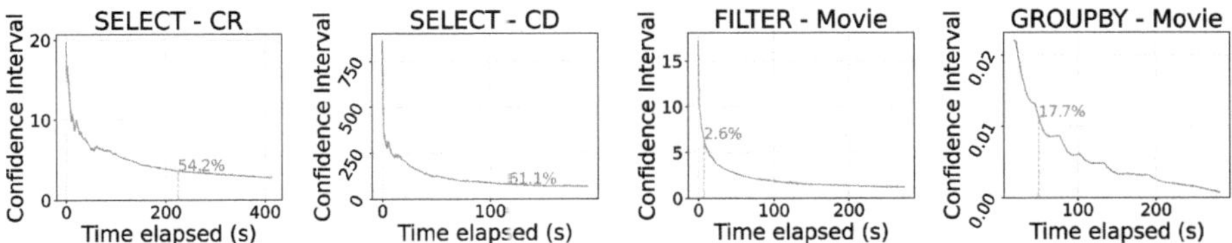

Fig. 5. Convergence of confidence intervals for representative query types. The red percentage is the time our method reaches a 5% error bound, divided by total batch execution time. Left to right: **SELECT**: Average *age* of resume on *Chinese Resume*(CR); **SELECT**: Average *total_price* of type *Invoices* on *Company Documents*(CD); **WHERE**: Average length of positive *Movie* reviews; **GROUP BY**: Proportion of Neutral *Movie* reviews.

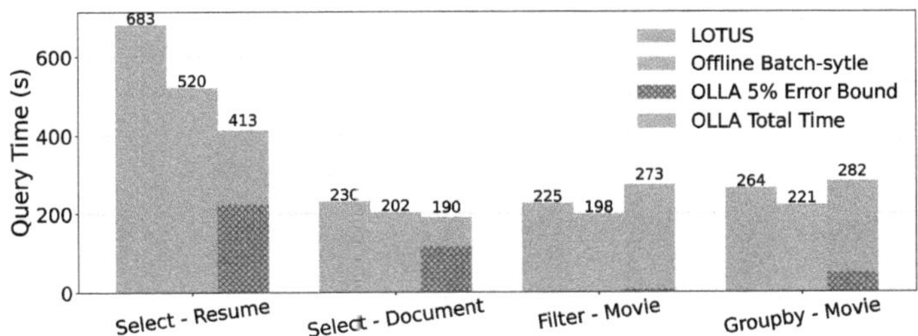

Fig. 6. Comparison of query execution times for LOTUS, a batch-style baseline, and OLLA for different query types. The hatched portion of the OLLA bar indicates the time to reach a 5% error bound, while the full bar represents its total execution time.

To validate the efficiency of OLLA, we conduct experiments across three representative query types illustrated in Fig. 2. In each case, we track the evolution of the confidence interval as the sampling progresses. This setup reflects a common use case where users seek to get early, progressively refined results without waiting for the full dataset to be processed.

Figure 5 presents the confidence interval trajectories for each query. We further evaluate the fraction of total execution time needed to reach a 5% error bound. It is approximately 61.1% (1.6× compared with scanning full data) of the total execution time for the **SELECT** scenario, 2.6% (38×) for the **WHERE** scenario, and 17.7% (5.6×) for the **GROUP BY** scenario.

We further conduct comparative analysis of OLLA against state-of-the-art system, LOTUS [22], and a naive offline baseline that processes the entire data via parallel random sampling. The results, presented in Fig. 6, highlight the trade-offs and core benefits of our approach. For the **SELECT** query, OLLA significantly reduces latency. For the **WHERE** and **GROUP BY** tasks, while our semantic stratified sampling introduces minor latency overhead, the primary strength of OLLA lies in its online nature. The time to reach the 5% error bound is substantially shorter than waiting for any offline batch-style method to complete.

5.4   Efficiency of Sampling Optimization

Online Filtering with Semantic Sampling. We evaluate the effectiveness of online Filtering with Semantic Sampling on two datasets: *Amazon Product* and *BBC News*. In each experiment, we compare our method against a random sampling baseline in terms of how quickly they accumulate informative samples and how fast the confidence intervals converge during online aggregation.

The results are presented in Fig. 7. The top two plots show the cumulative count of LLM-evaluated entries satisfying the predicate (i.e., labeled TRUE) as sampling steps. Our filter-aware strategy consistently accumulates valid samples more rapidly in the early phases of sampling. The bottom two plots show the width of the corresponding confidence intervals for the target aggregates, demonstrating that our method reaches the target error bound significantly faster than the baseline. Specifically, for the *Amazon* dataset, filter sampling reaches the 5% error bound after retrieving 5.28% of the full sampling process, while random sampling requires 6.44%. Similarly, for *BBC News*, filter sampling reaches this bound at 2.55% of the process, in contrast to 6.04% for random sampling, demonstrating that our method can provide tighter estimates with much lower sampling effort and, consequently, reduce the number of LLM calls in practice.

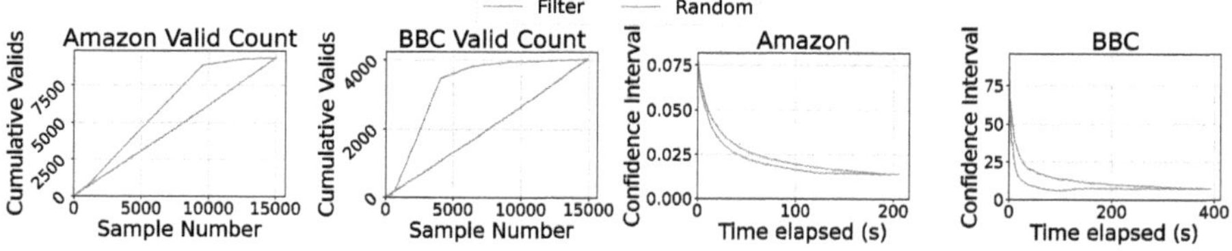

Fig. 7. Comparison of Online Filtering with Semantic Sampling ("Filter" in figure) and random sampling ("Random" in figure). Left: *verified_purchase* rate among reviews classified as positive by the LLM in *Amazon Product*; Right: average *like_count* of news articles identified by the LLM as sports-related in *BBC News*.

Group Aggregation with Semantic Stratified Sampling. We assess Semantic Stratified Sampling in group aggregation scenarios to estimate totals or proportions of LLM-identified groups. Using *Movie* (sentiment) and *BBC News* (topic) datasets, we compare our dynamic approach—which adjusts strata every 10% of steps—against random sampling and static stratification (UQE [8]).

Figure 8 shows that our *Adjust* strategy consistently yields faster confidence interval convergence. While static stratification (*No Adjust*) effectively leverages initial clustering on *BBC News*, it performs similarly to random sampling on the *Movie* dataset due to semantic misalignment. Quantitatively, we report the average time proportion relative to full processing needed to reach a 5% error bound. On the *Movie* dataset, *Adjust* requires only 13.75% of the time, significantly surpassing *No Adjust* (23.71%) and *Random* (21.80%). Likewise, for *BBC News*, *Adjust* (10.82%) proves more efficient than *No Adjust* (14.58%) and *Random* (39.40%).

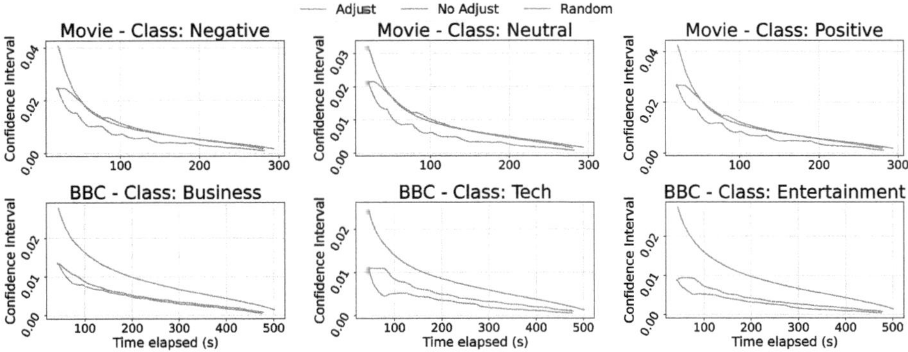

Fig. 8. Confidence interval evolution for group proportion estimates under different sampling strategies. Legend: "Adjust" – applies the Sampling-Recording- Adjustment workflow proposed in Sect. 4.3; "No Adjust" – performs sampling solely on the embedding vectors without adjusting the clustered strata proposed in UQE [8]; "Random" – draws samples by simple random selection. Top: ratio of *Movie* sentiment (Positive, Neutral, Negative); Bottom: ratio of *BBC News* topic (Business, Tech, Entertainment).

5.5   Impact of Hyper-parameters

In this section, we investigate the impact of key hyper-parameters on the performance of OLLA. We aim to provide empirical guidance for choosing appropriate hyper-parameter settings to maximize estimation accuracy.

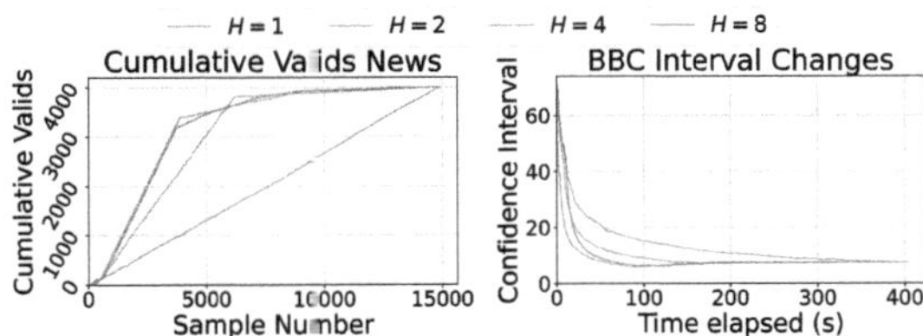

Fig. 9. Impact of the number of strata on convergence of Confidence Intervals in filtering scenario. The query is to filter items belong to the *sports* category in *BBC News*.

For the filtering scenario, we investigate the impact of the number of strata on the convergence of confidence intervals. The results, shown in Fig. 9, demonstrate that even with as few as two strata, semantic sampling already achieves a significant acceleration in convergence. This improvement can be attributed to the strong semantic separability of the *BBC News* dataset, which enables the clustering algorithm to effectively distinguish among different semantic groups. Furthermore, as the number of strata increases to four or five, reflecting the five major news categories present in the dataset, convergence further improves, yielding more accurate estimates with reduced sampling effort.

For group aggregation scenario, we further investigate the effects of key hyper-parameters on convergence behavior. This experiment is based on the

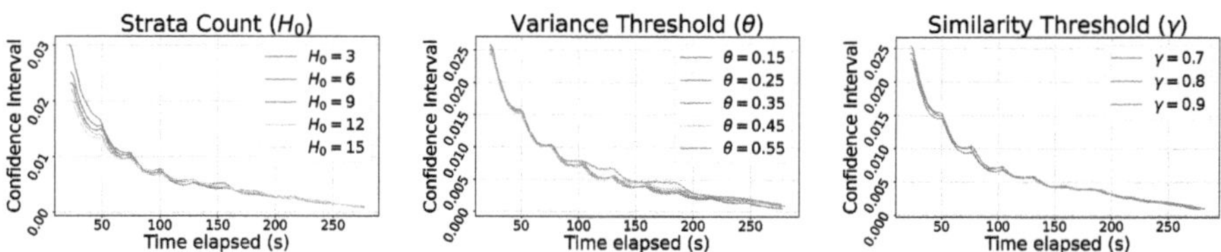

Fig. 10. Impact of hyper-parameter settings on convergence in group-by scenario. The query is designed to compute the ratio of different categories (Positive, Negative and Neutral) within the *Movie* dataset.

previously described *Movie* dataset, where the objective is to compute the proportion for each semantic group. The results, presented in Fig. 10, show that employing a larger number of initial strata typically contributes to faster convergence in early sampling phases. This observation can be explained by the enhanced semantic resolution that a greater number of strata provides, allowing for more accurate group-wise estimates. The second subplot illustrates the impact of the variance threshold on convergence: when this threshold is set too high (e.g., 0.55), it may prohibit finer-grained strata formation and consequently slow down convergence. In contrast, the third subplot reveals that similarity threshold exerts a relatively small influence on the overall convergence performance.

5.6   Scalability

We evaluate scalability of our system by increasing number of LLM serving instances to assess how performance improves. We deploy our system using LiteLLM [4] as a serving layer, scaling from 1 to 4 instances with identical hardware specifications, and measure the execution time required to reach the same error bound in two representative workloads: filtering and aggregation on *Amazon Product* and *BBC News*. Figure 11 illustrates the execution time with respect to the number of LLM serving instances. As shown in the figure, both workloads benefit from scaling, but the incremental gains diminish as additional instances are added. With 2, 3, and 4 serving instances, we observe speedups of $1.32\times$, $1.46\times$, and $1.52\times$ for the *Amazon Product* workload, and a more scalable $1.52\times$, $1.83\times$, and $2.03\times$ for the *BBC News* workload. These results confirm that our architecture scales effectively. The sub-linear trend arises mainly from LiteLLM's default routing and batching configuration, which is not yet tuned for our workload. With optimized settings, scalability could approach the linear ideal.

6   Conclusion

In this paper, we present OLLA, a novel framework that integrates large language models with online aggregation to enable scalable, interactive analytics over unstructured data. By transforming unstructured inputs into structured

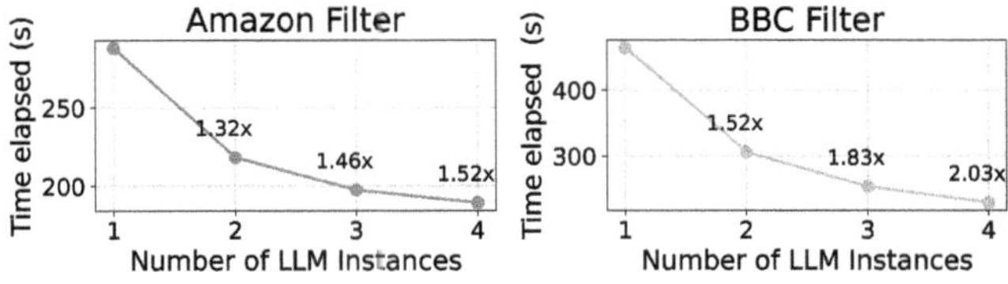

Fig. 11. System scalability evaluation.

representations through LLM-based extraction, embedding them for semantic indexing, and applying approximate query processing via online aggregation, OLLA addresses long-standing challenges in latency, scalability, and semantic ambiguity. Experimental results across diverse domains demonstrate that OLLA enables faster time-to-insight and higher-quality exploration compared to existing approaches.

Acknowledgments. This work was supported by BRAIN of RUC, the Fundamental Research Funds for the Central Universities and the Research Funds of RUC (24XNKJ22), the Beijing Natural Science Foundation (L247027), National Natural Science Foundation of China (No. 62272466, No. U2436209, and No. U24A20233). The computing resource was supported by PCC of RUC.

References

1. Agarwal, S., Mozafari, B., Panda, A., Milner, H., Madden, S., Stoica, I.: BlinkDB: queries with bounded errors and bounded response times on very large data. In: EuroSys 13, pp. 29–42 (2013)
2. Ahmed, M., Seraj, R., Islam, S.M.S.: The k-means algorithm: a comprehensive survey and performance evaluation. Electronics **9**(8), 1295 (2020)
3. Armbrust, M., et al.: Structured streaming: a declarative API for real-time applications in Apache spark. In: SIGMOD 18, pp. 601–613 (2018)
4. BerriAI: LiteLLM (2024). https://github.com/BerriAI/litellm. Accessed 16 June 2025
5. Bose, B.: BBC News classification (2019). https://kaggle.com/competitions/learn-ai-bbc. Kaggle
6. Chen, X., Jin, L., Zhu, Y., Luo, C., Wang, T.: Text recognition in the wild: a survey. ACM Comput. Surv. **54**(2) (2021)
7. Cherguelain, A.: Company documents dataset. Kaggle (2024). https://www.kaggle.com/datasets/ayoubcherguelaine/company-documents-dataset
8. Dai, H., et al: UQE: a query engine for unstructured databases. In: NeurIPS 2024, vol. 37, pp. 29807–29838 (2024)
9. Haas, P.J., Hellerstein, J.M.: Ripple joins for online aggregation. SIGMOD Rec. **28**(2), 287–298 (1999)
10. He, R., McAuley, J.: Ups and downs: modeling the visual evolution of fashion trends with one-class collaborative filtering. In: WWW 2016, pp. 507–517 (2016)
11. Hellerstein, J.M., Haas, P.J., Wang, H.J.: Online aggregation. In: SIGMOD 1997, pp. 171–182 (1997)

12. Johnson, J., Douze, M., Jégou, H.: Billion-scale similarity search with GPUs. IEEE Trans. Big Data **7**(3), 535–547 (2019)
13. Joshi, S., Jermaine, C.: Robust stratified sampling plans for low selectivity queries. In: ICDE 2008, pp. 199–208. IEEE (2008)
14. Kreps, J., Narkhede, N., Rao, J., et al.: Kafka: a distributed messaging system for log processing. In: NetDB 2011, pp. 1–7 (2011)
15. Kwon, W., et al.: Efficient memory management for large language model serving with pagedattention. In: SOSP 2023, pp. 611–626 (2023)
16. Li, Y., Wen, Y., Yuan, X.: Online aggregation: a review. In: Web Information Systems and Applications, pp. 103–114 (2018)
17. Liu, S., et al.: Optimizing LLM queries in relational workloads. In: MLSys 2025 (2025)
18. Neyman, J.: On the two different aspects of the representative method: the method of stratified sampling and the method of purposive selection. In: Kotz, S., Johnson, N.L. (eds) Breakthroughs in Statistics. Springer Series in Statistics, pp. 123–150. Springer, New York (1992). https://doi.org/10.1007/978-1-4612-4380-9_12
19. Olken, F., Rotem, D.: Random sampling from databases: a survey. Stat. Comput. **5**(1), 25–42 (1995)
20. Pang, B., Lee, L.: Seeing stars: exploiting class relationships for sentiment categorization with respect to rating scales. arXiv preprint cs/0506075 (2005)
21. Park, Y., Mozafari, B., Sorenson, J., Wang, J.: VerdictDB: universalizing approximate query processing. In: SIGMOD 2018, pp. 1461–1476 (2018)
22. Patel, L., et al.: Semantic operators and their optimization: towards AI-based data analytics with accuracy guarantees. PVLDB **18**(11), 4171–4184 (2025)
23. Reimers, N., Gurevych, I.: Sentence-BERT: sentence embeddings using Siamese BERT-networks. In: EMNLP 2019 (2019)
24. Su, Y., Zhang, J., Lu, J.: The resume corpus: a large dataset for research in information extraction systems. In: CIS 2019, pp. 375–378 (2019)
25. arXiv.org submitters: arXiv dataset (2024). htttps://www.kaggle.com/dsv/7548853
26. Wankhade, M., Rao, A.C.S., Kulkarni, C.: A survey on sentiment analysis methods, applications, and challenges. Artif. Intell. Rev. **55**(7), 5731–5780 (2022)

Cloud Data Management

ZTune: Model-Assisted Reinforcement Learning for Executor Tuning on Ad Hoc Spark SQL Query

Dejun Kong[1,2], Xiuqi Huang[3], and Xiaofeng Gao[1(✉)]

[1] Shanghai Jiao Tong University, Shanghai 200240, China
`kdjkdjkdj99@sjtu.edu.cn`, `gao-xf@cs.sjtu.edu.cn`
[2] University of New South Wales, Kensington, NSW 2052, Australia
[3] State Key Laboratory of CAD&CG, Zhejiang University, Hangzhou 310058, China
`huangxiuqi@zju.edu.cn`

Abstract. With the development of distributed computing systems, Spark has become widely used in big data parallel processing scenarios. To reduce costs and enhance efficiency, it is essential to make quick and precise adjustments to Spark configuration for each job. Ad hoc Spark SQL job tuning presents unique challenges due to its one-off execution nature, where existing iterative tuning methods for periodic jobs are inapplicable. In this paper, we focus on ad hoc Spark SQL job tuning within Spark executor parameters, which have the most significant influence on processing performance, aiming at improving the quality of service between service providers and clients. We propose a model-assisted reinforcement learning tuning framework, ZTune. ZTune integrates SQL plan representation, a novel query performance model (QPM), and a dual-phase tuning process: prediction and searching. In prediction phase, we combine the constructed model with historical data to better utilize observed patterns and sampling data for more robust performance prediction. In searching phase, we propose a Dual Sampling & Transfer Deep Q-Network to accelerate model training and improve configuration recommendations. Performance evaluation on a large-scale cluster illustrates that ZTune achieves significant performance results compared with baseline methods, including state-of-the-art and industrial approaches, with competitive execution overhead.

Keywords: Spark Tuning · Ad Hoc Query · Reinforcement Learning

1 Introduction

Along with the rapid development of big data processing and analysis, traditional computing on a single machine has encountered a series of problems, such as insufficient computing ability, low computing efficiency, and limited storage capacity [13]. To address these issues, distributed computing clusters are developed. Apache Spark has gradually risen in the past ten years, which is a memory-based distributed computing framework for easier data processing.

H. Jung et al. (Eds.): DASFAA 2026, LNCS 16536, pp. 597–612, 2026.
https://doi.org/10.1007/978-981-92-0366-6_36

Spark configuration is the most important factor influencing the quality of job processing performance. It is composed of multiple parameters involving all aspects of settings, while the importance of different parameters varies [9]. The value range of each parameter contributes to an enormous parameter space. Searching the best configuration group for a specific job can be extremely difficult. What's more, these parameters are to some extent coupled, meaning that modifying one parameter may change the optimal settings for others [8]. In this situation, it is hard to obtain the best configuration by searching each parameter individually. As a result, the configuration searching becomes more complicated. Hence, configuration tuning deserves an in-depth study.

Among all the sections of the Spark framework, the executor plays an important role in job processing, which is responsible for running tasks and storing data for Spark jobs [15]. Recent work implies the importance of the Spark executor parameters in Spark tuning [9]. Executor tuning makes a major contribution to the optimization with a percentage of over 50% [11]. Besides, Spark also supports dynamic allocation of executors, where executors can be added or removed dynamically. Dynamic allocation is also crucial for tuning.

In recent years, there are a number of works on Spark tuning, recommending parameters with various methods. However, there are still some limits as follows.

1. **Lack Attention to Ad Hoc Jobs:** Most recent works concentrate on periodic job tuning [16], which always requires multiple samplings on computing clusters to initialize or assist in optimizing the configuration. Such a tuning mode is suitable for periodic jobs since there are a number of attempts enabling progressive tuning, but not compatible with ad hoc jobs. 25% jobs are ad hoc in industrial scenario [17]. Existing methods rarely consider this.
2. **Limited Optimization Objective:** Most existing works only consider minimizing the processing time when conducting configuration tuning [14]. Some recent works improve the objectives and consider both processing time and resource consumption [11], but the motivation is still simple. Cloud service providers and clients need to balance processing time and resource costs. The processing time also shouldn't be much worse to avoid bad user experiences.
3. **Poor Transparency of Tuning:** A number of recent studies on tuning methods employ machine learning methods to study and predict the relationship between configuration and performance [4]. However, since the processing performance is easily affected by system fluctuation, especially online, it is challenging for machine learning models to distinguish real performance and disturbance. Poor transparency of machine learning cannot guarantee the performance of its mechanism, involving configuration recommendation risks.

In this work, we concentrate on configuration tuning of Spark executors for ad hoc SQL queries. In large-scale workloads, thousands of SQL queries run daily, dominating enterprise applications. Hence, improving the performance of SQL queries contributes to significant benefit improvement. We propose a novel tuning framework in Fig. 1, ZTune, to recommend configurations for ad hoc queries without pre execution, i.e., zero-shot. The whole framework consists of two phases, prediction phase and searching phase. With the arrival of ad hoc jobs and other history jobs, the tuning system first extracts the SQL plans from the jobs and obtains the representation of the SQL plans. In prediction phase, the framework first trains the regression model of history query performance with history performance data. With the history regression query performance model (rQPM) and SQL plan representation, the framework can train and predict the ad hoc query performance model (pQPM). In searching phase, the framework first selects the most similar history query and trains the Dual Sampling Deep Q-Network model (DS-DQN) with history query performance model and history data. Then the model is transferred to train Transfer DQN model (TDQN) for ad hoc query, recommending the optimal configuration.

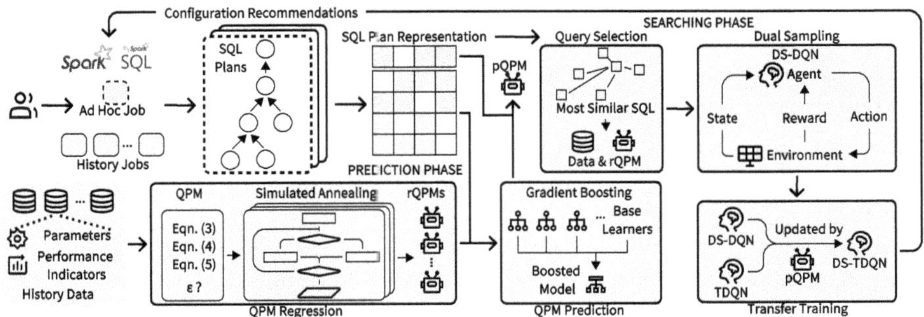

Fig. 1. ZTune: System Overview.

Our contributions are shown as follows to address the limits.

1. Our work fills a critical gap on the study of ad hoc job tuning. We propose a configuration tuning system, ZTune, incorporating multiple objective optimizations to balance processing time and costs with delay penalties.
2. We introduce a query performance model for modeling the relationship between Spark parameters and quality of service, which provides a white box relationship for tuning to enhance the transparency of tuning.
3. We propose a SQL plan representation scheme with three factors: operation number, parallelism, and data size.
4. We design a two-stage performance prediction method with QPM regression and prediction. We also design a Dual Sampling & Transfer Deep Q-Network (DS-TDQN) with query selection, dual sampling, and transfer learning for efficient and better configuration searching.
5. We conduct experiments to evaluate the performance from QPM effectiveness, configuration recommendation, and conditional time-resource balance.

2   Problem Definition

To process Spark jobs, a series of operations need to be completed. In the whole processing scheme, there are four aspects of processing characteristics.

With the whole physical cluster(s) consisting of a number of nodes, there is a group of computing resource features $R = \{r_1, r_2, \cdots, r_m\}$, including cluster node number, cluster core number, cluster memory scale and so on. These features characterize the distributed computing ability of the cluster. As for a Spark job j to be processed, there is a group of job features $J = \{j_1, j_2, \cdots, j_n\}$ to characterize the job, including codes, data size, DAG, physical plan and so on. To process job j, we need to set a series of Spark parameters $S^J = \{s_1^J, s_2^J, \cdots, s_p^J\}$ for processing like spark.executor.cores, spark.executor.memory, spark.dynamicAllocation.maxExecutors. For each Spark parameter $s_i^J, i \in [p]$, there is a value range $s_i^J \in D_i$. With all concerned Spark parameters, there is parameter space $S^J \in \mathbb{S} = D_1 \times D_2 \times \cdots \times D_p$ for parameter settings. With processing environment R, Spark job J and Spark configuration S^J, the job is processed with different performance indicators $U^J = \{u_1^J, u_2^J, \cdots, u_q^J\}$ indicating the performance of the processing procedure, including *app_duration*, *vcore_seconds*, *memory_mb_seconds* and so on. The definitions of Spark parameters and performance indicators are presented in Table 1.

Table 1. Definitions of Spark parameters and performance indicators

Parameter & Indicator	Definition	Unit
executor.cores	The number of cores to use on each executor.	[1,16]
executor.memory	Amount of memory to use per executor process.	[1,16]
initialExecutors	Dynamic Allocation: Initial number of executors to run.	1
minExecutors	Dynamic Allocation: Lower bound for executor number.	1
maxExecutors	Dynamic Allocation: Upper bound for executor number.	[1,16]
app_duration	The processing time of Spark application.	ms
vcore_seconds	The time integral of applied cores.	core · s
memory_mb_seconds	The time integral of applied memory.	MB · s

Given the above definitions, the goal is to find the best Spark configuration considering the processing performance. The objective function is defined as $\hat{S}^J = \arg\min_{S^J \in \mathbb{S}} F(U^J(R, J, S^J))$, where U^J is a function of R, J and S^J, F is a cost function w.r.t. all $u_i^J, i \in [q]$. Then we aim at finding the best Spark configuration in the parameter space to minimize the cost function F. To evaluate and optimize the overall performance comprehensively considering processing time and resource occupancy, the cost function F is designed as follows. As for a Spark job J, we take the default Spark configuration S_0^J and the corresponding performance U_0^J as a benchmark. Then the performance improvement ratio $r_i^J = \frac{u_i^J}{u_0^J}$ for each $i \in [q]$ is regarded as an important evaluation indicator to measure resource and time optimization. To balance the revenue of both, the cost function $F = r_t^J \times (\sum_{i \in [q]/t} w_i \times r_i^J)^\alpha$ is designed as a weighting function of each resource performance improvement ratio r_i^J and time improvement ratio r_t^J, where α is the weighted factor for the importance between resources and time, and w_i for $i \in [q]/t$ are the weighted factors for the importance among different resources. α and w_i can be determined by users on resources and time.

3   Modelling

3.1   Query Performance Model with Case Study

As for a Spark job, processing time, CPU usage, and memory usage are paid most attention to. Therefore, we focus on the optimization of the three performance indicators *app_duration*, *vcore_seconds* and *memory_mb_seconds* to balance the processing time and the computing cost. Besides, we select executor.cores, executor.memory, and dynamicAllocation.maxExecutors as tuning parameters to control the executor settings, which contribute most to the processing. We investigate the relationships between these parameters and performance indicators through comprehensive testing on a dedicated cluster using the TPC-DS benchmark. Analysis of the test results revealed consistent and explainable patterns regarding the impact of configuration on performance, as shown in Table 2.

Table 2. The variation pattern of performance indicators and Spark parameters

Indicator	Variation Pattern
app	decreases exponentially as executor.cores increases.
duration	decreases exponentially as maxExecutors increases.
vcore	decreases exponentially as executor.cores increases.
seconds	first decreases then increases or increases as maxExecutors increases.
	decreases exponentially as executor.cores increases.
memory	increases linearly as spark.executor.memory increases.
mb	first decreases then increases or increases as maxExecutors increases.
seconds	Insufficient memory: *app_duration* increases to a certain extent.
	Insufficient memory: too few executors may cause the job failing.

As there is strong regularity among the three Spark parameters and the three performance indicators, we propose a query performance model (QPM) to express it. QPM consists of three mathematical expressions as follows.

$$app\_duration = \epsilon_t^1 \times cores^{\epsilon_t^2} \times maxExecutors^{\epsilon_t^2} \tag{1}$$

$$+ \epsilon_t^3 \times \left(\frac{cores}{memory \times maxExecutors}\right)^{\epsilon_t^4} + \epsilon_t^5$$

$$vcore\_seconds = \epsilon_c^1 \times cores^{\epsilon_c^2} \times maxExecutors^{\epsilon_c^2} \tag{2}$$

$$\times app\_duration^{\epsilon_c^3} + \epsilon_c^4$$

$$mem\_seconds = \epsilon_m^1 \times memory^{\epsilon_m^2} \times maxExecutors^{\epsilon_m^2}$$

$$\times app\_duration + \epsilon_m^3 \tag{3}$$

Equation (1) implies the relationship of *app_duration* w. r. t. the three indicators. Equation (2) implies the relationship of *vcore_seconds* w. r. t. the three indicators. Equation (3) implies the relationship of *memory_mb_seconds*

w. r. t. the three indicators. All ϵs are hyper-parameters which are determined by the job J and processing environment R. The first term of the three equations represents the major relationship among performance indicators and Spark parameters with a power function. The second term of the *app_duration* equation illustrates the out-of-memory failure (OOM) as shown in Table 2 by setting the *app_duration* to a much larger value so that the OOM situation will not be preferred by ZTune. With QPM, the prediction accuracy of performance is guaranteed by two aspects, equation structure and model fitting.

3.2 SQL Plan Representation

The physical plan of SQL Query is an execution roadmap of a Spark SQL job, which is generated from the SQL code, as shown in Fig. 2. [10] indicate that operation content, SQL structure, and query data are three important aspects. Different operators have specific processing speeds and the number of each operator for a job is the most important information. Second, to represent the plan structure, we define a parallelism factor to describe the parallelism of each operation. First of all, the parallelism factor of the final operator in the directed acyclic graph (DAG) is set to 1. Then, from back to front, once there is more than one branch in the postorder one-for-one operator, the parallelism factor increases by 1. Otherwise, the parallelism factor remains the same as the postorder one. At last, the sizes of all the data tables involved in the physical plan are counted and collected, which decides the processing scale for each operator. With the three parts, we can outline the execution plan framework.

In summary, we denote three groups of encoded representation factors of the physical plan information. The first group includes the operation number factors $\mathcal{X} = \{x_1, x_2, \cdots, x_{op}\}$, in which each factor is the number of one operator and op is the number of operation types. The second group includes the operation parallelism factors $\mathcal{Y} = \{y_1, y_2, \cdots, y_{op}\}$, in which each factor is the sum of all the parallelism factors for one operator. The third group includes one data size factor $\mathcal{Z} = \{z_{size}\}$, where the factor is the sum of all data tables' sizes involved. An example is illustrated in Fig. 2. With the analysis of the submitted SQL query, the representation can be obtained before running the query.

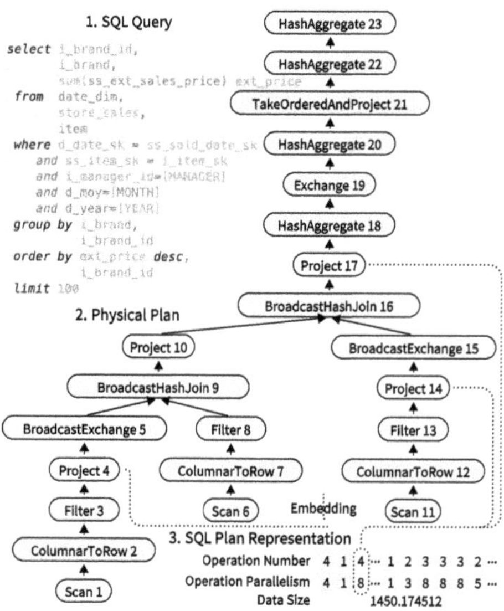

Fig. 2. Physical Plan Representation: Query 55

4   Prediction Phase

The prediction phase consists of a regression module and a prediction module. The regression module is designed for fitting QPM towards historical queries. With regressive QPM (rQPM) and SQL plan representation, the prediction module generates predicted QPM (pQPM) for the target query.

4.1   QPM Regression

In this section, we are going to obtain the QPM of every single historical query to illustrate the characteristics of job processing. For each query, the hyperparameters in the QPM need to be decided given the history processing data of one query. We employ a Simulated Annealing algorithm to search the best regression for the rQPM as shown in Algorithm 1.

Algorithm 1: Simulated Annealing for Regression

Input: Configuration and performance data for one query
Output: Solution ϵ_* with $* \in \{t, c, m\}$

1 Initialize solution ϵ_* randomly within configuration space;
2 Set initial temperature T, cooling rate α, stop criterion T';
3 **while** *stop criterion not met* **do**
4   **foreach** *neighbor ϵ'_* of ϵ_** **do**
5   Calculate difference $\Delta E = \mathrm{MSE}(\epsilon'_*) - \mathrm{MSE}(\epsilon_*)$;
6   **if** $\Delta E < 0$ **or** $rand(0,1) < e^{-\Delta E/T}$ **then**
7   Accept ϵ'_* as the new solution;
8      $T \leftarrow T \times \alpha$;   // cooled temperature
9 **return** ϵ_*;

With the history data of query processing, the algorithm starts with an initial random solution ϵ_* and a high temperature T that controls the probability of accepting worse solutions as it explores the configuration space. At each iteration, SA generates a neighboring solution ϵ_* and evaluates its quality by computing an evaluation function. We use mean squared error $\mathrm{MSE} = \frac{1}{N}\sum_{i=1}^{N}(* - \hat{*})^2$, where $*$ represents one of the three performance indicators computed by QPM with ϵ_*, $\hat{*}$ represents the true values from history data, and N represents the number of historical processing for one query. If the new solution is better, it is always accepted. If the new solution is worse, it may still be accepted with a probability that decreases as the temperature decreases. This probability is typically determined by the Metropolis criterion. Then the temperature gradually decreases with a rate of α following the iterations until the stopping criterion. With such a design, the algorithm is good at exploring the whole space at first. As temperature approaches the criterion, the algorithm becomes less likely to accept worse solutions and refine the current solution locally.

4.2 QPM Prediction

Following QPM Regression, we study the relationship between the SQL plan and regression model of the history queries to provide a most similar prediction for the target ad hoc query. We employ a gradient boosting algorithm to study rules and predict rQPM for target query as shown in Algorithm 2.

Algorithm 2: Gradient Boosting for Prediction

Input: Representation & QPM regression data for queries
Output: The boosted model: $F(\mathcal{R})$
1 Initialize dataset $(\mathcal{R}, \epsilon)$, base learner $h_0(x) = 0$;
2 Initialize number of boosting rounds M, learning rate η;
3 **for** $m = 1$ **to** M **do**
4   Compute the negative gradient: $r_i^m = - \left[\frac{\partial L(\epsilon_i, F(\mathcal{R}_i))}{\partial F(\mathcal{R}_i)} \right]_{F(\mathcal{R})=h_{m-1}(\mathcal{R})}$;
5   Fit a weak learner to the negative gradient:
6      $h_m(\mathcal{R}) = \arg\min_h \sum_{i=1}^{N} L(\epsilon_i, h_{m-1}(\mathcal{R}_i) + h(\mathcal{R}_i)) + \Omega(h)$;
7   Update base learner: $h_m(\mathcal{R}) = h_{m-1}(\mathcal{R}) + \eta \cdot h_m(\mathcal{R})$;
8 **return** $F(\mathcal{R}) = \sum_{m=1}^{M} \eta \cdot h_m(\mathcal{R})$;

The algorithm starts with initializing the dataset $(\mathcal{R}, \epsilon)$, where $\mathcal{R}$ represents the embedding vectors of SQL plan, ϵ is the regression target, along with an initialized base learner $h_0(x) = 0$. The number of boosting rounds M and the learning rate η are set at first. For each boosting round m from 1 to M, the negative gradient r_i^m is computed as the direction and magnitude of the current model's prediction error for each sample. A weak learner $h_m(\mathcal{R})$ is then fitted to these negative gradients by minimizing the loss function $L = \frac{1}{N}(r_i^m - (h_{m-1}(\mathcal{R}_i) + h(\mathcal{R}_i)))^2$ and a regularization term $\Omega = \gamma T + \frac{1}{2}\lambda \sum_{j=1}^{T} \beta_j^2$, where T is the node number of the decision tree, β_j is the weight of the j-th leaf node, and γ and λ are the regularization parameters. Then the base learner is updated by adding the scaled predictions of the weak learner. Finally, the boosted model $F(\mathcal{R})$ is produced by summing the contributions of all weak learners. With $F(\mathcal{R})$, the target query is applied to the model, and rQPM is predicted.

5 Searching Phase

In Sect. 4, we have preliminarily established the relationship among Spark configurations and performance indicators by prediction. However, searching for the best configuration of the target query is also a complicated circumstance due to the system fluctuations and accuracy of prediction. We try to utilize the history queries to recognize these cases and recommend the best configuration. In this section, we introduce Dual Sampling & Transfer Deep Q-Network (DS-TDQN) for parameter searching and recommendation. Reinforcement learning is used here for history learning and tuning experience learning.

$$r = \begin{cases} -\dfrac{app\_dur}{app\_dur_0} \times (w_1 \times \dfrac{mem\_sec}{mem\_sec_0} + w_2 \times \dfrac{vcore\_sec}{vcore\_sec_0})^\alpha, & app\_dur \leq const \times 2, \\ -100, & app\_dur > const \times 2. \end{cases} \tag{4}$$

5.1 Basic Model

The DS-TDQN model is shown in Fig. 3. The model is first trained on a similar history query towards the target query with history data and rQPM as dual environment. With the pre-trained agent, the model is trained on target query with pQPM to get the recommended Spark configuration.

State. The state space of the reinforcement learning model is a 6-dimension state space, consisting of the three Spark parameters and the three performance indicators as shown in Table 1. The three Spark parameters can be tuned directly by changing the state, while the three performance indicators are decided by the state of the three Spark parameters and environment.

Environment. The environment of reinforcement learning is based on QPM and historical data. The agent samples from the QPM to efficiently obtain the new state, including pQPM and rQPM, as well as historical data.

Action. The action of the reinforcement learning model is to change the value of the Spark parameters. Considering the tuning direction, the action includes increasing, decreasing, or unchanging the Spark parameters with different scales.

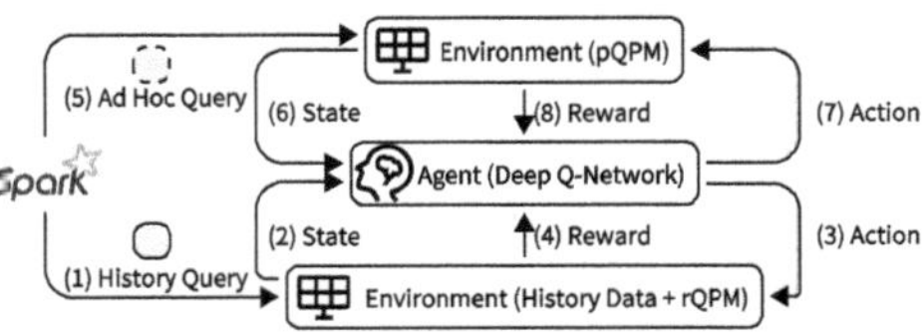

Fig. 3. Dual Sampling & Transfer Deep Q-Network

Reward. The reward function in Eq. (4) is designed based on the objective function $\hat{S}^J$ in Sect. 2 to satisfy the requirement of both the platform and job provider. In the proposed framework, we set a criterion in which the reward function gets a large penalty if the processing time is beyond two times of the benchmark processing time, making trade-off on resources and time. It ensures a sufficient buffer to accommodate processing uncertainties (e.g., workload fluctuations) and enforces a strict Service Level Agreement adherence.

5.2 Model Optimization

To better adapt the model to Spark configuration searching, we make improvements in three aspects, including query selection, dual sampling and transfer training. Query selection is introduced to select the most similar query for auxiliary configuration tuning. Dual sampling is designed for both considering the regression model and real history data. Transfer learning enables ad hoc query tuning to utilize the history data.

Algorithm 3: DS-TDQN Training

---

1 Initialize replay memory D to capacity N & exploration probability ϵ;
2 Initialize Q-network by transferred weights θ & target Q-network by $\theta^- = \theta$;
3 **for** *episode* $= 1$ ***to*** M **do**
4   Initialize state s_1;
5   **for** *timestep* $= 1$ ***to*** T **do**
6   Select random action a_t or $\arg\max_a Q(s_t, a; \theta)$;
7   Execute action a_t, observe reward r_t and state s_{t+1};
8   Store transition (s_t, a_t, r_t, s_{t+1}) in D;
9   Sample random minibatch of transitions from D;
10   Compute target values:

$$
y_i = \begin{cases} r_i & \text{if episode terminates at step } i+1 \\ r_i + \gamma \max_{a'} Q(s_{i+1}, a'; \theta^-) & \text{otherwise} \end{cases}
$$

        Update θ by minimizing loss: $\mathcal{L}(\theta) = \frac{1}{|B|} \sum_i \left(Q(s_i, a_i; \theta) - y_i \right)^2$;
11   **if** *every C steps* **then**
12   Update target Q-network: $\theta^- = \theta$;

Query Selection. Since we have no definite knowledge on the characteristics of the processing of the ad hoc query, involving a similar history query for pre-training and assistance is an inspiring idea. With the representation vector $\mathcal{R}$ of the SQL plans for all history queries and ad hoc queries, we calculate the Euclidean distance between each history query and ad hoc query, and then find out the minimum one and take the related history query as the similar query. The distance is calculated as $d_{rp} = (v_1 \cdot \sum_{i=1}^{op}(x_i^r - x_i^p)^2 + v_2 \cdot \sum_{j=1}^{op}(y_j^r - y_j^p)^2 + v_3 \cdot (z_{size}^r - z_{size}^p)^2)^{\frac{1}{2}}$, where different vectors can be weighted by v_1, v_2, v_3 to adjust the importance of different parts.

Dual Sampling. As for the history query, there exist errors with the rQPM, while the history data also might not be accurate due to the system fluctuation. To strengthen the sampling quality and obtain more accurate rewards, the model samples both from the rQPM and history data when training on history query. If there is a history job with the same Spark configuration, the model will obtain two groups of results for performance indicators. Then the final performance indicators are obtained by weighted sum as $\text{final}(*) = v_1 \cdot *(\text{from history}) + v_2 \cdot *(\text{from rQPM})$, where $*$ represents one of the three performance indicators. If no identical history job, model samples from rQPM only.

Transfer Learning. The transfer learning module is introduced to reduce search costs and improve recommendation quality when target job cannot be directly executed. To accelerate the convergence of training, we transfer the agent pre-trained from history query to initialize the agent for ad hoc query training, so that the experience on similar queries can be utilized for ad hoc query tuning. Specifically, the weights θ_r of the Q-network are transferred to the new training together with randomly initialized weights θ'. The initialized weights θ are illustrated as $\theta = u_1 \cdot \theta_r + u_2 \cdot \theta'$, where u_1 and u_2 are the weights with $u_1 + u_2 = 1$. This design aims to use the experience appropriately.

5.3   Model Training

With the design of the basic model and optimization, DS-TDQN is trained as shown in Algorithm 3. The training progress includes initialization, action selection, experience replay, and network update, helping to stabilize learning and avoid overfitting to recent experiences. In action selection, the next action is selected randomly with probability ϵ or $\arg\max_a Q$. In network update, the target value y_i is calculated by Bellman function. The network is updated by minimizing the mean square error of Q and y_i. DS-DQN can be pre-trained offline with history, then configuration can be obtained from TDQN for target query.

6   Experiment Setup

In this section, we evaluate the performance of ZTune from three aspects: (1) The ability of QPM model fitting - verify the effectiveness of QPM; (2) The ability of configuration recommendation - test the effectiveness of configuration recommendation towards baseline methods, including traditional, industrial and state-of-the-art methods; (3) The ability of conditional time-resource balance - verify the effectiveness of multi-objective optimization with time criterion.

6.1   Experiment Setting

Environment. We exclusively use 192GB memory and 720 CPU cores of a large-scale Spark cluster in an industrial environment. Tests are repeated five times.

Datasets. We utilize industrial datasets and benchmarks to evaluate our proposed methods, including TPC-DS 500G, which can better simulate massive query scenarios in industrial applications. 99 queries are collected from industrial scenarios, representing different business requirements. We randomly select 9 queries as ad hoc queries for tuning, while the rest are treated as history. Baseline methods are conducted directly interacting with our cluster.

Parameters. A number of hyperparameters used in ZTune are set to definite values in the experiment. α, w_1, w_2 (Reward) are set to $1, 0.5, 0.5$. *const* (Reward) is set to the *app_duration* of the case with moderate resource allocation, (cores, memory, maxExecutors) = $(2, 2, 4)$. T, T', α (SA) and γ, λ (GB) are set to $1, 10^{-9}, 0.1$ / $0.1, 1$. v_1, v_2, v_3 (RL) are set to 1, which means that data size is the most critical, then operation number and parallelism. u_1, u_2 (RL) are set to 0.5 to balance experience transfer and random initialization.

6.2   Baseline Methods

Default Settings. We set a group of default configurations with `cores` $= 1$, `memory` $= 1$, and `maxExecutors` $= 1$ as default input for a job.

Random Search [3]. Random Search recommends random configurations.

Rule-based Method. Rule-based method is based on expert knowledge of Spark. Such rules are usually set to tune several key parameters. In [16], a series of expert rules is introduced for tuning.

CherryPick [1]. The core idea of CherryPick is to utilize Bayesian inference to construct a surrogate model of the objective function and conduct the tuning.

Online Tuning [11]. This method proposes an adaptive multi-objective online tuning method based on Bayesian optimization, effective on periodic jobs. Approximation gradient descent is introduced to accelerate convergence.

ZTune with DQN. DS-TDQN is a variant of DQN under the circumstance of Spark Tuning. To evaluate the improvement of the DS-TDQN model, we test ZTune with the original DQN as the searching phase to analyze the variation.

7   Performance Evaluation

7.1   QPM Effectiveness Analysis

We first evaluate the effectiveness of the designed query performance model by observing the accuracy of model regression. With QPM regression method in Sect. 4.1, we obtain the rQPMs of the selected queries including the 12 ϵ hyperparameters. Then we calculate the mean square error (MSE) and the average (AVG) for each performance indicator, and the error rates (Error) of the models. The rQPM results of the 9 ad hoc queries are shown in Table 3.

Table 3. Performance Evaluation of QPM: rQPMs of queries (Partial, Error Unit: %)

Query	ϵ_t^1	ϵ_t^2	ϵ_t^3	ϵ_t^4	ϵ_t^5	MSE_t	AVG_t	$Error_t$	ϵ_c^1	ϵ_c^2	ϵ_c^3	ϵ_c^4	MSE_c	AVG_c	$Error_c$	ϵ_m^1	ϵ_m^2	ϵ_m^3	MSE_m	AVG_m	$Error_m$
27	69.78	−1.23	0.00	4.21	16.07	1.87	25.26	7.41	13.87	0.97	0.00	101.96	135.09	495.12	27.29	2174.95	0.81	−15829.67	74448.12	493850.72	15.08
30	50.16	−1.38	0.00	3.49	21.36	2.47	27.22	9.08	7.31	1.08	0.17	107.22	110.45	673.44	16.40	1851.33	0.86	5542.64	88990.67	582460.40	15.28
43	69.18	−1.19	0.00	4.98	14.90	1.83	24.36	7.52	3.60	1.07	0.28	122.82	95.86	464.52	20.64	2399.57	0.79	−26446.83	68712.70	470018.23	14.62
47	387.73	−1.13	3.88	0.55	35.04	5.75	94.62	6.07	4.33	1.24	0.25	577.16	186.98	1567.84	11.93	1685.77	0.89	41095.55	193904.97	1708132.60	11.35
68	31.11	−1.50	0.00	9.89	16.77	1.80	20.09	8.95	1.94	0.81	0.97	−4.27	120.05	455.13	26.38	2090.94	0.82	−6263.97	69473.27	423099.93	16.42
71	43.15	−1.35	17.49	0.00	0.33	1.54	22.96	6.72	13.07	0.95	0.08	79.28	134.29	505.75	26.55	2245.26	0.80	−12261.58	75774.73	467457.61	16.21
72	861.71	−1.13	0.00	5.85	69.33	17.95	198.70	9.03	0.21	1.71	0.54	1457.32	2434.40	3211.10	75.81	1611.87	0.91	140534.64	640742.45	3662199.91	17.50
79	63.21	−1.28	0.00	3.81	17.70	1.98	25.69	7.71	1.65	0.94	0.83	56.44	160.82	537.06	29.94	2106.18	0.82	−10606.34	78291.51	515810.31	15.18
89	59.60	−1.25	0.00	10.00	15.70	1.84	23.46	7.85	2.75	0.68	1.13	−174.13	126.16	463.65	27.21	2234.39	0.81	−20621.52	78009.63	463827.12	16.82

The results in Table 3 imply a good regression outcome on all the queries. The regression error for different queries and performance indicators is restricted to a satisfactory ratio. The regression results of processing time for different queries mostly have an error rate of no more than 10 percent, Based on the results of processing time, the regression results of executor cores and memory also have a good regression error with cores lower than 30 percent and memory lower than 20 percent. This implies that our query performance model is effective for performance regression and prediction. Although there are still regression errors on absolute values, the equation structures can ensure the trend of performance variation so that the prediction phase will not deviate too much from the optimal.

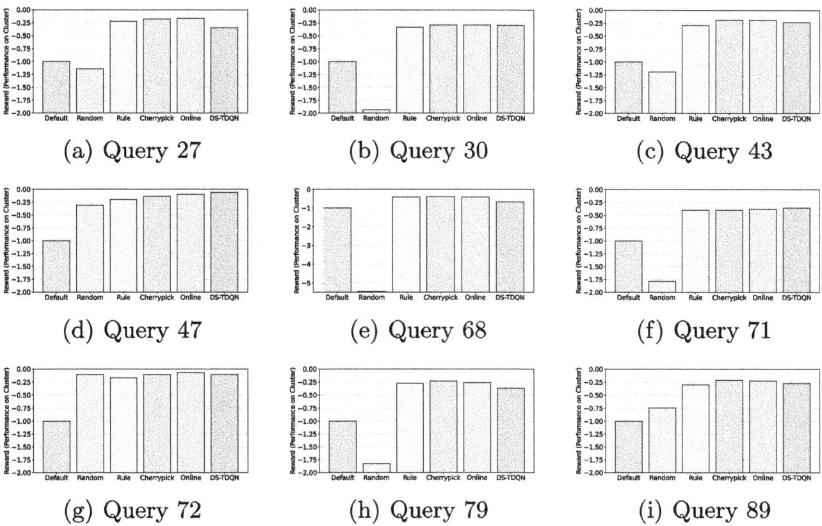

(a) Query 27   (b) Query 30   (c) Query 43

(d) Query 47   (e) Query 68   (f) Query 71

(g) Query 72   (h) Query 79   (i) Query 89

Fig. 4. Tuning Comparison on Rewards: ZTune towards Baseline Methods.

7.2 Configuration Recommendation Analysis

With effective QPM, we conduct comparison experiments on configuration recommendation evaluation, including proposed methods and other baseline methods introduced in Sect. 6.2. As shown in Fig. 4, Default is set as a criterion. Higher reward represents better comprehensive performance on processing time and resource occupancy. Random has large fluctuations on different queries, with mostly poor results due to the randomness. In contrast, Rule has better stability, results, and good iteration numbers on all queries since we modify the parameters of rules based on our recognition and experiment environment towards the SQL queries (QPM). Even more, CherryPick and Online have better rewards as adaptive tuning methods based on Bayesian optimization. Multiple iterations on real clusters enable them to tune more precisely so that a minor advantage towards DS-TDQN is achieved. In comparison, our proposed method DS-TDQN achieves balanced results on reward performance and interaction number. DS-TDQN has better results than Default and Random without interaction, while the results of DS-TDQN approach those of Rule, CherryPick, and Online within one run. DS-TDQN also has better results on queries 47, 71 than other methods.

7.3 Time-Resource Balance Analysis

We further analyze the performance control of processing time to evaluate the realization of the goal. Since the reward discussed above comes from weighted processing time and resources, it means that better resource saving contributes to worse time improvement. The processing time improvement ratio is the ratio of the processing time of different methods towards the processing time of default

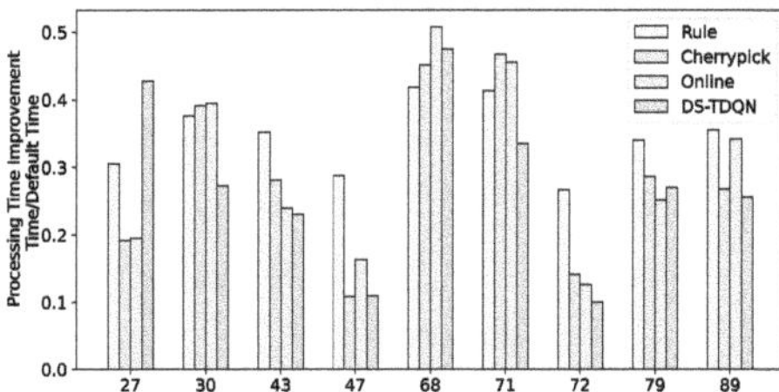

Fig. 5. The Improvement of Processing Time: The comparison of the processing time improvement ratios among Rule-based Method, CherryPick, Online Tuning, and ZTune (DS-TDQN).

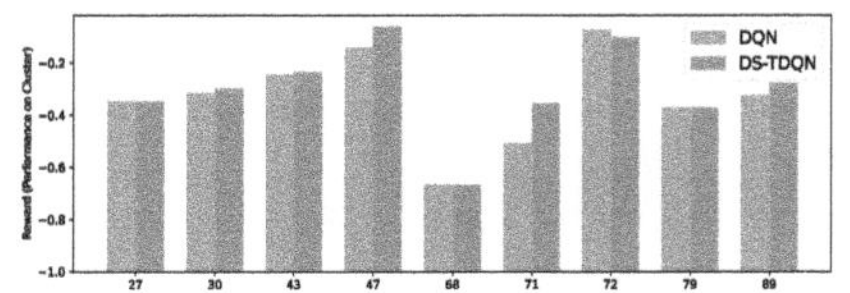

Fig. 6. Performance: DS-TDQN vs DQN

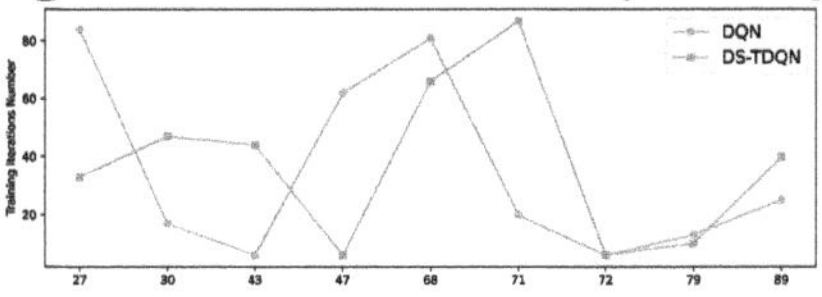

Fig. 7. Training Iters: DS-TDQN vs DQN

settings. Figure 5 shows that a lower ratio contributes to better processing time. From the results, we can find that ZTune (DS-TDQN) has the best performance on most cases, including queries 30, 43, 71, 72, 89. Performance on queries 47, 68, and 79 is also relatively better. Performance on query 27 is worse, but it is still acceptable, which means that the resource is saved much more in this case. In general, the conditional time-resource balance setting contributes to a better processing time compared to other methods due to a heavy penalty on time delay, simultaneously enabling suboptimal resource saving. The experiment results reach the expectation, i.e., time-resource balance with strict delay control.

7.4   Ablation Study

We conduct an ablation study to evaluate DS-TDQN's effectiveness, comparing DS-TDQN with DQN, which uses different methods in searching phase. From Fig. 6 we can find that DS-TDQN achieves better performance on queries 30, 43, 47, 71 and 89, while DS-TDQN and DQN have the same performance on queries 27, 68 and 79. DS-TDQN has worse performance on query 72 with little gap. Generally, DS-TDQN provides better rewards in most cases since dual sampling makes it possible to study more accurate tuning policy for transfer learning, so that a better Spark configuration can be recommended. The improvement of the algorithm is effective. Figure 7 presents the convergence of training DS-TDQN and DQN on different queries. DS-TDQN converges faster on queries 27, 47, 68, and 79, while DQN converges faster on queries 30, 43, 71, and 89. It depends on the probability of state transition and the transfer quality.

8   Related Work

With the development of distributed computing frameworks, a number of methods are proposed to optimize the processing performance. In general, these tuning methods are divided into six categories, including rule-based approach, cost modeling approach [2,7,12,15], simulated-based approach, experiment-driven approach, machine learning approach [10] and adaptive approach [11,16,18]. In these works, some methods are specially paid attention to and employed in Spark tuning in recent years. Iphicles [7] provides a DCN configuration framework that uses a GNN-based twin performance model for efficient and stable parameter tuning, improving flow completion time with minimal convergence time. [2] proposes a cost model for Spark SQL, providing an error of no less than 14 percents on processing time, while our results achieve a better result on processing time and extend it to resource occupancy index with an acceptable error. [5] proposes a multi-objective optimization model to recommend Spark configuration with a genetic algorithm for configuration updating and Adaboost for performance evaluation. [10] proposes a resource-aware deep learning model with attentional LSTM to predict the performance given the Spark SQL tasks. [6] studies data storage tuning on various hyper-parameters with deep learning and Bayesian optimization. In the most recent, online tuning has become a hotspot in studies. [11] proposes a new tuning framework based on Bayesian Optimization to solve function, cost, and efficiency problems. Some other works pay attention to specific scenarios to conduct the optimization. AutoExecutor [15] concentrates on tuning Spark executor with a machine learning model. Recent works make contribution in segmented fields, while there are still problems for application.

9   Conclusion

In this work, we concentrate on the problem of zero-shot executor tuning on ad hoc queries with balanced requirements of resource and time cost. We introduce ZTune, which integrates SQL plan representation, query performance model, model regression and prediction, and searching with DS-TDQN. ZTune enables high-quality configuration recommendations with SQL plan and history queries. The experiments imply that ZTune achieves better results towards baseline.

Acknowledgment. This work was supported by the National Key R&D Program of China [2024YFF 0617700], the National Natural Science Foundation of China [U23A20309, 62272302, 62502430, 62372296], and the ByteDance Research Project [CT20211123001686, CT20230320002435, CT20241217115379]. Dejun Kong and Xiaofeng Gao are with the Shanghai Key Laboratory of Scalable Computing and Systems. Xiuqi Huang finished most work at Shanghai Jiao Tong University. We also thank Tieying Zhang, Zuzhi Chen, Tiantian Wei, Ron Hu, Rong Kang, Binbin Chen, and Zhaowei Tan for their advice.

References

1. Alipourfard, O., Liu, H.H., Chen, J., Venkataraman, S., Yu, M., Zhang, M.: {CherryPick}: Adaptively unearthing the best cloud configurations for big data analytics. In: USENIX Symposium on Networked Systems Design and Implementation, pp. 469–482. USENIX Association (2017)

2. Baldacci, L., Golfarelli, M.: A cost model for spark sql. IEEE Trans. Knowl. Data Eng. **31**(5), 819–832 (2018)

3. Bergstra, J., Bengio, Y.: Random search for hyper-parameter optimization. J. Mach. Learn. Res. **13**(2), 281–305 (2012)

4. Chen, Y., Goetsch, P., Hoque, M.A., Lu, J., Tarkoma, S.: *d*-simplexed: adaptive delaunay triangulation for performance modeling and prediction on big data analytics. IEEE Transactions on Big Data **8**(2), 458–469 (2019)

5. Cheng, G., Ying, S., Wang, B.: Tuning configuration of apache spark on public clouds by combining multi-objective optimization and performance prediction model. J. Syst. Softw. **180**, 111028 (2021)

6. Dorier, M., et al.: Hpc storage service autotuning using variational-autoencoder-guided asynchronous bayesian optimization. In: 2022 IEEE International Conference on Cluster Computing, pp. 381–393. IEEE (2022)

7. Huang, S., Wang, M., Liu, Y., Liu, Z., Cui, Y.: Iphicles: tuning parameters of data center networks with differentiable performance model. In: 2024 IEEE/ACM 32nd International Symposium on Quality of Service, pp. 1–10. IEEE (2024)

8. Javaid, M.U., Kanoun, A.A., Demesmaeker, F., Ghrab, A., Skhiri, S.: A performance prediction model for spark applications. In: International Conference on Big Data, pp. 13–22. Springer (2020). https://doi.org/10.1007/978-3-030-59612-5_2

9. Kunjir, M., Babu, S.: Black or white? how to develop an autotuner for memory-based analytics. In: Proceedings of the 2020 ACM SIGMOD International Conference on Management of Data, pp. 1667–1683 (2020)

10. Li, Y., Wang, L., Wang, S., Sun, Y., Peng, Z.: A resource-aware deep cost model for big data query processing. In: IEEE International Conference on Data Engineering, pp. 885–897. IEEE (2022)

11. Li, Y., et al.: Towards general and efficient online tuning for spark. Proc. VLDB Endowment **16**(12), 3570–3583 (2023)

12. Li, Y., Lee, B.C.: Phronesis: efficient performance modeling for high-dimensional configuration tuning. ACM Trans. Archit. Code Optim. **19**(4), 1–26 (2022)

13. Qian, L., Luo, Z., Du, Y., Guo, L.: Cloud Computing: An Overview. In: Jaatun, M.G., Zhao, G., Rong, C. (eds.) CloudCom 2009. LNCS, vol. 5931, pp. 626–631. Springer, Heidelberg (2009). https://doi.org/10.1007/978-3-642-10665-1_63

14. Sagaama, H., Slimane, N.B., Marwani, M., Skhiri, S.: Automatic parameter tuning for big data pipelines with deep reinforcement learning. In: IEEE Symposium on Computers and Communications, pp. 1–7 (2021)

15. Sen, R., et al.: Autoexecutor: predictive parallelism for spark SQL queries. Proc. VLDB Endowment **14**(12), 2855–2858 (2021)

16. Shen, Y., et al.: Rover: An online spark sql tuning service via generalized transfer learning. In: ACM Conference on Knowledge Discovery and Data Mining, pp. 4800–4812 (2023)

17. Wu, Y., et al.: Towards resource efficiency: practical insights into large-scale spark workloads at bytedance. Proc. VLDB Endowment **17**(12), 3759–3771 (2024)

18. Xin, J., Hwang, K., Yu, Z.: Locat: Low-overhead online configuration auto-tuning of spark sql applications. In: Proceedings of the 2022 International Conference on Management of Data, pp. 674–684 (2022)

P-Raft: Distributed Consensus with Predictive Optimization Under Cross-Domain Sites

Yangyang Wang[1,2], Ziqian Cheng[1,2], Yucheng Ji[1,2], and Zichen Xu[1,2(✉)]

[1] School of Artificial Intelligence, Nanchang University, Nanchang, China
`yangyangwang@ncu.edu.cn`, `{Chengzq,JYC}@email.ncu.edu.cn`
[2] School of Mathematics and Computer Sciences, Nanchang University,
Nanchang, China
`xuz@ncu.edu.cn`

Abstract. As network scale and business concurrency continue to grow and data centers expand across multiple regions, cross-domain data processing has become the norm. In current distributed databases, leader-based consensus protocols face the challenge of high latency when processing cross-domain requests. To address this, we present P-Raft, a Raft-based protocol tailored for cross-domain sites that shortens the cross-domain commit's critical path and employs machine-learning–based, proactive leader migration to adapt to network topology and load shifts, thereby improving performance and reducing global latency. Under representative read/write-balanced workloads, P-Raft reduces average latency by 75.19% and 72.61% relative to Raft and EPaxos, respectively; compared with the state-of-the-art leader-management approach GeoLM, P-Raft reduces average latency by 63.89%.

Keywords: Big Data · Large-scale System · Optimization · Distributed System · Consensus

1 Introduction

Distributed databases, valued for their scalability and availability, are widely used in large-scale computing that combines data and task parallelism. Unfortunately, this typically requires frequent data synchronization across different devices and even across geographically separated sites to ensure strong consistency. This cross-domain synchronization imposes a simple yet stringent requirement on existing consensus protocols: to provide a more efficient consistency framework under long-distance conditions.

In leader-based consensus protocols, the commit phase requires the leader to obtain quorum acknowledgments from replicas geographically dispersed across domains. This wide-area commit path introduces long round-trip times (RTTs), making communication optimization critical for performance. Equally important, leader placement determines both the number and length of cross-domain

H. Jung et al. (Eds.): DASFAA 2026, LNCS 16536, pp. 613–628, 2026.
https://doi.org/10.1007/978-981-92-0366-6_37

round trips; thus, a principled leader-selection strategy is key to reducing global latency.

We introduce P-Raft, a Raft-based protocol designed for cross-domain sites. P-Raft combines Fast Commit, the Optimal Leader Domain Evaluation Model, and Prediction-Driven Leader Migration. First, Fast Commit compresses the commit path to a majority within the leader's domain, moving cross-domain replication and acknowledgment off the commit's critical path. Second, using a lightweight evaluation model over inter-domain latency and per-domain read/write load, P-Raft selects the optimal leader domain online. Third, leveraging machine-learning load predictions, P-Raft proactively drives leader migration to the evaluated optimal domain.

We have implemented the P-Raft prototype in a distributed database based on ETCD [1] and rigorously evaluated it using the YCSB [2] benchmark. Under a representative read/write-balanced workload, P-Raft reduces average latency by 75.19% and 72.61% compared to Raft [3,4] and EPaxos [5], respectively. Compared to GeoLM [6], it reduces average latency by 63.89%. These results show that P-Raft delivers substantial performance gains.

The main contributions of this paper are as follows:

(1) *We identify the core performance bottlenecks of leader-based distributed consensus in cross-domain sites.* During the commit phase, the leader must wait for cross-domain replication and quorum acknowledgment, and the current leader lacks dynamic awareness of network and load. The combination of the two results in significant end-to-end latency overhead.
(2) *We propose P-Raft, a protocol tailored for cross-domain sites.* P-Raft reduces global latency through Fast Commit, the Optimal Leader Domain Evaluation Model, and Prediction-Driven Leader Migration.
(3) *We implemented a prototype on ETCD and validated its performance through extensive experiments using the YCSB benchmark.* Under read/write-balanced workloads, P-Raft reduces average latency by 63.89%−75.19% compared with Raft, EPaxos, and GeoLM.

This paper is organized as follows. Section 2 introduces the background and motivation of our research. Section 3 presents the specific design of our proposed P-Raft. Section 4 describes the evaluation of P-Raft, and Sect. 5 introduces related work. Finally, Sect. 6 concludes this paper.

2   Background and Motivation

2.1   The Latency Bottleneck of Leader-Based Consensus Under Cross-Domain Sites

Leader-based consensus protocols are the dominant paradigm in engineering practice, with Raft [3,4] and Multi-Paxos [7] as canonical representatives. Raft is well known for its simplicity and ease of deployment compared with Paxos [8]; since its introduction in 2014, it has been adopted by many real-world systems such as ETCD [1], CockroachDB [9], Kudu [10], TiDB [11], LogCabin [12], Dragonboat [13], NebulaGraph [14] and PolarDB [15].

In such leader-based protocols, the leader serves as both the client-facing entry point and the coordinator of data replication, while the remaining nodes are followers that replicate data from the leader. Under cross-domain sites, Raft uses majority-quorum commit: the leader must collect acknowledgments from a majority of nodes before an entry can be committed. Because nodes are spread across geographic domains, cross-domain replication and acknowledgments lie on the commit's critical path, directly adding wide-area RTTs to the end-to-end latency.

As illustrated in Fig. 1, the client sends a write to the leader (Step ①). The leader appends the request to its local persistent log and forwards the log entry in parallel to all followers (Step ②). Followers append the entry and return acknowledgments (Step ③). Upon receiving acknowledgments from a majority, the leader deems the entry committed and replies to the client (Step ④). Under cross-domain sites, Steps ②–③ typically traverse cross-domain links and therefore dominate the overall latency.

Two direct observations follow: **(i)** the end-to-end latency of writes is primarily dominated by cross-domain link latency; **(ii)** the leader's domain determines both the number and the length of cross-domain paths, thereby fixing the effective number of RTTs required to reach a majority and the bottleneck link. This reveals two complementary levers for reducing global latency under cross-domain sites: shortening the commit's critical path and optimizing leader placement.

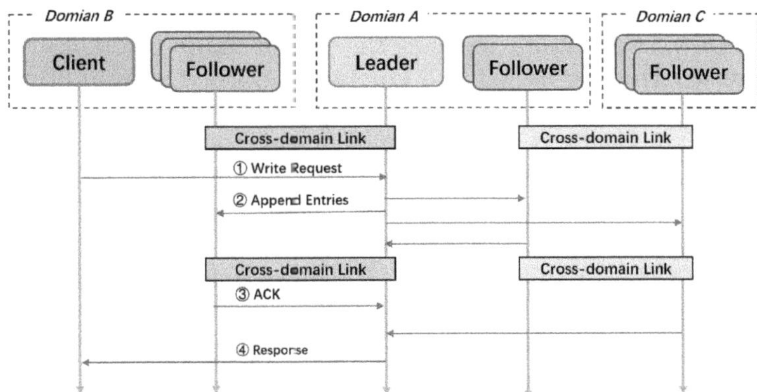

Fig. 1. Write request processing flow of the leader-based consensus protocol in cross-domain sites.

2.2 Leader-Based Consensus Protocols Lack Dynamic Awareness of Network and Request Load

Consensus operates over long-distance links under cross-domain sites, where both network conditions and workload patterns are non-stationary. Raft's leader is elected via timeout-triggered randomized elections and thus lacks dynamic awareness of network and request load, making it difficult to sustain high performance across domains.

At the same time, domain-level workloads (e.g., application intensity, business rhythms, holidays, and seasonal fluctuations) exhibit pronounced temporal correlation and periodic structure [16]. Based on this fact, we employ time-series regression to perform short-horizon predictions of per-domain read/write intensity, capturing imminent hotspot migrations and their intensity changes.

Inter-domain network latency is continuously probed online to ensure robust awareness of current link conditions. Leveraging the predicted load profile and the real-time, statistically maintained topology of inter-domain latencies, the system evaluates candidate leader domains online and proactively migrates the leader, achieving sustained adaptive optimization under dynamic network and load conditions.

2.3   Challenges

Challenge #1 How to Compress the Commit's Critical Path? To move cross-domain replication and acknowledgment off the commit's critical path, P-Raft only requires acknowledgment from a majority of nodes in the leader's domain to respond to the client; replicas in other domains catch up asynchronously. To ensure safety and availability, we constrain the election process to ensure that the newly elected leader must contain the committed log to avoid commit loss and rollback. Through the above strategies, the commit path collapses to a single-domain majority, while cross-domain links handle background replication only, substantially reducing the system end-to-end latency.

Challenge #2 How to Select the Optimal Leader Domain? Under cross-domain sites, the optimal leader domain depends on the latency topology and the load distribution. Accordingly, we build a lightweight model that evaluates candidate domains by combining the latency topology with the load distribution. The model selects the domain that minimizes the average latency of the system as the target leader domain.

Challenge #3 How to Control the Prediction Horizon? Prediction and leader migration both incur costs, and the overall cost increases with the prediction frequency. Thus, an overly long prediction horizon slows the prediction frequency and risks missing shifts in network and workload conditions; conversely, an overly short horizon raises the prediction frequency, amplifying computational overhead and potentially increasing the number of leader migrations. To balance these factors, we adopt the adaptive prediction horizon strategy: a variable prediction horizon serves as the look-ahead window, and the system automatically triggers a new prediction after the window expires. When short-term dynamics are pronounced, the horizon is shrunk to improve responsiveness; when conditions are stable, it is expanded to suppress unnecessary evaluations and migrations. This mechanism can continuously and adaptively match dynamic networks and loads at a controllable cost.

3   Design of P-Raft

This section introduces, in order: Fast Commit, the Optimal Leader Domain Evaluation Model, and Prediction-Driven Leader Migration.

3.1   Fast Commit

As discussed in Sect. 2.1, under cross-domain sites, Raft's majority-quorum commit must traverse cross-domain links to obtain a global majority; these cross-domain links lie on the commit's critical path and become the dominant performance bottleneck.

P-Raft introduces a Fast Commit mechanism. This mechanism compresses the commit quorum to a majority within the leader's domain. Once a majority of nodes in the leader domain have durably appended the entry, the leader can immediately respond to the client, while followers in other domains are replicated asynchronously.

Figure 2 illustrates a nine-node cross-domain database, used to explain the workflow of Fast Commit. The client sends a write to the leader (Step ①). After appending the log entry locally, the leader replicates it in parallel to all followers (Step ②). Once the leader receives acknowledgments from a majority of nodes within its domain (Step ③), it commits the entry and immediately responds to the client (Step ④). Data replication for followers in other domains proceeds asynchronously in the background (indicated by the dotted lines).

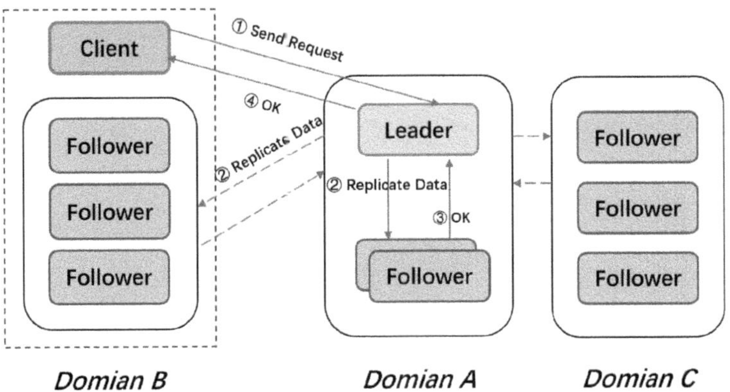

Fig. 2. The workflow of Fast Commit.

To ensure reliability and safety under Fast Commit, P-Raft constrains elections to the leader domain: initially, the system randomly selects a leader within one domain; subsequently, only followers in that domain may initiate and participate in elections, and candidates must receive votes from a majority of nodes within the domain to become leader. This safety constraint guarantees that any new leader necessarily contains all committed log entries.

In addition, P-Raft supports proactive leader migration: when the evaluation model indicates an optimal domain, that domain becomes the new target leader domain and the leader migrates to it.

Correctness. We developed a formal specification of the P-Raft protocol described in this section and proved its correctness. The specification is written in the TLA+ specification language, and we used the TLA+ proof system [17] to mechanically verify the correctness of P-Raft. The TLA+ specification is available at: https://github.com/Qranger-new/P-Raft-TLA.git.

3.2 Optimal Leader Domain Evaluation Model

Because both client request distribution and cross-domain latency vary among domains, leader domain determines the number and length of cross-domain round trips; therefore, the leader domain directly affects the overall cross-domain system latency. To determine the leader domain that minimizes future overall cross-domain system latency, we estimate the overall cross-domain system latency using the predicted client request load and the inter-domain latency matrix. Table 1 summarizes the symbols used in this model.

Table 1. Symbols used in the Optimal Leader Domain Evaluation Model.

Symbol	Description
l_{ij}	Latency between Domain i and Domain j
L_i	Cross-domain latency of operations in Domain i
W_i	Number of write requests from clients in Domain i
R_i	Number of read requests from clients in Domain i
$L_s(i)$	Overall cross-domain system latency when the leader resides in Domain i
N	Number of domains
z	Leader's domain
$\lambda(t_0, t_1)$	Predicted request statistics between (t_0, t_1)
Δ	Prediction horizon

Based on the predicted request volume of each domain (W_i, R_i) and the inter-domain latency (l_{ij}), we compute the overall cross-domain system latency under different domains and migrate the leader to the optimal domain to minimize system latency.

Assume the cluster consists of N domains, and the leader resides in domain z ($z \in [1, N]$). For any domain i ($i \in [1, N]$), the client's cross-domain latency L_i can be divided into two cases:

Case 1: The Client and the Leader are in the Same Domain ($i = z$). When the client and the leader are colocated, the leader commits write requests locally without cross-domain communication. Since read requests are also served locally, the cross-domain latency can be expressed as:

$$L_i(z) = 0. \tag{1}$$

Case 2: The Client and the Leader are in Different Domains ($i \neq z$). When the client and the leader are in different domains, the local domain node forwards the write request to the leader. The leader performs a Fast Commit within its domain and then immediately responds to the client. For reads, the request is sent to the leader; the leader serves the read from its local state and responds to the client. Consequently, since both reads and writes incur exactly one cross-domain communication, the latency for domain i becomes:

$$L_i(z) = 2 \cdot l_{iz} \cdot (W_i + R_i). \tag{2}$$

Overall Cross-Domain System Latency. The overall cross-domain system latency L_s can then be expressed as:

$$L_s(z) = \sum_{i=1}^{N} L_i(z), \quad i \in [1, N]. \tag{3}$$

By selecting the leader domain z that minimizes L_s, the total cross-domain communication cost is reduced, thereby improving system performance.

$$z_{new} = \arg \min_{z \in \{1,\dots,N\}} L_s(z). \tag{4}$$

3.3 Prediction-Driven Leader Migration

To adapt to dynamic and fluctuating loads, we employ machine-learning-based load prediction to guide proactive leader migration. Specifically, the leader continuously records the request statistics from each domain. At each prediction horizon Δ, the system takes the collected data as input and predicts the future request sequence $\lambda(t, t + \Delta)$.

The predicted distribution $\lambda(t, t + \Delta)$ is fed into the optimal leader domain evaluation model of Sect. 3.2, which directly selects the optimal domain z_{new}. If z_{new} differs from the current leader domain, the leader proactively migrates to domain z_{new}.

Adaptive Prediction Horizon. Both machine-learning prediction and leader migration introduce non-negligible overheads. A short prediction horizon adds extra computation and migration overhead, but a long prediction horizon delays the system's response to fluctuating loads (see Sect. 4.4).

To balance responsiveness and stability, we design an adaptive prediction horizon mechanism. Let the initial prediction horizon be Δ (significantly larger than a single migration cost). The horizon is always clipped within the range

$[\Delta_{\min}, \Delta_{\max}]$, where $\Delta_{\min}$ and $\Delta_{\max}$ are determined by the business logic. Formally, $\Delta \leftarrow \text{clip}(\Delta, \Delta_{\min}, \Delta_{\max})$.

After each prediction, the system compares the optimal leader domains derived from the first and second halves of the horizon ($\Delta/2$ each). If the two results differ, the prediction horizon is shortened to improve responsiveness according to:

$$\Delta_{\text{New}} = \max(\Delta/2, \Delta_{\min}).$$

Example 3.1. As shown in Fig. 3, the top part illustrates the case where the prediction horizon is shortened. The current horizon is $\Delta_{\text{Current}} = 2$. The prediction for the first half $\lambda(t, t + \Delta/2)$ selects Domain A as optimal, while the second half $\lambda(t + \Delta/2, t + \Delta)$ selects Domain C. Because the optimal domains of the two halves differ, the system shortens the next horizon, setting $\Delta_{\text{New}} = \Delta_{\text{Current}}/2 = 1$, and selects A as the next leader domain.

Otherwise, the system evaluates an additional sequence of length $\Delta/2$, i.e., $\lambda(t + \Delta, t + \Delta + \Delta/2)$. If the optimal domain remains consistent, the prediction horizon is extended to suppress unnecessary re-evaluations according to:

$$\Delta_{\text{New}} = \min(1.5 * \Delta, \Delta_{\max}).$$

Example 3.2. As shown in Fig. 3, the bottom part illustrates the case where the prediction horizon is extended. The two half-horizons $\lambda(t, t + \Delta/2)$ and $\lambda(t + \Delta/2, t + \Delta)$ select both domain B. So We evaluate one more half-horizon, $\lambda(t + \Delta, t + 1.5 * \Delta)$, and it still selects B. Since B remains optimal at $(t, t + 1.5 * \Delta)$, the system sets $\Delta_{\text{New}} = 1.5 * \Delta_{\text{Current}} = 3$ and selects B as the leader domain.

Otherwise, if the two half-horizons select the same leader but the additional half selects a different one, we maintain the current horizon and set:

$$\Delta_{\text{New}} = \Delta.$$

The adaptive prediction horizon mechanism dynamically controls the prediction horizon, achieving a balance between responsiveness and stability, and continuously drives the system toward lower overall cross-domain system latency.

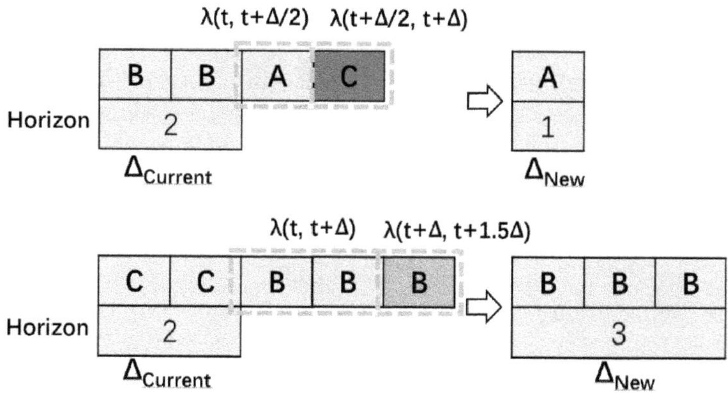

Fig. 3. Illustration of the Adaptive Prediction Horizon.

4   Implementation and Evaluation

4.1   System Implementation

We implemented a P-Raft prototype system based on ETCD, employing gRPC as the distributed communication framework. We integrate an LSTM-based time-series model [18] to predict cross-domain request load.

4.2   Experimental Setup

Environment and Parameters. Performance evaluation experiments for P-Raft were conducted on the Huawei Cloud platform [19]. The experimental cluster consisted of 12 nodes in total, including 9 server nodes and 3 client nodes. Each node was configured with 2 vCPUs, 4 GB of memory, and a 40 GB SSD, running CentOS 7.6 (64-bit). The experiment involved three geographical domains, each comprising three server nodes and one client node. To emulate cross-domain sites, we introduced network latency using the Linux Traffic Control (tc) utility with the netem module [20].

Workloads. We evaluated P-Raft across varied configurations with YCSB [2], carrying out comprehensive comparative experiments. The evaluated workloads include Load (insert-only), Workload A (50:50 read/update), Workload B (95:5 read/update), and Workload C (read-only); the workload definitions are listed in Table 2. Each key-value pair contains a 16 B key and a 1 KB value by default.

Table 2. The YCSB workloads used in evaluations.

Workload	Write Type	Query Type	Category
Load	Insert	/	Insert Only
A	Update	Point Query	50%update / 50%read
B	Update	Point Query	5%update / 95%read
C	/	Point Query	Read Only

Baselines. To comprehensively assess the improvements of P-Raft, we compared it against several representative consensus protocols:

- *Raft* [3,4]: a widely deployed leader-based consensus protocol ensuring strong consistency.
- *Mencius* [21]: a rotating-leader consensus protocol that balances load across replicas while maintaining strong consistency.
- *EPaxos* [5]: a leaderless consensus protocol that achieves low latency by executing non-conflicting requests in parallel.
- *GeoLM* [6]: a leader management strategy designed to improve system performance in cross-domain sites by adapting to network and load changes.
- *P-Raft*: a prediction-driven consensus protocol that migrates leaders dynamically across domains.

Experimental Configuration. Unless otherwise specified, all experiments were conducted under the three-domain configuration. The inter-domain RTTs were set to 30 ms, 40 ms, and 50 ms, respectively. Before each experiment, clients preloaded 1 million key–value pairs. To emulate cross-domain hotspot rotation, phased load scripts were used to impose peak pressure on different domains over time. The hotspot domain rotated every 10 s, and the read/write ratio was fixed at 1:1.

4.3   Overall Performance

Method. The overall performance experiment is conducted under the default three-domain configuration. We evaluate P-Raft and compare its average operation latency against Raft, Mencius, EPaxos, and GeoLM under the four workloads.

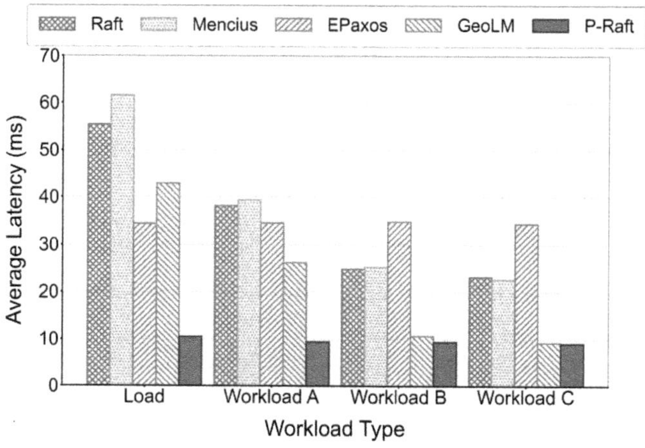

Fig. 4. Comparison of latencies among Raft, Mencius, EPaxos, GeoLM and P-Raft under YCSB workloads.

Results. Figure 4 shows the average latency of each protocol under the four workloads. As a rotating-leader protocol, Mencius achieves an average performance close to Raft's domain-averaged performance. EPaxos's average latency is similar across the four workloads. GeoLM shows similar read latency to P-Raft, but its write latency remains noticeably higher. P-Raft's read and write latency is significantly lower than all baselines across the four workloads.

Under heavy write loads (*Load* and *Workload A*), P-Raft consistently achieves the lowest average latency among all compared protocols. Under *Load*, relative to Raft, Mencius, EPaxos, and GeoLM, the average latency of P-Raft decreases by 81.08%, 82.97%, 69.58%, and 75.50%, respectively. Under *Workload A*, the corresponding reductions are 75.19%, 75.92%, 72.61%, and 63.89%, respectively.

Under read-dominant workloads (*Workload B* and *Workload C*), P-Raft achieves consistently lower latency. Under *Workload B*, relative to Raft, Mencius, EPaxos, and GeoLM, the average latency of P-Raft decreases by 61.88%, 62.62%, 72.93%, and 11.76%, respectively. Under *Workload C* (read-only), the corresponding reductions relative to Raft, Mencius, and EPaxos are 60.67%, 59.79%, and 73.56%, respectively.

Summary. Under the four workloads, P-Raft's advantages are most pronounced under write-heavy and mixed read/write loads; in read-intensive scenarios, it still achieves notable latency reductions. These findings align with our design goal: by predicting short-term workload shifts and combining them with the current inter-domain latency, P-Raft proactively migrates the leader to the predicted optimal domain and performs local commits, thereby removing cross-domain links from the commit path and reducing average latency.

4.4   Impact of Prediction Horizon

Method. Under the default three-domain configuration, we systematically evaluate the influence of the prediction horizon. After the model stabilizes, we measure the average system latency under prediction horizons of 5 s, 10 s, and 15 s, shown in Fig. 5.

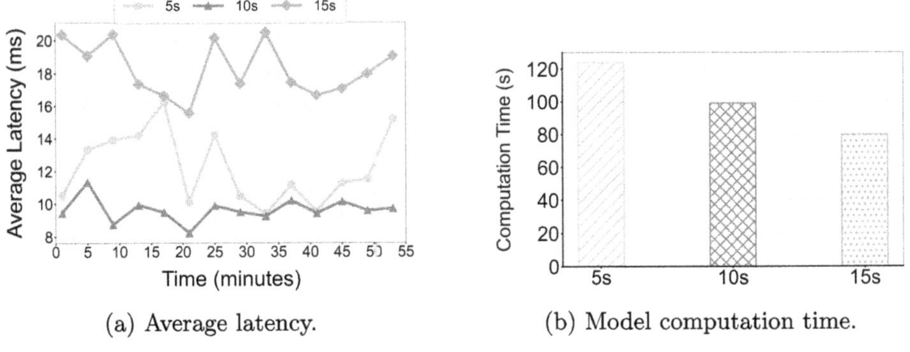

(a) Average latency.   (b) Model computation time.

Fig. 5. Impact of different prediction horizons (5 s, 10 s, 15 s) on system performance.

Results. As shown in Fig. 5(a), the system exhibits distinct latency variations under different prediction horizons. In the 5 s configuration, latency occasionally approaches that of the 10 s configuration but fluctuates significantly overall. The short prediction horizon causes frequent predictions and leader migrations, increasing model computation time and leading to instability. As shown in Fig. 5(b), its model computation time is about 25.1% higher than that of the 10 s configuration.

In the 15 s configuration, the horizon is overly long and fails to align with the hotspot migration frequency. As a result, the commit path often crosses inter-domain links for replication and confirmation.

In contrast, the 10 s configuration achieves the best overall performance and stability. The 10 s horizon matches the hotspot migration frequency, allowing the leader to react promptly to load shifts while avoiding redundant migrations and predictions. Statistically, the average latency of the 10 s configuration is 20.81% lower than that of 5 s and 47.20% lower than 15 s.

Summary. The prediction horizon should be close to the hotspot migration frequency. An excessively small horizon leads to instability, while an overly large one amplifies cross-domain latency. A horizon matched to the migration frequency maintains the lowest average latency, minimal fluctuation, and controllable model computation time under dynamic workloads.

4.5   Model Iteration

Method. We run this experiment under the standard three-domain configuration. The online model is updated every 5 min and uses an LSTM-based prediction model to predict future load. In each cycle, a prediction is counted as correct when its leader domain prediction equals the later validated optimum; accuracy is the fraction of correct predictions.

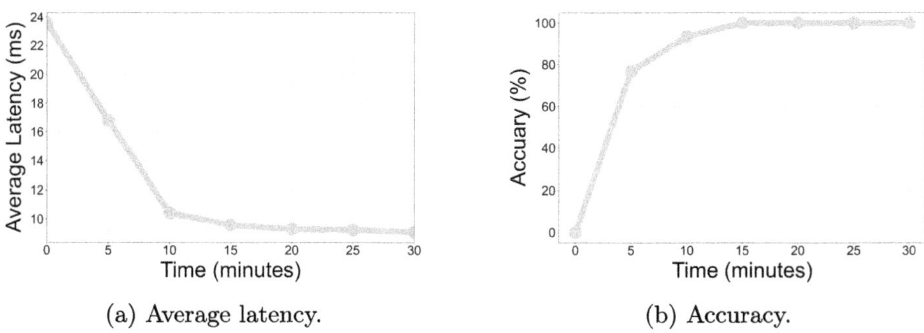

(a) Average latency.   (b) Accuracy.

Fig. 6. Model Iteration: Latency and Accuracy.

Results. Figure 6(a) shows the average system latency dropping from an initial 23 ms: it steps down markedly after the first update, enters a stable 9 ms band after the second update, and remains flat in subsequent cycles. The accuracy trend in Fig. 6(b) mirrors this decrease: it rises to 76% after the first update (5 min), approaches 93% after the second update (10 min), and then stabilizes at 100%.

Analysis. These outcomes indicate that machine-learning-based prediction is naturally effective under strongly periodic workloads. This is because the regular temporal structure of hotspot migration allows the model to reliably capture the recurring pattern, reaching stable performance within a short adaptation window.

4.6 Adaptive Prediction Horizon

Method. We evaluate four strategies under the default three-domain configuration: fixed 5 s, fixed 15 s, fixed 30 s prediction horizon, and adaptive prediction horizon. The system runs through three sequential phases in which the hotspot migration frequency is 10 s, 20 s, and 30 s, respectively; each phase lasts 20 min. Throughout the run, we record the time series of the average latency of the system (see Fig. 7).

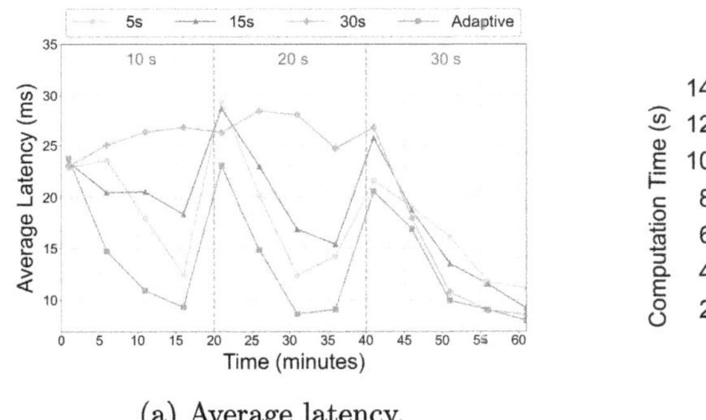

(a) Average latency.

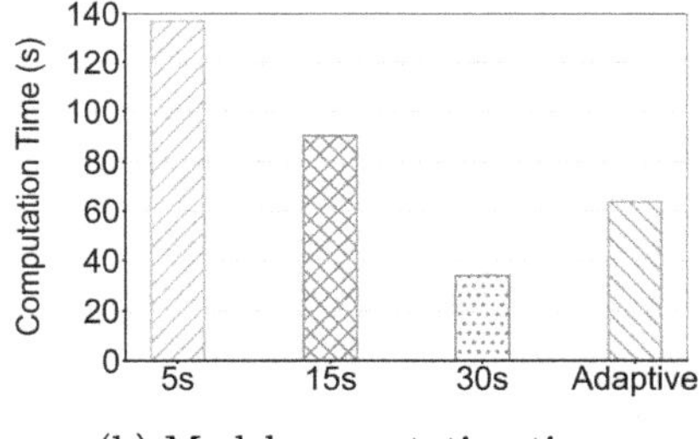

(b) Model computation time.

Fig. 7. Latency over time under four strategies (fixed 5 s vs. fixed 15 s vs. fixed 30 s vs. adaptive) across three hotspot-rotation phases (10 s, 20 s, 30 s).

Results. Under the three hotspot migration phases (10 s / 20 s / 30 s), the adaptive prediction horizon keeps the latency curve lower and smoother over the whole run. On average, Adaptive strategy reduces system latency by 22.95%, 27.59%, and 37.30% compared with the fixed 5 s, 15 s, and 30 s horizons, showing that the adaptive strategy consistently delivers lower average latency.

Fixed 5 s. Predictions and leader moves are triggered too frequently, disturbing the system and increasing model computation time. As Fig. 7(b) shows, its model computation time is about 2.14 times that of Adaptive strategy, which slows model convergence and degrades overall performance.

Fixed 15 s. As a middle-ground choice, the horizon is too long in the 10 s phase, so leader migration lags the hotspot and average latency rises. It improves at 20 s but is still sub-optimal; at 30 s—where the horizon equals half of the hotspot migration frequency—it performs relatively better.

Fixed 30 s. In the 10 s and 20 s phases, the horizon is too long, leading to very poor performance. Only in the 30 s phase, when it matches the hotspot migration frequency, does it converge quickly and remain stable.

Model Computation Time and Convergence. From Fig. 7(b), Adaptive strategy's model computation time is about 53.4% lower than the fixed 5 s strategy and 29.3% lower than the fixed 15 s strategy, though 87.5% higher than the fixed 30 s strategy. This shows that Adaptive strategy achieves a better trade-off: the fixed 30 s strategy has the lowest evaluation time but unacceptable performance, while the Adaptive strategy yields substantial latency gains at a low model computation cost. Meanwhile, Fig. 7(a) shows the model converges within two iterations and remains stable. This matches Experiment 4.5: an appropriate prediction horizon speeds up and stabilizes model iteration, so Adaptive strategy also outperforms fixed horizons in training quality.

Summary. In summary, the experimental results confirm the superiority of the adaptive prediction horizon mechanism in dynamic environments. It automatically adjusts the prediction horizon based on system feedback, allowing the system to maintain low latency and stable performance under multi-phase and non-uniform workload migrations.

5   Related Work

In cross-domain sites, existing research generally focuses on two main directions:

Quorum and Communication Path Optimization. WPaxos [22] and Flexible Paxos [23] relax quorum selection requirements, thereby reducing cross-domain communication costs in principle. Fast Paxos [24] minimizes latency by allowing clients to communicate directly with all replicas. Speculative Paxos [25] and Network-Ordered Paxos [26] rely on dedicated networks to ensure message ordering, which limits their general applicability. CURP [27] and EPaxos [5] determine dependencies among requests; when conflicts occur, they require additional communication to resolve them.

Leader-Oriented Optimization. PigPaxos [28] adopts a hierarchical structure to reduce cross-domain communication. Mencius [21] rotates the leader role to balance the workload. Raft-Plus [29] elects leaders based on network quality and node performance. DPaxos [30] mitigates cross-domain access frequency by placing data partitions close to the request sources. GeoLM [6] introduces performance-oriented leader management to adapt to dynamic network and workload conditions. However, in these protocols, cross-domain acknowledgments remain on the critical commit path and thus dominate overall latency.

P-Raft optimizes both aspects: it removes cross-domain replication and acknowledgment from the commit's critical path, and it dynamically migrates the leader to the optimal domain based on network topology and predicted workload dynamics, thereby systematically reducing end-to-end latency.

6   Conclusion

Geo-distributed databases span multiple domains. Cross-domain links are long, variable, and shared, and demand shifts across sites. In leader-based consensus, the commit phase still crosses domains to reach a quorum, so cross-domain communication and leader placement both matter for latency. Prior leader-management schemes react slowly and keep cross-domain links on the request commit path.

We present P-Raft, a Raft variant for cross-domain sites that combines Fast Commit, the Optimal Leader Domain Evaluation Model, and Prediction-Driven Leader Migration. Fast Commit keeps the commit path within the leader's domain. The evaluation model and migration mechanism collaboratively migrate the leader according to the predicted load and the current inter-domain latency. These components reduce the overall system latency while preserving Raft's safety and availability. We prove its correctness through a formal TLA+ specification.

Overall evaluation shows that P-Raft performs well across diverse workloads, consistent with its design goal of shortening the commit's critical path through Fast Commit, the Optimal Leader Domain Evaluation Model, and Prediction-Driven Leader Migration. Model Iteration experiments demonstrate that the model converges rapidly, enabling the system to maintain low, stable latency over time. Adaptive prediction horizon experiments further confirm that the mechanism adapts effectively under fluctuating network and load conditions.

Acknowledgements. This work is supported by the National Key R&D Program of China under Grant 2022YFB4501703, the Jiangxi Provincial Career-Early Young Scientists and Technologists Cultivation Project under Grant 20252BEJ730003, and the Jiangxi Provincial Natural Science Foundation under Grant 20252BAC200615.

References

1. ETCD (2025). https://github.com/etcd-io/etcd
2. Cooper, B.F., Silberstein, A., Tam, E., Ramakrishnan, R., Sears, R.: Benchmarking cloud serving systems with YCSB. In: Proceedings of the 1st ACM Symposium on Cloud Computing, pp. 143–154 (2010)
3. Ongaro, D., Ousterhout, J.: In search of an understandable consensus algorithm. In: 2014 USENIX Annual Technical Conference (USENIX ATC 14), pp. 305–319 (2014)
4. Ongaro, D.: Consensus: Bridging Theory and Practice. Stanford University (2014)
5. Moraru, I., Andersen, D.G., Kaminsky, M.: There is more consensus in egalitarian parliaments. In: Proceedings of the Twenty-Fourth ACM Symposium on Operating Systems Principles, pp. 358–372 (2013)
6. Xu, D., et al.: GeoLM: performance-oriented leader management for geo-distributed consensus protocol. In: IEEE INFOCOM 2025-IEEE Conference on Computer Communications. IEEE, pp. 1–10 (2025)
7. Lamport, L.: Paxos made simple. ACM SIGACT News (Distributed Computing Column) 32, 4 (Whole Number 121, December 2001), pp. 51–58 (2001)
8. Leslie, L.: The part-time parliament. ACM Trans. Comput. Syst. **16**, 133–169 (1998)
9. CockroachDB (2025). https://github.com/cockroachdb/cockroach

10. Kudu (2025). https://kudu.apache.org/
11. TiDB (2025). https://github.com/pingcap/tidb
12. Logcabin (2025). https://github.com/logcabin/logcabin
13. Dragonboat (2025). https://github.com/lni/dragonboat
14. NebulaGraph (2025). https://github.com/vesoft-inc/nebula
15. Cao, W., et al.: PolarDB serverless: a cloud native database for disaggregated data centers. In: Proceedings of the 2021 International Conference on Management of Data, pp. 2477–2489 (2021)
16. Aghili, R., Qin, Q., Li, H., Khomh, F.: Understanding web application workloads and their applications: systematic literature review and characterization. In: 2024 IEEE International Conference on Software Maintenance and Evolution (ICSME), pp. 474–486. IEEE (2024)
17. Cousineau, D., Doligez, D., Lamport, L., Merz, S., Ricketts, D., Vanzetto, H.: TLA+ Proofs. In: International Symposium on Formal Methods, pp. 147–154. Springer (2012)
18. Hochreiter, S., Schmidhuber, J.: Long short-term memory. Neural Comput. **9**(8), 1735–1780 (1997)
19. Huawei: Huawei cloud: Everything as a service (2025). https://www.huaweicloud.com/intl/en-us/
20. Traffic Control (TC) and Network Emulation (netem) (2025). https://man7.org/linux/man-pages/man8/tc.8.html
21. Mao, Y., Junqueira, F.P., Marzullo, K.: Mencius: building efficient replicated state machines for WANs (2008)
22. Ailijiang, A., Charapko, A., Demirbas, M., Kosar, T.: WPaxos: wide area network flexible consensus. IEEE Trans. Parallel Distrib. Syst. **31**(1), 211–223 (2019)
23. Howard, H., Malkhi, D., Spiegelman, A.: Flexible Paxos: quorum intersection revisited. arXiv preprint arXiv:1608.06696 (2016)
24. Lamport, L.: Fast Paxos. Distrib. Comput. **19**, 79–103 (2006)
25. Ports, D.R., Li, J., Liu, V., Sharma, N.K., Krishnamurthy, A.: Designing distributed systems using approximate synchrony in data center networks. In: 12th USENIX Symposium on Networked Systems Design and Implementation (NSDI 15), pp. 43–57 (2015)
26. Li, J., Michael, E., Sharma, N.K., Szekeres, A., Ports, D.R.: Just say {NO} to Paxos overhead: replacing consensus with network ordering. In: 12th USENIX Symposium on Operating Systems Design and Implementation (OSDI 16), pp. 467–483 (2016)
27. Park, S.J., Ousterhout, J.: Exploiting commutativity for practical fast replication. In: 16th USENIX Symposium on Networked Systems Design and Implementation (NSDI 19), pp. 47–64 (2019)
28. Charapko, A., Ailijiang, A., Demirbas, M.: PigPaxos: devouring the communication bottlenecks in distributed consensus. In: Proceedings of the 2021 International Conference on Management of Data, pp. 235–247 (2021)
29. Xu, J., Wang, W., Zeng, Y., Yan, Z., Li, H.: Raft-PLUS: improving raft by multi-policy based leader election with unprejudiced sorting. Symmetry **14**(6), 1122 (2022)
30. Nawab, F., Agrawal, D., El Abbadi, A.: DPaxos: managing data closer to users for low-latency and mobile applications. In: Proceedings of the 2018 International Conference on Management of Data, pp. 1221–1236 (2018)

CoPA–Fed: A Federated Reliability Auditing System Under Biased Client Participation

Raiha Tallat[1], Xu Xiaohua[1]([envelope]), Ammar Hawbani[2], and Xingfu Wang[1]

[1] University of Science and Technology of China (USTC), Hefei, China
raiha-tallat@mail.ustc.edu.cn, {xiaohuaxu,wangxfu}@ustc.edu.cn
[2] Shenyang Aerospace University, Shenyang, China
anmande@ustc.edu.cn

Abstract. Federated learning (FL) deployments in the wild often suffer from missing-not-at-random (MNAR) participation, where client availability depends on power, connectivity, and workload conditions. Such biased activation distorts global model calibration and fairness across heterogeneous devices. We introduce CoPA–Fed, a server-side reliability auditing framework that frames participation bias and coverage control as queryable reliability signals aligned with data-management principles. CoPA–Fed continuously estimates client participation propensities from lightweight telemetry (availability, latency, gradient, and loss signals) and instantiates classical inverse propensity weighting (IPW) within federated conformal calibration to approximately maintain target coverage under MNAR participation The framework produces two actionable diagnostics: a Γ-sensitivity ladder quantifying robustness to adversarial up-weighting and participation-fairness deciles measuring coverage across rarely to frequently active clients. CoPA–Fed is optimizer agnostic, plug-and-play, and imposes negligible overhead; clients only share small validation statistics. Experiments on HAR, EMNIST (Balanced), and PathMNIST under non-IID and MNAR conditions show that CoPA–Fed achieves near-target coverage ($\tau\approx0.90$), **improves worst-decile coverage relative to baselines, approaching target levels**, and yields robustness certificates around ($\Gamma^{\star}\approx3$).

Keywords: Federated Learning · Conformal Prediction · Reliability Auditing · Fairness · Data Systems

1 Introduction

Federated learning (FL) enables distributed model training across heterogeneous edge devices while keeping data local. In production deployments, however, devices participate only intermittently: availability fluctuates with diurnal cycles, battery and connectivity, foreground load, and even local training progress [5]. Consequently, the set of active clients at any round is rarely *missing completely*

H. Jung et al. (Eds.): DASFAA 2026, LNCS 16536, pp. 629–645, 2026.
https://doi.org/10.1007/978-981-92-0366-6_38

at random (MCAR). Participation is often *missing not at random (MNAR)*, so statistics computed on the "active" subset—such as calibration or coverage—can be biased relative to the full deployment population. This tension between partial participation and trustworthy uncertainty quantification remains under explored in FL, where prior work emphasizes communication efficiency, optimization under non-IID data, and scalable orchestration [8].

Classical missing-data theory (Rubin's Theorem 1 in [20]) establishes that MNAR mechanisms distort evaluation whenever the probability of missingness depends on unobserved or outcome-linked variables. In practice, *who shows up* in each FL round correlates with model loss, gradient signals, device scale, and time-of-day. Coverage measured on these participants can therefore appear optimistic for rarely available or systematically harder clients. Addressing this requires modeling participation propensities and correcting selection bias.

Conformal prediction (CP) provides distribution-free, finite-sample coverage guarantees by forming prediction sets from quantiles of nonconformity scores. Yet CP assumes exchangeability; under selection or covariate shift, it can miscalibrate. Recent work shows that importance-weighting by estimated propensities can restore nominal coverage when propensities are well behaved—a principle particularly relevant to FL, where "shift" arises from the participation mechanism rather than feature drift [14].

FL in the wild rarely enjoys uniformly available clients: mobile devices appear intermittently, and participation correlates with local dynamics [22]. While CP extensions for FL compute conformal sets either centrally or in distributed form, they typically assume representative calibration data and outcome-independent participation [16]. Our work targets exactly this gap: an auditing layer that learns participation propensities online and re-weights calibration so thatcoverage control is approximately maintained under realistic MNAR participation. We instantiate classical IPW within federated conformal calibration and expose robustness and fairness diagnostics useful for practitioners.

From a data-systems perspective, **CoPA–Fed** acts as a reliability-auditing layer atop federated data flows. It ingests client telemetry, performs conformal calibration *in situ*, and exports aggregated reliability statistics to a global dashboard. In a federated IoT deployment—e.g., hospital wearables or industrial sensors streaming to an edge gateway—devices join and leave asynchronously as users disconnect or machines undergo maintenance. Such non-random participation breaks calibration: global uncertainty estimates drift, and coverage guarantees fail for infrequent clients. CoPA–Fed addresses this by logging availability traces, computing inverse-propensity weights, and exposing coverage metrics through a structured auditing API.

Contributions.

- **System design.** A server-side reliability-auditing layer for FL under MNAR participation that models client propensities from lightweight telemetry and *instantiates classical IPW* within federated conformal calibration for coverage control.

- **Theory and diagnostics.** A theoretical coverage guarantee under MNAR participation and two actionable diagnostics: a Γ-sensitivity ladder for robustness certification and participation-fairness deciles for cross-client equity.
- **Implementation and evaluation.** A drop in implementation requiring no client-side code changes; clients export a small validation signal (< 1 KB/round) and are evaluated on HAR [17], EMNIST (Balanced) [4], and PathMNIST [23], achieving target coverage, improved worst-decile fairness, and low overhead.
- **Data-management alignment.** Reliability auditing exposed as structured, queryable logs with a minimal API, integrating with existing monitoring stacks (e.g., FLARE, Flower).

The rest of the paper is organized as follows. Section 2 reviews related work; Sect. 3 presents the system model and methodology; Sect. 4 details the CoPA–Fed architecture; Sect. 5 describes the experimental setup; Sect. 6 reports results; Sect. 7 discusses data-management implications; Sect. 8 outlines limitations; and Sect. 9 concludes.

2 Related Work

Research addressing participation challenges in federated learning (FL) spans four main directions.

Partial Participation. Early studies analyze stability and efficiency when only a fraction of clients are active [9–11,14]. These works refine aggregation or sampling strategies but typically assume participation is independent of client state, leaving selection bias uncorrected.

Fairness Objectives. A complementary stream enforces fairness during training [12,15,18], reweighting losses to mitigate performance gaps across clients. Such methods promote optimization equity yet do not guarantee calibrated uncertainty or reliable post-deployment coverage. Recent FL fairness-auditing works address training bias [19, 16] but seldom provide post-training reliability diagnostics, which CoPA–Fed targets.

Conformal Prediction in FL. Recent advances adapt conformal prediction (CP) to federated or decentralized settings [2,13,21], offering distribution-free coverage guarantees. However, most assume representative calibration data and outcome-independent participation. Under *missing-not-at-random (MNAR)* availability—where participation correlates with loss or device conditions—these approaches mis-estimate coverage and degrade reliability. Weighted conformal prediction has also been explored for covariate and selection shifts, where calibration points are reweighted by estimated density ratios or propensities (e.g., APS [21] and RAPS [2]). **CoPA–Fed** instantiates this principle within a federated setting, coupling inverse-propensity weighting (IPW) with online participation-propensity estimation to correct MNAR bias—an aspect unexplored in prior FL work.

System Frameworks and Availability Modeling. Modern FL infrastructures such as NVIDIA FLARE and Flower [19] enable production-grade deployments and motivate integrated reliability-auditing layers. Recent analyses model availability and scheduling effects on convergence [6,7], while surveys highlight participation bias as a first-class concern [3]. Yet these system efforts focus on training throughput rather than calibrated evaluation.

Distinctiveness. CoPA–Fed differs by coupling online propensity estimation with IPW-based conformal calibration to recover population-level coverage and to surface two interpretable diagnostics: a $\Gamma^\star$ robustness certificate and participation-fairness deciles for auditable reliability under MNAR participation.

3   Federated Reliability Model and Coverage Guarantee

We study federated multi-class classification under *MNAR* client participation. Let $\mathcal{D} = \{(x, y)\}$ denote the deployment distribution over inputs x and labels $y \in \{1, \ldots, K\}$. There are N clients, each holding a non-i.i.d. shard. At round t, a subset $\mathcal{S}_t \subseteq \{1, \ldots, N\}$ participates according to availability and system state, violating MCAR/MAR assumptions. A central server maintains parameters θ_t (e.g., FedAvg). Our contribution is a *server-side auditing layer* that restores deployment-calibrated coverage and exposes robustness and fairness diagnostics without altering client training; each client exports only a small validation signal.

3.1   Weighted Calibration and Quantile Estimation

Let $s(x, y; \theta) \in [0, 1]$ be a nonconformity score such as

$$s(x, y; \theta) = 1 - \text{softmax}\big(f_\theta(x)\big)_y.$$

For each calibration point (x_i, y_i) observed at time t_i, the auditor assigns a weight $w_i \propto 1/\hat{p}(x_i, y_i)$, where $\hat{p}$ estimates the probability that the client containing (x_i, y_i) participates. Weights are normalized so that $\sum_i w_i = 1$.

The weighted empirical CDF and the corresponding conformal quantile are

$$\hat{F}(q) = \frac{\sum_i w_i \, \mathbf{1}\{s_i \leq q\}}{\sum_i w_i}, \tag{1}$$

$$\hat{q}_\tau = \inf\{q : \hat{F}(q) \geq \tau\}, \tag{2}$$

yielding a prediction set $C_\tau(x) = \{y : \text{softmax}(f_\theta(x))_y > 1 - \hat{q}_\tau\}$ whose size controls efficiency. Here, $\hat{F}$ denotes the weighted empirical CDF of calibration scores, and $\hat{q}_\tau$ is its $(1 - \alpha)$ quantile satisfying $\hat{F}(\hat{q}_\tau) = \tau$.

3.2   Assumptions and Coverage Result

Assume the underlying population follows distribution $\mathcal{P}$, and calibration examples are observed through a missing-not-at-random (MNAR) mechanism with selection probability $p(x, y) \in (0, 1]$. After inverse-propensity weighting, the weighted empirical distribution $\widehat{F}_{\text{IPW}}(s) = \frac{1}{Z} \sum_j \frac{1}{p(x_j, y_j)} \mathbf{1}\{s_j \leq s\}$ (with $Z = \sum_j 1/p(x_j, y_j)$ the normalization constant) approximates the true $F(s)$ in expectation, thereby restoring unbiased coverage under MNAR participation. Let $\hat{p}$ be a consistent propensity estimator satisfying:

1. **Positivity:** $\exists\, \epsilon > 0$ such that $p(x, y) \geq \epsilon$ for all (x, y).
2. **Bounded weights:** $\max_i w_i \leq C < \infty$.
3. **Consistency:** $\hat{p}(x, y) \to p(x, y)$ in probability as $n \to \infty$.

Theorem 1 (Asymptotic Coverage under MNAR with IPW-CP).
Under Assumptions 1–3, the conformal predictor using $\hat{q}_\tau$ in (2) achieves population coverage

$$\Pr[y \in C_\tau(x)] = \tau \pm \mathcal{O}_p(n^{-1/2}) + \mathcal{B}(\hat{p}),$$

where $\mathcal{B}(\hat{p}) = \sup_{(x,y)} |\hat{p}(x, y) - p(x, y)|$ measures bias from propensity misestimation.

Remark. After inverse-propensity weighting, the weighted empirical distribution of calibration points converges to the target population distribution, effectively restoring an i.i.d. sample in expectation under standard MAR / MNAR correction assumptions.

Practical Regimes. For HAR and EMNIST, calibration points originate from MNAR-participating clients, and per-record IPW correction targets the deployment distribution $\mathcal{P}$ in expectation. For PathMNIST, calibration samples are i.i.d. from $\mathcal{P}$ and serve as a control condition. Thus, IPW-corrected calibration approximates population coverage in the MNAR case, while i.i.d. calibration provides a sanity check.

4   CoPA–Fed Architecture and Auditing Workflow

4.1   System Overview

CoPA–Fed is a lightweight, server-side reliability auditing layer integrated into standard FL orchestration. It consists of three components (Fig. 1): (i) a *Telemetry Collector* capturing client availability and proxy statistics, (ii) a *Conformal Auditor* maintaining per-client calibration buffers and computing inverse-propensity-weighted (IPW) quantiles, and (iii) a *Reliability Dashboard* exposing coverage, fairness, and robustness metrics via a REST/SQL-like API. This design requires no changes to client training or data pipelines, preserving privacy while enabling full audit traceability.

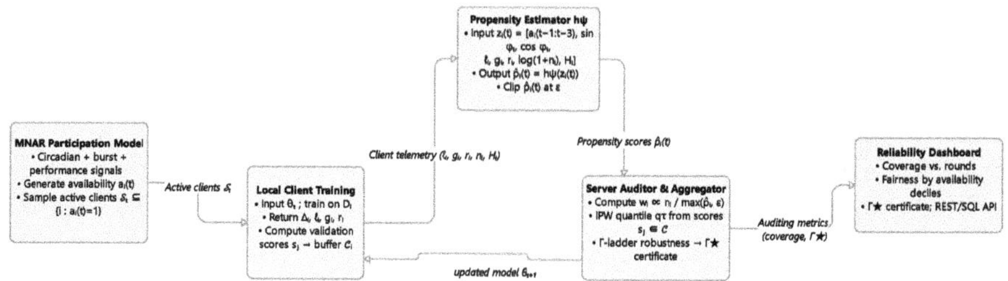

Fig. 1. Auditing workflow of CoPA–Fed under MNAR participation. Each round: availability sampling, local training, propensity estimation, IPW aggregation, and reliability reporting.

4.2   Telemetry Features and Propensity Estimation

Each client i reports lightweight telemetry summarizing recent availability and training dynamics. At round t, the feature vector is

$$z_i(t) = \big[a_i(t{-}1{:}t{-}3), \sin \phi_t, \cos \phi_t, g_i(t), \ell_i(t), r_i(t), \log(1{+}n_i), H_i\big],$$

where $a_i(\cdot) \in \{0,1\}$ encodes recent activity, ϕ_t denotes diurnal phase, (g_i, ℓ_i, r_i) are EMAs of gradient norm, loss, and latency, n_i the local sample count, and H_i label entropy. A compact logistic head $h_\psi : \mathbb{R}^d \to (0,1)$ trained online yields $\hat{p}_i(t) = h_\psi(z_i(t))$. In ablations, static descriptors $(\log(1{+}n_i), H_i)$ are removed $(d{=}8)$.

4.3   IPW Calibration and Coverage Control

The auditor attaches calibration weights

$$w_j \propto \frac{n_{i(j)}}{\max(\hat{p}_{i(j)}(t_j), \varepsilon)},$$

and computes the IPW conformal quantile $\hat{q}_\tau$ via Eq. (2). At inference, all labels with posterior probability $\geq 1{-}\hat{q}_\tau$ form the prediction set. Unweighted and uniform-p baselines are retained for ablation studies.

4.4   Diagnostics: Robustness and Fairness

Two interpretable diagnostics summarize calibration behavior:

- **Γ-sensitivity ladder:** Upweight the top 20% hardest calibration scores (largest non-conformity values), following robustness-sweep heuristics [1], via $w_j^{(\Gamma)} = \Gamma w_j$; sweep Γ upward until coverage drops below τ, and record the resulting $\Gamma^\star$ as the robustness tolerance index.

- **Participation-fairness deciles:** Compute mean availability as

$$\bar{a}_i = \tfrac{1}{3} \sum_{u=1}^{3} a_i(t-u),$$

bin clients into deciles, and report per-decile coverage to expose equity across participation strata.

Here, $\Gamma^\star \approx 3$ indicates that coverage remains at target even if the hardest 20% of calibration mass is effectively upweighted threefold, quantifying tolerance to adversarial weighting.

4.5 Complexity and Implementation Notes

Computation. The logistic head ($\sim 10^3$ parameters) adds $\mathcal{O}(Nd)$ cost per round ($d \leq 10$); weighted quantile evaluation is $\mathcal{O}(n)$. **Communication.** Telemetry adds only a few scalars ($< 1\,\mathrm{KE}/\mathrm{client}/\mathrm{round}$). **Memory.** Calibration buffers for 10^4 points occupy 1–2 MB (float32). **Implementation.** PyTorch 2.8 with `torch.compile`; default parameters $\tau{=}0.90$, $\rho{=}0.2$, $T{=}20$, $W{=}3$, $\varepsilon{=}10^{-3}$. Typical runtime per round: ~ 0.2 s (HAR) and ~ 1.0 s (PathMNIST).

5 Experimental Setup

This section details the experimental configuration used to evaluate CoPA–Fed across diverse federated settings. We describe datasets and partitions, participation regimes simulating missing-not-at-random (MNAR) behavior, baseline methods, metrics, and training protocols. All experiments assess both calibration reliability and system-level efficiency under realistic non-IID and partial participation conditions.

5.1 Datasets and Partitions

We evaluate CoPA–Fed on three standard benchmarks for federated calibration and uncertainty estimation:

- **UCI HAR:** Human Activity Recognition with six activity labels and 9-axis sensor features. Each client corresponds to one user; data are partitioned using Dirichlet($\alpha{=}0.3$) to induce non-IID behavior.
- **EMNIST Balanced:** A 47-class handwritten-character dataset with $N{=}50$ clients. Class-skewed shards are formed via Dirichlet($\alpha{=}0.3$) splits.
- **PathMNIST:** A 9-class medical imaging dataset of histopathology tiles. We simulate $N{=}100$ clients with moderate label imbalance to stress-test scalability.

Algorithm 1. CoPA–Fed: Participation-Aware Conformal Auditing under MNAR

Require: Target coverage $\tau \in (0,1)$; total rounds T; warm-up W; client fraction ρ; clip ε; MNAR configuration; model f_θ; propensity head h_ψ

Ensure: Per-round logs; robustness certificate $(\Gamma^\star, q_\tau^\star)$

1: **Initialization:** server model θ_0; exponential moving averages (EMAs) $\{\ell_i, g_i, r_i\} \leftarrow$ const; recent availability buffers $a_i(t-u) \leftarrow 0$ for $u \in \{1,2,3\}$; static features $(\log(1+n_i), H_i)$; CoPA buffer $\mathcal{C} \leftarrow \varnothing$

2: **for** $t = 0$ **to** $T-1$ **do**

3:     **MNAR availability:** compute $\pi_i(t)$ via circadian + burst + performance terms; sample $a_i(t) \sim \mathrm{Bern}(\pi_i(t))$

4:     Select $m = \lceil \rho N \rceil$ clients $\mathcal{S}_t \subseteq \{i : a_i(t) = 1\}$ uniformly at random

5:     **Feature extraction & propensity:**

6:         form $z_i(t) = [a_i(t-1{:}t-3), \sin \phi_t, \cos \phi_t, g_i(t), \ell_i(t), r_i(t), \log(1+n_i), H_i]$

7:         predict $\hat{p}_i(t) \leftarrow h_\psi(z_i(t))$ ▷ clip later at ε

8:     **Local update (each $i \in \mathcal{S}_t$):**

9:         broadcast θ_t; client trains one local epoch; returns update Δ_i and proxy stats (ℓ_i, g_i, r_i)

10:        update EMAs of ℓ_i, g_i, r_i; roll availability buffer $a_i(t) \leftarrow 1$

11:        compute logits on held-out validation; push $s_j = 1 - \mathrm{softmax}(f_{\theta_t}(x_j))_{y_j}$ to $\mathcal{C}$ tagged with $(n_i, \hat{p}_i(t))$

12:     Roll availability buffers for $i \notin \mathcal{S}_t$ with $a_i(t) \leftarrow 0$

13:     **Update propensity head:** set targets $y_i(t) := a_i(t)$; one step of SGD/Adam on $\sum_i \mathrm{CE}(h_\psi(z_i(t)), y_i(t))$

14:     **Server aggregation:**

15:     **if** $t < W$ **then**   ▷ warm-up via FedAvg weighted by n_i

16:        $\theta_{t+1} \leftarrow \theta_t + \sum_{i \in \mathcal{S}_t} \frac{n_i}{\sum_{k \in \mathcal{S}_t} n_k} \Delta_i$

17:     **else**   ▷ standard FedAvg; IPW applies only to calibration stage

18:        $\theta_{t+1} \leftarrow \theta_t + \sum_{i \in \mathcal{S}_t} \frac{n_i}{\sum_{k \in \mathcal{S}_t} n_k} \Delta_i$

19:     **end if**

20:     **IPW conformal quantiles:**

21:        set $\alpha = 1 - \tau$; compute calibration weights $u_j = \dfrac{n_{i(j)}}{\max(\hat{p}_{i(j)}(t_j), \varepsilon)}$

22:        compute q_τ as weighted $(1-\alpha)$-quantile of $\{s_j\}$ under $\{u_j\}$

23:        log unweighted and uniform-p variants for ablation

24:     **Evaluation:**

25:        on i.i.d. test set: coverage $\widehat{\mathrm{cov}} = \mathbb{P}(s(x,y) \leq q_\tau)$; mean set size $|\{k : \mathrm{softmax}_k \geq 1 - q_\tau\}|$

26:        compute worst-decile coverage by availability deciles

27: **end for**

28: **Robustness evaluation (Γ-ladder):** upweight the top 20% largest non-conformity scores s_j (i.e., the hardest calibration instances) via $u_j^{(\Gamma)} = \Gamma u_j$; sweep $\Gamma \uparrow$; recompute $q_\tau^{(\Gamma)}$ and coverage; output $\Gamma^\star = \max\{\Gamma : \mathrm{cov} \geq \tau\}$

29: **return** Logs and certificate $(\Gamma^\star, q_\tau^{(\Gamma^\star)})$

We evaluate two calibration regimes to test generality: MNAR-collected calibration with inverse-propensity weighting (HAR/EMNIST) and an i.i.d. control calibration (PathMNIST). This isolates the effectiveness of IPW under MNAR participation while providing a bias-free reference condition. Calibration examples for HAR and EMNIST are drawn from MNAR-participating clients each round, whereas PathMNIST uses a global i.i.d. calibration subset with a round-level propensity adjustment.

5.2   MNAR Participation Regimes

Client availability follows a stochastic process combining circadian and bursty dynamics:

$$\pi_i(t) = \mathrm{clip}(\beta_0 + \beta_{\mathrm{circ}} \sin \phi_t + s_i(t) - \beta_{\mathrm{perf}}\ell_i(t),\ \pi_{\min}, \pi_{\max}),$$
$$s_i(t) = \gamma s_i(t-1) + \mathrm{Bern}(\rho) \cdot \beta_{\mathrm{burst}}, \qquad a_i(t) \sim \mathrm{Bern}(\pi_i(t)). \tag{3}$$

Unless specified otherwise, two regimes are tested:

- **Severe MNAR:** strong performance coupling ($\beta_{\mathrm{perf}}{=}0.6$) and pronounced bursts.
- **Mild MNAR:** weaker coupling ($\beta_{\mathrm{perf}}{=}0.2$) with identical circadian base rate.

Warm-up rounds are $W{=}5$, sampling fraction $\rho{=}0.2$, and total rounds $T{=}20$.

5.3   Baselines

We benchmark CoPA–Fed against established conformal and participation-aware methods:

- **Vanilla:** unweighted conformal calibration.
- **APS:** Adaptive Prediction Sets [21].
- **RAPS:** Regularized APS [2], tuned over (τ, λ).
- **No-IPW** and **Uniform-p:** ablations using the same propensity model without inverse weighting.

We focus on *calibration-layer corrections* under MNAR participation. Training-time FL methods (e.g. client selection or aggregation strategies) are intentionally held fixed to isolate the effect of participation-aware calibration under MNAR.

5.4   Evaluation Metrics

Calibration and efficiency are quantified by:

1. Mean coverage ($\widehat{\mathrm{cov}}$),
2. Worst-decile coverage (by recent availability),
3. Mean prediction-set size (efficiency),
4. Robustness certificate ($\Gamma^\star$).

5.5   Training and Hardware Details

Each client trains for one epoch per round with batch size 64 and learning rate 0.01 (Adam optimizer). Propensity heads use cross-entropy loss with clipping at $\varepsilon=10^{-3}$. All experiments employ `PyTorch 2.8` with `torch.compile` on an NVIDIA T4 GPU (16 GB) or CPU fallback. Typical per-round runtime: $\sim$0.2 s for HAR and $\sim$1.0 s for PathMNIST. Full code and hyperparameters are included in the public artifact for reproducibility.

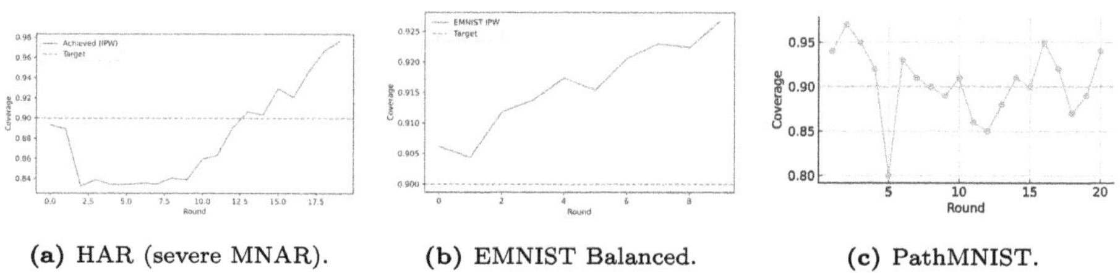

(a) HAR (severe MNAR).   (b) EMNIST Balanced.   (c) PathMNIST.

Fig. 2. Coverage vs. rounds at $\tau=0.90$. CoPA–Fed maintains near-target coverage under MNAR participation.

6   Results

We evaluate CoPA–Fed along five axes: (i) coverage control at target levels, (ii) empirical convergence of auditing statistics, (iii) prediction–set efficiency, (iv) fairness and robustness diagnostics, and (v) ablations. Unless noted, plots are post–warm-up ($t \geq W=5$). All experiments follow identical training and auditing protocols to ensure comparability across datasets and MNAR regimes.

6.1   Coverage Control

On **HAR** (severe MNAR, $\tau=0.90$),CoPA–Fed rapidly reaches near-target coverage shortly after warm-up and maintains it consistently thereafter (Fig. 2a); fluctuations are consistent with Theorem 1's $\mathcal{O}_p(n^{-1/2})$ rate. **EMNIST** exhibits comparable behavior despite a larger label space (Fig. 2b). On **PathMNIST** with $N=100$ clients, near-target coverage is sustained across all rounds (Fig. 2c). A cross-method comparison on PathMNIST (Table 3) shows that our framework achieves the highest coverage while preserving compact prediction sets.

6.2   Convergence Stability

Let $\{(s_j, u_j)\}_{j \in \mathcal{C}_t}$ be calibration scores and weights up to round t with $u_j \propto n_{i(j)}/\max(\hat{p}_{i(j)}, \varepsilon)$. Coverage $\widehat{\text{cov}}_t$ and mean set size $\hat{\kappa}_t$ are computed via Eqs. (1)–(2). We deem convergence when $(\epsilon_{\text{cov}}, \epsilon_q, \epsilon_\kappa)$ remain within $(0.01, 10^{-3}, 10^{-2})$ for five consecutive rounds. On **HAR**, coverage and set-size metrics approach stability shortly after the warm-up period, while **EMNIST** exhibits a longer transient phase due to its larger output space and higher label entropy.

6.3 Prediction–Set Efficiency

While meeting target coverage, the mean set size on **HAR** decreases to $\sim$2.7 by round 20 (Fig. 3a), confirming calibration is not achieved by oversized sets. On **PathMNIST**, set sizes stabilize in the 7–9 range (Fig. 3b), consistent with Table 3.

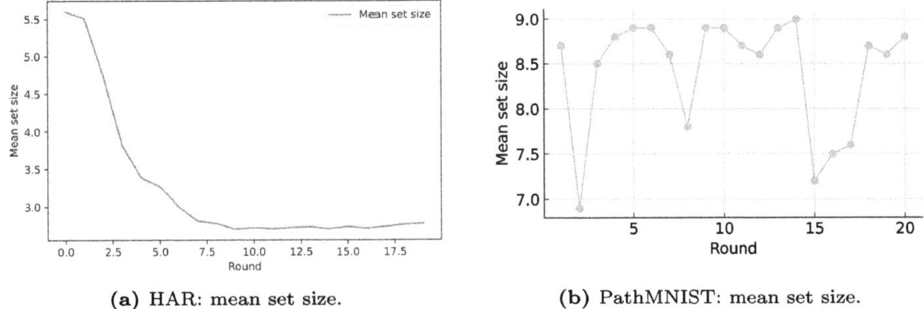

(a) HAR: mean set size.   (b) PathMNIST: mean set size.

Fig. 3. Efficiency under MNAR. CoPA–Fed reduces/stabilizes prediction–set width while preserving target coverage.

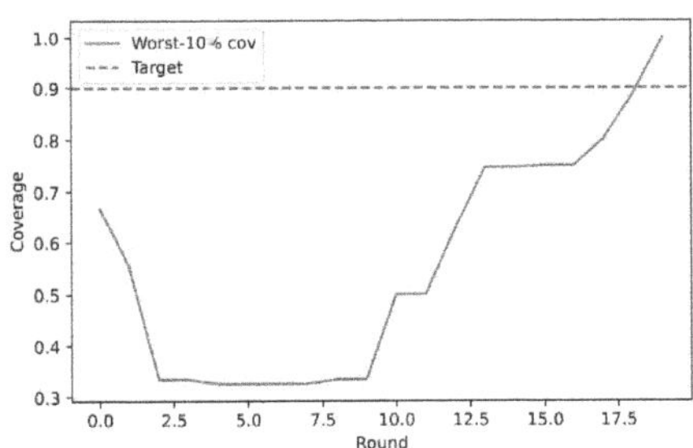

Fig. 4. Worst-decile coverage trajectory on HAR under MNAR participation. CoPA–Fed improves coverage for rarely available clients over rounds, approaching the target $\tau = 0.90$.

6.4 Fairness Under Participation Skew

Clients are binned into deciles by recent availability ($\bar{a}_i$ averaged over the past three rounds). Table 1 reports mean and worst-decile coverage, quantifying reliability across rarely to frequently active clients. The worst-decile trajectory in Fig. 4 shows that coverage for the least available clients progressively approaches the target level, indicating that CoPA–Fed mitigates participation bias without inflating prediction sets.

Table 1. Fairness summary on HAR and EMNIST under severe MNAR participation. The worst-decile coverage quantifies calibration quality for rarely active clients.

Dataset	Mean coverage	Worst-10% coverage	Mean set size
HAR (static feats)	0.890	0.834	2.79
HAR (feature-drop)	0.884	0.834	2.77
EMNIST (static feats)	0.923	0.832	40.7

6.5 PathMNIST: Calibration and Fairness Comparison

Figure 5 contrasts baselines on PathMNIST: (a) τ-mismatch (achieved minus target coverage) and (b) overall vs. worst-decile coverage. Table 3 corroborates that **CoPA–Fed** delivers the highest coverage and *the smallest mean set size (6.79)* while maintaining strong worst-decile coverage.

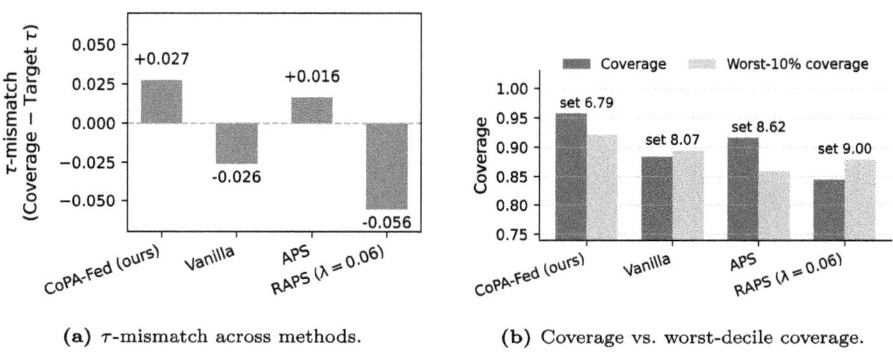

(a) τ-mismatch across methods.   (b) Coverage vs. worst-decile coverage.

Fig. 5. PathMNIST under MNAR participation.

6.6 Ablations and Robustness

We compare **IPW** (ours), **No-IPW** (unweighted), and **Uniform-p** (size-only) variants, each with/without static features. On **HAR**, full IPW tracks τ most closely; removing static features slightly degrades control (Fig. 6a). Robustness to calibration shift is quantified via the Γ-ladder: coverage remains above τ until $\Gamma^\star \approx 3.0$ (Fig. 6b).

6.7 Quantitative Summary

Table 2 consolidates the primary metrics—coverage, worst-decile coverage, efficiency, and $\Gamma^\star$—across datasets. On **HAR**, CoPA–Fed attains 0.97–0.98 mean coverage with mean set size ~ 2.7; on **EMNIST**, mean coverage is ~ 0.958 with $\Gamma^\star = 3.0$. For completeness, Table 3 provides the PathMNIST cross-method comparison, demonstrating consistent improvement in worst-decile coverage and efficiency relative to unweighted and regularized baselines.

7   System Integration and Data-Management Implications

CoPA–Fed maintains per-client calibration buffers $\mathcal{C}_i$ as distributed reliability logs. Each record stores a client identifier, timestamp, nonconformity scores, and audit outcomes. These logs can be queried or joined using standard data-processing primitives, effectively treating audit metadata as structured tables. A sliding-window retention policy bounds storage and ensures constant-time sampling for updates. This design enables direct integration with federated monitoring stacks and enterprise data-lake infrastructures.

Scalability. At large scale ($N > 10^4$ clients), calibration logs are sharded by client identifier or timestamp, guaranteeing bounded query latency and predictable storage growth—a property central to distributed data-management systems. CoPA–Fed thus operates as a lightweight reliability layer within FL orchestration frameworks such as NVIDIA FLARE or Flower.

Audit API. CoPA–Fed exposes reliability metrics through a minimal REST interface:

```
GET /audit/coverage?round=t   -> {cov, q_tau, gamma_star}
GET /audit/deciles?window=K   -> {(decile, cov)}
POST /audit/query             -> SQL-like query over audit_logs
```

These endpoints can be embedded into existing FL dashboards or reliability alerting pipelines, turning calibration diagnostics into first-class queryable data assets.

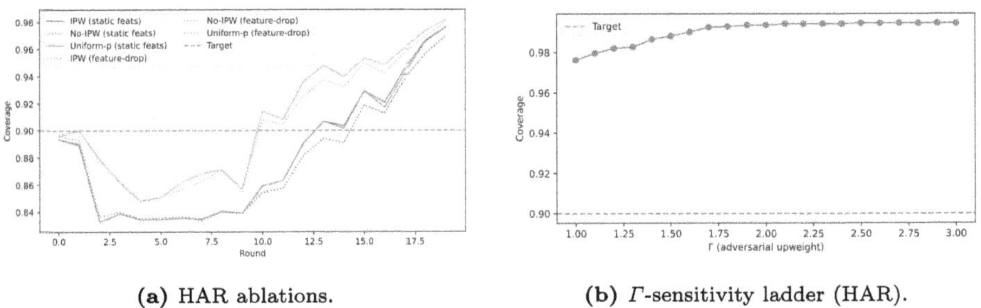

(a) HAR ablations.

(b) Γ-sensitivity ladder (HAR).

Fig. 6. Ablation and robustness analyses under MNAR participation.

Table 2. Summary of main metrics (coverage and fairness are primary; accuracy is reference-only).

Dataset	Mean cov. (post-burn)	Worst-10%	Mean size	$\Gamma^\star$
HAR	0.97–0.98	0.88–1.00	$\sim$2.7	3.0
EMNIST	$\sim$0.958	$\sim$0.91	$\sim$2.4–3.3	3.0

8   Discussion, Limitations, and Future Directions

CoPA–Fed, though effective under synchronous aggregation, assumes round-based coordination and bounded calibration memory. Extending it to asynchronous or streaming FL (e.g., FedAsync, FedBuff) will require redefining temporal consistency in conformal calibration.

Fairness auditing is currently per-round; continuous or adaptive fairness tracking remains open. Calibration drift under evolving client populations or concept shift (e.g., non-stationary IoT workloads) also warrants study.

Our evaluation uses simulated MNAR participation; applying CoPA–Fed to real-world telemetry (e.g., uptime or network logs) is a natural next step toward production-grade reliability auditing. Integration with federated observability systems could enable end-to-end accountability. Future work includes adopting alternative MNAR estimators (e.g., doubly-robust or stratified calibration) and extending the horizon ($T>50$) to confirm long-term convergence.

Table 3. PathMNIST ($N=100$). Conformal-set methods under IPW federated learning. Coverage on the test set; "Worst-10%" = lowest-participation clients over the last 10 rounds. *Note:* HAR and EMNIST use MNAR-collected calibration with per-example IPW correction, whereas PathMNIST uses an i.i.d. calibration subset as a control condition.

Method	τ	Coverage	Worst-10%	Mean set size	$\Gamma^\star$
CoPA–Fed (ours)	0.93	**0.957**	**0.922**	**6.79**	**2.8**
Vanilla	0.91	0.884	0.894	8.07	2.8
APS	0.90	0.916	0.860	8.62	2.8
RAPS ($\lambda=0.06$)	0.90	0.844	0.879	9.00	2.8

9   Conclusion

We presented CoPA–Fed, a participation-aware reliability auditing framework for federated learning under missing-not-at-random (MNAR) participation. Unlike prior work centered on optimization or aggregation, CoPA–Fed treats reliability as a first-class data-system concern. It estimates participation propensities from lightweight telemetry, applies inverse-propensity-weighted (IPW) conformal calibration, and exposes interpretable diagnostics for robustness and fairness.

Empirically, CoPA–Fed attains near-target coverage ($\tau\approx0.90$) on HAR, EMNIST, and PathMNIST, with compact prediction sets and robustness certificates around $\Gamma^\star\approx3$. It **improves worst-decile coverage relative to baselines, approaching target levels**, while requiring no client-side modification. Clients share only minimal validation signals, making CoPA–Fed readily deployable. From a data-management view, its calibration buffers act as auditable, queryable reliability logs bridging ML robustness and system accountability. Future extensions will address asynchronous FL, fairness drift, and integration with large-scale orchestration frameworks such as NVIDIA FLARE and Flower.

Acknowledgments. This work was supported by the National Natural Science Foundation of China (NSFC) under Grants No. 62172383 and No. 62231015, in part by Anhui Provincial Key R&D Program under Grant No. S202103a05020098, and in part by Research Launch Project of University of Science and Technology of China (USTC) under Grant No. KY0110000049.

Disclosure of Interests. The authors declare that they have no competing interests relevant to the content of this work.

References

1. Angelopoulos, A.N., Bates, S.: A gentle introduction to conformal prediction and distribution-free uncertainty quantification (2022). https://arxiv.org/abs/2107.07511
2. Barber, R.F., Candès, E.J., Ramdas, A., Tibshirani, R.J.: Conformal prediction beyond exchangeability. Ann. Stat. **51**(2), 816–845 (2023). https://doi.org/10.1214/23-AOS2276
3. Benarba, N., Bouchenak, S.: Bias in federated learning: a comprehensive survey. ACM Comput. Surv. **57**(11), 1–36 (2025). https://doi.org/10.1145/3735125
4. Cohen, G., Afshar, S., Tapson, J., Van Schaik, A.: EMNIST: extending MNIST to handwritten letters. In: 2017 International Joint Conference on Neural Networks (IJCNN), pp. 2921–2926. IEEE (2017)
5. Gao, X., Hou, L., Chen, B., Yao, X., Suo, Z.: Compressive-learning-based federated learning for intelligent IoT with cloud–edge collaboration. IEEE Internet Things J. **12**(2), 2291–2294 (2025). https://doi.org/10.1109/JIOT.2024.3505838
6. Garg, D., Sanyal, D., Lee, M., Tumanov, A., Gavrilovska, A.: Client availability in federated learning: It matters! In: Proceedings of the 5th Workshop on Machine Learning and Systems, pp. 114–121. EuroMLSys '25, Association for Computing Machinery, New York, NY, USA (2025). https://doi.org/10.1145/3721146.3721964
7. Jee Cho, Y., Wang, J., Joshi, G.: Towards understanding biased client selection in federated learning. In: Camps-Valls, G., Ruiz, F.J.R., Valera, I. (eds.) Proceedings of The 25th International Conference on Artificial Intelligence and Statistics. Proceedings of Machine Learning Research, vol. 151, pp. 10351–10375. PMLR (28–30 Mar 2022). https://proceedings.mlr.press/v151/jee-cho22a.html
8. Kairouz, P., et al.: Advances and open problems in federated learning. Found. Trends® Mach. Learn. **14**(1–2), 1–210 (2021). https://doi.org/10.1561/2200000083
9. Karimireddy, S.P., Kale, S., Mohri, M., Reddi, S., Stich, S., Suresh, A.T.: SCAFFOLD: stochastic controlled averaging for federated learning. In: III, H.D., Singh, A. (eds.) Proceedings of the 37th International Conference on Machine Learning. Proceedings of Machine Learning Research, vol. 119, pp. 5132–5143. PMLR (13–18 Jul 2020). https://proceedings.mlr.press/v119/karimireddy20a.html
10. Lai, F., Zhu, X., Madhyastha, H.V., Chowdhury, M.: Oort: Efficient federated learning via guided participant selection. In: 15th USENIX Symposium on Operating Systems Design and Implementation (OSDI 21), pp. 19–35. USENIX Association (2021). https://www.usenix.org/conference/osdi21/presentation/lai

11. Li, T., Sahu, A.K., Zaheer, M., Sanjabi, M., Talwalkar, A., Smith, V.: Federated optimization in heterogeneous networks. In: Dhillon, I., Papailiopoulos, D., Sze, V. (eds.) Proceedings of Machine Learning and Systems, vol. 2, pp. 429–450 (2020). https://proceedings.mlsys.org/paper_files/paper/2020/file/1f5fe83998a09396ebe6477d9475ba0c-Paper.pdf
12. Li, T., Sanjabi, M., Beirami, A., Smith, V.: Fair resource allocation in federated learning (2020). https://arxiv.org/abs/1905.10497
13. Lu, C., Yu, Y., Karimireddy, S.P., Jordan, M., Raskar, R.: Federated conformal predictors for distributed uncertainty quantification. In: Krause, A., Brunskill, E., Cho, K., Engelhardt, B., Sabato, S., Scarlett, J. (eds.) Proceedings of the 40th International Conference on Machine Learning. Proceedings of Machine Learning Research, vol. 202, pp. 22942–22964. PMLR (23–29 Jul 2023). https://proceedings.mlr.press/v202/lu23i.html
14. McMahan, B., Moore, E., Ramage, D., Hampson, S., Arcas, B.A.y.: Communication-efficient learning of deep networks from decentralized data. In: Singh, A., Zhu, J. (eds.) Proceedings of the 20th International Conference on Artificial Intelligence and Statistics. Proceedings of Machine Learning Research, vol. 54, pp. 1273–1282. PMLR (20–22 Apr 2017). https://proceedings.mlr.press/v54/mcmahan17a.html
15. Mohri, M., Sivek, G., Suresh, A.T.: Agnostic federated learning. In: Chaudhuri, K., Salakhutdinov, R. (eds.) Proceedings of the 36th International Conference on Machine Learning. Proceedings of Machine Learning Research, vol. 97, pp. 4615–4625. PMLR (09–15 Jun 2019). https://proceedings.mlr.press/v97/mohri19a.html
16. Plassier, V., Makni, M., Rubashevskii, A., Moulines, E., Panov, M.: Conformal prediction for federated uncertainty quantification under label shift. In: Krause, A., Brunskill, E., Cho, K., Engelhardt, B., Sabato, S., Scarlett, J. (eds.) Proceedings of the 40th International Conference on Machine Learning. Proceedings of Machine Learning Research, vol. 202, pp. 27907–27947. PMLR (23–29 Jul 2023). https://proceedings.mlr.press/v202/plassier23a.html
17. Reyes-Ortiz, Jorge, A.D.G.A.O.L., Parra, X.: Human Activity Recognition Using Smartphones. UCI Machine Learning Repository (2013). https://doi.org/10.24432/C54S4K
18. Ribero, M., Vikalo, H., de Veciana, G.: Federated learning under intermittent client availability and time-varying communication constraints. IEEE J. Sel. Top. Signal Process. **17**(1), 98–111 (2023). https://doi.org/10.1109/JSTSP.2022.3224590
19. Roth, H.R., et al.: Supercharging federated learning with flower and nvidia flare. In: Federated Learning in the Age of Foundation Models - FL 2024 International Workshops: FL@FM-WWW 2024, Singapore, May 14, 2024; FL@FM-ICME 2024, Niagara Falls, ON, Canada, July 15, 2024; FL@FM-IJCAI 2024, Jeju Island, South Korea, August 5, 2024; and FL@FM-NeurIPS 2024, Vancouver, BC, Canada, December 15, 2024, Revised Selected Papers, pp. 36–45. Springer-Verlag, Berlin, Heidelberg (2025). https://doi.org/10.1007/978-3-031-82240-7_3
20. Rubin, D.B.: Inference and missing data. Biometrika **63**(3), 581–592 (1976). https://doi.org/10.1093/biomet/63.3.581
21. Tibshirani, R.J., Foygel Barber, R., Candes, E., Ramdas, A.: Conformal prediction under covariate shift. In: Wallach, H., Larochelle, H., Beygelzimer, A., d'Alché-Buc, F., Fox, E., Garnett, R. (eds.) Advances in Neural Information Processing Systems, vol. 32. Curran Associates, Inc. (2019). https://proceedings.neurips.cc/paper_files/paper/2019/file/8fb21ee7a2207526da55a679f0332de2-Paper.pdf

22. Wang, S., Ji, M.: A unified analysis of federated learning with arbitrary client participation. Adv. Neural. Inf. Process. Syst. **35**, 19124–19137 (2022)
23. Yang, J., et al.: MedMNIST v2-a large-scale lightweight benchmark for 2D and 3D biomedical image classification. Sci. Data **10**(1), 41 (2023)

FSEM: Few-Shot Entity Matching Using Multi-loss Adversarial Training with Multi-attention Masking

Mengfei Xiong, Huayan Ma, Derong Shen$^{(\boxtimes)}$, Tiezheng Nie, and Yue Kou

School of Computer Science and Engineering, Northeastern University, Shenyang, China
`{xiongmf,mahy9}@mails.neu.edu.cn,`
`{shenderong,nietiezheng,kouyue}@cse.neu.edu.cn`

Abstract. Entity matching (EM) is a critical task for data integration. Existing methods based on Pre-trained Language Models (PLMs) have achieved promising performance with sufficient training samples, but exhibit significant performance drops in few-shot scenarios. To address the issue of poor performance in few-shot scenarios, we propose a novel RoBERTa-based framework called FSEM, which uses multi-loss adversarial training with multi-attention masking. The multi-attention masking module adopts four complementary masking mechanisms to capture multi-level semantic interactions, while the multi-loss adversarial training module, leveraging these multi-attention features, generates diverse adversarial samples by the Fast Gradient Method (FGM) to supplement scarce training data in few-shot scenarios and enriches gradient signals through multi-loss fusion to alleviate gradient sparsity caused by limited annotations. Experiments on nine benchmark datasets demonstrate that FSEM not only significantly outperforms baselines by an average of 14.94% in few-shot scenarios, but also remains competitive when data is sufficient. Furthermore, ablation studies validate the necessity of both modules.

Keywords: Entity Matching · Few-shot Scenarios · Pre-trained Language Model · RoBERTa · Adversarial Training · Multi-Attention Masking

1 Introduction

EM aims to determine whether two entity records from different data sources refer to the same real-world object, serving as a critical task in data integration [2,17,25,27,28]. In recent years, with the widespread application of PLMs [1,10], Transformer-based sequence-pair matching methods [3,6,13,26] have become mainstream. Leveraging PLMs' strong semantic modeling capabilities, these methods achieve promising performance but still face critical challenges in practice, especially low-resource scenarios with scarce labeled data.

© The Author(s), under exclusive license to Springer Nature Singapore Pte Ltd. 2026
H. Jung et al. (Eds.): DASFAA 2026, LNCS 16536, pp. 646–662, 2026.
https://doi.org/10.1007/978-981-92-0366-6_39

Input Entity Pair		Entity Matching		
	name + description + price	RoBERTa	FSEM	Ground Truth
RECORD 1	sanus 30 ' 58 ' visionmount flat panel tv black tilting wall mount lt25b1 sanus 30 ' 58 ' visionmount flat panel tv black tilting wall mount lt25b1 lateral shift adjustment virtual axis height and level adjustments clickstand clickfit system open wall plate black finish 199.0	match	non-match	non-match
RECORD 2	sanus visionmount tilting flat panel tv wall mount mt25-b1 steel 100 lb			

Fig. 1. An example of EM task. Match/non-match indicates whether two records refer to the same entity.

First, existing methods [9,13] rely on a single global token, while largely ignoring the internal structure of entities and fine-grained semantic interactions between entity pairs. Specifically, these methods directly use the [CLS] token for matching decisions, compressing complex entity descriptions into a single global representation and consequently losing fine-grained semantic features. This limitation makes it difficult for models to distinguish hard negative examples in EM. As shown in Fig. 1, this issue causes RoBERTa to classify RECORD1 and RECORD2 as matched samples, while their ground truth is non-match.

Second, performance degrades sharply when labeled samples are limited. Methods like JointBERT [13] and EMBA [26] rely heavily on sufficient labeled data. To alleviate the scarcity of labeled samples, Ditto [9] augments training samples by manipulating discrete tokens within individual attributes, including shuffling token order within attributes, deleting non-critical tokens, or replacing words with synonyms. However, such operations often distort core semantic associations, resulting in a significant degradation in sample quality. Adversarial training, particularly the Fast Gradient Method (FGM) [11], which perturbs continuous token embedding vectors directly instead of discrete text, has shown promise in few-shot tasks by generating adversarial samples to augment data diversity [16,23]. Even so, such perturbations still risk disrupting critical semantic structures in EM. In few-shot scenarios, where core semantic information is already sparse, such perturbations may introduce spurious signals rather than robust features. Moreover, the single classification loss in FGM yields sparse and noisy gradients, limiting its effectiveness in few-shot EM.

To address the two challenges, we propose a RoBERTa-based EM framework called FSEM, which uses multi-loss adversarial training with multi-attention masking. Instead of relying solely on a global token, the multi-attention masking module adopts four complementary masking mechanisms to capture multi-level fine-grained semantic interactions: the global mask handles cross-entity overall semantic alignment; the local mask models intra-attribute local consistency; the skip mask captures long-range non-contiguous dependencies; and the random mask randomly focuses on associations within entity-pair sequences. Meanwhile, the multi-loss adversarial training module, which is improved on FGM, generates diverse adversarial samples to augment limited training data by perturbing RoBERTa's word embedding layer parameters rather than input text, and enriches gradients through fusing main loss and four pseudo-losses to alleviate overfitting and boost generalization under limited labeled samples. We compare FSEM with six baselines on nine datasets. FSEM not only significantly outperforms baselines by an average of 14.94% in few-shot scenarios, but also remains competitive when samples are sufficient.

In conclusion, our main contributions are as follows:

- We propose FSEM, a RoBERTa-based framework that integrates multi-attention masking with multi-loss adversarial training to address the few-shot EM task.
- We propose a multi-attention masking module that employs four complementary masking mechanisms to capture multi-level fine-grained semantic representations.
- We propose a multi-loss adversarial training module improved upon FGM, which generates diverse adversarial samples to augment limited training data and enriches gradient signals via multi-loss fusion, thus improving generalization in few-shot scenarios.
- We evaluate FSEM against six baselines on nine datasets. Experimental results show that FSEM outperforms all baselines by an average of 14.94% in few-shot scenarios and is comparable to state-of-the-art (SOTA) method with sufficient samples.

2   Related Work

Early studies on EM primarily focused on rule-based approaches [4,15] and crowdsourcing-based methods [8,20]. With the advancement of PLMs, models exemplified by BERT and RoBERTa have become the cornerstone of mainstream EM approaches. They can automatically capture semantic correlations between entity attributes by deeply encoding entity attribute sequences. Representative PLM-based EM methods include Ditto [9] and JointBERT [13], which concatenate the attribute sequences of two entities into the format "[CLS] + Entity A + [SEP] + Entity B + [SEP]". After inputting this sequence into a Transformer encoder, they use the [CLS] token embedding to perform binary classification for matching decisions. To address the limitation that the [CLS] token lacks sufficient capacity to capture local semantics, JointMatcher [24] introduces a relevance-aware encoder and a numerically-aware encoder. These encoders enhance the model's sensitivity to semantically similar segments and number-containing segments, respectively, enabling JointMatcher to achieve SOTA performance on small- and medium-sized training sets. EMBA [26] introduces an Attention-over-Attention mechanism that dynamically aggregates discriminative semantic segments from contextual token representations. It achieves SOTA performance on large-scale datasets and significantly outperforms JointBERT, highlighting the importance of fine-grained semantic representations.

Although the aforementioned PLM-based methods have achieved strong performance, labeling such data requires substantial human and material resources. Thus, researchers have proposed various optimization strategies for few-shot EM. DADER [19], a domain adaptation approach, trains PLMs on a source domain with abundant labels and transfers EM knowledge to an unlabeled target domain by aligning feature distributions across domains, thereby alleviating domain shift. The self-supervised method CollaborEM [7] leverages PLMs to automatically construct pseudo-labels, achieving performance comparable to supervised models without requiring manual annotation. The CLER [22] framework introduces a collaborative learning paradigm between a blocker and a matcher, enabling mutual knowledge enhancement through pseudo-label transfer under limited labeling budgets.

Existing EM methods based on Large Language Models (LLMs) [5,14,21, 29] embed labeled entity pairs as demonstration examples within prompts and leverage LLMs' in-context learning capabilities to address the EM task in few-shot settings. However, due to their high computational overhead, expensive inference costs, and inferior performance to fine-tuned PLMs, we do not discuss these methods in detail.

3   The Framework of FSEM

The framework of FSEM, illustrated in Fig. 2, is built upon RoBERTa and integrates two core modules: a multi-attention masking module and a multi-loss adversarial training module. The framework takes a sequence pair containing two target entities as input and outputs a binary matching prediction.

Given two entities e_1 and e_2 with attribute value sets $\{V_{e_1}^1, \ldots, V_{e_1}^m\}$ and $\{V_{e_2}^1, \ldots, V_{e_2}^n\}$ respectively, we first concatenate all attribute values of each entity into a single string (separated by spaces) and then concatenate these two strings using RoBERTa's special tokens `<s>` and `</s>` to form the final entity-pair sequence for input to RoBERTa, in the following format:

$$\texttt{<s>}\{V_{e_1}^1, \ldots, V_{e_1}^m\}\texttt{</s></s>}\{V_{e_2}^1, \ldots, V_{e_2}^n\}\texttt{</s>}$$

RoBERTa converts this input sequence into corresponding token embeddings, and this set of embeddings is uniformly denoted as E in subsequent sections.

3.1   Multi-Attention Masking

The multi-attention masking module consists of four distinct masking mechanisms, which are designed to capture global, local, and long-range semantic correlations between tokens for entity pairs, thereby overcoming the limitations of existing methods that rely solely on global token embeddings for EM. As shown in Fig. 2, for the entity-pair sequence, the module takes the token embedding set E as input and processes it through the four masking mechanisms to generate four types of differentiated embedding features (f_1, f_2, f_3, f_4) as outputs. These four features are then concatenated with the global `<s>` token embedding to form fused features for EM classification.

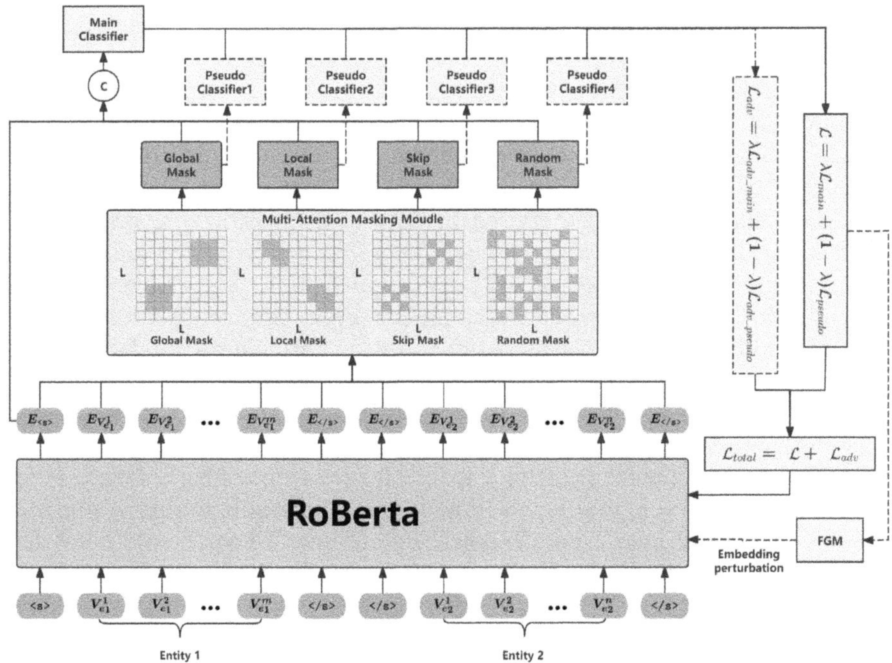

Fig. 2. The framework of FSEM.

Design of Four Masks. To clarify the design of these masking mechanisms, we define key parameters as follows: all masks are constructed based on the entity-pair token sequence, where L denotes the length of this sequence; G_1 and G_2 represent the index sets of tokens belonging to entity e_1 and e_2 respectively, with indices corresponding to positions in this sequence; and $G = G_1 \cup G_2$ represents all token indices of the entity pair.

Global Mask. This mask only allows attention interactions between the two entities (all tokens of e_1 attend to all tokens of e_2 and vice versa), while shielding token interactions within the same entity. By excluding interactions within a single entity, the model is forced to focus on semantic associations between the two entities, thereby enhancing its ability to capture their overall matching degree. The global mask matrix M is defined as Eq. 1:

$$M_{i,j} = \begin{cases} 1, & \text{if } (i \in G_1 \text{ and } j \in G_2) \text{ or } (i \in G_2 \text{ and } j \in G_1) \\ 0, & \text{otherwise} \end{cases} \tag{1}$$

where $M_{i,j} = 1$ means attention interaction is allowed and $M_{i,j} = 0$ means it is prohibited; attention interaction is allowed only for tokens from the two entities.

Local Mask. This mask limits attention interactions to a fixed local neighborhood range around the current token via a preset parameter w (local range threshold) and only allows such interactions within the same entity (tokens of e_1 only interact with their neighboring tokens, and the same for e_2). It prohibits attention interactions between the two entities, which blocks interference of irrelevant information across entities and reduces the impact of redundant noise on feature learning. The local mask matrix M is defined as Eq. 2:

$$M_{i,j} = \begin{cases} 1, & \text{if } |i - j| \leq w \text{ and } (i, j \in G_1 \text{ or } i, j \in G_2) \\ 0, & \text{otherwise} \end{cases} \tag{2}$$

where w is the local range threshold; attention interaction is allowed only for tokens from the same entity with a position difference $|i - j| \leq w$.

Skip Mask. By introducing a skip step d, this mask makes attention skip some intermediate tokens and only focus on non-consecutive token pairs between the two entities where their position difference is an integer multiple of d. It compensates for the limitations of the global mask (containing redundant interactions) and the local mask (only focusing on local neighborhoods), effectively capturing long-range semantic interactions in long-sequence entity pairs. The skip mask matrix M is defined as Eq. 3:

$$M_{i,j} = \begin{cases} 1, & \text{if } [(i \in G_1 \text{ and } j \in G_2) \text{ or } (i \in G_2 \text{ and } j \in G_1)] \\ & \quad \text{and } |i - j| \bmod d = 0 \\ 0, & \text{otherwise} \end{cases} \tag{3}$$

where d is the skip step; attention interaction is allowed only for tokens from the two entities with a position difference of an integer multiple of d.

Random Mask. This mask sparsifies attention interactions through a random strategy: with a fixed sparsity rate r, it randomly selects and retains some of all possible token interaction pairs (including those within a single entity and between the two entities), while masking the rest. The above three types of masks are all generated based on deterministic rules, while this mask introduces diverse perspectives through randomness, compensating for the coverage blind spots of rule-based masks in complex scenarios. It complements the other three rule-based masks, improving the model's adaptability to complex EM task. The random mask matrix M is defined as Eq. 4:

$$M_{i,j} = \begin{cases} 1, & \text{if } (i, j) \in S \subseteq G \times G \text{ and } |S| = \lfloor r \cdot |G|^2 \rfloor \\ 0, & \text{otherwise} \end{cases} \tag{4}$$

where r is the sparsity rate and S is a subset randomly selected from all interaction pairs in $G \times G$, with size $\lfloor r \cdot |G|^2 \rfloor$; attention interaction is allowed only for token pairs (i, j) from subset S.

Feature Generation and Fusion. After obtaining the four types of mask matrices, four different feature representations f_1, f_2, f_3, f_4 are generated respectively through Eq. 5 based on a unified attention computation paradigm:

$$f_i = \mathrm{AvgPool}\left(\mathrm{Softmax}\left(\frac{Q \cdot K^T}{\sqrt{H_d}} + (1 - M_i) \times -\infty\right) \cdot V\right) \tag{5}$$

Here, Q, K, and V represent the query matrix, key matrix, and value matrix respectively, generated from the input token embeddings E through linear transformation; H_d denotes the dimension of the hidden layer in the attention mechanism; M_1, M_2, M_3, M_4 denote the four types of mask matrices mentioned above; AvgPool($\cdot$) represents the average pooling operation on the sequence dimension, which is used to convert sequence features into fixed-dimensional vector features. The four features are generated through parallel computation, sharing Q, K, V, but applying mask constraints independently, ensuring the efficiency and diversity of feature extraction. Finally, the generated $f_1 - f_4$ are concatenated with the global token embedding output by RoBERTa to form fused features for EM classification, improving the accuracy of matching judgment through the integration of multi-dimensional semantic information.

3.2 Multi-Loss Adversarial Training

Although traditional FGM generates adversarial samples by perturbing token embeddings rather than discrete input text, perturbing each token independently still risks disrupting critical entity semantics and core associations. Moreover, limited samples often lead to sparse and unstable gradients, while relying on a single loss function fails to provide sufficiently stable optimization signals.

To address both issues, the proposed multi-loss adversarial training module shifts the perturbation target from token embeddings to the word embedding layer parameters of RoBERTa (denoted as θ_t), while fusing multiple losses to enrich gradient signals. First, since θ_t serves as the core weight matrix that maps input tokens to their corresponding semantic vectors, we fine-tune it along adversarial gradients. This strategy induces consistent global perturbations across all token embeddings, thereby generating adversarial samples in which semantically associated tokens retain strong correlations. Second, we design a multi-loss architecture that leverages features from the multi-attention masking module. Specifically, one main classifier takes the fused features as input to compute the main loss $\mathcal{L}_{\mathrm{main}}$, while four pseudo-classifiers each use a single-attention feature to compute the corresponding pseudo-losses $\mathcal{L}_{\mathrm{pseudo}\text{-}i}$ ($i = 1, 2, 3, 4$). All classifiers adopt binary cross-entropy loss, with heuristic weights set to 0.5 for $\mathcal{L}_{\mathrm{main}}$ and 0.125 for each $\mathcal{L}_{\mathrm{pseudo}\text{-}i}$, balancing the dominance of the core matching task and the gradient contributions from individual attention features. Compared with a single-loss objective, this multi-dimensional design not only provides more stable optimization directions but also enriches gradient sources, thereby alleviating gradient sparsity caused by insufficient annotations.

Algorithm 1 presents the specific implementation process of the multi-loss adversarial training module. First, we perform forward propagation on the input

---

Algorithm 1. Multi-loss Adversarial Training

Require: all model parameters θ, target parameters $\theta_t \subset \theta$, perturbation ϵ, loss function $\mathcal{L}$, input x, labels y, learning rate η.

1: **procedure** MLAT$(\theta, \theta_t, \epsilon, \mathcal{L}, x\ y, \eta)$
2:     **Compute Gradients**
3:     Forward pass: get main classifier output $\hat{y}_{\text{main}}$ and pseudo-classifier outputs $\hat{y}_{\text{pseudo-i}}$
4:     Total loss: $\mathcal{L}_{\text{total}}(\theta) = 0.5 \cdot \mathcal{L}(\hat{y}_{\text{main}}, y) + 0.125 \cdot \sum_{i=1}^{4} \mathcal{L}(\hat{y}_{\text{pseudo-i}}, y)$
5:     Gradients: $\nabla_\theta \mathcal{L}_{\text{total}}$
6:     **Perturb Target Parameters**
7:     Backup: $\theta_{\text{backup}} = \theta_t$
8:     Perturbation: $\Delta\theta_t = \epsilon \cdot \dfrac{\nabla_{\theta_t} \mathcal{L}_{\text{total}}}{\|\nabla_{\theta_t} \mathcal{L}_{\text{total}}\|_2}$
9:     Update: $\theta_t \leftarrow \theta_t + \Delta\theta_t$
10:     **Adversarial Training**
11:     Forward pass (perturbed θ_t): get adversarial outputs $\hat{y}_{\text{adv\_main}}$ and $\hat{y}_{\text{adv\_pseudo-i}}$
12:     Adversarial loss: $\mathcal{L}_{\text{adv}}(\theta) = 0.5 \cdot \mathcal{L}(\hat{y}_{\text{adv\_main}}, y) + 0.125 \cdot \sum_{i=1}^{4} \mathcal{L}(\hat{y}_{\text{adv\_pseudo-i}}, y)$
13:     Gradients: $\nabla_\theta \mathcal{L}_{\text{adv}}$
14:     Update: $\theta \leftarrow \theta - \eta \cdot (\nabla_\theta \mathcal{L}_{\text{total}} + \nabla_\theta \mathcal{L}_{\text{adv}})$
15:     **Restore Target Parameters**
16:     Restore: $\theta_t \leftarrow \theta_{\text{backup}}$
17: **end procedure**

---

data x: the input sequence is encoded by RoBERTa and then fed into the multi-attention masking module, which generates fused features and four differentiated single-attention features. These features are sent to their corresponding classifiers, yielding the main classifier output $\hat{y}_{\text{main}}$ and four pseudo-classifier outputs $\hat{y}_{\text{pseudo-i}}$. Next, we compute the original total loss $\mathcal{L}_{\text{total}}$ according to Eq. 6:

$$\mathcal{L}_{\text{total}}(\theta) = 0.5 \cdot \mathcal{L}(\hat{y}_{\text{main}}, y) + 0.125 \cdot \sum_{i=1}^{4} \mathcal{L}(\hat{y}_{\text{pseudo-i}}, y) \tag{6}$$

Then obtain the gradient of $\mathcal{L}_{\text{total}}$ with respect to the model's total parameters θ via backpropagation, which is denoted as $\nabla_\theta \mathcal{L}_{\text{total}}$. Subsequently, to avoid semantic damage, we first back up the word embedding layer parameters θ_t and then use the gradient of $\mathcal{L}_{\text{total}}$ with respect to θ_t (denoted as $\nabla_{\theta_t} \mathcal{L}_{\text{total}}$) to compute the increment $\Delta\theta_t$ according to the FGM perturbation formula, as shown in Eq. 7:

$$\Delta\theta_t = \epsilon \cdot \frac{\nabla_{\theta_t} \mathcal{L}_{\text{total}}}{\|\nabla_{\theta_t} \mathcal{L}_{\text{total}}\|_2} \tag{7}$$

Here, ϵ denotes the perturbation strength, and $\|\cdot\|_2$ is the L_2-norm for gradient normalization. The parameters θ_t are then updated with $\Delta\theta_t$ to introduce adversarial perturbations, which modifies the word embedding layer for adversarial training. To leverage these perturbations in training, we run forward propagation again on the same input x with the perturbed θ_t. Given this perturbation, RoBERTa's outputs and the multi-attention masking features acquire adversarial characteristics and thus are distinct from their original versions, serving as the adversarial samples for training. The classifiers then output $\hat{y}_{\text{adv\_main}}$ and $\hat{y}_{\text{adv\_pseudo-i}}$, with the adversarial total loss $\mathcal{L}_{\text{adv}}$ computed as in Eq. 8:

$$\mathcal{L}_{\text{adv}}(\theta) = 0.5 \cdot \mathcal{L}(\hat{y}_{\text{adv\_main}}, y) + 0.125 \cdot \sum_{i=1}^{4} \mathcal{L}(\hat{y}_{\text{adv\_pseudo-i}}, y) \tag{8}$$

This loss reflects the model's performance on adversarial features, providing critical signals for enhancing robustness. Based on these signals, we obtain the gradient of $\mathcal{L}_{\text{adv}}$ with respect to θ through backpropagation, which is denoted as $\nabla_\theta \mathcal{L}_{\text{adv}}$. Then sum the original gradient and the adversarial gradient to update all parameters θ of the model. Finally, we restore the perturbed word embedding layer parameters θ_t to their original backed-up values to ensure the independence of each round of adversarial training, avoid the cumulative impact of perturbations on subsequent training, maintain the stability of the training process, prompt the model to learn robust features insensitive to minor perturbations, and thus improve the generalization of the EM task.

4 Experiments

In this section, we conduct extensive experiments to evaluate the effectiveness of FSEM and verify the necessity of its two core modules.

4.1 Experimental Settings

Datasets and Metrics. We evaluate FSEM on nine benchmark datasets from diverse domains, covering both large-scale datasets (with $\geq$ 7k labeled samples) and few-shot datasets (with $\leq$ 500 labeled samples). Among them, three large-scale datasets are reused from the EMBA paper, while six few-shot datasets are included to assess performance under limited-label conditions. Detailed statistics of these datasets are presented in Table 1. For the WDC dataset[1], we adopt its official training/validation/test splits, where the training sets are provided in four sizes (small, medium, large, and xlarge), ranging from $\sim$2,000 to $\sim$70,000 product offer pairs. For the Abt-Buy, Dblp-Scholar, Beer, iTunes-Amazon$_1$ (IA$_1$), and iTunes-Amazon$_2$ (IA$_2$, dirty version of IA$_1$) datasets[2], we use their pre-processed versions and dataset splits as reported in DeepMatcher [12]. For the Baby-Products (BP), Bikes and Books datasets[3], we also adopt the same training/validation/test split ratio of 3:1:1 as used in DeepMatcher. Following previous EM research, we use the F1 score as the evaluation metric.

---

[1] https://webdatacommons.org/largescaleproductcorpus/v2/.
[2] https://github.com/anhaidgroup/deepmatcher/blob/master/Datasets.md.
[3] https://sites.google.com/site/anhaidgroup/useful-stuff/the-magellan-data-repository.

Table 1. Statistics of 9 benchmark datasets

Dataset	Size	# Pos. Pairs	# Neg. Pairs	# Test Set
WDC computers	xlarge	9690	58771	1100
	large	6146	27213	
	medium	1762	6332	
	small	722	2112	
WDC cameras	xlarge	7178	35099	1100
	large	3843	16193	
	medium	1108	4147	
	small	486	1400	
WDC watches	xlarge	9264	52305	1100
	large	5163	21864	
	medium	1418	4995	
	small	580	1675	
WDC shoes	xlarge	4141	38288	1100
	large	3482	19507	
	medium	1214	4591	
	small	530	1533	
Abt-Buy	default	822	6837	1916
Dblp-Scholar	default	4277	18688	5742
Baby-Products	default	86	234	80
Bikes	default	104	256	90
Books	default	74	244	79
Beer	default	54	305	91
iTunes-Amazon$_1$	default	105	325	109
iTunes-Amazon$_2$	default	105	325	109

Baselines. We compare FSEM with the following six representative baselines: DeepMatcher (DM) [12] is the SOTA deep learning-based method for EM before PLMs; Ditto [9] innovatively transforms the EM task into a sequence-pair classification task and achieves efficient performance with sufficient samples; Joint-Matcher (JM) [24] is equipped with a relevance-aware encoder and a numerically-aware encoder, and achieves strong performance on small- and medium-sized training sets; EMBA [26] optimizes JointBERT and achieves SOTA performance on large-scale datasets; DAME [18] is a domain adaptation method that supports zero-shot learning on the target domain; CollaborEM [7] is a self-supervised method specifically designed for few-shot scenarios. We also introduce MatchGPT [14], a representative LLM-based approach, as a supplementary few-shot baseline that leverages in-context learning with GPT-3.5-Turbo, GPT-4, and GPT-4o-mini. These baselines cover both general and few-shot scenarios, enabling a comprehensive evaluation of FSEM across different data scales.

Implementation Details. All experiments are conducted on four NVIDIA A5000 GPUs with 24GB VRAM. For the PLM, we selected RoBERTa, which comprises 12 layers and 768 dimensions, with the maximum token input length limited to 512. During training, we fixed the batch size at 32 and used the Adam optimizer to train FSEM for 30 epochs with a linearly decaying learning rate and a one-epoch warmup. A learning rate sweep was performed over the range [1e-5, 3e-5, 5e-5, 8e-5, 1e-4]. For the key parameters of the multi-attention masking module (local range threshold w, skip step d, random sparsity rate r), we performed a grid search on the WDC cameras dataset, with final optimized values: $w = 2$, $d = 2$, $r = 0.3$. For CollaborEM, FSEM, and its variants, we trained each model three times and reported the average results; for the remaining approaches, we adopted the optimal values reported in relevant papers [18, 24, 26, 28].

4.2 Results and Discussion

Table 2. Comparison of F1 scores between FSEM and baselines on large-scale datasets. The best value is bolded and the second best underlined. $\Delta F1$ denotes the F1 score difference between FSEM and the best baseline.

Dataset	Size	DM	Ditto	EMBA	JM	FSEM	$\Delta F1$
WDC computers	xlarge	88.95	96.53	**98.44**	95.73	96.39	−2.05
	large	84.32	93.81	**97.73**	94.03	95.95	−1.78
	medium	69.85	88.97	93.03	90.10	**94.67**	+1.64
	small	61.22	81.52	81.89	86.95	**90.51**	+3.56
WDCcameras	xlarge	84.88	94.74	**99.16**	93.57	95.51	−3.65
	large	82.16	94.41	**97.84**	92.00	95.26	−2.58
	medium	69.34	87.97	91.90	89.26	**94.81**	+2.91
	small	59.65	78.67	80.69	84.15	**90.37**	+6.22
WDCwatches	xlarge	88.34	97.05	**99.11**	96.61	96.66	−2.45
	large	86.03	97.17	**98.97**	95.89	96.79	−2.18
	medium	67.92	89.16	92.63	93.18	**94.94**	+1.76
	small	54.97	81.32	83.28	91.31	**92.15**	+0.84
WDCshoes	xlarge	86.74	93.28	**98.47**	90.22	90.79	−7.68
	large	83.17	90.07	**97.16**	89.01	91.38	−5.78
	medium	74.40	83.20	**88.47**	85.63	87.99	−0.48
	small	64.71	75.13	73.42	78.42	**84.49**	+6.07
Abt-Buy	default	62.80	82.11	84.81	-	**93.50**	+8.69
Dblp-Scholar	default	94.70	94.47	94.71	-	**96.23**	+1.52
mean	-	-	75.79	88.87	91.76	-	**93.24**	+1.48

Table 2 presents the F1 scores of FSEM and four baselines on three large-scale datasets. On the small and medium subsets of the WDC dataset (simulating few-shot scenarios), FSEM achieves the highest F1 score across all subsets except the medium subset of the WDC shoes dataset, outperforming the second-best baseline (JM or EMBA) by 0.84–6.22% while its average performance is 8.0% higher than Ditto and 25.98% higher than DM. On the large and xlarge subsets with sufficient samples, although EMBA takes the first place, FSEM still maintains a leading position in the second tier, with performance superior to DM and JM and even surpassing Ditto in most cases. Additionally, FSEM achieves the highest F1 scores on both the Abt-Buy and Dblp-Scholar datasets. In terms of overall performance, FSEM ranks first with an average F1 score of 93.24%, outperforming EMBA, Ditto, and DM by 1.48–17.45%.

Table 3. Comparison of F1 scores between FSEM and baselines on few-shot datasets.

	BP	Bikes	Books	Beer	IA$_1$	IA$_2$	mean
DAME	-	-	-	87.58	95.24	-	-
CollaborEM	55.00	73.50	70.55	90.32	92.50	92.50	79.06
GPT-3.5-Turbo	-	-	-	55.9	38.4	-	-
GPT-4o-mini	-	-	-	67.5	69.6	-	-
GPT-4	-	-	-	85.1	73.2	-	-
FSEM	**83.39**	**99.35**	**92.17**	**91.17**	**98.79**	**99.40**	**94.05**
$\Delta F1$	+28.39	+25.85	+21.62	+0.85	+3.55	+6.9	+14.94

To further evaluate FSEM in few-shot scenarios, we conduct additional experiments on six few-shot datasets and compare its performance with DAME, CollaborEM, and MatchGPT (with GPT-3.5-Turbo, GPT-4o-mini, GPT-4). As shown in Table 3, FSEM consistently outperforms all baselines across all datasets. Specifically, FSEM surpasses GPT-4 (the strongest model in MatchGPT) and achieves an average F1 improvement of 14.99% over CollaborEM. These results further demonstrate the superiority of FSEM in few-shot EM.

4.3   Ablation Experiments

We design three variants of FSEM to analyze the impact of the multi-attention masking module and the multi-loss adversarial training module. FSEM-MAM retains the multi-attention masking module but removes the multi-loss adversarial training module, whereas FSEM-MLAT retains the multi-loss adversarial training module but removes the multi-attention masking module. Finally, RoBERTa serves as a baseline variant that removes both modules and directly uses global token embeddings for classification.

Table 4. Comparison of F1 scores between FSEM and its 3 variants on WDC dataset.

Dataset	Size	RoBERTa	FSEM-MAM	FSEM-MLAT	FSEM
WDCcomputers	xlarge	94.73	<u>96.29</u>	96.15	**96.39**
	large	94.68	94.97	<u>95.46</u>	**95.95**
	medium	91.90	94.07	<u>94.47</u>	**94.67**
	small	86.37	89.06	**90.96**	<u>90.51</u>
WDCcameras	xlarge	94.39	<u>95.23</u>	94.31	**95.51**
	large	93.91	94.53	<u>95.03</u>	**95.26**
	medium	90.20	93.65	<u>94.33</u>	**94.81**
	small	85.74	<u>89.31</u>	89.14	**90.37**
WDCwatches	xlarge	94.87	96.03	<u>96.20</u>	**96.66**
	large	93.93	<u>96.68</u>	96.40	**96.79**
	medium	92.28	93.75	<u>94.04</u>	**94.94**
	small	87.16	89.99	<u>90.04</u>	**92.15**
WDCshoes	xlarge	88.88	**91.05**	90.00	<u>90.79</u>
	large	86.60	90.15	<u>91.26</u>	**91.38**
	medium	81.12	85.23	<u>86.88</u>	**87.99**
	small	80.29	<u>82.84</u>	81.26	**84.49**

The detailed results are presented in Table 4. On the small subsets of the WDC dataset, FSEM-MLAT achieves an average F1 score improvement of 2.96% compared to RoBERTa, which is higher than the average improvement of 1.60% on the large/xlarge subsets of the WDC dataset, and FSEM further outperforms FSEM-MLAT by 1.53%. These results clearly demonstrate the effectiveness of the multi-loss adversarial training module in alleviating data sparsity under few-shot settings, as well as the supporting role of the multi-attention masking module in enhancing the effectiveness of adversarial training. Furthermore, FSEM significantly outperforms its ablation variants, with the average F1 score dropping by 0.99% when removing the multi-loss adversarial training module and 0.8% when removing the multi-attention masking module, which fully confirms the necessity of both modules in the FSEM framework.

Table 5. Comparison of F1 scores between FSEM and its 4 mask-removed variants on WDC cameras dataset.

Size	FSEM-LSR	FSEM-GSR	FSEM-GLR	FSEM-GLS	FSEM
xlarge	95.06	95.13	<u>95.23</u>	94.95	**95.51**
large	<u>95.25</u>	94.75	95.22	95.00	**95.26**
medium	<u>94.38</u>	94.28	93.99	94.31	**94.81**
small	<u>90.31</u>	89.54	89.36	89.78	**90.37**

We further analyze the individual contributions of the four mask types in the multi-attention masking module by designing four variants, each removing one specific mask. Specifically, FSEM-LSR removes the global mask, FSEM-GSR removes the local mask, FSEM-GLR removes the skip mask, and FSEM-GLS removes the random mask, while retaining the other three masks. As shown in Table 5, removing any mask causes performance degradation to varying degrees across different data scales on the WDC cameras dataset, confirming that each mask is indispensable to the multi-attention masking module.

4.4 Analysis of the Perturbation Parameter

The multi-loss adversarial training module generates adversarial samples by introducing perturbations to the word embedding layer parameters, thereby expanding semantic diversity and enhancing the model's robustness to noise. Perturbation parameter ϵ determines the strength of perturbations: an overly small value fails to generate effective adversarial signals, while an overly large one introduces excessive noise that distorts the original semantic features of entities.

Table 6. F1 scores of FSEM under different perturbation parameters ϵ on WDC cameras dataset.

Size	$\epsilon=0$	$\epsilon=0.5$	$\epsilon=1.0$	$\epsilon=1.5$	$\epsilon=2.0$	$\epsilon=2.5$	$\epsilon=3.0$
xlarge	94.39	95.17	<u>95.47</u>	**95.51**	95.07	94.93	94.98
large	93.91	94.60	95.03	<u>95.26</u>	**95.37**	94.57	94.78
medium	90.20	93.48	<u>94.34</u>	**94.81**	94.22	93.97	94.18
small	85.74	89.63	89.89	**90.37**	<u>90.02</u>	89.58	89.05

To find the optimal perturbation parameter ϵ, we conduct experiments on the WDC cameras dataset to test 7 perturbation values (0, 0.5, 1.0, 1.5, 2.0, 2.5, 3.0). Table 6 presents the F1 scores of FSEM under different ϵ values. The model performance first improves and then decreases as ϵ increases on all subsets, reaching the optimal result when $\epsilon = 1.5$.

5   Conclusion

In this paper, we identify critical limitations of existing EM methods in few-shot scenarios and propose a novel FSEM framework integrating multi-attention masking and multi-loss adversarial training to address these challenges. The multi-attention masking module provides fine-grained semantic features for the multi-loss adversarial training module through four complementary masking mechanisms, enabling the generation of more diverse adversarial samples and effectively alleviating the limited generalization ability of models in few-shot scenarios. Extensive experiments on nine benchmark datasets demonstrate that FSEM significantly outperforms existing few-shot EM methods by an average of 14.94%, confirming its status as a new SOTA solution for few-shot EM. Future work will explore extending FSEM to cross-domain few-shot EM scenarios and further optimize the mask design within the multi-attention masking module.

Acknowledgments. This work is supported by National Natural Science Foundation of China (62172082, 62072084, 62072086), and the Fundamental Research Funds for the Central Universities (N2116008).

Disclosure of Interests. The authors have no competing interests to declare that are relevant to the content of this article.

References

1. Devlin, J., Chang, M.W., Lee, K., Toutanova, K.: BERT: pre-training of deep bidirectional transformers for language understanding. In: Proceedings of the 2019 Conference of the North American Chapter of the Association for Computational Linguistics: Human Language Technologies, vol. 1 (Long and Short Papers), pp. 4171–4186 (2019)
2. Dou, W., et al.: Enhancing deep entity resolution with integrated blocker-matcher training: balancing consensus and discrepancy. In: Proceedings of the 33rd ACM International Conference on Information and Knowledge Management, pp. 508–518 (2024)
3. Dou, W., et al.: Soft target-enhanced matching framework for deep entity matching. In: Proceedings of the AAAI Conference on Artificial Intelligence, pp. 4259–4266 (2023)
4. Elmagarmid, A., Ilyas, I.F., Ouzzani, M., Quiané-Ruiz, J.A., Tang, N., Yin, S.: NADEEF/ER: generic and interactive entity resolution. In: Proceedings of the 2014 ACM SIGMOD International Conference on Management of Data, pp. 1071–1074 (2014)
5. Fan, M., et al.: Cost-effective in-context learning for entity resolution: a design space exploration. In: 2024 IEEE 40th International Conference on Data Engineering (ICDE), pp. 3696–3709. IEEE (2024)
6. Fang, L., Li, L., Liu, Y., Torvik, V.I., Ludäscher, B.: KAER: a knowledge augmented pre-trained language model for entity resolution. arXiv preprint arXiv:2301.04770 (2023)
7. Ge, C., Wang, P., Chen, L., Liu, X., Zheng, B., Gao, Y.: CollaborEM: a self-supervised entity matching framework using multi-features collaboration. IEEE Trans. Knowl. Data Eng. **35**(12), 12139–12152 (2023)
8. Gokhale, C., Das, S., Doan, A., Naughton, J.F., Rampalli, N., Shavlik, J., Zhu, X.: Corleone: Hands-off crowdsourcing for entity matching. In: Proceedings of the 2014 ACM SIGMOD International Conference on Management of Data, pp. 601–612 (2014)
9. Li, Y., Li, J., Suhara, Y., Doan, A., Tan, W.: Deep entity matching with pre-trained language models. Proc. VLDB Endowment **14**(1), 50–60 (2020)
10. Liu, Y., et al.: RoBERTa: a robustly optimized BERT pretraining approach. arXiv preprint arXiv:1907.11692 (2019)

11. Miyato, T., Dai, A.M., Goodfellow, I.: Adversarial training methods for semi-supervised text classification. In: International Conference on Learning Representations (2017)
12. Mudgal, S., et al.: Deep learning for entity matching: a design space exploration. In: Proceedings of the 2018 International Conference on Management of Data, pp. 19–34 (2018)
13. Peeters, R., Bizer, C.: Dual-objective fine-tuning of BERT for entity matching. Proc. VLDB Endowment **14**, 1913–1921 (2021)
14. Peeters, R., Steiner, A., Bizer, C.: Entity matching using large language models. In: Proceedings 28th International Conference on Extending Database Technology, EDBT 2025, Barcelona, Spain, March 25–28, 2025. pp. 529–541. OpenProceedings.org (2025)
15. Singh, R., et al.: Synthesizing entity matching rules by examples. Proc. VLDB Endowment **11**(2), 189–202 (2017)
16. Su, S., Shao, D., Ma, L., Yi, S., Yang, Z.: ADCL: an attention feature enhancement network based on adversarial contrastive learning for short text classification. Adv. Eng. Inform. **65**, 103202 (2025)
17. Tang, J., Dou, W., Shen, D., Nie, T., Kou, Y.: Towards long-text entity resolution with chain-of-thought knowledge augmentation from large language models. In: International Conference on Database Systems for Advanced Applications, pp. 322–336. Springer (2024)
18. Trabelsi, M., Heflin, J., Cao, J.: DAME: domain adaptation for matching entities. In: Proceedings of the Fifteenth ACM International Conference on Web Search and Data Mining, pp. 1016–1024 (2022)
19. Tu, J., et al.: Domain adaptation for deep entity resolution. In: Proceedings of the 2022 International Conference on Management of Data, pp. 443–457 (2022)
20. Wang, J., Kraska, T., Franklin, M.J., Feng, J.: CrowdER: crowdsourcing entity resolution. Proc. VLDB Endowment **5**(11), 1483–1494 (2012)
21. Wang, T., et al.: Match, compare, or select? An investigation of large language models for entity matching. In: Proceedings of the 31st International Conference on Computational Linguistics, COLING 2025, Abu Dhabi, UAE, January 19-24, 2025, pp. 96–109. Association for Computational Linguistics (2025)
22. Wu, S., Wu, Q., Dong, H., Hua, W., Zhou, X.: Blocker and matcher can mutually benefit: a co-learning framework for low-resource entity resolution. Proc. VLDB Endow. **17**(3), 292–304 (2023)
23. Wu, T., Xin, Z., Chen, S., Zou, Y., You, X.: Adversarial feature training for few-shot object detection. IEEE Trans. Circuits Syst. Video Technol. **35**(9), 9324–9336 (2025)
24. Ye, C., et al.: JointMatcher: numerically-aware entity matching using pre-trained language models with attention concentration. Knowl.-Based Syst. **251**, 109033 (2022)
25. Zeakis, A., Papadakis, G., Skoutas, D., Koubarakis, M.: Pre-trained embeddings for entity resolution: an experimental analysis. Proc. VLDB Endowment **16**(9), 2225–2238 (2023)
26. Zhang, J., Sun, H., Ho, J.C.: EMBA: entity matching using multi-task learning of BERT with attention-over-attention. In: Proceedings of the 27th International Conference on Extending Database Technology, pp. 281–293 (2024)
27. Zhang, Z., Groth, P., Calixto, I., Schelter, S.: Anymatch–efficient zero-shot entity matching with a small language model. arXiv preprint arXiv:2409.04073 (2024)

28. Zhang, Z., Groth, P., Calixto, I., Schelter, S.: A deep dive into cross-dataset entity matching with large and small language models. In: Proceedings 28th International Conference on Extending Database Technology, EDBT 2025, Barcelona, Spain, March 25-28, 2025, pp. 922–934. OpenProceedings.org (2025)
29. Zhang, Z., Zeng, W., Tang, J., Huang, H., Zhao, X.: Active in-context learning for cross-domain entity resolution. Inf. Fus. **117**, 102816 (2025)

DPC-Net: A Decouple-Predict-Correct Framework for Long-Term Time Series Forecasting

Xiangyu Su[1], Zhihong Cui[2], Hengyu Liu[3], Tiancheng Zhang[1(✉)], and Minghe Yu[4]

[1] School of Computer Science and Engineering, Northeastern University, Shenyang 110819, China
tczhang@mail.neu.edu.cn
[2] Department of Informatics, The University of Oslo, Oslo, Norway
[3] Department of Computer Science, Aalborg University, Aalborg, Denmark
[4] Software College, Northeastern University, Shenyang, China

Abstract. Long-term time series forecasting (LTSF) is essential for applications requiring accurate future insights in domains such as energy management, traffic planning, and financial prediction. While Transformer architectures have recently achieved remarkable performance on LTSF tasks, they often come with high model complexity, large parameter counts, and costly inference latency, which limits their practicality in real-time or resource-constrained environments. To overcome these limitations, we propose DPC-Net, a lightweight and efficient dual-branch framework that explicitly decouples input time series into trend and seasonal branches. The trend branch captures global low-frequency patterns through simple yet effective channel-wise linear projections, enhancing interpretability and computational efficiency. The seasonal branch incorporates a novel Temporal Multi-Scale Block (TMB), which employs multi-branch convolutional filters with different receptive fields to model fine-grained seasonal structures across diverse frequencies. Furthermore, a GateCorrector adaptively fuses and refines the outputs of both branches, effectively reducing residual prediction errors. Extensive experiments on six benchmarks demonstrate that DPC-Net consistently outperforms state-of-the-art models in both accuracy and efficiency.

Keywords: Long-term Time Series Forecasting · Dual-branch Neural Networks · Multi-branch Convolution · Lightweight Neural Architecture

1 Introduction

Long-term time series forecasting (LTSF) plays an increasingly critical role across a wide range of application domains, including energy load management, traffic control, and financial market prediction [7]. The ability to forecast future values over extended horizons enables decision-makers to proactively respond to

H. Jung et al. (Eds.): DASFAA 2026, LNCS 16536, pp. 663–678, 2026.
https://doi.org/10.1007/978-981-92-0363-6_40

anticipated trends and anomalies. However, real-world time series often exhibit a complex mixture of global trends and seasonal variations, posing significant challenges for accurate modeling [11].

Time series forecasting has evolved from classical statistical methods to modern deep learning approaches, with each paradigm addressing specific limitations of its predecessors. Traditional models such as ARIMA and exponential smoothing offer interpretable ways to capture trend and seasonality but struggle with nonlinear, non-stationary, and multivariate dynamics over extended horizons [14,20]. Machine learning methods, including random forests and support vector regression (SVR), improve nonlinear modeling but fail to effectively capture temporal dependencies [3]. Recurrent neural networks (RNNs) introduced memory mechanisms to model sequential patterns but suffer from vanishing gradients and limited capacity for long-range dependency modeling [19]. More recently, Transformer-based architectures have demonstrated strong performance in long-term forecasting by leveraging self-attention to capture global dependencies [15]. Variants such as Informer, Autoformer, and PatchTST [13,26,31] further improve accuracy via trend–seasonality decomposition and sparse attention mechanisms. Despite their improvement, these models often involve large parameter sizes, high computational costs, and limited interpretability, making them less suitable for real-time and resource-constrained deployment scenarios. These considerations motivate lightweight, structurally simple alternatives that preserve forecasting accuracy while improving efficiency and interpretability. However, two core challenges still hinder deployment-ready time-series forecasting.

Challenge 1: How to reduce model inefficiency and resource overhead? In recent years, time series forecasting has been increasingly dominated by complex architectures, including Transformers [22], Graph Neural Networks (GNNs) [27], and Latent ODE-based models [16]. These models are designed to capture intricate temporal and cross-variable dependencies through mechanisms such as global self-attention, graph structure modeling, and continuous latent dynamics, achieving state-of-the-art performance in long-term forecasting tasks. However, these capabilities come at a high computational cost. For instance, Transformer-based models like Informer and Autoformer [26,31] incorporate trend–seasonality decomposition and sparse attention to enhance temporal modeling, but at the expense of increased architectural complexity. As a result, these models often exhibit high memory usage, slow inference, and prolonged training times, limiting their applicability in real-time or edge deployment scenarios. In addition to computational overhead, model transparency remains a major concern. The use of deep attention layers and latent-variable components often obscures the decision-making process, making it difficult for practitioners to interpret predictions or diagnose unexpected behaviors. Therefore, it is essential to design a new forecasting model that minimizes unnecessary architectural complexity while retaining the core capability for accurate long-range prediction.

Challenge 2: How to uniformly model diverse and conflicting temporal patterns? Currently, many time-series forecasting methods treat the sequence as a uniform signal, by applying the same modeling operations to all parts of the time series. For example, Transformer-based models such as Autoformer and FEDformer [26,32] introduce frequency-aware or autocorrelation based mechanisms to capture long-term dependencies. However, we found that in the real world, time series data often consist of multiple temporal components rather than signal one. As shown in Fig. 1, time series data includes global trends and seasonal cycles. These components differ not only in frequency but also in statistical characteristics—trends evolve slowly and irregularly, while seasonal components repeat rapidly and regularly. As the anomaly point shown in 1, even a simple temperature series exhibits complex dynamics, including an upward global trend from winter to summer and seasonal oscillations reflecting daily or weekly cycles. In addition, anomaly points highlight deviations that cannot be explained by trend or seasonality alone, requiring careful detection. Notably, the seasonal component varies across different seasons in both frequency and amplitude, making it difficult to capture using fixed cycle models. Recently, many methods consider utilizes adaptive or multi-scale decomposition approaches to accurately model and detect anomalies in real-world time series. For instance, ETSformer [24] employs a level–trend–seasonal decomposition inspired by classical exponential smoothing, while TimesNet [25] leverages multiscale convolutions to capture features across varying temporal scales. However, these methods are computationally expensive. In this end, it is desired for a lightweight and structurally simple alternatives that can retain strong forecasting performance while improving both computational efficiency and model interpretability.

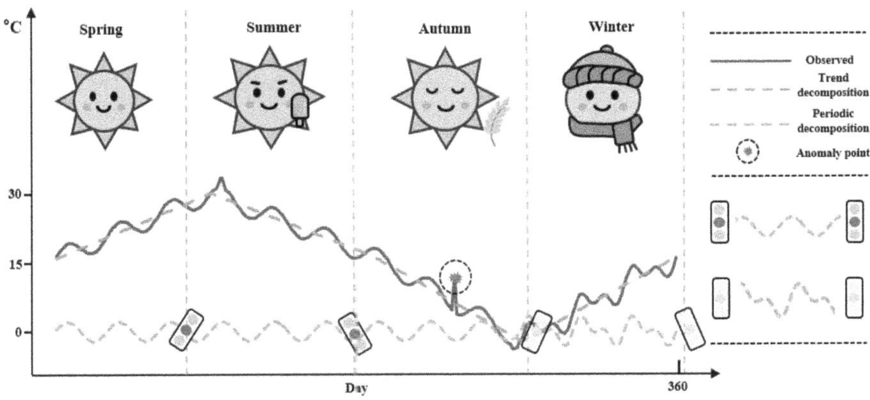

Fig. 1. Illustration of time series decomposition on seasonal temperature data. The original signal (blue) is separated into a global trend component (orange) capturing long-term variations and a seasonal component (green) representing seasonal fluctuations across spring, summer, autumn, and winter. (Color figure online)

To address these challenges, we propose a lightweight dual-branch forecasting framework DPC-Net that explicitly disentangles time series signals into trend and seasonal branches. **To address Challenge 1**, the trend branch is designed to model slowly evolving, global patterns using a lightweight linear projection module. This module applies a channel-specific linear transformation to the entire trend sequence, enabling it to summarize long-range temporal dynamics. By processing the full historical window jointly, it captures low-frequency global trends without relying on recurrent or attention-based mechanisms. This ensures high computational efficiency and interpretability, directly overcoming the limitations of resource-intensive and opaque architectures, and making the model suitable for real-time and resource-constrained deployments. **To address Challenge 2**, the seasonal branch captures fine-grained, high-frequency fluctuations using a nonlinear modeling module. Specifically, we employ multi-branch convolutions with diverse kernel sizes to extract patterns across different temporal resolutions, enabling effective modeling of short-term seasonal signals. This architectural separation results in a model that balances predictive performance with interpretability and computational cost, without relying on deep attention stacks or handcrafted priors. The main contributions are summarized as follows:

- We propose a dual-branch forecasting architecture that explicitly disentangles trend and seasonal branches. The framework leverages linear modeling for global patterns and nonlinear modules for fine-grained seasonal variations, enabling interpretable and targeted temporal representation.
- We propose a Temporal Multi-Scale Block (TMB) as a core component of the seasonal branch. It uses multi-branch convolutions with diverse receptive fields to efficiently capture fine-grained seasonal components. This lightweight module enhances frequency-aware modeling while maintaining low computational cost, supporting real-time deployment.
- We demonstrate superior forecasting performance on multiple long-term time series benchmarks. Our model consistently outperforms strong Transformer-based baselines in both accuracy and inference efficiency, with lower memory usage and fewer parameters.

2 Related Work

2.1 From Traditional to Deep Models

Time series forecasting initially relied on classical statistical methods such as ARIMA, VAR, and Exponential Smoothing, which are favored for their interpretability and robustness in capturing short-term linear patterns [21]. However, their reliance on assumptions like stationarity limits their ability to model nonlinear dynamics, long-range dependencies, and multivariate interactions [6,18]. With the advent of deep learning, Recurrent Neural Networks (RNNs) [17], along with their variants such as Long Short Term Memory (LSTM) [5] and Gated Recurrent Unit (GRU) [2], introduced the capacity to model nonlinear temporal dependencies, leading to significant improvements in diverse forecasting tasks [9,30]. Despite their effectiveness, these models are hindered by sequential training bottlenecks, gradient vanishing or explosion, and limited scalability, which restrict their deployment in large-scale, real-time applications.

2.2   Modeling Long-Term Dependencies

The introduction of the Transformer architecture has transformed time series forecasting by enabling parallel computation and global context modeling via self-attention [8]. Building on this, models such as Informer, Autoformer, and FEDformer [26,31,32] have incorporated sparse attention mechanisms, series decomposition, and frequency-domain transformations to enhance efficiency and capture long-term seasonal structures. For instance, Autoformer explicitly separates seasonal and trend components to improve forecasting accuracy, while FEDformer applies spectral transformations to reduce attention complexity. Despite their effectiveness, these Transformer-based models typically involve high model complexity, large parameter sizes, and substantial training costs [28]. Moreover, their ability to consistently capture seasonal components remains under debate, leading to increased interest in lightweight, interpretable models.

2.3   Period-Aware and Efficient Modeling for Time Series Forecasting

Recent research on time series forecasting has increasingly focused on two primary directions: enhancing seasonal pattern modeling and improving model efficiency. In the realm of seasonal modeling, TimesNet [25] employs Fourier-based spectral transformation to project time series into 2D representations, enabling the learning of intra- and inter-seasonal variations. PatchTST [13] introduces patch-based learning with a channel-independent structure to extract seasonal subsequences efficiently. Concurrently, significant efforts have been made to enhance the efficiency and lightweight nature of forecasting models. DLinear [29] achieves accurate modeling of trend and seasonal components with minimal complexity through a simple single-layer linear projection. TSMixer [10] further extends this paradigm by introducing multi-layer perceptron (MLP) blocks to improve representation capacity. These approaches demonstrate that competitive performance can be achieved with reduced computational cost, promoting the development of interpretable and resource-efficient forecasting frameworks [4,12].

3   Preliminaries

In this study, we focus on the problem of Multivariate Long-term Time Series Forecasting (MLTSF). The input time series is defined as:

$$\mathbf{X} = \left\{ \left(x_1^t, x_2^t, \ldots, x_c^t \right) \right\}_{t=1}^{L} \in \mathbb{R}^{L \times C}, \tag{1}$$

where c denotes the number of variables, L is the length of the observation window, and x_i^t represents the value of the i-th variable at time step t.

The forecasting objective is to predict the values of all variables for the next H time steps based on the historical data of length L:

$$\hat{\mathbf{Y}} = \left\{ \left(\hat{y}_1^t, \hat{y}_2^t, \ldots, \hat{y}_c^t \right) \right\}_{t=L+1}^{L+H} \in \mathbb{R}^{H \times C}, \tag{2}$$

where $\hat{y}_c^t$ denotes the predicted value of the c-th variable at time step t, H is the forecasting horizon, L is the length of the observation window, and C is the total number of variables.

The ground-truth labels are defined as:

$$\mathbf{Y} = \left\{ \left(y_1^t, y_2^t, \ldots, y_c^t \right) \right\}_{t=L+1}^{L+H}. \tag{3}$$

4   Methodology

4.1   Overview

The overall framework of DPC-Net is illustrated in Fig. 2. It consists of four modules: (1) Decouple Module, (2) Trend Path, (3) Seasonal Path, and (4) Residual Fusion and Gated Correction.

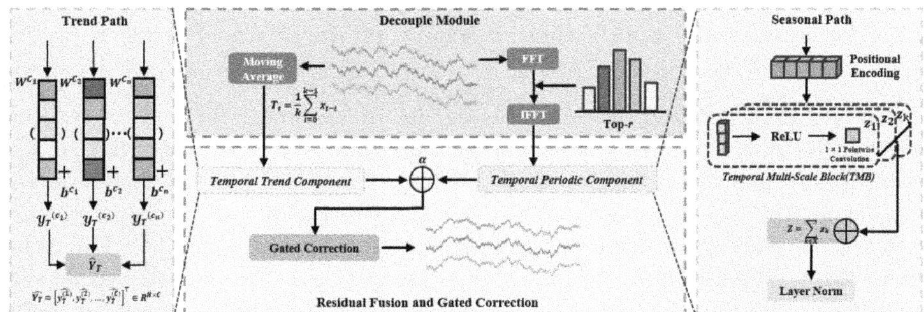

Fig. 2. Architecture of DPC-Net.

4.2   Decouple Module

To better capture the trend and seasonal components in time series data, we introduce a decoupling mechanism into the input sequence by decomposing it into trend and seasonal branches. Specifically, we employ a Moving Average operation to extract long-term, slowly-varying trends, while applying the Fast Fourier Transform (FFT) to identify prominent seasonal components. The trend branch captures low-frequency variations in the sequence, whereas the seasonal branch preserves finer-grained seasonal structures and local fluctuations.

Trend Extraction. We employ the Moving Average as a trend extractor to smooth the sequence fluctuations and retain the long-term slowly-varying trend component. Given an input sequence $x \in \mathbb{R}^{L \times C}$ (where L is the length of the historical window and C is the number of channels), the moving average at time step t is defined as:

$$\mathbf{T_t} = \frac{1}{k} \sum_{i=0}^{k-1} x_{t-i}, \tag{4}$$

where $\mathbf{T}_t$ is the smoothed trend at time step t, $\mathbf{x}_{t-i}$ is the multivariate observation at time $t - i$, and k is the size of the moving average window.

Seasonal Component Extraction. To extract significant seasonal variations in the sequence, we apply the FFT to convert the time-domain signal into the frequency domain. For each channel c, we perform the transformation on its corresponding time series $x^{(c)} \in \mathbb{R}^L$:

$$F(x^{(c)}) = \text{FFT}(x^{(c)}). \tag{5}$$

We then select the top r frequency components with the highest amplitudes:

$$f_1, f_2, \ldots, f_r = \text{Top}_r(|\mathcal{F}(x^{(c)})|). \tag{6}$$

Finally, we apply the Inverse Fast Fourier Transform (IFFT) to these selected frequencies to obtain the seasonal component:

$$p^{(c)} = \text{IFFT}(f_1, \ldots, f_r), \tag{7}$$

where $p^{(c)}$ represents the reconstructed seasonal component of the c-th variable, and $f_1, \ldots, f_r$ are the selected frequency components used in IFFT.

The original sequence can be decomposed into the following form:

$$x = T + P + \varepsilon, \tag{8}$$

where T denotes the trend branch that captures long-term stable variations, P represents the seasonal branch that accounts for seasonal fluctuations, and ε corresponds to high-frequency residuals, typically interpreted as noise or fine-grained discrepancies.

4.3   Trend Path: Structured Modeling of Temporal Trend Component

Following the decomposition of the input sequence into trend and seasonal components, the trend signal $T \in \mathbb{R}^{L \times C}$ which captures the low-frequency dynamics is processed by a structured linear projection module. This component is tailored to model global temporal evolution patterns independently for each variable channel. To preserve channel-wise independence and avoid interference across variables, we adopt a channel-specific modeling strategy. Specifically, for each channel $c \in \{1, \ldots, C\}$, the historical trend sequence $T^{(c)} \in \mathbb{R}^L$ is mapped to the forecasted output via a parametric transformation:

$$\hat{\mathbf{y}}_{\mathbf{T}}^{(\mathbf{c})} = f_{\text{trend}}^{(c)}(T^{(c)}) = W^{(c)}T^{(c)} + b^{(c)}, \tag{9}$$

where $W^{(c)} \in \mathbb{R}^{H \times L}$ and $b^{(c)} \in \mathbb{R}^H$ are the learnable weights and biases. To obtain the final trend forecast, all channel-wise projections are performed in parallel, leading to the following output:

$$\hat{\mathbf{Y}}_{\mathbf{T}} = \left[\hat{y}_T^{(1)}, \hat{y}_T^{(2)}, \ldots, \hat{y}_T^{(C)}\right]^\top \in \mathbb{R}^{H \times C}. \tag{10}$$

4.4 Seasonal Path: Structured Modeling of Temporal Seasonal Component

To more effectively capture the seasonal variations embedded in the input sequence, we introduce a lightweight multi-branch convolutional module, termed the **Temporal Multi-Scale Block** (TMB). Instead of explicit frequency transformation, TMB uses multiple convolution branches with different receptive fields to simulate multi-frequency behaviors in the time domain. This module is designed to simulate multiple frequency modes in the time domain and enhance the capacity of the model to represent complex seasonal components. Specifically, TMB adopts a parallel multi-branch architecture, where each branch corresponds to a predefined convolution kernel size $k \in \mathcal{K}$, intended to extract features under different receptive-field frequencies. Each branch consists of a one-dimensional convolution with kernel size k, followed by a ReLU activation, and a 1×1 pointwise convolution for feature transformation and compression. The operation of each branch can be formulated as follows:

$$\mathbf{z}_k = \mathrm{Conv}_{1 \times 1}\left(\mathrm{ReLU}\left(\mathrm{Conv}_k(\mathbf{x})\right)\right), \quad \forall k \in \mathcal{K}, \tag{11}$$

where $\mathbf{x}$ is the input, Conv_k denotes a 1D convolution with kernel size k, and $\mathrm{Conv}_{1 \times 1}$ is a point-wise convolution for transformation and compression.

Subsequently, the outputs from all branches are aggregated as:

$$\mathbf{z} = \sum_{k \in \mathcal{K}} \mathbf{z}_k. \tag{12}$$

By leveraging multi-scale convolutions with diverse receptive fields and combining them through residual learning, TMB effectively captures localized seasonal fluctuations across multiple frequency bands while maintaining computational efficiency and model compactness.

Finally, to enhance numerical stability and improve feature generalization, we apply a LayerNorm operation to the aggregated output:

$$\mathbf{Out} = \mathrm{LayerNorm}(\mathbf{z}). \tag{13}$$

4.5 Residual Fusion and Gated Correction

To refine the fused prediction and improve robustness, we introduce a lightweight residual correction module with a learnable gating mechanism, termed GateCorrector. This module adjusts the final output by modulating each timestep and channel through dynamic gating and biasing.

Let $\hat{y}_0$ denote the fused prediction from the trend and seasonal branches:

$$\hat{\mathbf{y}}_0 = \alpha \cdot \mathrm{TrendOut} + (1 - \alpha) \cdot \mathrm{SeasonOut}. \tag{14}$$

Finally, We apply the Gatecorrector as follows:

$$\hat{\mathbf{y}}_{\mathrm{final}} = \hat{y}_0 \odot \sigma(G) + B. \tag{15}$$

Here, $G \in \mathbb{R}^{H \times C}$ and $B \in \mathbb{R}^{H \times C}$ are learnable parameters denoting the gate and bias respectively, $\sigma(\cdot)$ denotes the sigmoid activation, and $\odot$ represents element-wise multiplication.

5   Experiments

5.1   Datasets

We conduct extensive experiments on six widely adopted multivariate time series forecasting benchmarks, including Weather, Electricity, and four variants of the ETT dataset: ETTh1, ETTh2, ETTm1, and ETTm2. They are commonly utilized in recent literature as standardized benchmarks for evaluating the performance of long-term forecasting models. Comprehensive dataset statistics, including sequence length, number of variables, sampling intervals, and forecasting settings, are provided in Table 1.

Table 1. Statistics of popular dataset of benchmark.

Dataset	Features	Timesteps	Frequency
Weather	21	52,696	10 min
Electricity	321	26,304	1 h
ETTh1	7	17,420	1 h
ETTh2	7	17,420	1 h
ETTm1	7	69,680	15 min
ETTm2	7	69,680	15 min

5.2   Baselines

To comprehensively evaluate the effectiveness of our proposed method, we benchmark it against a suite of state-of-the-art models for MLTSF. The selected baselines span three representative architecture families: Transformer-based, MLP-based, and convolutional models.

Transformer-based models: We include *Autoformer* [26], *FEDformer* [32], and *PatchTST* [13] as representative attention-driven or token-based architectures. These models exemplify recent advances in modeling long-range temporal dependencies through attention mechanisms and spectral decomposition.

MLP-based models: We incorporate *TSMixer* [10], a recent model that leverages MLP-based spatial-temporal token mixing. It replaces self-attention with linear or convolutional projections, offering a favorable trade-off between accuracy and efficiency for long-horizon forecasting.

CNN-based models: We include *TimesNet* [25], *MICN* [23], and *TCN* [1], three classic convolutional approach that models sequential dependencies via dilated causal convolutions, known for its efficiency and robustness in time-series modeling.

5.3 Implementation Details

All baseline and proposed models are implemented following the official configurations specified in their respective papers to ensure a fair comparison. We adopt four widely used prediction lengths, set as $H \in \{96, 192, 336, 720\}$, and the look-back input length is uniformly set to $L = 720$ for all experiments in the multivariate LTSF task. The data split strategy (train/validation/test) and pre-processing pipeline remain consistent with prior works to maintain comparability.For performance evaluation, we utilize two commonly adopted error metrics: Mean Squared Error (MSE) and Mean Absolute Error (MAE).

5.4 Comparison with Baselines

Table 2 presents the performance comparison between DPC-Net and seven competitive baselines on six multivariate time series datasets under four forecast horizons. Across datasets and horizons, DPC-Net achieves lower average MSE and MAE than existing methods. Specifically, on the Weather dataset, DPC-Net achieves the lowest MSE and MAE across all forecast lengths, demonstrating its strong capability in modeling high-variability meteorological data. For

Table 2. Performance comparison (MSE / MAE) on eight models across six datasets and four forecast lengths.

Models		DPC-Net		Autoformer		FEDformer		PatchTST		MICN		TSMixer		TimesNet		TCN	
Datasets	Metrics	MSE	MAE	MSE	MAE	MSE	MAE	MSE	MAE	MSE	MAE	MSE	MAE	MSE	MAE	MSE	MAE
Weather	96	**0.150**	**0.195**	0.265	0.331	0.218	0.297	<u>0.152</u>	<u>0.197</u>	0.181	0.232	0.158	0.201	0.195	0.236	0.296	0.321
	192	**0.192**	**0.224**	0.307	0.367	0.278	0.335	<u>0.203</u>	0.249	0.237	0.268	0.199	<u>0.245</u>	0.215	0.268	0.526	0.473
	336	**0.241**	**0.281**	0.359	0.395	0.341	0.382	<u>0.253</u>	<u>0.283</u>	0.289	0.315	0.265	0.302	0.269	0.299	0.635	0.526
	720	**0.304**	**0.334**	0.416	0.425	0.402	0.416	<u>0.310</u>	0.354	0.364	<u>0.343</u>	0.321	<u>0.343</u>	0.345	0.631	0.422	0.395
Electricity	96	**0.164**	**0.273**	0.201	0.317	0.191	0.308	0.186	<u>0.276</u>	<u>0.182</u>	0.288	0.198	0.286	0.185	0.291	0.333	0.389
	192	**0.149**	**0.236**	0.221	0.334	0.201	0.315	0.159	0.249	0.154	<u>0.248</u>	<u>0.152</u>	0.254	0.194	0.289	0.331	0.392
	336	**0.162**	<u>0.261</u>	0.230	0.339	0.215	0.317	<u>0.165</u>	0.276	0.171	0.269	0.169	**0.258**	0.201	0.299	0.298	0.372
	720	**0.190**	**0.294**	0.253	0.365	0.243	0.355	0.199	0.299	<u>0.193</u>	<u>0.295</u>	0.213	0.297	0.216	0.315	0.369	0.412
ETTh1	96	**0.451**	**0.442**	0.476	0.459	0.459	0.458	0.482	<u>0.443</u>	0.499	0.476	0.485	0.452	<u>0.456</u>	0.454	0.815	0.674
	192	0.458	<u>0.431</u>	0.451	0.448	0.501	0.479	0.458	0.443	<u>0.431</u>	0.453	**0.423**	**0.429**	0.551	0.512	0.862	0.701
	336	**0.438**	**0.423**	0.459	0.466	0.521	0.496	<u>0.439</u>	0.455	0.478	0.489	0.443	<u>0.451</u>	0.496	<u>0.451</u>	1.126	0.801
	720	**0.456**	**0.501**	0.506	<u>0.507</u>	0.514	0.513	<u>0.488</u>	0.511	0.701	0.612	0.493	0.515	0.650	0.587	1.092	0.856
ETTh2	96	**0.316**	**0.359**	0.346	0.388	0.358	0.397	0.385	0.391	<u>0.318</u>	<u>0.363</u>	0.380	0.376	0.404	0.434	3.324	1.691
	192	**0.326**	**0.356**	0.429	0.439	0.456	0.452	0.351	0.380	0.396	0.414	<u>0.346</u>	0.375	<u>0.431</u>	0.463	3.786	1.740
	336	<u>0.365</u>	**0.398**	0.496	0.488	0.482	0.486	0.367	<u>0.405</u>	0.415	0.438	**0.357**	0.410	0.501	0.499	2.819	1.563
	720	0.409	<u>0.441</u>	0.463	0.474	0.515	0.511	**0.396**	**0.439**	0.423	0.510	<u>0.391</u>	<u>0.441</u>	0.456	0.448	2.999	1.325
ETTm1	96	**0.293**	**0.333**	0.379	0.419	0.505	0.475	0.295	0.343	0.316	0.365	<u>0.294</u>	0.348	0.392	<u>0.341</u>	0.613	0.496
	192	<u>0.344</u>	**0.346**	0.423	0.440	0.521	0.469	0.345	0.349	0.349	0.388	**0.325**	<u>0.348</u>	0.465	0.435	0.542	0.632
	336	0.368	**0.391**	0.454	0.431	0.598	0.499	**0.361**	<u>0.392</u>	0.380	0.410	<u>0.362</u>	0.395	0.416	0.426	1.153	0.812
	720	**0.421**	**0.423**	0.543	0.459	0.701	0.559	0.426	0.439	0.449	0.475	<u>0.423</u>	<u>0.425</u>	0.488	0.469	1.123	0.736
ETTm2	96	**0.162**	**0.250**	0.203	0.279	0.255	0.399	<u>0.172</u>	0.254	0.181	0.275	0.175	<u>0.253</u>	0.199	0.289	0.704	0.615
	192	**0.211**	**0.261**	0.256	0.319	0.281	0.342	<u>0.218</u>	<u>0.295</u>	0.238	0.312	0.230	0.312	0.309	0.355	0.956	0.800
	336	**0.272**	0.346	0.325	0.365	0.339	0.372	0.292	<u>0.336</u>	0.297	0.349	<u>0.291</u>	**0.335**	0.362	0.381	2.112	1.203
	720	**0.351**	**0.378**	0.423	0.415	0.431	0.432	<u>0.366</u>	<u>0.381</u>	0.381	0.409	0.370	0.389	0.399	0.413	5.201	2.130

the Electricity dataset, which exhibits stable seasonal components, DPC-Net again surpasses almost baselines with significant margins. On more challenging datasets such as ETTh1 and ETTm1, DPC-Net achieves comparable or superior results, especially at longer horizons (e.g., 0.421 MSE and 0.423 MAE on ETTm1 at horizon 720, outperforming all Transformer-based models). Notably, on the ETTm2 dataset, which is highly non-stationary, DPC-Net delivers robust predictions with clear improvements over methods like PatchTST and TimesNet (e.g., 0.351 vs. 0.366 vs. 0.399 at horizon 720).

These results demonstrate the effectiveness and generalization ability of DPC-Net across both smooth and complex time series domains. The dual-branch decomposition and gated refinement mechanism enable accurate forecasting with fewer parameters and improved efficiency.

5.5 Ablation Study

To better understand the contribution of each branch in DPC-Net, we conduct ablation experiments by removing or modifying individual modules, including the trend branch, seasonal branch, and the GateCorrector refinement module. All experiments are performed in the Weather and ETTm1 datasets with a prediction length of 96 and 336, and the results are shown in Fig. 3.

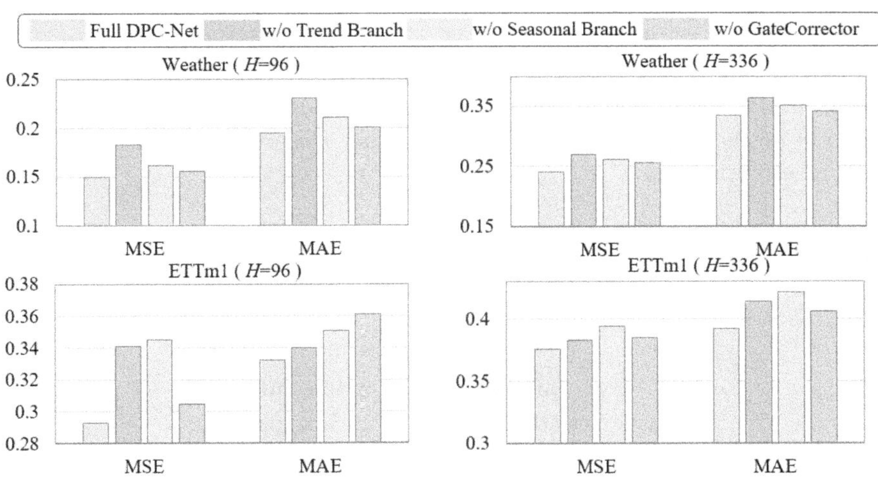

Fig. 3. Ablation Study on Model branches across Forecast Lengths (96/336).

- **w/o Trend Branch:** Removing the trend branch leads to a noticeable performance drop, especially on datasets with strong global shifts, indicating the importance of long-term trend capture.
- **w/o Seasonal Branch:** Eliminating the seasonal path significantly degrades prediction accuracy, especially on seasonal datasets like ETTm1, confirming the necessity of frequency-aware modeling.

- **w/o GateCorrector:** Without the residual refinement, the model exhibits increased error accumulation, particularly at longer prediction horizons, showing the effectiveness of the multi-stage correction.

As shown in Fig. 3, the full model achieves the best performance in terms of both MSE and MAE across all settings, which validates the effectiveness of the overall architecture. When the trend branch is removed, performance drops considerably, particularly on the Weather dataset at $H = 96$, indicating that modeling the global trend is essential for capturing long-term dependencies. The absence of the seasonal branch leads to an even more pronounced degradation, especially in MAE, highlighting the critical role of seasonal component modeling for precise forecasting. Moreover, excluding the GateCorrector module results in slightly worse performance, demonstrating its utility in refining residual errors and enhancing prediction stability. These results confirm that trend modeling, seasonal decomposition, and residual correction each contribute positively and complementarily to the overall performance of DPC-Net, enabling it to better capture multiscale temporal dynamics in multivariate long-term forecasting.

5.6   Model Efficiency

In addition to predictive accuracy, computational efficiency is a crucial factor for the practical deployment of long-term time series forecasting models, especially in resource-constrained environments. We therefore conduct a comprehensive comparison of training efficiency and memory consumption between our proposed DPC-Net and representative baseline models. Table 3 summarizes the results in terms of epoch time and peak memory usage across multiple datasets and forecast lengths.

Table 3. Comparison of practical efficiency of LTSF models under $L = 96$ and $H = 720$ on the Electricity dataset. MACs are the number of multiply-accumulate operations.

Models	MACs	Parameter	Time	Memory
DPC-Net	**0.08G**	**152.5K**	**0.4ms**	**691MiB**
Transformer	4.05G	15.61M	29.8ms	6231MiB
Informer	4.13G	15.38M	51.9ms	4269MiB
Autoformer	4.61G	17.93M	184.1ms	7972MiB
FEDformer	4.94G	22.86M	45.5ms	4243MiB

The results strongly highlight the lightweight nature and computational efficiency of DPC-Net. On average, DPC-Net achieves a remarkable 194.56× acceleration in training time and an 8.22× reduction in memory consumption compared to competitive Transformer-based baselines. These substantial improvements stem from our efficient architectural design, which eliminates redundant computations in both temporal and channel dimensions, while avoiding the quadratic complexity inherent in Transformer architectures.

Such efficiency gains are particularly valuable for large-scale or real-time forecasting applications, where computational resources and latency constraints are critical. By maintaining competitive forecasting performance while drastically reducing computational cost, DPC-Net achieves an exceptional balance between accuracy and efficiency, enabling broader deployment on resource-limited hardware platforms without compromising predictive capability.

5.7   Parameter Sensitivity Analysis

To assess the sensitivity of DPC-Net to the number of retained frequency components (r) in the FFT-based seasonal decomposition, we conduct experiments on datasets with varying seasonal characteristics.

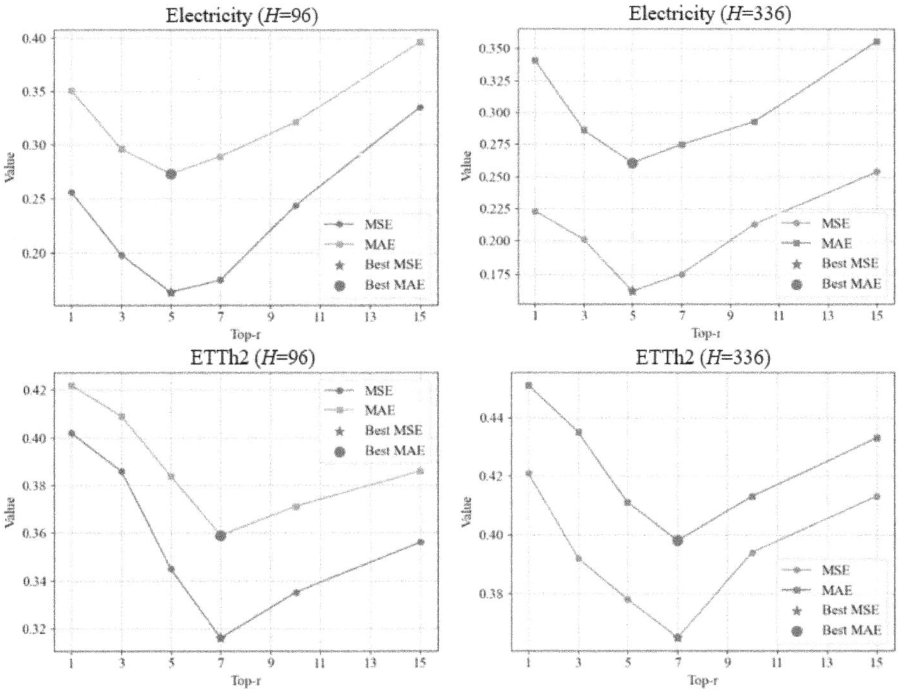

Fig. 4. Effect of top-r frequency components on forecasting performance.

Specifically, we evaluate on two representative datasets: (1) **Electricity**, which exhibits strong and stable seasonal components; (2) **ETTh2**, characterized by weaker seasonality and higher noise levels. For each dataset, the input length is fixed to $L = 720$, and we consider two forecasting horizons $H \in \{96, 336\}$.

The number of retained frequency components is varied as $r \in \{1, 3, 5, 7, 10, 15\}$, while all other hyperparameters remain unchanged.

As illustrated in Fig. 4, DPC-Net exhibits a U-shaped performance curve with respect to the number of retained frequency components (r) in the seasonal decomposition. Across both the Electricity and ETTh2 datasets and under forecast horizons $H = 96$ and $H = 336$, we observe that overly small r (e.g., $r \leq 3$) leads to insufficient seasonal information, while overly large r (e.g., $r \geq 10$) introduces high-frequency noise, both resulting in degraded performance. The optimal r consistently falls within a narrow range, typically $r \in \{5, 7\}$, suggesting that a heuristic choice in this range can yield strong results across datasets with different seasonal characteristics. Furthermore, the consistency between short- and long-horizon settings highlights the robustness of DPC-Net with respect to forecast length. These results confirm that DPC-Net achieves stable and accurate forecasting without requiring exhaustive hyperparameter tuning, offering a practical balance between performance, efficiency, and adaptability for real-world time series forecasting scenarios.

6   Conclusion

We proposed DPC-Net, a compact and efficient dual-branch architecture for multi-scale time series forecasting. By explicitly decoupling trend and seasonal branches, the model employs channel-wise linear projections for long-term trend modeling and a multi-branch convolutional module for capturing fine-grained seasonal components. Additionally, a lightweight residual correction GateCorrector is introduced to refine predictions with minimal overhead.

Extensive experiments on six LTSF benchmarks show that DPC-Net consistently outperforms Transformer-based baselines in both accuracy and efficiency, achieving lower latency, fewer parameters, and reduced memory consumption. Ablation studies further validate the effectiveness of the proposed decomposition and correction mechanisms.

Benefiting from its strong predictive capability and structural simplicity, DPC-Net is well-suited for real-time and resource-constrained forecasting scenarios. As future work, we aim to extend DPC-Net to handle irregularly sampled and multi-modal temporal data, enabling broader applicability to more complex real-world forecasting tasks.

Acknowledgments. This research was funded by the National Natural Science Foundation of China (Nos. 62272093, 62137001).

References

1. Bai, S., Kolter, J.Z., Koltun, V.: An empirical evaluation of generic convolutional and recurrent networks for sequence modeling. arXiv preprint arXiv:1803.01271 (2018)

2. Cho, K., et al.: Learning phrase representations using RNN encoder–decoder for statistical machine translation. In: Proceedings of the 2014 Conference on Empirical Methods in Natural Language Processing (EMNLP), pp. 1724–1734 (2014)
3. Gonçalves, P., Pereira, T.: A survey on machine learning methods for time series forecasting. Expert Syst. Appl. **224** (2025)
4. Goyal, R., Singh, A., Mehta, S.: Lightweight and interpretable models for time series forecasting: a review. IEEE Trans. Neural Networks and Learning Syst. (2024)
5. Hochreiter, S., Schmidhuber, J.: Long short-term memory. Neural Comput. **9**(8), 1735–1780 (1997)
6. Kim, Y., Lim, J., Kang, S.: Time series forecasting: a review of recent advances. Inf. Fus. **90**, 123–147 (2023)
7. Lee, S., Hong, J., Liu, L., Choi, W.: TS-Fastformer: fast transformer for time-series forecasting. ACM Trans. Intell. Syst. Technol. **15**(3), 1–24 (2024)
8. Li, X., Zhu, J., Lin, C.: A survey of transformer-based time series models. Inf. Fus. **93**, 68–86 (2023)
9. Lim, B., Zohren, S.: Time series forecasting: recent advances and future challenges. IEEE Trans. Neural Networks Learn. Syst. (2024)
10. Liu, M., Zeng, A., Xu, Z., Zhang, Q.: TSMixer: an all-MLP architecture for time series forecasting. arXiv preprint arXiv:2303.06053 (2023)
11. Liu, X., Wang, W.: Deep time series forecasting models: a comprehensive survey. Mathematics **12**(10), 1504 (2024)
12. Luo, X., Yin, T., Zhang, L.: Recent advances in lightweight transformer architectures for time series forecasting. Knowl.-Based Syst. **285** (2024)
13. Nie, Y., Nguyen, N.H., Sinthong, P., Kalagnanam, J.: A time series is worth 64 words: long-term forecasting with transformers. In: International Conference on Learning Representations (ICLR) (2023)
14. Oliveira, J.M., Ramos, P.: Evaluating the effectiveness of time series transformers for demand forecasting in retail. Mathematics **12**(17), 2728 (2024)
15. Patro, S., Meher, S.: A comprehensive review of transformer-based architectures for time series forecasting. J. Big Data **11**(1), 78 (2024)
16. Rubanova, Y., Chen, R.T.Q., Duvenaud, D.: Latent odes for irregularly-sampled time series. In: Advances in Neural Information Processing Systems (NeurIPS), vol. 32, pp. 5320–5330 (2019)
17. Rumelhart, D.E., Hinton, G.E., Williams, R.J.: Learning representations by back-propagating errors. Nature **323**(6088), 533–536 (1986)
18. Salinas, D., Flunkert, V., Gasthaus, J., Januschowski, T.: DeepAR: probabilistic forecasting with autoregressive recurrent networks. Int. J. Forecast. **38**(3), 1191–1201 (2022)
19. dos Santos, R.P., Matos-Carvalho, J.P., Rocha, A.M.A.C.: Deep learning in time series forecasting with transformer models and RNNs. PeerJ Comput. Sci. **11**, e3001 (2025)
20. Song, X., Deng, L., Wang, H., Zhang, Y., He, Y.: Deep learning-based time series forecasting: a survey. Artif. Intell. Rev. **57**(6), 9275–9334 (2024)
21. Ullah, I., Ameen, U., Lee, M.: A comprehensive review of time series forecasting using deep learning and classical models. J. King Saud Univ.-Comput. Inf. Sci. **35**, 101685 (2023)
22. Vaswani, A., et al.: Attention is all you need. In: Advances in Neural Information Processing Systems (NeurIPS), vol. 30, pp. 5998–6008 (2017)

23. Wang, H., Peng, J., Huang, F., Wang, J., Chen, J., Xiao, Y.: MICN: multi-scale local and global context modeling for long-term series forecasting. In: The Eleventh International Conference on Learning Representations (ICLR) (2023)

24. Woo, J., Lee, J., Bae, J.: ETSformer: exponential smoothing inspired transformer for time-series forecasting. In: Proceedings of the International Conference on Learning Representations (ICLR) (2023)D

25. Wu, H., Hu, T., Liu, Y., Zhou, H., Wang, J., Long, M.: TimesNet: temporal 2D-variation modeling for general time series analysis. In: International Conference on Learning Representations (ICLR) (2023)

26. Wu, H., Xu, J., Wang, J., Long, M.: AutoFormer: decomposition transformers with auto-correlation for long-term series forecasting. In: Advances in Neural Information Processing Systems (NeurIPS), vol. 34, pp. 22419–22430 (2021)

27. Wu, Z., Pan, S., Chen, F., Long, G., Zhang, C., Yu, P.S.: A comprehensive survey on graph neural networks. IEEE Trans. Neural Networks Learn. Syst. **32**(1), 4–24 (2021)

28. Yuan, X., Huang, Y., Li, C.: On the interpretability and complexity of transformer models for time series forecasting. Expert Syst. Appl. **222**, 119495 (2023)

29. Zeng, A., Dao, T., Zhang, K., Fu, A., Ré, C.: DLinear: exponentially improving linear models for time series forecasting. arXiv preprint arXiv:2210.06518 (2023)

30. Zhang, Y., Li, X., Tang, J.: A comprehensive survey on time series forecasting using deep learning models. Neurocomputing **531**, 36–70 (2023)

31. Zhou, H., et al.: Informer: Beyond efficient transformer for long sequence time-series forecasting. In: Proceedings of the AAAI Conference on Artificial Intelligence, vol. 35, pp. 11106–11115 (2021)

32. Zhou, T., Ma, Z., Wen, Q., Wang, X., Sun, L., Jin, R.: FEDformer: frequency enhanced decomposed transformer for long-term series forecasting. In: International Conference on Machine Learning (ICML), pp. 27268–27286 (2022)

Author Index

GPSR Compliance
The European Union's (EU) General Product Safety Regulation (GPSR) is a set of rules that requires consumer products to be safe and our obligations to ensure this.

If you have any concerns about our products, you can contact us on

ProductSafety@springernature.com

In case Publisher is established outside the EU, the EU authorized representative is:

Springer Nature Customer Service Center GmbH
Europaplatz 3
69115 Heidelberg, Germany

www.ingramcontent.com/pod-product-compliance
Ingram Content Group UK Ltd.
Pitfield, Milton Keynes, MK11 3LW, UK
UKHW020813080726
473059UK00007B/2218